Magnetic Interactions and Spin Transport

Magnetic Interactions and Spin Transport

Edited by

Almadena Chtchelkanova
Strategic Analysis, Inc.
Arlington, Virginia

Stuart Wolf
Naval Research Laboratory
Materials Science & Technology Division
Washington, D.C.

Defense Advanced Research Projects Agency
Arlington, Virginia

and

Yves Idzerda
Montana State University
Bozeman, Montana

Kluwer Academic/Plenum Publishers
New York, Boston, Dordrecht, London, Moscow

ISBN: 0-306-47352-6

©2003 Kluwer Academic / Plenum Publishers, New York
233 Spring Street, New York, New York 10013

http://www.wkap.nl/

10 9 8 7 6 5 4 3 2 1

A C.I.P. record for this book is available from the Library of Congress

All rights reserved

No part of this book may be reproduced, stored in a retrieval system, or transmitted in any form or by any means, electronic, mechanical, photocopying, microfilming, recording, or otherwise, without written permission from the Publisher, with the exception of any material supplied specifically for the purpose of being entered and executed on a computer system, for exclusive use by the purchaser of the work

Printed in the United States of America

Preface

Stuart Wolf

This book originated as a series of lectures that were given as part of a Summer School on Spintronics in the end of August, 1998 at Lake Tahoe, Nevada. It has taken some time to get these lectures in a form suitable for this book and so the process has been an iterative one to provide current information on the topics that are covered. There are some topics that have developed in the intervening years and we have tried to at least alert the readers to them in the Introduction where a rather complete set of references is provided to the current state of the art. The field of magnetism, once thought to be dead or dying, has seen a remarkable rebirth in the last decade and promises to get even more important as we enter the new millennium. This rebirth is due to some very new insight into how the spin degree of freedom of both electrons and nucleons can play a role in a new type of electronics that utilizes the spin in addition to or in place of the charge. For this new field to mature and prosper, it is important that students and postdoctoral fellows have access to the appropriate literature that can give them a sound basis in the fundamentals of this new field and I hope that this book is a very good start in this direction.

The Chapter 1 covers the physical concepts related to the magnetic and electrical properties of transition metal oxides. These materials are important for some of the existing as well as future applications of spintronics.

Chapter 2 provides a fundamental description of the origins of magneto-crystalline anisotropy resulting from spin–orbit interactions and magnetism. It also deals with the magneto-optic properties of materials such as the Kerr and Faraday effects which are important tools for understanding magnetic properties of materials.

Chapter 3 addresses many of the theoretical aspects of spin transport, mostly in metallic systems.

Chapter 4 provides an overview of experimental results on spin transport and covers such topics as GMR and spin dependent tunneling.

Chapter 5 describes in detail methods for quantitative measurements of the magnetization of materials. This chapter provides a detailed description of magnetic units and their connection to the measurements of magnetic properties. Instruments described in this chapter are the vibrating sample magnetometer, the SQUID magnetometer, Mossbauer spectrometers and NMR spectrometers.

Chapter 6 reviews experimental techniques for looking at surface magnetism and involves Kerr microscopy, scanning electron microscopy with polarization analysis, magnetic force microscopy, x-ray dichroism, etc.

Chapter 7 discusses the origins of magnetic noise that in some sense is the limiting characteristic for many state of the art magnetic devices including magnetic disk storage.

Chapter 8 describes the methods for preparing thin films of magnetic materials that include molecular beam epitaxy (MBE), sputtering and pulsed laser deposition.

Chapter 9 is a description of magnetic sensors, Chapter 10 is a detailed description of magneto-resistive memories and Chapter 11 is a description of hybrid magnetic devices.

I hope that this book will provide an important introduction to the very exciting and potentially revolutionary field of spin electronics.

Contents

Introduction xi
    Almadena Chtchelkanova

Chapter 1. Electron Spins in Ionic Molecular Structures
    Gerald F. Dionne
  1.1. Introduction 1
  1.2. Magnetic Cations 2
  1.3. Cation Sites in Anion Lattices 8
  1.4. Electron Energy Levels in Crystal Fields 12
  1.5. Magnetocrystalline Anisotropy 33
  1.6. Spin Exchange Between Cations 54
  1.7. Spin Ordering in Ferrimagnetic Oxides 75
  1.8. Spin Transport in Oxides 92
  Bibliography 125
  References 126

Chapter 2. Secondary Magnetic Properties
    James MacLaren
  2.1. Introduction 131
  2.2. Magneto-Optic Properties 134
  2.3. The Kubo Formula 142
  2.4. *Ab Initio* Calculations of the Kerr Effect 149
  2.5. Magneto-Crystalline Anisotropy 160
  2.6. First-Principle Calculations 168
  Appendix A 177
  Appendix B 180
  References 181

Chapter 3. Spin-Dependent Transport in Magnetic Multilayers
    William Butler
  3.1. Introduction 185
  3.2. Kubo–Greenwood Formula for Conductivity 185
  3.3. Free Electrons with Random Point Scatterers 187
  3.4. Boltzmann Equation 194
  3.5. Landauer Formula for Conductance 211
  3.6. Spin-Dependent Tunneling 212

Appendix 215
References 216

Chapter 4. Magnetotransport (Experimental)
Jack Bass

4.1. Introduction and Overview 219
4.2. Electronic Transport in F-Metals and F-Based Alloys 221
4.3. Spin-Dependent Tunneling and "Polarization" 227
4.4. Spin-Injection Studies 232
4.5. Giant Magnetoresistance in F/N Multilayers and Granular Alloys . 235
4.6. Granular Magnetoresistance 275
4.7. Tunneling Magnetoresistance 278
4.8. Colossal MR 292
4.9. Miscellaneous Phenomena 301
References 308

Chapter 5. Magnetic Characterization of Materials
Chia-Ling Chien

5.1. Introduction 313
5.2. Units 313
5.3. Magnetic Moment, Magnetization, and Susceptibility 314
5.4. Diamagnetism and Paramagnetism 319
5.5. Magnetic Ordering 323
5.6. Magnetometry 328
5.7. Magnetic Characterizations through Hyperfine Interactions 331
5.8. Concluding Remarks 339
Bibliography 333

Chapter 6. Magnetic Domain Imaging of Spintronic Devices
Robert J. Celotta, John Unguris, and Daniel T. Pierce

6.1. Introduction 341
6.2. Scanning Electron Microscopy with Polarization Analysis 343
6.3. Magnetic Force Microscopy 349
6.4. Magneto-Optic Imaging 352
6.5. Transmission Electron Microscopy 357
6.6. Magnetic Imaging with X-Ray Dichroism 366
6.7. Conclusions 371
References 372

Chapter 7. Domain Dynamics and Magnetic Noise
Heidi Hardner

7.1. Introduction 375
7.2. Overview of Magnetic Noise in Device Applications 375
7.3. Magnetic Domains and Magnetization Reversal 377
7.4. Electrical Noise 384
7.5. From Magnetic Fluctuations to Electrical Noise 388
7.6. Stabilization of Magnetic Sensors 398
7.7. Media Noise in Magnetic Recording 405

Bibliography 410
References 410

Chapter 8. Deposition Techniques for Magnetic Thin Films and Multilayers
David Keavney and Charles Falco
- 8.1. Introduction 413
- 8.2. Thermal Evaporation 414
- 8.3. Sputtering 430
- 8.4. Pulsed Laser Deposition 439
- 8.5. Film Thickness Measurement and Control Techniques 441
- 8.6. Characterization Techniques 442
- 8.7. Summary 445
- References 446

Chapter 9. Magnetic Sensors
Jim Daughton and Carl Smith
- 9.1. Introduction 449
- 9.2. Why We Sense Magnetic Fields 449
- 9.3. Magnetic Sensing Technologies 454
- 9.4. Spin-Dependent Magnetoresistive Materials 457
- 9.5. Magnetoresistive Sensor Design 460
- 9.6. Anisotropic Magnetoresistive Sensors 461
- 9.7. Giant Magnetoresistive Sensor Design 463
- 9.8. Spin-Dependent Tunneling Sensor Design 468
- 9.9. Conclusions 474
- References 474

Chapter 10. High Speed Magnetoresistive Memories
Arthur V. Pohm
- 10.1. Introduction 477
- 10.2. Memory Fundamentals 478
- 10.3. Plated Wire and Planar Film Memories, Precursors 481
- 10.4. Magnetoresistive Memory Design Factors 484
- 10.5. Honeywell's Original MRAM 488
- 10.6. Pseudo Spin Valve, Giant Magnetoresistive Memories 497
- 10.7. Memories with Spin Valve Materials 508
- 10.8. Spin Dependent Tunneling Memory Cells 510
- 10.9. Summary 513
- References 513

Chapter 11. Hybrid Devices
Mark Johnson
- 11.1. Introduction 515
- 11.2. Hybrid Ferromagnet–Superconducting Devices 530
- 11.3. Hybrid Ferromagnet–Semiconductor Devices 545
- References 562

Index 565

Introduction

Almadena Chtchelkanova

INTRODUCTION TO SPINTRONICS

The word "spintronics" originated in 1994, when the Defense Advanced Research Projects Agency (DARPA) started a new program, named Spintronics (**spin tr**ansport electr**onics**), to create a new generation of electronic devices where the spin of the carriers would play a crucial role in addition to or in place of the charge. The main goal of the program was to create new magnetic field sensors and magnetic random access memory (MRAM) based on the giant magnetoresistance (GMR) effect and spin dependent tunneling, see Chapters 9 and 10. Over time, the program expanded to include semiconductor magnetoelectronics and semiconductor quantum spin electronics.

Spintronics research includes electronically and photonically controlled magnetism in semiconductors, theoretical and experimental investigation of ferromagnetic semiconductors, optical properties of quantum dot molecules, electronic spin injection, manipulation and detections in heterostructures, combinatorial search for new materials for spintronics applications. Technological advances in fabrication and characterization of hybrid nanostructures have expanded spin studies into low dimensional structures, such as quantum wells, quantum wires and quantum dots. Understanding of spin transport through hetero-interfaces is important for modeling and building spintronic devices. Spintronics has become a multidisciplinary effort bringing together specialists in semiconductors, magnetism and optical electronics studying spin dynamics and transport in semiconductors, metals, superconductors and heterostructures.[1,2]

This volume resulted from lectures given at Summer School at Lake Tahoe, Nevada in 1998 by prominent scientists in the field of magnetism and spin transport. The authors had a chance to update their respective chapters to bring the information up-to-date because the field of spintronics is changing very quickly. Originally, spintronics was concerned mostly with all-metal devices,[3,4] and this volume reflects this view. A second volume is planned to cover topics such as semiconductor magnetoelectronics, semiconductor spin electronics and spin studies in low dimensions. The goal of this introduction is to mention the spintronics issues not covered in this volume and provide references to reviews and original articles for future in-depth reading.

HIGH DENSITY DISKS AND GMR READ HEADS

The charge of the electron in semiconductors is used primarily in integrated circuits and spin of electron in magnetic materials is used for data storage. The spin has been an important part of magnetic high density data storage technology for many years. Hard disk drives (HDD) store information as tiny magnetized regions along concentric tracks. The recent progress in the increasing storage areal density (expressed as billions of bits per square inch of disk surface area, Gb/in.^{2}), is due to high sensitivity read heads made possible after the discovery of the GMR effect. The first commercially available disk drive using a GMR sensor head was the 1998 IBM Deskstar 16GP disk drive that had 2.69 Gb/in.^{2} areal density. The commercially available areal density is now up to 40 Gb/in.^{2} Fujitsu and Seagate recently reported laboratory demonstration of 110 Gb/in.^{2} using longitudinal magnetic recording technology. To achieve high density magnetic recording, it is very important to suppress media noise (see Chapter 7). This requires decreasing the media grain size and improvement in grain uniformity and magnetic grain isolation. Continuing this trend has two major concerns. First, as areal densities are approaching 100 Gb/in.^{2}, the width of the sensor strip is approaching 0.1 μm and the current density is becoming high. It is unclear if the conventional spin valve (SV) read head can be extended to those levels, or if a new form of read head will have to be introduced. Second, further reduction in the media grain size will eventually compromise thermal stability and limit the areal density due to super-paramagnetic decay. New emerging technologies, such as perpendicular magnetic recording (PMR), heat-assisted magnetic recording (HAMR) and patterned media recording, might be required to achieve areal densities significantly beyond 110 Gb/in.^{2}

A GMR structure widely used in HDD read head is a SV originally proposed by IBM[5] in 1994. A SV has two ferromagnetic layers sandwiching a thin non-magnetic metal (usually copper), with one of the two magnetic layers being "pinned", i.e., the magnetization in that layer is relatively insensitive to moderate magnetic fields.[6] This can be achieved by using a layer of an antiferromagnetic material. The other magnetic layer is called the "free" layer, and its magnetization can be changed by application of a relatively small magnetic field. In SV read heads for high density recording, the magnetic moment of the pinned layer is fixed along the transverse direction by exchange coupling with an antiferromagnetic layer (FeMn), while the magnetic moment of the free layer rotates in response to signal fields. When a weak magnetic field, such as from a bit on a hard disk, passes beneath the read head, the magnetic orientation of the free layer changes its direction relative to the pinned layer, generating a change in electrical resistance due to the GMR effect. As the alignment of the magnetizations in the two layers change from parallel to antiparallel, the resistance of the SV typically rises from 5 to 10%. The original SV configuration was modified to achieve a higher change in manetoresistance. In one approach, the simple "pinned" layer is replaced with a synthetic antiferromagnet — two magnetic layers separated by a very thin (~10 Å) non-magnetic conductor.[7] The magnetizations in the two magnetic layers are strongly antiparallel coupled, and are thus effectively immune to outside magnetic fields. In the second approach, a nano-oxide layer (NOL) is formed at the outer surface of the soft magnetic film. This layer reduces resistance due to surface scattering,[8] thus reducing background resistance and thereby increasing the percentage change in magnetoresistance of the structure. The magnetoresistance of SV has increased dramatically from about 5% in early heads to about 15–20% today, using synthetic antiferromagnets and NOLs. In general, all-metal devices are not integrated with the silicon circuits, have low impedance and high currents. The use of SV for GMR sensors is detailed in Chapter 9. Chapter 10 addresses use of GMR sandwiches (SV without pinning layers) or pseudo-SVs for GMR memories. The need for spintronic devices combining magnetic and semiconductor materials is covered in Chapter 11.

SPIN INJECTION AND SPINTRONIC DEVICES

Conventional electronic devices are based on charge transport. Prospective spintronic devices utilize the direction and coupling of the spin of the electron in addition to or in place of the charge. To be commercially useful, a spintronic device has to work at room temperature and be compatible with existing semiconductor-based electronics. Almost every imaginable spintronic device should have means of spin injection, manipulation and detection. Well-known sources of spin-polarized electrons are ferromagnetic materials (FMs). The magnetic field of the FM interacts with the spins of electrons and as a result, the majority of electrons are in the states such that their spins are aligned with the local magnetization. Spin-polarized current injection was achieved from FM into superconductors,[9] from FM into normal metals,[10] between two FM separated by a thin insulating film,[11] from a normal metal using magnetic semiconductors as an aligner into non-magnetic semiconductors,[12–14] and hole injection from a p-type FM into a non-magnetic semiconductor.[15] For in-depth review of spin injection issues see Chapter 11. The main obstacle to effective electrical injection is the conductivity mismatch between the FM electrode and the semiconductor.[16] The effectiveness of the spin-injection depends on the ratio of the (spin-dependent) conductivities of the FM and non-ferromagnet (NFM) electrodes, σ_F and σ_N, respectively. When $\sigma_F \leq \sigma_N$ as in the case of a typical metal, then efficient and substantial spin injection can occur, but when the NFM electrode is a semiconductor, $\sigma_F \gg \sigma_N$ and the spin-injection efficiency will be very low. Only for a ferromagnet where the conduction electrons are 100% spin-polarized, can efficient spin injection be expected in diffusive transport.

Insertion of a tunnel contact (T) at an FM-to-normal conductor interface can be a solution of the conductivity mismatch problem.[17] Two possible configurations considered were an FM–T–semiconductor and a Schottky barrier diode. A 2% room temperature tunneling spin injection was achieved from Fe into GaAs in the Schottky diode configuration.[18] The Schottky barrier formed at the Fe/GaAs interface provides a natural tunnel barrier for injecting spin-polarized electrons under reverse bias. A 30% injection efficiency was achieved from an Fe contact into a semiconductor light emitting diode structure ($T = 4.5$ K) and persisted to almost room temperature[19] (4% at $T = 240$ K). Measurements of spin-polarization of the current transmission across an FM–insulator–2DEG junction yield 40%, with little dependence over the range 4–295 K.[20]

Tunnel injection of "hot" (energy much greater than E_F) electrons into a ferromagnet can be used to create spin-polarized currents. Because inelastic mean free paths for majority and minority electrons differ significantly, hot electron passage through a 3-nm Co layer can result in 90% spin-polarized currents.[21–23] The highly polarized current then can be used for further injection into the semiconductor. The disadvantage of hot electron injection is that the overall efficiency is low.

NEW MATERIALS FOR SPINTRONIC DEVICES

Progress in new materials engineering and research is very important because, as mentioned above, one can expect efficient spin injection if the source of spin-polarized electrons is 100% spin-polarized. Measurements in a variety of metals, half-metals, metallic binary oxides, Heusler alloys and other compounds have shown a 35–90% range of spin-polarization.[24–27]

Materials combining ferromagnetic and semiconducting properties, magnetic semiconductors, can be a very attractive option as a source of spin-polarized carriers because there is no interface problem or conductivity mismatch. To achieve large spin-polarization in semiconductors, the Zeeman splitting of the conduction (valence) band must be greater than the Fermi energy, E_F, of the electrons (holes). In concentrated materials, this occurs easily because the net magnetization upon ordering is proportional to the concentration of magnetic species. If the concentration of magnetic species is low, ~5% or below, to produce large polarization large externally applied magnetic fields and low temperatures are required.

Mixed-valence manganese perovskites with the general formula $A_{1-x}B_xMnO_3$ (where A = La, Nd or Pr, and B = Ca, Ba or Sr) have been extensively studied[28] for their magnetic and transport properties resulting in "colossal magnetoresistance" (CMR) observed at the vicinity of the Curie temperature, T_C. Applying external magnetic fields and lowering the temperature promotes spin ordering. Magnetoresistance values of more than 400% at low temperatures were measured in all-oxide SV in which $La_{0.7}Sr_{0.3}MnO_3$ electrodes were separated by a thin insulating layer, and spin-polarization was estimated to be at least 83% at the Fermi level.[29] For many of the mixed-valence manganese perovskites, T_C is above room temperature, but photoemission data show that the spin-polarization decreases to almost zero at room temperature.

Europium chalcogenides with the general formula EuX (X = O, S, Se, Te) in which the magnetic ion Eu^{2+} resides on every lattice site, have low T_C (~80 K).[30]

Diluted magnetic semiconductors (DMSs) are alloys in which atoms of a group-II element of a II–VI compound (CdTe or ZnS) semiconductor are randomly replaced by magnetic atoms (e.g., Mn). However, II–VI-based DMS have been very difficult to dope to create p- and n-type semiconductors used in electronic applications. At high magnetic atom concentration and low temperature, anti-ferromagnetic interaction between magnetic ions creates a magnetically ordered phase. Below T_C, DMS exhibit ferromagnetic behavior. So far for II–VI semiconductors (e.g., CdMnTe), the ferromagnetic phase was observed only below 2 K.[31] The non-equilibrium growth conditions of low-temperature molecular beam epitaxy allowed the successful preparation of (In, Mn)As[32] for which hole-induced ferromagnetic ordering was detected below 35 K.[33] A GaAs-based DMS was fabricated in 1996[34] and T_C around 110 K was reported for some samples in 1998.[35] For a detailed review of the properties of ferromagnetic III–V semiconductors see Ohno.[36] GaMnAs heterostructures can be epitaxially grown with abrupt interfaces and with atomically controlled layered thickness.[37] Tanaka and Higo reported large tunneling magnetoresistance in GaMnAs/AlAs/GaMnAs ferromagnetic semiconductor tunneling junctions.[38] Although the total change in the magnetoresistance was 44% at 4.2 K the TMR ratio due to the SV effect was estimated to be only 15–19%.[39] From a technological point of view the use of semiconducting magnetic elements may reduce the large currents required in all-metal SVs. The current state of research in the area of ferromagnetic semiconductors is given in report.[40] So far Mn is the only successfully used ion for doping DMS. Some groups stared to use Cr ions to dope GaAs to grow new DMS $Ga_{1-x}Cr_xAs$ by MBE.[41]

A thin film magnetic system consisting of nanoscale $Mn_{11}Ge_8$ ferromagnetic clusters ($T_C \sim$ room temperature) embedded into a Mn_xGe_{1-x} DMS semiconductor matrix, exhibits magnetoresisitance at low fields (2 kOe) and at low temperatures (22 K).[42] A group IV DMS Mn_xGe_{1-x} was successfully grown using MBE in which the Curie temperature was found linearly dependent on the Mn concentration from 25 to 116 K.[43] Theoretical calculations predict even higher T_C than was measured. T_C calculations[44] predicting above room temperature ferromagnetism prompted active interest in DMS GaMnN.[45] Ferromagnetic behavior was recently confirmed at 300 K[46,47] and T_C is estimated to be 940 K.[48]

Half-metallic ferromagnets (HMF) are potential sources of spin-polarized electrons. The Fermi level of HMF intersects the majority spin electron band while the minority band has an energy gap near the Fermi level. The HMFs have simultaneously both metallic and semiconducting characteristics, and theory predicts that the conduction electrons are 100% spin-polarized.[49] As a result, the magnetoresistance in magnetic multilayers or tunneling junctions is expected to be significantly higher than with conventional FMs.

Heusler alloys have T_C slightly above room temperature[50] and spin-polarization of 58%.[25]

Half-metallic oxides were predicted[51] to be potential sources of fully spin-polarized electrons. Spin-resolved photoemission from polycrystalline CrO_2 films have shown[52] a spin-polarization of nearly +100% with confirmation by point contact Andreev reflection measurements.[53,54] Large TMR effects were observed for magnetite Fe_3O_4[55] ($T_C \sim$ 860 K) at helium temperatures, but at room temperature the effect is less than 1%.

Transition metal pnictides MnAs, MnSb, CrAs, CrSb can be easily incorporated into existing semiconductor technology if proven to be useful as spin-polarized sources. For magnetooptical device applications, such as an optical isolator, high optical transmission and a large magnetooptic effect must be realized at room temperature. The GaAs/MnAs nanocluster system was fabricated and characterized by Shimizu and Tanaka.[56] Zinc blende structure CrAs thin films were synthesized on GaAs (001) substrates by MBE, and show a ferromagnetic behavior at room temperature. Calculations predict a highly spin-polarized electronic band structure.[57] Thin films of CrSb grown by MBE on GaAs, (Al, Ga)Sb and GaSb are found to have a zinc blende structure and $T_C > 400$.[58] A thousand fold magnetoresistance effect was discovered in granular MnSb films[59] at room temperature with low magnetic fields (less than 0.5 T). A 20% positive, photoinduced magnetoresistance effect was observed in GaAs with inclusions of MnSb nanomagnets when irradiated with photons with energies above the band gap of GaAs. This effect is presumably due to an enhancement of the tunneling probability between MnSb islands by photogenerated carriers in the GaAs matrix.[60]

Epitaxial thin films of the DMS oxide, Mn-doped ZnO, fabricated by pulse-laser deposition showed considerable magnetoresistance at low temperature.[61] Following the prediction that ZnO has a T_C above room temperature[44] and would become ferromagnetic by doping with $3d$ transition elements, intensive combinatorial work began in Japan on dilute magnetic oxides. Room temperature ferromagnetism was reported in transparent Co-doped (up to 8%) Ti_2O anatase films.[62] This material is transparent to visible light and might be of great importance to optoelectronics.

Electronic structure calculations for La-doped CaB_6[63] ($T_C \sim 900$ K) showed that it has a semiconductor band structure and can be considered as a new semiconducting material for spin electronics.

OPTICALLY AND ELECTRICALLY CONTROLLED FERROMAGNETISM

Optically and electrically controlled ferromagnetism is now a low-temperature effect, but extension to higher temperatures may have important implications in areas ranging from optical storage to photonically and electrically driven micromechanical elements.

Ferromagnetism was induced by photogenerated carriers in an MBE-grown p-(In, Mn)As/GaSb film at temperatures below 35 K.[64] The order was preserved even after the light was switched off and recovered to the original paramagnetic condition above 35 K. The results were explained in terms of hole transfer from GaSb to InMaAs which then enhanced the ferromagnetic spin exchange between Mn ions in the heterostructures.

Electric-field control of hole-induced ferromagnetism was demonstrated in (In, Mn)As using an insulating-gate field effect transistor (FET) at temperatures below 20 K[65] in 2000. Manganese substitutes indium and provides a localized magnetic moment and a hole. These holes mediate magnetic interactions resulting in ferromagnetism. Changing the hole concentration by applying a gate voltage modifies the ferromagnetic properties in the DMS below the transition temperature. It was also found that the new group IV DMS, Mn_xGe_{1-x}, allows control over ferromagnetic order by applying a 0.5 V gate voltage.[43]

CURRENT-INDUCED MAGNETIZATION SWITCHING AND SPIN WAVE GENERATION

Current-induced magnetization switching and spin wave generation are recently discovered effects allowing manipulation of the magnetization of FM layers in heterostructures. The change in scattering of the electrons traversing alternating ferromagnetic and non-magnetic multilayers depends on the relative orientation of the magnetization. The scattering of the electrons within the alternating layers of FM and regular metals can affect the moments in the magnets. Theoretical calculations indicate[66-69] that spin-polarized currents perpendicular to the layers can transfer angular momentum between layers, causing torque on the magnetic moments, and as a result, causing rotation and possibly even precession of the layer magnetization and high frequency switching. A few independent experiments have demonstrated current-induced magnetization rotation.[70-76] Current-induced magnetization switching has potential applications to high speed, high density GMR-based MRAM as a convenient writing process.[77] Other possible applications might include spin-filter devices and spin-wave emitting diodes.[78] In the presence of a large magnetic field, the spin rotation can become a high frequency coherent precession of the moments. This effect has been labeled spin amplification by simulated emission of radiation (SWASER), was predicted by Berger,[66] and the feasibility of it was studied by Tsoi et al.[76]

OPTICAL MANIPULATIONS OF SPINS IN SEMICONDUCTORS

Spin transport in bulk semiconductors and across junctions in heterostructures has been studied by optoelectronic manipulation of spins allowing spatial selectivity and temporal resolution.[79] Optically excited coherent spin states in semiconductors[80-83] have been shown to exist for more than 100 ns. Optical pulses are used both to create a superposition of the basis spin states defined by an applied magnetic field, and to follow the phase, amplitude, and location of the resulting electronic spin precession (coherence) in bulk semiconductors, heterostructures, and quantum dots. Data show that spin coherence across heterojunctions[84] can be preserved when a "pool" of coherent spins is crossing the interface between two semiconductors. A phase shift of spins on opposite sides of the interface can be set by the difference in electron g-factors between the two materials, and can be controlled by utilizing epitaxial growth techniques. More recent measurements have established an increase in spin injection efficiency with bias-driven transport: relative increases of up to 500% in electrically biased structures, and 4000% in p–n junctions with intrinsic bias have been observed[85] relative to the unbiased interfaces. Another important aspect for the development of spin-based electronics is the effect of defects on spin-coherence. Studies of electron spin coherence times in GaN epilayers, a III–V semiconductor, having eight orders of magnitude higher concentration of charged dislocations as in GaAs, reach ~20 ns at $T = 5$ K, with observable coherent precession at room temperature.[1,86]

POTENTIAL SPINTRONIC DEVICES

A spin-light emitting diode (spin-LED) is a device used primarily to measure the effectiveness of the injection of spin-polarized currents into semiconductor heterostructures. In a spin-LED, recombination of spin-polarized carriers results in the emission of right (σ^-) or left (σ^+) circularly polarized light in the direction normal to the surface according to selection rules.[87] Polarization analysis of the resulting electroluminescence (EL) provides quantitative measurement of the injection efficiency. A spin-LED was used to measure the electrical spin injection of electrons[13,14] into a GaAs quantum well at low temperatures (4.2 K) and with an external magnetic field up to 8 T using a DMS spin-aligner concept.[12]

The spin-polarized field effect transistor (spin-FET) was proposed by Datta and Das,[88] see Chapter 11 for in-depth description.

Introduction of FMs into resonant tunneling diodes (spin-RTDs) can greatly enhance their functionality. The RTD is a vertical tunneling diode with the vertical dimensions produced by growth and the lateral dimensions produced by lithography. If a quantum well is placed between two thin barriers, the tunneling probability is greatly enhanced when the energy level in the quantum well coincides with the Fermi energy (resonant tunneling). The typical dimensions in the tunneling direction are a few atomic layers, and this determines the current and power dissipation. To produce lower power devices, smaller dimensions are required and issues of control of the uniformity of the tunneling layer become very important because the tunneling current depends exponentially on the thickness of the tunnel barrier. Magnetization controlled resonance tunneling in GaAs/ErAs[89] RTDs shows splitting and enhancement of the resonant channels, which depend on the orientation of the external magnetic field with respect to the interface. Theoretical calculations predict that polarization of the transmitted beam can achieve 50%[90] or even higher.[91] Spin-RTDs can be used both as spin filters and energy filters. Current–voltage characteristics of AlAs/GaAs/AlAs double-barrier RTDs with ferromagnetic p-type (Ga, Mn)As on one side and p-type GaAs on the other, have been studied.[35,92] A series of resonant peaks have been observed in both polarities, i.e., injecting holes from p-type GaAs and from (Ga, Mn)As. When holes are injected from the (Ga, Mn)As side, spontaneous resonant peak splitting has been observed below the ferromagnetic transition temperature of (Ga, Mn)As without magnetic field. The temperature dependence of the splitting is explained by the spontaneous spin splitting in the valence band of ferromagnetic (Ga, Mn)As. Introducing a ferromagnetic quantum well in a ferromagnetic junction is shown to greatly enhance the tunneling magnetoresistance effect,[93] due to spin filtering as well as energy filtering.

Some other new spintronic devices are highly speculative and have not been implemented yet. One proposal[94] considers the possibility of constructing unipolar electronic devices by utilizing ferromagnetic semiconductor materials with variable magnetization directions. Such devices should behave very similarly to p–n diodes and bipolar transistors, and they could be applicable for magnetic sensing, memory and logic. Another theoretical device[95] is a spin-polarized p–n junction with spin-polarization induced either optically (in which case it is a solar cell) or electronically to produce majority or minority carriers. It is suggested that spin-polarization can be injected through the depletion layer by both minority and majority carriers, making semiconductor devices such as spin-polarized solar cells and bipolar transistors feasible. Spin-polarized p–n junctions allow for spin-polarized current generation, spin amplification, voltage control of spin-polarization, and a significant extension of the spin diffusion range. Spin filters can possibly be constructed by carefully engineering[96] ordered interfaces between FM and S, and a ballistic spin-filter transistor[97] can be added to the growing list of possible spintronic devices.

QUANTUM COMPUTATION AND SPINTRONICS

The introduction of coherent spins into ferromagnetic structures could lead to quantum spintronics. As an example, some proposed quantum computation schemes rely on the controllable interaction of coherent spins with FMs to produce quantum logic operations.[98] There is also experimental evidence that FMs can be used to imprint nuclear spins in semiconductors, offering a way of manipulating and storing information at the atomic scale.[99] Semiconductor-based quantum spin electronics is focused on developing a solid-state quantum information processing device. Spin lifetimes of nuclear spins[100–102] exceed those of electrons by at least several orders of magnitude and have been proposed as candidates for storing both classical and quantum information. It has been proposed that the spin of an electron confined to quantum dot is a promising candidate for quantum bits, and that arrays of quantum dots can be used in principle to implement a large scale quantum computer.[103,104] All proposals briefly mentioned above are very far from being implemented.

To summarize, future spintronic devices can be used as magnetic memories, magnetic field sensors, spin-based switches, modulators, isolators, transistors, diodes, and perhaps some novel devices without conventional analogues and that can perform logic functions in novel ways. As always, the key question is whether any potential benefit of such technology will be worth the production costs. Hopefully by the time the second volume will be published some of these questions remaining today will be answered, and may be a prototype of a spin-based quantum computer will be demonstrated...

ACKNOWLEDGMENTS

This Introduction is loosely based on the review article.[1] I would like to express my gratitude to Stuart Wolf, David Awschalom, Robert Buhrman, Jim Daughton, Stephan von Molnár, and Daryl Treger for their contributions to the original paper. Some of the material in this Introduction has been previously included in Ref. [105].

REFERENCES

1. S. Wolf, et al., Spintronics: A spin-based electronics vision for the future. *Science* **294**, 1488 (2001).
2. H. Ohno, F. Matsukura, and Y. Ohno, Semiconductor spin electronics., *JSAP Int.*, **5**, 4 (2002).
3. G. Prinz, Magnetoelectronics. *Science* **282**, 1660 (1998).
4. G. Prinz, Magnetoelectronics applications. *J. Magn. Magn. Mater.* **200**, 57 (1999).
5. C. Tsang, et al., Design, fabrication and testing of spin-valve read heads for high density recording. *IEEE Trans. Magn.* **30**, 3801 (1994).
6. B. Dieny, et al., *J. Appl. Phys.* **69**, 4774 (1991).
7. S. S. P. Parkin and D. Mauri, *Phys. Rev. B* **44**, 7131 (1991).
8. W. F. Egelhoff, Jr., et al., *J. Vac. Sci. Technol. B* **17**, 1702 (1999).
9. R. Meservey, P. M. Tedrow, and P. Fulde, Magnetic field splitting of the quasiparticle states in superconducting aluminium films. *Phys. Rev. Lett.* **25**, 1270 (1970).
10. M. Johnson and R. Silsbee, Interfacial charge-spin coupling: Injection and detection of spin magnetization in metals. *Phys. Rev. Lett.* **55**, 1790 (1985).

11. M. Jullière, Tunneling between ferromagnetic films. *Phys. Lett. A* **54**, 225 (1975).
12. M. Oestreich, *et al.*, Spin injection into semiconductors. *Appl. Phys. Lett.* **74**, 1251 (1999).
13. R. Fiederling, *et al.*, Injection and detection of a spin-polarized current in a light-emitting diode. *Nature* **402**, 787 (1999).
14. B.T. Jonker, *et al.*, Robust electrical spin injection into a semiconductor heterostructures. *Phys. Rev. B* **62**, 8180 (2000).
15. Y. Ohno, *et al.*, Electrical spin injection in a ferromagnetic semiconductor heterostructure. *Nature* **402**, 790–792 (1999).
16. G. Schmidt, *et al.*, Fundamental obstacle for electrical spin injection from a ferromagnetic metal into a diffusive semiconductor. *Phys. Rev. B* **62**, R4790 (2000).
17. E. I. Rashba, Theory of electrical spin injection: Tunnel contacts as a solution of the conductivity mismatch problem. *Phys. Rev. B* **62**, R16267 (2000).
18. H. J. Zhu, *et al.*, Room-temperature spin injection from Fe in to GaAs. *Phys. Rev. Lett.* **87**, 16601 (2001).
19. A. Hanbiki, *et al.*, Electrical spin injection from a magnetic metal/tunnel barrier contact into a semiconductor. *Appl. Phys. Lett.* **80**, 83 (2002).
20. P. R. Hammar and M. Johnson, Spin-dependent current transmission across a ferromagnet–insulator–two-dimensional gas junction. *Appl. Phys. Lett.* **79**, 2591 (2001).
21. D. J. Monsma, J. C. Lodder, Th. J. A. Popma, and B. Dieny, Perpendicular hot electron spin-valve effect in a new magnetic field sensor: The spin-valve transistor. *Phys. Rev. Lett.* **74**, 5260 (1995).
22. W. H. Rippard and R. A. Buhrman, Spin-dependent hot electron transport in Co/Cu thin films. *Phys. Rev. Lett.* **84**, 971 (2000).
23. R. Jansen, *et al.*, The spin-valve transistor: Fabrication, characterization, and physics. *J. Appl. Phys.* **89**, 7431 (2001).
24. J.W. Dong, *et al.*, Spin-polarized quasiparticle injection devices using $Au/Yba_2Cu_3O_7/LaAlO_3/Nd_{0.7}Sr_{0.3}MnO_3$ heterostructures. *Appl. Phys. Lett.* **71**, 1718 (1997).
25. R. J. Soulen, *et al.*, Measuring the spin polarization of a metal with a superconducting point contact. *Science* **282**, 85 (1998).
26. D. C. Worledge and T. H. Geballe, Spin-polarized tunneling in $La_{0.67}Sr_{0.33}MnO_3$. *Appl. Phys. Lett.* **76**, 900 (2000).
27. D. J. Monsma and S. S. P. Parkin, Spin polarization of tunneling current from ferromagnet/Al_2O_3 interfaces using copper-doped aluminum superconducting films. *Appl. Phys. Lett.* **77**, 720 (2000).
28. J. M. Coey, M. Viret, and S. von Molnár, Mixed-valence manganites. *Adv. Phys.*, **48**, 167 (1999).
29. M.Viret, *et al.*, Spin polarized tunneling as a probe of half metallic ferromagnetism in mixed-valence manganites. *J. Magn. Magn. Mater.* **198–199**, 1 (1999).
30. F. Holtzberg, T. McGuire, S. Methfessel, and J. Suits, Effect of electron concentration on magnet exchange interactions in rare earth chalcogenides. *Phys. Rev. Lett.* **13**, 18 (1964).
31. A.Haury, *et al.*, Observation of a ferromagnetic transition induced by two-dimensional hole gas in modulation-doped CdMnTe quantum wells. *Phys. Rev. Lett.* **79**, 511 (1997).
32. H. Munekata, H. Ohno, S. von Molnär, A. Segmüller, L.Chang, and L. Esaki, Diluted magnetic III–V semiconductors. *Phys. Rev. Lett.* **63**, 1849 (1989).
33. H. Ohno, H. Munekata, S. von Molnär, and L. Chang, Magnetotransport properties of p-type (in, Mn)As diluted magnetic semiconductors. *Phys. Rev. Lett.* **68**, 2664 (1992).
34. H. Ohno, *et al.*, (Ga,Mn)As: A new diluted magnetic semiconductor based on GaAs. *Appl. Phys. Lett.* **69**, 363–365 (1996).
35. H. Ohno, Making nonmagnetic semiconductor ferromagnetic. *Science*, **281**, 951 (1998).
36. H. Ohno, Properties of ferromagnetic III–V semiconductors. *J. Magn. Magn. Mater.* **200**, 100 (1999).
37. M. Tanaka, *J. Vac. Sci. Technol. B* **16**, 2267 (1998); M. Tanaka, *J. Vac. Sci. Technol. A* **18**, 1247 (2000).
38. M. Tanaka and Y. Higo, Large tunneling magnetoresistance in GaMnAs/AlAs/GaMnAs ferromagnetic semiconductor tunneling junctions. *Phys. Rev. Lett.* **87**, 26602 (2001).
39. T.Hayashi *et al.*, *J. Cryst. Growth* **201/202**, 660 (1999).
40. H. Ohno, F. Matsukura, and Y. Ohno, Semiconductor spin electronics. *JSAP Int.*, **5**, 4 (2002).
41. H. Saito, *et al.*, Transport properties of a III–V $Ga_{1-x}Cr_x$ As Diluted Magnetic Semiconductor. In Spintech1 International Conference, Maui, Hawaii, May 2001.

42. Y. D. Park, *et al.*, Magnetoresistance of Mn:Ge ferromagnetic nanoclusters in a diluted magnetic semiconductor matrix. *Appl. Phys. Lett.* **78**, 2739 (2001).
43. Y. D. Park, *et al.*, A group IV ferromagnetic semiconductor Mn_xGe_{1-x}. *Science* **295**, 651 (2002).
44. T. Dietl, F. Matsukura, J. Cibert, and D. Ferrand, Zener model description of ferromagnetism in zinc-blende magnetic semiconductors. *Science* **287**, 1019 (2000).
45. M. Overberg, *et al.*, Epitaxial growth of dilute magnetic semiconductors: GaMnN and GaMnP. *Mater. Res. Soc. Symp. Proc.* **674**, T6.5.1 (2001).
46. M.L. Reed *et al.*, Room temperature magnetic (Ga, Mn)N: A new material for spin electronic devices. *Mater. Lett.* **51**, 500 (2001).
47. M. L. Reed, *et al.*, Room temperature ferromagnetic properties of (Ga, Mn)N. *Appl. Phys. Lett.* **79**, 3473 (2001).
48. S. Sonoda, *et al.*, Molecular beam epitaxy of wurtzite (Ga, Mn)N films on sapphire (0001) showing the ferromagnetic behaviour at room temperature. *J. Cryst. Growth* **237–239**, 1358 (2002).
49. R. A. de Groot, *et al.*, New class of materials: Half-metallic ferromagnets. *Phys. Rev. Lett.* **50**, 2024 (1983).
50. J. W. Dong, *et al.*, MBE growth of ferromagnetic single crystal Heusler alloys on (001) (001) $Ga_{1-x}In_xAs$. *Physica E* **10**, 428 (2001).
51. K. Schwartz, CrO_2 predicted as a half-metallic ferromagnet. *J. Phys. F* **16**, L211 (1986).
52. K. P. Kaemper, *et al.*, CrO_2 — A new half-metallic ferromagnet. *Phys. Rev. Lett.* **59**, 2788 (1988).
53. R. J. Soulen, *et al.*, Measuring the spin polarization of a metal with a superconducting point contact. *Science* **282**, 85 (1998).
54. Y. Ji, *et al.*, Determination of the spin polarization of half-metallic CrO_2 by point contact Andreev reflection. *Phys. Rev. Lett.* **86**, 5585 (2001).
55. J. J. Versluijs, M. A. Bari, and J. M. D. Coey, Magnetoresistance of half-metallic oxide nanocontacts. *Phys. Rev. Lett.* **87**, 26601 (2001).
56. H. Shimizu and M. Tanaka, Magneto-optical properties of semiconductor-based superlattices having GaAs with MnAs nanoclusters. *J. Appl. Phys.* **89**, 7281 (2001).
57. H. Akinaga, T. Manago, and M. Shirai, Material design of half-metallic zinc-blende CrAs and the synthesis by molecular beam epitaxy. *Jpn. J. Appl. Phys.* **39**, L1120 (2000).
58. J. Zhao, *et al.*, Room-temperature ferromagnetism in zincblende CrSb grown by molecular beam epitaxy. *Appl. Phys. Lett.* **79**, 2776 (2001).
59. H. Akinaga, M. Mizuguchi, K. Ono, and M. Oshima, Room-temperature thousandfold magnetoresistance change in MnSb granular films: Magnetoresistive switch effect. *Appl. Phys. Lett.* **76**, 357 (2000).
60. H. Akinaga, M. Mizuguchi, K. Ono, and M. Oshima, Room-temperature photoinduced magnetoresistance effect in GaAs including MnSb nanomagnets. *Appl. Phys. Lett.* **76**, 2600 (2000).
61. T. Fukumura, Z. Jin, A. Ohtomo, H. Koinuma, and M. Kawasaki, An oxide-diluted magnetic semiconductor Mn-doped ZnO. *Appl. Phys. Lett.* **75**, 3366 (2000).
62. Y. Matsumoto, *et al.*, Room-temperature ferromagnetism in transparent transition metal-doped titanium dioxide. *Science* **291**, 854 (2001).
63. H. J. Tromp, *et al.*, CaB_6: A new semiconducting material for spin electronics. *Phys. Rev. Lett.* **87**, 16401 (2001).
64. S. Koshihara, *et al.*, Ferromagnetic order induced by photogenerated carriers in magnetic III–V semiconductor heterostructures of (In, Mn)As/GaSb. *Phys. Rev. Lett.* **78**, 4617 (1997).
65. H. Ohno, *et al.*, *Nature* **408**, 944 (2000).
66. L. Berger, Emission of spin waves by a magnetic multiplayer traversed by a current. *Phys. Rev. B* **54**, 9353 (1996).
67. J. C. Slonczewski, Current-driven excitation of magnetic multilayers. *J. Magn. Magn. Mater.* **159**, L1 (1996).
68. J. C. Slonczewski, Excitation of spin waves by an electric current. *J. Magn. Magn. Mater.* **195**, L261 (1999).
69. Ya. B. Bazaliy, B. Jones, and S.-C. Zhang, Modification of the Landau-Lifshitz equation in the presence of a spin-polarized current in colossal- and giant-magnetoresistive materials. *Phys. Rev. B* **57**, R3213 (1998).
70. M. Tsoi, *et al.*, Excitation of a magnetic multiplayer by an electric current. *Phys. Rev. Lett.* **80**, 4281 (1998); **81**, 493 (1998) (E).

71. E. B. Myers, D. C. Ralph, J. A. Katine, R. N. Louie, and R. A. Buhrman, Current-induced switching of domains in magnetic multiplayer devices. *Science* **285**, 867 (1999).
72. J. E. Wegrowe, *et al.*, Current-induced magnetization reversal in magnetic nanowires. *Europhys. Lett.* **45**, 626 (1999).
73. J. Z. Sun, Current-driven magnetic switching in manganite trilayer junctions. *J. Magn. Magn. Mater.* **202**, 157 (1999).
74. S. M. Rezende, *et al.*, Magnon excitation by spin injection in thin Fe/Cr/Fe films. *Phys. Rev. Lett.* **84**, 4212 (2000).
75. J. A. Katine, *et al.*, Current-driven magnetization reversal and spin-wave excitation in Co/Cu/Co pillars. *Phys. Rev. Lett.* **84**, 3149 (2000).
76. M. Tsoi, A. G. M. Jansen, J. Bass, W.-C. Chiang, V. Tsoi, and P. Wyder, Generation and detection of phase-coherent current-driven magnons in magnetic multilayers. *Nature* **406**, 46–48 (2000).
77. J. Slonszewski, U.S. Patent 5,695,864. December, 9, 1997.
78. L. Berger, Multilayer as a spin-wave emitting diode. *J. Appl. Phys.* **81**, 4880 (1997).
79. D. D. Awschalom and J. M. Kikkawa, Electron spin and optical coherence in semiconductors. *Phys. Today* **52**, 33 (1999).
80. J. M. Kikkawa, D. D. Awschalom, I. P. Smorchkova, and N. Samarth, *Science* **277**, 1284 (1997).
81. J. M. Kikkawa and D. D. Awschalom, *Phys. Rev. Lett.* **80**, 4313 (1998).
82. J. M. Kikkawa and D. D. Awschalom, Lateral drag of spin coherence in gallium arsenide. *Nature* **397**, 139 (1999).
83. J. A. Gupta, X. Peng, A. P. Alivisatos, and D. D. Awschalom, *Phys. Rev. B* **59**, R10421 (1999).
84. I. Malajovich, *et al.*, *Phys. Rev. Lett.* **84**, 1015 (2000).
85. I. Malajovich, *et al.*, *Nature* **411**, 770 (2001).
86. B. Beschoten, *et al.*, *Phys. Rev. B* **63**, R121202 (2001).
87. B. T. Jonker, U.S. Patent 5,874,749.
88. S. Datta and B. Das, Electronic analog of the electro-optic modulator. *Appl. Phys. Lett.* **56**, 665 (1990).
89. D. E. Brenner, *et al.*, *Appl. Phys. Lett.* **67**, 1268 (1995).
90. E. A. de Andrada e Silva and G. C. La Rocca, Electron spin polarization by resonant tunneling, *Phys. Rev. B* **59**, R15583 (1999).
91. A. G. Petukhov, D. O. Demchenko, and A. N. Chantis, Spin-dependent resonant tunneling in double-barrier magnetic heterostructures. *J. Vac Sci.Technol. B* **18**, 2109 (2000).
92. H. Ohno, N. Akiba, F. Matsukura, A. Shen, K. Ohtani, and Y. Ohno, Spontaneous splitting of ferromagnetic (Ga, Mn)As valence band observed by resonant tunneling spectroscopy. *Appl. Phys. Lett.* **73**, 363 (1998).
93. T. Hayashi, M. Tanaka, and A.Asamitsu, Tunneling magnetoresistance of a GaMnAs-based double barrier ferromagnetic junction. *J. Appl. Phys.* **87**, 4673 (2000).
94. M. E. Flatte and G. Vignale, Unipolar spin diodes and transistors. *Appl. Phys. Lett.* **78**, 1273–1275 (2001).
95. I. Žutić, *et al.*, Spin injection through the depletion layer: A theory of spin-polarized $p-n$ junctions and solar cells. *Phys. Rev. B* **64**, 121201R (2001).
96. G. Kirczenow, Ideal spin filters: A theoretical study of electron transmission through ordered and disordered interfaces between ferromagnetic metals and semiconductors. *Phys. Rev. B* **63**, 54422 (2001).
97. D. Grundler, Ballistic spin-filter transistor. *Phys. Rev. B* **63**, R161307 (2001).
98. D. P. DiVincenzo, *et al.*, *Nature* **408**, 339 (2000).
99. R. K. Kawakami, *et al.*, *Science* **294**, 131 (2001).
100. B. Kane, *Nature* **393**, 133 (1998).
101. R. Vrijen, *et al.*, Electron spin-resonance transistors for quantum computing in silicon–germanium heterostructures. *Phys. Rev. A* **62**, 12306, (2000).
102. T. Ladd, J. Goldman, F. Yamaguchi, Y. Yamamoto. E. Abe, and K. Itoh, An all silicon quantum computer, quant-ph/0109039.
103. D. Loss and D. P. DiVincenzo, *Phys. Rev. A* **57**, 120 (1998).
104. G. Burkard, H. Engel, and D. Loss, *Fortschr. Phys.* **48**, 965 (2000).
105. S. Wolf, A. Chtchelkanova, and D. Treger, in *The Handbook on Nanoscience, Engineering, and Technology*, W. Goddard III, D. Brenner, S. Lyshevski, and G. Iafrate, eds., CRC Press, Boca Raton, FL, 2002, (in press).

1

Electron Spins in Ionic Molecular Structures

Gerald F. Dionne*

1.1. INTRODUCTION

Magnetism in ionic compounds has been investigated from the earliest decades of the twentieth century, culminating during the middle years with a plethora of research results from a variety of academic, government, and industrial laboratories. Motivation behind these efforts came mainly from two directions: microwave spectroscopy of individual magnetic ions in paramagnetic crystals that served as the foundation of amplification devices, e.g., the *maser*, that were intended for advanced radar and communications applications, and the multidiscipline investigations of ferrimagnetic oxide materials (ferrites) for high frequency and microwave applications, as well as for the burgeoning field of information storage and the more conventional functions requiring high permeability combined with electrical insulating properties for electromagnet and inductor cores. The scientific endeavors over these years were staggering in scope as is evidenced by the volumes of literature produced during the period following World War II and extending into the 1970s. More recently three advances that involve the transport of charge carriers has re-energized the field of magnetic oxides. Among these initiatives is high-temperature superconductivity, magnetoresistance, and polarized-spin transport found in select transition-metal oxide compounds that offer the promise of integration for the enabling of future electronic technologies.

In this chapter, the physical concepts that underlie the magnetic and electrical behaviors of transition-metal oxides will be described with the intention of providing starting points for the informed scientist and research engineer seeking to acquire a working understanding of oxide magnetism and spin-transport phenomena in oxides and similar compounds. The review will begin with the atomic origins of the positive magnetic ions (cations) as located in the Periodic table, and the determination of their angular momentum ground states. A sampling of lattice sites formed by negatively charged ligands (anions) with commonly occurring symmetries are then explained. When a cation resides in a lattice site, the electric field (crystal field) of the negative anions couples to the orbital component of the angular momentum and can cause a lifting of its orbital

---

*Lincoln Laboratory, Massachusetts Institute of Technology, Cambridge, Massachusetts.

degeneracy through a Stark effect. This effect is termed "orbital quenching." The degree to which the orbital degeneracy is removed is reflected by the extent to which its excited states are split from the ground state. For complete quenching, the ground state is an orbital singlet, and the energies of excitation to the higher states would tend to infinity. In the presence of a magnetic field (which can also be simulated by internally generated by exchange interactions), spin–orbit coupling that tends to bind the directions of the orbital moment and the spin moment can have major effects on the properties of the ion in its site. In situations where the magnetic-field coupling to the spin competes with the crystal-field coupling to the orbital moment, local destabilization can have directional features that give rise to magnetic anisotropy and magnetostriction, as well as influence high-frequency electromagnetic properties.

The discussion will proceed to the examination of systems where high densities of magnetic ions interact through chemical bonding with their anion neighbors to establish superexchange coupling, i.e., the basis for spontaneous magnetism. In the case of ferromagnetic parallel spin alignment, the opportunity for polarized spin transport is also available. Examples of magnetic oxide systems that have achieved utilitarian value are spinel and garnet ferrites. Promising materials for exploiting spin transport in magnetoresistance and superconductivity are the manganite and cuprate perovskites, respectively. Partially polarized spin transport at room temperature can also be achieved in select oxides with metallic conduction properties.

As the title of the chapter indicates, the underlying physical concepts that determine the properties of these materials are described from the standpoint of local sites of the individual magnetic ions rather than from traditional periodic lattice band theory. This approach is prompted by the need to understand the role of random substitutions into these materials that cannot easily be analyzed by momentum–space formalisms. Furthermore, with the trend in technology being driven towards increasingly smaller dimensionality, the local-site molecular orbital approximations could become more representative of the physical systems on which future magnetic materials engineering will be based.

1.2. MAGNETIC CATIONS

To establish the pattern of this chapter, a logical first step is to review the Periodic table of chemical elements, to identify the magnetic groups within it, and to explain the local influences of the chemical bonding environment in which the various ions reside in a crystal lattice. The magnetic ions are formed from those elements for which the inner shells remain unfilled while electrons occupying outer shells participate in chemical bonding. As a consequence, the electrons of the unfilled inner shells are responsible for a variety of magnetic properties because of the magnetic moments carried by their unpaired spins.

1.2.1. Periodic Table and Atomic Structure

To preserve electrical charge neutrality, the number of electrons occupying the orbital states of any atom is equal to the number of protons in its nucleus, i.e., the atomic number. To understand the actual arrangements of the electrons in the orbital states, one must recall the details of atomic structure that are compiled in the Periodic table of chemical elements.

In general, the n shells of the Bohr atomic model are segregated according to rows ($n = 1, 2, 3, 4$, etc.) and the orbital angular momentum (l) states within each n shell are designated by columns. According to the standard nomenclature of atomic physics, the orbital angular momentum of an electron bound to its atom is identified according to s ($l = 0$), p ($l = 1$), d ($l = 2$), f ($l = 3$), g ($l = 4$), etc. For each l value there are $2l + 1$ quantum states. Because there are also two spin states for each orbital state, there are correspondingly 2s electrons, 6p electrons, 10d electrons, 14f electrons, and 18g electrons. As a general rule, all inner orbital shells are complete, i.e., the only unfilled orbital states are in the outermost shell. Certain important departures from the standard order of atomic state occupancies exist, however, and it is from these elements that most magnetic effects originate. They are called the *transition-metal* groups and are found in the central columns of the Periodic table.

Figure 1.1 is the Mendeleyev Periodic table that has been modified to highlight the transition element groups. In the building of the table, there is an orderly progression of state occupations from hydrogen with a single 1s electron all the way out to the first transition series that begins with a single electron in the 10-state 3d shell of scandium Sc (atomic number 21) outside a completed argon (Ar) core, but inside a completed two-state 4s shell, as sketched in Fig. 1.2(a). The series ends at number 29 with copper (Cu) when the 10 electrons fill the 3d shell (designated $3d^{10}$). This series is commonly referred to as the "iron group" and will be the central feature of this chapter.

In a similar manner, the second series is built on a krypton core (Kr) with a filled outer $4p^6$. It begins at number 39 with yttrium (Y) and ends with number 47, silver (Ag), but notably features a completely unfilled 4f shell, shown in Fig. 1.2(b). Next in importance to the 3d shell, the "rare-earth" or lanthanide transition series is defined by the filling of the 4f shell and begins with number 57 lanthanum (La), continuing to number 70, ytterbium (Yb), with a completed $4f^{14}$. In Fig. 1.2(c), it is shown schematically that the rare-earth series is built on an xenon (Xe) core, even though the 4f shell is shown pictorially inside the Xe $5s^2$ and $5p^6$ filled shells (not shown). Beyond the Xe core, there are also empty 5d, 5f, and 5g shells before the outermost $6s^2$ shell completes the electronic configuration of these elements. The transition series shown in Fig. 1.2(d) begins with number 71 lutetium (Lu) and ends with number 79, gold (Au), and is fashioned from the completion of the 5d shell outside of the Xe core (now with the 4f shell filled), still inside the $6s^2$ shell. A final transition series may be considered as beginning with number 89, actinium (Ac), and ending with number 101, mendelevium (Md), formed on the radon (Rn) core, but with an incomplete 5f shell inside the radon core, analogous to the configurations of the rare-earth series.

Among the four groups of interest, only certain members become interesting for their magnetic properties in the ionic states, and only the iron group and rare-earth group have so far made a significant impact on the practical properties of oxide compounds.

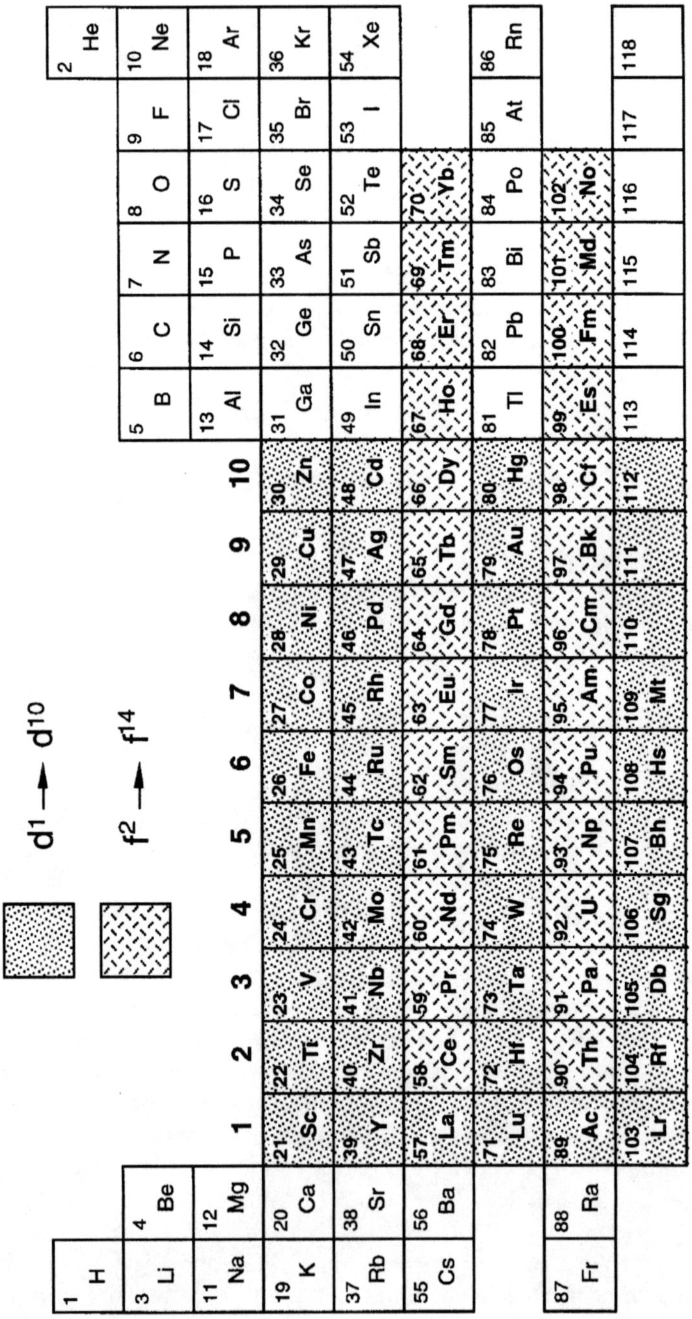

Figure 1.1. Periodic table configured to highlight the transition element groups.

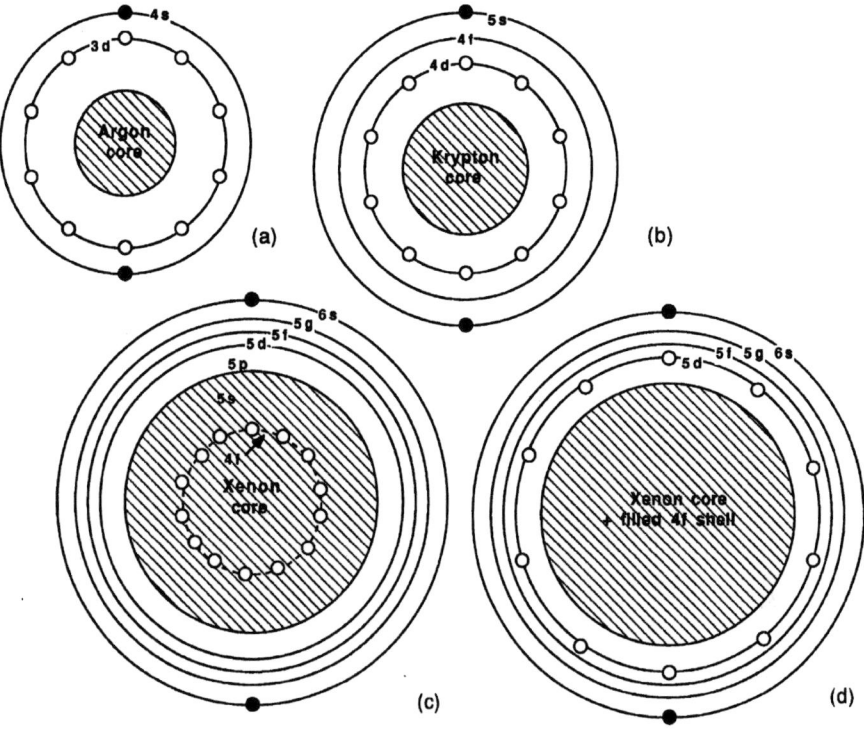

Figure 1.2. Electron shell diagrams of the transition elements: (a) $3d^n$; (b) $4d^n$; (c) $4f^n$; (d) $5d^n$.

1.2.2. Transition Elements

In the Periodic table of Fig. 1.1, the elements with ions that produce magnetic effects in combination with oxygen are distinguished by shading. In general, the atoms surrender their outer s electrons to form ionic bonds with atoms that accept them to complete their own unfilled shells. In an analogy to electron tubes with a cathode (emitter) and anode (plate), the metal atom that donates electrons is called a cation, while the "anode" atom becomes an anion. Because oxygen normally has an anion valence of $2-$, the cations of the transition metal oxides usually have valence charges of at least $2+$, particularly in the compounds that produce the magnetic properties of practical interest. One immediate observation is that once the 4s electrons are stripped from the $3d^n$ elements when they form ionic bonds, the partially filled 3d shell is exposed to the molecular environment, which will be a crystal lattice of specific symmetry and comprising electric and magnetic fields of its own.

The ions of the lower-third of the group, $3d^1$, $3d^2$, and $3d^3$ are generally paramagnetic and exhibit few collective properties, causing only perturbing effects on the remainder of the series. The magnetic moments of ions from $3d^4$ to $3d^9$ are capable of producing strong spontaneous magnetism when in sufficient densities to allow an exchange coupling to align spins as ferro-, antiferro- or ferrimagnets. In all cases where the ion is chemically bonded in an anion lattice, the combined orbital angular momentum vector L of all the electrons in the shell is uncoupled from the corresponding spin vector S by the electrostatic fields of the lattice and the spin moments that dominate the magnetic properties. Where this occurs the total angular momentum $J \neq L + S$ and is no longer a meaningful quantum operator. Consequently, the g factor is approximately equal to the free electron value of 2 except for special situations. The relevant parameter values for the iron group are summarized in Table 1.1.

The second most important transition group has an unfilled $4f^n$ shell. These elements are commonly referred to as the rare earths or lanthanides because La is the first member of the series. The other groups with unfilled $4d^n$ and $5d^n$ shells also have magnetic properties. As mentioned above, an important distinction between the $4f^n$ ions and those of the $3d^n$ iron group is the shielding of the 4f shell inside the $5s^2$ and $5p^6$ outer shells of the Xe core. In other words, the magnetically active electrons are buried inside the Xe core and are therefore shielded from electrostatic fields of the molecular environment. As a result, the ions act largely independent of one another, even in high concentrations, and are generally paramagnetic because the multiple lobes of the 4f orbital wave functions do not extend far enough for covalent bonding and magnetic exchange to be significant, as will be discussed later.

Table 1.1. Parameters of the Iron Group $3d^n$ Ion Series

Electrons	Ion	Ground State				Radius[a] (Å)	O^{2-}	Remarks
		Term	L	S	J			
$3d^0$	Sc^{3+}	$^1S_0$	0	0	0	0.73	6	Diamagnetic ion
$3d^1$	Ti^{3+}	$^2D_{3/2}$	2	1/2	3/2	0.67	6	Metallic with Ti^{4+}, forms sapphire in Al_2O_3
$3d^2$	V^{3+}	$^3F_2$	3	1	2	0.64	6	
	Ti^{2+}					0.86	6	Large ion
$3d^3$	Cr^{3+}	$^4F_{3/2}$	3	3/2	3/2	0.61	6	Forms ruby in Al_2O_3
	Mn^{4+}					0.54	6	
$3d^4$	Cr^{2+}	$^5D_0$	2	2	0	0.82	6	J–T ion, magnetostrictive
	Mn^{3+}					0.65	6	Metallic with Mn^{4+}
$3d^5$	Mn^{2+}	$^6S_{5/2}$	0	5/2	5/2	0.82	6	S-state ion,
	Fe^{3+}					0.64,49	6,4	low-spin $S=1/2$
$3d^6$	Fe^{2+}	$^5D_4$	2	2	4	0.77	6	Magnetostrictive,
	Co^{3+}					0.61	6	low-spin $S=0$
$3d^7$	Co^{2+}	$^4F_{9/2}$	3	3/2	9/2	0.73	6	Spin–orbit stabilized,
	Ni^{3+}					0.60	6	highly anisotropic, fast-relaxing, low-spin $S=1/2$
$3d^8$	Ni^{2+}	$^3F_4$	3	1	4	0.70	6	Magnetostrictive
	Cu^{3+}					–	6	low-spin $S=0$
$3d^9$	Cu^{2+}	$^2D_{5/2}$	2	1/2	5/2	0.73	6	J–T ion, magnetostrictive, metallic conductor with $S=0$ Cu^{3+} [b]
$3d^{10}$	Cu^{1+}	$^1S_0$	0	0	0	0.96	6	Large diamagnetic ion, metallic with Cu^{2+}

[a] Based on the radius of divalent oxygen of 1.40 Å.
[b] J. B. Goodenough et al.[134]

Upon inspection of Table 1.2, it will be noticed that the J values of the ions are divided into a lower and upper group, based on whether L is larger or smaller than S. For the lower-half from Ce^{3+} ($4f^1$) to Eu^{3+} ($4f^7$) and $J = |L - S|$; the upper-half from Gd^{3+} ($4f^8$) to Yb^{3+} ($4f^{13}$) features larger spin values, $J = |L + S|$. The J of the rare earths is the important angular momentum parameter because of strong spin–orbit coupling energies. Consequently, the g factors also are heavily dependent on J through the unquenched L contribution. The standard relation for g in the lanthanide rare earths was derived by Van Vleck:[1]

$$g = 1 + \frac{J(J+1) + S(S+1) - L(L+1)}{2J(J+1)} \quad (1.2.1)$$

Rare-earth ions contribute a number of important effects that include the tailoring of magnetization versus temperature behavior in magnetic garnets, the control of high power properties of microwave ferrites, and the Faraday rotation of magnetic garnets and other compounds for optical applications.

The unfilled 4d and 5d shells of the other two transition series have ions that resemble the 3d series in magnetic properties and can be used as alternatives for them in certain cases. Tetravalent ruthenium (Ru^{4+}) with a $4d^4$ configuration, e.g., has magnetoelastic properties similar to those of important trivalent manganese (Mn^{3+}) with a $3d^4$ occupancy. In general, these ions have not attracted much interest for their magnetic properties because the compounds formed from them do not exhibit

Table 1.2. Parameters of the Rare-Earth $4f^n$ Ion Series

Electrons	Ion	Term	L	S	J	g	Radius[a] (Å)	Remarks
$4f^0$	La^{3+}	$^1S_0$	0	0	0	—	1.18	Diamagnetic ions
	Ce^{4+}						0.97	
$4f^1$	Ce^{3+}	$^2F_{5/2}$	3	1/2	5/2	6/7	1.14	Strong magneto-optical properties
	Pr^{4+}						0.99	
$4f^2$	Pr^{3+}	$^3H_4$	5	1	4	4/5	1.14	Strong magneto-optical properties
$4f^3$	Nd^{3+}	$^4I_{9/2}$	6	2	4	8/11	1.12	Strong magneto-optical properties
$4f^4$	Pm^{3+}	$^5I_4$	6	2	4	3/5	0.98	Synthetic element
$4f^5$	Sm^{3+}	$^6H_{5/2}$	5	5/2	5/2	2/7	1.09	—
$4f^6$	Eu^{3+}	$^7F_0$	3	3	0	—	1.07	Diamagnetic ion
$4f^7$	Eu^{2+}	$^8S_{7/2}$	0	7/2	2	2	1.25	S-state ions
	Gd^{3+}						1.06	
$4f^8$	Tb^{3+}	$^7F_6$	3	3	6	3/2	1.04	Fast-relaxing ion
$4f^9$	Dy^{3+}	$^6H_{15/2}$	5	5/2	15/2	4/3	1.03	Fast-relaxing ion
$4f^{10}$	Ho^{3+}	$^5I_8$	6	2	8	5/4	1.02	Fast-relaxing ion
$4f^{11}$	Er^{3+}	$^4I_{15/2}$	6	3/2	15/2	6/5	1.00	Fast-relaxing ion
$4f^{12}$	Tm^{3+}	$^3H_6$	5	1	6	7/6	0.99	Fast-relaxing ion
$4f^{13}$	Yb^{3+}	$^2F_{7/2}$	3	1/2	7/2	8/7	0.98	Fast-relaxing ion
$4f^{14}$	$Lu^{3+\,b}$	$^1S_0$	0	0	0	—	0.97	Fast-relaxing ion

[a] Based on an oxygen coordination of 8.
[b] Lutetium is included here to complete the 4f shell. It also represents the beginning of the $5d^n$ series.

strong spontaneous magnetic properties. Moreover, several of them, such as rhodium (Rh), palladium (Pd), osmium (Os), iridium (Ir), and platinum (Pt) are not available in sufficient abundance to be considered for low-cost applications.

1.3. CATION SITES IN ANION LATTICES

Transition-metal ions are chemically reactive and are naturally bonded to anions of the seventh or eighth columns in compounds with distinct crystal structures. Such compounds can be viewed as comprising separate metal (cation) and anion lattices. In the discussions to follow, we focus our attention on the immediate surroundings of the metal ions, specifically the disposition of the oxygen or ligand coordinations relative to the transition ion.

1.3.1. Crystal Systems and Point Groups

A familiarity of crystallographic symmetry is essential for understanding of the concepts that determine the properties of the transition ions in oxygen coordinations. For a thorough treatment of the subject, the reader is referred to a standard text.[2] In this review, however, the topic will be focused on the concepts that form the basis for the crystal-field and molecular-orbital theories from which most of the properties of interest are examined.

Table 1.3 lists the seven systems that comprise all the naturally occurring types of crystalline structures. Corresponding cell sketches are presented in Fig. 1.3. These systems are in turn composed of 32 crystallographic "point groups" or crystal classes which describe the basic symmetry of crystallographic building blocks or unit cells. A point group therefore consists of a collection of symmetry operators that serve to define a particular crystal structure. There are three such basic operations: rotation, whereby the structure repeats itself upon rotation around a particular direction or axis, e.g., fourfold meaning that it repeats the image of its projection along the axis

Table 1.3. Crystal Systems and Their Symmetry Elements

System	Generic Point Group	Unit Cell	Symmetry Elements
Triclinic	—	$a \neq b \neq c; \alpha \neq \beta \neq \gamma \neq 90°$	No axes, no planes
Monoclinic	C_{2h}	$a \neq b \neq c; \alpha = \beta = 90° \neq \gamma$	One twofold axis or one plane
Orthorhombic	D_{2h}	$a \neq b \neq c; \alpha = \beta = \gamma = 90°$	Three orthogonal twofold axes; two planes intersecting a twofold axis
Tetragonal	D_{4h}	$a = b \neq c; \alpha = \beta = \gamma = 90°$	One fourfold axis or a fourfold inversion axis
Rhombohedral (trigonal)	D_{3d}	$a = b = c; \alpha = \beta = \gamma \neq 90°$	One threefold axis
Hexagonal	D_{6h}	Three axes, a in x–y plane at $\alpha = 120°$; $c \neq a$	One sixfold axis
Cubic (isometric)	O_h	$a = b = c; \alpha = \beta = \gamma = 90°$	Four threefold axes

ELECTRON SPINS IN IONIC MOLECULAR STRUCTURES

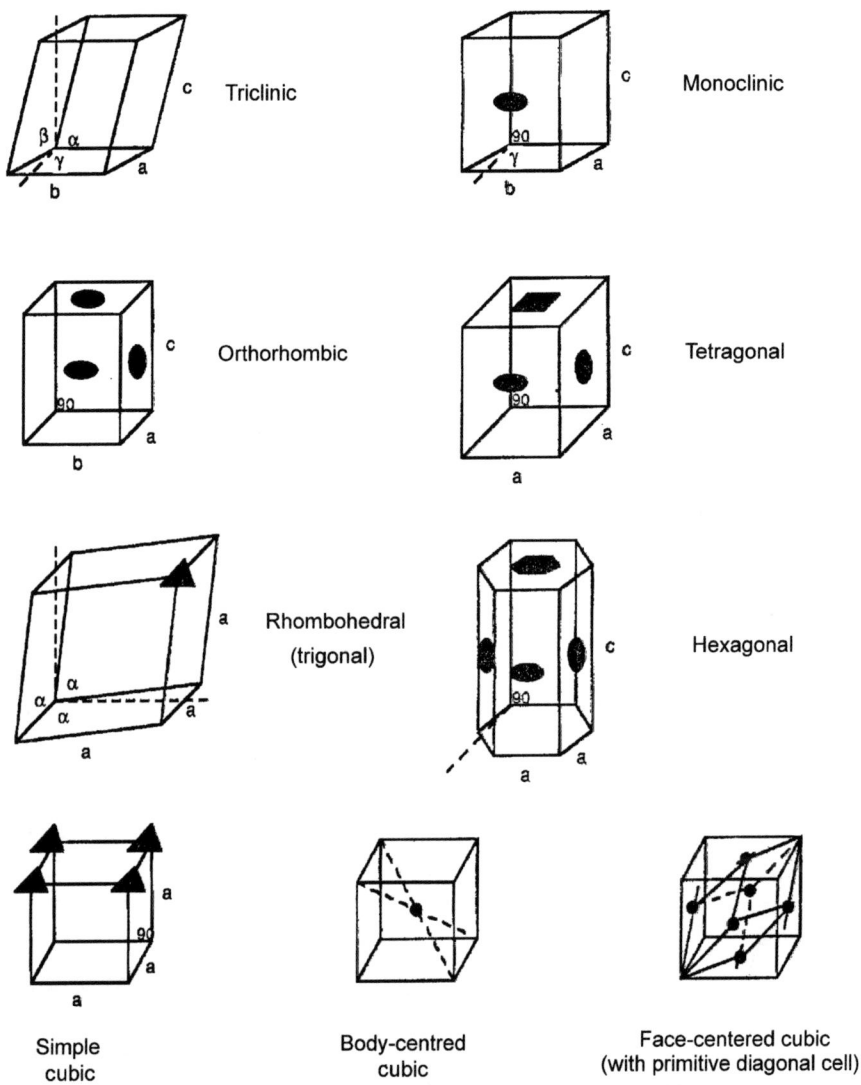

Figure 1.3. Crystal system diagrams.

every 90°, reflection about a plane, and inversion, meaning that the crystal retains its appearance after undergoing reversal of each of the x, y, and z coordinates. A further refinement categorizes the point groups into 240 "space groups", but this level of detail will not be necessary for the present purposes.

1.3.2. Cubic Symmetry Sites

In the magnetic oxides of interest, the cations usually reside in sites comprising oxygen arrangements referred to as coordinations. The basic building block is generally of cubic (isometric) symmetry or one that is derived from it. In Fig. 1.4, the most common situations are sketched in relation to Cartesian axes and are of

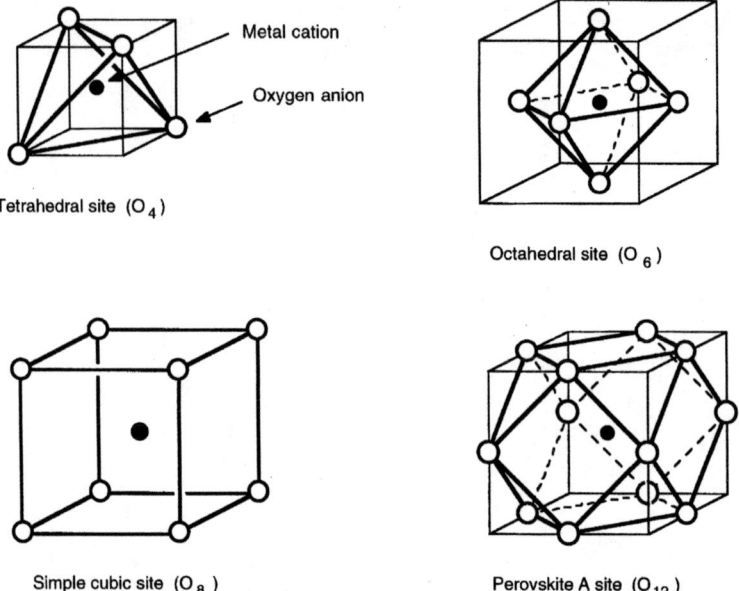

Figure 1.4. Cation sites with ligand coordinations of cubic symmetry.

four types: tetrahedral, with four sides and four anions on alternate corners as shown; octahedral with eight sides and six anions on the corners of an octahedron formed with anions at the cube face centers; simple cubic with 12 sides and eight anions at the cube corners; and a site that houses the larger rare-earth or alkaline-earth ions in the A site of the perovskite family with 12 anions located at the midpoints of the 12 cube sides. Another large site (not shown) is called dodecahedral, with 12 sides and 14 anions located at the eight corners of a half-sized cube and six more at the face centers of the full cube. This latter structure does not actually occur in the magnetic oxides of interest here. In reality, the "dodecahedral" site of the garnet lattice is a 12-sided cell formed from a twisted cube with only eight corner anions.

Each of these coordinations can have the required four threefold symmetry axes directed along body diagonals, referred to as the $\langle 111 \rangle$ axes family according to the convention of the Miller indices.* The corresponding indices for the three fourfold cubic axes (x, y, and z directions) are the $\langle 100 \rangle$ family, and for the face diagonals, the $\langle 110 \rangle$ family. The most common oxygen coordination is octahedral, here labeled O_6 for the six O^{2-} anions. The octahedral site is of paramount importance in spinels, garnets and the various perovskite-related compounds. The tetrahedral coordination, designated O_4, is the alternate cation site in the spinels and garnets and is generally occupied by smaller metal ions. Because the smaller

---

*The Miller indices were developed to identify the various planes in a crystallographic lattice. The system is based on the values of the three intercepts of the plane with the x, y, and z axes expressed as the lowest integer values. The labeling convention for family of planes is $\{hkl\}$ and an individual plane is (hkl). Alternatively, the normal axes to the planes are labeled $\langle hkl \rangle$ for the family and $[hkl]$ for an individual axis.

separation between anions leads to higher mutual repulsive forces, the higher coordination numbers result in larger site volumes to minimize the bonding energies. For this reason the larger ions, such as the lanthanide rare earths, usually occupy O_8 as in the cubic perovskites, or the dodecahedrally distorted O_8 sites in the garnets.

1.3.3. Lower Symmetry Sites

In all but some of the simplest oxides, the cubic symmetry is usually reduced by lattice distortions that are often large enough to be considered as part of a phase transition to another point group even if the effect is local and involves only an isolated cation complex. Even when the distortions are subtle, their effects on the magnetic properties can be significant. The topic of lattice distortions and symmetry changes will recur regularly throughout this text.

Although departures from cubic symmetry can vary widely in extent, there is no immediate need to discuss more than the ones depicted in Fig. 1.5. The most convenient vehicle to examine these distortions is the octahedral site, which may undergo extensions or compressions along any of the principal axes of symmetry. In most cases, the $\langle 111 \rangle$ and $\langle 100 \rangle$ groups are the ones of concern. If the distortion is along a $\langle 111 \rangle$ axis, the symmetry is reduced from cubic O_h to trigonal or rhombohedral D_{3d} and the immediate environment of the cation has one threefold symmetry axis. Figure 1.5 also indicates the effects of a tetragonal D_{4h} compression along the [001] axis and the addition of a second distortion, this time an extension along the [010] axis to create a lower orthorhombic D_{2h} symmetry. Combinations of trigonal with tetragonal distortions that can occur through spontaneous local distortions called Jahn–Teller (J–T) effects can reduce the symmetry even further, as will be discussed in Section 1.5.

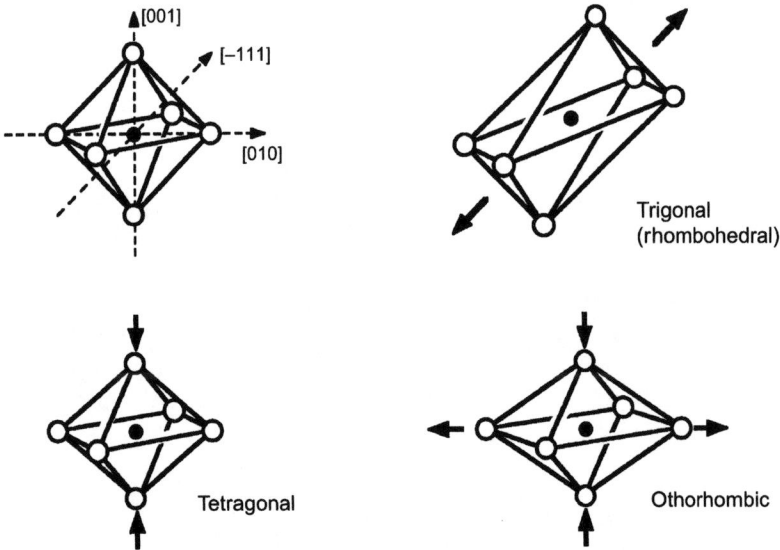

Figure 1.5. Cubic cation sites with simple distortions.

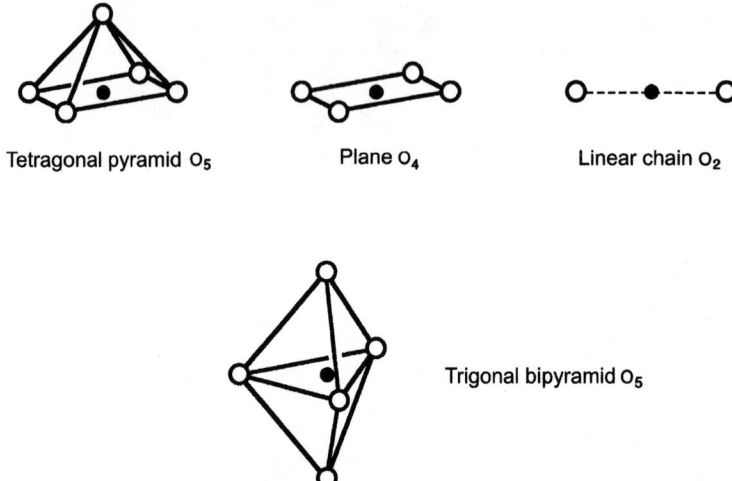

Figure 1.6. Other ligand coordinations in oxides.

Four other situations commonly encountered in magnetic oxides are shown in Fig. 1.6. A tetragonal pyramid, which occurs when one of the z-axis oxygen ions is missing, provides an O_5 coordination of C_4 symmetry (no reflection plane in this case); a square or rectangular planar configuration is formed by the removal of both apical anions and provides the ultimate tetragonal or orthorhombic D_{4h} or D_{2h} symmetry; and a one-dimensional chain that remains after two opposite planar oxygens are removed gives pure cylindrical symmetry. In recent years, these abbreviated octahedra have been detected in the superconducting layered perovskite-type compounds. A fourth cation site takes the form of a trigonal bipyramid and occurs in the hexagonal structure originally called "magnetoplumbite." This site is only one of 13 in the type-M hexagonal ferrite compounds, and is a strong contributor to the highly anisotropic magnetic properties of these important compounds.

1.4. ELECTRON ENERGY LEVELS IN CRYSTAL FIELDS

In a crystal lattice, the metal ion is bonded electrostatically to its neighbors through either simple cation–anion Coulomb attractive forces, covalent (electron sharing) molecular bonds, or collective electron stabilization typical of metal elements and alloys. For oxide systems, the bonding is principally ionic (Coulomb), but features a smaller covalent component that is critically important for the electronic and magnetic properties of compounds with cations of a d^n transition series. To analyze states of an ion for which the d electrons are exposed to the negative charges of the anion coordination, point-charge crystal-field theory is applied first.

1.4.1. Angular Momentum of Free Ions

To introduce the quantum mechanical effects of Stark splittings to the free ion orbital angular momentum states, it is instructive to review the formation of the multi-electron orbital terms that are usually governed by Hund's rules, which state that the lowest energy term has:

1. the maximum possible combined spin value S;
2. within the maximum S manifold, the maximum combined L.

These rules originate from the Pauli exclusion principle and the quantum mechanical preference for spins to align parallel when dispersed among the set of orthogonal orbital wave functions by mutual electrostatic repulsion. To help visualize the "laddering" exercise, Fig. 1.7 is included to depict the situation among the states of a d^n series. The rows of stacked boxes represent an orbital angular momentum value m_l of operator l_z. Each box can hold two electrons, one for each up or down spin orientation as required by the Pauli exclusion principle. Beginning with the lower-half of the series from d^1 to d^5, the electrons are added sequentially, obeying the spin polarization requirement to fill the first five up-spin compartments and produce a half-filled set of orbitals with the maximum spin value of $S = 5/2$ when the d^5 limit is reached.

From an energy standpoint, the half-filled shell is most stable because each d electron occupies a separate orthogonal orbital state, and the destabilizing effect of the mutual repulsion is minimum. This ordered spatial dispersal of the polarized spins beyond a random distribution reduces the screening of the nucleus and stabilizes the spins in proportion to their numbers (or their combined S). As the upper-half of the shell begins to fill, the sixth electron must now share an orbit with its spin antiparallel to the net spin of the lower-half in order to satisfy the Pauli principle. The natural consequence is an abrupt increase in energy that is the direct

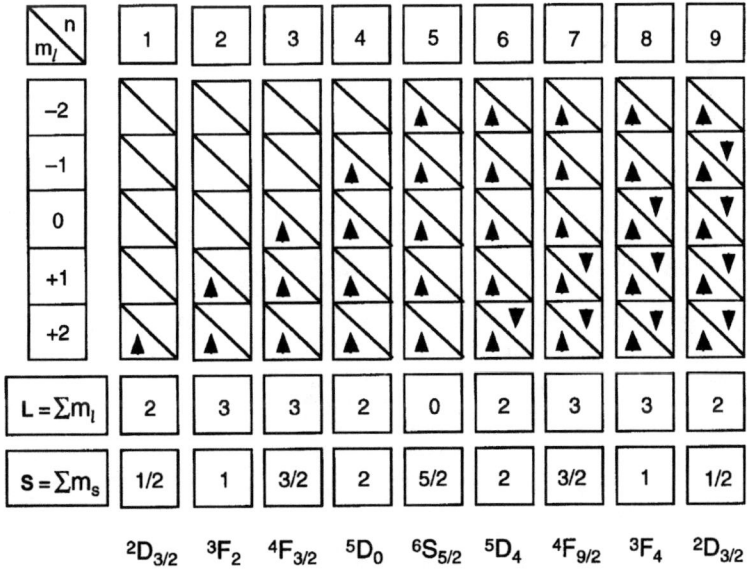

Figure 1.7. Application of Hund's rule in the formation of angular momentum ground terms in the $3d^n$ shell. Terms are either D ($L=2$), F ($L=3$), or in the $3d^5$ case, S ($L=0$).

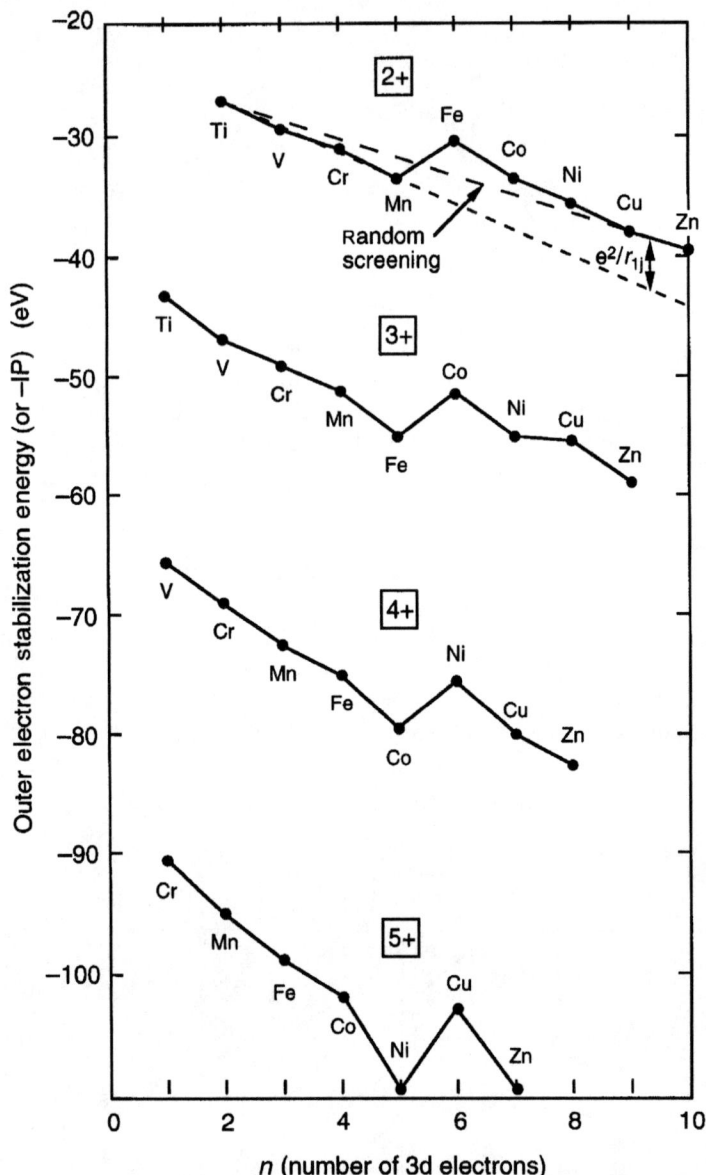

Figure 1.8. Electron stabilization energies of the d-shell ions, plotted as reverse ionization potentials. Note effects of Hund's rule spin polarization and the Pauli spin pairing, particularly as the upper-half of shell begins to fill. The half-filled shell d^5 electron is the most stable, and the d^6 electron is the least stable, which explains why Fe^{2+} ions frequently act as electron "donors" in charge transfer phenomena. To suppress conduction electrons (or create holes), d^4 Mn^{3+} acts as an "acceptor." Data are from C.E. Moore, NSRDS-NBS 34, Office of Standard Reference Data, National Bureau of Standards, Washington, DC.

result of the mutual repulsion. From d^6 to d^{10}, the process is repeated with a positive energy increment until the d shell is filled and the net spin returns to $S = 0$, at which point the d shell can be considered part of the electron core. The effect on energy from the filling of the d-orbital shell by spin pairing following Hund's rule can be seen in the ionization potentials (IP) (plotted negative relative to free space) in Fig. 1.8 as a function of n for different ionic valences across the transition series. Note that the departures from the baseline for random dispersal is consistent with the concept that the internal alignment (intra-exchange) energy is proportional to the net spin value of the ion, and that the maximum destabilization between up and down spins is on the order of 3 eV. In band theories, where the Fermi level is used as the zero reference energy for electrical conduction, it has been found phenomenologically convenient to split the d shell into separate bands for up and down spin bands in order to account for the variations in ionization potential between the lower and upper halves.

More important for the immediate discussion, however, are the combined values of L which can be calculated by straightforward additions of m_l in each column. The resultant designations for the ground terms of each free ion $^{2S+1}L_J$ indicates that only three orbital degeneracies occur in the d-electron transition series: S, D, and F. Since the S represents an $L = 0$ state, d^5 automatically becomes a spin-only magnetic entity, which greatly simplifies analysis of magnetic properties, at least to first-order theory.

1.4.2. Orbital Quenching by Crystal Fields

When a positive magnetic ion is subjected to the electric field of the negative anion charges, the lobes of the electron orbital wave functions react to repulsive forces that either stabilize or destabilize the different orbitals depending on their relative proximity to the orbital lobes of the ligands, e.g., $2p_{x,y,z}$ orbital functions of oxygen. In the broadest context, the orbital momentum is captured by the crystal field, and in the process is decoupled from the magnetic moments of the electron spins. A simple sketch of this quenching effect is given in Fig. 1.9 for an uniaxial crystal field that separates the L and S vectors aligned through spin-orbit interaction when it is not collinear with a magnetic field vector H. Crystal-field theory is applied through quantum mechanical methods to determine the energy level structures of the resultant orbital states and their associated eigenfunctions.

Before considering the effects of the crystalline environment on the cation energy states, we must review briefly the Hamiltonian of the free or unperturbed ion. A more complete discussion may be found in other texts[3-5] that have emanated from the treatise by Condon and Shortley.[6] The Hamiltonian for an ion with angular momentum vectors L and S coupled by $\mathcal{H}_{LS}$ is

$$\mathcal{H} = \left[\mathcal{H}_{\text{Coul}} + \mathcal{H}_{\text{rep}}\right] + \mathcal{H}_{LS} \tag{1.4.1}$$

where the bracketed terms are the free-ion energies comprising

$$\mathcal{H}_{\text{Coul}} = -\frac{\hbar^2}{2m_e}\sum_i \nabla_i^2 - \sum_i \frac{Ze^2}{r_i} \tag{1.4.2a}$$

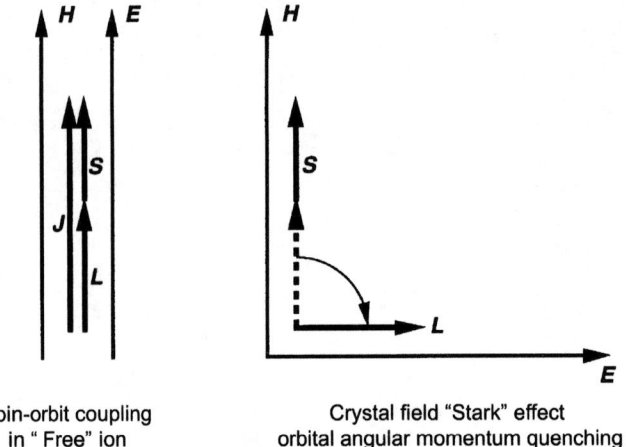

Spin-orbit coupling
in "Free" ion

Crystal field "Stark" effect
orbital angular momentum quenching

Figure 1.9. Orbital angular momentum quenching in a crystal field. Where the electric field E of the anion charges is not collinear with the magnetic field H, the Stark effect disrupts spin–orbit coupling.

which is the basic relation containing the Coulomb attractive energy between the Z electrons and the nuclear charge separated by r_i, and

$$\mathcal{H}_{\text{rep}} = \sum_{i>j} \frac{e^2}{r_{ij}} \quad (1.4.2b)$$

which is the energy of mutual repulsion between orbiting electrons i and j separated by r_{ij}. In combination with the crystal field, $\mathcal{H}_{\text{rep}}$ is responsible for the distribution of electrons among the various orbital states of the unfilled shell and therefore the ordering and separation of the energy terms which have been computed and thoroughly documented in the literature of atomic physics.[6] It is also responsible for the parallel spin alignments that produce the high-spin states associated with Hund's rule. The eigenfunctions of this free-ion Hamiltonian are the familiar solutions of the Schrödinger equation in the form of exponentially decaying radially symmetric functions $R(r)$ combined with spherical harmonics that shape the various wave function lobes, according to $\Psi_l^{m_l} = R(r) Y_l^{m_l}$

The $R(r)$ function common to all orbitals within a main Bohr n shell. Within a given shell, such as the transition series 3d, it can be treated as a scale factor. The angular parts of the eigenfunctions of the 2p states of oxygen and the d and f states of the transition group are listed as[7]

$$Y_1^{-1} = p_{-1} = \left(\frac{3}{8\pi}\right)^{1/2} \frac{x - iy}{r}$$

$$Y_1^0 = p_0 = \left(\frac{3}{4\pi}\right)^{1/2} \frac{z}{r} \quad \text{(P state)} \quad (1.4.3)$$

$$Y_1^1 = p_1 = \left(\frac{3}{8\pi}\right)^{1/2} \frac{x + iy}{r}$$

$$Y_2^{-2} = d_{-2} = \left(\frac{5}{4\pi}\right)^{1/2}\left(\frac{3}{8}\right)^{1/2}\frac{(x-iy)^2}{r^2}$$

$$Y_2^{-1} = d_{-1} = \left(\frac{5}{4\pi}\right)^{1/2}\left(\frac{3}{2}\right)^{1/2}\frac{z(x-iy)}{r^2}$$

$$Y_2^0 = d_0 = \left(\frac{5}{4\pi}\right)^{1/2}\left(\frac{1}{4}\right)^{1/2}\frac{3z^2-r^2}{r^2} \qquad \text{(D state)} \qquad (1.4.4)$$

$$Y_2^1 = d_1 = -\left(\frac{5}{4\pi}\right)^{1/2}\left(\frac{3}{2}\right)^{1/2}\frac{z(x+iy)}{r^2}$$

$$Y_2^2 = d_2 = \left(\frac{5}{4\pi}\right)^{1/2}\left(\frac{3}{8}\right)^{1/2}\frac{(x+iy)^2}{r^2}$$

$$Y_3^{-3} = f_{-3} = \left(\frac{35}{64\pi}\right)^{1/2}\frac{(x-iy)^3}{r^3}$$

$$Y_3^{-2} = f_{-2} = \left(\frac{75}{32\pi}\right)^{1/2}\frac{z(x-iy)^2}{r^3}$$

$$Y_3^{-1} = f_{-1} = \left(\frac{21}{64\pi}\right)^{1/2}\frac{(x-iy)(5z^2-r^2)}{r^3}$$

$$Y_3^0 = f_0 = \left(\frac{7}{16\pi}\right)^{1/2}\frac{z(5z^2-3r^2)}{r^3} \qquad \text{(F state)} \qquad (1.4.5)$$

$$Y_3^1 = f_1 = -\left(\frac{21}{64\pi}\right)^{1/2}\frac{(x+iy)(5z^2-r^2)}{r^3}$$

$$Y_3^2 = f_2 = \left(\frac{75}{32\pi}\right)^{1/2}\frac{z(x+iy)^2}{r^3}$$

$$Y_3^3 = f_3 = -\left(\frac{35}{64\pi}\right)^{1/2}\frac{(x+iy)^3}{r^3}$$

The various terms formed from the spherical harmonics tend to be ordered energetically according to Hund's ^{2S+1}L rule, which states that the lower energies favor first the highest multiplicities $2S+1$ and then the highest L within each $2S+1$ group. For the $3d^n$ series, the $3d^1$ case with $L=2$ and $S=1/2$, ^2D is the only orbital term because the $\mathcal{H}_{\text{rep}}$ mutual repulsion energy is zero.

The solutions for the multiple-electron cases, which are sorted according to the influence of $\mathcal{H}_{\text{rep}}$, are listed in Table 1.4. An example of the important five-electron case $3d^5$ corresponding to Fe^{3+} is shown in Fig. 1.10, with the ^6S ground term and the first excited term ^4G with its subsequent multiplet splittings and eventual Zeeman splittings in a magnetic field. A physical picture of the ^4G state would have the $m_l = -2$ electron shown in the occupancy diagram of Fig. 1.7 reversing its spin sense to form a pair in the $m_l = 2$ orbit and provide a resultant $L=4$ with an $S=3/2$. The spin–orbit coupling energy term in Eq. (1.4.2), $\mathcal{H}_{LS} = \sum_i \xi_i(r) l_i \cdot s_i$

Table 1.4. 3d^n (Iron-Group) Free-Ion Energy Terms (Lowest 5)

d^1 (d^9)	d^2 (d^8)	d^3 (d^7)	d^4 (d^6)	d^5[a]
—	^{1}S	^{2}F	^{3}D	^{4}F
—	^{1}G	^{2}G	^{3}F[b]	^{4}D
—	^{3}P	^{2}H	^{3}G	^{4}P
—	^{1}D	^{4}P	^{3}H	^{4}G
^{2}D	^{3}F	^{4}F	^{5}D	^{6}S

[a] For this case, in particular, the order of the term energies does not follow the approximation of Hund's rule. This is a characteristic of the higher energy terms in configurations with greater numbers of d electrons.
[b] There are two values for this term.

($\sim 10^2$ cm^{-1} for the iron group and 10^3 cm^{-1} for the rare-earth group), is the perturbation that produces the multiplet structure observed in atomic spectra. Where the coupling functions $\xi_i(r)$ are effectively invariant among the states, they are treated as a semiempirical spin–orbit coupling constant λ.

To account for the repulsive interaction between the electrons of the cation and the electric field of the anion neighbors, a crystal-field energy term $\mathcal{H}_{cf}$ must be inserted before $\mathcal{H}_{LS}$ in Eq. (1.4.2),

$$\mathcal{H} = [\mathcal{H}_{\text{Coul}} + \mathcal{H}_{\text{rep}}] + \mathcal{H}_{cf} + \mathcal{H}_{LS} \qquad (1.4.6)$$

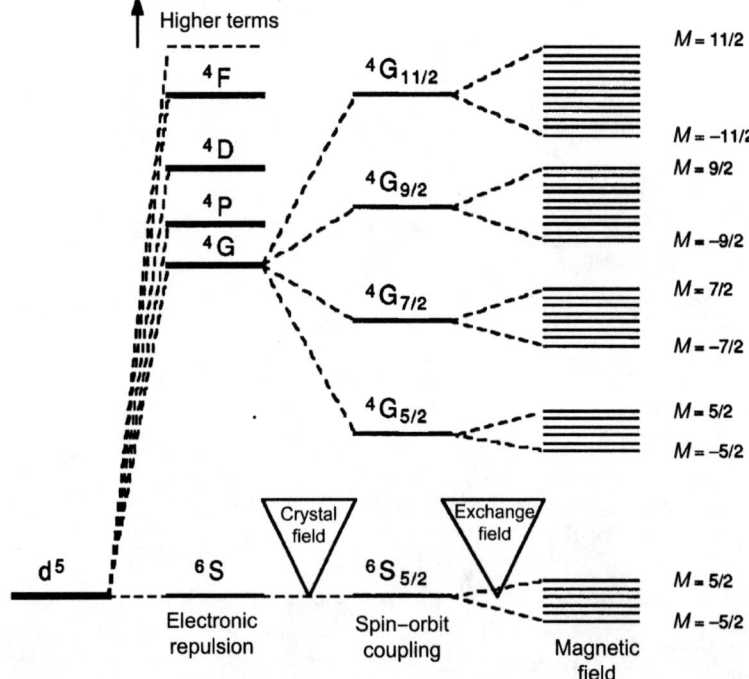

Figure 1.10. Generic model of energy-level structure of five-d electron (d^5) configuration, typical of the Fe^{3+} ^{6}S-state ion.

In the simplest approximation, the source of $\mathcal{H}_{cf}$ is represented as point charges fixed at the locations of the particular ligands surrounding the cation. The purpose is to simulate a Stark effect coupling between the orbital angular momentum L and the electric field E_{cf} that disrupts the spin–orbit coupling by freeing S from L, as depicted in Fig. 1.9. Therefore, the effect on paramagnetism is to render the magnetic properties dependent on S rather than J. This crystal-field quenching of the orbital magnetism results in $g \approx 2$ when the quenching is nearly complete.

1.4.3. Perturbation Hierarchy

As suggested by the order of terms in Eq. (1.4.4), $\mathcal{H}_{LS}$ is usually smaller than the larger lattice-related perturbation terms. However, that is not always the case. At this point it becomes both convenient and instructive to define three crystal-field regimes:[4,5]

$$\begin{aligned}
\mathcal{H}_{rep} > \mathcal{H}_{cf} > \mathcal{H}_{LS} \quad & (3d^n \text{ series}) \\
\mathcal{H}_{cf} \geq \mathcal{H}_{rep} \quad & (4d^n \text{ and } 5d^n \text{ series}) \\
\mathcal{H}_{cf} < \mathcal{H}_{LS} \quad & (4f^n \text{ series})
\end{aligned} \quad (1.4.7)$$

A summary of these schemes and their ranges of applicability is given in Table 1.5. The first is the one of principal interest because it applies to the most commonly encountered iron group $3d^n$ series. In this "weak-field" case, the crystal field is smaller than the energy term separations due the $\mathcal{H}_{rep}$ repulsive energy of Eq. (1.4.1) listed in Table 1.4. Consequently, the starting free-ion terms in a perturbation calculation are not mixed, only their degeneracies are lifted by the electrostatic coupling of the crystal fields. For interpreting most magnetic effects, only the ground terms need to be considered.

The second or "strong-field" case is also important, perhaps more for the high-energy transitions of magneto-optical properties. It is analytically more challenging than the "weak-field" case because the $\mathcal{H}_{cf}$ magnitudes are equal or greater than the free ion term splittings set by $\mathcal{H}_{rep}$, and are therefore strong enough to mix the starting orbital terms prior to the removal of their degeneracies. As a result, the various possible electron distributions among the individual d orbital states, i.e., the excited states, must be included as separate energy levels prior to application of the symmetry constraints imposed by the $\mathcal{H}_{cf}$ operator. The strong-field situation is sometimes referred to as the covalent limit because the $\mathcal{H}_{cf}$ potential energy is

Table 1.5. Perturbation Hamiltonians

Weak Field (3d)	Strong Field (4d, 5d)	Rare Earth (4f)
$\mathcal{H}_{rep} \rightarrow e^2/r_{ij}$ (<3 eV)	$\mathcal{H}_{rep} + \mathcal{H}_{cf} \rightarrow e^2/r_{ij} + V_{cf}$ (<6 eV)	$\mathcal{H}_{rep} \rightarrow e^2/r_{ij}$ (<3 eV)
$\mathcal{H}_{cf} \rightarrow V_{cf}$ (<3 eV)	—	$\mathcal{H}_{LS} \rightarrow \lambda L \cdot S$ (>1 eV)
$\mathcal{H}_{LS} \rightarrow \lambda L \cdot S$ (<1 eV)	$\mathcal{H}_{LS} \rightarrow \lambda L \cdot S$ (<1 eV)	$\mathcal{H}_{cf} + \mathcal{H}_{ex} + \mathcal{H}_{mag} \rightarrow$
		$V_{cf} + JS \cdot S + g_\perp m_B J \cdot H$
		(<<1 eV)
$\mathcal{H}_{ex} \rightarrow JS \cdot S$ (<1 eV)	$\mathcal{H}_{ex} \rightarrow JS \cdot S$ (<1 eV)	—
$\mathcal{H}_{mag} \rightarrow g_{eff} m_B S \cdot H$ (<<1 eV)	$\mathcal{H}_{mag} \rightarrow g_{eff} m_B S \cdot H$ (<<1 eV)	—

enhanced by the overlap of the cation and anion orbital lobes. It is more common among the $4d^n$ and $5d^n$ transition series ions with larger ionic radii, but can also apply in the $3d^n$ series when the anion complex provides a locally stronger crystal field than that of the standard O^{2-} coordinations.

The third case occurs with the rare-earth $4f^n$ series, where the Stark effect of the crystal field is not great enough to decouple L from S because of the shielding by the filled $5s^2$ and $5p^6$ shells. In this series, $\lambda \boldsymbol{L} \cdot \boldsymbol{S}$ remains a constant of the motion and the $\mathcal{H}_{LS}$ operation creates the various multiplet terms now identified by $^{2S+1}L_J$, where L represents the particular orbital term designated by S, P, D, F, G, etc., with respective values of L being 0, 1, 2, 3, 4. For the rare earths, it is the total angular momentum J and its associated g value as defined by Eq. (1.2.1), rather than S with its fixed $g = 2$, that determines the individual ion contributions to the magnetic properties.

1.4.4. Weak-Field Solutions for D and F Ions

The subject of crystal-field theory has been presented in many excellent texts.[3-5] Historically, the seminal work was carried out by Kramers,[8] Van Vleck,[9] and Penney and Schlapp,[10] who treated the electric fields from the anions surrounding a given cation site as resulting from negative point charges. Because the potential energy $\mathcal{V}_{cf}$ at the cation site from the assembly of neighboring charges satisfies Laplace's equation $\nabla^2 \mathcal{V}_{cf} = 0$, $\mathcal{H}_{cf}$ $(= \mathcal{V}_{cf})$ may be expressed as an expansion of generalized Legendre polynomials which take the same familiar form of spherical harmonics comprising Eq. (1.4.2). The problem of applying quantum perturbation theory to determine the electronic states of the cation in a particular crystal field is then reduced to setting up a perturbation matrix and solving a secular equation,

$$\mathcal{H}_{cf}^k = |\mathcal{H}_{cfij} - E_{cf}^k \delta_{ij}| = 0 \tag{1.4.8}$$

where $\mathcal{H}_{cfij} = \langle \Psi_i | \mathcal{H}_{cf} | \Psi_j \rangle$, i and j are integers that run from 1 to k. E_{cf}^k are the k eigenvalue solutions of the matrix, each representing a perturbed energy state depending on the extent of the degeneracy removal. In this problem, it is the degeneracies of the orbital angular momentum represented by the spherical harmonic parts of the free ion wave functions of Eq. (1.4.2) that determine the order of the splittings.

To illustrate the method, the example of a single d electron will be reviewed. The spherical harmonic functions (also expressed in the d_{m_l} abbreviations) for the ^2D term are given in Eq. (1.4.4). For an octahedral (O_6) site, the crystal field potential energy with x, y, and z axes coincident with the $\langle 100 \rangle$ cubic axes is given by[11]

$$\mathcal{V}_{cf}^{oct} = D_4[x^4 + y^4 + z^4 - (3/5)r^4] + \text{ higher order terms} \tag{1.4.9}$$

Expressed in spherical harmonics, Eq. (1.4.10) becomes

$$\mathcal{V}_{cf}^{oct} = \frac{7}{2}D_4\left[Y_4^0 + \left(\frac{5}{14}\right)^{1/2}(Y_4^4 + Y_4^{-4})\right] + \frac{3}{4}D_6\left[Y_6^0 - \left(\frac{7}{2}\right)^{1/2}(Y_6^6 + Y_6^{-6})\right] \tag{1.4.10}$$

where $D_4 = (35/4)Ze^2/a^6$ and $D_6 = (21/2)Ze^2/a^7$ and a is the cation to anion distance. The relevant harmonics are given by

$$Y_2^0 = \left(\frac{5}{4\pi}\right)^{1/2} \left(\frac{1}{4}\right)^{1/2} \frac{3z^2 - r^2}{r^2}$$

$$Y_4^0 = \left(\frac{9}{4\pi}\right)^{1/2} \left(\frac{1}{64}\right)^{1/2} \frac{35z^4 - 30z^2r^2 + 3r^4}{r^4}$$

$$Y_4^4 + Y_4^{-4} = \frac{3}{16}\left(\frac{70}{\pi}\right)^{1/2} \frac{x^4 - 6x^2y^2 + y^4}{r^4} \tag{1.4.11}$$

$$Y_6^0 = \frac{1}{32}\left(\frac{13}{\pi}\right)^{1/2} \frac{231z^6 - 315z^4r^2 + 105z^2r^4 - 5r^6}{r^6}$$

$$Y_6^6 + Y_6^{-6} = \frac{231}{64}\left(\frac{54}{231\pi}\right)^{1/2} \frac{x^6 - 15x^4y^2 + 15x^2y^4 - y^6}{r^6} \frac{x-iy}{r}$$

For the tetrahedral (O_4) and cubic (O_8) coordinations only the Y_4 terms of Eq. (1.4.10) enter the calculation. Their crystal field energies scale according to

$$V_{cf}^{tet} = -\left(\frac{4}{9}\right)V_{cf}^{oct} \qquad V_{cf}^{cub} = -\left(\frac{8}{9}\right)V_{cf}^{oct} \tag{1.4.12}$$

If the radial part of Eq. (1.4.2) is folded into the scale factor of the matrix elements within the $n = 3$ shell, we may work only with the $Y_l^{m_l}$ functions of Eq. (1.4.11) to set up a 5×5 matrix based on Eq. (1.4.8). Following this step, diagonalization simplified by rotational symmetry insights from group theory or by solution of the secular equation will separate the 5×5 matrix into a 2×2 and a 3×3 matrix expressed in standard group theory designations e_g and t_{2g}, according to:

$$e_g^b = \frac{1}{(2)^{1/2}}(d_2 + d_{-2})$$

$$e_g^a = d_0$$

$$t_{2g}^+ = d_1 \tag{1.4.13}$$

$$t_{2g}^- = d_{-1}$$

$$t_{2g}^0 = \frac{1}{(2)^{1/2}}(d_2 - d_{-2})$$

The degree of crystal-field quenching of the orbital angular momentum about the [001] axis can be checked by a straightforward application of the l_z operator to show that in addition to e_g^a, the e_g^b and t_{2g}^0 states now have zero z-axis angular momentum, while the $t_{2g}^\pm$ states retain $m_l = \pm 1$.

If the function set of Eq. (1.4.13) are grouped into the set of linear combinations in real form sketched in Fig. 1.11, they are expressed as

$$e_g \begin{cases} d_{x^2-y^2} = \left(\frac{3}{2}\right)^{1/2}(d_2 + d_{-2}) = \left(\frac{3}{2}\right)^{1/2}(Y_2^2 + Y_2^{-2}) = \frac{(3)^{1/2}}{2}(x^2 - y^2) \\ d_{z^2} = d_0 = Y_2^0 = \frac{1}{2}(3z^2 - r^2) \end{cases}$$

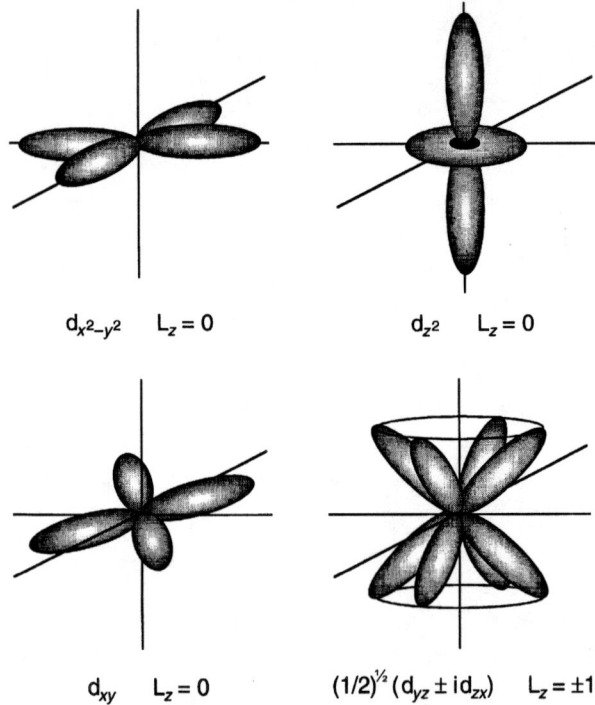

$d_{x^2-y^2}$ $L_z = 0$ $\qquad$ d_{z^2} $L_z = 0$

d_{xy} $L_z = 0$ $\qquad$ $(1/2)^{1/2}(d_{yz} \pm id_{zx})$ $L_z = \pm 1$

Figure 1.11. The orbital eigenfunctions of the d-electron shell.

$$t_{2g} \begin{cases} d_{xy} = \dfrac{(3)^{1/2}}{i(2)^{1/2}}(d_2 - d_{-2}) = \dfrac{(3)^{1/2}}{i(2)^{1/2}}(Y_2^2 - Y_2^{-2}) = (3)^{1/2}xy \\ d_{xz} = -\left(\dfrac{3}{2}\right)^{1/2}(d_1 - d_{-1}) = -\left(\dfrac{3}{2}\right)^{1/2}(Y_2^1 - Y_2^{-1}) = (3)^{1/2}xz \\ d_{yz} = -\dfrac{(3)^{1/2}}{i(2)^{1/2}}(d_1 + d_{-1}) = -\dfrac{(3)^{1/2}}{i(2)^{1/2}}(Y_2^1 + Y_2^{-1}) = (3)^{1/2}yz \end{cases} \quad (1.4.14)$$

where the radial factor $R(r)$ and other common factors have been dropped for convenience. A more general nomenclature for these states that is used where multiple electrons are involved, is A_{1g}, A_{2g} for singlets (also B_{1g}, B_{2g} in lower symmetry refinements), E_g for doublets and T_{1g}, T_{2g} for triplets. With this set of wave functions, the matrix is diagonal, so that the energy eigenvalues of Eq. (1.4.8) become

$$E_g \to \begin{cases} \varepsilon_0 + \langle d_{x^2-y^2} | \mathcal{V}_{cf}^{oct} | d_{x^2-y^2} \rangle = \varepsilon_0 + \varepsilon_1 \\ \varepsilon_0 + \langle d_{z^2} | \mathcal{V}_{cf}^{oct} | d_{z^2} \rangle = \varepsilon_0 + \varepsilon_1 \end{cases}$$

$$T_{2g} \to \begin{cases} \varepsilon_0 + \langle d_{xy} | \mathcal{V}_{cf}^{oct} | d_{xy} \rangle = \varepsilon_0 + \varepsilon_2 \\ \varepsilon_0 + \langle d_{xz} | \mathcal{V}_{cf}^{oct} | d_{xz} \rangle = \varepsilon_0 + \varepsilon_2 \\ \varepsilon_0 + \langle d_{yz} | \mathcal{V}_{cf}^{oct} | d_{yz} \rangle = \varepsilon_0 + \varepsilon_2 \end{cases} \quad (1.4.15)$$

where V_{cf}^{oct} is given by Eq. (1.4.11) for z parallel to [001] axis. Equation (1.4.15) indicates that the fivefold degeneracy of the D term is split into an upper doublet and lower triplet as sketched in Fig. 1.12. For the case of a single empty d state or "hole" (d^9), the unpaired spin occupies a state in the E_g doublet, which therefore becomes the ground term shown as part of the inverted D level structure of Fig. 1.12. The correspondence between d^n and d^{10-n} configurations can be reasoned by recognizing that d^n features electrons and d^{10-n} holes, which become distinguishable under the influence of the ligand charges. For this reason, level inversion occurs between the d^n and d^{10-n} ions, i.e., equal numbers of unpaired electron spins versus "hole" spins. A further convention is to label the overall splitting equal to 10Dq, where Dq > 0. Then,

$$\varepsilon_1 - \varepsilon_2 = 10\text{Dq} \tag{1.4.16}$$

Since diagonal elements remain unchanged before and after the perturbation is applied, for the five ^{2}D orbital states

$$2(\varepsilon_0 + \varepsilon_1) + 3(\varepsilon_0 + \varepsilon_2) = 5\varepsilon_0 \tag{1.4.17}$$

and it follows that, if ε_0 is arbitrarily set to zero,

$$\begin{aligned} ^2E_g &\rightarrow \varepsilon_1 = 6\text{Dq} \\ ^2T_{2g} &\rightarrow \varepsilon_2 = -4\text{Dq} \end{aligned} \tag{1.4.18}$$

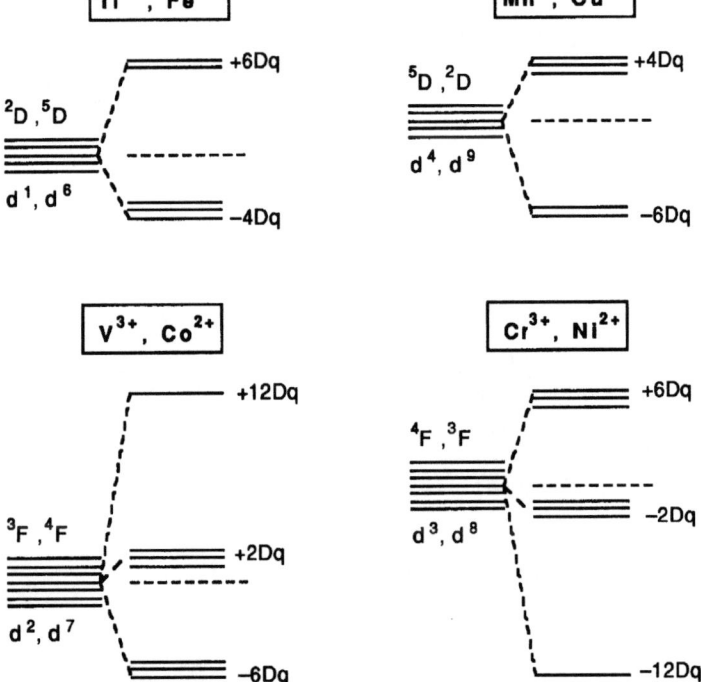

Figure 1.12. Multiple electron crystal-field energy levels for D and F terms indicating correspondence between members of the lower- and upper-halves of the $3d^n$ transition series.

The value of Dq is determined semi-empirically, i.e., by experiment, but an expression for it can be obtained analytically.[12,13]

$$Dq = \pm\left(\frac{2}{63}\right)D_4\langle r^4\rangle \quad \text{for D states}$$

$$= \mp\left(\frac{2}{315}\right)D_4\langle r^4\rangle \quad \text{for F states} \tag{1.4.19}$$

where the lower signs apply to ions of the lower-half of the d^n series.

From the eigenfunctions listed as Eq. (1.4.14) that are sketched in Fig. 1.11, the ordering of the energy levels can be determined by inspection of the relative positions of the negatively charged lobes in relation to the negative ligand point charges. Note that in an octahedral site the e_g orbitals are directed towards the negatively charged ligands and therefore will assume the higher energy states.

From these orbital sketches, the existence of an unquenched l_z angular momentum component may also be discerned. A test for deciding whether an orbital momentum about an axis can still be present is whether the eigenstate can be transformed into another eigenstate within its degenerate manifold by a rotation about that axis. In this case of the [001] as axis of quantization, it can be seen by visual inspection that $d_{x^2-y^2}$ rotates into d_{xy} by a 45° rotation about the z axis, and that the same applies to d_{xz} and d_{yz}. Only the latter pair are eigenstates in a cubic field, however, which means that the remaining three states have their l_z fully quenched, including the degenerate e_g orbitals. This latter condition is important for the discussion of the J–T effect.

The second case to be discussed is the three-electron ^4F term, which is of greater historical importance than the one-electron case because it was the basis for the invention of the maser. For this situation, the orbital spherical harmonics are the Y_3 group,[7]

$$Y_3^{-3} = \left(\frac{7}{4\pi}\right)^{1/2}\left(\frac{3}{8}\right)^{1/2}\frac{(x-iy)^3}{r^3}$$

$$Y_3^{-2} = \left(\frac{7}{4\pi}\right)^{1/2}\left(\frac{15}{8}\right)^{1/2}\frac{z(x-iy)^2}{r^3}$$

$$Y_3^{-1} = \left(\frac{7}{4\pi}\right)^{1/2}\left(\frac{3}{16}\right)^{1/2}\frac{(x-iy)(5z^2-r^2)}{r^3}$$

$$Y_3^0 = \left(\frac{7}{4\pi}\right)^{1/2}\left(\frac{1}{4}\right)^{1/2}\frac{z(5z^2-r^2)}{r^3} \tag{1.4.20}$$

$$Y_3^1 = -\left(\frac{7}{4\pi}\right)^{1/2}\left(\frac{3}{16}\right)^{1/2}\frac{(x+iy)(5z^2-r^2)}{r^3}$$

$$Y_3^2 = \left(\frac{7}{4\pi}\right)^{1/2}\left(\frac{15}{8}\right)^{1/2}\frac{z(x+iy)^2}{r^3}$$

$$Y_3^3 = -\left(\frac{7}{4\pi}\right)^{1/2}\left(\frac{5}{16}\right)^{1/2}\frac{(x+iy)^3}{r^3}$$

For the F states, the Y_6 terms of Eq. (1.4.11) must also be part of the calculation. Solutions of the resulting secular equation are a singlet ground state A_{2g} and two excited triplets T_{1g} and T_{2g}, with corresponding eigenfunctions

$$T_{1g} \begin{cases} \left(\dfrac{3}{8}\right)^{1/2} Y_3^1 + \left(\dfrac{5}{8}\right)^{1/2} Y_3^{-3} \\ \left(\dfrac{3}{8}\right)^{1/2} Y_3^{-1} + \left(\dfrac{5}{8}\right)^{1/2} Y_3^3 \\ Y_3^0 \end{cases}$$

$$T_{2g} \begin{cases} \left(\dfrac{3}{8}\right)^{1/2} Y_3^1 - \left(\dfrac{5}{8}\right)^{1/2} Y_3^{-3} \\ \left(\dfrac{3}{8}\right)^{1/2} Y_3^{-1} - \left(\dfrac{5}{8}\right)^{1/2} Y_3^3 \\ \dfrac{1}{(2)^{1/2}} (Y_3^2 + Y_3^{-2}) \end{cases} \quad (1.4.21)$$

$$A_{2g} \left\{ \dfrac{1}{(2)^{1/2}} (Y_3^2 - Y_3^{-2}) \right\}$$

The energy level schemes for Eq. (1.4.21) and the inverted arrangement are sketched in Fig. 1.12(c) and (d). Following the reasoning leading up to Eq. (1.4.19), for ^4F of d^3 and ^4F of d^7 the term energies are

$$\begin{aligned} T_{1g} &\to 6Dq \\ T_{2g} &\to -2Dq \\ A_{2g} &\to -12Dq \end{aligned} \quad (1.4.22)$$

The eigenfunctions of Eq. (1.4.21) can serve as an instructional example of orbital angular momentum quenching. The ground state is a linear combination of two spherical harmonics of the $L=2$ manifold that yield the singlet A_{2g} term under the influence of the cubic crystal field. As such, it has the symmetry properties of a singlet s orbital and therefore would carry the properties of $L=0$; its only contribution to the magnetic moment must come from the ion spin. Conversely, if the orbital levels are inverted in energy, the triplet T_{2g} becomes the ground state and the orbital angular momentum would have the characteristics of a triplet p state with $L=1$. In this case, L is reduced from the d state value of 2 down to 1, and the result is only partial quenching because a threefold degeneracy remains in the ground state. This residual degeneracy is an important factor in the properties of certain $3d^n$ transition ions, e.g., Co^{2+}.

It can be pointed out that where the crystal-field splitting parameter Dq is small enough to allow the influence of the upper terms in a subsequent perturbation calculation, appropriate orbital contributions to the magnetic moment will enter the eigenfunctions after spin–orbit and magnetic field perturbations are applied. Recalling the ground state terms of the $3d^n$ series listed in Table 1.4, we can now mention that the two example solutions for ^2D and ^4F will apply equally to the ^5D

Table 1.6. 3d^n States and Energies in Weak Octahedral Fields

d Electrons	Orbital Ground State	States and Energies in Dq^a
d^1	^{2}D	^{2}E$_g$ (+6) ^{2}T$_{2g}$ (−4)
d^2	^{3}F	^{3}A$_{2g}$ (+12) ^{3}T$_{2g}$ (+2) ^{3}T$_{1g}$ (−6)
d^3	^{4}F	^{4}T$_{1g}$ (+6) ^{4}T$_{2g}$ (−2) ^{4}A$_{2g}$ (−12)
d^4	^{5}D	^{5}T$_{2g}$ (+4) ^{5}E$_g$ (−6)
d^5	^{6}S	^{6}A$_{1g}$ (0)
d^6	^{5}D	^{5}E$_g$ (+6) ^{5}T$_{2g}$ (−4)
d^7	^{4}F	^{4}A$_{2g}$ (+12) ^{4}T$_{2g}$ (+2) ^{4}T$_{1g}$ (−6)
d^8	^{3}F	^{3}T$_{1g}$ (+6) ^{3}T$_{2g}$ (−2) ^{3}A$_{2g}$ (−12)
d^9	^{2}D	^{2}T$_{2g}$ (+4) ^{2}E$_g$ (−6)

^{a}For tetrahedral (O_4) coordinations, multiply Dq by −4/9; for cubic (O_8), multiply by −8/9.

and ^{3}F cases. Table 1.6 lists these crystal field terms for the series in units of Dq, with their signs adjusted to take into account the sign reversal for the upper-half of the series.

The point charge calculation may also be approached by another powerful technique called "operator equivalents" developed by Stevens.[13] This method consists of the replacement of the Cartesian operator functions of the V_{cf} potential energy with the equivalent L or J (whichever is applicable) angular momentum operators. Expressed in operator equivalents, the first part of the D_4 term of the octahedral field given by Eq. (1.4.9):

$$V_{cf}^{oct} = D_4[x^4 + y^4 + z^4 - (3/5)r^4]$$
$$= D_4\{L_x^4 + L_y^4 + L_z^4 - (1/5)L(L+1)[3L(L+1) - 1]\} \quad (1.4.23)$$

Since the eigenfunctions of these angular momentum operators are linear combinations of the spherical harmonics, the calculation of matrix elements is straightforward. Operator equivalents can be very useful for quantitative calculations of more complex symmetries that involve higher order terms in V_{cf} and also for cases of higher L (or J) values. An introduction to these techniques is given in Ballhausen[4] and Low[11] and a more comprehensive discussion including many tables of matrix elements may be found in Hutchings.[14]

1.4.5. Group Theory and Lower Symmetry

Conventional perturbation calculations to determine crystal-field states can become arduous for more complicated systems. The solutions for the simple cases outlined in Section 1.4.4 will prove almost sufficient for our discussion of the various magnetic properties. To cope with the frequently encountered trigonal (rhombohedral), tetragonal, and orthorhombic distortions of the cubic coordinations, however, the solutions are found by a shortcut that is derived from symmetry considerations. In the point charge calculations, diagonalization of matrices by solutions of higher order secular determinants may be accomplished by applying group theory to determine not only the best linear combinations of wave functions but also the degeneracies of the different eigenstates, e.g., the A_{2g}, E_g, T_{1g}, and T_{2g} terms of the type defined by Eq. (1.4.21).

Unfortunately, the scope of this text will not permit a detailed exposition of group theory. The interested reader is directed towards any number of excellent treatments of this subject, including those cited earlier.[3-5] For the purposes at hand, we need to recognize that the diagonalization process involves the construction of wave function combinations that conform to the symmetry of the perturbation operator. Group theory provides a method for predetermining the correct eigenfunction combinations for a particular perturbation problem and is useful in solving for the eigenfunctions of lower symmetry fields.

For the simple case of a descent in symmetry from cubic $O_h \rightarrow$ tetragonal $D_{4h} \rightarrow$ orthorhombic D_{2h}, relation of the basis vector lobes to the changing locations of the negatively charged ligands along the x, y, and z axes can be visualized in Fig. 1.11. For a tetragonal distortion shown in Fig. 1.5, T_{2g} (d_{xy}, d_{xz}, and d_{yz}) splits in the same way as the trigonal case, but the upper E_g doublet is also now split because of the relation of the $d_{x^2-y^2}$ and d_{z^2} lobes to the octahedral ligands. The orthorhombic distortion D_{2h} of Fig. 1.5 will remove the final degeneracy and split the d_{xz} and d_{yz} states. The splittings of the D orbital term in these fields are presented in Fig. 1.13.

The energy level structure for the special cases of the tetragonal or orthorhombic distortions that occur with pyramidal O_5 and planar O_4 coordinations shown in Fig. 1.6 may be inferred by extrapolating the results for the weak-field solutions. The descent in symmetry from cubic to the planar structure is particularly important in the cuprate superconductors. To obtain a quantitative sense of the influence of a strong tetragonal field, we return briefly to the point charge calculation and examine the effects of a tetragonal component V_{cf}^T added to the octahedral crystal-field energy of Eq. (1.4.10) to obtain[4]

$$V_{cf}^{oct+T} = V_{cf}^{oct} + V_{cf}^T \qquad (1.4.24a)$$

where

$$V_{cf}^T = A_T f_2(r) R(r) Y_2^0 + B_T f_4(r) R(r) Y_4^0 \qquad (1.4.24b)$$

and $f_2(r)$ and $f_4(r)$ are radially dependent coefficients.[15] With the eigenfunction set of Eq. (1.4.14), diagonal matrix elements may be obtained by straightforward integral computations. Part of this process involves the defining of two additional splitting parameters representing the integrals of the radial components of the respective matrix elements over space, according to

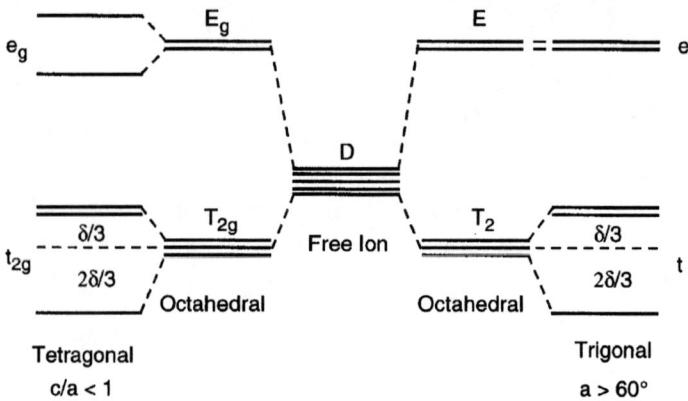

Figure 1.13. Comparison of d-electron energy levels in crystal fields of tetragonal ($c/a < 1$) and trigonal ($\alpha > 60°$) symmetries.

$$\text{Ds} = (3/2) \int f_2(r)[R(r)]^2 d\tau$$
$$\text{Dt} = (3/2) \int f_4(r)[R(r)]^2 d\tau \tag{1.4.25}$$

From these definitions, it can be shown that the energy states of the tetragonal perturbation follow directly from the diagonal matrix elements. The splitting of the upper doublet E_g of the $O_h + D_{4h}$ group is reflected in the matrix elements

$$\langle d^*_{x^2-y^2} | V^{oct}_{cf} + V^T_{oct} | d_{x^2-y^2} \rangle = 6\text{Dq} + 2\text{Ds} - \text{Dt}$$
$$\langle d^*_{z^2} | V^{oct}_{cf} + V^T_{oct} | d_{z^2} \rangle = 6\text{Dq} - 2\text{Ds} - 6\text{Dt} \tag{1.4.26a}$$

and that of the lower T_{2g} triplet is

$$\langle d^*_{x^2-y^2} | V^{oct}_{cf} + V^T_{cf} | d_{x^2-y^2} \rangle = 6\text{Dq} + 2\text{Ds} - \text{Dt}$$
$$\langle d^*_{z^2} | V^{oct}_{cf} + V^T_{cf} | d_{z^2} \rangle = 6\text{Dq} - 2\text{Ds} - 6\text{Dt} \tag{1.4.26b}$$

where the lowest d_{xz} and d_{yz} orbitals retain their degeneracy. For Ds and Dt > 0, the order of energy levels is shown in Fig. 1.14. As drawn, the structure is shown with the doublet as the ground state, but this may not necessarily be the case since the relative individual values of Ds and Dt would determine the correct order. Note that the uppermost state is still d_{z^2} and that it reaches a maximum separation of 10Dq from the next highest state, which is now d_{xy} instead of d_{z^2}. The crossover point where this upper state splitting becomes equal to 10Dq can be attained with a large tetragonal distortion, but may not necessarily require a complete removal of the two apical ligands along the z-axis that would leave only an O_4 planar coordination.

Another demonstration of energy level determinations by group theory is realized by the descent in symmetry from cubic $O_h \rightarrow$ trigonal D_{3d} that is commonly encountered in magnetic oxides. If the z-axis of quantization is taken as the [111] direction of threefold symmetry, the appropriate basis vectors may be constructed

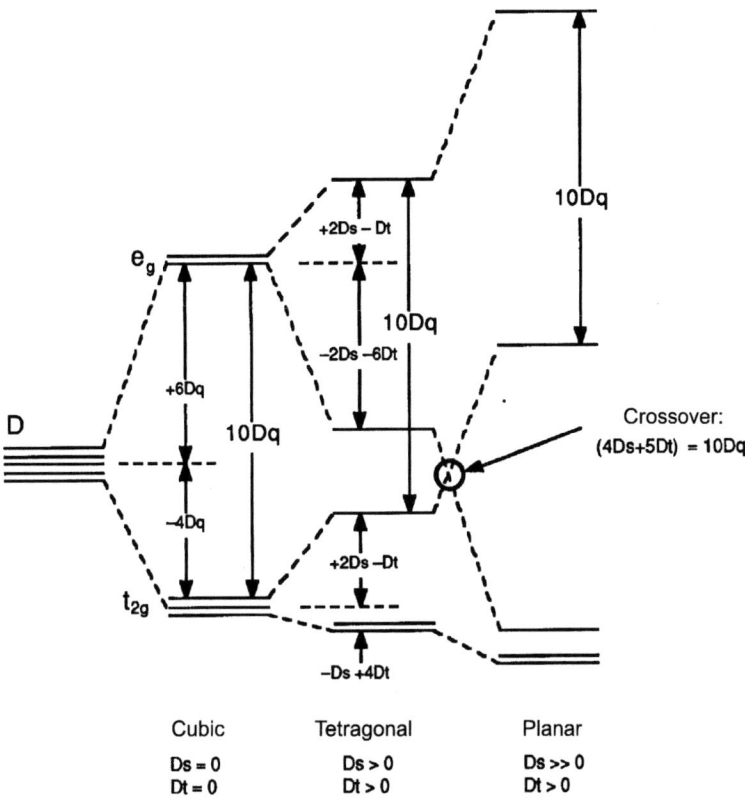

Figure 1.14. Details of orbital energy-level splittings as a tetragonal crystal evolves from cubic to planar, showing the 10Dq destabilization of the highest e_g level. Diagram is based on Fig. A.47 of ref. 5.

from the set of Eq. (1.4.14) to obtain eigenfunctions of V_{cf}^{oct} expressed in this frame of reference:[4]

$$e = \begin{cases} \frac{1}{(3)^{1/2}} d_{x^2-y^2} + \left(\frac{2}{3}\right)^{1/2} d_{xz} = e^+ \\ \frac{1}{(3)^{1/2}} d_{xy} - \left(\frac{2}{3}\right)^{1/2} d_{yz} = e^- \end{cases}$$

$$t_2 = \begin{cases} \left(\frac{2}{3}\right)^{1/2} d_{x^2-y^2} - \frac{1}{3^{1/2}} d_{xz} = t_2^+ \\ \left(\frac{2}{3}\right)^{1/2} d_{xy} + \frac{1}{3^{1/2}} d_{yz} = t_2^- \\ d_{z^2} = t_2^0 \end{cases} \quad (1.4.27)$$

As with the earlier eigenfunction set with the z-axis along the [001] direction, a straightforward application of the l_z operator along the [111] direction will verify that t_2^0 and the two e states have zero angular momentum, while the remaining two t_2 states retain $m_l = \pm 1$. If an analytical problem that involved a trigonal (rhombohedral) perturbation along the [111] axis were to be solved with this combination of basis vectors expressed in the regular cubic coordinate system with x, y, and z transformed back into the x', y', and z' coordinates set up coincident with the (001) family of axes, eigenfunctions for this purpose have been reported by Pryce and Runciman[16] and Dionne and Palm.[17]

For this set, the appropriate octahedral crystal-field potential energy is given by a relation analogous to Eq. (1.4.10), abbreviated and without normalizing factors[14]

$$V_{cf}^{oct} = Y_4^0 + \left(\frac{10}{7}\right)^{1/2}(Y_4^3 - Y_4^{-3}) \quad \text{with } z \text{ along [111] axis} \qquad (1.4.28)$$

where

$$Y_4^3 - Y_4^{-3} = -\left(\frac{9}{4\pi}\right)^{1/2}\left(\frac{35}{4}\right)^{1/2}\frac{z(x^3 - 3xy^2)}{r^4} \qquad (1.4.29)$$

The resulting eigenfunctions for a trigonal distortion of an octahedral site reveal that the lower triplet T_{2g} is split into a doublet and a singlet, while the degeneracy of the upper doublet E_g is unchanged, as shown in Fig. 1.13. The survival of the upper doublet is an important feature that distinguishes the trigonal from the tetragonal case. For a trigonal distortion, the crystal field potential energy is

$$V_{cf}^{oct+\tau} = V_{cf}^{oct} + V_{cf}^{\tau} \qquad (1.4.30)$$

The trigonal crystal field operator $V_{cf}^{\tau} = A_\tau f_2(r)R(r)Y_2^0 + B_\tau f_4(r)R(r)Y_4^0$ is applied in a fashion similar to that of the tetragonal field component V_{cf}^{T} given by Eq. (1.4.24), except that the set of orbital functions that it operates on are the group of Eq. (1.4.27).[18] Upon application of this perturbation, the matrix elements are

$$\begin{aligned}\langle e^\pm | V_{cf}^{oct} + V_{cf}^\tau | e^\pm \rangle &= 6Dq + (7/3)D\tau \\ \langle t_2^\pm | V_{cf}^{oct} + V_{cf}^\tau | t_2^\pm \rangle &= -4Dq + D\sigma + (2/3)D\tau \\ \langle t_2^0 | V_{cf}^{oct} + V_{cf}^\tau | t_2^0 \rangle &= -4Dq - 2D\sigma - 6D\tau \\ \langle t_2^\pm | V_{cf}^{oct} + V_{cf}^\tau | e^\pm \rangle &= -(2)^{1/2}D\sigma - (5(2)^{1/2}/3)D\tau \end{aligned} \qquad (1.4.31)$$

where $D\sigma$ and $D\tau$ are defined analogously to Ds and Dt of Eq. (1.4.25).

At this point, it is instructive to compare Eqs. (1.4.26) and (1.4.31). The tetragonal and trigonal cases are similar in that the T_{2g} (and T_2) group is split into a singlet and doublet, but as illustrated earlier in Fig. 1.13, the E_g (and E) term remains degenerate in the trigonal field. Moreover, we now see that the $t_2^\pm$ and $e^\pm$ states mix under the V_{oct}^τ perturbation. Pryce and Runciman[16] have studied this question in detail, but the problem can be simplified if we assume that $D\sigma \approx (5/3)D\tau$, and that the off-diagonal elements can be ignored.

1.4.6. Ground-State Occupancy Diagrams

When cations are placed in a lattice, the crystal field from the anion charges is the basis for the ionic part of the chemical bonding. Within the point charge approximation, the resultant electrostatic potential between neighboring ions (often approximated by a Madelung energy calculation) and the ionization potentials of the various ionic species gives the bonding energy of the system. For the 3d^n-electron series, the crystal field provides additional binding energy by lowering the ground state energy as part of the splitting of the orbital degeneracy that was reviewed in the previous section. The ligand-field stabilization is therefore an electronic component of the overall chemical bond, which may be further enhanced by a covalent component.

In the weak-field regime, a one-electron model may be constructed for the 3d^n series as a general qualitative approximation to provide an occupancy map of the ground state electronic structures sketched. The energy levels of the T_{2g} and E_g cubic field terms from Fig. 1.13 of the single d-electron energy level structure are used as a "floor plan" for the distributions of all the d electrons in the ground states diagrammed in Figs. 1.15 and 1.16.[19] In this abbreviation, the "Aufbau principle" is applied by adding d electrons sequentially beginning with the lowest state, while observing Hund's rule of spin polarization for the lower-half of the d shell and the Pauli principle of spin pairing for the upper-half. The stabilization energy of each configuration varies according to the degree of intra-orbital electron repulsion energy e^2/r_{ij} and is absorbed into the ionization potential. It must therefore be pointed out here that this approximation can apply with rigor only to the d^1 or d^9 cases. Where additional perturbations such as spin–orbit coupling, magnetic exchange, and Zeeman effects in an external magnetic field are involved, multi-electron systems with their corresponding excited terms would have to be taken into account.

For the weak-field case where $\mathcal{H}_{rep}$ exceeds $\mathcal{H}_{cf}$, this procedure establishes a ground state with the maximum available spin number. The electrons are distributed among the orbital states with spin pairing permitted only after each of the five orbitals are half filled (Hund's rule). Since the energy distribution also changes with

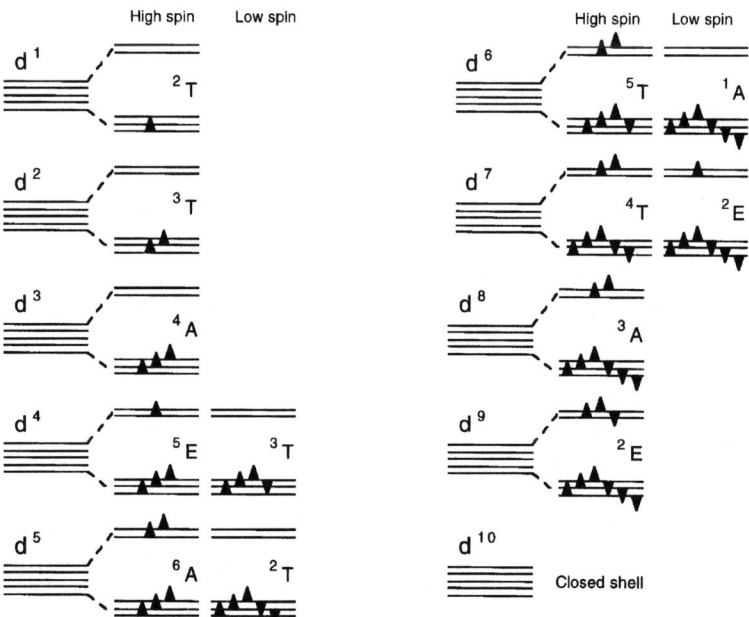

Figure 1.15. One-electron d-orbital occupancy diagrams: octahedral site.

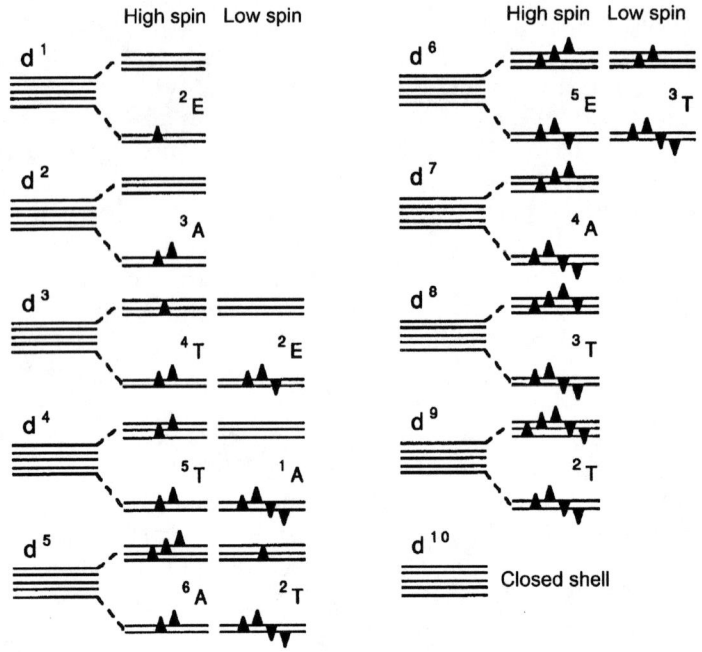

Figure 1.16. One-electron d-orbital occupancy diagrams: tetrahedral site.

the assignment of electrons, one of the main insights gained from this approximation is the relative magnitudes of the cation site stabilization energies. In Figs. 1.15 and 1.16, the one d-electron ground-state configurations for the octahedral and tetrahedral sites include the "low-spin" states that occur in strong crystal fields (violation of Hund's rule when the destabilization from the splitting overrides the e^2/r_{ij} spin polarization effects).

In the weak-field limit, the spins are polarized to the maximum extent and the result is logically called the "high-spin" state. Although these spin alignments are the usual situations in the $3d^n$ series, low-spin occupancies may also occur in selected instances where the cubic 10Dq splitting is large enough to breakdown the Hund's rule spin polarization and allow the lower shell to fill before the upper one. Many magnetic and electronic properties can be explained by the nature of these different electron distributions. In these "floor plan" diagrams of the ground state, the low-spin configurations are shown for comparison where they differ from the high-spin case.

For standard divalent or trivalent cations of the $3d^n$ series, low-spin states do not usually occur in tetrahedral sites because of the 4/9 reduction factor in the value of Dq, but the hypothetical configurations are included here for completeness. The respective configurations for octahedral and tetrahedral sites are listed in Tables 1.7 and 1.8, respectively, and their corresponding d-electron diagrams are shown in Figs. 1.15 and 1.16 together with stabilization energies in units of Dq. Comparison of these estimates of site stabilization energy can be used to decide the likely preference for ions in lattices such as spinels and garnets where both octahedral and tetrahedral sites are present. For the specific example of Cr^{3+}, a site preference energy of 4Dq can be deduced from the stabilization energies listed in the tables. The validity of this

Table 1.7. High- and Low-Spin d-Electron Stabilization Energies in an Octahedral Coordination

d Electrons	High-Spin Configuration	Stabilization Energy	Low-Spin Configuration	Stabilization Energy
1	t_{2g}^1	4Dq	t_{2g}^1	4Dq
2	t_{2g}^2	8Dq	t_{2g}^2	8Dq
3	t_{2g}^3	12Dq	t_{2g}^3	12Dq
4	$t_{2g}^3 e_g^1$	6Dq	t_{2g}^4	16Dq
5	$t_{2g}^3 e_g^2$	0	t_{2g}^5	20Dq
6	$t_{2g}^4 e_g^2$	4Dq	t_{2g}^6	24Dq
7	$t_{2g}^5 e_g^2$	8Dq	$t_{2g}^6 e_g^1$	18Dq
8	$t_{2g}^6 e_g^2$	12Dq	$t_{2g}^6 e_g^2$	12Dq
9	$t_{2g}^6 e_g^3$	6Dq	$t_{2g}^6 e_g^3$	6Dq
10	$t_{2g}^6 e_g^4$	0	$t_{2g}^6 e_g^4$	0

Table 1.8. High- and Low-Spin d-Electron Stabilization Energies in a Tetrahedral Coordination

d Electrons	High-Spin Configuration	Stabilizaton Energy	Low-Spin Configuration	Stabilizaton Energy
1	e_g^1	6Dq	e_g^1	4Dq
2	e_g^2	12Dq	e_g^2	8Dq
3	$e_g^2 t_{2g}^1$	8Dq	e_g^3	18Dq
4	$e_g^2 t_{2g}^2$	4Dq	e_g^4	24Dq
5	$e_g^2 t_{2g}^3$	0	$e_g^4 t_{2g}^1$	20Dq
6	$e_g^3 t_{2g}^3$	6Dq	$e_g^4 t_{2g}^2$	16Dq
7	$e_g^4 t_{2g}^3$	12Dq	$e_g^4 t_{2g}^3$	12Dq
8	$e_g^4 t_{2g}^4$	8Dq	$e_g^4 t_{2g}^4$	8Dq
9	$e_g^4 t_{2g}^5$	4Dq	$e_g^4 t_{2g}^5$	4Dq
10	$e_g^4 t_{2g}^6$	0	$e_g^4 t_{2g}^6$	0

model can be observed from heat of hydration data[20] for which the excess energy attributed to the octahedral ligand-field stabilization is seen to follow the values predicted in Table 1.7 for the high-spin case.

In the tables and figures presented so far, it is evident that the cubic fields alone fail to remove all of the ground-state degeneracy. Further stabilization can occur and there are two mechanisms by which this can take place locally: (i) orbit–lattice coupling manifested in the now-celebrated J–T spontaneous distortion effects; and (ii) the more conventional spin–orbit (S–O) coupling that can override and reverse the J–T distortion in select cases. Both of these phenomena are responsible for important magnetic and electronic behavior.

1.5. MAGNETOCRYSTALLINE ANISOTROPY

Since the effects of $\mathcal{H}_{cf}$ on the orbital states have been reviewed for D and F terms of the iron group, we now continue with an examination of the next perturbations for this series: spin–orbit coupling $\lambda L \cdot S$ and the combined effect of the orbital and spin moments in a magnetic field $m_B(L+2S) \cdot H$. To this end, we can examine the effects of the remaining electron interactions that influence the magnetic moment of an isolated ion. For the individual ions of the $3d^n$ transition series, the Zeeman operator $\mathcal{H}_Z$ is added to Eq. (1.4.6) according to

$$\begin{aligned}\mathcal{H} &= [\mathcal{H}_{\text{Coul}} + \mathcal{H}_{\text{rep}}] + \mathcal{H}_{\text{cf}} + \mathcal{H}_{LS} + \mathcal{H}_Z \\ &= [\mathcal{H}_{\text{Coul}} + \mathcal{H}_{\text{rep}}] + \mathcal{H}_{\text{cf}} + \lambda L \cdot S + m_B(L+g_e S) \cdot H \\ &= [\mathcal{H}_{\text{Coul}} + \mathcal{H}_{\text{rep}}] + \mathcal{H}_{\text{cf}} + \lambda L \cdot S + m_B g_{ij} S_i H_j\end{aligned} \qquad (1.5.1)$$

where m_B is the Bohr magneton, $g_e = 2$ is the g factor for an electron, and g_{ij} is its tensor element that reflects orbital quenching and anisotropy from the crystal field. In the context of magnetocrystalline anisotropy, the extent to which the crystal field lifts the orbital degeneracy determines the magnetoelastic properties. To this end, we will review the various electronic structures of the transition metal ions in ligand fields of initially cubic symmetry.

Analysis of any of these electronic configurations is solvable by conventional degenerate perturbation theory with the use of high-speed digital computation once the appropriate Hamiltonian matrices are set up. To demonstrate the origin of single-ion anisotropy as reflected in the variation of the g factor with magnetic field direction, the simplest case can introduce the reader to spin–lattice interactions.

1.5.1. Single-Ion Anisotropy

For the general case of a single d electron in an octahedral crystal field with an orthorhombic distortion dominated by a z-axis extension, the energy level structure is shown in Fig. 1.17. Note that for cases where the free-ion term is a $^{2S+1}D_J$, the ground state occupancy diagram conforms to the actual energy states of the ion. The ground term is indicated by $^2D_{5/2}$ and the eigenvectors of the crystal-field states listed as $|0\rangle, |1\rangle, |2\rangle, |3\rangle$, and $|4\rangle$, where from Eq. (1.4.14):

$$\begin{aligned}|4\rangle &= d_{z^2} = \frac{1}{2}(3z^2 - r^2) \\ |3\rangle &= d_{x^2-y^2} = \frac{3^{1/2}}{2}(x^2 - y^2) \\ |2\rangle &= d_{yz} = 3^{1/2} \; yz \\ |1\rangle &= d_{xz} = 3^{1/2} \; xz \\ |0\rangle &= d_{xy} = 3^{1/2} \; xy\end{aligned} \qquad (1.5.2)$$

The solutions of these perturbation effects are carried out in two steps: first, the eigenstates of $\mathcal{H}_{LS}$ term are determined and a new ground state found; then, the Zeeman term is applied to determine the anisotropy of the magnetic moment that is reflected in the differing values of the g_{ij} tensor elements. The problem was analyzed in terms of the Ti^{3+}-substituted hydrated alum salt $Rb^{1+}(Al^{3+}, Ti^{3+})(SO_4)_2^{2-} \cdot 12H_2O$ from three approaches,[21] beginning with the most general.* First, degenerate theory

---

*The octahedron of H_2O molecules simulates O^{2-} ligands in these compounds because the O part of the water dipole is closest to the Ti^{3+} cation.

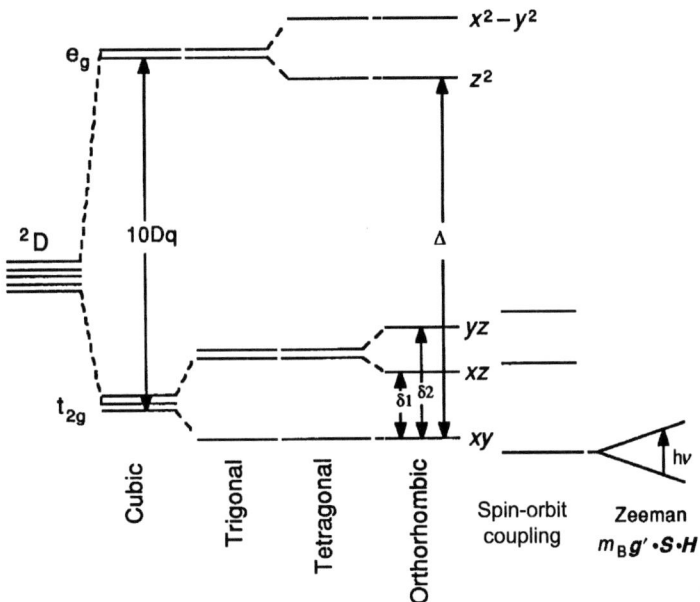

Figure 1.17. 3d¹ energy-level structure with complete removal of orbital degeneracy in an orthorhombic crystal field, showing spin stabilization in a Zeeman splitting of the ground state Kramers doublet, with $g' \neq 2$.

was applied whereby a full perturbation matrix is established for the elements $\langle n|\mathcal{H}_{cf} + \lambda \mathbf{L} \cdot \mathbf{S}|k\rangle$, where $\langle n|$ and $|k\rangle$ are eigenfunctions from Eq. (1.5.2) with the spin states included. For the spin–orbit contributions the conventional operator is $\mathbf{L} \cdot \mathbf{S} = L_z S_z + (1/2)(L_+ S_- + L_- S_+)$. Since each orbital state has a twofold spin degeneracy ($m_s = \pm 1/2$) this exercise involves a total of 10 wave functions, requiring the diagonalization of a 10×10 matrix.

The secular equation constructed from Eq. (1.4.8) is separable into two identical functions that yield degenerate energy states, each representing a twofold spin degeneracy (Kramers doublets, which occur for odd-electron cases). The resulting ground state eigenfunctions

$$|0+\rangle = a|0,\tfrac{1}{2}\rangle + b|1,-\tfrac{1}{2}\rangle + c|2,-\tfrac{1}{2}\rangle + d|3,\tfrac{1}{2}\rangle$$
$$|0-\rangle = a|0,-\tfrac{1}{2}\rangle + b|1,\tfrac{1}{2}\rangle - c|2,\tfrac{1}{2}\rangle - d|3,-\tfrac{1}{2}\rangle \qquad (1.5.3)$$

where the highest energy $|4\rangle$ level has been dropped to simplify the calculation with only a small loss in accuracy, are then used for the 2×2 ground-state Zeeman matrix according to $\langle n|m_B(\mathbf{L} + 2\mathbf{S}) \cdot \mathbf{H}|k\rangle$. To compute the Zeeman splittings for a particular direction of $\mathbf{H}$, we determine the matrix elements by applying the appropriate operators, recalling that $L_x = (1/2)(L_+ + L_-)$ and $L_y = -(i/2)(L_+ - L_-)$, and likewise for the spin operators. In each case, the Zeeman matrix is diagonalized and the expressions for the energies of the split spin doublet are subtracted to give $\Delta E_i = g_i m_B H_i$, from which expressions for g_z, g_x, and g_y in terms of the a, b, c, and d coefficients can be deduced. To solve for the g factors, values are assigned to the crystal-field splittings δ_1, δ_2, and Δ shown in Fig. 1.17. A value for λ must also be assigned, usually the free-ion value. Conversely, if g factors are found from measurement, e.g., paramagnetic resonance (EPR) at microwave frequencies in many cases, the reverse procedure could be followed to find values for δ_1, δ_2, and Δ.

If we assume a lower symmetry crystal field that produces an orbital singlet ground state, we can also treat the problem as a conventional nondegenerate case whereby the matrix elements of interest become the off-diagonal $\langle k|\lambda L \cdot S|n\rangle$ for use in the standard relation for the perturbed eigenfunction

$$|n'\rangle = |n\rangle - \sum_{k>n} |k\rangle \frac{\langle k|\lambda L \cdot S|n\rangle}{E_k - E_n} \tag{1.5.4}$$

where $|n\rangle$ refers to $|0, +1/2\rangle$ and $|0, -1/2\rangle$ and $|k\rangle$ is any of the other excited orbital eigenfunctions in Eq. (1.5.2) with the $|+1/2\rangle$ and $|-1/2\rangle$ spin functions added. After solving the secular equation, we obtain for the ground states

$$\begin{aligned}|0+\rangle &= |0,\tfrac{1}{2}\rangle - i(\lambda/2\delta_1)|1,-\tfrac{1}{2}\rangle - (\lambda/2\delta_2)|2,-\tfrac{1}{2}\rangle + i(\lambda/\Delta)|3,\tfrac{1}{2}\rangle \\ |0-\rangle &= |0,-\tfrac{1}{2}\rangle - i(\lambda/2\delta_1)|1,\tfrac{1}{2}\rangle + (\lambda/2\delta_2)|2,\tfrac{1}{2}\rangle - i(\lambda/\Delta)|3,-\tfrac{1}{2}\rangle\end{aligned} \tag{1.5.5}$$

To calculate the Zeeman energy splitting of the Kramers doublet $|0\pm\rangle$ we follow the same procedure for the degenerate solution above to produce matrix elements for the z direction

Table 1.9. Anisotropic g Factors and Crystal Field Splittings of Ti^{3+} (d^1) in Octahedral Sites ($\lambda = 154$ cm^{-1})

Ti^{3+} Host	Radius (Å)	Theory	g_z	g_x	g_y	δ_1 (cm^{-1})	δ_2 (cm^{-1})	Δ^a (cm^{-1})	Ref.
Rb alum	1.48	Spin Hamiltonian	1.895	1.715	1.767	1070	1310	11,500	21
		Nondegenerate	1.895	1.715	1.767	1050	1320	17,000	21
		Degenerate	1.895	1.715	1.767	1050	1320	20,300	21
K alum	1.33	Degenerate	1.975	1.828	1.897	1780	2950	20,300	23
Tl alum	1.40	Degenerate	1.938	1.790	1.834	1462	1843	20,300	22
Na alum	0.95	Degenerate	~2.00	~1.86	~1.86	> 2000	> 2000	20,300	23

			$g_\parallel$	$g_\perp$	δ (cm^{-1})	Δ (cm^{-1})	
CsTi alum	1.69	Degenerate	1.24	0.93	~200	20,300	24
		Degenerate	1.19	0.70	~200	20,300	24
		Degenerate	1.17	0.23	~200	20,300	24
Cs alum	1.69	Degenerate	1.25	1.14	~300	20,300	25
Al_2O_3	–	–	1.067	< 0.1			26

[a] The value $\Delta = 20,300$ cm^{-1} was measured by optical absorption[27] in an aqueous solution, where the ligands are an octahedron of water molecules. This situation is close to that of oxygen ligands because of the relative location of the O in the H_2O molecule. To obtain the measured value of $g_z = 1.895$ by the degenerate method, $\Delta = 22,000$ cm^{-1}.

$$\mathcal{H}_{11} = \left[1 - \frac{4\lambda}{\Delta} - \frac{\lambda^2}{2\delta_1\delta_2} - \frac{\lambda^2}{4}\left(\frac{1}{\delta_1^2} + \frac{1}{\delta_2^2}\right) + \frac{\lambda^2}{\Delta^2}\right] m_B H_z \qquad (1.5.6)$$

$$\mathcal{H}_{22} = -\mathcal{H}_{11}$$

$$\mathcal{H}_{12} = \mathcal{H}_{21} = 0$$

By solving the secular equation and relating $\Delta E = 2\mathcal{H}_{11} = g_z m_B H_z$, and then repeating the procedure for the x and y directions, the diagonal elements of the g tensor can be expressed as

$$g_z = 2 - \frac{8\lambda}{\Delta} - \frac{\lambda^2}{\delta_1\delta_2} - \frac{\lambda^2}{2}\left(\frac{1}{\delta_1^2} + \frac{1}{\delta_2^2}\right) + \frac{2\lambda^2}{\Delta^2}$$

$$g_x = 2 - \frac{2\lambda}{\delta_1} + \frac{\lambda^2}{2}\left(\frac{1}{\delta_1^2} - \frac{1}{\delta_2^2}\right) + \frac{2\lambda^2}{\delta_2\Delta} - \frac{2\lambda^2}{\Delta^2} \qquad (1.5.7)$$

$$g_y = 2 - \frac{2\lambda}{\delta_2} + \frac{\lambda^2}{2}\left(\frac{1}{\delta_2^2} - \frac{1}{\delta_1^2}\right) + \frac{2\lambda^2}{\delta_1\Delta} - \frac{2\lambda^2}{\Delta^2}$$

As listed in Table 1.9 for Rb alum, splitting energies for the measured $g_z = 1.895$, $g_x = 1.715$, and $g_y = 1.767$ are $\delta_1 = 1050$, $\delta_2 = 1320$, and $\Delta = 17{,}000$ cm^{-1}.[21] Parameter values for other members of the alum family are also listed.[22-27]

To complete this instructional example, we introduce the "spin" Hamiltonian $\mathcal{H}_s$ approximation to the perturbation terms of Eq. (1.5.1) that was developed by Pryce and Abragam.[28] This method was used to analyze the paramagnetic resonance spectrum of ions with an orbital *singlet* ground state, in particular the ^{4}F-state Cr^{3+} ion that was the key to the early ruby maser and laser demonstrations. If the excited orbital states of the crystal-field splittings are high enough to be ignored in the following calculations, the approach differs from those of the two conventional methods in that the perturbation matrix elements are computed for combined spin–orbit and Zeeman perturbations, $\langle k|\lambda L \cdot S + m_B(L+2S)\cdot H|n\rangle$, with the result that the spin S is treated as the only angular momentum quantum number and the contribution of L becomes absorbed in the g factors, consistent with Eq. (1.5.7) but with less precision. Since the $|0\rangle$ state is an orbital singlet that carries no orbital angular momentum, i.e., S-state behavior, the only first-order nonzero elements will be $\langle 0|2m_B S \cdot H|0\rangle$ and the second-order correction will simplify to

$$\langle k|\mathcal{H}_s|0\rangle = -\sum_{k>0} \frac{|\langle k|\lambda L \cdot S + m_B H \cdot L|0\rangle|^2}{E_k - E_0} \qquad (1.5.8)$$

If only the first- and second-order terms that are linear in $\lambda m_B H$ are included, Eq. (1.5.8) can be reduced to

$$\langle k|\mathcal{H}_s|0\rangle = 2m_B(\delta_{ij} - \lambda\Lambda_{ij})H_i S_j \qquad (1.5.9)$$

where

$$\Lambda_{ij} = \sum_{k>0} \frac{\langle 0|L_i|k\rangle\langle k|L_j|0\rangle}{E_k - E_0} \qquad (1.5.10)$$

From Eq. (1.5.8) we note that the matrix elements used are only those linked to the ground state, thereby substantially simplifying the solution. With this approximation only the terms to first-order in λ appear, and Eq. (1.5.7) reduces to

$$g_z = 2 - \frac{8\lambda}{\Delta}$$
$$g_x = 2 - \frac{2\lambda}{\delta_1} \quad (1.5.11)$$
$$g_y = 2 - \frac{2\lambda}{\delta_2}$$

For Rb alum measured g factors are fitted to splitting energies of $\delta_1 = 1070$, $\delta_2 = 1310$, and $\Delta = 11{,}500$ cm^{-1}, as listed in Table 1.9. Note that only the larger cubic field component Δ deviates significantly from the measured reference value of 20,300 cm^{-1}.[27]

For ions with more complicated spin multiplets, i.e., $S > 1/2$, such as Cr^{3+} and Ni^{2+}, higher order terms in λ become important for spectral analysis and a more general form of Eq. (1.5.9) can be expressed as[28]

$$\langle k|\mathcal{H}_s|0\rangle = 2m_B\left(\delta_{ij} - \lambda\Lambda_{ij}\right)H_iS_j - \lambda^2\Lambda_{ij}S_iS_j \quad (1.5.12)$$

leading to the general form of the spin Hamiltonian expressed to second-order in λ as

$$\mathcal{H}_s = m_B \mathbf{g} \cdot \mathbf{H} \cdot \mathbf{S} + DS_z^2 + E\left[S_x^2 - S_y^2\right] + \frac{1}{6}a\left[S_z^4 + S_x^4 + S_y^4\right] \quad (1.5.13)$$

where the anisotropic g is now expressed in tensor format. The zero-field splitting parameters D and E (both functions of Λ_{ij}) can be determined from experiment and can be used to deduce information about the electronic structure of the ions in their crystal field. A sketch of the resulting energy level structure of $3d^3$ is presented in Fig. 1.18. Included in this version of the spin Hamiltonian is the fourth-order term characterized by the a parameter which becomes significant in S-state ions $3d^5$ (Fe^{3+}, Mn^{2+}) and $4f^7$ (Gd^{3+}, Eu^{2+}) for which the first-order orbital angular momentum contributions are absent. For this case, the small anisotropy effects are caused by the spin–orbit coupling contributions from the excited orbital terms that form hybrids with the ^6S ground state., e.g., ^4G. In Fig. 1.10, the structure is sketched with the expected location of relevant crystal field and exchange field perturbations indicated.*

Equations (1.5.7) and (1.5.11) expose the role of spin–orbit coupling in determining the magnitude and anisotropy of the magnetic moments in these transition-metal ions. The ratios of the constant λ to the various orbital level splittings can dictate not only the degree of anisotropy of the magnetic moment as the magnetic field is applied in different directions, but also whether the moment increases or decreases, as in this case of the Ti^{3+} ion. According to Table 1.10, the λ constants increase in magnitude and reverse sign in the upper-half of the $3d^n$ series, thereby alerting us to the fact that g factors greater than 2 as well as increased anisotropy are to be expected from the $3d^n$ ions with $n > 5$.

---

*Where applicable, the exchange field is treated as an external magnetic field acting on the spin moment and causing a (large) Zeeman-type splitting. The magnitude of spin degeneracy splitting from an exchange field can be on the order of 10^3 cm^{-1} in oxides such as ferrites.

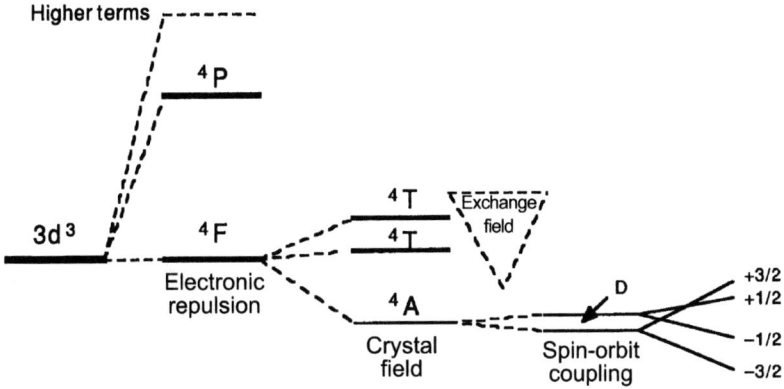

Figure 1.18. Generic model of energy-level structure of $3d^3$ configuration, typical of the Cr^{3+} ^4F-state ion.

Before advancing to some of the more esoteric aspects of the electronic level structures of the $3d^n$ series, it is instructive to compare the $3d^3$ series case, e.g., Cr^{3+}, with its rare-earth counterpart $4f^3$, usually Nd^{3+}, sketched in Fig. 1.19. Contrasting this diagram with Fig. 1.18, one observes how the crystal-field and spin–orbit coupling transpose their insertion points in the perturbation heirarchy, consistent with Table 1.5. Magnetocrystalline anisotropy is also derived from the combination of these interactions, but in reverse order. Characteristic of the entire rare-earth series, J is largely preserved even into the exchange interactions that couple only S vectors. Without significant L quenching, J remains a "good" quantum number and will follow the S directions to provide the magnetic moment.

Table 1.10. $3d^n$ Transition Group Data[a]

	d^1	d^2	d^3	d^4	d^5	d^6	d^7	d^8	d^9
Free ion	Ti^{3+}	V^{3+} Ti^{2+}	Cr^{3+} Cr^{2+}	Mn^{3+} Cr^{2+}	Fe^{3+} Mn^{2+}	Fe^{2+} Co^{3+}	Co^{2+} Ni^{3+}	Ni^{2+} Cu^{3+}	Cu^{2+}
λ (cm^{-1})	154	104	87 55	85 57	—	-100	-180	-335	-852
Hund's term	$^2D_{3/2}$	$^3F_2$	$^4F_{3/2}$	$^5D_0$	$^6S_{5/2}$	$^5D_4$	$^5F_{9/2}$	$^3F_4$	$^5D_{3/2}$
High spin	—	—	—	e_g^1 t_{2g}^3	e_g^2 t_{2g}^3	e_g^2 t_{2g}^4	e_g^2 t_{2g}^5	e_g^2 t_{2g}^6	e_g^3 t_{2g}^6
	t_{2g}^1	t_{2g}^2	t_{2g}^3						
Low spin	—	—	—	—	—	—	e_g^1 t_{2g}^6	e_g^{2b} t_{2g}^6	e_g^{3b} t_{2g}^6
	t_{2g}^1	t_{2g}^2	t_{2g}^3	t_{2g}^4	t_{2g}^5	t_{2g}^6			
S (high spin) (low spin)	1/2	1	3/2	2 1	5/2 1/2	2 0	3/2 1/2	1 0^b	1/2
g	0–~1.9	~1.9	1.9–2	1.9–2	~2	>3–~7	>4	>2.25	2–2.5

[a] Data obtained from W. Low, Ref. 11, Table XIX; and J. S. Griffith, Ref. 3, Appendix 6.
[b] Low-spin states can occur in lower symmetry crystal fields that split the E_g degeneracy sufficiently to cause a violation of Hund's rule by creating a spin pair in the lower of the split e_g states.

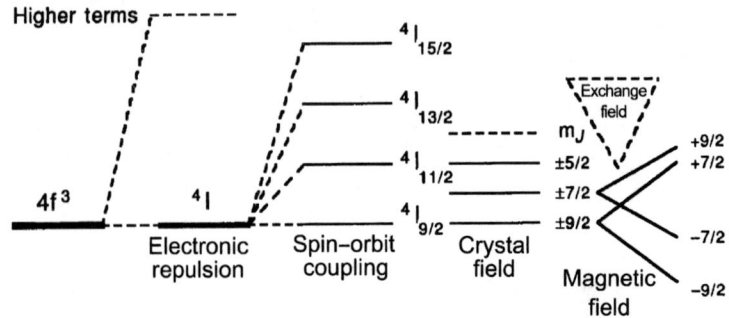

Figure 1.19. Generic model of energy-level structure of $4f^3$ configuration typical of the rare-earth Nd^{3+} ^{4}I-state ion.

1.5.2. Jahn–Teller and Spin–Orbit Stabilizations

In Figs. 1.15 and 1.16, it is evident that in many cases not all of the degeneracy is removed by quenching of the orbital angular momentum in cubic fields. Group theory dictates that lower symmetry fields of the oxygen ligands can lift the remaining degeneracies and eliminate all vestiges of unquenched orbital angular momentum in the ground states. Except for situations where the lattice itself furnishes the lower symmetry through its own point group, the sources for such lower symmetry field components are not always apparent. In two particular situations of great importance in determining many magnetic and magnetoelastic properties, the distortion may occur spontaneously at the cation site because of orbit–lattice or spin–orbit–lattice interactions peculiar to the electron configuration of the transition ion.

In analyzing the paramagnetic resonance behavior of Cu^{2+} ($3d^9$) ions in hydrated salts, Abragam and Pryce[28] explained the occurrence of an unexpected tetragonal component to the cubic crystal field by invoking a theorem deduced by Jahn and Teller:[29] "If a molecule has a degenerate orbital angular momentum state, the immediate environment of the site will spontaneously adjust to create lower symmetry if such a distortion will lift the degeneracy to provide a state of lower energy for the electrons that occupy the degenerate state." In its elemental definition, the J–T effect produces a singlet orbital ground state, which may still retain its spin degeneracy, e.g., a Kramers doublet. Such J–T distortions can be static when the orbit–lattice interaction is much stronger than spin–orbit coupling and can determine the lattice symmetry through cooperative involvements of many sites. If the J–T interaction is weaker, it can manifest a dynamic behavior caused by random lattice vibrations at higher temperatures. In special cases, coherent vibronic actions of ligand normal modes can influence spin exchange phenomena.

The J–T effect in its essential form may be seen from the schematic picture of Fig. 1.20. Here, the example is an e_g orbital doublet that is split by a tetragonal crystal field component from a $\langle 100 \rangle$ axis distortion to create a stabilization energy $\Delta E_{el} = E_{cf}|z - z_0|$ that is linearly proportional to an elastic distortion along the apical (z-axis). Because the lattice energy may be approximated as quadratic in the form of Hooke's law, $\Delta E_{latt} = (1/2)k(z - z_0)^2$, an equilibrium energy can be established for

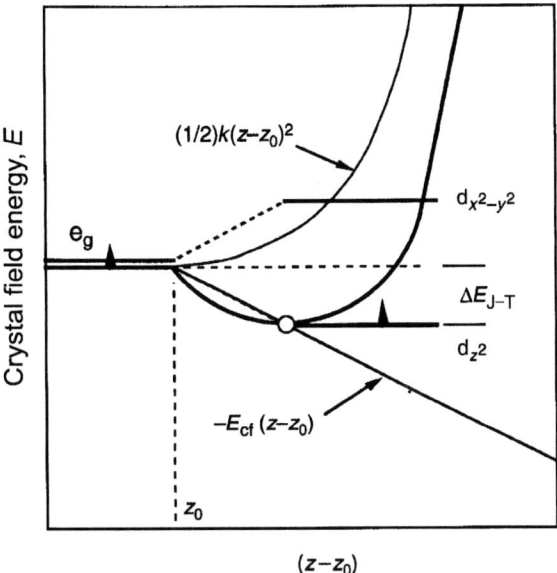

Figure 1.20. Generic model of the J–T effect.

any *unpaired* electron that occupies the doublet. By minimizing $E_{tot} = \Delta E_{el} + \Delta E_{latt}$ by differentiation, we obtain the equilibrium value of $(z-z_0) = \pm E_{cf}/k$, and the net reduction in E_{tot} is the J–T stabilization energy:

$$\Delta E_{J-T} = \tfrac{1}{2} E_{cf}^2/k \qquad (1.5.14)$$

where E_{cf} is the crystal-field stabilization energy and k is the lattice equivalent of the force constant, both of which are parameters that vary with each cation–ligand–lattice combination. In elasticity theory, Eq. (1.5.21) would be expressed in terms of an elastic constant c ($\sim 10^{12}$–10^{13} dyn/cm^3) and the square of the strain ε^2 ($\sim 10^{-10}$–10^{-12}).

The double-valued solution for z refers to the option of extension or compression along the z-axis that would allow the order of the $d_{x^2-y^2}$ and d_{z^2} levels to reverse. For this reason the J–T effect can be dynamic, responding to the overall lattice phonon spectrum or vibronic, responding to the specific normal modes of vibration with the local anion coordination.[30] In cases of static cooperative J–T effects, the sign of the distortion is determined by higher-order factors that influence the total energy. What distinguishes the pure J–T effect from other spontaneous lattice distortions is that the $d_{x^2-y^2}$ and d_{z^2} states involved in the splitting each have zero orbital angular momentum, i.e., $m_l = 0$. For J–T splitting of the e_g doublet, there are four situations among the high-spin d^n electron transition groups where this can occur in an octahedral site. As diagrammed in Fig. 1.21, they are d^4 and d^9 in octahedral and d^1 and d^6 in tetrahedral sites. If the low-spin configurations are included, d^3 and d^7 ions in tetrahedral sites are also J–T configurations.

In reviewing the occupancy diagrams for the ground state in Figs. 1.15 and 1.16, one observes that certain t_{2g} electron configurations are degenerate because of partially

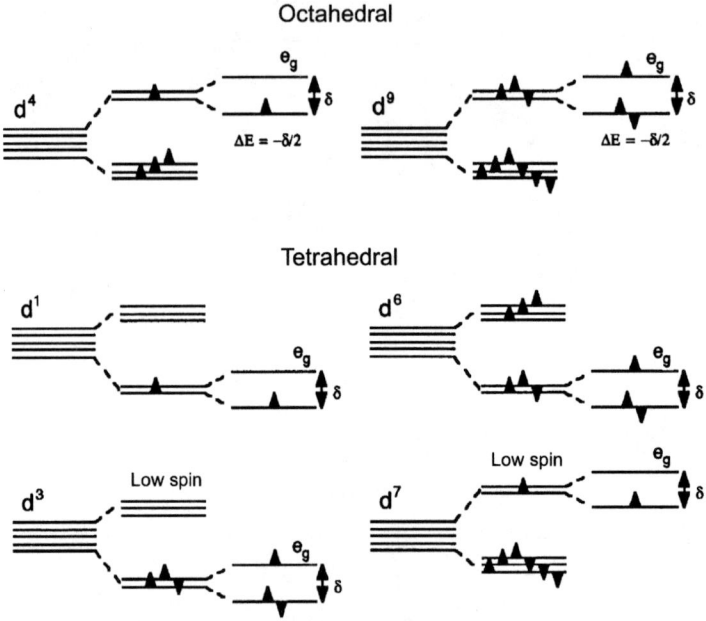

Figure 1.21. One-electron model of J–T stabilizations of the e_g shell.

filled states. Unlike the e_g states with $m_l = 0$ that can undergo pure J–T splittings independent of spin–orbit interactions, the t_{2g} triplet contains $d_{yz,xz}$ with nonzero l_z values ($m_l = \pm 1$) because they are hybrids formed from $t_{2g}^+(d_1)$ and $t_{2g}^-(d_{-1})$ of Eq. (1.4.13). By definition, a pure J–T distortion quenches L by stabilizing a nondegenerate ground state configuration involving only orbitals with zero measurable l_z, i.e., $m_l = 0$. The stabilization of a nondegenerate electron configuration in the ground state can also occur in the t_{2g} shell through a splitting of the triplet into a $[1/(2)^{1/2}](d_{yz} \pm id_{xz})$ doublet ($m_l = \pm 1$) and d_{xy} singlet ($m_l = 0$) with appropriate distributions of spins. Unlike the e_g stabilization that requires a tetragonal or orthorhombic distortions along $\langle 100 \rangle$ axes, the t_{2g} splitting can occur by either $\langle 100 \rangle$ or trigonal axis $\langle 111 \rangle$ distortions, as indicated by Fig. 1.13. Sketches of the degenerate orbital lobes in these two situations are shown in Fig. 1.22. The preference for one type of distortion over the other lies in the peculiarities of the particular combination of cation and ligand in their lattice environment.

Where exchange fields (H_{ex}) impose ordering of the spins, an alternative to the singlet J–T ground state can be manifested as a collective distortion of the opposite sign. Spin–orbit coupling can stabilize the ($m_l = \pm 1$) doublets in concert with an exchange energy, creating additional perturbation terms $\lambda L \cdot S + g m_B H_{ex} \cdot S$. Because the splitting energy of the doublet from the t_{2g} triplet is only half that of the singlet in order to preserve the t_{2g} total energy (see Fig. 1.13), a minimum requirement for S–O dominance is $\lambda L \cdot S \geq \delta/3$, which is the energy preference for the stabilization of the J–T singlet over the doublet when spin–orbit stabilization is not dominant.[31] The actual J–T stabilization energy $\Delta E_{J-T} = 2\delta/3$. A recent examination of this effect in an $S = 1/2$ system produced a more explicit dependence on H_{ex}.[32]

ELECTRON SPINS IN IONIC MOLECULAR STRUCTURES 43

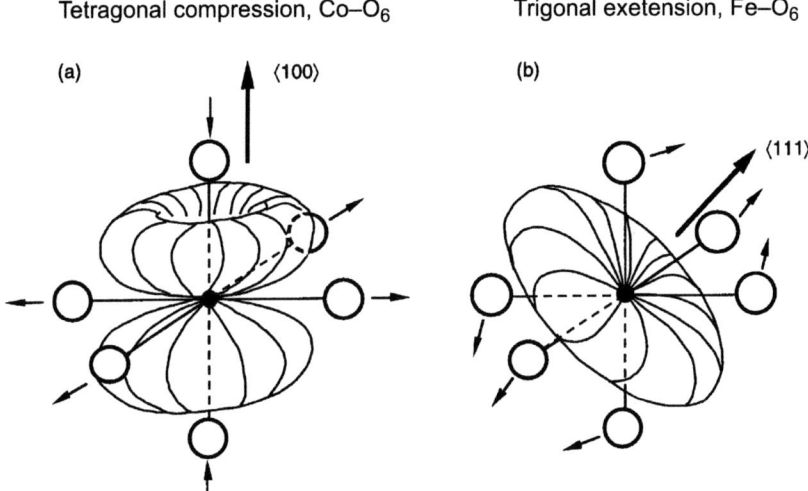

Figure 1.22. Octahedral-site deformations from stabilization of spin–orbit doublet create ground states in the partially quenched t_{2g} shell: (a) a single electron in degenerate hybrid $d_{yz} \pm id_{xz}$ causes a $\langle 100 \rangle$ tetragonal compression (characteristic of CoO); and (b) a single electron in degenerate hybrid of a trigonal field produces a $\langle 111 \rangle$ extension (characteristic of FeO). Based on Figs. 44 and 53 of ref. 31.

$$\frac{\lambda}{\delta} \geq \frac{2}{3}\left(1 + \frac{2}{3}\frac{\delta}{gm_B H_{ex}}\right) \quad (1.5.15)$$

These rudimentary concepts are indicated throughout the various t_{2g} spin occupancy situations shown in Figs. 1.23 and 1.24. The sign of the lattice deformations can be a guide to which effect is occurring in a particular situation. A simple test for the J–T effect is its insensitivity to a breakdown in spin ordering at the Curie temperature.

1.5.3. Mn^{3+}, Cu^{2+}, and Co^{2+} Ions

In Fig. 1.25, two common examples of the J–T and S–O stabilizations (d^4 and d^7 in octahedral sites) are compared in relation to observed ligand displacements. Note that the z-axis ligand distortion of the S–O case is in the expected opposite sense to create a degenerate ground state stabilized by the $m_l = \pm 1$ orbital angular momenta. There are many instances where potential spin–orbit–lattice stabilizations could take place. A summary of these possibilities is given in Table 1.11 with known occurrences highlighted.

The distinction between these two spontaneous stabilization effects should be restated because it can have important influence on the properties of magnetic oxides. Because the pure J–T effect is independent of spin interactions, cooperative manifestations of it can appear independently of any spin ordering condition. However, the S–O effect requires collinearity of spins for cooperative distortions to occur and is therefore expected only in spin-ordered systems as expressed in Eq. (1.5.15). The attendant local distortions of either sign can produce magnetostriction

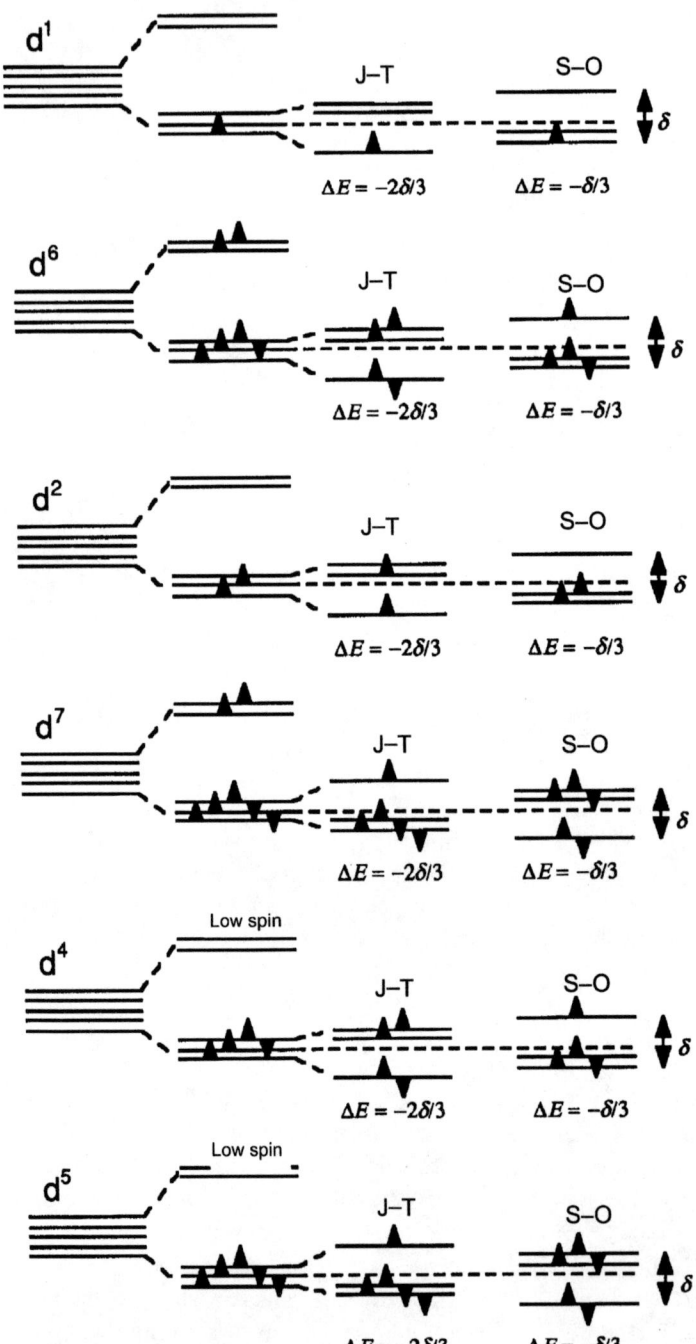

Figure 1.23. One-electron model of J–T/S–O stabilizations of the t_{2g} shell: octahedral site.

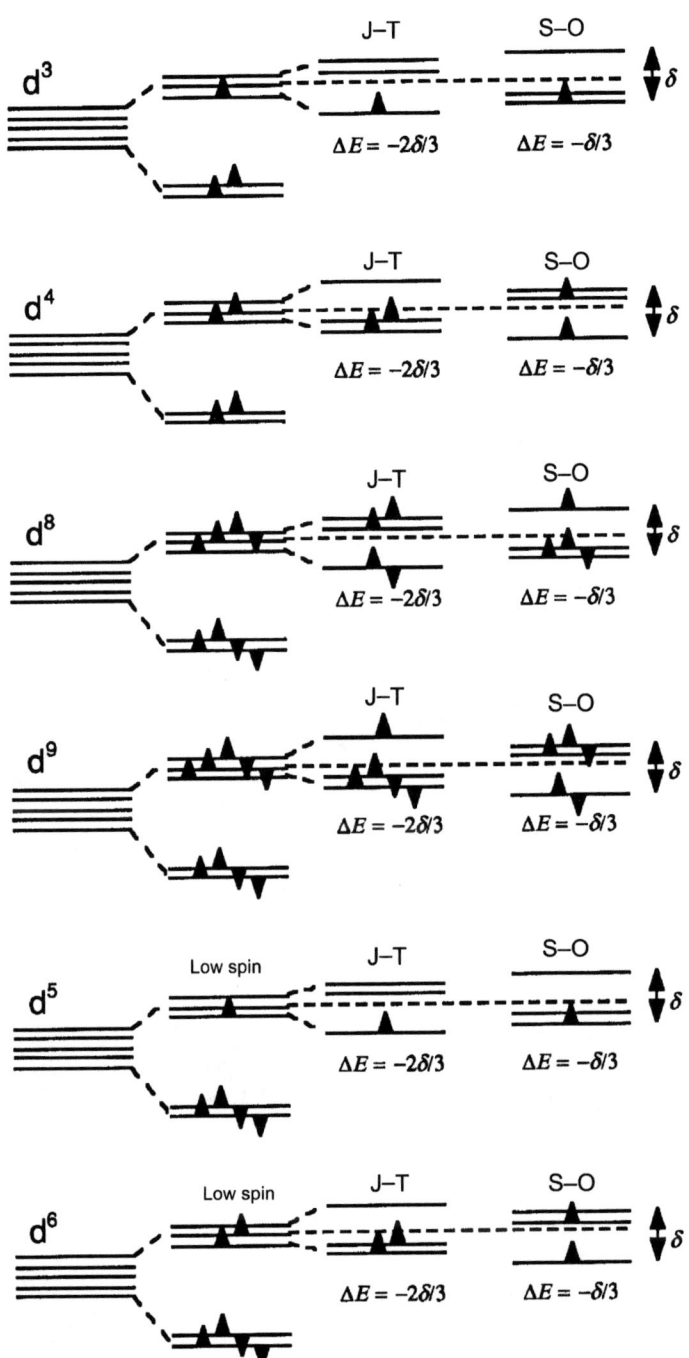

Figure 1.24. One-electron model of J–T/S–O stabilizations of the t_{2g} shell: tetrahedral site.

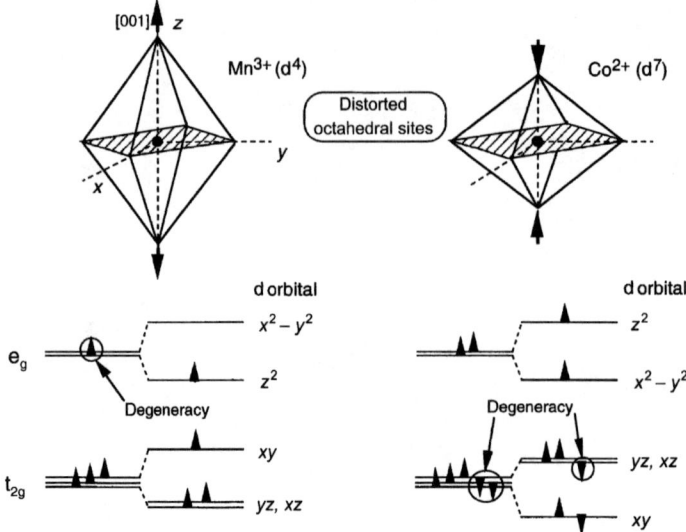

Figure 1.25. Examples of J–T (d^4, Mn^{3+}) and S–O (d^7, Co^{2+}) stabilizations.

effects in spin-ordered systems, one through direct spin–orbit–lattice coupling, the other by indirect influence of the local lattice environment. If the concentration of J–T or S–O stabilized ions is high enough, the effect can become cooperative and a crystallographic phase transition can take place.

In addition to the lattice distortions, spin–orbit stabilization also brings strong spin–lattice effects that manifest themselves in magnetoelastic properties of ferrites, particularly with Co^{2+} (d^7) ions octahedral sites. It should also be remarked that any lower symmetry crystal field, regardless of origin, that leaves an unquenched orbital ground state can have great influence on spin–lattice relaxation properties.

Table 1.11. Jahn–Teller and Spin–Orbit Stabilized d^n Ions[a]

Pure J–T Effect e_g Shell		S–O or J–T Effect t_{2g} Shell	
Octahedral	Tetrahedral	Octahedral	Tetrahedral
—	d^1	d^1	—
—	—	d^2	—
—	d^3 (low spin)	—	d^3
d^4	—	d^4 (low spin)	d^4
—	—	d^5 (low spin)	d^5 (low spin)
—	d^6	d^6	d^6 (low spin)
d^7 (low spin)	—	d^7	—
—	—	—	d^8
d^9	—	—	d^9 (low spin)

[a] Bold type indicates known cases.

Because the combination of a weak crystal field and a stronger spin–orbit coupling that leaves the total angular momentum J largely unperturbed, these phenomena can be even more intriguing with ions of the $4f^n$ rare-earth series. In some compounds, e.g., paramagnetic $TmPO_4$ with a degenerate E_g doublet ground term from the partial quenching by the cubic crystal field, the distinction between J–T and S–O stabilization is somewhat blurred. Many observed spin–orbit–lattice effects such as structural phase transitions induced by a high magnetic field are viewed as part of a broad generic J–T category,[33] although spin–orbit effects remain dominant in other properties such as spin–lattice relaxation.

1.5.4. Summary of $3d^n$ Series Ions

In Section 1.5.1, the $3d^1$ configuration was analyzed to illustrate some of the theoretical tools available for determining the anisotropy of the single-ion magnetic moment in a magnetic field. We now offer an overview of the $3d^n$ group.

1.5.4.1. $3d^1$ and $3d^6$ D-State Triplet: d^1 (Ti^{3+}) and d^6 (Fe^{2+}).

The Ti^{3+} (d^1) and Fe^{2+} (d^6) ions are described by the one-electron ground state diagrams and the corresponding multiple electron energy level structure of Fig. 1.26. For the d^6 case, it is convenient to view the degeneracy as arising from the single electron that begins the second-half of the d shell, recalling that the first five are locked into collinear spin polarization dictated by Hund's rule with $L=0$ that therefore contribute nothing to the orbital degeneracy. (A similar approach can be used for the

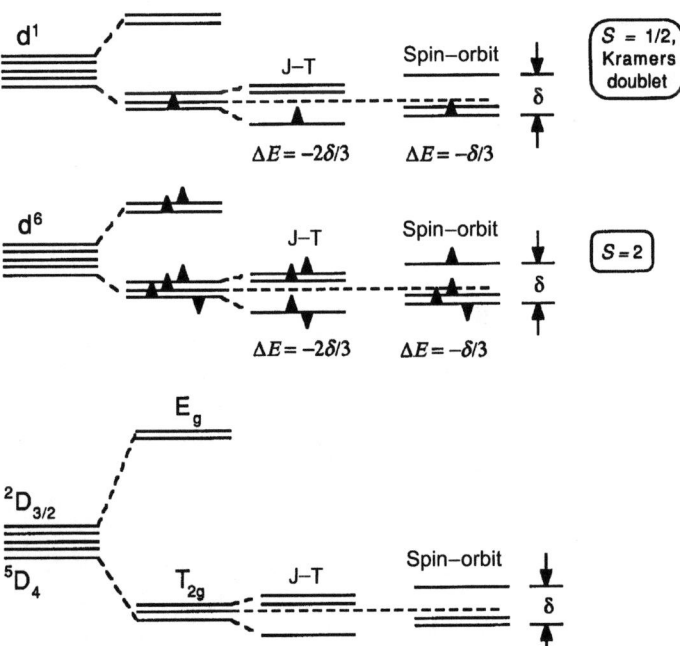

Figure 1.26. Electronic structure of d^1 and d^6 configurations: one-electron and combined electron models.

case of two electrons in the T_{2g} state (d^2 or d^7) by treating the source of degeneracy as spin vacancies or "holes".) For both d^1 and d^6 ions, the triplet can be further stabilized as a singlet (analyzed in Section 1.4) or as a doublet, which can have important influence on all magnetoelastic properties because the remaining degeneracy carries with it an orbital angular momentum and hence a spin–orbit link to the crystal lattice.

Despite the success in interpreting the data for the orthorhombic crystal field of Ti^{3+} in the Rb alum salt, the Cs alum case remains a mystery in some respects because of the uncertainty in the nature of the lower symmetry crystal field and its origins.[23] The primary question is whether the orbital ground state is a singlet originating from the dispositions of the $(SO_4)^{2-}$ radicals in the Na, K, Rb, and Tl members of the family, or whether it can be a doublet due to fields from axially arranged neighbor ions or possible spin–orbit coupling in magnetically aligned situations. The unquenched orbital state of the doublet would lead to strong magnetoelastic effects that would cause paramagnetic resonance line broadening. Later studies have suggested that J–T influences within the T_{2g} manifold could account for some of the observed peculiarities in the CsTi case.[34,35]

The principal $3d^6$ ion of interest is Fe^{2+}, usually occupying an octahedral site. Despite the similarities of the ground state, this configuration differs from the d^1 case in various ways: with an even number of spins, there are no Kramers doublets to enable microwave spectroscopy; the value of Dq is smaller because of the lower cation valence charge; the sign of the multi-electron spin–orbit coupling constant λ is reversed for ions in the upper half of the d^n shell. This latter effect causes the g factor to become greater than 2 in the perturbation analysis given in Eqs. (1.5.7) and (1.5.11), therefore causing an increased magnetic moment and larger paramagnetic anisotropy. Despite the absence of Kramers doublets in this even-electron system, microwave resonance measurements were carried out on Fe^{2+} in MgO because the negative λ combined with the smaller δ splittings of the T_{2g} term likely caused by local random J–T effects produced two isotropic transitions with respective g factors of 3.47 and 6.83.[36]

1.5.4.2. $3d^4$ and $3d^9$ D-State Doublet: d^4 (Mn^{3+}) and d^9 (Cu^{2+}).

For the classic J–T configurations, the g factors are not strongly influenced by the ligand environment for the simple reason that the ground E_g doublet described by the diagrams in Fig. 1.27 contains no orbital angular momentum. The influence of the upper T_{2g} level therefore occurs through mixing by the $\lambda \mathbf{L} \cdot \mathbf{S}$ operator, but as the perturbation theory dictates, the effect on g will be reduced to terms on the order of $\lambda \mathbf{L} \cdot \mathbf{S}/10Dq \sim 10^{-2}$ in most situations—significant but not substantial for spin–lattice interaction considerations. Moreover, the d^4 configuration does not yield Kramers doublets for convenient microwave spin resonance analysis of the g factors, although anisotropic factors $g_\parallel = 1.95$ and $g_\perp = 1.99$ were found for Cr^{2+} in rhombic $CrSO_4 \cdot 5H_2O$ with a large zero-field splitting parameter $D = 2.24$ cm^{-1}.[37] In this case the lower symmetry field of the host probably pre-empted the major part of a J–T effect. The most common example of this configuration is Mn^{3+} (although Fe^{4+} makes an occasional appearance in nonstoichiometric garnets). It almost universally occupies octahedral sites in oxides and therefore brings with it structural perturbations associated with the attendant

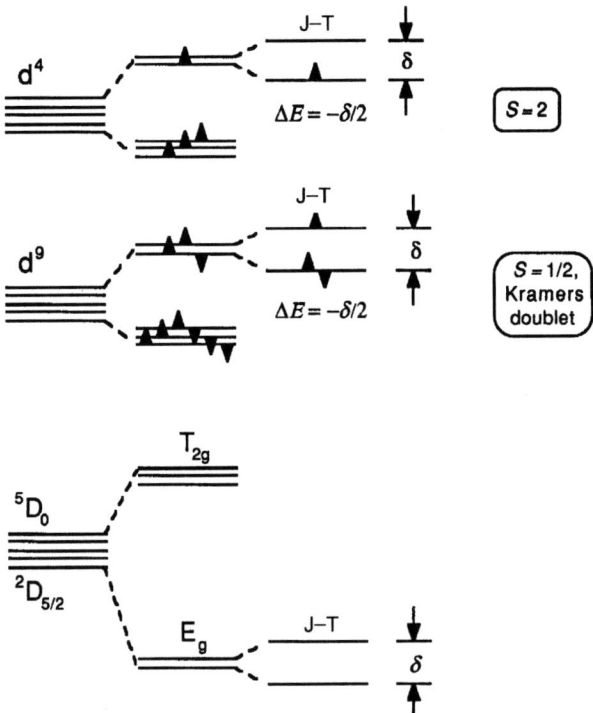

Figure 1.27. Electronic structure of d^4 and d^9 configurations: one-electron and combined electron models.

J–T tetragonal distortions of the ligands that manifest themselves as reversible magnetostriction effects or actual lattice phase changes when they become cooperative. In mixed-valence situations with Mn^{2+} or Mn^{4+}, charge transfer can provide interesting and sometimes anomalous electrical conductivity behavior, as examined in Section 1.8.

The paramagnetism of d^9 is seen most commonly in Cu^{2+} which served as the vehicle for the discovery of the J–T effect and in that respect parallels the magnetoelastic behavior of the d^4 configuration. In early electron paramagnetic resonance EPR work with the cupric salts, g factors in the range of 2–2.5 were found, consistent with the expectation for $\lambda < 0$.[38] The importance of this ion, however, may reside more in its polaronic charge transfer capability with Cu^{3+} and Cu^{1+} oxidation states for high-temperature superconductivity than in its fundamental magnetism. The low-spin value contributes little to magnetic systems and although, like Mn^{3+}, its magnetoelastic capability can assist in tailoring magnetostriction in magnetic oxides, cooperative J–T effects on lattice structure can sometimes be more of an irritant than an asset.

1.5.4.3. $3d^2$ and $3d^7$ F-State Triplet: d^2 (V^{3+}) and d^7 (Co^{2+}).

Although the two d-electron case produces a ground-state triplet similar to that of the d^1 and d^6 cases, and can be treated in a similar manner by considering the T_{2g} triplet as occupied by a single spin vacancy or hole for simple approximations, an important distinction must be made. The total L value for these configurations is 3 instead of 2 (see Fig. 1.7). This means that the free-ion term is F, not D, and there are seven orbital states instead of only five, as displayed in Fig. 1.12. The distinction becomes particularly of interest for optical transitions because of the different orbital splitting energies. For F states, the first expected orbital term above a ground triplet is the other triplet which is separated by an energy 8Dq. Nonetheless, the ground state can still be treated by the single-electron model for our present purposes. Inspection of the diagrams of Fig. 1.28 indicate that J–T and S–O stabilizations can take place in a manner similar to the d^1 and d^6 cases, but with inverted electronic energy structures. The splitting of the lower triplet, however, will contribute significantly to the single-ion anisotropic g factors because the δ splittings are small enough to render denominator of the spin Hamiltonian Λ_{ij} factor in Eq. (1.5.10) small enough to cause a significant departure of the g factor from the isotropic spin-only value of 2. For this reason, g factors for V^{3+} (d^2) in Al_2O_3 and Co^{2+} (d^7) in MgO have been measured, respectively, as 1.92 with a positive $\lambda = 104 \text{ cm}^{-1}$,[39] and 4.28 with a negative $\lambda = -180 \text{ cm}^{-1}$.[40]

When the d^7 case is analyzed in the multi-electron ^4F term format, the degenerate T_{1g} state is lowest and its threefold degeneracy can be approximated by a pure P state with $L=1$.[41] This approach simplifies the perturbation calculation by the adoption of an effective spin–orbit coupling operator to produce a set of orbital

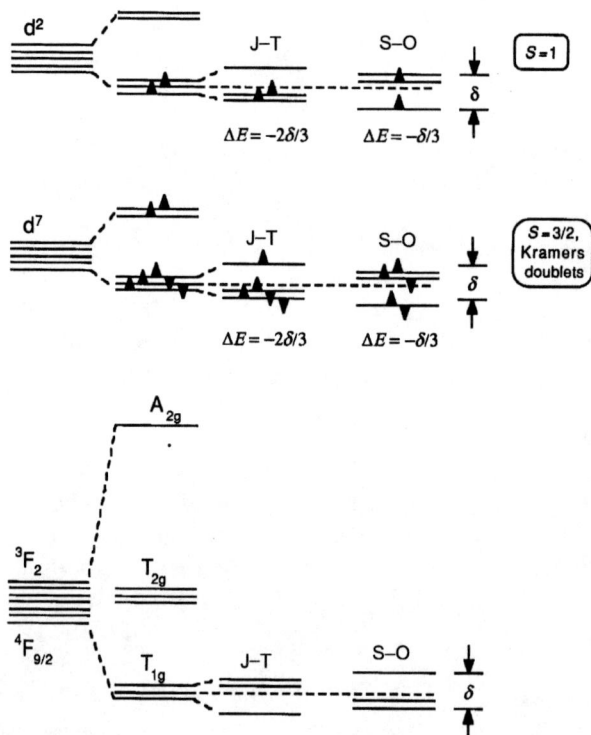

Figure 1.28. Electronic structure of d^2 and d^7 configurations: one-electron and combined electron models.

ELECTRON SPINS IN IONIC MOLECULAR STRUCTURES

states with Kramers spin degeneracies included. For our purposes, however, the one-electron model is sufficient to gain a physical understanding of how spin–orbit coupling can further stabilize an unquenched orbital doublet by enhancing the effects of an exchange field. This effect can account for the large Co^{2+} anisotropy contributions in magnetically ordered systems. It can also help to explain the source of the small orbital contribution to the paramagnetic anisotropy and magneto-optical effects of the S-state Fe^{3+} ion in cubic crystal fields typical of spinel and garnet ferrimagnets.

1.5.4.4. $3d^3$ and $3d^8$ F-State Singlet: d^3 (Cr^{3+}) and d^8 (Ni^{2+}).

The influence of orbital angular momentum on the paramagnetism and anisotropy of the g factors is felt less in the F-state cases where the A_{1g} singlet in cubic field is the lowest, as shown in Fig. 1.29 for d^3 and d^8 configurations. With the first excited orbital state separated in energy from the ground state by 10Dq, a smaller $8\lambda/10Dq$ correction to g in Eq. (1.5.11) can usually be expected. The most studied ion of this type in an octahedral site is Cr^{3+} (d^3), with a spin $S = 3/2$ that provides three Kramers doublets. Its EPR spectrum has been studied in alum lattices[42] and others including corundum Al_2O_3 (ruby),[43] where it not only served as the vehicle

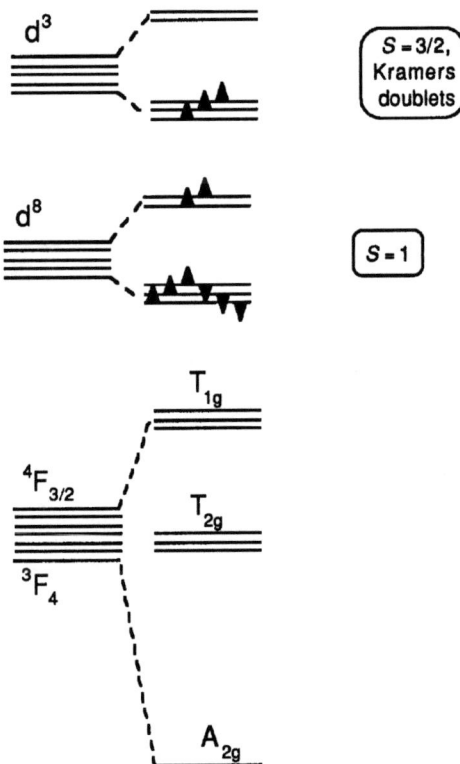

Figure 1.29. Electronic structure of d^3 and d^8 configurations: one-electron and combined electron models.

for the invention of the maser but also the earliest solid-state laser utilizing the $\sim 17{,}000$ cm^{-1} crystal-field transition. To analyze these singlet ground states, the expanded spin Hamiltonian of Eq. (1.5.13) is employed, with emphasis on the second-order D and E terms. Neither Cr^{3+} nor its d^8 counterpart Ni^{2+} are expected to be active in a spin–lattice sense because of the cubic-field stabilized singlet.

However, despite the absence of J–T or S–O stabilization possibilities in the octahedral sites that they almost always occupy, the Ni^{2+} ions have exhibited an isotropic $g \approx 2.22$ in MgO.[44] The absence of anisotropy is consistent with the first-order cubic crystal field, and the origin of the apparent larger spin–lattice contribution can be found in a consideration of the relative magnitudes of λ and Dq in the first-order approximation given by Eq. (1.5.11), $g = 2 - 8\lambda/10\text{Dq}$. If we apply $\lambda = -335$ cm^{-1} from Table 1.10 and assume $\text{Dq} \approx 900$ cm^{-1} for divalent cations taken from optical absorption measurements of $Ni(H_2O)_6^{2+}$,[45] we arrive at $g \approx 2.9$ without even taking into account the influence of covalent bonding on reducing orbital contributions. For Cr^{3+} with $\lambda = 87$ cm^{-1} and $\text{Dq} \approx 1700$ cm^{-1} for trivalent cations, this calculation yields $g \approx 1.96$ in good agreement with the measured value. However, the EPR spectrum of this ion features large angular variations because of the D and E terms of the spin Hamiltonian.

1.5.4.5. 3d^5 S-State Singlet: d^5 (Fe^{3+}, Mn^{2+}).

There remains to be discussed the d^5 configuration exemplified most commonly by Fe^{3+} and Mn^{2+}, which is the least sensitive to its ligand environment while providing the strongest superexchange interaction fields of all the transition-metal ions. To explain this apparent paradox, we must begin by pointing out that although five d electrons in the high-spin state produce an $S = 5/2$ total spin, the addition of orbital angular momentum leaves an $L = 0$ state and a $^6S_{5/2}$ free-ion ground term. Small contributions to the anisotropy are introduced through the spin–orbit coupling of excited free-ion terms. In this one case, where the issue of orbital quenching is rendered moot because of the $L = 0$ condition, any orbital angular momentum influence must be derived from coupling between the ground ^6S term and the excited terms from the Russell–Saunders coupling listed in Table 1.4. As indicated by the sketches in Fig. 1.30, ^4G, ^4P, ^4D, ^4F, and other levels are sequentially raised above the ground ^6A singlet, each with its own set of degeneracies. The states that are sensitive to the cubic crystal field parameter are shown with limiting-case electron occupancies. These energy levels were computed by Tanabe and Sugano[46] and Orgel,[47] and the complexities are discussed by Griffith.[3]

In a cubic field, the first excited state above the singlet $^6A_{1g}$ ground state that has orbital angular momentum is the triplet $^4T_{1g}$ from the ^4G term for both octahedral and tetrahedral sites. Orbital influence felt in the ground state through spin–orbit coupling that links the two separate terms, instead of states within a J manifold, can be analyzed as a generic hybrid orbital ground state φ_0 i.e., a mixture of the lower singlet and upper triplet combined through nondegenerate perturbation theory, according to

$$|\varphi_0\rangle \approx |^6A_{1g}, m_s\rangle + \varepsilon|^4T_{1g}, m_s\rangle \quad (1.5.16)$$

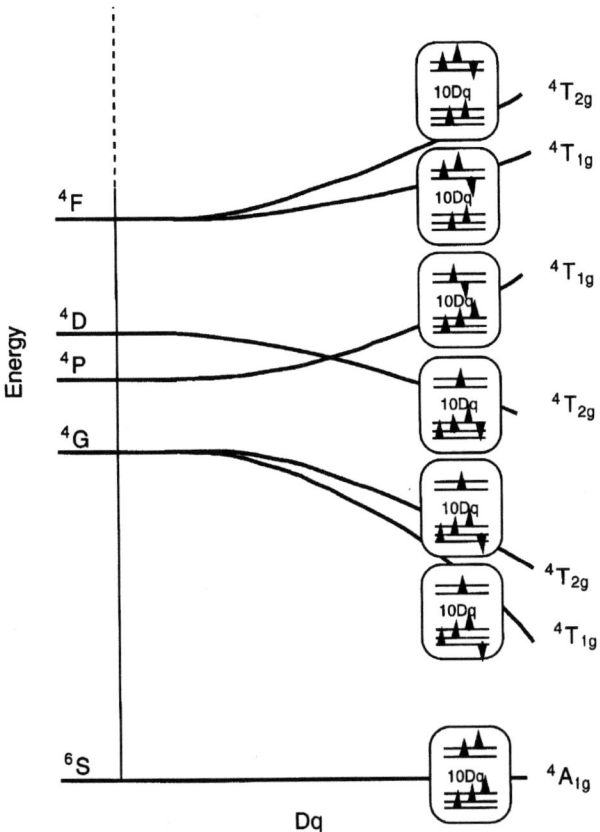

Figure 1.30. Variation of d⁵ energy terms as a function of cubic field strength 10Dq, showing proposed electron configurations.

where

$$\varepsilon \approx \frac{\sum_n \langle {}^6A_{1g}, m_s | \lambda L \cdot S | {}^4T_{1g,n}, m_s \rangle}{|E_{T_{1g,n}} - E_{A_{1g}}|}$$

Here n corresponds to any of the three orbital states within the $^4T_{1g}$ triplet (analogous to a P state with $m_l = 1, 0, -1$). The influence of the cubic crystal field can be seen in Fig. 1.31 where the inverse dependence of ΔE_{oct} and ΔE_{tet} on Dq is illustrated. As a consequence, the degree of mixing of the excited state into the ground singlet would be greater for an octahedral site than a tetrahedral site because $Dq_{tet} = -(4/9)Dq_{oct}$. A short review of this particular problem can be found in Low.[48]

Paramagnetic anisotropy of single d⁵ ions therefore is a higher-order effect than for the other members of the series. For almost all purposes, g factors are assumed to be equal to the pure spin value of 2. However, paramagnetic resonance measurements reveal anisotropic effects in the angular dependence of the microwave spectra that are attributed to the *a* parameter. These data were interpreted by

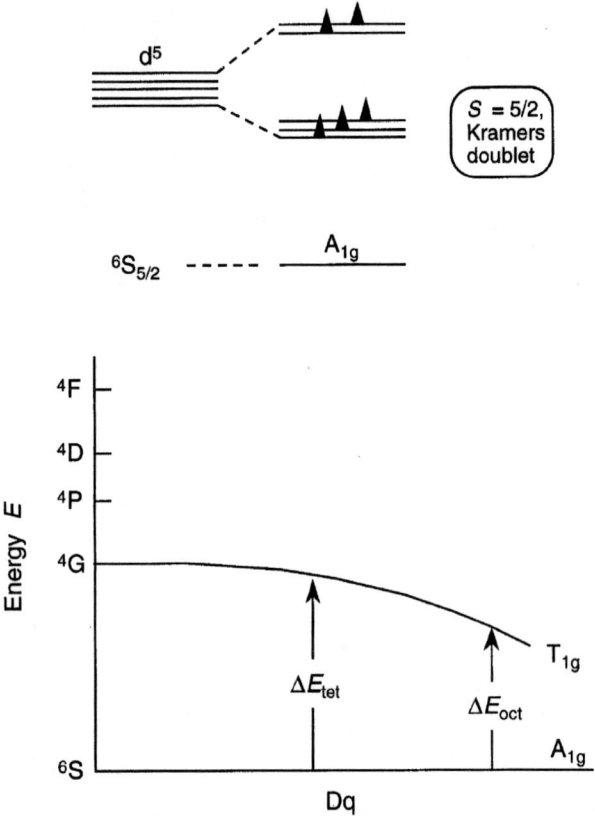

Figure 1.31. Electronic structure of d^5 configurations: one-electron and combined electron models, plus the more complete diagram showing the relation of the A_{1g} ground state and the ^4G term.

the spin Hamiltonian of the $^6S_{5/2}$ state in a cubic crystal field with lower symmetry components.[48]

1.6. SPIN EXCHANGE BETWEEN CATIONS

Quantum theory has proven to be a powerful tool for analyzing electronic structures and properties of crystalline systems. In general, the basic philosophies have evolved from the two extremes of chemical bonding: (i) the collective approach in which all of the s, p, and d electrons outside of the filled core interact as "community property" without assignment to any specific nucleus, and (ii) the local model in which the electrons retain an allegiance to their original nuclei and participate in covalent sharing as a perturbation. The factor that determines which model should apply is the relative magnitudes of the interionic electron–nuclear and electron–electron energies. Where the e^2/r_{ij} term is dominant in the Hamiltonian, a collective (band) model is applied to explain ferromagnetism in the manner of Hund's rule parallel spin alignment, usually in systems with metallic conduction properties. If the molecular bonds are principally ionic, however, the transferred s and p electrons can now be treated as fully localized on the their respective nuclei, and the system usually becomes a d-electron antiferromagnetic insulator that fits the local approximation.

To describe the electronic origins of spontaneous magnetism in transition-metal compounds where the d electrons remain primarily local to their nuclei, an extension of the crystal-field model called molecular orbital theory provides a convenient introduction. In this approach, point charges of the crystal field are replaced by perturbations from overlapping orbital lobes of the actual anions, and their p-orbital wave functions are combined with the d orbitals of the cations to provide hybrid states, i.e., linear combinations of atomic orbitals (LCAO), that reflect the energy stabilization due to covalent bonding. In this section, single-electron molecular orbital functions are used to estimate magnetic exchange stabilization energies of transition-metal oxides. The two-electron analysis that includes mutual electron repulsion e^2/r_{ij} term is based on the valence-bond method that was introduced by Heitler and London[49] in their analysis of the H_2 molecule, and will be mentioned in relation to ferromagnetic effects in select metallic oxides.

1.6.1. Covalent Stabilization and Molecular Orbitals

All chemical bonding is electrostatic, arising from nuclear and electron charge interaction. Molecular orbital theory is a method to describe covalent bonding mathematically, and is formalized in terms of hybrid wave functions of the participating ions. For transition-metal oxides, the method is a direct extension of crystal-field theory, employing the one-electron eigenfunctions, with the interaction Hamiltonian approximated by the averaged resultant of the various electrostatic terms that determine the stabilization of each bonding electron.

To account for interactions among electrons of adjacent ions occupying individual one-electron orbital states, hybrid states between cations and anions belonging to the overall nuclear skeleton can be constructed from the individual orbitals of the $3d^1$ crystal field eigenfunctions and the 2p orbitals of the oxygen ligand. In this method, the combined orbitals are assigned to the molecule rather than to the individual atoms or ions. The available bonding electrons among the ions can then be distributed among these hybrid orbitals with a maximum of two spin-paired electrons per state following the dictates of the Pauli principle and Hund's rule, analogous to the formation of electronic configurations in atomic structure. Once the molecular orbital scheme is established, the Aufbau principle that was applied to map the electron distributions in the crystal-field ground states of the transition-metal ions in Figs. 1.15 and 1.16 can be adopted for the molecule.

For a generic heteronuclear diatomic molecule with ions a and b, the molecular orbital wave functions are hybridized into LCAO from the corresponding one-electron orbital functions φ_a and φ_b according to

$$\varphi_- = N_-(\lambda_-\varphi_a - \varphi_b) \quad \text{(antibonding state)} \tag{1.6.1a}$$

and

$$\varphi_+ = N_+(\varphi_a + \lambda_+\varphi_b) \quad \text{(bonding state)} \tag{1.6.1b}$$

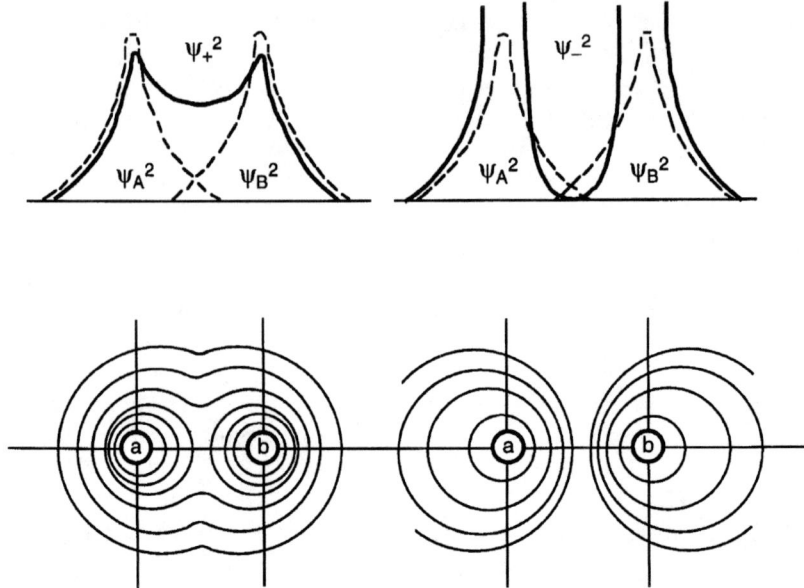

Figure 1.32. Schematic of bonding and antibonding states for spherically symmetric s-electron wave functions of two similar atoms A and B, showing the probability distributions of electron charge between them. Except in ferromagnetic cases with partially filled d shells, the more stable is called the bonding state with greater charge density between the nuclei.

where $N_+ = (1 + \lambda_+^2 + \lambda_+ S)^{-1/2}$ and $N_- = (1 + \lambda_-^2 - \lambda_- S)^{-1/2}$ and $S = \langle \varphi_a | \varphi_b \rangle$ is the orbital overlap integral that emerges from the natural process of normalization whereby $\langle \varphi_i | \varphi_j \rangle = \delta_{ij}$. Thus, the electron transfer or ionic character of the covalence is retained through the parameter $\lambda_{+,-} < 1$ as part of the sharing of orbital states. (If the molecule is homonuclear, such as H_2, $\lambda_{+,-} = 1$). The spatial probability distributions of the functions φ_+ and φ_- are represented by the sketches in Fig. 1.32. Where the electrons accumulate between the nuclei (usually with spins aligned antiparallel in observance of the Pauli exclusion principle), the energy is lower because the electrons serve to screen the repulsive forces acting between the two positively charged nuclei. For this reason, φ_+ is called the "bonding" state; the opposite condition occurs in the "antibonding" state φ_- with spin directions parallel because wave function charge clouds that repel each other tend to have maximum combined spin (consistent with Hund's rule of maximum spin polarization).

Estimates of the molecular orbital functions for a diatomic molecule can be obtained by solving two-level perturbation problems in the conventional way. For the example of overlapping orbitals of ions separated by r_{ab} with atomic numbers Z_a and Z_b and respective radii r_a and r_b, the Hamiltonian can be written as

$$\mathcal{H} = \mathcal{H}_a + \mathcal{H}_b + \frac{Z_a Z_b e^2}{r_{ab}} \tag{1.6.2}$$

ELECTRON SPINS IN IONIC MOLECULAR STRUCTURES

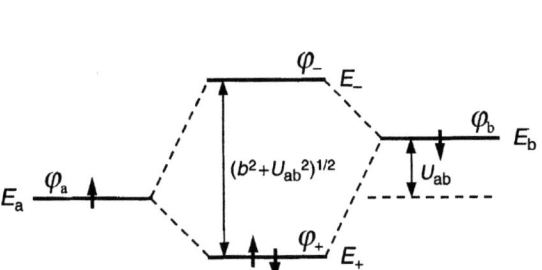

Figure 1.33. Basic two-ion molecular orbital diagram, with one electron per ion. Both electrons are stabilized into the lower energy bonding state φ_+.

with

$$\mathcal{H}_a = \left[-\left(\frac{\hbar^2}{2m}\right)\nabla_a^2 - \frac{Z_a e^2}{r_a}\right] + V_b$$

$$\mathcal{H}_b = \left[-\left(\frac{\hbar^2}{2m}\right)\nabla_b^2 - \frac{Z_b e^2}{r_b}\right] + V_a$$

where the bracketed terms contain the free-ion kinetic and potential energies, and the attractive energies $Z_a e^2/r_b$, $Z_b e^2/r_a$ from the neighboring ions are included in the lattice Madelung potentials V_a and V_b. The inter-nuclear repulsion term $Z_a Z_b e^2/r_{ab}$ is also absorbed into the resultant lattice Madelung energy. The stabilization energy and coefficients may be estimated by means of a self-consistent approach developed by Wolfsberg and Helmholtz.[50] The perturbation matrix for the diatomic molecule becomes

$$\begin{vmatrix} \mathcal{H}_{aa}-E & \mathcal{H}_{ab}-ES \\ \mathcal{H}_{ba}-ES & \mathcal{H}_{bb}-E \end{vmatrix} = \begin{vmatrix} E_a-E & b-ES \\ b-ES & E_b-E \end{vmatrix} = 0 \quad (1.6.3)$$

and the secular equation is $|\mathcal{H}_{ab}-ES_{ab}|=0$, where $S_{aa}=S_{bb}=1$ and $S_{ab}=S_{ba}=S$, and the matrix element $\mathcal{H}_{ab}=b=\langle\varphi_a|\mathcal{H}|\varphi_b\rangle$. To implement the solution of this equation when applied to the molecular orbital problem, it must first be recognized that φ_a and φ_b are not part of an orthonormal set. Since the different energies of the φ_a and φ_b states appear on the diagonal of the determinant in Eq. (1.6.3), the corresponding energies can be represented directly by E_a and E_b. In this approximate calculation, E_a can be estimated from the ionization potential of the outer cation electron destabilized by the repulsive field of the negative anion, and E_b from the electron affinity of the outer anion electron stabilized by the attractive field of the positive cation. To be Hermitian, however, $\mathcal{H}_{ba}$ must equal $\mathcal{H}_{ab}$. Because these elements represent the one-electron cation–anion transfer integral b, the solution requires a singular value of b. Relations for the matrix elements can be approximated by[50]

$$\mathcal{H}_{ab} = \langle \varphi_a|\mathcal{H}|\varphi_a\rangle = E_a\langle \varphi_a|\varphi_a\rangle = E_a$$
$$\mathcal{H}_{bb} = \langle \varphi_b|\mathcal{H}|\varphi_b\rangle = E_b\langle \varphi_b|\varphi_b\rangle = E_b \qquad (1.6.4)$$
$$\mathcal{H}_{ab} = b \approx \zeta\langle \varphi_a|\tfrac{1}{2}(E_a + E_b)|\varphi_b\rangle \approx (E_a + E_b)\langle \varphi_a|\varphi_b\rangle = (E_a + E_b)S$$

where $\zeta \approx 2$.* This expression for $\mathcal{H}_{ab}$ is based on the assumption that $\mathcal{H}$ of Eq. (1.6.2) is nearly uniform within the orbital overlap region and the factor.

The values of E_a and E_b are defined graphically in the basic two-body molecular orbital diagram of Fig. 1.33 as the stabilization energies of the outer electrons on the respective atoms (or ions). Note that the values of E_a and E_b are negative, with E_a chosen to be of lower energy. Regardless of the exact expression for $\mathcal{H}_{ab}$, the above exercise points out that the magnitude of the transfer integral is jointly dependent on the energy of the electron states of the free ions involved in covalent bond and the magnitude of the overlap integral $S = \langle \varphi_a|\varphi_b\rangle \leq 1$.

Solutions of Eq. (1.6.3) for the bonding and antibonding states are:

$$E_{\pm} = \frac{(E_a+E_b-2bS) \pm [(E_a+E_b-2bS)^2 - 4E_aE_b - b^2]^{1/2}}{2(1-S^2)} \qquad (1.6.5)$$

where $E_+ = (E_a + E_b)/2$ is the mean energy of the free ion states, and the corresponding energy difference $U_{ab} = E_a - E_b$. If we use the relation $b = \langle \varphi_a|\mathcal{H}|\varphi_b\rangle \approx (E_a + E_b)S$, Eq. (1.6.5) can be simplified to

$$E_{\pm} = \frac{(E_a+E_b)(1-2S^2) \pm [(E_a-E_b)^2(1-S^2) + b^2]^{1/2}}{2(1-S^2)} \qquad (1.6.6)$$

Note that signs of U_{ab} and b must be considered before actual computations are attempted. To conform with metal–ligand case, we assume $U_{ab} < 0$; for metal–metal energies, U_{ab} can be positive. The transfer integral b has also been designated as a negative quantity consistent with E_a and E_b. However, in Section 1.8 the magnitudes of these parameters are used to avoid confusion in reading into the physical effect they influence. To determine the value of S by computation, a Wolfsberg–Helmholtz procedure can be followed,[50] but in practice S (< 1) is typically treated as a semi-empirical parameter to be determined from experiment.

If $b^2 \ll U_{ab}^2$ and $S^2 \ll 1$ in Eq. (1.6.6) a binomial approximation yields

$$E_{\pm} \approx \frac{(E_a+E_b)}{2} \pm \frac{1}{2}[(E_a-E_b)^2 + b^2]^{1/2}$$
$$\approx E_F \pm \frac{U_{ab}}{2}[1 + b^2/(E_a-E_b)^2]^{1/2} \approx E_F \pm \frac{U_{ab}}{2} \pm \frac{b^2}{4U_{ab}} \qquad (1.6.7)$$

Philosophically, the $b^2/4U_{ab}$ term represents the amount by which the hybrid energy is stabilized by the b transfer integral.

---

*A curiosity in the computation can be pointed out here. If the self-consistent term ES is not included in the diagonal elements of Eq. (1.6.3), the result will be identical to that obtained if $\zeta = 1$ with the Wolfsberg–Helmholtz model. For the simpler model, b would be only half as large and all multpliers of it would then be double.

For $b^2 \gg U_{ab}^2$ and $S^2 \ll 1$,

$$E_\pm \approx \frac{(E_a+E_b)}{2} \pm \frac{1}{2}[(E_a-E_b)^2+b^2]^{1/2}$$

$$\approx E_F \pm \frac{b}{2}[1+(E_a-E_b)^2/b^2]^{1/2} \approx E_F \pm \frac{b}{2} \pm \frac{U_{ab}^2}{4b} \quad (1.6.8)$$

In this limit, $U_{ab}^2/4b$ is the additional stabilization of the single spin in the bonding state. It can be readily recognized that if $U_{ab}=0$, there is no trap barrier for the transfer of the spin between φ_a and φ_b because their coefficients in the hybrid orbital state are equal. Therefore, $U_{ab}^2/4b$ is bonding state energy that must be gained to eliminate the trap. As a consequence, $U_{ab}^2/4b$ is an effective activation energy for polaronic charge transport.

Eigenfunctions of the E_+ and E_- levels can now be determined by forming the linear hybrid combinations

$$\begin{aligned}\varphi_- &= N_-(c_{ba}\varphi_a - c_{bb}\varphi_b) \quad \text{(antibonding)}\\ \varphi_+ &= N_+(c_{aa}\varphi_a + c_{ab}\varphi_b) \quad \text{(bonding)}\end{aligned} \quad (1.6.9)$$

and $N_- = (c_{ba}^2 + c_{bb}^2 - 2c_{ba}c_{bb}S)^{-1/2}$ and $N_+ = (c_{aa}^2 + c_{ab}^2 + 2c_{aa}c_{ab}S)^{-1/2}$. Expressed in terms of the corresponding LCAO coefficients, $c_{ab} = c_{ba} = \lambda_{+,-}$. The c_{ij} coefficients are then determined from the standard relation $\sum_j (\mathcal{H}_{ij} - ES_{ij})c_{ij} = 0$ for each solution value of E. Accordingly, we can write

$$\begin{aligned}(b-E_-S)c_{ba} + (E_b - E_-)c_{bb} &= 0\\ (E_a - E_+)c_{aa} + (b - E_+S)c_{ab} &= 0\end{aligned} \quad (1.6.10)$$

and after recognizing the normalizing conditions $N_-^2(c_{ba}^2 = c_{bb}^2) = 1$ and $N_+^2(c_{aa}^2 + c_{ab}^2) = 1$, we obtain the general relations

$$\begin{aligned}N_-^2 c_{ba}^2 &= \left[\frac{(E_b-E_-)^2}{(E_b-E_-)^2+(b-E_-S)^2}\right]\\ N_-^2 c_{bb}^2 &= \left[\frac{(b-E_-S)^2}{(E_b-E_-)^2+(b-E_-S)^2}\right]\\ N_+^2 c_{aa}^2 &= \left[\frac{(b-E_+S)^2}{(E_a-E_+)^2+(b-E_+S)^2}\right]\\ N_+^2 c_{ab}^2 &= \left[\frac{(E_a-E_+)^2}{(E_a-E_+)^2+(b-E_+S)^2}\right]\end{aligned} \quad (1.6.11)$$

For $b^2 \ll U_{ab}^2$, Eq. (1.6.9) can be then reduced to the un-normalized expressions

$$\begin{aligned}\varphi_- &\approx \left(\frac{E_F}{E_a}\frac{b}{2U_{ab}}\right)\varphi_a - \varphi_b\\ \varphi_+ &\approx \varphi_a + \left(\frac{E_F}{E_b}\frac{b}{2U_{ab}}\right)\varphi_b\end{aligned} \quad (1.6.12)$$

The square of coefficient $(E_F/E_{a,b})(b/2U_{ab})$ approximates the lesser fractional share of the ionic state φ_a and φ_b states in their respective hybrids.

Discussion of the symmetry contributions to covalent bonding will not be attempted in this text, but an appreciation of its importance can be introduced through the definition of the primary types of overlap interactions. For the transition-metal oxides, we need only to consider the influence of an unfilled d electron shell and its interaction with the 2p shell of the O^{2-} ligands. Although the electrons in the outer p and s shells of the cation are principally responsible for the chemical bond, the magnetic properties enter through the d shell. In Fig. 1.34, four examples are sketched for the d–p orbital lobe arrangements characteristic of transition-metal ions in octahedral oxygen coordinations.

The two principal factors that determine the magnitude of S are: (i) the amount of spatial overlap of the cation d and ligand p orbital lobes, designated according to whether the ligand lobe is directed along the axis joining the two nuclei (σ bonds) or at a significant angle to it (π bonds); and (ii) the relative signs of the lobes. A rule often followed is that *a nonzero S will occur if both orbitals have the same symmetry about the axis joining the two nuclei.* Upon inspection of the examples in Fig. 1.34, it is seen that $e_g(d_{x^2-y^2})-p_{xy}\sigma$ provides the largest S and strongest bond while $t_{2g}(d_{xy})-p_{x,y}\pi$ can also contribute, but through a significantly weaker interaction. Similar constructions can also be made for the $e_g(d_{z^2}) - p_{xy}\sigma, \pi$ couplings. For a

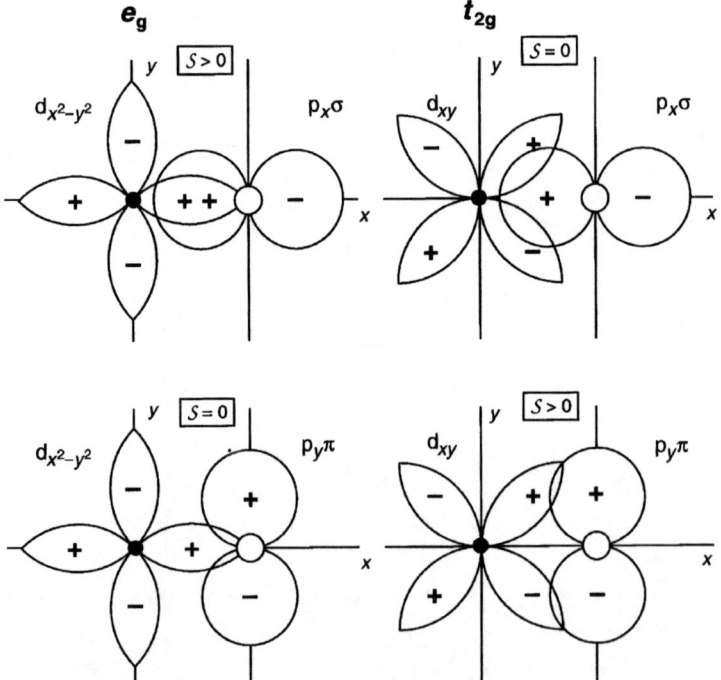

Figure 1.34. Illustrations of the difference between σ (direct overlap with oxyen p orbital lobes directed at the cation nucleus) and π (oblique overlap with oxygen p orbital lobes directed orthogonal to the cation nucleus). Example is given for the *x–y* plane.

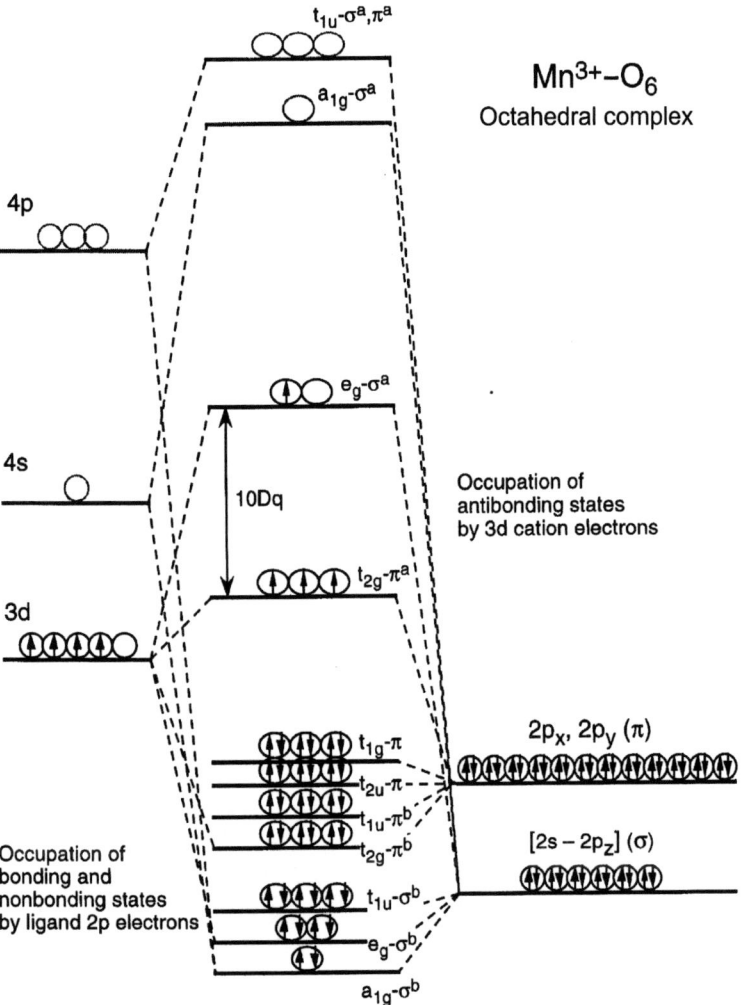

Figure 1.35. Detailed molecular orbital energy diagram for an $(Mn^{3+}-O_6)^{3-}$ complex. Note that the Mn^{3+} four d electrons occupy mainly the nonbonding (t_{2g}-π) and antibonding (e_g-σ) states.

first-order estimate of covalent energy in octahedral complexes, only e_g–$p\sigma$ is generally taken into account in determining the bonding and antibonding states, while the t_{2g} orbitals in octahedral sites are treated as inactive or "nonbonding."

In Fig. 1.35, a full molecular orbital energy level diagram is shown for a 3d transition-metal ion and oxygen ligands in an octahedral coordination. To a first approximation, the diagram could be constructed by computing bonding and antibonding states for each of the metal levels linked covalently to the oxygen 2p states in the manner of Fig. 1.33. Once this diagram is formed, the individual molecular orbital states are then assigned electrons from the two participating ions according Hund's rule and the Aufbau principle that was employed previously in the one-electron ground state models of the crystal-field theory. As indicated, the low-energy bonding states are filled mainly with spin-paired electrons from the oxygen ligands in the fractional amounts $N_+^2 c_{aa}^2$ for each hybrid orbital combination while the antibonding distributions would favor the 3d electrons of the σ-bonding e_g states in the corresponding fractions $N_-^2 c_{bb}^2$. Also, note that the weaker π-bonding t_{2g} electrons are designated as nonbonding and are considered to remain in their initial cation orbital states. To place the crystal-field orbital angular momentum quenching in the context of the full molecular orbital scheme, the 3d and 2p spin occupancies of an Mn^{3+} (d^4) ion in an oxygen octahedral complex are included in Fig. 1.35. Note how the 10Dq splitting between the t_{2g} and e_g states is retained, although relabeled as antibonding t_{2g}-π^a and e_g-σ^a to conform to the expanded nomenclature.

1.6.2. Molecular Orbital Superexchange Approximation

In the simplest context, the spontaneous magnetism can take two forms: collinear spins (ferromagnetism) and opposing or paired spins (antiferromagnetism). These two extremes are the equivalent of high- and low-spin states of the individual

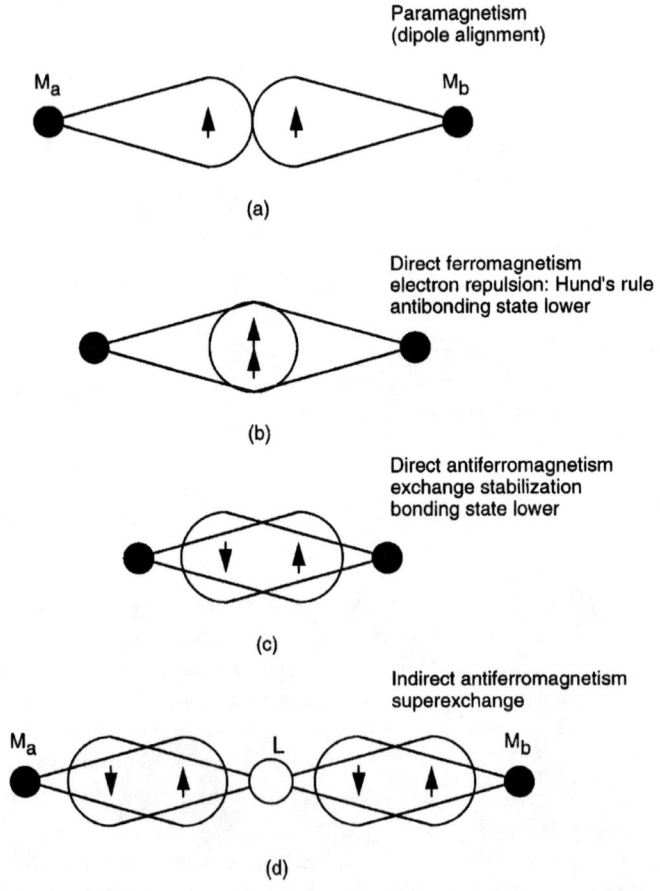

Figure 1.36. Orbital lobe diagrams for the different bonding situations.

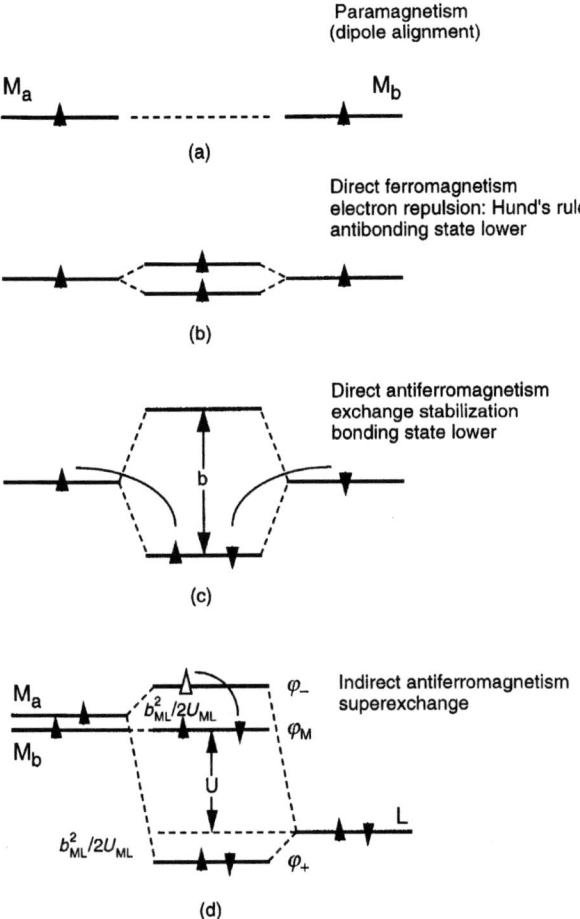

Figure 1.37. Molecular orbital diagrams corresponding to the orbital lobe configurations in Fig. 1.36.

ions and the interactions that select one in preference to the other are virtually the same. In Fig. 1.36, diagrams show how the different spin–spin interactions evolve as two magnetic cations are brought together in a direct chemical bond, beginning with (a) paramagnetism when the spins interact at a distance through inverse square dependence on separation, and (b) passing from direct exchange ferromagnetism within the smaller orbital overlap limit for which Hund's rule of orthogonality (the maximum spin requirement) applies, on to (c) antiferromagnetism as the stabilization from b becomes dominant in the covalent bond. Note that in the case (d) of the metal–oxygen–metal bond the antiferromagnetic ordering of the two metal spins results from a pair of orbital overlaps. These models can serve as a starting point for the discussion of magnetic exchange.

Before specific magnetic ion combinations are considered, it is useful to demonstrate how the molecular orbital approach can also be used to illustrate spin alignment stabilization. In Fig. 1.37, we consider the molecular orbital diagrams that correspond to the simple models similar metal ions M_a and M_b illustrated in

Fig. 1.36. For the paramagnetic case of Fig. 1.37(a), the nuclei are too far apart to permit the orbital lobes to overlap. Consequently, the two spins can interact only through their dipolar fields. As the two nuclei are moved closer together in Fig. 1.37(b), the overlap integral S increases and bonding/antibonding states are formed from the one-electron wave functions separated in energy by the transfer integral b. Initially, the ground state distributions are parallel, representing a sustaining of the Hund's rule $S = 1$ high-spin state.* As b increases and the splitting between the bonding and antibonding states becomes larger, the orbital energy stabilization shown in Fig. 1.37(c) favors the low-spin $S = 0$ antiparallel configuration, which represents the onset of the domination by the interionic electron–nuclear attraction that is responsible for the antiferromagnetic spin alignment. Analogous to the one d-electron model used to display the spin occupancies of the crystal-field ground states determined by the Aufbau principle, these molecular orbital diagrams can map the spin occupancies of the bonding and antibonding levels of the multi-electron ground state.

These concepts can now be carried forward to include metal–ligand bonds of basically ionic compounds, but with the added consideration of the excitation energy U_{ML}. In this approach, U_{ML} represents the difference in energy of an electron on the cation and one on the ligand or anion in the crystal lattice as determined by the various electrostatic fields. For the superexchange case that involves three ions, we begin with a linear molecule M_a–L–M_b, where M_a and M_b represent the two metal ions and L is the common ligand, e.g., oxygen, depicted in Figs. 1.36(d) and 1.37(d). The secular determinant equation was extended from the two-body version of Eq. (1.6.3) by Ballhausen,[51] according to

$$\begin{vmatrix} E_a - E & b_a - ES_a & 0 \\ b_a - ES_a & E_L - E & b_b - ES_b \\ 0 & b_b - ES_b & E_b - E \end{vmatrix} = 0 \quad (1.6.13)$$

If M_a and M_b are identical, then $E_a = E_b = E_M$, $b_a = b_b = b$, $S_a = S_b = S$, and we define $U_{ML} = E_L - E_M$ as the difference between the unperturbed ligand and metal ground state energies. The three solutions for the case of $b_{ML}^2/U_{ML}^2 \ll 1$ then become

$$\begin{aligned} E_- &\approx E_M - \frac{b_{ML}^2}{2U_{ML}} \\ E_M &= E_a = E_b \\ E_+ &\approx E_L + \frac{b_{ML}^2}{2U_{ML}} \end{aligned} \quad (1.6.14)$$

with corresponding eigenfunctions (un-normalized) approximated by

$$\varphi_- \approx \frac{1}{(2)^{1/2}} (\varphi_a - \varphi_b) - \left(\frac{E_F}{E_L} \frac{b_{ML}}{U_{ML}}\right) \varphi_L$$

---

*This initial ferromagnetic bias arises from the electrostatic repulsion of two electrons that is not included in the one-electron molecular orbital Hamiltonian and must be presumed. It is convenient to treat it as a high-spin state analogous to the d^1 electron models that retain parallel spin alignments until the stabilization made possible by stronger crystal fields (or larger b integrals) overcome the mutual repulsion and cause antiparallel spin alignment.

$$\varphi_M = \frac{1}{(2)^{1/2}}(\varphi_a + \varphi_b) \tag{1.6.15}$$

$$\varphi_+ \approx \frac{1}{(2)^{1/2}}\left(\frac{E_F}{E_M}\frac{b_{ML}}{U_{ML}}\right)(\varphi_a - \varphi_b) + \varphi_L$$

A molecular orbital diagram can now be constructed for these energy levels to show how the energy stabilization arises from magnetic superexchange. In the energy-level diagram of Fig. 1.37(d) for the three-body case, the bonding state φ_+ will be filled with two spin-paired electrons from the ligand ion according to the Pauli principle. The justification for this can be seen in the eigenfunctions of Eq. (1.6.15), where the weighting of φ_L dominates the bonding state φ_+ for small values of b_{ML}/U_{ML}, while the single electrons of the separated half-filled d orbitals of φ_a and φ_b will form either ferromagnetic or antiferromagnetic spin pairs in the φ_M molecular orbital state which has an energy ΔE_{ML} below the antibonding state φ_-. It is this stabilization that distinguishes antiferromagnetism from the ferromagnetism of the mutual electron repulsion responsible for the orthogonality of the orbital eigenfunctions (Hund's rule). As indicated in Fig. 1.37(d), formation of the ferromagnetic state would require one of the d electrons to occupy the φ_- orbital, thereby destabilizing the molecule relative to the antiferromagnetic state by

$$\Delta E_{ML} \approx b_{MLM} = \frac{b_{ML}^2}{2U_{ML}} \tag{1.6.16}$$

To ensure the ferromagnetic alignment where ΔE_{ML} is small, the missing contribution from the e^2/r_{ij} repulsion would have to be introduced by a two-electron exchange calculation, which would make a positive addition to the transfer element and lead to a reversal in the cation–cation bonding/antibonding states φ_M and φ_-. It is important to recognize that although these parameters are not the same as those used in the more formal analysis of superexchange, at a qualitative level the model provides helpful insight as well as providing an estimate of the effective transfer integral b_{MLM} and allows conventional exchange theory to be applied to direct cation–cation interactions. Note also that since b_{ML} varies directly the first power of the overlap integral S_{ML}, from Eq. (1.6.16) the quasi-direct transfer integral b_{MLM} would be proportional to S_{ML}^2.

1.6.3. Two-Electron Valence-Bond Exchange

Computation of bonding energies for specific structures can be carried out by an approximation that allows the introduction of multi-electron wave functions in a form referred to as the valence-bond method. For bonding energy calculations, each atom is assigned a valence charge after a specific lattice structure is defined. In most simple cases, however, the lattices are treated as either ionic or covalent, according to the location of the atomic components in the Periodic table. Solutions to bonding problems are then carried out by constructing a multi-electron Hamiltonian that retains the nuclear and electronic interactions, but adds electron–electron repulsion terms. The effects of crystal fields on the partially filled d-electron states can be included if the appropriate molecular orbitals are employed as the starting wave functions. For the analysis of direct magnetic

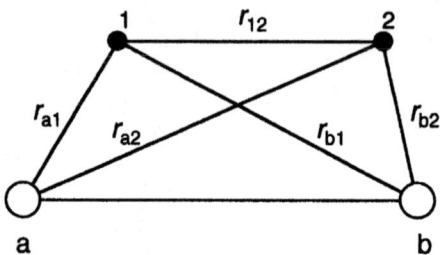

Figure 1.38. Two electrons 1 and 2 in a two-ion a and b coordinate system.

exchange (applied originally to the H_2 molecule[49]), the valence-bond method is used to compute the orbital interaction energies that dictate the high-spin (parallel) and low-spin (antiparallel) states based on the indistinguishability requirement of the two electrons sharing hybrid orbital states of both nuclei. For this purpose, the Madelung terms of Eq. (1.6.2) are replaced by the actual interionic electron–nuclear attraction and electron–electron mutual repulsion terms similar to obtain a two-electron Hamiltonian $\mathcal{H} = \mathcal{H}_0 + \mathcal{H}_1$, where

$$\mathcal{H}_0 = -\left(\frac{\hbar^2}{2m}\right)(\nabla_a^2 + \nabla_b^2) - \frac{Z_a e^2}{r_{a1}} - \frac{Z_b e^2}{r_{b2}} + \frac{Z_a Z_b e^2}{r_{ab}}$$

$$\mathcal{H}_1 = -\frac{Z_a e^2}{r_{a2}} - \frac{Z_b e^2}{r_{b1}} + \frac{e^2}{r_{12}}$$

(1.6.17)

As defined in Fig. 1.38, the distances r_{a2} and r_{b1} are the electron–nuclear separation of the exchanged electrons labeled 1 and 2, and r_{12} is the separation between the electrons themselves. Two-electron degenerate eigenfunctions $\varphi_{a1}\varphi_{b2}$ (initial occupancy) and $\varphi_{a2}\varphi_{b1}$ (exchanged occupancy) of this Hamiltonian are constructed from products of single-electron functions for computation of the elements $K_{12} = \langle \varphi_{a1}\varphi_{b2}|\mathcal{H}_1|\varphi_{a1}\varphi_{b2}\rangle$ and $J_{12} = \langle \varphi_{a1}\varphi_{b2}|\mathcal{H}_1|\varphi_{a2}\varphi_{b1}\rangle$ of the 2 × 2 Slater determinant. The solutions of the secular equation give bonding and antibonding states energies according to $K_{12} \pm J_{12}$. Consequently, the sign of J_{12} determines whether the spins in the ground state are parallel ($S=1$) for $J_{12} > 0$, or antiparallel ($S=0$) for $J_{12} < 0$. Where molecular-orbital solutions provide the starting functions, $\mathcal{H}_0$ can absorb all of the nuclear terms, which then leaves a simplified $\mathcal{H}_1 = e^2/r_{12}$. The details of the quantum mechanical solution, including the issue of electron indistinguishability, are discussed in many standard texts.[51–56] The eventual result is the expression for the net stabilization energy summed over an array of spins which had been deduced phenomenologically by Heisenberg:[57]

$$E_{ex} = -2\sum_{i>j} J_{ij} S_i \cdot S_j$$

(1.6.18)

where J_{ij} is the exchange constant between spins now designated i and j instead of 1 and 2.

From the valence-bond analysis the following guidelines for estimating spin alignment can be inferred by inspection of Fig. 1.36:

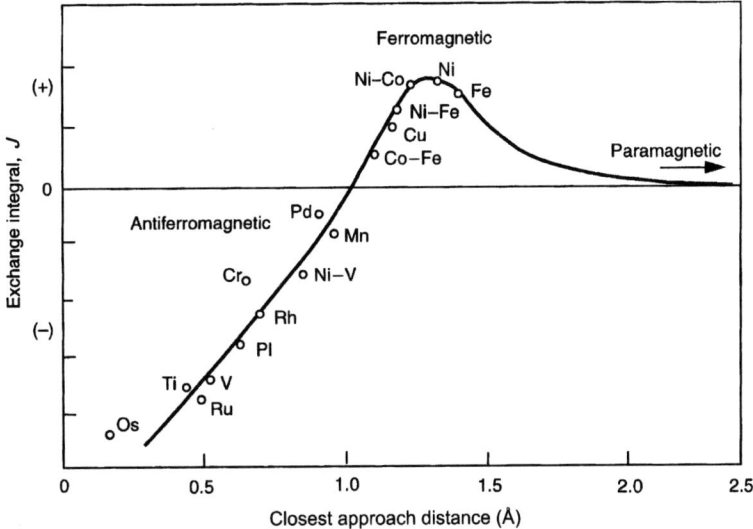

Figure 1.39. Relative value of the exchange constant J as a function of the distance of closest approach of electrons of neighboring atoms. Based on Lax and Button.[58]

1. Ferromagnetism (parallel spins) occurs when the electron repulsion term of Eq. (1.6.17) is dominant. This is expected when the nuclear separation r_{ab} is large enough that the electron orbits overlap enough to disperse the spins according to Hund's rule without being unduly influenced by the attraction fields of their opposite nucleus (antibonding state is lowest). To comply with Eq. (1.6.18), this condition requires that $J > 0$.
2. Antiferromagnetism (opposing spins) occurs when the inter-ionic electron–nuclear terms of Eq. (1.6.17) are dominant. This is expected when the nuclear separation is small enough that the electron charge clouds overlap greatly ($r_{ab} \ll r_{a2}, r_{b1}$) and the dispositions of the spins are controlled by the attractive fields of the opposite nuclei (bonding state is lowest). For this case, the electrons can be more stable by filling states as spin pairs according to the Pauli principle, and $J < 0$. Experimental results that support these predictions are shown in Fig. 1.39 for a variety of direct exchange coupled compounds.[58]

As a consequence, both the origin of Hund's rule and the existence of collective ferromagnetism depend on the presence of mutual electron repulsion.[59] Since the maximum S ground term of a free ion arises from the dispersal of spins among the set of orthogonal orbital states, the condition can be characterized by overlap integrals $S = 0$. Thus, the destabilization caused by the repulsion is minimized if the number of filled orbital states is also a minimum, thereby dictating that a maximum of half-filled states will be created to produce the highest net S. For a collective electron case, it can be reasoned that the spin alignment after hybridization will also depend on the degree of orthogonality, i.e., on the effective value of S.

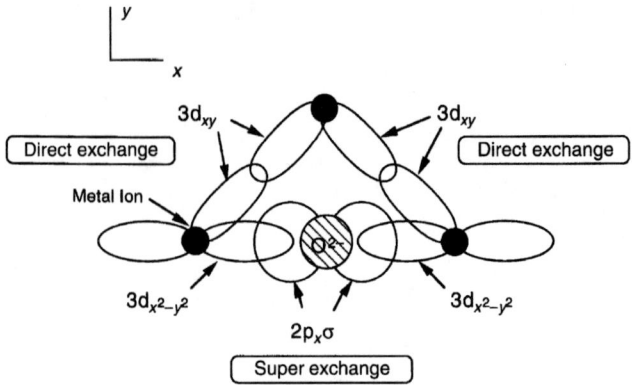

Figure 1.40. Schematic of metal–oxygen exchange bonding showing the comparison between direct and indirect or superexchange. Superexchange stabilizes antiferromagnetism; direct exchange is commonly ferromagnetic.

For a high-spin alignment ($J > 0$) in a partially filled d shell, the electrons in the internuclear space will be driven into states with small orbital overlap to minimize the repulsive energy, i.e., both r_{12} and S_{12} are small. Such a situation can be likened to the antibonding picture of Fig. 1.32 in which parallel spins would accompany the antisymmetric hybrid orbital state. If, however, the opposite is true and the electron repulsion is overcome by the nuclear attractive forces, the electrons will then stabilize as pairs in states with larger r_{12} and S_{12}, so that the ground state has paired spins favoring a low-spin antiferromagnetic alignment. In either case, the energy stabilization that results from the exchange interactions can be represented in the Hamiltonian as a quasi-magnetic field $H_{ex} \simeq JS$.

In oxides and other ionic compounds, the magnetic ions are not usually nearest neighbors. Therefore, any direct exchange effects should be weak. If the collective-electron formalism that is successful for metals is applied to oxides, r_{a2} and r_{b1} are seen as too large to create a negative J, thereby predicting that a weak ferromagnetic spin alignment might be expected. However, in most instances this conclusion is not supported by experiment. The cation lattice–spin ordering is most often antiferromagnetic. Following the early suggestion of Kramers[60] that *indirect* exchange must occur, the general concept sketched in Fig. 1.40 was adopted. Indirect or *superexchange* is the coupling of cations spins through the medium of nearest-neighbor anion ligands. Ferromagnetism can also be produced by direct exchange ($J > 0$) between next-nearest neighbors, however, shown as $3d_{xy}$ orbitals overlapping across the diagonals of a two-dimensional model, e.g., a cube face.

With initial degenerate wave functions already separated by the crystal field, the lattice-related electrostatic terms can be absorbed in the unperturbed Hamiltonian. The main concerns are then the magnitudes of the excitation energy U_{MM} (now more likely a positive quantity in contrast to U_{ML}) associated with a virtual charge transfer $2M^{n+} \rightarrow M^{(n+1)+} + M^{(n-1)+} + U_{MM}$ and the effective transfer integral between the orbital states of the coupled cations. For a pair of similar 3d orbital states of cations linked by a 2p orbital of an oxygen ligand, perturbation theory was applied to compute the energy of the stabilized ground state occupied by two spin-paired electrons. Therefore, we can consider the exchange stabilization of a magnetic oxide as resulting from a combination of both direct exchange and indirect or superexchange. The mathematical formalization of superexchange which is characterized by the exchange integral J_{MM} ($\propto b_{MLM}^2/U_{MM}$) was developed by Anderson by the method of second quantization. Discussion of this elegant work is beyond the scope of this text, and the reader is encouraged to consult his writings on this subject.[61]

In generalized format for n orbital levels linking two similar cations with respective spins S_i and S_j with net cation ionization energy for charge transfer to a virtual excited state U_{ij}, the resultant exchange constant in the presence of direct exchange with $J_{ij}^{\text{direct}} > 0$ is expressed as[56,62]

$$J_{ij} = J_{ij}^{\text{direct}} + J_{ij}^{\text{super}} = \frac{1}{2S_iS_j}\sum_n \left(J_{ij}^{n\,\text{direct}} - \frac{b_{ij}^{n2}}{U_{ij}^n}\right) \quad (1.6.19)$$

where n is a particular orbital state, $b_{ij} \sim b_{MLM}$ for a M—L—M interaction as defined by Eq. (1.6.16) and $U_{ij} = U_{MM} \sim 5\text{--}10$ eV. Quantitatively, $b_{ML} \sim -5$ eV and $U_{ML} \sim -15$ eV thereby yielding $b_{MLM} \sim -0.5$ to -1 eV from Eq. (1.6.16), in general agreement with Anderson's estimates.[61] In nonmetals, the ferromagnetic term J_{ij}^{direct} contains contributions from the metal—ligand—metal charge transfer states and any direct exchange from other orbitals that might link the two cations directly, as depicted in Fig. 1.40. Where charge transfer is absent the b_{ij}^2/U_{ij} term usually exceeds the ferromagnetic stabilization energy from the direct interaction between cations, resulting in $J_{ij} < 0$ and antiferromagnetism.

In typical situations the spin directions are not precisely aligned, but form an angle θ_{ij} with each other. To introduce this angular dependence into Anderson's result, the use of the exact relation for the scalar product $S_i \cdot S_j$ of Eq. (1.6.18) is not convenient because of the quantum correction $S(S+1)$ in the angular momentum addition. If the Ising approximation is adopted, whereby only the z components of S are retained in the vector product, $S_z = S\cos(\theta_{ij}/2)$ and the exchange energy relations for the parallel case can be expressed as

$$E_{ex}^{\text{ferro}} = -2J_{ij}S_i\cdot S_j \approx -2J_{ij}S_iS_j \cos^2\left(\frac{\theta_{ij}}{2}\right)$$
$$= -J_{ij}S_iS_j[1 + \cos\theta_{ij}] \quad (1.6.20a)$$

and for the antiparallel case,

$$E_{ex}^{\text{antiferro}} = -2J_{ij}S_iS_j \sin^2\left(\frac{\theta_{ij}}{2}\right)$$
$$= -J_{ij}S_iS_j[1 - \cos\theta_{ij}] \quad (1.6.20b)$$

From these general results, we can discuss the various spin ordering situations that occur in transition-metal oxides.

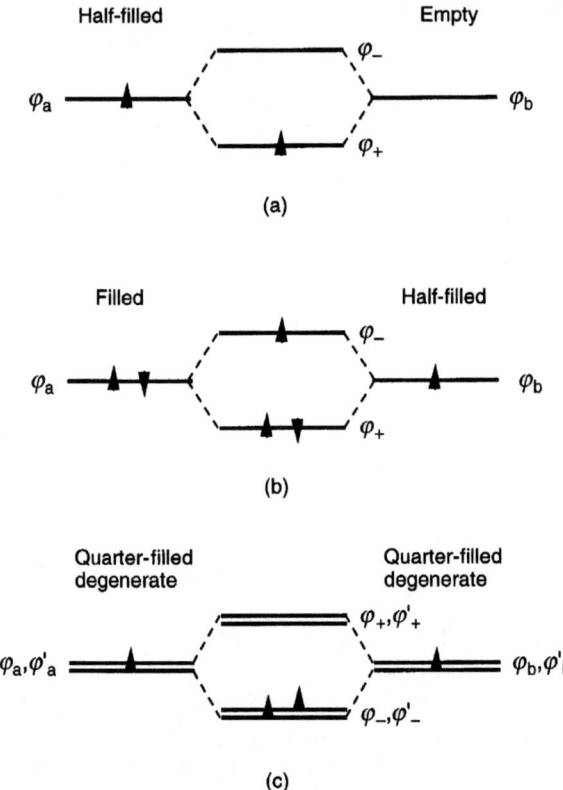

Figure 1.41. Molecular orbital diagrams of ferromagnetic superexchange: (a) delocalization from half-filled to empty; (b) delocalization from filled to half-filled; and (c) Hund's rule ferromagnetism in quarter-filled degenerate orthogonal metal–ligand states.

1.6.4. Magnetic Exchange in the $3d^n$ Oxides

In addition to the mechanism where like cations with half-filled orbitals couple through common ligands by *correlated* virtual charge transfer of two electrons to produce antiferromagnetic spin alignments, *delocalization* by real charge transfer can induce ferromagnetism by the sharing of a single electron. Figure 1.41 illustrates two delocalization conditions: (a) half-filled to empty and (b) filled to half-filled orbital transfers.[62] A special case of delocalization exchange occurs between neighboring cations of the same atomic element but with different valence charges. When this mechanism operates in an antiferromagnetic or ferrimagnetic system in which opposing sublattices are present, the transfers are intra-sublattice and the resulting energy stabilization is called *double* exchange.[63–65]

A somewhat related situation can occur in crystal fields (cubic, trigonal, or tetragonal) that allow orbital degeneracy to survive unquenched. In Fig. 1.41(c), a single electron occupying a doublet, for example the t_{2g} spin–orbit stabilized case of d^2 or d^7 in a $c/a < 1$ tetragonal site, presents a degenerate antibonding state that allows

Hund's rule to be sustained and provides metallic ferromagnetic superexchange. This effect is discussed further in Section 1.8.

A fourth type of indirect exchange is referred to as *polarization* whereby the spins couple to each other through electrical polarization of the "charge clouds" of their mutual environment rather than by a formally defined covalent bond. This mechanism, called RKKY after its collective authors Ruderman–Kittel–Kasuya–Yosida and described in most standard texts on magnetism, is generally associated with the rare-earth or lanthanide series for which the electrons of the shielded partially filled $4f^n$ inner shell interact with the crystal field and its magnetic neighbors by perturbing the charge distributions of their own outer shells and those of the relevant ligands. This mechanism likely contributes to the indirect exchange effects of rare-earth ions in the c sublattice of the magnetic garnets.

Although less important from a purely magnetic standpoint, the delocalization case of *real* transfer is of interest because of the electrical conductivity implications. In analyzing the magnetism created by a single mobile spin, it must be recognized from the outset that there can be no spin-dependent stabilization associated solely with the transfer between the two orbital states. However, if the two cations each have a net spin or reside in a cluster of lattice spins that would provide an exchange field to dictate the orientation of the transferring spin, the net spins of the cations involved in the transfer would likely be ferromagnetically aligned. Otherwise the transfer electron would undergo a spin reversal in order to obey Hund's first rule, but would then violate the $\Delta S = 0$ requirement. If the spin retains its orientation in opposition to the polarization dictates of Hund's rule, its energy of the e^2/r_{ij} repulsion represents a destabilization that would offset any antiparallel alignments that are favored by unpaired spins occupying other orbital states on the same ion. The result of an antiparallel alignment between the cations would be the discouragement of the spin transfers through the creation of a spin trap or magnetic activation energy that would result in a loss of the energy that would otherwise be gained by the sharing of the spin.

As a consequence, in cases where only one of the orbital states is half-filled, the Pauli principle is no longer the main concern because only one electron can be involved in the coupling, and the two-electron exchange effects need not apply. Because the second orbital would be either empty or filled with no net spin in either case, correlation is not involved in the charge transfer. For practical purposes, it can be convenient to view these two orbital combinations as analogous, with electron transfer into the empty orbital and "hole" transfer into the filled orbital. Except for the extremes of d^0, d^{10} or certain low-spin configurations that are diamagnetic because of low-spin $S = 0$ configurations, Hund's rule is followed for ions with $S \neq 0$ to prevent the loss of transfer stabilization energy that would occur through the formation of excited states within the d-electron shell.

The various exchange interactions can be examined qualitatively. Three basic situations can be expressed in chemical ionic notation as follows:

1. Correlation superexchange between ions with the same d-shell occupancy (virtual spin transfer between half-filled orbitals):

$$M_a^{n+} + M_b^{n+} \rightarrow M_a^{(n+1)+} + M_b^{(n-1)+} + U$$

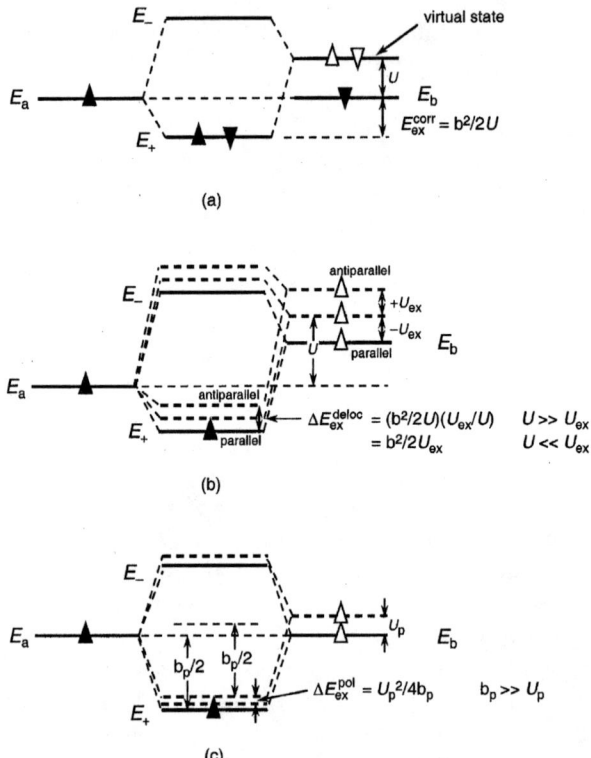

Figure 1.42. Schematic molecular orbital diagrams indicating the origin of the exchange stabilization for: (a) Aufbau model of two-electron correlation antiferromagnetism; (b) one-electron delocalization ferromagnetism (general case); and (c) one-electron delocalization ferromagnetism, special case of mixed-valence charge transfer.

2. Delocalization (or kinetic) superexchange between ions of different elements M and M' in the same lattice (or the same element in different lattices) with different d-shell occupancy (real spin transfer between half-filled/empty or filled/half-filled orbitals):

$$M_a^{n+} + M_b'^{n+} \rightarrow M_a^{(n+1)+} + M_b'^{(n-1)+} + U \pm U_{ex}$$

3. Delocalization (or kinetic) superexchange ions of the same element M in the same lattice but with different valence charges (real spin transfer between a half-filled/empty or filled/half-filled orbitals):

$$M_a^{n+} + M_b^{(n+1)+} \rightarrow M_a^{(n+1)+} + M_b^{n+} + U_{ex}$$

where $U = U_{MM}$ and U_{ex} originates from Hund's of spin polarization. When not specified, the MM subscript will be assumed for the balance of this discussion.

For the correlation case 1, a one-electron molecular orbital model is depicted in Fig. 1.42(a). The covalent stabilization determined for each electron can approximate the two-electron solution with the result in Eq. (1.6.7). Referenced to E_a, the magnitude of the b-dependent term

$$E_{ex}^{corr} \approx \frac{2b^2}{4U} = \frac{b^2}{2U} \quad \text{(antiferromagnetic system)} \qquad (1.6.21)$$

which is consistent with Anderson's result[61] in Eq. (1.6.19).

The general delocalization (case 2) where the ions have different electron configurations results from the transfer of a single electron as described by Eq. (1.6.8). This situation can occur with ions of different atomic elements, e.g., $Mn^{3+}(d^4)$ and $Cr^{3+}(d^3)$ in octahedral ligand coordinations. The energy reduction from the sharing of an electron between a half-filled/empty hybrid state is also given by Eq. (1.6.21), but with half of the energy because of one electron instead of two involved in the exchange. Since there is no *a priori* requirement for a spin polarization, no electron repulsion within the hybrid wave function, the amount of energy contributed by the kinetic action of the single spin will be influenced by the resultant spin alignment of the ions involved. However, a spin flip could be necessary to complete the transfer in obedience of Hund's rule, and the additional U_{ex} would be added to U to lower the transfer probability as diagrammed in Fig. 1.42(b). The magnitudes of the respective exchange energies are given by

$$E_{ex}^{deloc} \approx \frac{b^2}{4(U - U_{ex})} \quad \text{(ferromagnetic system)}$$

$$E_{ex}^{deloc} \approx \frac{b^2}{4(U + U_{ex})} \quad \text{(antiferromagnetic system)} \qquad (1.6.22)$$

$$\Delta E_{ex}^{deloc} \approx \frac{b^2}{4(U - U_{ex})} - \frac{b^2}{4(U + U_{ex})} \approx \frac{b^2}{2U}\left(\frac{U_{ex}}{U}\right)$$

in agreement with Goodenough's result.[62] As discussed in connection with Eq. (1.6.8), ΔE_{ex}^{deloc} represents a trap energy for the itinerant electron. Typical values of the relevant parameters are $b \sim -1$ eV, $U \sim 5$–10 eV, and $U_{ex} < 3$ eV (Slater integral), which permit the $b^2/U^2 \ll 1$ requirement to be satisfied in many cases.

The third situation shown in Fig. 1.42(c) is a special case of delocalization often called "double exchange" in which spin transfer between mixed-valence ions are of the same atomic element, e.g., $Mn^{3+}(d^4)$ and $Mn^{4+}(d^3)$, does not produce a net energy change because the initial and final states are identical, i.e., $U = 0$. However, there remains the carrier or polaron trap energy that results from the local electrostatic (elastic) environment and more importantly, from the presence of U_{ex} in a system without perfect ferromagnetic spin order. For this case, we assign a generic polaron trap U_p and introduce a polaron tranfer integral b_p that is temperature dependent through the influence of lattice elastic energy. Because the relative magnitudes of b_p and U_p can range from $b_p^2 \ll U_p^2$ to $b_p^2 \gg U_p^2$, mathematical computation becomes awkward. For the latter case, which is of interest for polarized spin transport, the magnitudes of the contributions can then be deduced from the one-electron molecular orbital solution for double exchange from Eq. (1.6.8), according to

$$E_{\text{ex}}^{\text{pol}} \approx \frac{b_{\text{p}}}{2} \quad \text{(ferromagnetic system)}$$

$$E_{\text{ex}}^{\text{pol}} \approx \frac{1}{2}\left(b_{\text{p}}^2 + U_{\text{p}}^2\right)^{1/2} \quad \text{(antiferromagnetic system)}$$

$$\Delta E_{\text{ex}}^{\text{pol}} \approx \frac{1}{2}\left(b_{\text{p}}^2 + U_{\text{p}}^2\right)^{1/2} - \frac{b_{\text{p}}}{2}$$

$$\approx \frac{U_{\text{p}}^2}{4 b_{\text{p}}} \quad \text{(for } U_{\text{p}}^2/b_{\text{p}}^2 \ll 1\text{)}$$

(1.6.23)

where $\Delta E_{\text{ex}}^{\text{pol}}$ is the polaron trap energy U_{p} reduced in proportion to the covalence ratio $U_{\text{p}}/b_{\text{p}}$. The implications of this effect on electrical conduction are examined in Section 1.8.

Based on these results for the exchange energies of individual molecular orbital states, relations for the total ionic exchange constants can be deduced by applying Eq. (1.6.19) to combine the contributions from every occupied state. To this end, we define a resultant exchange constant between ions i and j:

$$J_{ij} = -\frac{1}{2S_i S_j} E_{\text{ex}} = -\frac{1}{2S_i S_j}\left(\sum E_{\text{ex}}^{\text{corr}} - \sum E_{\text{ex}}^{\text{deloc}}\right), \quad (1.6.24)$$

where $E_{\text{ex}}^{\text{corr}}$ and $E_{\text{ex}}^{\text{deloc}}$ are positive quantities by definition.

For any superexchange combination M–L–M, estimates of the J_{ij} constant can be made if values can be assigned to the respective E_{ex} for each participating orbital state. To make calculations, it must first be realized that three of the five d orbitals are t_{2g}-$p\pi$ and the other two are e_g-$p\sigma$. The distinction is important because the overlap integral S_π is significantly less than S_σ. Estimates of these contributions based on the Néel temperature data suggest that for two-electron correlation exchange $E_{\text{ex}}^\pi : E_{\text{ex}}^\sigma \sim 0.1$.[61]

In most cases, the ferromagnetic influence of the orbital states linked by the single-electron transfers competes with the antiferromagnetism of any lower energy half-filled orbitals. In octahedral sites, where t_{2g} orbitals form weaker π bonds and e_g stronger σ bonds, the resultant magnetic alignment will depend on the particular ion combination. A more typical example is that of a d^4–d^3 coupling between ions of the same atomic element. Here, the ferromagnetic stabilization of the single electron in the e_g-$p\sigma$ state is offset by the antiferromagnetic energies of the six t_{2g}-$p\pi$ electrons. As will be discussed later in connection with the magnetoresistance properties of manganite compounds, the resultant spin alignments of these particular ions can assume parallel or antiparallel configurations depending on a variety of factors.

By pursuing this reasoning to evaluate the probable spin alignments of various combinations favored by particular metal—ligand—metal bonds, a catalog of superexchange interactions among members of the $3d^n$ series can be created including relative strengths of the expected J_{ij} constants. Such endeavors produced the Goodenough–Kanamori rules that have become valuable guidelines for the chemical design of magnetic compounds and are summarized in Anderson's review chapter.[61] In Table 1.12, a compilation of superexchange interactions based on Kanamori's review is reproduced with some additions.[66]

A more comprehensive study of cation combinations including those involving low-spin configurations was done by Goodenough[67] and Goodenough and Loeb.[68]

Table 1.12. Summary of Spin Alignment Estimates of d^n Ion 180° Superexchange Interactions in Octahedral Sites (after Kanamori[66])

d^n–d^n Combinations	Cations	Bond Mechanism	Spin Alignment	Probable Net Result
d^3–d^3	Mn^{4+}–Mn^{4+}	σ-, π-bonds		
	Cr^{3+}–Cr^{3+}	A, G, A–H, S	↑↓	↑↓
d^8–d^8	Ni^{2+}–Ni^{2+}	σ-bonds		
		A, G, A–H, S	↑↓	↑↓
d^5–d^5	Mn^{2+}–Mn^{2+}	σ-bonds		
	Fe^{3+}–Fe^{3+}	A, G, A–H, S	↑↓	
		π-bonds		
		G, A–H, S	↑↓ (weak)	↑↓
		π-bonds		
		A	Uncertain (weak)	
d^8–d^3	Ni^{2+}–V^{2+}	σ-, π-bonds		
	Ni^{2+}–Cr^{2+}	A, G, A–H, S	↑↑	↑↑
d^5–d^3	Fe^{3+}–Cr^{3+}	σ-bonds		
	Mn^{2+}–V^{2+}	A, G, A–H, S	↑↑	
	Fe^{3+}–V^{2+}	π-bonds		
	Mn^{2+}–Cr^{3+}	G, A–H	↑↓ (weak)	↑↑
		π-bonds		
		A, S	Uncertain (weak)	
d^4–d^4	Mn^{3+}–Mn^{3+}	dependent on		
	Mn^{3+}–Fe^{4+}	bond angle		
d^6–d^6	Fe^{2+}–Fe^{2+}	σ-bonds		
	(e.g., FeO)	A, G, A–H, S	↑↓	
		π-bonds		↑↓
	Co^{3+}–Co^{3+}		Uncertain	
	(e.g., Co_2O_3)[a]		(weak)	
d^7–d^7	Co^{2+}–Co^{2+}	σ-bonds		
		A, G, A–H, S	↑↓	
	(e.g., CoO)	π-bonds	Dependent on bond angle	↑↓

A, Anderson mechanism;[61] G, Goodenough mechanism;[67,68] A–H, Anderson–Hasegawa mechanism;[65] S, Slater mechanism.[52]
[a]Not 180° bond angles.

In Table 1.13, estimates of probable spin ordering for 180° bonds between cations in octahedral sites are summarized. It is important, however, to recognize that the spin ordering temperatures could be meaningful only where a homogeneous magnetic lattice is created. In most of these situations, particularly those involving combinations far from the diagonal or of widely differing cation valences, the actual materials structures may not be thermodynamically stable and can be synthesized only as metastable phases.

1.7. SPIN ORDERING IN FERRIMAGNETIC OXIDES

To this point in the description of transition metal ions in insulating structures, the emphasis has been on the fundamentals of local or isolated-ion phenomena in crystal fields leading to covalent interactions and their relations to

Table 1.13. Spin Ordering of d^n Ions in Octahedral Sites (High-Spin States, 180° Bonds)

		d^0	d^1	d^2	d^3	d^4	d^5	d^6	d^7	d^8	d^9
	d^1	$S=0$ CET	NEGL	↑↓ W	↑↓ W	↑↑ ~	↑↑ M	↑↑ M	↑↑ M	↑↑ M	NEGL
	d^2			↑↓ W	↑↓ W	↑↑ ~	↑↑ M	↑↑ M	↑↑ M	↑↑ M	↑↑ W
	d^3				↑↓ W	↑↑ ~	↑↑ M	↑↑ M	↑↑ M	↑↑ M	↑↑ W
	d^4					↑↑ ~M	↑↓ ~	↑↓ ~	↑↓ ~	↑↑ ~	↑↓ ~
	d^5						↑↓ S	↑↓ S	↑↓ S	↑↓ S	↑↓
	d^6							↑↓ S	↑↓ S	↑↓ S	↑↓
	d^7								↑↓ S	↑↓ S	↑↓
	d^8									↑↓ S	↑↓
	d^9										↑↓
	d^{10}										$S=0$ CET

S = Strong   W = Weak   NEGL = Negligible
M = Moderate   ~ = Quasi-static   CET = Covalent Electron Transfer

magnetic superexchange. For the balance of this discussion, the influence of these magnetic ion–ligand complexes on the properties of materials will be examined. There are two important areas where the foregoing physical concepts can be applied: (i) ferrimagnetism in which correlation superexchange is the dominant mechanism that creates a broad class of practical magnetic oxides with electrical insulating properties; and (ii) delocalization exchange which shares a cause-and-effect relation with electrical conduction in materials systems that are the media for colossal magnetoresistance (CMR) and high critical temperature superconductivity (HTS). In both cases, the exchange or molecular field anchors the macroscopic theory on which collective magnetic and electrical properties are based. We begin with a review of the origins of spontaneous magnetism.

1.7.1. Molecular-Field Theory

Magnetic exchange effects in spontaneous magnetism are most readily observed in the macroscopic behavior of ferro- and antiferromagnetic materials. For this text, we begin the discussion with the application of the Brillouin theory of paramagnetism to the case where exchange effects provide an effective magnetic field that causes spin ordering in the lattice.

For paramagnetism of a collection of n spins S per unit volume in a fixed magnetic field H, the ratio of $M(T, H)$ to the saturation magnetization $M(0, H)$

$$\frac{M(T,H)}{M(0,H)} = \mathcal{B}_S(a) = \frac{2S+1}{2S}\coth\left(\frac{2S+1}{2S}\right)a - \frac{1}{2S}\coth\frac{a}{2S} \qquad (1.7.1)$$

where $M(0,H) = ngm_BS$ (assuming complete orbital quenching leaves the individual ionic moments equal to gm_BS), $\mathcal{B}_S(a)$ is the Brillouin function, and $a = gm_BSH/kT$.

The variation of magnetization with temperature and magnetic field in the ferromagnetic state may be computed by adding to the applied field a molecular field or exchange field H_{ex} that is proportional to the magnetization. Accordingly, we can construct the Brillouin–Weiss variable

$$a_{\text{eff}}(T, H) = \frac{gm_BS}{kT}[H + H_{ex}] = \frac{gm_BS}{kT}[H + N_W M(T,H)] \qquad (1.7.2)$$

where N_W is the Weiss molecular field coefficient. Since M is also a function of T and H from Eq. (1.7.1), Eq. (1.7.2) can be expressed as

$$a_{\text{eff}}(T, H) = \frac{gm_BS}{kT}[H + N_W M(0, H)\mathcal{B}_S(a_{\text{eff}})] \qquad (1.7.3)$$

Because a_{eff} is now a function of $\mathcal{B}_S$, Eq. (1.7.1) cannot be solved in closed form. Prior to the advent of high-speed digital computers, graphical solutions were used. There are, however, two useful analytical results that emerge from manipulations of the above equations. For $H = 0$ and $a_{\text{eff}} \ll 1$, it can be shown that the Curie temperature at which spontaneous magnetism vanishes is given by

$$T_C = \frac{ng^2 m_B^2 S(S+1)}{3k} N_W \qquad (1.7.4)$$

It should be noted that this relation for a single value of T_C can apply only if $H = 0$, as will be seen in Section 1.8 in the discussion of magnetoresistance. In addition, a relation for $\mathcal{B}_S$ as a function of T/T_C can be constructed according to

$$\mathcal{B}_S\left(\frac{T}{T_C}\right) = \left(\frac{S+1}{3S}\right) a_{\text{eff}}\left(\frac{T}{T_C}\right) \qquad (1.7.5)$$

Universal curves of the Brillouin–Weiss theory are shown in Fig. 1.43 where $\mathcal{B}_S$ is presented for different values of S. The curve for $S \to \infty$ is the Langevin limit where $[S(S+1)]^{1/2} \to S$.

To relate the molecular-field coefficient to the exchange constant, it is again convenient to adopt the Ising model for the exchange energy. According to the

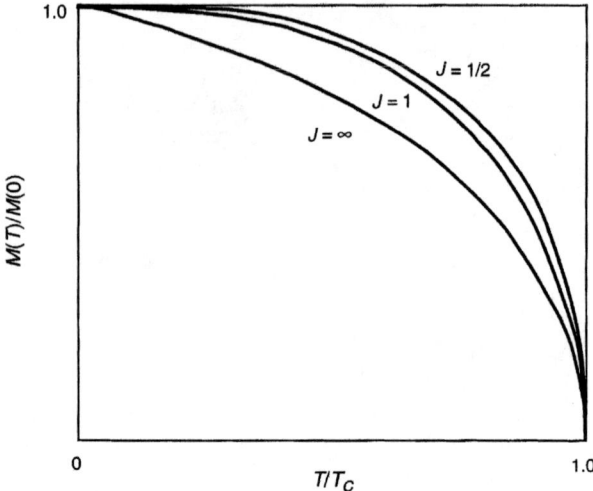

Figure 1.43. Universal Brillouin–Weiss curves for $J = 1/2$, 1, and ∞, with $H_{ex} \gg H$.

theory of Larmor precession, the spin vectors will precess about a magnetic field, applied or molecular (exchange), directed along the z-axis of quantization. As a consequence, the time averages of the x and y spin components are assumed to be zero, and

$$E_{ex} \approx -2zJ_{ij}S_{zi}S_{zj} \quad (1.7.6)$$

where z is the number of nearest-neighbor S_j spins.* Since we consider only the z component of the magnetization M_z $(= ngm_B S_{zj})$ and Eq. (1.7.6) can be written as

$$E_{ex} \approx -2\left(\frac{z}{n}\right)\frac{J_{ij}S_{zi}M_z}{g_i m_B} \quad (1.7.7)$$

For $H = 0$, E_{ex} can also be expressed in terms of the Weiss molecular field by equating it to the numerator of a_{eff} in Eq. (1.7.2), according to

$$E_{ex} \approx -g_i S_{zi} m_B N_W M_z \quad (1.7.8)$$

after combining Eqs. (1.7.7) and (1.7.8), a relation between the exchange integral J_{ij} and the Weiss molecular-field coefficient N_W can be derived:

$$N_W = \left(\frac{z}{n}\right)\frac{2J_{ij}}{g_i^2 m_B^2} \quad (1.7.9)$$

and we have for the Curie temperature from Eq. (1.7.4), after Eq. (1.7.9) is substituted,

---

*When the bonding energy is divided among multiple equivalent neighbors instead of the single ligand used in the foregoing simple molecular models, care must be taken to lower J_{ij} and related energy parameters accordingly to maintain the correct E_{ex}.

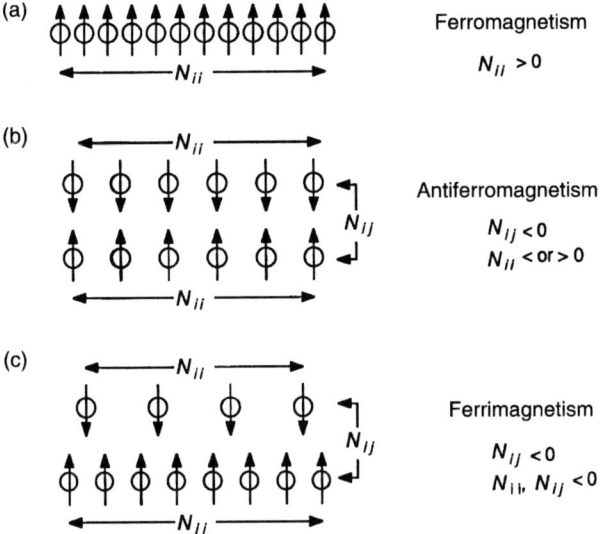

Figure 1.44. One-dimensional models of spontaneous spin alignment. (a) Ferromagnetism: $N_{ii}>0$; (b) Antiferromagnetism: $N_{ij}<0$, $N_{ii} \langle \text{or} \rangle\, 0$; (c) Ferrimagnetism: $N_{ij}<0$, $N_{ii}, N_{ij}<0$.

$$T_C = \frac{2zJ_{ij}S(S+1)}{3k} \tag{1.7.10}$$

The relations for N_W and T_C are important in the models for thermomagnetization of ferrites and magnetoresistance of perovskites.

1.7.2. Magnetic Sublattices and Ferrimagnetism

One of the most important and successful applications of the molecular-field theory is computation of the spontaneous magnetization characteristic of a ferrimagnetic composition as a function of temperature (thermomagnetization). The seminal work was reported by Néel for the case of two magnetic sublattices modeled as one-dimensional rows in Fig. 1.44, where the spin ordering represented as two antiparallel arrays with unequal populations[69] is contrasted with the standard ferro- and antiferromagnetic spin configurations. In three-dimensional reality, the sublattices comprise different ligand sites (usually octahedral and tetrahedral) that alternate through the crystal lattice. Recalling the previous discussion of correlation superexchange that produces an antiferromagnetic exchange stabilization between like magnetic ions, we recognize that there must be three molecular-field coefficients, one for each sublattice (intra-sublattice N_{ii} and N_{jj}) and one for the exchange between the sublattices (inter-sublattice N_{ij}). As expected from the molecular-field theory, the antiferromagnetic alignment of the spin in i and j sites occurs because $N_{ij} > N_{ii}$ and N_{jj}. The resultant magnetization or magnetic moment per mole, depending on the application, is given by

$$M(T) = |M_i(T) - M_j(T)| \tag{1.7.11}$$

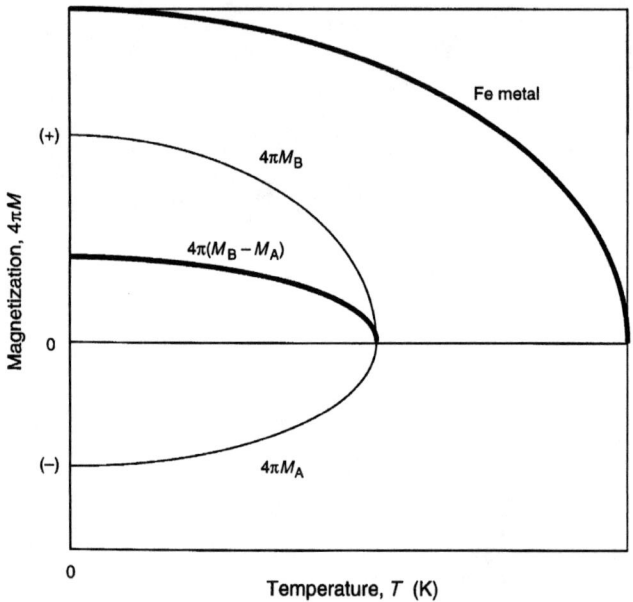

Figure 1.45. Schematic models of the thermomagnetization of a metal and a ferrite.

The procedure once again involves the Brillouin–Weiss function, which is applied to each sublattice according to

$$M_i(T) = M_i(0)\mathcal{B}_{S_i}(T)$$
$$M_j(T) = M_j(0)\mathcal{B}_{S_j}(T) \quad (1.7.12)$$

where

$$\mathcal{B}_{S_i}(T) = \frac{m_i H_{ex}^{(i)}}{kT} = \frac{g_i m_B S_i}{kT}(N_{ii}M_i + N_{ij}M_j)$$

$$\mathcal{B}_{S_j}(T) = \frac{m_j H_{ex}^{(j)}}{kT} = \frac{g_j m_B S_j}{kT}(N_{ji}M_i + N_{jj}M_j) \quad (1.7.13)$$

and $M_i(0) = n_i g_i m_B S_i N_A$, $M_j(0) = n_j g_j m_B S_j N_A$ (expressed in emu/mol), and N_A is Avogadro's number. Solution of Eq. (1.7.11) cannot be carried out in closed form, but can be accomplished by self-consistent iteration procedures involving multiple sublattices simultaneously[70–72] through the use of high-speed digital computers.* A sketch of a typical thermomagnetization computation is given in Fig. 1.45. The usefulness of the molecular-field model has proven to be enormous over the past four decades, particularly the refined versions of Dionne that have made possible the explanation and prediction of thermomagnetism behavior of compounds in which magnetic ions have been replaced by nonmagnetic substitutes for the purpose of tailoring magnetic properties to specific applications.[70–74] In most of these situations, dilution of

---

*These reports are published and can be obtained from the U.S. National Technical Information Service or the U.S. Defense Technical Information Center.

the magnetic sublattices has been accompanied by departures from ideal magnetic spin alignments, commonly referred to as "spin canting." These reductions in the effective magnetic moments occur beyond the normal disruptions of the magnetic ordering arising from the thermal randomization accounted for in the application of the Brillouin–Weiss function and are the result of antiferromagnetic frustration.[75]

1.7.3. Spinel Ferrites

In Section 1.7.2, some of the basic concepts peculiar to ferrimagnetism were described. In particular, spontaneous magnetism and its relation to the magnetic exchange was explained. However, the actual magnetic properties of the various ferrimagnetic systems are important in themselves. The following sections will sketch some features of ferrimagnetic spinels and garnets—the most frequently encountered in electronic applications. Magnetocrystalline anisotropy and magnetostriction effects, designated respectively by the conventional symbols K_1 (first-order constant) and λ_s (isotropic average of the λ_{111} and λ_{100} constants), are included because of their relevance to hysteresis and microwave properties. A major part of these effects originate from the spin–orbit–lattice coupling between single ions and their ligands that was described previously.

The 2 : 1 octahedral B to tetrahedral A site ratio in the spinel lattice is determined by the crystal lattice structure sketched in Fig. 1.46. A cursory inspection of this structure will reveal that the Fe–O–Fe bond angles for either octahedral B or tetrahedral A sites are not 180°, which was presumed in the foregoing discussion of theory. As a general rule, the exchange interaction is greatest for the highest bond angles. This situation also applies to the garnet structure. Unlike the garnets,

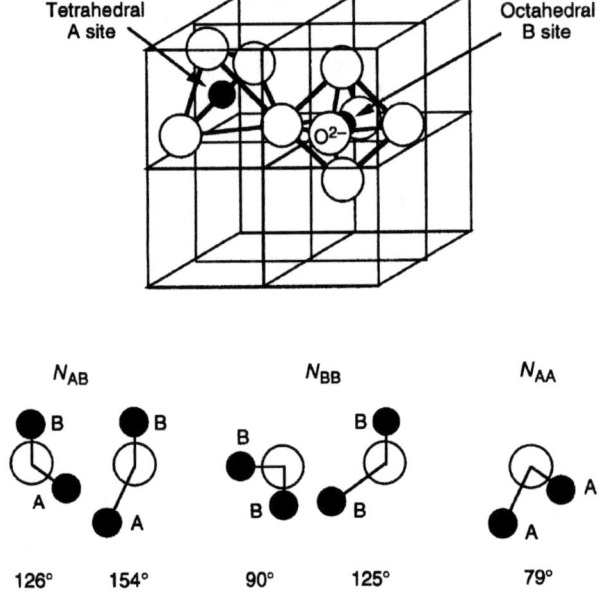

Figure 1.46. Spinel crystal structure with bond angle diagrams.

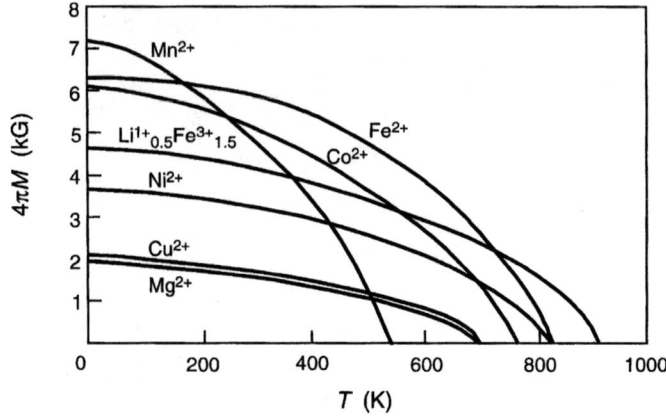

Figure 1.47. Approximate thermomagnetization curves of basic spinel ferrites. Format was inspired by Smit and Wijn.[76]

however, the spinel cell features congruity between the overall crystal symmetry axes to those of the individual sites. In its generic form the spinel can be: (i) normal, with the divalent ions occupying only B sites, i.e., $A^{3+}[B^{3+}B^{2+}]O_4$; or (ii) inverse, in which the divalent ion resides in A sites. In most practical cases, the ferrite is a mixture between normal and inverse. A number of excellent references document the properties of ferrites.[76-83] The magnetization properties of common spinel families are described in the following paragraphs based on the site distribution estimates and magnetic moment values listed in Tables 1.14 and 1.15, based on values from Smit and Wijn.[76] Thermomagnetization properties are shown in Fig. 1.47.

Magnetite is a naturally occurring normal spinel of chemical formula $Fe^{3+}[Fe^{3+}Fe^{2+}]O_4$ that serves as a crystallographic and chemical template for the design of the various spinel ferrites. Because the electrical conductivity introduced by the full complement of octahedral-site Fe^{2+} ions that can transfer electrons among

Table 1.14. Saturation Magnetic Moments per Formula Unit of Common Spinel Ferrites at $T = 0$ K

Ferrite	A-Site Ions	B-Site Ions	A Moment (Bohr Magnetons)	B Moment (Bohr Magnetons)	Net Moment[a] (Bohr Magnetons)
Fe_3O_4	Fe^{3+}	$Fe^{2+} + Fe^{3+}$	5	4 + 5	5
$Li_{0.5}Fe_{2.5}O_4$	Fe^{3+}	$Li^{3+}_{0.5} + Fe^{3+}_{1.5}$	5	0 + 7.5	2.5
$NiFe_2O_4$	Fe^{3+}	$Ni^{2+} + Fe^{3+}$	5	1 + 5	2
$MnFe_2O_4$	$Fe^{2+}_{0.2} + Mn^{2+}_{0.8}$	$Mn^{2+}_{0.2} + Fe^{3+}_{1.8}$	5	5 + 5	5
$MgFe_2O_4$	Fe^{3+}	$Mg^{2+} + Fe^{3+}$	5	0 + 5	0
$CoFe_2O_4$	Fe^{3+}	$Co^{2+} + Fe^{3+}$	5	3 + 5	3
$CuFe_2O_4$	Fe^{3+}	$Cu^{2+} + Fe^{3+}$	5	1 + 5	1

[a]These values are first-order approximations, but are reasonably accurate, with the possible exception of Mg ferrite which shows a net moment of 1.1. This discrepancy is attributed to some of the Mg ions occupying A sites.

Table 1.15. Room-Temperature Magnetizations and Curie Temperatures of Common Spinel Ferrites

Ferrite	$4\pi M_s$ at $T = 0$ K (G)	$4\pi M_s$ at $T = 300$ K (G)	T_C (K)
Fe_3O_4	6400	6000	858
$Li_{0.5}Fe_{2.5}O_4$	4200	3900	943
$NiFe_2O_4$	3800	3400	858
$MnFe_2O_4$	7000	5000	573
$MgFe_2O_4$	1800	1500	713
$CoFe_2O_4$	6000	5300	793
$CuFe_2O_4$	2000	1700	728

the Fe^{3+} neighbors within the B sublattice, magnetite is more a vehicle for studying the electrical properties of oxides than a practical magnetic material.[84] It is also one of the best hosts for studying the magnetic contribution of the Fe^{2+} ion, which is responsible for important magnetoelastic and dielectric effects, as well as polarized spin transport by the double exchange mechanism.

Lithium ferrite contains the most iron ions next to magnetite and is the ferrite compound with the largest Curie temperature ($T_C \approx 940$ K). As can be recognized immediately from the chemical formula $Fe[Li_{0.5}Fe_{1.5}]O_4$, 25% of the B sublattice is diluted by the Li^{1+} ions. Nonetheless the net number of Bohr magnetons is 2.5, which is sufficient to produce a room-temperature $4\pi M \sim 3500$ G. To lower the moment, magnetic dilution can be accomplished by direct substitution of Al^{3+} in the B sublattice or more commonly by substitutions of $0.5Li^{1+} + Ti^{4+} \rightarrow 1.5Fe^{3+}$ with chemical formula $Fe_{1-t/2}Li_{t/2}[Li_{0.5}Fe_{1.5-t}Ti_t]O_4$. As the site distributions indicate, when the B sublattice is diluted with Ti^{4+} ions the extra Li^{1+} ions required for electrical charge neutralization replace Fe^{3+} ions in the A sites. To increase the moment, $Zn^{2+} \rightarrow 0.5Fe^{3+} + 0.5Li^{1+}$ are substituted into the A sublattice according to $Fe_{1-z}Zn_z[Li_{0.5-z/2}Fe_{1.5+z/2}]O_4$. This modification also increases the Fe^{3+} content of the B sublattice, but it also causes a reduction in T_C because of the decreased net number of Fe_A^{3+}–Fe_B^{3+} interactions. The results of a molecular-field analysis have been published[74] and the relevant parameters are listed in Table 1.16.

The uses of Li ferrite have become mainly in microwave applications because of its high T_C made possible by the larger amounts of Fe^{3+} in both sublattices and generally good dielectric properties in the microwave bands. Its higher $4\pi M$ capabilities make it particularly attractive in the millimeter-wave bands (above 35-GHz frequencies). Further experimental investigations of the magnetic properties of this family was carried out by a team led by White and Patton.[85]

Nickel ferrite $Fe^{2+}[Ni^{2+}Fe^{3+}]O_4$ is a fully inverted and chemically well-behaved spinel with Ni^{2+} occupying half of the octahedral B sublattice. Because Ni^{2+} carries a spin value $S = 1$ in addition to an effective g factor greater than 2, nickel ferrite can offer higher magnetizations. Since Ni^{3+} occurs only occasionally the incidence of Fe^{2+} is usually negligible, and dielectric properties are excellent. Because the e_g orbitals are half-filled, this ion presents a significant covalent transfer integral b and consequently, a strong superexchange coupling to neighbors of its own kind and to the adjacent Fe^{3+} ions of both sublattices. Similar to Li ferrite, its magnetic moment can be reduced, in this case typically by direct substitutions of Al^{3+} for Fe^{3+}

Table 1.16. Molecular and Exchange Parameters of Lithium Spinel and Yttrium Iron Garnet Ferrite

	$z_{ij} : n_j$	M_j (emu/mol × 10^4)	N_{ij} with M_j in emu/mol (mol/cm³)	N_{ij} with M_j in emu/cm³ (× 4π)	J_{ij} (ergs × 10^{-15})	$H_{ex}^{(i)}$ at $T = 0$ K (Tesla)
$Li_{0.5}Fe_2.5O_4$						
$Fe_B \rightarrow \begin{cases} Fe_B \\ Fe_A \end{cases}$	$4.5^a : 1.5^a$ / $6 : 1$	4.2 / 2.8	-60 / 273	-212 / 969	-2.1 / -4.8	512
$Fe_A \rightarrow \begin{cases} Fe_A \\ Fe_B \end{cases}$	$4 : 1$ / $9^b : 1.5^b$	2.8 / 4.2	-150 / 273	-533 / 969	-3.8 / -4.8	722
$Y_3Fe_5O_{12}$						
$Fe_d \rightarrow \begin{cases} Fe_d \\ Fe_a \end{cases}$	$4 : 3$ / $4 : 2$	8.4 / 5.6	-30.4 / 97.0	-347 / 1107	-2.4 / -4.9	288
$Fe_a \rightarrow \begin{cases} Fe_a \\ Fe_d \end{cases}$	$6 : 2$ / $6 : 3$	5.6 / 8.4	-65.0 / 97.0	-742 / 1107	-2.3 / -4.9	451

[a] In the determination of these values, the lattice parameter of the garnet and spinel $a_0 \approx 12.4$ and 8.4 Å, respectively.
[b] These values reflect the 25% initial dilution of the B sublattice by Li^{1+} ions.

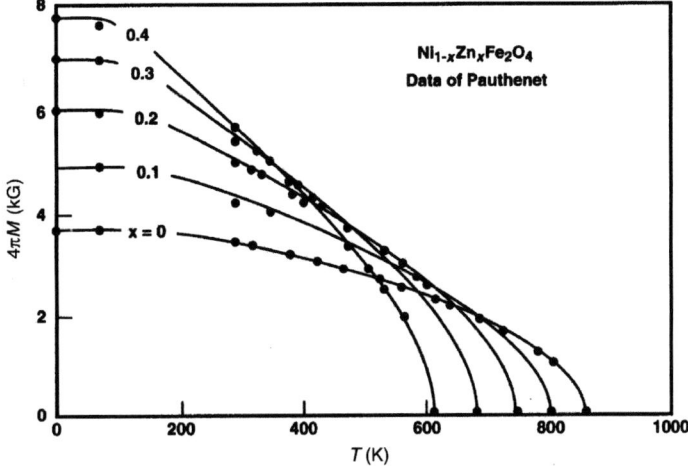

Figure 1.48. Thermomagnetization curves of nickel–zinc spinel ferrite, showing the effects of Zn^{2+} dilution of the minority tetrahedral A sites. Note decrease in Curie temperatures. Data are from Pauthenet.[87]

ions which initially favor B-site dilution according to the theoretical analysis reported by Borghese.[86] The magnetic moment can also be raised by the Zn^{2+} ions diluting the A sublattice, with the attendant increase in the sensitivity of $4\pi M$ to temperature resulting from the decrease in T_C, as shown by Pauthenet's data[87] in Fig. 1.48. Nickel ferrite can also feature large magnetostriction that renders the anisotropy and hysteresis-loop properties of the material highly stress sensitive. Manganese in the 3+ state helps to compensate magnetostriction through its J–T property to control stress sensitivity.[88]

Manganese ferrites are mainly normal with site distributions indicated by the formula $Mn_{0.8}^{2+}Fe_{0.9}^{3+}[Mn_{0.2}^{2+}Fe_{1.8}^{3+}]O_4$. The Mn^{2+} ion has a larger radius than most of the ions of the $3d^n$ series, 0.80 Å instead of ~0.65 Å. To accommodate this larger cation, the lattice parameter is increased with the result that some of the b^2/U factors of the covalent bonding are reduced, thereby leading to a Curie temperature that is the lowest of all of the basic spinel ferrite families, as shown in the data of Fig. 1.47. This effect is reflected in the values of the molecular-field coefficients N_{ij} which reveal weaker interactions that involve Mn^{2+}, where large reductions occur in the N_{AB} coefficients for Fe–Mn and Mn–Mn.[89] These results support the conclusions of Simsa and Brabers[90] that a canting angle of 54° exists among the Mn^{2+} ions in the B sublattice that accounts for an initial deficiency in the magnetic moment at $T = 0$ K.

For microwave applications, dielectric losses due to poor dielectric properties stem from the charge transfer mechanism that stabilizes Fe^{3+} in B sites, i.e., $Fe^{2+} + Mn^{3+} \rightarrow Fe^{3+} + Mn^{2+}$. There are also two possible polaronic conduction opportunities here: $Fe^{2+} \leftrightarrow Fe^{3+} + e^-$ and $Mn^{3+} \leftrightarrow Mn^{2+} - e^-$. In general, the Curie temperatures of this family are unacceptably low for applications with power dissipation concerns that would likely increase the operating temperature. Similar to Ni ferrite, T_C is further reduced with the addition of Zn^{2+} to increase $4\pi M$.

Magnesium ferrite is mainly inverted with a typical formula $Mg_{0.1}^{2+}Fe_{0.9}^{3+}$ $[Mg_{0.9}^{2+}Fe_{1.1}^{3+}]O_4$. The magnetization and permeability are too low for most low-frequency applications, but the family has proven to be useful for certain microwave regimes in situations where the high-temperature sensitivity is tolerable. In these cases, manganese is added for the purpose of suppressing any formation of Fe^{2+} through the charge transfer equation $Fe^{2+} + Mn^{3+} \rightarrow Fe^{3+} + Mn^{2+}$. As a consequence, this family is usually referred to as the *magnesium–manganese ferrites*. The principal virtues of this hybrid family are its square hysteresis loops and insensitivity to stress made possible by the presence of Mn^{3+} ions.

Cobalt ferrite is also an inverted spinel of formula $Fe^{2+}[Co^{2+}Fe^{3+}]O_4$, with Co^{2+} occupying the B sites. For reasons suggested by its strong spin–orbit–lattice interaction, this compound is of interest primarily for its magnetoelastic properties from which the phenomena of magnetocrystalline anisotropy and magnetostriction arise. Like Ni^{2+}, it can also feature a g factor greater than 2 in octahedral sites. Divalent cobalt in a B site is a classic spin–orbit stabilized ion that favors a compressive distortion along the c-axis of the ligand octahedron. Unfortunately, these properties are generally undesirable for many conventional applications. Moreover, the strong coupling of Co^{2+} spins to the lattice is responsible for large power losses when this ferrite is used as a microwave propagation medium. In small amounts, however, Co^{2+} ions can be beneficial for adjusting anisotropy and magnetostriction, and in controlling relaxation effects that can raise high-power limits.

Copper ferrite is another spinel that features a magnetoelastically active constituent Cu^{2+}, which like Mn^{3+} is a classic J–T ion in an octahedral site. The formula is generally inverted, in the form of $Fe^{3+}[Cu^{2+}Fe^{3+}]O_4$, but can in theory assume a monovalent state in $Cu_{0.5}^{1+}Fe_{2.5}^{3+}O_4$ analogous to Li ferrite, but with part of the Cu^{1+} ions occupying tetrahedral A sites.[91] As in the case of Ni^{2+} and Co^{2+} in octahedral sites, the g factor of the Cu^{2+} ion can be greater than 2, but because of its low spin value $S = 1/2$. Furthermore, the strong J–T tendencies of the Cu^{2+} ion can drive the lattice symmetry tetragonal, as it does in $Fe^{3+}[Cu^{2+}Fe^{3+}]O_4$ (and also the superconducting cuprate perovskite host compounds such as $La_2^{3+}Cu^{2+}O_4$), and will also produce strong positive magnetostrictive effects when used as an additive to cubic lattices. Another drawback is the polaronic conduction mechanisms similar to those of Mn ferrite, $Cu^{2+} \leftrightarrow Cu^{3+} + e^-$ and $Cu^{2+} \leftrightarrow Cu^{1+} - e^-$. An interesting possibility is the use of monovalent copper in A sites to produce higher magnetization compounds for microwave applications.[92,93]

The effects of various ions when used as substitutes for iron to dilute the magnetic sublattices or to alter the magnetoelastic properties of spinel ferrites are summarized in Table 1.17.

1.7.4. Yttrium Garnet Ferrites

From chemical, structural, magnetic, and dielectric standpoints, the garnet family is the best behaved of all ferrimagnetic oxides. A thorough compilation of data with relevant discussion of the various magnetic properties was authored by Winkler.[94] The first careful analysis of the crystal structure was carried out by Geller and Gilleo.[95–97] The three cation sites dodecahedral {c}, octahedral [a], and tetrahedral (d) are shown in relation to a section of crystallographic unit cell in Fig. 1.49,

Table 1.17. Initial Effects of Dilutant Ions in Inverted Spinel Ferrite $A^{3+}(B^{3+})_1(B^{2+})_1O_4$

Ions	Site	M	$|K_1|$	$|K_1|/M$	$|\lambda_s/K_1|$	Footnotes
Al^{3+}, Ga^{3+}	B(A)	⇓	⇓	—	—	a
Mg^{2+}	B	⇓	⇓	—	—	b
$Li^{1+} + 2Ti^{4+}$	A + 2B	⇓	⇓	—	—	c
Zn^{2+}	A	⇑	—	↓	—	d
Co^{2+}	B	—	⇓⇒⇑	⇓⇒⇑	⇑⇒⇓	d
Mn^{3+}	B	—	—	—	⇓⇒⇑	e
Fe^{2+}	B	—	⇑⇑⇑	⇑⇑⇑	—	f

[a] Al^{3+} and Ga^{3+} occupy both octahedral and tetrahedral sites, but opposite to the garnets, they initially have strong preference for the octahedral sites in the inverted spinels.
[b] Mg^{2+} is the divalent-ion basis for the magnesium ferrite family.
[c] $Li^{1+} + 2Ti^{4+}$ combination is used in Li ferrite. The added Li^{1+} replaces Fe^{3+} in the tetrahedral sites, but the net result is a lowering of M by the $2Ti^{4+}$ ions diluting the octahedral sublattice.
[d] Co^{2+} is used in very small amounts to compensate K_1. It will cause K_1 to pass from negative to positive at about 0.01–0.02 ions per formula unit. Because it is spin–orbit stabilized, it is also a fast-relaxing ion.
[e] Mn^{3+} is a J–T ion that influences magnetostriction by adding a positive contribution to the large λ_{100} constant that makes the average value λ_σ pass through zero at a concentration of about 0.2 ions per formula unit.
[f] Fe^{2+} is a spin–orbit stabilized ion that is a natural constituent of spinel ferrites. It enhances the negative anisotropy constant by favoring the $\langle 111 \rangle$ axis extensions, and thereby increases the λ_{111} magnetostriction constant. In combination with host Fe^{3+} ions, it can provide a hopping electron conduction mechanism that is detrimental to high-frequency applications.

where it can be readily seen that the principal symmetry axes of the a and d sites do not conform to the cubic cell axes $\langle 100 \rangle$, $\langle 111 \rangle$, and $\langle 110 \rangle$, in marked contrast to the spinels which have the A and B sites neatly packed into the cubic cell. Only the octahedral site has one of its trigonal axes aligned with a lattice $\langle 111 \rangle$ direction.

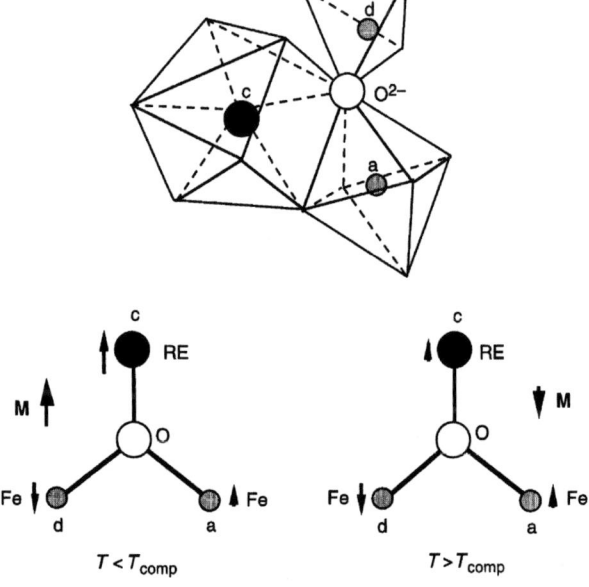

Figure 1.49. Garnet crystal structure with bond angle diagrams, below and above the compensation temperature T_comp that occurs when the c-sublattice is occupied by heavy rare-earth ions of the $4f^n$ series.

An even more important difference is that the tetrahedral sites are more numerous than the octahedral sites in the ratio of 3 : 2, thereby making the d sublattice dominant in contributing to the magnetization. This feature allows the magnetoelastic properties to be varied over a wider range.

Because all of the cations are naturally in the 3+ state, thereby affording less opportunity for Fe^{2+} to form, dielectric properties of the garnets are generally superior to those of the spinels. Magnetocrystalline anisotropy and magnetostriction are also manageable quantities. Moreover, there are a plethora of substitutions available for all three sites that may be used to tailor magnetic, microwave, and optical properties. Because the system is more refractory than the spinels, chemical stability is generally higher among conventional compounds, i.e., those based on yttrium or lanthanum series ions in the c sublattice.

Yttrium iron garnet is the basic magnetic host of this family, and can be altered by a variety of substitutions. Al^{3+} and Ga^{3+} ions are used to reduce the magnetization by forming solid solutions of $Y_3Fe_5O_{12}$ (YIG) with either $Y_3Al_5O_{12}$ (YAG) or $Y_3Ga_5O_{12}$ (YGG). With these two ions, the dilution begins in the d sublattice and gradually spills over to the a sites, as was determined by Mössbauer measurements.[98] For this reason Al^{3+} and Ga^{3+} are used to reduce the magnetization, as shown in Gilleo and Geller's data[99] of Fig. 1.50. To increase magnetization (at the expense of lowering the Curie temperature), the larger ions In^{3+} or Sc^{3+} can be used up to a concentration level of about 0.5 per formula unit because they occupy a sites almost exclusively. With the greater reduction in the number of J_{ad} exchange couplings by a-site dilution, however, the Curie

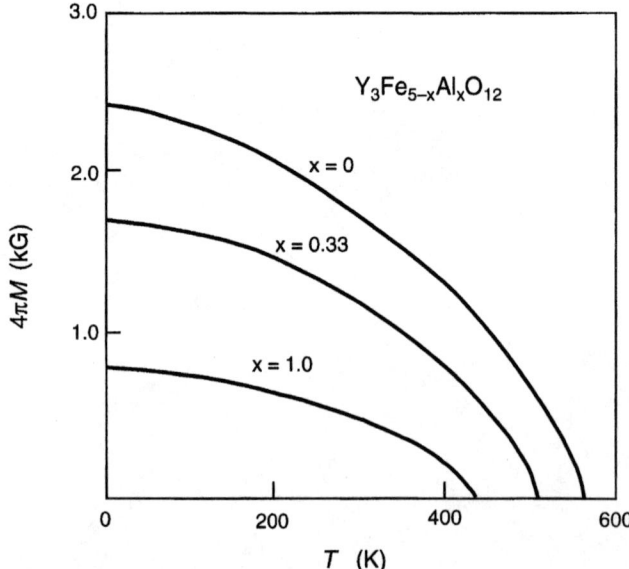

Figure 1.50. Thermomagnetization curves of yttrium aluminum iron garnet ferrite, showing the effects of Al^{3+} dilution of the majority tetrahedral d sites (and to a lesser extent the a sublattice). Note the contrast with Figs. 1.48 and 1.51. Data are from Gilleo and Geller.[99]

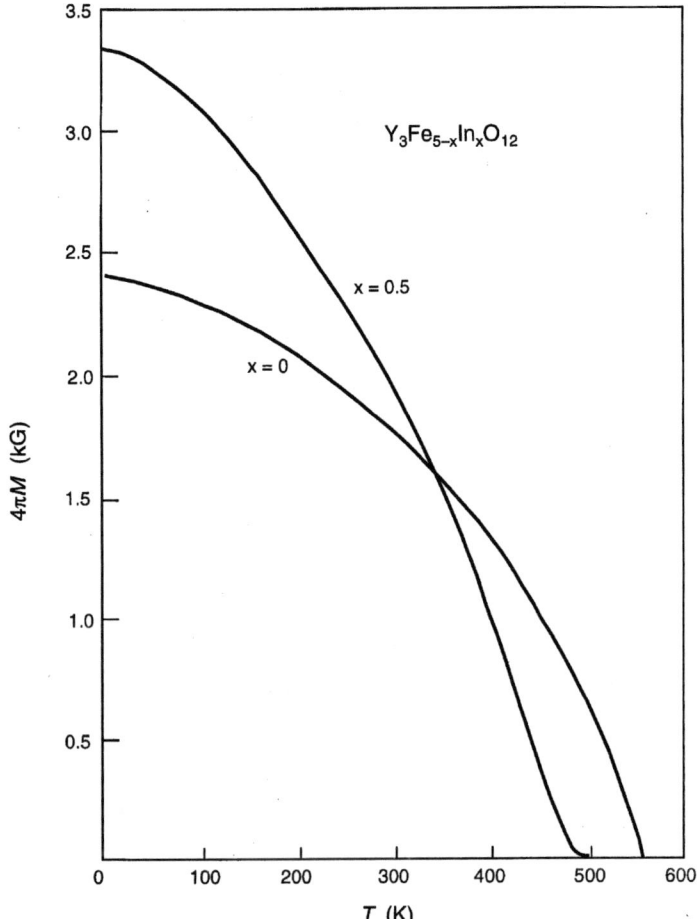

Figure 1.51. Thermomagnetization curves of yttrium indium iron garnet ferrite, showing the effects of In^{3+} dilution of the minority octahedral a sites. Compare with Figs. 1.48 and 1.50. Data are from Gilleo and Geller.[99]

temperature drops quickly with dilution, similar to the case of Zn^{2+} in the A sites of spinels. An example of these effects are shown in the data for In^{3+} substitutions of Fig. 1.51. For application to Eq. (1.7.13), relations for the molecular-field coefficients as a function of dilution fractions k_d and k_a were determined by Dionne:[73]

$$N_{dd} \approx N_{dd}^0 (1 - 0.87k_a)$$
$$N_{aa} \approx N_{aa}^0 (1 - 1.26k_d) \qquad (1.7.14)$$
$$N_{da} = N_{ad} \approx N_{da}^0 (1 - 0.25k_a - 0.38k_d)$$

In units of mol/cm^3, the values for the undiluted molecular field coefficients are $N_{dd}^0 = -30.4$, $N_{aa}^0 = -65.0$, $N_{ad}^0 = +97.0$. A summary of the results of this analysis is given in Table 1.16.

Table 1.18. Initial Effects of Dilutant Ions in Garnet Ferrites $\{c^{3+}\}_3[a^{3+}]_2(d^{3+})_3O_{12}$

Ions	Site	M	$|K_1|$	$|K_1|/M$	$|\lambda_s/K_1|$	Footnotes
Al^{3+}, Ga^{3+}	d(a)	⇓	—	⇓,⇑	—	a
$2Ca^{2+} + V^{5+}$	2c + d	⇓	—	⇑	—	b
$Ca^{2+} + Ge^{4+}$	c + d	⇓	—	⇑	—	b
$Ca^{2+} + Si^{4+}$	c + d	⇓	—	⇑	—	b
In^{3+}, Sc^{3+}	a	⇑	⇓	⇓⇓	⇑	b
$Ca^{2+} + Zr^{4+}$	c + a	⇑	⇓	⇓⇓	⇑	b
$Ca^{2+} + Sn^{4+}$	c + a	⇑	⇓	⇓⇓	⇑	b
$Co^{2+} + Si^{4+}$	a + d	—	⇓⇒⇑	⇓⇒⇑	⇑⇒⇓	c
Mn^{3+}	a	—	—	—	⇓⇒⇑	d
Bi^{3+}, Pb^{2+}	c	—	—	—	—	e
Fe^{2+}	a	—	⇑⇑⇑	⇑⇑⇑	—	f

[a] Al^{3+} and Ga^{3+} occupy both octahedral and tetrahedral sites, but initially have strong preference for the tetrahedral sites in the garnets.
[b] Sr^{2+} can be used instead of Ca^{2+} in c sites, as well as La^{3+} instead of Y^{3+} for lattice parameter adjustments.
[c] Co^{2+} is used in very small amounts to compensate K_1. It will cause K_1 to pass from negative to positive at about 0.02 ions per formula unit. Because it is spin–orbit stabilized, it is also a fast-relaxing ion.
[d] Mn^{3+} causes a J–T distortion that adds a positive contribution to the magnetostriction constants and makes the average value pass through zero at a concentration between 0.05 and 0.1 ions per garnet formula unit.
[e] Bi^{3+} and Pb^{2+} are important for magneto-optical applications but can increase microwave losses.
[f] Fe^{2+} is a spin–orbit stabilized ion that occurs as a natural impurity. It enhances the negative anisotropy constant by favoring the ⟨111⟩ axis extensions, and thereby increases the λ_{111} magnetostriction constant. In combination with host Fe^{3+} ions, it can provide a hopping electron conduction mechanism that is detrimental to high-frequency applications.

Calcium–vanadium–zirconium iron garnet is a common alternative to the basic YIG composition that has been considered for use in special microwave and magneto-optical applications. Dilution of the iron sublattices is achieved with vanadium and zirconium ions. The advantages of this family are twofold: (i) V^{5+} and Zr^{4+} occupy exclusively their respective d and a sites, which means that the adjustments in magnetization can be achieved without unwanted dilution of the opposite sublattice that would cause a further decrease in Curie temperature; and (ii) calcium, vanadium, and zirconium combinations are less expensive than the alternative yttrium, lanthanum, with indium or scandium, and aluminum or gallium. A further benefit of the use of calcium actually comes from the reduced amounts of yttrium or lanthanum because these rare-earth elements can contain fast-relaxing impurities that increase microwave absorption, particularly at low temperatures. For enhanced magneto-optical effects, Bi^{3+} and Pb^{2+} are substituted into c sites because of their large spin–orbit coupling. This property, however, can be deleterious to microwave performance because it reduces spin–lattice relaxation times. Table 1.18 presents a qualitative summary of magnetic and magneto-elastic properties of YIG-based garnets similar to that given for the spinels in Table 1.16.

Other ions of the lanthanide group can be used in the c sublattice to form a distinct set of ferrimagnetic compounds. Because the ionic spins that are introduced to the c sublattice exchange couple to the iron in the d and a sublattices, a wide variety of thermomagnetic, magnetoelastic, microwave, and magneto-optical properties can be created in these rare-earth iron garnets.

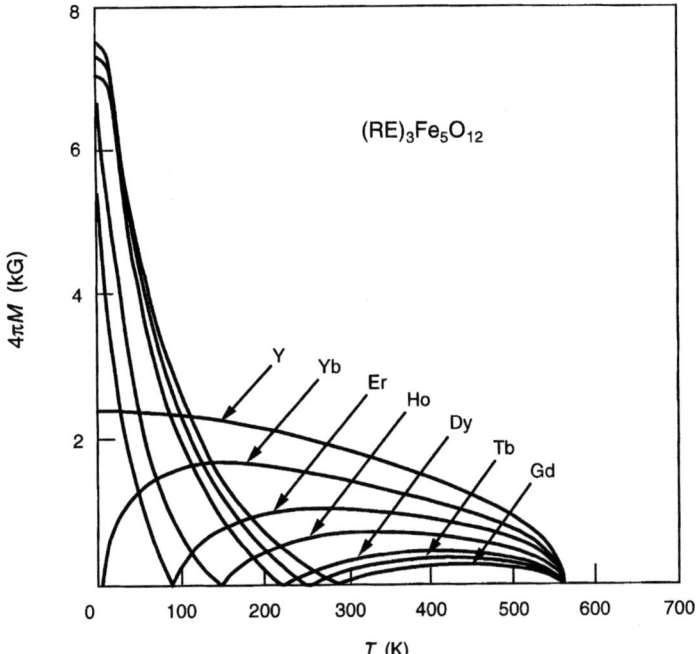

Figure 1.52. Thermomagnetization curves of the rare-earth series occupying c sites in the iron garnet system $(RE)_3Fe_5O_{12}$.

1.7.5. Rare-Earth Garnet Ferrites

When rare-earth ions with unfilled $4f^n$ shell are substituted into the c sublattice to form solid solutions of $(RE)_3Fe_5O_{12}$ and $Y_3Fe_5O_{12}$, the magnetic properties of the iron garnets take on a remarkably different character. The earliest studies on the magnetization behavior of this family were reported first by Pauthenet[100] whose data are replotted in Fig. 1.52 for the series from $Gd_3Fe_5O_{12}$ to $Yb_3Fe_5O_{12}$, with Y^{3+} and Lu^{3+} included as diamagnetic reference ions in the c sites. The most distinctive feature of these curves is the compensation temperature that occurs when the magnetic moments of the c sublattice cancel the net moments of the opposing d and a sublattices (see three-ion model in Fig. 1.53) by aligning with the a-site spins, i.e., $M = |M_d - M_a - M_c|$.

To explain this effect analytically the Néel theory must be expanded to three or more sublattices with the number of molecular field coefficients increasing from four to nine or more. In the relations involving the c sublattice, the magnetic moment is determined by the total angular momentum J_c instead of only the spin component S_c. Because the rare-earth ions have the unpaired spins in the $4f^n$ shell, the orbital angular momentum is largely unquenched by the crystal field. Unlike the $3d^n$ series where the crystal-field strength $10Dq \sim 1$ eV, the crystal fields to which the unpaired 4f electrons are exposed are only $\sim 10^{-2}$ eV (~ 100 cm^{-1}). This means that the sum J_c remains a "good" quantum number.

The earliest attempt to determine the molecular-field coefficients for the rare-earth series was reported by Aléonard from reduction of paramagnetic susceptibility

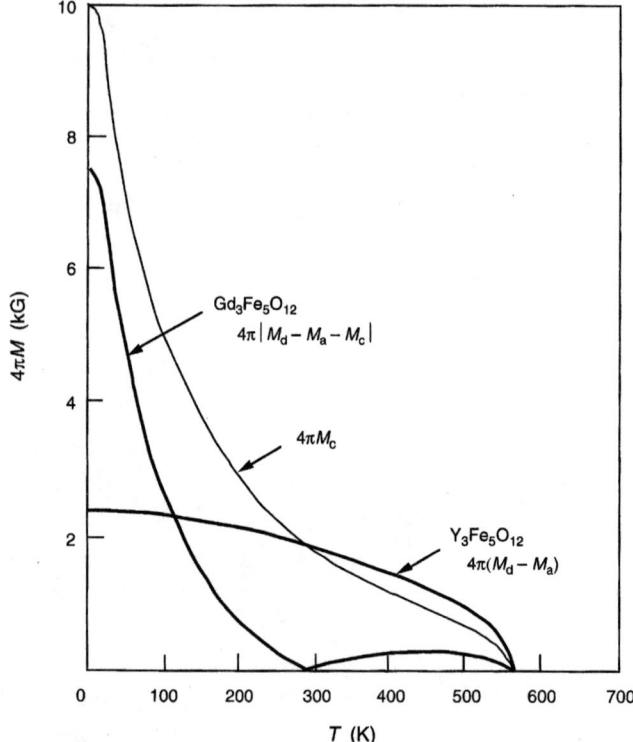

Figure 1.53. Thermomagnetization curves of gadolinium iron garnet ($Gd_3Fe_5O_{12}$) and yttrium iron garnet ($Y_3Fe_5O_{12}$), showing the contribution of the Gd^{3+} sublattice and indicating the formation of the compensation temperature T_{comp}. The curve for the Gd sublattice was obtained by a simple arithmetic subtraction, which is not strictly accurate but adequate for illustrative purposes.

data above the Curie temperature.[101] Although efforts to apply these results to fit the magnetization versus temperature measurement data below the Curie temperature were unsuccessful, Aléonard's values served as useful starting points for Dionne's solution[73,102] that was carried out in the manner described previously for diluting the $Y_3Fe_5O_{12}$ host system. In Table 1.19 molecular field values are listed, including a set reported by Brandle and Blank[103] who also used the approach of Dionne. A comprehensive review of these properties for the entire magnetic garnet system is given by Hansen.[104]

1.8. SPIN TRANSPORT IN OXIDES

Electrical conductivity in solids is traditionally associated with collective electron metals, intermetallic compounds and alloys. With outer shell s and p electrons unbound in the sense of a Sommerfeld gas, the analysis of electrical properties usually takes the form of a density of states calculation from a theory that assumes a periodic lattice potential and applies Fermi statistics. The exercise leads to the creation of broadened energy states (bands) and the definition of a Fermi level to serve as the zero energy reference. For transition metals with unfilled "inner" d shells,

Table 1.19. Molecular Field Coefficients of Heavy Rare-Earth Iron Garnets in mol/cm^3

c-Site Ion	g_c	J_c Free-Ion	$J_c'^a$ In-Site	N_{ac} (Dionne)	N_{cd} (Dionne)	N_{ac} (Aléonard)	N_{cd} (Aléonard)	N_{ac} (Brandle)	N_{cd} (Brandle)
Gd^{3+}	2	3.5	3.5	−3.44	6.0	−1.2	3.4	—	—
Tb^{3+}	3/2	6	4.6	−4.2	6.5	−4.4	4.6	−1.80	3.40
Dy^{3+}	4/3	7.5	5.3	−4.0	6.0	−3.2	3.6	−3.35	3.95
Ho^{3+}	5/4	8	4.98	−2.1	4.0	−4.0	2.4	−0.75	1.50
Er^{3+}	6/5	7.5	4.62	−0.2	2.2	−0.6	1.0	−0.75	1.25
Tm^{3+}	7/6	6	1.085	−1.0	17.0	0	0	−1.00	8.00
Yb^{3+}	8/7	3.5	1.49	−4.0	8.0	−1.0	8.8	−1.70	2.00

^{a} Reduced values of the effective total angular momentum is attributed to canting of the spin vectors.

complications arise from the hybridization of "free" electron s states with states of unpaired spins in the d shell that forms the collective electron band structure. The first successful attempt to model ferromagnetism in metal elements was developed by Stoner employing a phenomenological theory that explained the net magnetization by introducing an exchange field that separated the spins into up and down polarizations with unequal populations.[53] This exchange field represented the effects of electron repulsion energy e^2/r_{ij} (appearing as U_{ex} in Section 1.6), and accounted for the variation in the d-electron ionization potentials plotted in Fig. 1.8.

Regrettably, the basic assumption of a periodic potential with Bloch-type functions forces the abandonment of important local interactions that are of critical importance in ionic compounds. These include randomly spaced effects such as lower symmetry crystal-field splittings from electron–lattice vibronic interactions at individual sites, variations in spin–orbit coupling, single-ion magnetoelastic effects, spin–lattice relaxation that controls spin resonance line shapes and spin-wave propagation, and to some extent, the superexchange interactions that determine the type and stability of magnetic ordering. In select cases where metallic properties have been observed in homogeneous ferromagnetic oxides, the electrical properties have been interpreted by Hartree–Fock tight-binding approximations that include the spin-density-wave model of Overhauser[105] to split the d shell into majority and minority spin bands with the Fermi energy set by the highest occupied state, in the manner of Stoner. However, in most cases where electronic conduction in oxides occurs with spontaneous magnetism, the charge transfer is from randomly dispersed polarons.

Charge transfer by electron hopping between mixed-valence cations of the same atomic element was mentioned previously as a mechanism of electrical conduction in ferrites, particularly $Fe^{2+} \leftrightarrow Fe^{3+} + e^-$ in the octahedral sublattice. The conductivity that results from these random events generally has the temperature characteristic of an insulator or semiconductor at room temperature. In other systems, however, charge transfer can be metallic in temperature dependence and sometimes sufficiently coherent that superconductivity can exist to temperatures greater than 130 K. These phenomena fall under the general class of polaronic motion, which is strongly influenced by the state of spin ordering in the lattice. The general study of polarons was pioneered by a number of workers, prominent among them being Fröhlich[106] and Mott.[107] For metallic oxides, however, the molecular crystal approach of Holstein[108] and the insightful review by Goodenough[109] are also appropriate to the present discussion because of their conformance with the molecular orbital framework outlined in the previous sections. Much of the following sections has been developed by the author[110–116] inspired by this earlier work.

1.8.1. Polarons in Spin Systems

A polaron is a charge carrier that resides in an electrostatic potential trap created by the interaction between the charge and its local crystalline surroundings in the manner of a dipole with one of its charges mobile. The trap consists of lattice distortions that are induced by the electric field of the charge and the associated differences in the size of the ion that carries the charge. Another term that is used to describe a polaron is a "dressed" carrier. In ionic compounds, polarons occur as a result of cation chemistry and mixed valence induced by departures from stoichio-

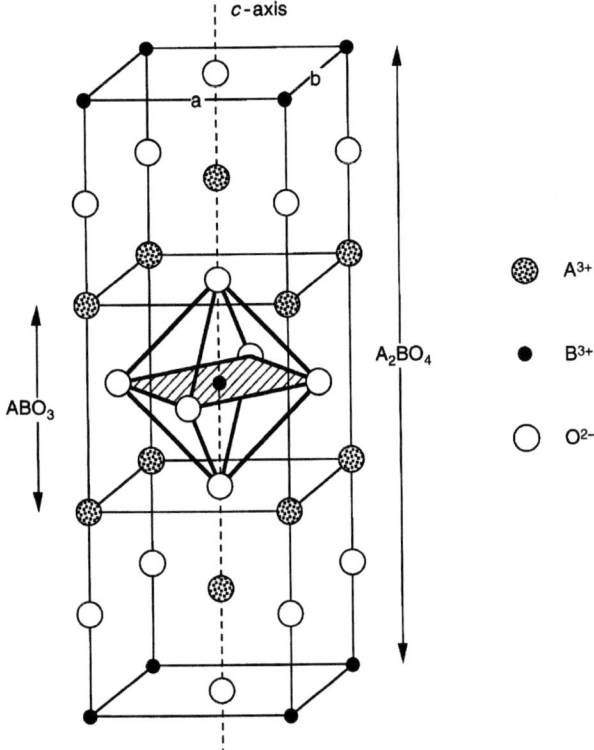

Figure 1.54. Perovskite unit cells cubic ABO_3 and tetragonal A_2BO_4, highlighting the octahedral B site.

metry, in contrast to the *excitation* necessary to create an electron (or hole) in a doped conventional band-gap semiconductor. For electrical conduction, the carrier part of the dipole is not excited, but rather released from its trap by phonons. The participation of the polaron in an electrical current is usually described as an *activation* of its mobility. Convenient vehicles for studying these conduction mechanisms are the cubic ABO_3 and tetragonal A_2BO_4 perovskites shown in Fig. 1.54, emphasizing the octahedral B site at which the polaron is centered. An important feature of this crystallographic family in relation to spin exchange and transport are the 180° B–O–B bond angles that form B–O_2 chains along the z (or c)-axis and B–O_4 planar layers in the x–y (frequently called the a–b) plane.

Consider the case of a mixed-valence complex transition metal perovskite $A^{3+}B^{3+}O_3^{2-}$ in which different ions of smaller valence charge A'^{2+} are substituted. To restore electrical neutrality: (i) either a corresponding number of the O^{2-} ions are converted to O^{1-} to create a partial peroxide $A^{3+}_{1-x}A'^{2+}_{x}B^{3+}O^{1-}_{x}O^{2-}_{3-x}$; or (ii) where possible a corresponding number of B^{3+} are converted to B^{4+} to produce the mixed-valence cations in the B sublattice of $A^{3+}_{1-x}A'^{2+}_{x}B^{3+}_{1-x}B^{4+}_{x}O_3^{2-}$. Because larger ionic charges usually produce greater ionic bonding energies and higher stability, once the ionization potential and electron affinity tradeoffs are made between cations and anions, the latter arrangement would likely result in a lower Madelung energy. Relative to the neutral background, A'^{2+} is a stationary fixed negative charge and

Charge balance: $[A_{1-x}A'^-{}_x](B_{1-x}B^+{}_x)O_3$

Charge transfer: $B \Leftrightarrow B^+ + e^-$

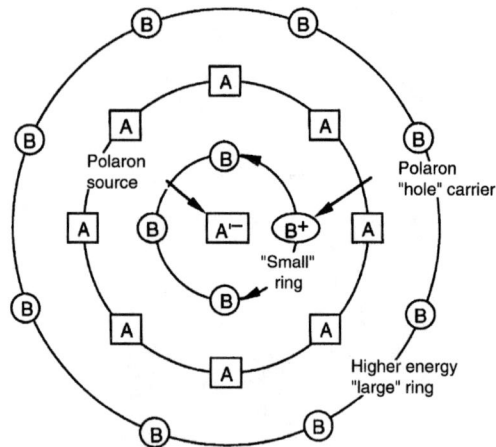

Figure 1.55. Positive polaron (hole) formation in mixed-valence ABO$_3$ and the concentric ring model centered about a fixed polaron source. Inner ring is the radius of a small polaron, outer rings of increasingly higher electrostatic potential energy are regions of large polarons.

B^{4+} forms the positive half of a dipole, as depicted in the two-dimensional sketch of Fig. 1.55. Since the charge at the electronic hole labeled as B^+ is capable of transferring to equivalent sites surrounding the A' fixed charge, in this case without a net change in energy, it is called a polaron and A' is the polaron source.

If the model of Fig. 1.55 is followed, we can identify two parts to the problem of analyzing charge movement in these materials: first, the transfer of electrons among sites of equivalent energy, depicted as locations around the rings, and second, the transport of charges to sites of higher relative energy, depicted by the transitions to an outer ring. In the first case, which is that of a *small* polaron, the gain in binding energy is derived directly from the actual transfer of charge, and therefore favors a large covalent exchange energy integral to stabilize the transfer ground state. The stabilization energy associated with the transfer is critically related to the states of spin polarization of the transfer ions. In the second part, which determines the itinerancy of the charges that leads to the formation of *large* polarons, the stabilization results from a standard Coulomb potential of the dipole attraction and can be examined from the standpoint of polaron theory independent of spin ordering issues.

1.8.1.1. Transfer among Equivalent Sites (Small Polarons).

When an atom is ionized in free space, the removal (or addition) of an electron is accomplished at the cost of an ionization energy of many electron volts. In a crystal lattice, this energy (e.g., the electronic work function) is reduced to a few electron volts because of the lower electrostatic fields due to polarizability of the dielectric medium. In cases where covalent bonding is significant, however, a less energy-expensive dynamic excitation takes place by charge transfer between cations through the covalent interaction with intermediary anions. Where selection rules for energy-free transfer are satisfied, i.e., $\Delta S = 0$, the transfer can take place spontaneously, and when there is no net energy transferred to the lattice, the transfer can be termed "adiabatic." The activation energy from spin exchange is therefore dependent on the angle θ_{ij} of Eq. (1.6.20). Where $\theta_{ij} > 0$, spin flips are required to satisfy any intra-orbital exchange (Hund's rule) requirements on the receptor ion, and the spontaneous excitation-free sharing of the transfer spin is easily allowed because activation energy must be supplied to restore the charge transfer state stabilization energy given by

$$E_{ex}(\theta_{ij}) = -2zJ_{ij}S_iS_j \cos^2\left(\frac{\theta_{ij}}{2}\right)$$
$$= -zJ_{ij}S_iS_j(1 + \cos\theta_{ij}) , \qquad (1.8.1)$$

where J_{ij} is defined by Eq. (1.6.24), and the angular dependence is taken from Eq. (1.6.20a).

When $\theta_{ij} = 0$, E_{ex} has a maximum value for z neighboring S_j spins that are collinear with S_i to satisfy the $\Delta S = 0$ requirement for the charge transfer. If $\theta_{ij} > 0$, E_{ex} decreases and the energy needed to restore the spin alignment takes the form of a magnetic trap of energy equal to the loss in stabilization energy

$$E_{hop}^{ex} = E_{ex}(\theta_{ij}) - E_{ex}(0) = zJ_{ij}S_iS_j(1 - \cos\theta_{ij}) \qquad (1.8.2)$$

which is also the activation energy necessary to effect the charge transfer between the two cation sites i and j. It should be noted that θ_{ij} values in the range $0 - \pi$ are allowed by Eq. (1.8.2), and that the theoretical maximum hopping activation energy from exchange is actually $2zJ_{ij}S_iS_j$ when the spins are antiferromagnetic ($\theta_{ij} = \pi$). For the ferromagnetic to paramagnetic transition that occurs at the Curie temperature, where the average angle between spins becomes $\pi/2$.* The Boltzmann transfer probability for small polarons among equivalent sites relative to their sources can be written as $p_{ex} = \exp(-E_{hop}^{ex}/kT)$.

1.8.1.2. Transport to Higher Energy Sites (Large Polarons).

In the simplest case, there are no spin polarization limitations and the only restriction on the polaron mobility is the electrostatic forces binding it to its source in the lattice. It is therefore assumed that the trap is of elastic (hence electrostatic) origin, with the polaron carrier tethered to its nearby fixed charge of opposite sign. The solution is worked out in terms of the difference in electrostatic energy ΔE_p between the two lattice metal ions M_a and M_b by following the procedures used previously in the derivation of Eq. (1.6.8). In the model defined in Fig. 1.56 (similar to Fig. 1.42(c), but with $U_p = \Delta E_p$), we again assume that the effective polaron integral b_p for direct transfer between M_a and M_b is greater than the energy ΔE_p with the approximate

---

*Note that Eq. (1.8.2) is consistent with Eq. (1.6.22) if the antiferromagnetic spin alignment factor $\sin^2(\theta_{ij}/2)$ from Eq. (1.6.20b) is included, and $2zJ_{ij}S_iS_j = \Delta E_{ex}^{deloc}$ from Fig. 1.42(b).

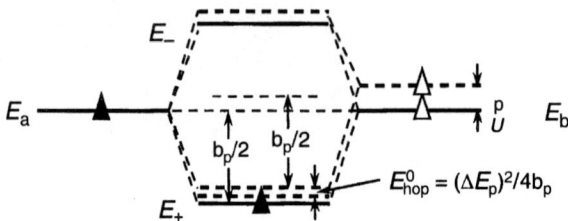

Figure 1.56. Two-cation molecular orbital diagram for a polaron in a trap of energy ΔE_p. Polaron transfer integral is defined as $b_p = |b_{ML}^2/2U_{ML}|$.

solution for the elastic trap energy equal to the U_{ab}-dependent term of Eq. (1.6.8), now defined as a spin-independent case of Eq. (1.6.23),

$$\Delta E_{ex}^{pol} = E_{hop}^0 = \frac{\Delta E_p^2}{4b_p} \quad (1.8.3)$$

(In the previous work, the simpler molecular orbital model was used,[116] so that the multiplier of b_p was 8 instead of 4.) The trap energy E_{hop}^0 that emerges from the calculation also represents the activation energy between polaron cells required for charge transport to a site on an outer ring in Fig. 1.55.

Since the individual activation energies influence respective Boltzmann probability functions defined by p_{ex} and p_p for local and itinerant polarons, a probability for the spin-polarized charge transport can be expressed as

$$p = p_p p_{ex} = \exp\left(\frac{E_{hop}^0}{kT}\right) \exp\left(\frac{E_{hop}^{ex}}{kT}\right) \quad (1.8.4)$$

and the combined activation energy required for transfer in a spin-order dependent exchange field can be expressed as

$$E_{hop} = E_{hop}^0 + E_{hop}^{ex} = \frac{\Delta E_p^2}{4b_p} + zJ_{ij}S_iS_j(1 - \cos\theta_{ij}) \quad (1.8.5)$$

Equation (1.8.5) governs incoherent electrical conduction by thermal hopping, which is dominant at high temperatures or when E_{hop} is small. It also provides an opportunity to define a *ferromagnetic* polaron. In simple terms, when neighboring spins of the trapped carrier are aligned parallel, the local environment of the carrier is ferromagnetic and $\cos\theta = 1$, thereby *removing* the exchange trap. The carrier would then be constrained only by the dielectric (elastic) trap E_{hop}^0 and its size or range (if we think in terms of mobility) would be limited magnetically only by the degree of canting of the ferromagnetic spins surrounding the carrier. The ferromagnetic polaron therefore has the transport properties of a conventional dielectric polaron.

1.8.1.3. Covalent Tunneling Transport.

At low temperatures, a more efficient transfer mechanism can produce remarkable effects when the right conditions of chemical bonding and spin ordering are present. In the context of molecular-orbital states charge transfer can also be viewed as a quantum mechanical probability stemming directly from the covalent bond, i.e., through a "covalent transfer" mechanism. This concept can be readily appreciated by imagining the concentric ring diagram in Fig. 1.55 as analogous to a hydrogen atom (with reversed polarity in this case), where the rings would represent a series of higher energy states. The wave function of the mobile polaron charge could then be viewed as an s orbital wave function and its probability of occupancy at a distance r from the source would then be determined by its contribution to the molecular orbital state as the carrier moves towards the outer rings. Such an exercise can be approximated from the eigenfunctions of Eq. (1.6.12) after being modified to account for the polaron trap energy $\Delta E_p(r)$ and sketched in the energy level diagram of Fig. 1.56.

For the large polaron limit of $b_p \gg \Delta E_p$, the φ_M eigenfunction for a bonding state would be[112]

$$\varphi_M^{large} \approx \frac{1}{(2)^{1/2}} \left[c_{aa} \varphi_{M_a} + c_{ab} \varphi_{M_b} \right]$$

$$\varphi_M^{large} \approx \frac{1}{(2)^{1/2}} \left[\left(1 + \frac{\Delta E_p}{b_p}\right)^{1/2} \varphi_{M_a} + \left(1 - \frac{\Delta E_p}{b_p}\right)^{1/2} \varphi_{M_b} \right] \quad (1.8.6)$$

and the transfer probability of the polaron carrier from site a to site b can be defined as the square of the coefficient c_{ab} of φ_{M_b}. If we define a transfer efficiency as the ratio of the two probabilities

$$\eta(r)_{large} \approx \frac{c_{ab}^2}{c_{aa}^2} \approx \frac{1 - (\Delta E_p/b_p)}{1 + (\Delta E_p/b_p)} \approx 1 - \frac{2\Delta E_p}{b_p} \quad (1.8.7)$$

For the small polaron limit of $b_p \ll \Delta E_p$, Eqs. (1.8.6) and (1.8.7) are replaced by

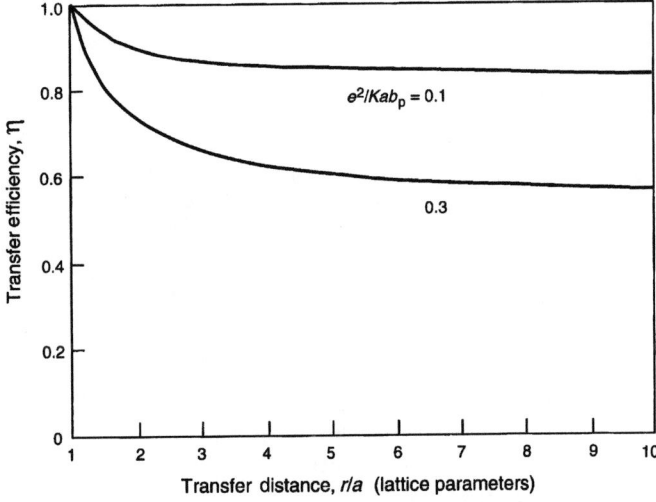

Figure 1.57. Polaron transfer efficiency as a function of reduced lattice length for two ratios of electrostatic energy to polaron half-bandwidth.

$$\varphi_M^{small} \approx \left(1 - \left(\frac{b_p}{2\Delta E_p}\right)^2\right)^{1/2} \varphi_{M_a} + \left(\frac{b_p}{2\Delta E_p}\right)\varphi_{M_b} \quad (1.8.8)$$

and

$$\eta(r)_{small} \approx \frac{(b_p/2\Delta E_p)^2}{1 - (b_p/2\Delta E_p)^2} \approx \left(\frac{b_p}{2\Delta E_p}\right)^2 \quad (1.8.9)$$

Recalling that $b_p = |b_{ML}^2/2U_{ML}|$ and $b_{ML} = (E_M + E_L)S_{ML}$, we recognize that the effective covalent interaction between cations through the mediation of the common ligand is a function of S_{ML}^4.

From the elementary electrostatic attraction between two oppositely charged particles (the hydrogen atom in classical terms) in a medium of dielectric constant K and M_a–L–M_b dimension a,

$$\Delta E_p(r) = \frac{e^2}{K}\left(\frac{1}{a} - \frac{1}{r}\right) = \frac{e^2}{Ka}\left(1 - \frac{a}{r}\right) \quad (1.8.10)$$

and we can express the charge transfer probabilities by quantum mechanical tunneling to a distance r from the polaron source by substituting Eq. (1.8.10) (or other appropriate function) into Eqs. (1.8.7) and (1.8.9). Figure 1.57 compares the η efficiency parameter variation with r/a for large polarons. A somewhat more rigorous approach would consider consecutive discrete transfers of the polaron carrier from its source.[112]

1.8.1.4. The Holstein Polaron Theory.

In his quantum mechanical analysis of polaron motion,[108] Holstein defined three stages of conductivity that were determined by the relative magnitudes of two energy parameters: (i) the polaronic charge transfer energy b_p, which establishes the lifetime τ_p of the polaron carrier in its site according to $\tau_p \sim \hbar/b_p$, and thereby presents b_p as the width of the polaron energy band which is the product of the electronic transfer integral b and a lattice vibronic coupling factor; and (ii) the energy ΔE_p of the electrostatic potential well in which the carrier resides and which is shaped by the charge of the carrier and a neighboring charge of opposite sign that created it, i.e., the other half of an extendible dipole. The temperature dependence of b_p is determined by the number of vibrational modes (phonons) available to interact with the carriers. The vibrational overlap integral curiously is a decreasing function of T and its contribution to b_p is maximal at absolute zero, diminishing rapidly with rising temperatures.

Based on these concepts, Holstein defined the criterion for a large polaron (one in which the carrier extends beyond its immediate environment by spontaneous charge transfer) as $b_p \gg E_{hop}^0$, and that of a small polaron (which is limited to its immediate neighboring sites) as $b_p < E_{hop}^0$. These transfer mechanisms are often referred to as coherent or adiabatic because of their energy conserving nature. In the final stage which sets in at higher temperatures, the metallic conductivity begins to break down and semiconduction by incoherent nonadiabatic electron hopping becomes dominant. With increasing temperatures the large polaron condition begins to fail because higher frequency lattice vibrations allow the carriers to stabilize

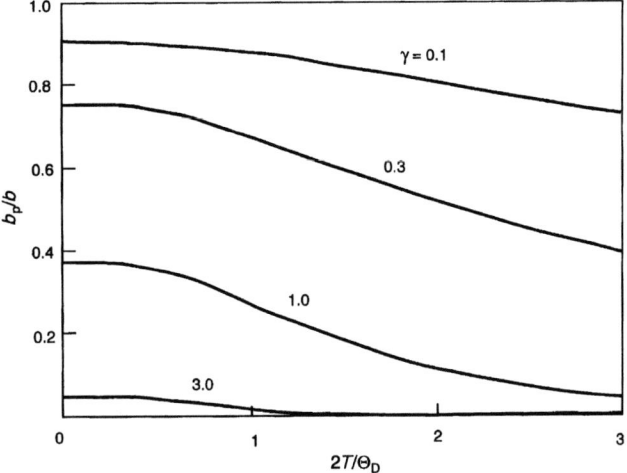

Figure 1.58. Reduced polaron bandwidth as a function of reduced temperature for different values of γ.

in their traps by permitting elastic adjustments to occur more quickly, thereby lengthening the polaron lifetime of τ_p. This action leads to a decrease in b_p which begins to set in as the temperature approaches the Debye temperature Θ_D.*

The polaron bandwidth b_p is reduced by vibronic (orbit–lattice) coupling that produces a temperature-dependent narrowing given by[117]

$$b_p(T) = b \exp\left[-\gamma \coth\left(\frac{\Theta_D}{2T}\right)\right] \qquad (1.8.11)$$

that ranges from $b_p(T) = b \exp(-\gamma)$ at $T = 0$ to the simplification

$$b_p(T) = b \exp\left[-\gamma\left(\frac{2T}{\Theta_D}\right)\right] \qquad (1.8.12)$$

when $T = \Theta_D/2$.

Although not readily discernible in this brief summary, another facet of Holstein's model is that the threshold where thermal activation becomes a significant competitor to tunneling is $T \approx \Theta_D/2$. As seen in the complete function Eq. (1.8.11) plotted in Fig. 1.58, the polaron bandwidth is still a sizable fraction of $b \exp(-\gamma)$ at this point. The parameter $\gamma \sim 50/K_{opt}^2$ would be ~ 0.1–1 for transition-metal oxides with an optical dielectric constant K_{opt} in the 5 – 25 range. As a consequence, the decrease in b_p will eventually serve to reduce the carrier population according to a probability approximated by the Boltzmann relation $p_p = \exp(-E_{hop}^0/kT)$ in the usual diffusion relation for electron hopping by thermal activation. The transport of a polaronic

---

*The relation between the Debye energy $k\Theta_D$ and ΔE_p in determining the actual activation energy E_{hop} can be appreciated if one recognizes that the electrostatic potential of the polaron dipole field has an intimate tie to the elastic distortion that dresses the polaron charge site. The distortion of the lattice can be viewed as a reaction to the dipolar field and both energies involve the polarizability of the lattice through the dielectric constant K. As origins of the trap energy, they could be considered to have some equivalence.

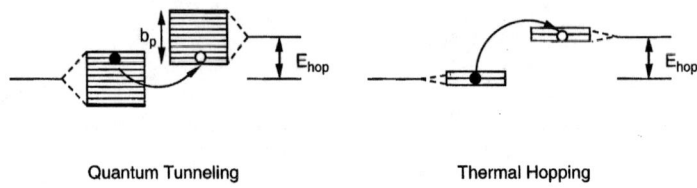

$b_p > E_{hop}$	large polarons in broad bands; spontaneous conduction by covalent transfer
$b_p \lesssim E_{hop}$	small polarons in thermally narrowed bands; local conduction among equivalent neighbors
$b_p \ll E_{hop}$	hopping electrons from deep in traps; random conduction by thermal activation

Figure 1.59. Diagram of polaron stages based on the relative magnitudes of b_p and E_{hop}. Note that the $b_p > E_{hop}$ condition for a large polaron is consistent with the $b^2 \gg \Delta E_p^2$ limit of Eq. (1.6.23).

carrier, therefore, requires that an increase in energy E_{hop}^0 be supplied to remove it from its trap. A qualitative summary of the polaron stages is presented in Fig. 1.59.

1.8.2. Metallic Oxides

For electronic conduction in solids, two conditions must be satisfied: (i) there must be charge carriers; and (ii) there must be a mechanism to enable their transport. Carriers can be "free," as in collective-electron metals, excited as in band-model semiconductors, or activated from polaron traps in ionic compounds. The second requirement involves the environment of the destination sites and what lies in between, i.e., the mobility. The temperature dependence of these variables is what generally defines whether the material is a metal (carrier density decreasing with T) or insulator (both carrier density and mobility increasing with T). There is, however, another significant distinction that applies particularly to magnetically ordered materials.

Charge carriers in metals are drawn mainly from s and p states, despite the formation of hybrids with d states in the collective intermingling. As a result, the d-electron magnetic carriers represent only a fraction of the population that produces current. In transition-metal oxides, the s and p electrons of the cations are transferred to the oxygen to establish the anion lattice, and have only negligible hybrid mixing with the d shell. Therefore, in the rare situations where the d electrons become itinerant, the current has the spin polarization of the transfer states, which is usually ferromagnetic in contrast to the more random situation in a ferromagnetic metal. Since the vast majority of the carriers are of only one spin orientation, metallic compounds of the transition-metal series are frequently called "half-metals." The possible origins of polarized spin transport in select metallic oxides can be reviewed in the context of molecular orbital theory.

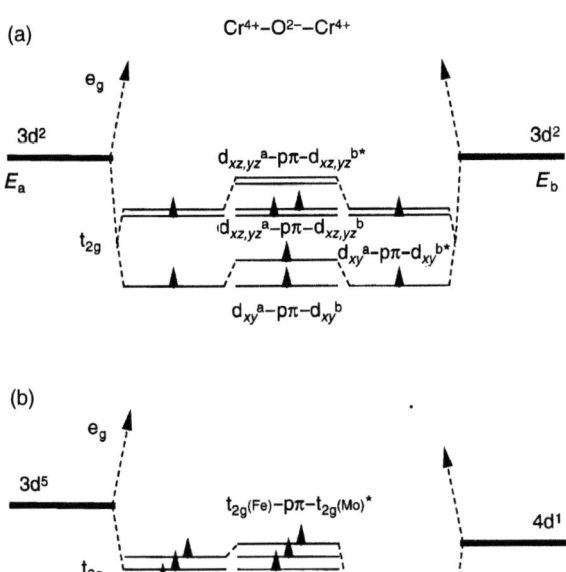

Figure 1.60. Molecular orbital approximations for two special cases of spin polarized metallic charge transfer: (a) $Cr^{2+}O_2$ where a ferromagnetic moment occurs because of the survival of a spin–orbit $d_{xz,\,yz}$ doublet that forms a "band" with partially filled orthogonal states allowing Hund's rule to apply; and (b) $Sr^{2+}{}_2[Fe^{3+}Mo^{5+}]O_6$, which is a paramagnet/unbalanced antiferromagnet (quasi-ferrite) that occurs by delocalization superexchange involving the sharing of the single Mo^{5+} electron. Oxygen ligand states are not shown.

In systems where only the t_{2g} orbital states are occupied, i.e., the lighter members of a $3d^n$ transition series, metallic conduction can occur if the t_{2g} levels retain a degeneracy in a lower symmetry crystal field. The most common occurrences are found with d^1 and d^2 configurations in simple monocation-site compounds. Because the cation sites are principally octahedral (but frequently distorted), the t_{2g} electrons usually form weak π bonds to the O^{2-} anions and therefore feature small exchange stabilization that would normally be expected to provide low antiferromagnetic Néel temperatures T_N. At room temperature, even with dynamic J–T or S–O splittings ($\sim 10^{-2}$ eV) of the triplet, the t_{2g} states would then form a partially filled t_{2g}–$p\pi$–t_{2g} states resulting from the localized metal–ligand–metal superexchange depicted in Fig. 1.37(d).

In cases such as TiO and TiO_2, single d-electron transfer can occur as a result of hopping electrons between mixed-valence states of the Ti ions when the compound is off stoichiometry, e.g., from O^{2-} vacancies. Magnetic ordering is not involved here because the spin density is too small to produce anything but paramagnetism. A more interesting situation occurs with stoichiometric $CrO_2(3d^2)$. The crystal structure is rutile (tetragonal cation site symmetry D_{4h} with $c/a < 1$), which splits the t_{2g} triplet into a lower d_{xy} singlet and an upper $d_{xz,\,yz}$ doublet, as sketched in Fig. 1.60(a). The molecular orbital hybrid states would then have a single electron in each level, but with competing spin alignment tendencies. In the lower state, the half-filled d_{xy}–$p\pi$–d_{xy} hybrid would be expected to follow the rules of antiferromagnetic superexchange (Table 1.13). In the upper $d_{xz,yz}$–$p\pi$–$d_{xz,yz}$ doublet, however, Hund's rule high-spin alignment can exist within the two orthogonal orbital states [see Fig. 1.41(c)]. The net result would be determined by the configuration that produces the lowest net energy, according to Eq. (1.6.22) applied here in reverse. Consequently, the antibonding half of d_{xy}–$p\pi$–d_{xy} is stabilized to accommodate the electrons with the lower energy $S=1$ parallel spin alignment. The ferromagnetic order gives the full $2m_B$ magnetic moment per cation (~100% polarization) at low temperatures and coexists with metallic conductivity up to a Curie temperature of 400 K.[118, 119] A more in-depth discussion of the various orbital interactions that lead to the magnetic order and electrical conductivity of these compounds is given by Goodenough.[120]

Another uncommon situation occurs in mixed (or double) perovskites (Fig. 1.54) in which ordered dissimilar octahedral-site cations can produce metallic conduction and "ferromagnetic" spin alignments that reach above room temperature. The prime example is $Sr_2^{2+}[Fe^{3+}Mo^{5+}]O_6$.[121] The charge transfer and magnetic properties originate in the B sublattice from the half-filled/half-filled correlation superexchange (see Table 1.13) between the two lowest states of Fe^{3+} (d^5) and Mo^{5+} (d^1) that form the d_{xy}(Fe)–$p\pi$–d_{xy}(Mo) hybrid. Antiferromagnetism is expected in this case, but with a net ferromagnetic moment associated with the remaining four unpaired spins of the Fe^{3+} ion. As a consequence, the compound is a quasi-ferrimagnet with only one crystallographic sublattice and only one molecular-field coefficient. If this material was an ideal uncanted ferrite, the magnetic moment per formula unit would be $4m_B$. Although measurements indicate that the actual value at low temperatures is closer to $3m_B$, the conductivity is still metallic. One possible explanation can be gleaned from the molecular-orbital model in Fig. 1.60(b). Beginning with the conductivity issue, we can reason that the single spin resident on the Mo ion is shared with the Fe ion. Since the Fe^{2+} (d^6) ion is the least stable of the iron group, with the second d_{xy} spin loosely bound according to the ionization energy plotted in Fig. 1.8, we can argue that two ionic configurations can exist simultaneously through electron transfer reactions

$$Fe^{3+} + Mo^{5+} \Leftrightarrow Fe^{2+} + Mo^{6+} + U_{eff}$$

$$t_{2g}^3 e_g^2 + t_{2g}^1 e_g^0 \Leftrightarrow t_{2g}^4 e_g^2 + t_{2g}^0 e_g^0 + U_{eff}$$

The degree of sharing and hence the mobility of the Mo^{5+} spin will once again depend on the value U_{eff}^2/b, where U_{eff} is the difference in the binding energy between the initial and final states of the transfer, and b is the effective $b_{MLM'}$ transfer integral between the Fe and Mo ions mediated by the O^{2-} ligand [not shown in Fig. 1.60(b)]. If the Mo spin becomes itinerant in this sense, spin-polarized transport could take place in the narrow half-filled band constructed from the $d_{xy}(Fe)$–$p\pi$–$d_{xy}(Mo)$ hybrid. Inspection of the above reactions reveals that the left-hand side represents an antiferromagnet (or quasi-ferrite) and the right-hand side a paramagnet. Therefore, the probability of obtaining the maximum $4m_B$ is reduced by the probability of paramagnetism that occurs when Mo is missing its spin. If a paramagnetic component is present, experiments at very high magnetic fields might help to clarify the situation.

In a series of compositions with Cr, Mn, and Co used in place of Fe,[123] ferromagnetism was found only with Cr. Since it assumes its 3+ valence with a highly stable electronic configuration $t_{2g}^3 e_g^0$, d^3–d^1 correlation superexchange is again the likely result. There was no real evidence of metallic conduction, however, which suggests that the value of U_{eff} is significantly greater than that of the Fe case. The other ions also showed insulating properties, but no magnetic ordering. The prospects of antiferromagnetism are reduced because the most stable forms of Mn and Co are divalent, which would force the Mo valence to be raised to the diamagnetic 6+ state, thereby lowering the chance of spontaneous magnetism. Similar analysis can be applied to other combinations of transition metals, such as Ti, W, and Re, but the prospects for strong superexchange with spin polarized metallic conductivity between the alternating B-site cations probably remains highest with a d^5–d^1 combination.

Ferromagnetism in metallic oxides also occurs with spins in the e_g shell of the mixed-valence perovskite $(La_{1-x}^{3+}Ca_x^{2+})[Mn_{1-x}^{3+}Mn_x^{4+}]O_3$, for temperatures reaching above 300 K. The origins of this unexpected ferromagnetism are complex and are discussed in Section 1.8.5. Normally when the e_g shell is occupied, strong σ bonds are formed that produce antiferromagnetic stabilizations that can survive to well above room temperature. Simple oxides of the heavier $3d^n$ elements, MnO through CuO, all feature antiferromagnetism and usually p-type semiconduction when mixed valence occurs, despite the ferromagnetic tendencies that can still exist with the t_{2g} shell partially filled.

Returning to the discussion that produced Eq. (1.8.5), we can now examine the question of conductivity in polaronic oxides. Most transition-metal oxides are termed mobility-activated semiconductors in contrast to more conventional band model version with the usual description of holes and electrons dictated by Fermi statistics. The electrical resistivity of mixed-valence oxides obeys the standard relation for mobility-activated semiconduction in which the polaron trap energy E_{hop} and the activation energy are equivalent in the temperature regime where random hopping is dominant,[123–125]

$$\rho = \left[\frac{x_{\text{eff}}}{V} e \left(\frac{eD}{kT}\right)\right]^{-1} \exp(E_{\text{hop}}/kT) \qquad (1.8.13)$$

where the carrier concentration per chemical formula unit, $x_{\text{eff}} = n_{\text{eff}}/V$, the ratio of the volume carrier density to the volume of a formula unit V, the factor (eD/kT) is the Einstein diffusion mobility, and D the diffusion constant that equals the ratio of the square of the mean-free path to the carrier lifetime $d_{\text{hop}}^2/\tau_{\text{hop}}$. To account for the probability of a transfer being completed, an effective carrier concentration is defined by $x_{\text{eff}} = xP$, where P is a polaron dispersal probability equal to $(1-x)$ for a random distribution. Further dependencies of relevance are that $\tau_{\text{hop}} = v_{\text{hop}}^{-1}$, which is on the order of the Debye frequency and $d \approx a/(1-x)$ to account for the average hop distance as a function of concentration.[112] Upon substituting appropriately into Eq. (1.8.13), a working expression for the resistivity becomes

$$\rho = \frac{C(1-x)kT}{x} \exp(E_{hop}/kT)$$
$$= \frac{C(1-x)kT}{x} \quad \text{for } kT \gg E_{hop}$$
(1.8.14)

where $C = V/e^2 a^2 v_{hop}$. As can be seen by inspection of Eq. (1.8.14), ρ plotted as a function of T will show insulating behavior (decreasing) at low temperatures and pass through a minimum at $T = E_{hop}/k$. As T increases further, the curve will approach a metallic straight line of (increasing) slope $C(1-x)k/x$. Note that the asymptote of the curve will pass through the origin at $T = 0$, which is a result that is not expected in experiment because residual resistivity effects would appear.

The temperature of metal–insulator transition therefore resides in the value of E_{hop}, which from Eq. (1.8.5) is controlled greatly by the magnetic state of the spin system. There are two prominent examples where E_{hop} is reduced to the basic elastic polaron limit E_{hop}^0 by the absence or removal of E_{hop}^{ex}. If the spins are collinear, $\cos\theta = 1$ and $E_{hop}^{ex} \approx 0$. In oxides, this condition is uncommon because superexchange usually dominates to produce antiferromagnetism. An important exception can occur with Mn ions in certain ferromagnetic perovskites compounds that feature metallic properties but also display large magnetoresistance effects at the Curie temperature.

1.8.3. Colossal Magnetoresistance

In the generic $(RE^{3+}A^{2+})MnO_3$ perovskite system, the conditions for metallic conduction in the Mn sublattice can occur in a ferromagnetic phase that appears at lower temperatures. When the ferromagnetism is dissipated at the Curie temperature, a metal-insulator transition takes place. Application of magnetic fields large enough to influence the Curie temperature has been shown to produce dramatic

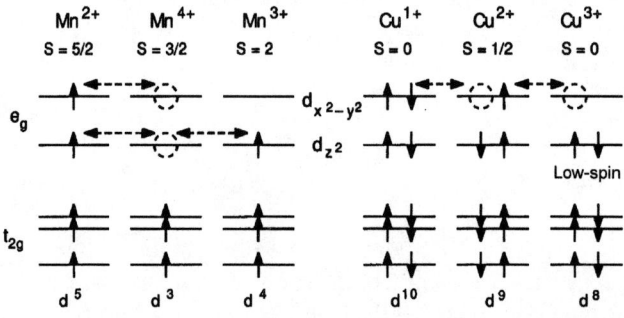

Figure 1.61. Schematic diagram of delocalization exchange in the e_g shell, comparing the various cases of mixed-valence Mn and Cu.

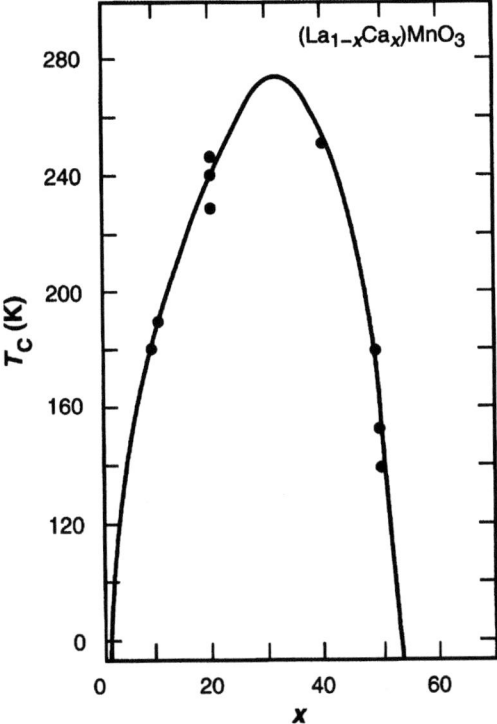

Figure 1.62. Characteristic plot of the Curie temperature versus x for $(La^{3+}_{1-x}Ca^{2+}_{x})Mn^{3+}_{1-x}Mn^{4+}_{x}O_3$. Data are from Jonker and Van Santen.[128]

magnetoresistance effects—CMR.[126] To explain the origin of this phenomenon, we first examine the source of the ferromagnetic spin alignment.

For the basically cubic octahedral oxygen coordination, crystal-field effects dictate that the t_{2g} orbital states are of lower energy and are half-filled to satisfy the Hund's rule spin polarization requirement for both Mn^{3+} ($3d^4$) and Mn^{4+} ($3d^3$). For each combination of exchange linkages, the t_{2g} electrons produce weak antiferromagnetism via covalent π bonding through the O^{2-} anions, e.g., $Mn^{4+}-O^{2-}-Mn^{4+}$. For the $Mn^{3+}-O^{2-}-Mn^{3+}$ combinations, the larger transfer integrals are the result of the stronger 180° σ-bonding in the e_g states. A single electron in the e_g shell (Mn^{3+} case) can be stabilized by a static J–T distortion that splits energy levels as shown in Fig. 1.61. Where the distortions are cooperative, a tetragonal/orthorhombic phase will appear (with site axis ratio $c/a, b > 1$) and the half-filled d_{z^2} orbital is stabilized relative to the empty $d_{x^2-y^2}$ state. Superexchange spin ordering possibilities for this system were examined in a seminal paper by Goodenough.[127]

From the Curie temperature data of Jonker and Van Santen[128] for $(La^{3+}_{1-x}Ca^{2+}_{x})[Mn^{3+}_{1-x}Mn^{4+}_{x}]O_3$ presented in Fig. 1.62, the variations in exchange field with Mn^{4+} concentration may be analyzed on the basis of changes in the nature of the J–T effect. At $x=0$, the J–T effect should be mainly static and cooperative, favoring an orthorhombic distortion and antiferromagnetic order in most cases.

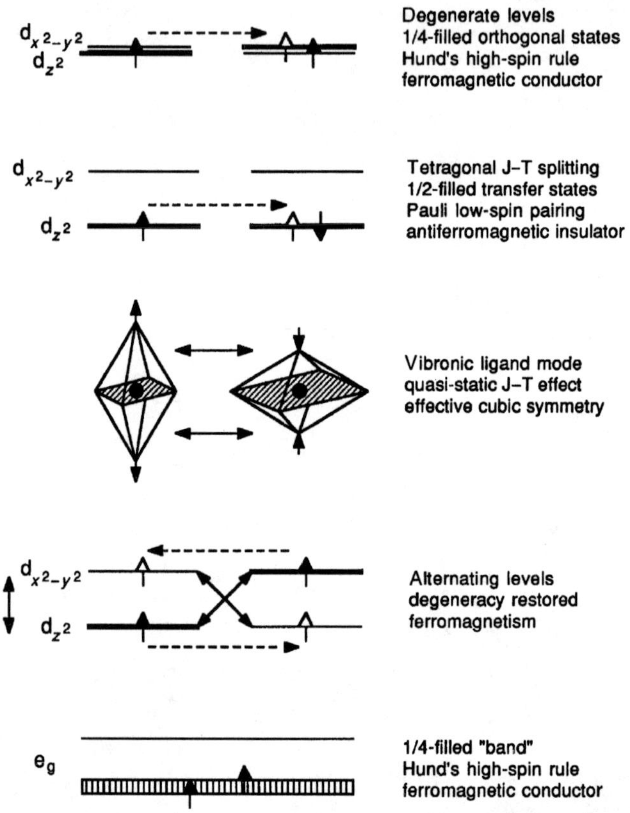

$d_{x^2-y^2}$
d_{z^2}
Degenerate levels
1/4-filled orthogonal states
Hund's high-spin rule
ferromagnetic conductor
$d_{x^2-y^2}$
d_{z^2}
Tetragonal J–T splitting
1/2-filled transfer states
Pauli low-spin pairing
antiferromagnetic insulator
Vibronic ligand mode
quasi-static J–T effect
effective cubic symmetry
$d_{x^2-y^2}$
d_{z^2}
Alternating levels
degeneracy restored
ferromagnetism
e_g
1/4-filled "band"
Hund's high-spin rule
ferromagnetic conductor

Figure 1.63. Schematic diagrams of J–T effects, including the quasi-static case in which carrier transfers into empty e_g orbital states that are mixed by vibronic modes may cause ferromagnetism.

Table 1.20. Magnetic Exchange in Mixed-Valence Manganites[a]

Ion Pair	d_{xy}, d_{xz}, d_{yz}	$d_{x^2-y^2}$ *a–b* axes	d_{z^2} *c* axis	Net
$d^5 \leftrightarrow d^5$	↑↓π(wk)	↑↓σ(str)	↑↓σ(str)	↑↓(str)
$d^5 \leftrightarrow d^4$	↑↓π(wk)	↑↑σ(str)	—	↑↑(str)
—	—	—	↑↓σ(str)	↑↓(str)
$d^4 \leftrightarrow d^{4b}$	↑↓π(wk)	↑↓σ(str)	—	↑↓(str)
$d^5 \leftrightarrow d^{3c}$	↑↓π(wk)	↑↑σ(mod)	↑↑σ(mod)	↑↑(mod)
$d^4 \leftrightarrow d^{3d}$	↑↓π(wk)	↑↑σ(str)	—	↑↑(mod/str)
$d^3 \leftrightarrow d^3$	↑↓π(wk)	—	—	↑↓(wk)

wk, weak; mod, moderate; str, strong.
[a] *Based on rules developed by J. B. Goodenough.*[67,68]
[b] *J–T splitting of* e_g *orbitals producing* $c/a > 1$ *distortion.*
[c] *Proposed quasi-static J–T version of* $d^4 \leftrightarrow d^4$ *that causes ferromagnetism through charge transfer into empty orbital states. It occurs in a rhombohedral structural phase that exists when the static cooperative distortions that cause orthorhombic phases are absent.*
[d] *Conditions similar to those of footnote b, but feature charge transfer superexchange that promotes ferromagnetism.*

With increasing Mn^{4+} concentration, ferromagnetism dominates in the regime up to $x = 0.5$, with a peak near $x = 0.3$. As Mn^{4+} ions are introduced, parallel spin alignments result from a combination of factors: (i) the $Mn^{3+}-O^{2-}-Mn^{4+}$ couplings contribute ferromagnetism by charge transfer among half-filled/empty orbital combinations as studied by Zener[63] and de Gennes;[64] and (ii) the anticipated antiferromagnetism from the $Mn^{3+}-O^{2-}-Mn^{3+}$ couplings in a static J–T effect is converted to ferromagnetism possibly by vibronic-induced J–T effects proposed by Goodenough et al.[128] that alternates the order of the e_g–$p\sigma$ antibonding levels (bands) and allows the two e_g electrons to be stabilized with parallel spin alignments in separate molecular-orbital states.

Ferromagnetism can therefore occur because of the absence of static tetragonal deformation that leaves the e_g states degenerate and removes the necessity for Pauli spin pairing, similar to the t_{2g} degeneracy in the case of CrO_2 discussed previously. As illustrated in Fig. 1.63, in cases where the $Mn^{3+}-O^{2-}-Mn^{3+}$ couplings dominate and the electronic bandwidth is broad enough for the interactions to be collective, a ligand vibronic mode may cause the d_{z^2} and $d_{x^2-y^2}$ states of adjacent Mn cations to oscillate out of phase and form a quarter-filled e_g shell. This situation allows Hund's rule to apply and gives rise to ferromagnetic order and spin-polarized metallic conductivity. The quasi-static J–T effect is consistent with the absence of the static orthorhombic distortion in the regime of the observed ferromagnetism that would normally be expected to stabilize the e_g electron of the Mn^{3+} ions in the lower of the split e_g states. This condition would tend to deny tunneling transfer by increasing the spin-dependent part E_{hop}^{ex} of the polaron trap energy. The anticipated couplings for the various combinations are summarized in Table 1.20.

Above $x \approx 0.1$, the vibronic actions can influence a change in crystallographic phase from orthorhombic to cubic or rhombohedral (trigonal), thereby restoring the degeneracy of the e_g levels in the crystal field (see Fig. 1.13), and remaining such until x approaches 0.5. In some cases, the vibronic effect on e_g can occur in the presence of an orthorhombic bias. With half the Mn ions in the 4+ state, the quasi-static behavior breaks down as the lattice symmetry returns to orthorhombic. At this point, the e_g levels are split, and the remaining $Mn^{3+}-O^{2-}-Mn^{3+}$ couplings revert to antiferromagnetism, combining with the existing antiferromagnetic $Mn^{4+}-O^{2-}-Mn^{4+}$ couplings to produce various cation charge order and antiferromagnetic configurations in the range from $0.5 < x < 1.0$. It should also be pointed out that in the ferromagnetic region the maximum available m_B per formula unit is apparent at cryogenic temperatures, confirming the anticipation of complete uncanted spin polarization similar to that of CrO_2.

To account for the magnetic exchange effects that produce the observed ferromagnetism, a single magnetic lattice is assumed. Because both the carrier charges are among the d electrons that provide the magnetic moments, the disposition of spins cannot be static. Valence-charge ordering may occur coincident with spin ordering in spatially variable phases, which could explain the reported observation of a metal/insulator mosaic that probably corresponds to ferro/antiferromagnetic domain patterns.[130] To describe this system by traditional analytical methods is a formidable challenge. Nonetheless, a model based on a random distribution of Mn^{3+} and Mn^{4+} cations carrying spins S_3 and S_4 can be fashioned. For this exercise, three exchange interactions are defined: $J_{33}S_3 \cdot S_3$, $J_{44}S_4 \cdot S_4$, and $J_{34}S_3 \cdot S_4$

(or $J_{43}S_4 \cdot S_3$), from which an average exchange energy is constructed for use with the Brillouin–Weiss theory.

For an individual charge transfer between Mn^{3+} and Mn^{4+} ions, Eq. (1.8.2) can be applied to express the activation energy as

$$E_{ex}(\theta_{34}) = zJ_{34}S_3 \cdot S_4 \approx zJ_{34}S_3S_4(1 + \cos\theta_{34}) \qquad (1.8.15)$$

where z is the number of nearest neighbors and θ_{34} is the average angle between the Mn^{3+} and Mn^{4+} spins which will be assumed to be simply θ, the average angle between adjacent spins within the entire system. When $\theta = 0$, E_{34} is a maximum, the spins are collinear, and the $\Delta S = 0$ requirement for the charge transfer is satisfied. If $\theta > 0$, E_{34} decreases and energy must be provided to restore the spin alignment and maintain $\Delta S = 0$. According to Eq. (1.8.2), the additional energy has the effect of a magnetic trap of depth

$$E_{hop}^{ex} = E_{34}(\theta_{34}) - E_{34}(0) = zJ_{34}S_3S_4(1 - \cos\theta_{34}) \qquad (1.8.16)$$

and E_{hop}^{ex} is also the activation energy necessary to effect the charge transfer between the two cation sites. It should be noted that θ_{34} values in the range 0 to π are allowed by Eq. (1.8.16), and that the theoretical maximum activation energy from exchange is actually $2E_{hop}^{ex}$ when the spins are antiferromagnetic. For the present problem, however, the regime of interest is the ferromagnetic to paramagnetic transition that occurs at the Curie temperature where the average angle between spins becomes $\pi/2$.

Since the Brillouin–Weiss function B_S represents the average z-axis projection of spins within a cone of half-angle θ, it also represents the average angle between a spin and the direction of the exchange field in which it resides. Consequently, $\cos\theta$ may be represented by B_S Eq. (1.8.16) and the spin canting effect on the binding energy can now be expressed as a function of temperature and magnetic field.

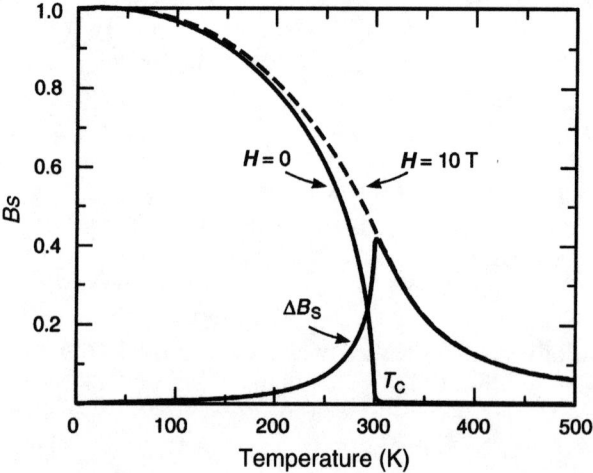

Figure 1.64. Calculated plots B_S and ΔB_S versus T for $T_C = 300$ K with $H = 0$ and 10 T.

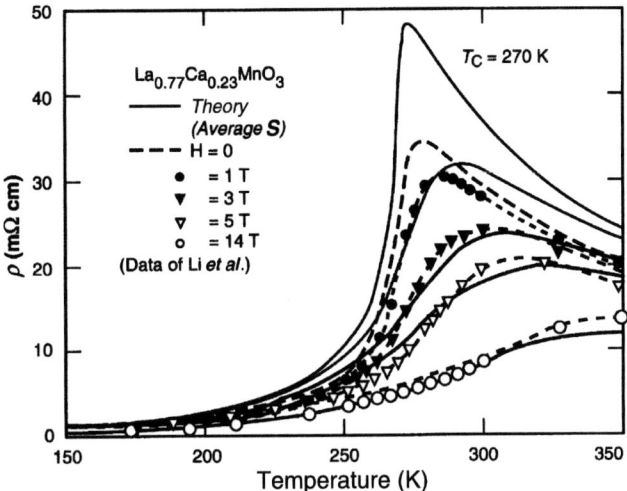

Figure 1.65. Comparison of B_S approximation theory with experiment for ρ versus T with $H = 0, 1, 3, 5,$ and 14 T. Data are from Li et al.[131]

From the discussion in Section 1.7.1, the total activation energy may then be expressed in terms of molecular field theory according to

$$E_{\text{hop}} = E_{\text{hop}}^0 + E_{\text{hop}}^{\text{ex}}[1 - B_S(T, H)] \tag{1.8.17}$$

where E_{hop}^0 is the polaron trap energy in the absence of spin polarization constraints (chosen as 0.004 eV), and $E_{\text{hop}}^{\text{ex}}$ (~0.1 eV) is the magnetic exchange contribution that reaches its full value when the spins become disordered at $T > T_C$.[116]

In Fig. 1.64, the Brillouin–Weiss function is plotted as a function of T for an average molecular-field coefficient $N = 114$ mol/cm^3 (derived from the J_{33}, J_{34}, and J_{44} parameters of the randomly dispersed Mn^{3+} and Mn^{4+} ions with $x = 0.23$), resulting in a Curie temperature of 300 K.[117] The effect of an external field $H = 10$ T in extending the ordered ferromagnetic region above T_C is shown together with the corresponding ΔB_S that occurs when the field is applied. From Eq. (1.8.17), it is seen that ΔB_S reflects the change in $E_{\text{hop}}^{\text{ex}}$ due to the applied field that causes the magnetoresistance effect.

By combining Eqs. (1.8.17) and (1.8.13), magnetoresistance curves can be computed for any set of material parameters or external field values. In Fig. 1.65, ρ versus T data of Li et al.[131] for a composition estimated as (La$_{0.77}^{3+}$Ca$_{0.23}^{2+}$)MnO$_3$ subjected to H fields of 0, 1, 3, 5, and 14 T are fitted by curves generated from Eq. (1.8.14) in combination Eq. (1.8.17) using the above values for E_{hop}^0 and $E_{\text{hop}}^{\text{ex}}$, $x = 0.23$, and $C = 6$ mΩ cm/eV. Except for the $H = 0$ curve, which did not reach its full peak probably due to the inhomogeneously broadened tail of the M versus T curve and the possibility that the specimen was not magnetically saturated, and the one for 14 T, which may exceed the range of validity of the approximations, theory and data are in reasonably good agreement. Because of the magnetocrystalline

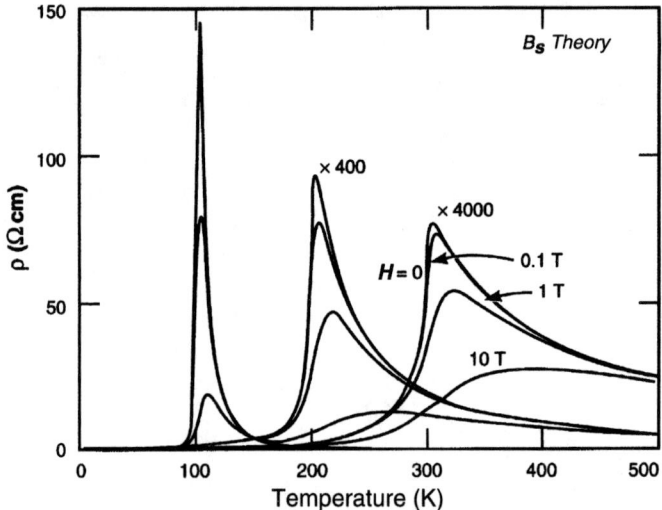

Figure 1.66. Predicted plots of ρ versus T by the $\mathcal{B}_S$ approximation for T_C values of 100, 200, and 300 K. Magnetic field strengths are $H = 0$, 0.1, 1, and 10 T.

anisotropy fields, which can probably reach beyond fields of 0.1 T, the presence of domains of varying size and disposition should be expected at low fields and temperatures approaching T_C.

Part of the discrepancy between theory and experiment is the result of the molecular-field approximation which represents the z-axis projection of the combined magnetic moment from all of the spins that occupy a cone of average half-angle θ. Any difference between θ and the average canting angle between neighboring S_3 and S_4 spins could account for the relatively small disagreement between theory and measurement in the metallic region below T_C. It has also been assumed that the polaron charges are randomly dispersed providing a net molecular-field coefficient that is constant with temperature when in fact it probably changes as the various exchange couplings compete for dominance as polaron charges shift about to maintain the lowest lattice energy.

Another consideration is the role of the polaron bandwidth (or inverse lifetime) which narrows with increasing temperature. As indicated by Eqs. (1.8.11) and (1.8.12), carrier transport is likely to be by tunneling at the lowest temperatures, but the coherence could dissipate and give way to random thermal hopping well before the Curie temperature is reached. Since the Debye temperatures of these compounds can be less than 200 K, the onset of nonadiabatic hopping could begin at temperatures below the liquid nitrogen range. This tunneling temperature regime would fall into the range of high-temperature superconductivity found in perovskite cuprate lattices.

In Fig. 1.66, example curves of ρ from the $\mathcal{B}_S$ approximation are plotted as functions of T with H values of 0, 0.1, 1, and 10 T for Curie temperatures at 100, 200, and 300 K. Since E_{hop}^{ex} is a constant of the transfer ions, the peaks of ρ at $H = 0$ should theoretically touch the insulator-phase envelope (given by a ρ calculation using the full

$E_{hop} \approx 0.1$ eV) at each of the T_C values. From these results, the magnitude of the anomalous increase in ρ at T_C is shown to increase by almost four orders of magnitude between 300 and 100 K.*

Some almost obvious conclusions can be drawn from this analysis: (i) the metal–insulator transition occurs at the Curie temperature, which is an intrinsic property of the exchange field and therefore the chemical bonding of the material; (ii) the metallic property defined by the positive slope of the ρ versus T curve is the direct result of the negative slope of the M versus T curve; (iii) the magnetoresistance is the result of enhancement of the intrinsic exchange field by an external magnetic field; (iv) the magnitude of the external field needed to cause significant changes in resistivity must be on the same scale as the exchange field, i.e., greater than 10 T; and (v) the peak resistivity and magnitude of the magnetoresistance decrease with rising temperatures.

The model used above is derived from the notion that spin directions undergo canting as the temperature increases. This is the Boltzmann statistical basis of the Brillouin–Weiss theory that was applied in a direct fashion to the carrier trap energy which is allowed to vary as a continuous function of temperature and magnetic field. A more simplified view of the CMR effect could be constructed by treating the carriers as comprising two groups, each with fixed activation energies: those free to be transported with minimum activation energy (E_{hop}^0) permitted by ferromagnetic ordering, and those from a paramagnetic phase with trap energy E_{hop} from Eq. (1.8.5). Partitioning of the carrier populations could be determined by separating the magnetic components according to $\mathcal{B}_S$ (for ferromagnetic spins) and $(1-\mathcal{B}_S)$ (for paramagnetic spins), each weighted by their respective activation probabilities $\exp(E_{hop}^0/kT)$ and $\exp(E_{hop}/kT)$. Such a reasoning would lead to an alternative version of Eq. (1.8.13):

$$\rho = \left[n'_{eff} e \left(\frac{eD}{kT} \right) \right]^{-1} \qquad (1.8.18)$$

where

$$n'_{eff} = n_{eff} \left[\mathcal{B}_S \exp\left(-E_{hop}^0/kT\right) + (1 - \mathcal{B}_S)\left(-E_{hop}/kT\right) \right]$$

By inspecting Eq. (1.8.1), we see that ρ approaches Eq. (1.8.13) in the high temperature limit. In the regime below $T = T_C$, a ρ versus T curve can be constructed directly from Fig. 1.64. However, this approach could be useful in the immediate vicinity of T_C where the approximation of a two-phase magnetic system might reasonably represent the breaking down of spin ordering. Where magnetic ordering is not a direct issue in determining the degree of carrier availability, a model based on two distinct carrier trap energies can also produce interesting results when applied to the case of high-temperature superconductivity.

---

*Note that if the curves in Fig. 1.66 were extended to $T = 0$, ρ would begin to rise sharply at $T \approx 40$ K because of the elastic trap energy $E_{hop}^0 = 4$ meV. In reality, the thermal hopping mechanism may be dominated by polaronic tunneling at these lowest temperatures and the metallic region would theoretically reach $T = 0$, where ρ would also approach some residual value.

At higher temperatures, changes in the cation charge distribution could enable the rhombohedral phase to extend beyond $x = 0.5$, giving rise to the peculiar antiferromagnetic/ferromagnetic transition first reported by Jonker and Van Santen[128] for $(La_{0.3}^{3+}Sr_{0.7}^{2+})MnO_3$. In the regime where the ferromagnetic stabilization can no longer dominate, a variety of antiferromagnetic ordering configurations can appear labeled as types A, C, and CE, where mixed Mn^{3+} and Mn^{4+} ions compete for spin alignments, and G for the stable antiferromagnetic end member $x = 1$ with only Mn^{4+} ions. The phenomenon of magnetoresistance can still occur when a large enough applied field is able to upset the net exchange field and reverse the sign of the resultant J constant. Structural symmetries react to the magnetic and charge order, as the relative disposition of the d_{z^2} and $d_{x^2-y^2}$ orbitals continue to determine the nature of the spin ordering and the anisotropy of charge transfer, whether along respective z-axis chains or within x–y planes. Detailed low-temperature phase diagrams for the manganite systems that correlate magnetic, crystallographic, and phases can be found in Goodenough and Longo[132] and Goodenough.[133]

A brief comment on the conduction properties of magnetite $Fe^{3+}[Fe^{2+}Fe^{3+}]O_4$ is in order. Although metallic conduction can be attributed to the polaronic charge transfer via $Fe^{2+} \leftrightarrow Fe^{3+} + e^-$, above the charge ordering Verwey temperature (120 K), where the trap energy drops from 0.15 to 0.04–0.06 eV,[84] there will always remain the antiferromagnetic contribution to the trap energy from the A sublattice. As a result the prospects of achieving a high degree of polarized spin transport in a true ferrimagnetic seem remote due to the frustration tendencies intrinsic to the competing sublattice system. The relatively small E_{hop} at room temperature, however, does lend credence to the suggestion that the value of U_{eff} in the $Sr_2^{2+}[Fe^{3+}Mo^{5+}]O_6$ compound is also small.

1.8.4. High-Temperature Superconductivity

Another situation where $b_p > E_{hop}$ and metallic behavior is observed occurs where the spin alignment and the $\Delta S = 0$ issues are moot altogether because one of the spins is zero and the lattice environment is without long range antiferromagnetic order. The most readily available examples of this situation are the $Cu^{2+} \leftrightarrow Cu^{3+}$ (low-spin $S = 0$) $+ e^-$ and $Cu^{1+} \leftrightarrow Cu^{2+} + e^-$ charge transfers of the high-temperature superconducting cuprates. Other compounds where polaronic conductivity without magnetic constraints condenses into superconductivity are the mixed-valence Ti, V, and Bi oxides.

1.8.4.1. Zero-Spin Polaron Ions.
Metallic spin transport in transition-metal oxides in which significant magnetic exchange energy can be present may also occur in select situations where the minority transfer ion is in a zero-spin state ($S = 0$). A remarkable feature of this situation is seen in p-type $Cu^{2+}(d^9)–O^{2-}–Cu^{3+}(d^8$, low-spin) and n-type $Cu^{2+}(d^9)–O^{2-}–Cu^{1+}(d^{10})$ configurations, also in 180° bonds of the perovskite structure. For the Cu^{3+} ion, the $S = 0$ state arises from a low-spin d^8 configuration in the e_g shell that occurs because of a large splitting of the e_g doublet first reported in $LaSr Cu^{3+}O_4$,[134] leaving the upper $d_{x^2-y^2}$ orbital empty and available to accept a transferred spin. It should be noted that this arrangement dictates a two-dimensional property and differs from the charge transfer situation in the manganites, which can occur in either e_g orbital, depending on the sign of the crystal-field

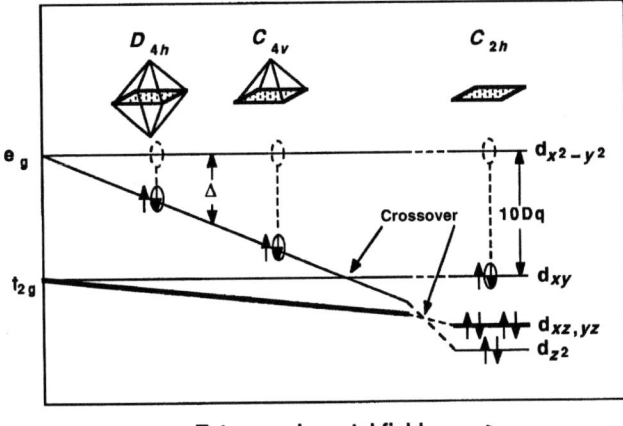

Figure 1.67. Growth of e_g doublet splitting and stabilization of the d^8 low-spin state as the tetragonal crystal field component increases from z-axis distorted octahedron (D_{4h}) to pyramidal (C_{3v}) to planar (C_{2h}). Compare this diagram with Fig. 1.14.

distortion. Here, the splitting of $d_{x^2-y^2}$ and d_{z^2} exists naturally as a result of the tetragonal/orthorhombic symmetry of the layered-type of perovskites and not necessarily from a J–T effect required to create the e_g splitting in the cubic or rhombohedral manganites.

The crystal-field origin of this effect is shown in Fig. 1.14 and its dependence on the degree of tetragonal distortion as it proceeds from c-axis extension, to the formation of a pyramid and finally to a planar ligand arrangement is diagrammed in Fig. 1.67. Unlike the case where ferromagnetism induced as a byproduct of spin-polarized electron transfer in the $Mn^{3+(4+)}$ combinations eliminates the exchange contribution E^{ex}_{hop} to the activation energy, the involvement of $S=0$ ions removes the internal polarization exchange energy and therefore renders moot any Hund's rule considerations. In addition, charge transfer from these mobile nonmagnetic ions causes local breakdowns of the antiferromagnetic couplings and can eventually reduce the Néel temperature T_N to zero by causing the spin alignment frustration to spread throughout the entire lattice.[112,124]

As sketched in Fig. 1.68, the polaronic hole carriers created by zero-spin Cu^{3+} ions of concentration $x < 0.08$ in $La_{2-x}Sr_xCuO_4$ would be more than sufficient to eliminate the magnetic order among the sites immediately surrounding the fixed negative charge represented by the Sr^{2+} dopant by establishing a ring of mobile carriers (as in Fig. 1.55), and then to frustrate through spin canting the regions beyond these lowest energy polaron sites. Where the charge transfer condenses into a superconducting state at critical temperature T_c, the threshold for frustration could be reduced further. Experimental evidence of this phenomenon is shown in Fig. 1.69 for two common cuprate superconductors.[135,136] Other possible $S=0$ candidates among the $3d^n$ series with occupied transfer orbitals $d_{x^2-y^2}$ and d_{z^2} are diagrammed schematically in Fig. 1.70.[112]

The influence of magnetic exchange on the $Cu^{2+(3+)}$ conductivity in the presence of rare-earth ions occupying the A sublattice is shown dramatically

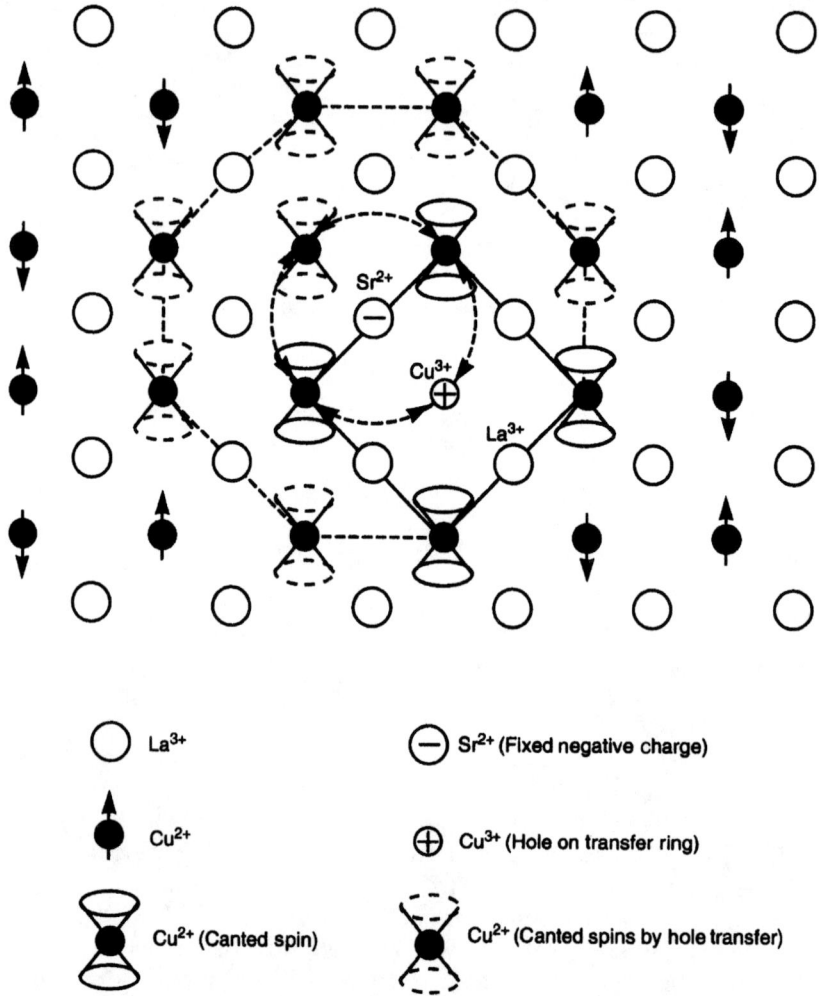

Figure 1.68. Origin of antiferromagnetic frustration caused by zero-spin Cu^{3+} ions in $La_{2-x}Sr_xCuO_4$. Breakdown in ordering can occur with only 1 out of 12 ($x \approx 0.08$) $S = 0$ sites in the Cu^{2+} sublattice in the region surrounding an Sr^{2+} "impurity" that acts as a fixed negative charge. Note that the Cu^{3+} hole can transfer around the inner ring in the manner of the model in Fig. 1.55, carrying the canted spins of its four Cu^{2+} neighbors with it.

in Fig. 1.71, where the data of George *et al.*[138] reveal the virtual absence of an activation energy for the nonmagnetic La compound. In all of the others, antiferromagnetic interactions deter the formation of large polarons that are necessary for the coherent activationless charge transfer of superconductivity. As shown in Fig. 1.69, only about 5% zero-spin Cu^{3+} could be sufficient to frustrate the long-range antiferromagnetic ordering in $La_{2-x}Sr_xCuO_4$ and create the metallic phase that allows the onset of superconductivity (the T_N, $T_c = 0$ condition). When one considers that less than one-third of the copper cations supply charge carriers, it would not be surprising to find the superconducting state to consist of spontaneously

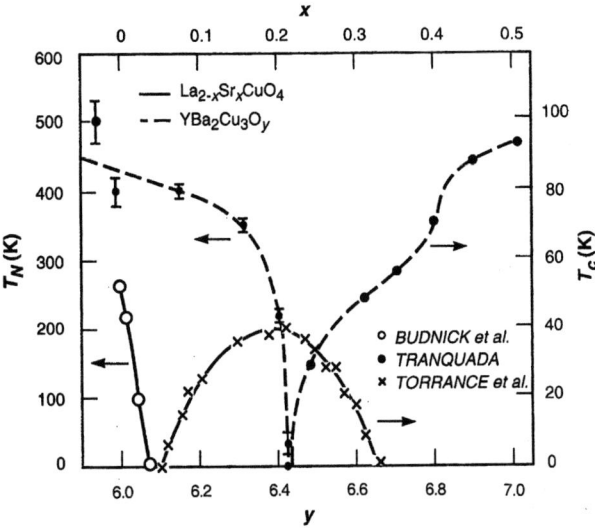

Figure 1.69. Experimental verification of the magnetic frustration requirement prior to the onset of superconductivity in $La_{2-x}Sr_xCuO_4$ and $YBa_2Cu_3O_y$ systems. Data from Budnick et al.,[135] Tranquada,[136] and Torrance et al.[137]

organized domains of metallic diamagnetism interspersed with regions of insulating or semiconducting antiferromagnetism.[139] Such a result is not unlike the room temperature formation of metallic and insulator regions in the manganites.[129]

1.8.4.2. Large-Polaron Superconductivity (Coherent Tunneling).

Metallic conduction properties in oxides have been demonstrated to condense into superconductivity at temperatures far above those of conventional metal and intermetallic compounds, reaching above 130 K.[140–142] To understand how covalent electron transfer (CET) through large polarons can be the agent of high-temperature superconductivity (HTS), we consider the nature of isolated larger polarons in diamagnetic or magnetically frustrated lattices in the context of superconductivity

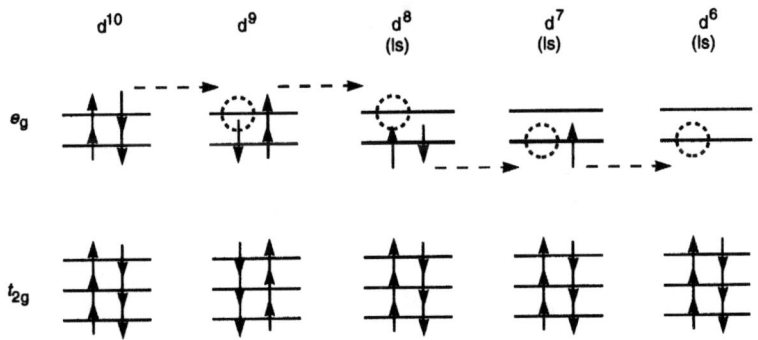

Figure 1.70. Schematic diagrams of $S = 1/2 \rightarrow S = 0$ transfers among $3d^n$ ions.

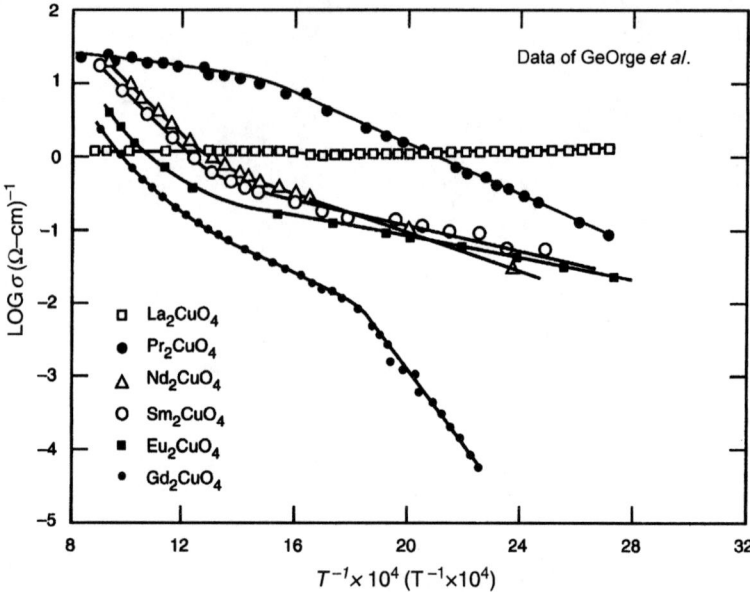

Figure 1.71. Conductivity σ data as a function of temperature, showing the influence on the activation energies from rare-earth ion exchange interactions with Cu^{2+} ions in a $(RE)_2CuO_4$ series. Note the absence of E_{hop}^{ex} in the La_2CuO_4 case. Data are from George et al.[138]

phenomenology.[143] A review of the London theory and the implications of the London equations as complements to those of Maxwell will not be included here. For this text, the discussion will focus on a primary condition that evolves from the London equations when applied to a system of supercarriers of volume density n_s: in the real-space ideal case, supercarriers must be spatially ordered, thereby rendering their transport coherent, i.e., $\nabla n_s = 0$.[112]

In collective electron systems this spatial order condition can be treated by the ensemble wave function ψ_{so} with spatial decay according Pippard's proposal

$$\psi_s = \psi_{so} \exp\left(-\frac{r}{\xi}\right) \tag{1.8.19}$$

where ξ is the coherence length. Because n_s is represented by the quantum mechanical probability density $|\psi_s|^2$, the gradient expressed from Eq. (1.8.19) can be written as

$$\nabla |\psi_s(r)|^2 = -\frac{2}{\xi} |\psi_{so}|^2 \exp\left(-\frac{2r}{\xi}\right) \tag{1.8.20}$$

If $\xi \to \infty$, the London requirement is satisfied. The magnitude of the coherence length serves as a measure of the purity of the superconducting state.[144]

For the polaron case, the equivalent of $|\psi_s|^2$ would be the transfer efficiency $\eta(r)$ of Eq. (1.8.7), which regrettably cannot be expressed as an exponential function in this model because of the central electrostatic field of the polaron trap given by Eq. (1.8.10). To extend the comparison to Eq. (1.8.20), the notion of coherence would be defined by the stabilization energy of the covalent bond according to

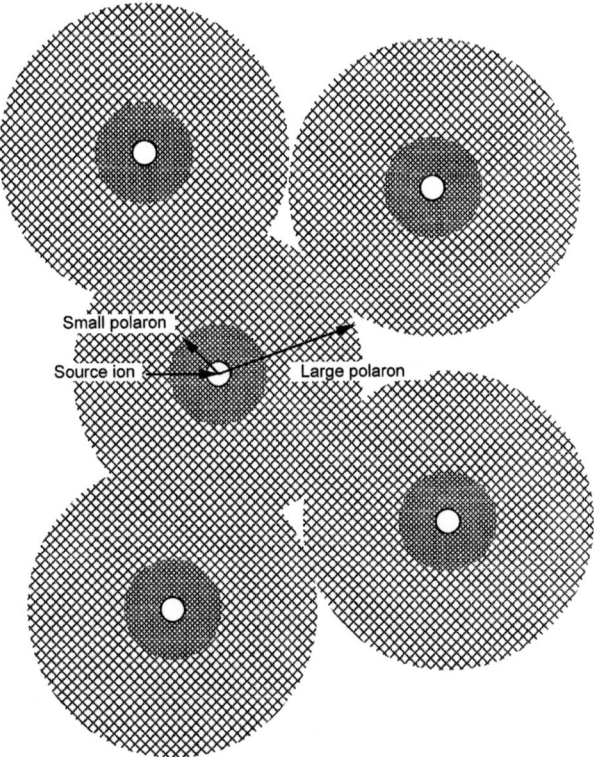

Figure 1.72. Two-dimensional diagram of large polaron cells amidst fixed polaron ionic sources.

$$\nabla \eta(r) \approx \frac{\partial}{\partial r}\left(1 - \frac{2\Delta E_p}{b_p}\right) = -\frac{1}{a}\frac{2e^2}{Kab_p}\left(\frac{a}{r}\right)^2 \qquad (1.8.21)$$

If $(a/r)^2$ is treated as the decaying exponential factor of Eq. (1.8.20), a real-space coherence length in the sense of ξ can be defined in the form of a large-polaron radius

$$r_p = a\left(\frac{Kab_p}{e^2}\right) \qquad (1.8.22)$$

which is the same as the result for the large polaron radius derived by Holstein,[108] except for a numerical multiplication factor. Consequently, the use of the polaron radius as a measure of coherence of the superconducting state can be adopted with the restriction that polaron dispersal must be achieved by real-space positioning of the polaron sources. This limitation does not exist with collective electron superconductivity formalized in momentum-space.

Within an individual polaron cell, the carrier ranges spontaneously and coherently through correlated transfers to a boundary defined by the covalent strength of the orbital bonds in which it resides and the polarizability of the lattice. In terms of Eq. (1.8.22), the "quality" of the superconductor is defined by the product of the dielectric constant K and the exchange integral b_p. When the concentration reaches a percolation threshold where cells begin to overlap, the dispersed polaron carriers and

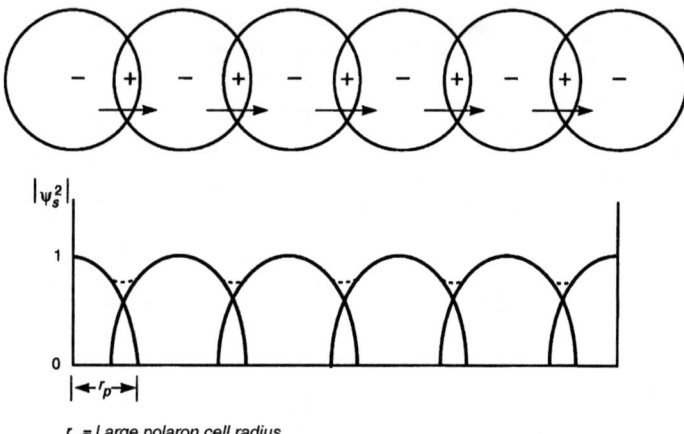

Figure 1.73. One-dimensional model of a large polaron chain, indicating the merger of carrier density functions $|\psi_s|^2$ to establish a continuous molecular orbital state.

their fixed source ions sketched in Fig. 1.72 condense into a string or chain of dipoles and a dynamic ferroelectric state of coherently tunneling carriers can form as in Fig. 1.73, still with one carrier per polaron cell. The smallest concentration threshold would then be determined by r_p, subject to the attendant requirement that antiferromagnetic frustration must also occur.

Following the concepts of local versus collective charge transfer, coherent tunneling would survive only under the conditions: (i) that polaron cell overlap be small enough that Pauli scattering would be moot and Fermi statistics not be a concern; and (ii) that thermal energy would not produce random hopping events to disrupt the correlated charge flow in the superconducting chains. The first condition would be concentration dependent, whereby the orbital levels (e_g-σ^a in the case of perovskites) would merge into a partially filled band at high concentrations, producing normal metallic properties typical of the t_{2g}-band metals such as CrO_2 mentioned in Section 1.8.4.1. A second limitation, which is reflected in Holstein's conclusion that hopping would dominate over tunneling where $T > \Theta_D/2$, could show its effect at even lower temperatures because of the added requirement that the coherent chains of polaron carriers be continuous.

1.8.4.3. Normal Resistivity and Critical Temperature.

From the above qualitative concepts, a "two-fluid" model can be constructed to account for the resistivity characteristics of the high-T_c cuprate superconductors. This can be accomplished in a straightforward manner; the total carrier concentration x is divided into normal and superconducting fractions by means of the thermal activation probability function. The basic premise of this large-polaron or CET theory is that carriers that are not activated by random lattice vibrations can be transported through the bonding, and the formation of a coherent tunneling state among this portion of the polaron population can then be established.

To begin the analysis, we define the normal carrier concentration as

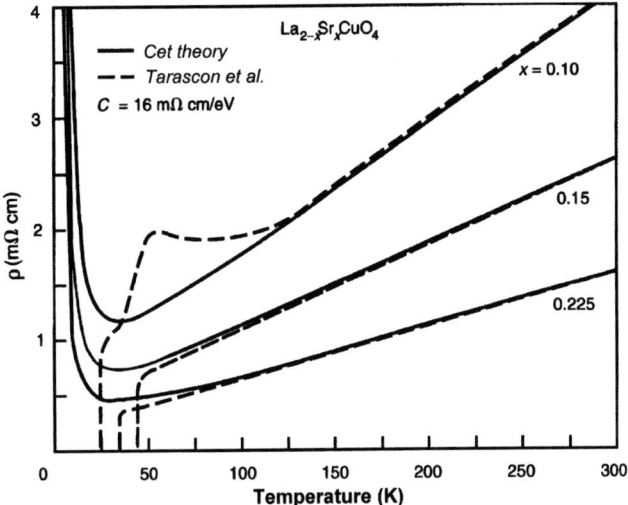

Figure 1.74. Comparison of theory with measured ρ versus T for the $La_{2-x}Sr_xCuO_4$ system case. Data are from Tarascon et al.[145]

$$x_n = x \exp\left(-\frac{E_{hop}}{kT}\right) \quad (1.8.23)$$

where E_{hop} is the basic polaron trap energy defined previously as E_{hop}^0 in our discussion of the manganites. As an example of the normal resistivity behavior of polaronic oxide compounds, Eq. (1.8.23) can be introduced to Eq. (1.8.24) to compute ρ versus T curves. Above the critical temperature T_c, the resistivity behavior of $La_{2-x}Sr_xCuO_4$ superconductors is metallic, as plotted in Fig. 1.74. For

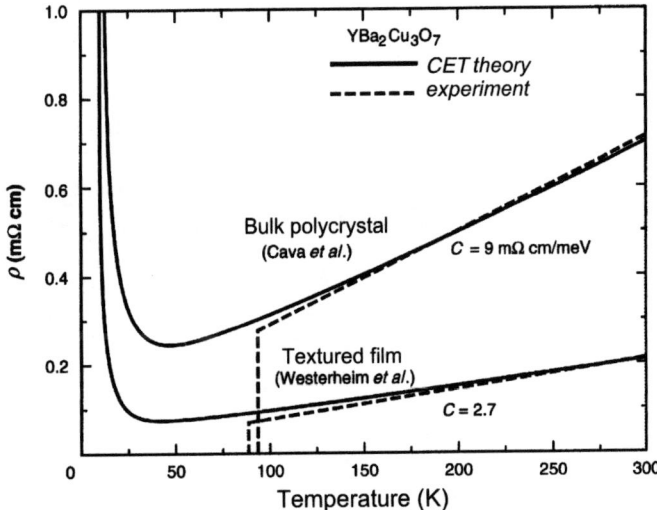

Figure 1.75. Comparison of theory with measured ρ versus T for bulk polycrystalline and oriented film $YBa_2Cu_3O_7$. Respective data are from Cava et al.[146] and Westerheim et al.[147]

the fitting of these data,[145] $C = 16$ mΩ cm/eV, larger than that for La$_{1-x}$Ca$_x$MnO$_3$ partly because of 50% higher lattice volume fraction occupied by octahedral B sites and the large angle grain boundaries in these bulk ceramic specimens. In these calculations $E_{hop} = 4$ meV for each value of $x = 0.10, 0.15$, and 0.225. Note that the onset of superconductivity occurs at the resistivity minimum, which suggests that the polaron trap could be the key to determining the value of the critical temperature. A second comparison between theory and experiment is presented in Fig. 1.75 for data from the most commonly studied compound YBa$_2$Cu$_3$O$_7$ (YBCO) in bulk ceramic[146] and quasi-single crystal film[147] forms. In this latter case, the extrapolated straight-line asymptote appears to reach the origin at $T = 0$.

If x_n is now subtracted from x, the concentration fraction, i.e., available for tunneling is the result. From this group, however, the limitations of transfer efficiency or probability and polaron dispersal must be taken into account. To this end, the supercarrier density is expressed as

$$x_s = \eta P(x - x_n) = \eta Px\left[1 - \exp\left(-\frac{E_{hop}}{kT}\right)\right] \quad (1.8.24)$$

where the transfer efficiency is reduced by a dispersal factor $P = (1 - 2\beta x)$ which represents the probability that a receptor site is adjacent to a carrier.[112] For these purposes, a dispersal parameter $0 \leq \beta \leq 1$ is defined whereby $\beta = 0$ for perfect ordering and 0.5 for random ordering; the higher values represent various degrees of clustering. Because the transfer involves adjacent pairs of mixed-valence ions, the maximum value of x is 0.5 for single transfers. Arguments can be made to support a double transfer as the minimum event (real-space pair transfer). This contention was discussed previously,[110,112] and would be consistent with the suggested presence of spin waves.[148] If the coherence of the condensed state requires a two-carrier transfer as the minimum event, the factor P would be applied twice as $P^2 \approx (1 - 4\beta x)$ and the limit of x would become 0.33.

From a concentration x_s of dispersed large polarons depicted in Figs. 1.72 and 1.73, a percolation threshold defined as x_t would be reached at the maximum temperature for which coherent tunneling can exist, i.e., the critical temperature T_c. Above this temperature, there would not be enough tunneling carriers to sustain the condensed state; below it, there would be excess carriers to provide supercurrent necessary for the "perfect" diamagnetism and other properties associated with the superconducting state. Accordingly, from Eq. (1.8.24) T_c can be related to x_t according to

$$T_c = \frac{E_{hop}}{k} W \quad (1.8.25)$$

where

$$W = \ln\left(1 - \frac{x_t}{\eta Px}\right)^{-1}$$

In Fig. 1.76, the T_c versus x results of interpretation of experiment by means of Eq. (1.8.25) are presented for hole-carrier (p-type) (La$_{2-x}^{3+}$Sr$_x^{2+}$)Cu$_{1-x}^{2+}$Cu$_x^{3+}$O$_4$ and YBa$_2$Cu$_3$O$_y$ (YBCO) compounds.[137,145,149,150] To standardize the T_c results in terms

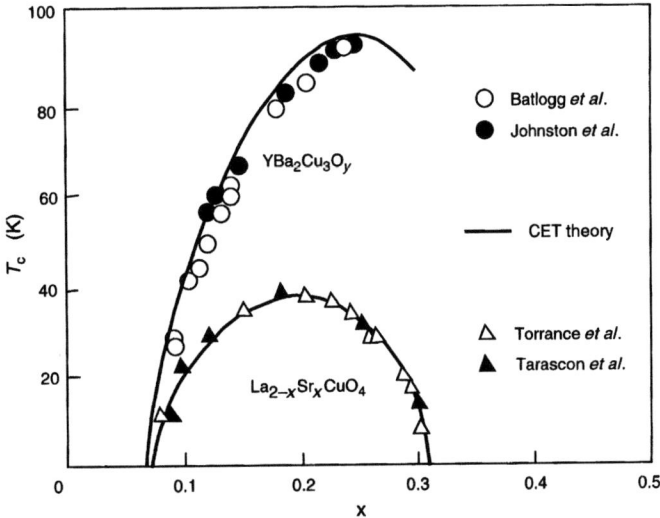

Figure 1.76. Critical temperature T_c versus x for $La_{2-x}Sr_xCuO_4$, and $YBa_2Cu_3O_y$ (for which y has been converted to x using the linear relation $y = 0.25x - 1.5$). Data are from Torrance et al.,[137] Tarascon et al.,[145] Johnson et al.,[149] and Batlogg et al.[150]

of Cu^{3+} polaron ion concentration in YBCO, x has been extracted from the charge–balance relation $y = 0.25x - 1.5$ for each value of y. The parabolic character of the curves arises from the introduction of the dispersal factor P to the W parameter. There are two readily discernible differences between the results for the two compounds: (i) the peak in T_c is greater for YBCO, suggesting that the value of β is

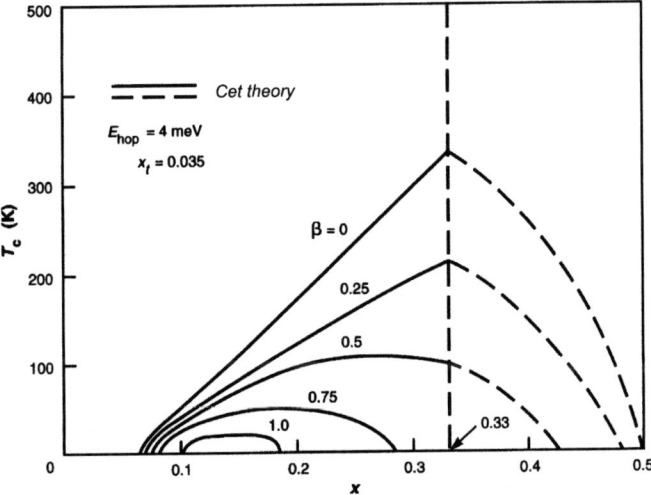

Figure 1.77. Projected T_c versus x curves over the range of $0 \leq \beta \leq 2$ for individual carriers and $0 \leq \beta \leq 1$ for pairs. Dashed curves indicate that the "real-space" pair model does not apply beyond $x = 0.33$.

reduced (from 0.7 to 0.57) because of better spatial ordering of the polaron sources;*
and (ii) the entire curve is higher, suggesting that E_{hop}^0 is increased (from 2.5 to
4 meV). The implications of this latter possibility are complicated because an
increase in the polaron trap energy would result from a decrease in b_p, which in turn
would mean that U_{eff} could have been increased by the D_{4h} field as suggested by
the pyramidal-to-planar descent depicted in Fig. 1.67. If the transfer integral is
affected, it would also mean that the transfer probability η and the polaron radius
that determines x_t would also change. As an example of the effects of the polaron
dispersal variation, Fig. 1.77 is offered for the YBCO values of $E_{hop}^0 = 4$ meV,
$x_t = 0.035$, where it is shown that ideal ordering could theoretically produce T_c values
approaching room temperature.

A condensed state of large polaron carriers with n-type properties can
theoretically occur for compositions in which there are more Cu^{3+} than Cu^{2+} ions,
i.e., $x > 0.5$. Other compounds such as $(La_{2-x}^{3+}Ce_x^{4+})Cu_{1-x}^{2+}Cu_x^{1+}O_4$ are n-type
superconductors, the implications of which were discussed elsewhere.[114]

From this model, it is possible to construct two-fluid functions to analyze
penetration depths, critical magnetic fields, critical current densities, and other properties as a function of temperature. The key variable in each case is the supercarrier
density n_s.[112] One central result of this approach to the superconducting state that
arises from the Boltzmann-type partitioning of the carriers, however, is that the
supercarrier population will decrease accordingly with temperature until the density
of supercarriers falls below the threshold value of n_t and the superconducting state is
extinguished. As a consequence, high supercurrent densities would only be expected
at the lower temperatures regardless of how great the critical temperature can be
engineered.

In summation, the mixed-valence manganites and cuprates are metallic if the
polaron trap energy from exchange is eliminated. $La_{1-x}Ca_xMnO_3$ is metallic for
$x < 0.5$ because of the ferromagnetism that results from vibronic J–T effects and
delocalization exchange between partially filled e_g orbital states. Above $x = 0.5$, the
crystallographic structures are influenced by a complexity of factors that include J–T
distortions that in turn control the splitting of the e_g states. Magnetic order can then
be established in a variety of antiferromagnetic configurations, often tenuous and
subject to insulator–metal transitions when high magnetic fields reverse the sign of
the resultant exchange field to revisit the ferromagnetism and create and produce a
significant magnetoresistance effect. The internal field from a spontaneous magnetic
moment, however, can preclude the possibility of superconductivity. $La_{2-x}Sr_xCuO_4$
and other high-T_c cuprates are metallic because mobile zero-spin Cu^{3+} or Cu^{1+} ions
remove the possibility of magnetic exchange trapping and frustrate any antiferromagnetic order above a concentration threshold of only a few percent. In both
materials systems, large polarons with itinerant properties can form at low temperatures, in one case contributing to a condition from which a magnetoresistance
anomaly may occur at the Curie temperature, and in the other, making possible the
unimpeded covalent charge transfer between $S = 1/2$ and $S = 0$ ions that is a basis
for high-T_c superconductivity.

---

*In these layered structures, the mixed valence in the Cu cation lattice caused by the charges of the substitutional ions or oxygen vacancies occurs in the crystallographic layer of the Cu–O_4 planes.

From the foregoing discussions, some observations concerning polarized spin transport can be made. The subjects reviewed were confined to those involving electrons bound within molecular structures that form the building blocks of crystal lattices. Because of the dictates of Hund's rule and the Pauli exclusion principle that set the $\Delta S = 0$ transition rule, spin alignment should always be preserved locally. To the extent that magnetic exchange ordering of spins influences the ground state energy, polarized transfer can be long-range and even coherent.

Elements from the d^n transition shells can fulfill this promise, but with important limitations. Metals feature impressive ferromagnetic and electrical conduction properties that are of less certain origin due to corruption by hybrid mixtures of free s and partially bound d electrons. Ionic compounds, particularly oxides, offer the possibility of long-range d-electron ferromagnetic spin transport, but only in select situations where partially filled degenerate states survive crystal-field quenching (in symmetries higher than orthorhombic), or where a real-charge mechanism such as polaron tunneling can effect the transfer. Regardless of the magnetic system, however, ideal polarization can only be expected far below the Curie temperature.

BIBLIOGRAPHY

General Background

A. H. Morrish, *The Physical Principles of Magnetism* (John Wiley, New York, 1965).

S. Chikazumi, *Physics of Magnetism* (John Wiley, New York, 1964).

K. Yosida, *Theory of Magnetism* (Springer, New York, 1996).

R. M. White, *Quantum Theory of Magnetism* (Springer, New York, 1983).

C. J. Ballhausen, *Molecular Electronic Structures of Transition Metal Complexes* (W & J Mackay, Chatham, England, 1979).

J. S. Griffith, *The Theory of Transition-Metal Ions* (Cambridge University Press, London, 1961).

M. M. Schieber, *Experimental Magnetochemistry* (John Wiley, New York, 1967).

C. J. Ballhausen and H. B. Gray, *Molecular Orbital Theory* (W.A. Benjamin, New York, 1964).

Crystal Fields and Oxide Chemistry

L. E. Orgel, *An Introduction to Transition-Metal Chemistry: Ligand Field Theory* (John Wiley, New York, 1960).

C. J. Ballhausen, *Introduction to Ligand Field Theory* (McGraw-Hill, New York, 1962).

H. L. Schläfer and G. Gliemann, *Basic Principles of Ligand Field Theory* (Wiley-Interscience, New York, 1969).

W. Low, *Paramagnetic Resonance in Solids* (Academic Press, New York, 1960).

G. E. Pake, *Paramagnetic Resonance* (W. A. Benjamin, New York, 1962).

D. S. McClure, Electronic spectra of molecules and ions in crystals. Part II. Spectra of ions in crystals, *Solid State Phys.* **9**, 399 (1959).

M. T. Hutchings, Point-charge calculations of energy levels of magnetic ions in crystalline electric fields, *Solid State Phys.* **16**, 227 (1964).

G. F. Dionne, Calculation of crystal-field energy level splittings of the Ti^{3+} ion in $RbAl(SO_4) \cdot 12H_2O$, *Phys. Rev. A* **137**, 743 (1965).

Covalence and Superexchange

C. J. Ballhausen and H. B. Gray, *Molecular Electronic Structures* (Benjamin/Cummings, New York, 1980).

P. W. Anderson, New approach to the theory of superexchange interactions, *Phys. Rev.* **115**, 2 (1959).

J. Kanamori, Superexchange interaction and symmetry properties of electron orbitals, *J. Chem. Phys. Solids* **10**, 87 (1959).

P. W Anderson, Theory of magnetic exchange interactions: exchange in insulators and semiconductors, *Solid State Phys.* **14**, 99 (1963).

J. B. Goodenough, *Magnetism and the Chemical Bond* (Interscience Publishers, John Wiley, New York, 1963).

J. B. Goodenough and J. M Longo, Crystallographic and magnetic properties of perovskites and perovskite-related compounds, in: *Landolt-Bornstein Tabellen*; New Series III/4a, edited by K. Hellwege (Springer, Berlin, 1970) p. 231.

Ferrimagnetism

J. Smit and H. P. J. Wijn, *Ferrites* (John Wiley, New York, 1959).

E. P. Wohlfarth (Ed.), *Ferromagnetic Materials*, vol. 2 (North Holland, New York, 1980). Chapters by M. A. Gilleo (magnetic garnets), J. Nicolas (microwave ferrites), and P. I. Slick (nonmicrowave ferrites).

G. Winkler, *Magnetic Garnets* (Braunschweig, Wiesbaden, Vieweg, 1981).

S. Geller, Crystal and static magnetic properties of garnets, *Proceedings of the International School of Physics "Enrico Fermi,"* Course LXX, Varenna, Italy 1977 (North Holland, New York, 1978).

G. F. Dionne, Molecular-field coefficients of substituted yttrium iron garnets, *J. Appl. Phys.* **41**, 4874 (1970).

W. P. Wolf, Effect of crystalline electric fields on ferromagnetic anisotropy, *Phys. Rev.* **108**, 1152 (1957).

G. F. Dionne, Origin of the magnetostriction effects from Mn^{3+}, Co^{2+}, and Fe^{2+} ions in ferrimagnetic spinels and garnets, *J. Appl. Phys.* **50**, 4263 (1979).

B. Lax and K. J. Button, *Microwave Ferrites and Ferrimagnetics* (McGraw Hill, New York, 1962).

W. H. Von Aulock, *Handbook of Microwave Ferrite Materials* (Academic Press, New York, 1965).

E. Schloemann, Microwave ferrite materials, in: *Encyclopedia of Electrical and Electronics Engineering*, vol. 13 (John Wiley, New York, 1999), p. 90.

Charge Transfer

R. J. D. Tilley, *Defect Crystal Chemistry and Its Applications* (Blackie, Glasgow and London, 1987), Chapter 6.

J. B. Goodenough, Metallic oxides, *Prog. Solid State Chem.* **5**, 145 (1972).

T. Holstein, Studies of polaron motion, *Ann. Phys.* **8**, 325 (1959).

N. F. Mott, *Metal–Insulator Transitions* (Taylor and Francis, London, 1990).

J. Appel, Polarons, *Solid State Phys.* **21**, 193 (1968).

J. B. Goodenough, A. Wold, N. Menyuk, and R. J. Arnott, Relationship between crystal symmetry and magnetic properties of ionic compounds containing Mn^{3+}, *Phys. Rev.* **124**, 373 (1961).

C. Zener, Interaction between the d-shells in the transition metals. II. Ferromagnetic compounds of manganese with perovskite structure, *Phys. Rev.* **82**, 403 (1951).

P. -G. de Gennes, Effects of double exchange in magnetic oxides, *Phys. Rev.* **118**, 141 (1960).

Proceedings of the International Workshop on the Colossal Magnetoresistance Effect, edited by N. Witte and J. B. Goodenough, *Aust. J. Phys.* **52** (1999).

G. F. Dionne, Magnetic exchange and charge transfer in mixed valence manganites and cuprates, *J. Appl. Phys.* **79**, 5172 (1996); Anomalous magnetoresistance in the lanthanide manganites and its relation to high-T_c superconductivity, MIT Lincoln Laboratory Technical Report 1029, (1996). AD-A309080

R. R. Heikes and W. D. Johnston, Mechanisms of conduction in Li-substituted transition metal oxides, *J. Chem. Phys.* **26**, 582 (1957).

G. F. Dionne, Magnetic frustration in high-T_c superconductors, *J. Appl. Phys.* **69**, 5194 (1991).

G. F. Dionne, Covalent electron transfer theory of superconductivity, MIT Lincoln Laboratory Technical Report 885 (1992). AD-A253975

REFERENCES

1. J. H. Van Vleck, *The Theory of Electric and Magnetic Susceptibilities* (Oxford University Press, London, 1932).
2. W. Borchardt-Ott, *Crystallography* (Springer, New York, 1995); M. J. Buerger, *Elementary Crystallography* (John Wiley, New York, 1956).
3. J. S. Griffith, *The Theory of Transition-Metal Ions* (Cambridge University Press, London, 1961).

4. C. J. Ballhausen, *Introduction to Ligand Field Theory* (McGraw-Hill, New York, 1962).
5. H. L. Schläfer and G. Gliemann, *Basic Principles of Ligand Field Theory* (Wiley-Interscience, New York, 1969).
6. E. U. Condon and G. H. Shortley, *The Theory of Atomic Spectra* (Cambridge University Press, London, 1963).
7. C. J. Ballhausen, *Introduction to Ligand Field Theory* (McGraw-Hill, New York, 1962), Appendix I, p. 93.
8. H. B. Kramers, *Proc. Amsterdam Acad. Sci.* **32**, 1176 (1929).
9. J. H. Van Vleck, *Phys. Rev.* **41**, 208 (1932).
10. W. G. Penney and R. Schlapp, *Phys. Rev.* **41**, 194 (1932); R. Schlapp and W. G. Penney, *Phys. Rev.* **42**, 666 (1932).
11. W. Low, *Paramagnetic Resonance in Solids* (Academic Press, New York, 1960), p. 15.
12. W. Low, *Paramagnetic Resonance in Solids* (Academic Press, New York, 1960), p. 22.
13. K. W. H. Stevens, *Proc. Phys. Soc. A* **65**, 209 (1952).
14. M. T. Hutchings, *Solid State Phys.* **16**, 227 (1964).
15. C. J. Ballhausen, *Introduction to Ligand Field Theory* (McGraw-Hill, New York, 1962), p. 99.
16. M. H. L. Pryce and W. A. Runciman, *Discuss. Faraday Soc.* **26**, 34 (1958).
17. G. F. Dionne and B. J. Palm, *J. Magn. Reson.* **68**, 355 (1986).
18. C. J. Ballhausen, *Introduction to Ligand Field Theory* (McGraw-Hill, New York, 1962), p. 103.
19. L. E. Orgel, *Introduction to Transition-Metal Chemistry: Ligand-Field Theory* (John Wiley, New York, 1959).
20. O. G. Holmes and D. S. McClure, *J. Chem Phys.* **26**, 1686 (1957).
21. G. F. Dionne, *Phys. Rev. A* **137**, 743 (1965); G. F. Dionne *Can. J. Phys.* **42**, 2419 (1964).
22. J. A. MacKinnon and G. F. Dionne, *Can. J. Phys.* **44**, 2329 (1966).
23. G. F. Dionne and J. A. MacKinnon, *Phys. Rev.* **172**, 325 (1968).
24. G. A. Woonton and J. A. MacKinnon, *Can. J. Phys.* **46**, 59 (1968).
25. B. Bleaney, G. S. Bogle, A. H. Cooke, R. J. Duffus, M. C. M. O'Brien, and K. W. H. Stevens, *Proc. Phys. Soc. A* **68**, 57 (1955).
26. L. S. Kornienko and A. M. Prokhorov, *Sov. Phys. JETP* **11**, 1189 (1960).
27. H. Hartmann and H. L. Schlafer, *Z. Phys. Chem. Leipzig* **197**, 116 (1951).
28. M. H. L. Pryce, *Proc. Phys. Soc. A* **63**, 25 (1951); A. Abragam and M. H. L. Pryce, *Proc. Roy. Soc. A* **205**, 135 (1951).
29. H. A. Jahn and E. Teller, *Proc. R. Soc. A* **161**, 220 (1937); H. A. Jahn, *Proc. R. Soc. A* **164**, 117 (1938).
30. J. H. Van Vleck, *Phys. Rev.* **57**, 426 (1940).
31. J. B. Goodenough, *Magnetism and the Chemical Bond* (Interscience Publishers, John Wiley, New York, 1963), pp. 192, 213.
32. G. F. Dionne, *J. Appl. Phys.* **91**, 7367 (2002).
33. M. Kaplan and B. Vekhter, *Cooperative Phenomena in Jahn–Teller Crystals* (Plenum Publishing Corporation, New York, 1995).
34. L. Dubicki and M. J. Riley, *J. Chem. Phys.* **106**, 1669 (1997).
35. P. L. W. Tregenna-Piggott, M. C. M. O'Brien, J. R. Pilbrow, H. U. Güdel, S. P. Best, and C. Noble, *J. Chem. Phys.* **107**, 8275 (1997).
36. W. Low, *Phys. Rev.* **101**, 1827 (1956).
37. K. Ono, S. Koide, S. Sekiyama, and H. Abe, *Phys. Rev.* **96**, 38 (1954).
38. B. Bleaney, K. D. Bowers, and R. J. Trenam, *Proc. R. Soc. A* **228**, 157 (1955).
39. G. M. Zverev and A. M. Prokhorov, *J. Exp. Phys. USSR* **34**, 1023 (1958).
40. W. Low, *Paramagnetic Resonance in Solids* (Academic Press, New York, 1960), p. 91.
41. C. J. Ballhausen, *Introduction to Ligand Field Theory* (McGraw-Hill, New York, 1962), p. 124 (on work of Abragam and Pryce).
42. K. D. Bowers and J. Owen, *Rep. Prog. Phys.* **18**, 310 (1955).
43. J. W. Orton, *Rep. Prog. Phys.* **22**, 204 (1959).
44. W. Low, *Phys. Rev.* **109**, 247 (1958).
45. D. S. McClure, *J. Chem. Solids* **3**, 311 (1957).
46. Y. Tanabe and S. Sugano, *J. Phys. Soc. Jpn.* **9**, 753 (1954).
47. L. E. Orgel, *J. Chem. Phys.* **23**, 1004 (1955).
48. W. Low, *Paramagnetic Resonance in Solids* (Academic Press, New York, 1960), p. 120.

49. W. Heitler and F. London, *Z. Phys.* **44**, 455 (1927).
50. M. Wolfsberg and L. Helmholz, *J. Chem. Phys.* **20**, 837 (1952); C. J. Ballhausen, *Introduction to Ligand Field Theory* (McGraw-Hill, New York, 1962), p. 161; C. J. Ballhausen and H. B. Gray, *Molecular Electronic Structures: An Introduction* (Benjamin/Cummings Reading, MA, 1980).
51. C. J. Ballhausen, *Molecular Electronic Structures of Transition Metal Complexes* (McGraw-Hill International, Chatham, Great Britain, 1979), pp. 84–89.
52. J. C. Slater, *Phys. Rev.* **35**, 509 (1930).
53. E. C. Stoner, *Proc. Leeds Philos. Soc.* **2**, 391 (1933).
54. A. Sommerfeld and H. Bethe, *Handbuch der Physik*, XXIV/2, (J. Springer, Berlin 1933), p. 596.
55. A. H. Morrish, *The Physical Principles of Magnetism* (John Wiley, New York, 1965), p. 279.
56. K. Yosida, *Theory of Magnetism* (Springer, New York, 1996), p. 54.
57. W. Heisenberg, *Z. Phys.* **49**, 619 (1928).
58. B. Lax and K. J. Button, *Microwave Ferrites and Ferrimagnetics*, Lincoln Laboratory Publication, (McGraw-Hill, New York, 1962), p. 65.
59. J. B. Goodenough, *Magnetism and the Chemical Bond* (Interscience Publishers, John Wiley, New York, 1963), p. 213, Chapter II.
60. H. B. Kramers, *Proc. Amsterdam Acad. Sci.* **33**, 959 (1930); H. B. Kramers *Physica* **1**, 182 (1934).
61. P. W. Anderson, *Phys. Rev.* **79**, 350 (1950); P. W. Anderson, *Phys. Rev.* **115**, 2 (1959); P. W. Anderson, *Solid State Phys.* **14**, 99 (1969). Review chapter.
62. J. B. Goodenough, *Magnetism and the Chemical Bond* (Interscience Publishers, John Wiley, New York, 1963), p. 167–171.
63. C. Zener, *Phys. Rev.* **82**, 403 (1951).
64. P. G. de Gennes, *Phys. Rev.* **118**, 141 (1960).
65. P. W. Anderson and H. Hasegawa, *Phys. Rev.* **100**, 675 (1955); T. Nagamiya, K. Yosida, and R. Kubo, *Adv. Phys.* **4**, 1 (1955).
66. J. Kanamori, *J. Phys. Chem. Solids* **10**, 87 (1959).
67. J. B. Goodenough, *Phys. Rev.* **117**, 1442 (1960); J. B. Goodenough, *Magnetism and the Chemical Bond* (Interscience Publishers, John Wiley, New York, 1963), p. 213, Chapter III.
68. J. B. Goodenough and A. L. Loeb, *Phys. Rev.* **8**, 391 (1955).
69. L. Néel, *Ann. Phys. Paris* **3**, 137 (1948).
70. G. F. Dionne, *Magnetic Moment versus Temperature Curves of Ferrimagnetic Garnet Materials* (MIT Lincoln Laboratory Technical Report 480, 1970), AD-A715284.
71. G. F. Dionne, *Magnetic Moment versus Temperature Curves of LiZnTi Ferrites* (MIT Lincoln Laboratory Technical Report 502, 1974), AD-A782421/2.
72. G. F. Dionne, *Magnetic Moment versus Temperature Curves of Rare-Earth Iron Garnets* (MIT Lincoln Laboratory Technical Report 588, 1981), AD-A107898/9.
73. G. F. Dionne, *J. Appl. Phys.* **41**, 4874 (1970).
74. G. F. Dionne, *J. Appl. Phys.* **45**, 3621 (1974).
75. G. F. Dionne, *J. Appl. Phys.* **85**, 4627 (1999).
76. J. Smit and H. P. J. Wijn, *Ferrites* (John Wiley & Sons, New York, 1959), p. 156.
77. W. H. von Aulock, *Handbook of Microwave Ferrite Materials* (Academic Press, New York, 1965).
78. E. W. Gorter, *Philips Res. Rep.* **9**, 295 (1954).
79. G. Blasse, *Philips Res. Rep.* Suppl. no. 3 (1964).
80. A. Broese van Groenou, P. F. Bongers, and A. L. Stuijts, *Mater. Sci. Eng.* **3**, 317 (1968).
81. *Landolt-Börnstein Tabellen*, New Series III/4b, (Springer, New York, 1970).
82. A. P. Greifer, *IEEE Trans. Magn.* **5**, 774 (1969).
83. M. M. Schieber, *Experimental Magnetochemistry* (North-Holland, Amsterdam, 1967).
84. M. I. Klinger and A. A. Samokhvalov, *Phys. Stat. Sol. (b)* **79**, 9 (1977).
85. G. O. White and C. E. Patton, *J. Magn. Magn. Mater.* **9**, 299 (1978).
86. C. Borghese, *J. Phys. Chem. Solids* **28**, 2225 (1967).
87. R. Pauthenet, *Ann. Phys.* **7**, 710 (1952).
88. G. F. Dionne and R. G. West, *Appl. Phys. Lett.* **48**, 1488 (1986); G. F. Dionne and R. G. West, *J. Appl. Phys.* **61**, 3868 (1987).
89. G. F. Dionne, *J. Appl. Phys.* **63**, 3777 (1988).
90. Z. Simsa and V. A. M. Brabers, *IEEE Trans. Magn.* **11**, 1303 (1975).
91. M. M. Schieber, *Experimental Magnetochemistry* (North-Holland, Amsterdam, 1967), p. 188.

92. G. F. Dionne, *J. Appl. Phys.* **61**, 3865 (1987).
93. K. E. Kuehn, D. Sriram, S. S. Bayya, J. J. Simmins, and R. L. Snyder, *J. Mater. Res.* **15**, 1635 (2000).
94. G. Winkler, *Magnetic Garnets* (Friedr. Vieweg & Sohn, Braunschweig/Wiesbaden, 1981).
95. S. Geller, *Physics of Magnetic Garnets*, Proceedings of the International School of Physics, Course LXX (North-Holland, New York, 1978), p. 1.
96. M. A. Gilleo, In: *Ferromagnetic Materials*, vol. 2, edited by W. P. Wohlfarth (North-Holland, New York, 1980), Chapter 1.
97. S. Geller and M. A. Gilleo, *J. Phys. Chem. Solids.* **3**, 30 (1957); S. Geller and M. A. Gilleo, *J. Phys. Chem. Solids* **9**, 235 (1959).
98. E. R. Czerlinsky and R. A. MacMillan, *Phys. Stat. Sol.* **41**, 333 (1970); E. R. Czerlinsky, *Phys. Stat. Sol.* **34**, 483 (1969).
99. M. A. Gilleo and S. Geller, *Phys. Rev.* **110**, 73 (1958).
100. R. Pauthenet, *Compt. Rendus* **243**, 1499, 1737 (1956); R. Pauthenet, *Ann. Phys.* **3**, 424 (1958).
101. R. Aléonard, *J. Phys. Chem. Solids* **15**, 167 (1960).
102. G. F. Dionne, *J. Appl. Phys.* **46**, 3347 (1976); G. F. Dionne and P. L. Tumelty, *J. Appl. Phys.* **50**, 8257 (1979).
103. C. D. Brandle and S. L. Blank, *IEEE Trans. Magn.* **12**, 14 (1976).
104. P. Hansen, *Physics of Magnetic Garnets* Proceedings of the International School of Physics, Course LXX (North-Holland, New York, 1978) p. 56.
105. A. W. Overhauser, *Phys. Rev.* **128**, 1437 (1962).
106. H. Fröhlich, *Adv. Phys.* **3**, 325 (1954).
107. N. F. Mott, *Metal–Insulator Transitions*, (Taylor and Francis, New York, 1990).
108. T. Holstein, *Ann. Phys.* **8**, 325 (1959); L. Friedman and T. Holstein, *Ann Phys.* **21**, 494 (1963).
109. J. B. Goodenough, *Prog. Solid State Chem.* **5**, 145 (1972).
110. G. F. Dionne, *Transition-Metal Oxide Superconductivity*, MIT Lincoln Laboratory Technical Report 802 (1988). AD-A197069
111. G. F. Dionne, *J. Appl. Phys.* **67**, 4561 (1990).
112. G. F. Dionne, *Covalent Electron Transfer Theory of Superconductivity*, MIT Lincoln Laboratory Technical Report 885 (1992). AD-A253975
113. G. F. Dionne, *J. Appl. Phys.* **69**, 5194 (1991).
114. G. F. Dionne, *A Strategy for Higher Temperature Superconductors*, MIT Lincoln Laboratory Technical Report 1021 (1995). AD-A296885
115. G. F. Dionne, *Temperature Dependence of Large Polaron Superconductivity*, MIT Lincoln Laboratory Technical Report 1024 (1995). AD-A297287
116. G. F. Dionne, *Anomalous Magnetoresistance in the Lanthanide Manganites and its Relation to High-T_c Superconductivity*, MIT Lincoln Laboratory Technical Report 1029 (1996). AD- A309080; G. F. Dionne, *J. Appl. Phys.* **79**, 5172 (1996).
117. L. Friedman and T. Holstein, *Ann Phys.* **21**, 494 (1963).
118. K. Siratori and S. Iida, *J. Phys. Soc. Jpn.* **15**, 210 (1960); P. Hansen, *Physics of Magnetic Garnets* Proceedings of the International School of Physics, Course LXX (North-Holland, New York, 1978) p. 56, Table 1.9.
119. S. M. Watts, S. Wirth, S. von Molnar, A. Barry, and J. M. Coey, *Phys. Rev. B*, **61**, 9621 (2000).
120. J. B. Goodenough, *Prog. Solid State Chem.* **5**, 359 (1972).
121. K. I. Kobayashi, T. Kimura, H. Sawada, K. Terakura, and Y. Tokura, *Nature* **395**, 677 (1998).
122. Y. Moritomo, Sh. Xu, A Machida, T. Akimoto, E. Nishibori, M. Takata, and M. Sakata, *Phys. Rev. B* **61**, R7827 (2000).
123. R. R. Heikes and W. D. Johnston, *J. Chem. Phys.* **26**, 582 (1957).
124. R. J. D. Tilley, *Defect Crystal Chemistry and its Applications* (Blackie, Glasgow, London, 1987), Chapter 6.
125. G. F. Dionne, *J. Appl. Phys.* **69**, 5194 (1991).
126. S. Jin, T. H Tiefel, M. McCormack, R. A. Fastnacht, R. Ramesh, and L. H. Chen, *Science* **264**, 413 (1994).
127. J. B. Goodenough, *Phys. Rev.* **100**, 564 (1955).
128. G. H. Jonker and J. H. Van Santen, *Physica* **XVI**, 337 (1950).
129. J. B. Goodenough, A. Wold, N. Menyuk, and R. J. Arnott, *Phys. Rev.* **124**, 373 (1961).
130. M. Fäth, S. Freisem, A. A. Menovsky, Y. Tomioka, J. Aarts, and J. A. Mydosh, *Science* **285** (1999).

131. Y. -Q. Li, J. Zhang, S. Pombrik, S. DiMascio, W. Stevens, Y. F. Yan, and N. P. Ong, *J. Mater. Res.* **10**, 2166 (1995).
132. J. B. Goodenough and J. M. Longo, Crystallographic and magnetic properties of perovskites and perovskite-related compounds, in: *Landolt-Bornstein Tabellen*, New Series III/4a, edited by K. Hellwege (Springer, Berlin, 1970) p. 231.
133. J. B. Goodenough, *Aust. J. Phys.* **52**, 155 (1999).
134. J. B. Goodenough, G. Demazeau, M. Pouchard, and P. Hagenmüller, *Solid State Chem.* **8**, 325 (1973).
135. J. I. Budnick, B. Chamberland, D. P. Yang, Ch. Neidermayer, A. Golnik, E. Recknagel, M. Rossmanith, and A. Weidinger, *Europhys. Lett.* **5**, 651 (1988).
136. J. M. Tranquada, *J. Appl. Phys.* **64**, 6071 (1988).
137. J. B. Torrance, Y. Tokura, A. I. Nazzal, A. Bezinge, T. C Huang, and S. S. P. Parkin, *Phys. Rev. Lett.* **61**, 1127 (1988).
138. A. M. George, I. K. Gopalakrishnan, and M. D. Karkhanavla, *Mater. Res. Bull.* **9**, 721 (1974).
139. J. M. Tranquada, B. J. Sternlieb, J. D. Axe, Y. Nakamura, and S. Uchida, *Nature* **375**, 661 (1995).
140. J. G. Bednorz and K. A. Müller, *Z. Phys. B* **64**, 189 (1986); J. G. Bednorz, K. A. Müller, and M. Takashige, *Science* **236**, 73 (1987).
141. M. K. Wu, J. R. Ashburn, C. J. Torng, P. H. Hor, R. L. Meng, L. Gao, Z. J. Huang, Y. Q. Wang, and C. W. Chu, *Phys. Rev. Lett.* **58**, 908 (1987).
142. M. A. Subramanian, C. C. Torardi, J. C. Calabrese, J. Gopalakrishnan, K. J. Morrissey, T. R. Askew, R. B. Flippen, U. Chowdhry, and A. W. Sleight, *Science* **239**, 1015 (1988).
143. F. London and H. London, *Physica* **2**, 341 (1935).
144. A. B. Pippard, *Proc. R. Soc. (London) A* **216**, 547 (1953); M. Tinkham, *Introduction to Superconductivity* (Robert E. Krieger, Malabar, FL, 1980).
145. J. M. Tarascon, L. H. Greene, W. R. McKinnon, G. W. Hull, and T. H. Geballe, *Science* **235**, 1373 (1987).
146. R. J. Cava, B. Batlogg, R. B. van Dover, D. W. Murphy, S. Sunshine, T. Siegrist, J. P. Remeika, E. A. Reitman, S. Zahurak, and G. P. Espinosa, *Phys. Rev. Lett.* **58**, 1676 (1987).
147. A. C. Westerheim, L. S. Yu-Jahnes, and A. C. Anderson, *IEEE Trans. Magn.* **27**, 1001 (1991).
148. G. Shirane, Y. Endoh, R. J. Birgeneau, M. A. Kastner, Y. Hidaka, M. Oda, M. Suzuki, and T. Murakami, *Phys. Rev. Lett.* **59**, 1613 (1987).
149. D. C. Johnston, A. J. Jacobson, J. M. Newsam, J. T. Lewandowski, D. P. Goshorn, D. Xie, and W. B. Yelon, in: *Chemistry of High-Temperature Superconductors*, edited by D. L. Nelson, M. S. Whittingham, and T. F. George (American Chemical Society, Washington, DC, 1987), p. 136.
150. B. Batlogg, R. J. Cava, C. H. Chen, G. Kourouklis, W. Weber, A. Jayaraman, A. E. White, K. T. Short, E. A. Rietman, L. W. Rupp, D. Werder, and S. M. Zahurak, in: *Novel Superconductivity*, edited by S. A. Wolf and V. Z. Kresin (Plenum Press, New York, 1987), p. 653.

2

Secondary Magnetic Properties

James MacLaren*

2.1. INTRODUCTION

Magnetic phenomena have their origins in a quantum mechanical description of electronic structure. An introduction to the quantum mechanical origins of magnetism has been given in an earlier chapter of this book. In particular, the exchange coupling between spins, which follows from the Coulomb interaction between the electrons, and the Pauli principle, gives rise to properties such as ferromagnetism and antiferromagnetism, and can be understood within the framework of non-relativistic quantum mechanics. Other magnetic properties, such as the magneto-crystalline anisotropy, which results in a preferred orientation for the magnetization with respect to the crystal axes, are relativistic effects resulting from the combined effects of spin–orbit coupling and magnetism. Other secondary magnetic phenomena including magneto-optics and magneto-elastic coupling also arise from a relativistic description of magnetism.

Spin–orbit coupling is a relativistic effect, which arises naturally in the Dirac equation of the electron (an excellent reference is the book by Rose[1]). However, it can be treated in a non-relativistic theory as a perturbation when the Dirac equation is reduced to a two-component Pauli equation. The physical origin of spin–orbit coupling is discussed in most text books on quantum mechanics, and hence will only be briefly reviewed here for completeness. Spin–orbit coupling can be understood from the following argument. In the frame of reference in which the electron is at rest, the moving nucleus produces both magnetic and electric fields. The magnitude of this magnetic field is $\mathbf{v} \times \mathbf{E}/c$ (Gaussian units), where $\mathbf{v}$ is the velocity of the electron and $\mathbf{E}$ the electric field from the nucleus. The spin of the electron $(e/mc)\mathbf{S}$ interacts with this magnetic field. The interaction energy, or spin–orbit coupling is given by $-(e/mc^2)\,\mathbf{S}\cdot\mathbf{v}\times\mathbf{E}$, which leads to the familiar expression in the case of a central potential of

$$H_{so} = \frac{1}{m^2 c^2 r}\frac{dV}{dr}\mathbf{L}\cdot\mathbf{S} \qquad (2.1.1)$$

where V is the potential energy $(e\mathbf{E} = -dV/dr)$, $\mathbf{L}$ and $\mathbf{S}$ are the orbital and spin angular momenta, respectively. A detailed calculation, which accounts for the fact that the frame of reference in which the electron is at rest is accelerating introduces an extra factor of 2 (the Thomas precession), leading to the well-known expression

---

*Associate Provost, Tulane University, New Orleans, Louisiana.

$$H_{so} = \frac{1}{2m^2c^2r}\frac{dV}{dr}\mathbf{L}\cdot\mathbf{S} \qquad (2.1.2)$$

A derivation is given in an appendix of Jackson's book "Classical Electrodynamics".[2] Despite the small size of the spin–orbit coupling term, when compared with the Coulomb energy, its consequences are important for determining many of the magnetic properties important to applications.

In the limit where the Dirac equation reduces to a two-component Pauli equation, $V/mc^2 \ll 1$, the correct form of the spin–orbit coupling, including the Thomas precession factor, occurs naturally. The many-electron version of this Pauli equation is quite complicated. Slater[3] derived a form which in addition to the sum of single-particle equations, that contain the usual spin–orbit coupling, there are interactions between pairs of electrons such as spin–other orbit coupling. The reader is referred to a recent review article by Gay and Richter[4] where a discussion of all the terms in the many-electron Pauli equation is given.

Quantitative calculations of electronic and magnetic properties are almost exclusively based upon density functional theory. This theory has turned out to be a remarkable success at predicting ground state properties of solids such as crystal structures, equilibrium lattice constants and elastic constants, magnetic moments, and magneto-crystalline anisotropies. Despite being a ground state theory, it has also provided a framework for interpreting excited state phenomena such as optical properties.

Density functional theory, while developed first for non-relativistic and non-magnetic systems,[5,6] has been extended to both magnetic and relativistic systems.[7,8] The formalism results in a mean-field theory of the electronic structure in which the wave function of each electron is solved independently in an effective potential that includes the external potential from the nuclei, a Hartree repulsion resulting from the electron density, and a term known as the exchange correlation potential which arises from the mapping of the interacting many-electron system into the single-particle model. This formalism is reviewed briefly in appendices. More information can be found in Ref. 9 and references therein.

Combining relativity and magnetism in the Dirac equation is quite complicated and leads to an infinite set of coupled differential equations since the magnetization (given by an operator proportional to S_z) no longer commutes with the Hamiltonian. Those band structure methods that solve the radial Dirac equation in the atomic spheres, such as the Korringa–Kohn–Rostoker methods,[10–13] always truncate the set of equations at the lowest order, leading to a coupling between states of the same ℓ only. In principle, basis set approaches such as the linearized muffin–tin orbital method (LMTO)[14,15] or full potential linearized augmented plane wave method (FLAPW)[9,15] can use these solutions to compute matrix elements of the Dirac Hamiltonian. A fully relativistic LMTO program, developed by Ebert,[16] has been used to study secondary magnetic properties such as anisotropy[17] and magneto-optics.[18]

However, most of the relativistic effects, Darwin and mass velocity terms, can be treated simply within the scalar relativistic approach.[19] This is the theory that results from converting the Dirac equation for a central potential into a second-order differential equation and then neglecting the spin–orbit coupling. The result is a Schrödinger-like equation which can be used with a spin-dependent potential.

Spin–orbit coupling can be added back in later as a perturbation. Because of its simplicity and accuracy, this approach is widely used in electronic structure calculations and appears to be quite accurate for all but the heaviest elements.

In this chapter, the secondary magnetic properties which result from the combined effects of spin–orbit coupling and magnetism will be discussed. In the case of magneto-optics, first a phenomenological theory will be presented, which involves solving Maxwell's equations for a material which has off-diagonal elements in the dielectric tensor. These off-diagonal elements are first order in the spin–orbit coupling. The philosophy involves first finding the normal modes for propagation of electromagnetic waves, and then matching the fields at the interface between vacuum and the magnetic material by using the appropriate boundary conditions. In the absence of free charges and surface currents, these are continuous perpendicular components of **D** and **B**, and continuous parallel components of **E** and **H**. At this level, the theory is based on classical physics.

The two common magnetic–optic phenomena, the Kerr and Faraday effects, are proportional to the off-diagonal elements in the dielectric tensor. These effects describe the rotation of the plane of polarization of reflected and transmitted plane-polarized light, respectively. The Faraday effect was first observed by Faraday in a piece of glass placed in an external magnetic field.[20] Absorption in a metal rapidly attenuates the Faraday rotation signal in metals in all but the thinnest of samples. Hence, the Faraday effect is not as useful for studying metallic magnetic materials. The Kerr effect is observed in light reflected from the surface of a ferromagnetic material.[21,22] The size of the rotation scales with the magnetization, and thus the Kerr effect is often used to map out hysteretic behavior. In general, the analysis required to compute the rotation angle is quite involved, and can only be done simply for certain special orientations of the magnetization and wave vector of the incident light. In this chapter, we will only derive expressions for the Kerr effect for these simpler situations. For the more complex geometries, we will quote the general expressions for the Kerr rotation and outline the principle steps in the derivation. There are several Refs. 23–25 where algebraic details are given, and the reader is referred to these papers.

The symmetry of the dielectric tensor follows from the Lorenz force, $\mathbf{F} = e(\mathbf{E} + \mathbf{v} \times \mathbf{B})$, which describes the motion of the electrons in the presence of both electric and magnetic fields. This classical picture is extended to a proper quantum mechanical treatment, in which the dielectric tensor is found from linear response theory by using the Kubo formula. An outline of the derivation of the Kubo formula is presented. A practical formula for both computation and interpretation, based upon approximate many body wave functions, derived from single-particle wave functions, such as output from a first principle's local spin density band structure technique, is presented. This approach will be illustrated with some first-principle calculations for some magnetic materials. An alternate derivation of the absorptive part of the conductivity tensor, based upon Fermi's golden rule, has been given by several authors,[26–28] leading to expressions for the elements of the conductivity tensor that are equivalent to those obtained from the Kubo formula when the sharp limit is taken. The dispersive parts can be found from a Kramers Kronig transformation.

The origin of the magneto-crystalline anisotropy and related effects are also a result of both magnetism and spin–orbit coupling. One of the first mechanisms proposed to explain magneto-crystalline anisotropy was based upon dipolar interactions. However, it is easy to show that in a cubic crystal such an interaction vanishes. At finite temperatures, the dipoles are not perfectly aligned leading to effective quadrupolar interactions that resulted in an anisotropy even in a cubic crystal. The size of the anisotropy, from purely dipolar forces is too small to account for experiment (see, e.g., the appendix in Ref. 29). An alternative mechanism was based upon spin–orbit coupling since this term couples the spin to the lattice though the spin and orbital angular momentum. Several calculations showed that this model leads to anisotropies of the correct size.[30–33] The magneto-crystalline anisotropy will be discussed in terms of simple models such as the Neel model,[34] as well as more detailed *ab initio* electronic structure calculations which are based upon the local spin-density approximation (LSDA) to density functional theory. Several studies of magneto-crystalline anisotropy using a tight-binding description of the underlying electronic structure have been made. Including studies on surfaces and monolayers by Bennett and Cooper,[35] Takayama *et al.*,[36] and more recently by Bruno[37] who also studied the quenching of the orbital moment. Cinal *et al.*,[38] extended the studies of monolayers to investigate the role of film thickness. The tight binding picture describes the d bands in transition metals quite accurately and as such can also describe magneto-crystalline anisotropy too.

The Neel theory, which can successfully describe many of the features of magneto-crystalline anisotropy, will be discussed in some detail. Comparisons to experiment will be made both for elemental magnetic materials and technologically important ultra-thin films. For the latter, the shape anisotropy will also be discussed. Its origin is not in the spin–orbit coupling, but rather in the dipolar forces between magnetic moments. In cubic materials, these dipolar forces produce no anisotropy; however, for ultra-thin films, this energy plays an important role in determining the direction of the magnetization. Most of the discussions will be aimed at transition metals and alloys. The issue of magneto-crystalline anisotropy in the rare-earth compounds is described elsewhere in this volume. The general Refs. 39–41 have good introductory discussions of magneto-crystalline anisotropy and magneto-elastic coupling.

2.2. MAGNETO-OPTIC PROPERTIES

2.2.1. Introduction

The discovery of magneto-optic effect dates back to the middle of the nineteenth century. Using a beam of linearly polarized light propagating through a piece of glass placed between the poles of a magnet, Faraday observed that the plane of polarization of the light was rotated.[20] The observed rotation of the plane of polarization can be understood in a phenomenological manner in terms of different refractive indexes for the left and right circularly polarized waves. A linearly polarized beam of light can be decomposed into a sum of left and right circularly polarized light. The difference in refractive index implies that the speed of light will be different for left and right circularly polarized light, and hence when the beam exits the sample, the two circularly polarized beams will have rotated by different amounts. Recombining these two beams results in one which is elliptically

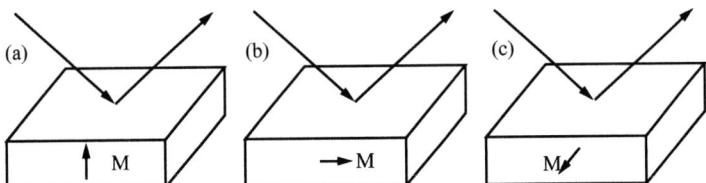

Figure 2.1. The three common geometries for the Kerr effect: (a) the polar Kerr effect, where the magnetization is perpendicular to the film, (b) the longitudinal Kerr effect where the magnetization lies in the plane of the film and the plane formed by the surface normal and the wave vector of the incident light, and (c) the equatorial Kerr effect where the magnetization lies in the plane of the film and is perpendicular to the plane formed by the surface normal and the wave vector of the incident light.

polarized, with the major axis of the ellipse rotated by the Faraday rotation angle with respect to the incident beam.

The magneto-optical Kerr (MOKE) effect is observed in light reflected from a magnetic surface. As in the Faraday effect, the plane of polarization of light reflected is rotated. Kerr[21,22] discovered the effect in ferromagnetic materials. Expressions for the Faraday and Kerr rotations in terms of refractive indexes can be derived from the Fresnel formula for transmission and reflection of electromagnetic waves from an interface along with a dielectric tensor of the correct symmetry.

Three different geometries for the Kerr effect are usually considered, these are distinguished from one another by the relative orientation of the sample magnetization and the wave vector of the incident light. These are shown schematically in Fig. 2.1. The derivation of the Kerr angle and ellipticity is quite straight forward for the polar geometry when the light is incident along the surface normal and parallel to the magnetization, since for this geometry, circularly polarized **D** and **E** fields are the correct normal modes. However, the computation is much more complex for angles other than for normal incidence, as well as for the other common orientations of the magnetizations: the longitudinal and equatorial Kerr effects. Expressions, including derivations, correct to first-order in the spin–orbit coupling (or equivalently the off-diagonal elements of the dielectric tensor) have been given in several papers.[23–25,42–44]

All of the derivations start from an assumption of the form of the dielectric tensor, dictated by symmetry. If the magnetization is assumed to be aligned along the z direction, then the dielectric tensor takes the form

$$\epsilon = \begin{pmatrix} \kappa_1 & \kappa_2 & 0 \\ -\kappa_2 & \kappa_1 & 0 \\ 0 & 0 & \kappa_3 \end{pmatrix} \quad (2.2.1)$$

Both the symmetry of the dielectric tensor and the rotation of the plane of polarization of the reflected light can also be understood from the symmetry of the Lorenz force. As an example, consider the case of the polar Kerr effect when the light is incident along the surface normal, the z direction. The electric field vector can be taken to be along the y axis. The electric field will cause an oscillatory excitation of the electrons along the y axis. The part of the Lorenz force resulting from the magnetic field is at right angles to both the magnetization and electric field, i.e., along the x axis. There are, therefore, components of the acceleration, and electronic motion, along both x and y directions. Hence, the electric field due to the radiation from the accelerated electrons have both x and y components. Thus, the reflected light is composed of an electric field along the y direction along with a small electric field along the x direction. The antisymmetric nature of ϵ follows directly from the cross-product in the Lorenz force.

Since to the lowest order in the magneto-optic coupling, $\kappa_3 = \kappa_1$,* we can write the dielectric tensor in a particular compact form (see Landau and Lifshitz, "Electrodynamics of Continuous Media"[45]):

$$\epsilon_{ij} = \kappa_1 \delta_{ij} + \kappa_2 \epsilon_{ijk} u_k \tag{2.2.2}$$

where ϵ_{ijk} is the usual symbol for the third rank antisymmetric tensor, and **u** is the unit vector parallel to the magnetization. When the magnetization is parallel to the z axis, this reproduces the standard form of Eq. (2.2.2), though because it is formed from objects that transform properly under rotations, it is the appropriate expression for an arbitrary orientation of the magnetization.

The origin of both the Faraday and Kerr effects has been known for some time, and a phenomenological theory based upon the symmetry of the dielectric tensor can successfully explain the effect. The quantum mechanical origin of magneto-optic effects lies in the spin–orbit coupling which aligns the magnetic moments along a particular crystal direction. The magnetic moments act as axial vectors and thus interact differently with left and right handed circularly polarized light. The dielectric tensor, which was used to provide an explanation of magneto-optic effects can be calculated from the linear response theory. Recently, there have been several calculations based upon this approach, in which the Kubo formula has been evaluated using wave functions and band structures derived from first-principle LSDA electronic structure calculations. While this theory is formally a ground state theory, and optical properties require a knowledge of the excited states, in practice the optical properties obtained from such a one-electron calculation are often in close agreement with experiment.

The Kerr effect finds applications in optical data storage, where the rotation is used to detect stored data, and is often used to map out hysteresis loops in magnetic thin films, since the signal is proportional to the magnetic induction.

2.2.2. Phenomenological Theory

A phenomenological description of magneto-optic effects proceeds as follows. In the media, normal modes for the propagation of waves are found from a solution of Maxwell's equations, and the dielectric tensor given by Eq. (2.2.2). To the lowest order in the magneto-optic coupling, or spin–orbit coupling, $\kappa_3 = \kappa_1$. In this case, the normal modes for the **D** field are left and right circularly polarized waves which are transverse ($\nabla \cdot \mathbf{D} = 0$). Note that the electric fields (**E**) are not generally transverse, as they would be in a material with a diagonal dielectric tensor, though they can be found from the relation $\mathbf{E} = \epsilon^{-1} \mathbf{D}$. For each of the two modes of

---

*Of course, the assumption of equality between κ_1 and κ_3 is not essential, it merely simplifies the derivations, and since the final results are typically only needed to first order in the magneto-optic coupling, it makes sense to start at that level of approximation.

propagation, there are different refractive indexes, which relate the wave speed, or wave vector, to that in vacuum.

Expressions for the Kerr rotation and ellipticity are obtained as follows. First, one solves for the electric and magnetic fields in vacuum which match, using the appropriate boundary conditions, to each normal mode in the material. Then, an appropriate linear combination of these solutions which represents an incident plane-polarized wave is taken. The corresponding reflected wave will in general be elliptically polarized and will determine the Kerr rotation and ellipticity. This requires transforming the fields between two bases: one in which the x and y axes lie in the plane whose normal is along the direction of propagation, and a second where the x and y axes lie in the interface plane.[23]

If the dielectric tensor was diagonal, then the normal modes can be taken as plane-polarized waves. The relations connecting reflected and transmitted fields to the incident fields are the Fresnel coefficients which are given in standard texts on electricity and magnetism.[2] The normal modes for propagation in the magnetic media are found from Maxwell's equations, which are given by

$$\nabla \cdot \mathbf{D} = 4\pi\rho \qquad \nabla \times \mathbf{H} = \frac{4\pi}{c}\mathbf{J} + \frac{1}{c}\frac{\partial \mathbf{D}}{\partial t}$$
$$\nabla \cdot \mathbf{B} = 0 \qquad \nabla \times \mathbf{E} = -\frac{1}{c}\frac{\partial \mathbf{B}}{\partial t} \qquad (2.2.3)$$

where for a linear media,

$$\mathbf{D} = \epsilon\mathbf{E}, \qquad \mathbf{J} = \sigma\mathbf{E}, \qquad \mathbf{B} = \mu\mathbf{H} \qquad (2.2.4)$$

In the case of metals at optical frequencies, the permeability $\mu \approx 1$. In addition, for metals, the permittivity $\epsilon \approx 1$, since the core electrons are tightly bound and not polarizable while the valence and conduction electrons are free to move.

Consider an incident plane wave for both $\mathbf{E}$ and $\mathbf{H}$:

$$\mathbf{E}(\mathbf{r}, t) = \mathbf{E}_0 \exp[i(\mathbf{k} \cdot \mathbf{r} - \omega t)]$$
$$\mathbf{H}(\mathbf{r}, t) = \mathbf{H}_0 \exp[i(\mathbf{k} \cdot \mathbf{r} - \omega t)] \qquad (2.2.5)$$

substituting into Maxwell's equations using these plane wave expressions for $\mathbf{E}$ and $\mathbf{H}$ leads to the following:

$$\nabla \times \mathbf{H} = \frac{1}{c}\epsilon(\omega)\frac{\partial \mathbf{E}}{\partial t} \qquad (2.2.6)$$

where the frequency dependent dielectric constant is redefined to include the contributions from the conductivity σ, i.e.,

$$\epsilon(\omega) = 1 + \frac{i4\pi\sigma(\omega)}{\omega} \qquad (2.2.7)$$

Taking the curl of Faraday's law, and using Eq. (2.2.3) leads to linear equations satisfied by the components of the $\mathbf{E}$ and $\mathbf{D}$ fields in the media:

$$k_0^2 \mathbf{D} = k^2 \mathbf{E} - (\mathbf{k} \cdot \mathbf{E})\mathbf{k} \qquad (2.2.8)$$

where $k_0 = \omega/c$.

Using the fact that the **D** fields are transverse, allows Eq. (2.2.7) to be written for these transverse components as

$$\mathbf{D} = n^2 \epsilon^{-1} \mathbf{D} \tag{2.2.9}$$

where $n = k/k_0$ is the refractive index. This equation can be solved quite simply, since to the lowest order in the magneto-optic coupling

$$\epsilon_{ij}^{-1} = 1/\kappa_1 \delta_{ij} - \kappa_2/\kappa_1^2 \epsilon_{ijk} u_k \tag{2.2.10}$$

Taking a transverse basis which defines x and y axes that are perpendicular to **k** and the z axis parallel to **k**, leads to

$$1/n^2 = 1/\kappa_1 \pm i\kappa_2/\kappa_1^2 u_z \tag{2.2.11}$$

for the refractive index of each mode. The corresponding eigenvectors are

$$\mathbf{D} = \hat{x} \pm i\hat{y} \tag{2.2.12}$$

which correspond to circularly polarized modes.

Since the Kerr angle describes the rotation of the plane of plane-polarized incident light, which will in general couple to both modes inside the magnetic material, we have to consider the polarization of the electric (and magnetic) field vectors of these incident plane waves. There are two cases, either the polarization of the electric field vector is perpendicular (labeled s) or in the plane of incidence (labeled p). The plane of incidence is defined by the wave vector $\mathbf{k}_0$ and the surface normal. This is shown in Fig. 2.2.

2.2.3. The Polar Kerr Effect

2.2.3.1. The Polar Kerr Effect at Normal Incidence.

As an example, consider the polar Kerr geometry in which the light is incident normal to the surface. The refractive indexes, obtained from Eq. (2.2.11) for the two normal modes are $n_{\pm}^2 = \kappa_1 \pm i\kappa_2$, since $\mathbf{u} = (0, 0, 1)$. In this geometry, the normal modes for the electric field are also circularly polarized waves. Normal incidence has one further simplification, namely that in the absence of magneto-optic coupling, the Fresnel reflection coefficients for light incident normally are the same for both s- and p-polarized light, and are given by

$$r_{ss} = r_{pp} = \frac{1-n}{1+n}$$

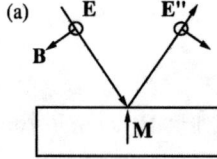 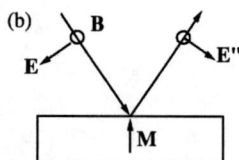

Figure 2.2. Illustration of the two polarizations of the electric field vectors for plane-polarized radiation: Polarization of **E** (a) perpendicular to plane of incidence (s) (b) parallel to plane of incidence (P).

SECONDARY MAGNETIC PROPERTIES

Hence, the circularly polarized beams, formed from a linear combination of s- and p-polarized beams have reflection coefficients given by $(1-n_+)/(1+n_+)$. The electric field vector parallel for the incident plane wave is chosen to be parallel to the x axis. This vector is then resolved into the left and right circular polarized components, i.e.,

$$\mathbf{E} = 2E_0 \qquad \hat{\mathbf{x}} = \mathbf{E}_+ + \mathbf{E}_- \qquad (2.2.13)$$

where $\mathbf{E}_\pm = E_0(\hat{\mathbf{x}} \pm i\hat{\mathbf{y}})$.

The reflected light is obtained from these Fresnel coefficients, i.e.,

$$\begin{aligned}\mathbf{E}^r &= \frac{1-n_+}{1+n_+}\mathbf{E}_+ + \frac{1-n_-}{1+n_-}\mathbf{E}_- \\ &= 2E_0 \frac{[(1-n_+n_-)\hat{\mathbf{x}} + i(n_- - n_+)\hat{\mathbf{y}}]}{(1+n_+)(1+n_-)}\end{aligned} \qquad (2.2.14)$$

Hence, the rotation of the reflected light's polarization is given by

$$\tan\phi = \frac{i(n_- - n_+)}{1 - n_+n_-} \qquad (2.2.15)$$

Since $\kappa_2 \ll \kappa_1$, we can write $\tan\phi \approx \phi$, and hence

$$\phi = \frac{\kappa_2}{(\kappa_1)^{1/2}(1-\kappa_1)} = \frac{\epsilon_{xy}}{(\epsilon_{xx})^{1/2}(1-\epsilon_{xx})} = \frac{-\sigma_{xy}}{\sigma_{xx}(1+i\frac{4\pi}{\omega}\sigma_{xx})^{1/2}} \qquad (2.2.16)$$

The real part θ_k provides the rotation of the polarization plane whereas the imaginary part ϵ_k gives the ellipticity of the reflected light.

2.2.3.2. The Polar Kerr Effect at General Angle of Incidence.

The situation becomes much more complicated when the light is incident at a general angle, since in the absence of magneto-optic coupling the reflection coefficients r_{ss} and r_{pp} differ. The two refractive indexes are $n^2 = \kappa_1 \pm i\gamma'\kappa_2$, where γ' is cosine of the angle the light in the magnetic material makes to the surface normal or z direction. (For this geometry, γ' is the cosine of the angle of refraction.) These are also found from Eq. (2.2.11). The reflection of an incident plane wave is described by a matrix with coefficients r_{ss}, r_{pp}, r_{sp}, and r_{ps}, depending on the polarization of incident and reflected waves. Formulae for these reflection coefficients for the polar Kerr effect at a general angle of incidence are

$$r_{pp} = \frac{n_0\gamma - \gamma'}{n_0\gamma + \gamma'}$$

$$r_{ss} = \frac{\gamma - n_0\gamma'}{\gamma + n_0\gamma'} \qquad (2.2.17)$$

$$r_{sp} = r_{ps} = \frac{\kappa_2/n_0}{(\gamma + n_0\gamma')(n_0\gamma + \gamma')}$$

where $n_0^2 = \kappa_1$. γ is the cosine of the angle between the light and the surface normal in vacuum, or for the polar geometry, the cosine of the angle of incidence. In terms of the matrix of Fresnel coefficients, the polar Kerr angle is given by r_{sp}/r_{ss} or

r_{sp}/r_{ss} depending on the polarization of the incident beam. For the case where $\theta_i = 0$, then $\gamma = \gamma' = 1$ and the Kerr angle is independent of the incident polarization as expected:

$$\phi = \frac{r_{sp}}{r_{ss}} = \frac{r_{sp}}{r_{pp}} = \frac{\kappa_2}{n_0(1-n_0^2)} = \frac{\epsilon_{xy}}{(\epsilon_{xx})^{1/2}(1-\epsilon_{xx})} \qquad (2.2.18)$$

2.2.4. The Longitudinal Kerr Effect

In this geometry, the refractive indexes for the two normal modes are the same as those in the polar geometry, i.e., $n^2 = \kappa_1 + i\gamma'\kappa_2$, where the coordinate system is set up such that the plane of incidence is the y–z plane. The matrix of Fresnel coefficients are

$$r_{pp} = \frac{n\beta - \beta'}{n_0\beta + \beta'}$$

$$r_{ss} = \frac{\beta - n_0\beta'}{\beta + n_0\beta'} \qquad (2.2.19)$$

$$r_{ps} = -r_{sp} = \frac{\gamma\beta\kappa_2/n_0^2}{\beta'(\beta + n_0\beta')(n_0\beta + \beta')}$$

In this geometry, $\beta = \cos(\theta_i)$, $\gamma = \sin(\theta_i)$, $\beta' = \cos(\theta_r)$, and $\gamma' = \sin(\theta_r)$. Where θ_i and θ_r are the angles of incidence and refraction, respectively.

2.2.5. The Equatorial Kerr Effect

In this geometry, the plane of incidence is perpendicular to the magnetization, and the refractive indexes are given by $n^2 = \kappa_1$ to first order in the magneto-optic coupling κ_2 (u_z is zero). In this case, the normal modes for the **D** and **E** fields can be taken to be s- and p–polarized waves, and the matrix of Fresnel coefficients can be shown to be

$$r_{pp} = \frac{n_0\gamma - \gamma'}{n_0\gamma + \gamma'}\left[1 + \frac{\sin(2\theta_i)\kappa_2/n_0^2}{n_0^2\gamma^2 - \gamma'^2}\right]$$

$$r_{ss} = \frac{\gamma - n_0\gamma'}{\gamma + n_0\gamma'} \qquad (2.2.20)$$

$$r_{sp} = r_{ps} = 0$$

Note that the magneto-optic coupling appears to be first order in r_{pp}, implying that the reflected intensity of unpolarized light would change upon reversal of the magnetization, and since s- and p-polarized waves are the appropriate normal modes, we have that $r_{sp} = r_{ps} = 0$.

2.2.6. Concluding Remarks about the Kerr Effect

One important application of the Kerr effect is in the study of the hysteretic behavior of a magnetic thin film, since the Kerr signal is proportional to components of the magnetization. Of many examples in the literature, we will just discuss two cases. In the first example, Philip's group studied epitaxial Co/Pt[46] and Co/Pd[47] trilayers. The magnitude of the polar Kerr effect was used to measure the component of the magnetization perpendicular to the surface, and hence a hysteresis loop. The epitaxial samples were grown in such a way that a wedge of varying Co thickness was deposited on either a Pt (111) or Pd (111) substrate. The Co wedge was then capped with Pt or Pd. This geometry allowed magnetic properties as a function of Co thickness to be studied all within a single sample. Fitting the tail of the hysteresis loop, which was obtained from the Kerr spectrum, where magnetization rotation was assumed to occur allowed the anisotropy constants to be obtained. The orientation of the applied field with respect to the sample was chosen to ensure that magnetization rotation, rather than domain wall motion, occurred. Anisotropy versus Co thickness allowed both volume and interface contributions to be determined.

In the second example, Florczak and Dahlberg[48] also used the Kerr effect to study the magnetic properties of a thin film. They considered the case where the easy axis of the magnetization lay in plane of the film. The applied field was also in the plane of the sample, and under the action of the field, the magnetization points along a different direction, one which minimizes the energy. Thus, the Kerr geometry is a mix between the equatorial and longitudinal geometries. Expressions for the intensities of reflected s and p polarized light in this more general case have been given by the authors. The technique allows both in-plane components of the magnetization to be measured simultaneously.

The analysis of the Kerr effect has also been generalized from the case of a single magnetic thin film to the case of a magnetic multilayer, allowing the Kerr signal to be computed for a stack of magnetic and non-magnetic layers. The approach is one of a transfer matrix which propagates and matches the fields at each interface. The result, for a superlattice, has to be obtained numerically and is obtained by multiplying 4×4 matrices together. More details can be found in the following papers.[42-44]

2.2.7. The Faraday Effect

The Faraday effect, or rotation of the plane of polarization in transmitted light, can be analyzed in a similar manner. For example, when the magnetization is perpendicular to the film and the light is incident normally, the effect can be conveniently described by viewing linearly polarized light as a sum of left- and right-handed circularly polarized components as was done for the polar Kerr effect. In magnetized medium, the refractive indexes for the two circular components are slightly different. This means that the two components travel at different speeds inside the medium, and hence the times to travel a distance L through the material are $t_\pm = L/v_\pm = Ln_\pm/c$. After traveling a distance L, the left circular component will rotate an angle of $\theta_+ = \omega t_+ = \omega L n_+/c$, and the right circular component will rotate by $\theta_- = \omega t_- = \omega L n_-/c$, but in the opposite sense. Ignoring the difference in attenuation between the two components, the rotation θ_F of the composite vector of left and right circular components is related to θ_+ and θ_- by

$$\theta_+ - \theta_F = \theta_- + \theta_F \qquad (2.2.21)$$

The Faraday rotation of linearly polarized light per unit distance traveled inside a magnetized medium is given by

$$\theta_F/L = \frac{1}{2}(\theta_+ - \theta_-)/L$$
$$= \frac{\omega}{2c}(n_+ - n_-) \qquad (2.2.22)$$

2.3. THE KUBO FORMULA

The Kubo formula provides an expression for the optical conductivity tensor elements in the linear response regime in terms of the microscopic current operator and the many-body wave functions of the system in the absence of any applied fields. This formalism can be approximated with many-body wave functions constructed out of single-particle wave functions, such as those found in a LSDA electronic structure calculation. This allows not only an often quite accurate computational evaluation of magneto-optic properties, but also the framework of the underlying band structure in which to interpret the results. Wang and Callaway[49,50] undertook one of the first studies of optical conductivity based upon this approach. Since then several groups have studied magneto-optics from this point of view.[18,51–61]

Linear response means that the signal is directly proportional to the external perturbation. This assumption is usually valid when the magnitude of the perturbing field is small. The final expression for the optical conductivity will be expressed in terms of single-particle Bloch states and energy eigenvalues. The physical picture will involve electronic transitions between occupied and unoccupied band states separated by the photon energy. Since we will only be interested in the processes involving photons of energies between 0 and 12 eV, and the wavelength of a 12 eV photon is about 1000 Å which is large compared to a typical lattice constant, hence, the spatial variation of the electric field over a unit cell can be neglected. The wave vector of the light is effectively zero when compared to the momentum of the electron.

The derivation and notation of the Kubo formula follows most closely the work of Mahan.[62] Although the derivation is standard, and is given in most books on many-body theory. These references, however usually stop with a formal expression in terms of the current–current correlation function evaluated for many-body wave functions. An approximate expression based upon single-particle Kohn–Sham wave functions is derived and it is this wave function, i.e., used in numerical calculations of the Kerr effect.

The conductivity which we wish to evaluate is the current response of the system to the local electric field. This local electric field may differ from the applied external electric field since the induced current will also set up fields. The local electric field $(E_\beta(\mathbf{r}',t'))$ is simply the sum of the external and internal fields:

$$J_\alpha(\mathbf{r},t) = \sum_\beta \int d\mathbf{r}' \int_{-\infty}^{t} dt' \sigma_{\alpha\beta}(\mathbf{r}-\mathbf{r}', t-t') E_\beta(\mathbf{r}',t') \qquad (2.3.1)$$

SECONDARY MAGNETIC PROPERTIES

In a periodic solid, the conductivity will not be homogeneous in space. When the applied field is simply a plane wave with a single frequency and wave vector, the response will have in addition to the wave vector $\mathbf{q}$ of the applied field, wave vectors of the form $\mathbf{q}+\mathbf{G}$, where $\mathbf{G}$ is a reciprocal lattice vector. These Umklapp processes are neglected. Using the convolution theorem, we find

$$J_\alpha(\mathbf{q},\omega) = \sum_\beta \sigma_{\alpha\beta}(\mathbf{q},\omega) E_\beta(\mathbf{q},\omega) \tag{2.3.2}$$

where σ has been defined to be zero when $t-t'<0$.

For the particular choice of a plane-polarized time-dependent electric field, given by

$$E_\alpha(\mathbf{r},t) = E_0 \exp[i(\mathbf{q}\cdot\mathbf{r}-\omega t)] \tag{2.3.3}$$

where $\alpha = x, y,$ or z, one of the spatial directions, the induced current is given by

$$J_\alpha(\mathbf{r},t) = \sum_\beta \sigma_{\alpha\beta}(\mathbf{q},\omega) E_\beta(\mathbf{r},t) \tag{2.3.4}$$

This is the fundamental definition of the microscopic conductivity. It is only necessary to find the response to a single frequency, since different frequencies act independently. The total current is the summation of responses at different frequencies.

The Hamiltonian in the presence of a perturbing photon field can be expressed as

$$H = \sum_i \frac{1}{2m}\left[\mathbf{p}_i - \frac{e}{c}\mathbf{A}(\mathbf{r}_i)\right]^2 + V(\mathbf{r}) \tag{2.3.5}$$

This form of the Hamiltonian, known as minimal coupling, assumes the Coulomb gauge, in which $\nabla\cdot\mathbf{A} = 0$ is chosen ($\mathbf{A}$ is the magnetic vector potential, $\mathbf{B} = \nabla\times\mathbf{A}$). The Hamiltonian is separated as $H = H_0 + H'$, where H_0 is the Hamiltonian when the applied field $\mathbf{A}$ is zero. Neglecting the higher order term in A^2, leads to H' given by

$$H' = -\frac{e}{2mc}\sum_i [\mathbf{p}_i\cdot\mathbf{A}(\mathbf{r}_i) + \mathbf{A}(\mathbf{r}_i)\cdot\mathbf{p}_i] \tag{2.3.6}$$

With the choice of electric field given by Eq. (2.3.3), the magnetic vector potential $\mathbf{A}$ has the form

$$A_\alpha = A_0 \exp[i(\mathbf{q}\cdot\mathbf{r}-\omega t)] \tag{2.3.7}$$

which leads to

$$H' = -\frac{e}{2mc} A_0 \sum_i [p_{\alpha i}\exp(i\mathbf{q}\cdot\mathbf{r}_i) + \exp(i\mathbf{q}\cdot\mathbf{r}_i)p_{\alpha i}]\exp(-i\omega t) \tag{2.3.8}$$

The perturbing Hamiltonian can be re-expressed in terms of the current operator, since the current operator can be defined by

$$j_\alpha(\mathbf{r}) = \frac{1}{2m}\sum_i e_i[p_i\delta(\mathbf{r}-\mathbf{r}_i) + \delta(\mathbf{r}-\mathbf{r}_i)p_i]_\alpha \tag{2.3.9}$$

Note that this form is equivalent to the usual quantum mechanical expression for the particle flux. In the second quantized form, the current operator is given by

$$\mathbf{j}(\mathbf{r}) = \frac{e}{2m} \int \psi^\dagger(\mathbf{r})[\mathbf{p}(\mathbf{r}')\delta(\mathbf{r}-\mathbf{r}') + \delta(\mathbf{r}-\mathbf{r}')\mathbf{p}(\mathbf{r}')]\psi(\mathbf{r})d\mathbf{r}' \qquad (2.3.10)$$

where ψ and $\psi^\dagger$ are the usual field operators. Integrating by parts and appealing once again to the Coulomb gauge, implies

$$\mathbf{j}(\mathbf{r}) = \frac{e}{2m}[\psi^\dagger(\mathbf{r})\mathbf{p}\psi(\mathbf{r}) - \psi(\mathbf{r})\mathbf{p}\psi^\dagger(\mathbf{r})]$$

$$= \frac{e\hbar}{2mi}[\psi^\dagger(\mathbf{r})\nabla\psi(\mathbf{r}) - \psi(\mathbf{r})\nabla\psi^\dagger(\mathbf{r})] \qquad (2.3.11)$$

It is convenient to introduce the Fourier transform of the current operator since we will be concerned with the response to a single plane wave external field. This can be achieved by using Cauchy's integral theorem

$$\delta(\mathbf{r}-\mathbf{r}_i) = \frac{1}{(2\pi)^3}\int d\mathbf{q}\,\exp[-i\mathbf{q}\cdot(\mathbf{r}-\mathbf{r}_i)] \qquad (2.3.12)$$

Substituting into Eq. (2.3.9) results in

$$\mathbf{j}(\mathbf{r}) = \frac{1}{(2\pi)^3}\int d\mathbf{q}\,\exp(-i\mathbf{q})\cdot\mathbf{r}\left\{\frac{e}{2m}\sum_i[\mathbf{p}_i\exp(i\mathbf{q}\cdot\mathbf{r}_i) + \exp(i\mathbf{q}\cdot\mathbf{r}_i)\mathbf{p}_i]\right\} \qquad (2.3.13)$$

where we can identify the term in square parentheses as the Fourier transform of the current operator ($\mathbf{j}(\mathbf{q})$). Comparing this expression with the form of the perturbation given in Eq. (2.3.8), it follows that

$$H' = -\frac{1}{c}j_\alpha(\mathbf{q})A_0\exp(-i\omega t) \qquad (2.3.14)$$

At this stage, it is more useful to express the magnetic vector potential in terms of the electric field. For plane wave fields, this can be easily done by using the Maxwell equation

$$\nabla\times\mathbf{E} = -\frac{1}{c}\frac{\partial\mathbf{B}}{\partial t} \qquad (2.3.15)$$

and the definition of $\mathbf{A}$

$$\mathbf{B} = \nabla\times\mathbf{A} \qquad (2.3.16)$$

Using these equations, the perturbation can be written as

$$H' = \frac{I}{\omega}j_\alpha(\mathbf{q})E_0\exp(-i\omega t) \qquad (2.3.17)$$

The actual induced current measured in an experiment is the average value of the velocity of the particles in the system multiplied with the charge. To distinguish it from the quantum mechanical current operator, we will label it as $\mathbf{J}$ (as compared to $\mathbf{j}$). The measured current is therefore given by

SECONDARY MAGNETIC PROPERTIES

$$J_\alpha(\mathbf{r}, t) = \frac{e}{\Omega}\left\langle \sum_i v_{i\alpha} \delta(\mathbf{r} - \mathbf{r}_i)\right\rangle \tag{2.3.18}$$

where the average is taken over the sample whose volume is Ω. In the presence of the external field, quantization of the velocity implies that the velocity is given by the momentum minus the vector potential:

$$\mathbf{v}_i = \frac{1}{m}\left[\mathbf{p}_i - \frac{e}{c}\mathbf{A}(\mathbf{r}_i)\right] \tag{2.3.19}$$

thus

$$J_\alpha(\mathbf{r}, t) = \frac{e}{m\Omega}\sum_i \langle p_{i\alpha}\rangle - \frac{e^2}{mc\Omega}\sum_i A_\alpha(\mathbf{r}_i, t)$$

$$= \langle j_\alpha(\mathbf{r}, t)\rangle + i\frac{ne^2}{m\omega} E_\alpha(\mathbf{r}, t) \tag{2.3.20}$$

where $\mathbf{j} = e\mathbf{p}/m$. The electric field is assumed to be constant over the volume Ω; hence, the summation over particles divided by this volume is just the density n. In order to evaluate the first term, which will result in interband contributions to the conductivity, it is easiest to work in the interaction representation. In this picture, the operators and wave functions are time dependent and are labeled with a caret:

$$\langle j_\alpha(\mathbf{r}, t)\rangle = \langle \hat{\psi}(t) | \hat{j}_\alpha(\mathbf{r}, t) | \hat{\psi}(t)\rangle \tag{2.3.21}$$

where

$$\hat{j}_\alpha(\mathbf{r}, t) = \exp(iH_0 t/\hbar) j_\alpha(\mathbf{r}) \exp(-iH_0 t/\hbar) \tag{2.3.22}$$

and

$$\hat{\psi}(t) = \exp(iH_0 t/\hbar)\exp(-iHt/\hbar)\psi \tag{2.3.23}$$

In the above equation, $H = H_0 + H'$. The operator in front of the wave function ψ is the time evolution operator $U(t)$ which can be shown to satisfy

$$\frac{\partial U}{\partial t} = -\frac{i}{\hbar}\hat{H}'(t) U(t) \tag{2.3.24}$$

Thus, keeping the lowest order in H', we find

$$U(t) \approx 1 - \frac{i}{\hbar}\int_{-\infty}^{t} dt' \hat{H}'(t') \tag{2.3.25}$$

The lower limit of the integration is $-\infty$ since the perturbation is assumed to be switched on adiabatically and hence is zero as $t \to -\infty$. Thus, $\langle j_\alpha(\mathbf{r}, t)\rangle$ is given by

$$\langle j_\alpha(\mathbf{r}, t)\rangle = \langle \psi | \left[1 + \frac{i}{\hbar}\int_{-\infty}^{t} dt' \hat{H}'(t')\right] \hat{j}_\alpha(\mathbf{r}, t) \times \left[1 - \frac{i}{\hbar}\int_{-\infty}^{t} dt' \hat{H}'(t')\right] |\psi\rangle$$

$$= \langle \psi | \hat{j}_\alpha(\mathbf{r}, t)|\psi\rangle - \frac{i}{\hbar}\langle \psi |\left\{\int_{-\infty}^{t} dt'\left[\hat{j}_\alpha(\mathbf{r}, t), \hat{H}'(t')\right]\right\}|\psi\rangle \tag{2.3.26}$$

The first term vanishes since there is no current in the absence of an applied field. Substituting for $\hat{H}'$ given by Eq. (2.3.17), leads to the following expression for $\langle j_\alpha(\mathbf{r},t)\rangle$:

$$\langle j_\alpha(\mathbf{r},t)\rangle = \frac{1}{\hbar\omega}\int_{-\infty}^{t} dt'\langle\psi|[\hat{j}_\alpha(\mathbf{r},t),\hat{j}_\beta(\mathbf{r},t')]|\psi\rangle\exp(-i\omega t')E_\beta^0$$

$$= \frac{1}{\hbar\omega}\exp(-i\mathbf{q}\cdot\mathbf{r})\int_{-\infty}^{t} dt'\exp(i\omega(t-t'))\langle\psi|[\hat{j}_\alpha(\mathbf{r},t),\hat{j}_\beta(\mathbf{r},t')]|\psi\rangle E_\beta(\mathbf{r},t)$$

(2.3.27)

where the factor of $\exp[(-i\mathbf{q})\cdot\mathbf{r}]$ comes from the definition of the electric field $E_\beta(\mathbf{r},t)$ (see Eq. (2.3.3)). Substituting Eq. (2.3.27) into the expression for the current given by Eq. (2.3.20), and comparing to our definition of $\sigma_{\alpha\beta}$, we arrive at the following for the elements of the conductivity tensor

$$\sigma_{\alpha\beta} = i\frac{ne^2}{m\omega}\delta_{\alpha\beta} + \frac{1}{\hbar\omega}\exp(-i\mathbf{q}\cdot\mathbf{r})\int_{-\infty}^{t}dt'\exp[(i\omega(t-t'))]$$
$$\times\langle\psi|[\hat{j}_\alpha(\mathbf{r},t),\hat{j}_\beta(\mathbf{q},t')]|\psi\rangle \quad (2.3.28)$$

A final average over the volume Ω is taken to average out atomic fluctuations, leading to

$$\sigma_{\alpha\beta} = i\frac{ne^2}{m\omega}\delta_{\alpha\beta} + \frac{1}{\hbar\omega\Omega}\int_{-\infty}^{t}dt'\exp(i\omega(t-t'))\langle\psi|[\hat{j}_\alpha(-\mathbf{q},t),\hat{j}_\beta(\mathbf{q},t')]|\psi\rangle \quad (2.3.29)$$

Finally, we observe that the conductivity is only a function of $t-t'$. Choosing this as the integration variable leads to the Kubo formula for $\sigma_{\alpha\beta}$:

$$\sigma_{\alpha\beta} = i\frac{ne^2}{m\omega}\delta_{\alpha\beta} + \frac{1}{\hbar\omega\Omega}\int_{0}^{\infty}dt'\exp(i\omega t)\langle\psi|[\hat{j}_\alpha(-\mathbf{q},t),\hat{j}_\beta(\mathbf{q},0)]|\psi\rangle \quad (2.3.30)$$

As mentioned earlier, we are only interested in the $\mathbf{q}\to 0$ limit of the Kubo formula since the electric field is slowly varying over the unit cell. Thus, we need to evaluate

$$\sigma_{\alpha\beta} = i\frac{ne^2}{m\omega}\delta_{\alpha\beta} + \frac{1}{\hbar\omega\Omega}\int_{0}^{\infty}dt'\exp(i\omega t)\langle\psi|[\hat{j}_\alpha(\mathbf{0},t),\hat{j}_\beta(\mathbf{0},0)]|\psi\rangle \quad (2.3.31)$$

In order to have a practical implementation of the Kubo formula, we must approximate the ground state many-body wave function ψ in terms of single-particle states computed by an LSDA electronic structure calculation. In order to do this, it is convenient to use a second quantized form. First, the commutator is expanded, and then a complete set of many-body states is inserted between the two current operators. The states are labeled by $|n\rangle$ where the ground state ψ will be denoted by $|0\rangle$. In terms of the occupancies of single-particle states, we can write

$$|n\rangle = |n_1,n_2,\ldots,n_n\rangle \quad (2.3.32)$$

where n_i is the occupancy of the ith single-particle state. Also we have $\sum_i n_i = N$, where N is the total number of electrons. The occupancies are either 1 or 0 according to the Pauli principle, and the states are ordered in increasing value of the single-particle energies E_i. The wave function $|n\rangle$ is a single Slater determinant formed from the N occupied orbitals. In this notation,

$$\langle\psi|[\hat{j}_\alpha(t'),\hat{j}_\beta(0)]|\psi\rangle = \sum_n \langle 0|\hat{j}_\alpha(t')|n\rangle\langle n|\hat{j}_\beta(0)|0\rangle - \langle 0|\hat{j}_\beta(0)|n\rangle\langle n|\hat{j}_\alpha(t')|0\rangle \quad (2.3.33)$$

converting the current operator from the interaction to the Schrödinger representation leads to

$$\langle\psi|[\hat{j}_\alpha(t'),j_\beta(0)]|\psi\rangle =$$
$$\sum_n \langle 0|\exp(iH_0 t')j_\alpha\exp(-iH_0 t')|n\rangle\langle n|j_\beta(0)|0\rangle$$
$$- \langle 0|j_\beta(0)|n\rangle\langle n|\exp(iH_0 t')j_\alpha\exp(iH_0 t')|0\rangle \quad (2.3.34)$$

H_0 is a sum of single-particle Hamiltonians which are identical except for the coordinates

$$H_0 = \sum_i H_0^{sp}(\mathbf{r}_i) \quad (2.3.35)$$

Since the single-particle orbitals forming the state $|n\rangle$ are eigenfunctions of the single-particle Hamiltonian H_0^{sp}, we have

$$\exp(-iH_0 t'/\hbar)|n\rangle = \exp(-i\sum_j n_j E_j t'/\hbar)|n\rangle$$
$$\langle n|\exp(iH_0 t'/\hbar) = \langle n|\exp(i\sum_j n_i E_j t'/\hbar) \quad (2.3.36)$$

If $|n\rangle = |0\rangle$ then clearly the two terms cancel. If $|n\rangle \neq |0\rangle$, then we can write $|n\rangle$ as

$$|n\rangle = |1,1,\ldots,0,\ldots,1,0,0,\ldots,0,1,0,\ldots,0\rangle \quad (2.3.37)$$

where an orbital with energy E_i ($E_i < E_f$) is unoccupied, and an orbital with energy E_j ($E_j > E_f$) is occupied. For this choice of $|n\rangle$, we have

$$\langle 0|\hat{j}_\alpha(t')|n\rangle = \exp[i(E_i - E_j)t'/\hbar]\langle 0|j_\alpha|n\rangle \quad (2.3.38)$$

The matrix element on the right-hand side of this equation can be evaluated by using second quantized notation, where

$$j_\alpha = \sum_{l,m} j_{\alpha,lm}^{sp} c_l^\dagger c_m \quad (2.3.39)$$

The matrix elements $j_{\alpha,lm}^{sp}$ involve the single-particle states ϕ_i and are given by

$$j_{\alpha,lm}^{sp} = \int d\mathbf{r}\,\phi_l^*(\mathbf{r}) j_\alpha^{sp}(\mathbf{r}) \phi_m(\mathbf{r}) \quad (2.3.40)$$

From now on, all operators are single-particle operators, and the superscript sp will be dropped. The subscript α on the matrix elements of the single-particle current operator will, for the ease of reading, be placed as a superscript, hence we have

$$j_{lm}^\alpha = \int d\mathbf{r}\,\phi_l^*(\mathbf{r}) j_\alpha^{sp}(\mathbf{r}) \phi_m(\mathbf{r}) \quad (2.3.41)$$

The operators $c_l^\dagger$ and c_m are the usual creation and annihilation operators for single-particle states l and m, respectively. In order for the matrix element in Eq. (2.3.39) to be non-zero, $c_l^\dagger c_m$ must convert $|n\rangle$ to $|0\rangle$. Thus, state j must be destroyed and state i created. Hence, only one term from the double summation is non-zero that where $l = i$ and $m = j$. Then, $\langle 0|c_l^\dagger c_m|n\rangle = \delta_{il}\delta_{mj}$, and

$$\langle 0|\hat{j}_\alpha(t')|n\rangle = \exp[i(E_i - E_j)t'/\hbar]j_{ij}^\alpha \tag{2.3.42}$$

Similar arguments apply to the second term in Eq. (2.3.34), and the final result for the commutator is

$$\sum_i{}' \sum_j{}'' \left\{ \exp[i(E_i - E_j)t'/\hbar]j_{ij}^\alpha j_{ji}^\beta - \exp[i(E_j - E_i)t'/\hbar]j_{ij}^\beta j_{ji}^\alpha \right\} \tag{2.3.43}$$

The summation over the complete set of states n becomes a double sum over all occupied single-particle states i and all unoccupied single-particle states j. The sum over occupied states is labeled with a prime and the sum over unoccupied states by a double prime. The conductivity is given by

$$\sigma_{\alpha\beta} = i\frac{ne^2}{m\omega}\delta_{\alpha\beta} + \frac{1}{\hbar\omega\Omega}\int_0^\infty dt' \exp(i\omega t)\sum_i{}'\sum_j{}''\left\{\exp[i(E_i - E_j)t'/\hbar]j_{ij}^\alpha j_{ji}^\beta\right.$$
$$\left. - \exp[i(E_j - E_i)t'/\hbar]j_{ij}^\beta j_{ji}^\alpha \right\} \tag{2.3.44}$$

In order to converge the time integration, ω is given a small imaginary part δ:

$$\sigma_{\alpha\beta} = i\frac{ne^2}{m\omega}\delta_{\alpha\beta} + \frac{i}{\omega\Omega}\sum_i{}'\sum_j{}''\left[\frac{j_{ij}^\alpha j_{ji}^\beta}{E_i - E_j + \hbar\omega + i\hbar\delta} - \frac{j_{ij}^\beta j_{ji}^\alpha}{E_j - E_i + \hbar\omega + i\hbar\delta}\right] \tag{2.3.45}$$

The $\omega \to 0$ part of interband conductivity is combined with the intraband term leading to a Drude expression which includes the effective mass tensor m^*:

$$\sigma_{\alpha\beta} = i\frac{n_0 e^2}{(\omega + i/\tau_D)}[m^*]^{-1}$$
$$+ \frac{i}{\Omega}\sum_i{}'\sum_j{}''\frac{\hbar}{E_j - E_i}\left[\frac{j_{ij}^\alpha j_{ji}^\beta}{\hbar\omega - E_j + E_i + i\hbar\delta} - \frac{j_{ij}^\beta j_{ji}^\alpha}{E_j - E_i + \hbar\omega + i\hbar\delta}\right] \tag{2.3.46}$$

It is convenient to use the momentum rather than the current operator. Π is used instead of p since in principle there are relativistic spin-flip terms that are included in Π. In practice, these terms are negligible. In which case, we arrive at

$$\sigma_{\alpha\beta} = i\frac{n_0 e^2}{(\omega + i/\tau_D)}[m^*]^{-1}$$
$$+ \frac{ie^2}{m^2\hbar\Omega}\sum_k\sum_{l\sigma}{}'\sum_{l'\sigma'}{}''\frac{1}{\omega_{l'\sigma',l\sigma}}\left[\frac{\Pi^\alpha_{l\sigma l'\sigma'}\Pi^\beta_{l'\sigma' l\sigma}}{\omega - \omega_{l'\sigma',l\sigma} + i\delta} - \frac{\Pi^\beta_{l\sigma l'\sigma'}\Pi^\alpha_{l'\sigma' l\sigma}}{\omega + \omega_{l'\sigma',l\sigma} + i\delta}\right] \tag{2.3.47}$$

where the notation has been converted into one better suited for a periodic solid. In this case, the summation over states is divided into one over bands at a particular k-point and a summation over all k-points in the first Brillouin zone. The band states

SECONDARY MAGNETIC PROPERTIES

are labeled by a band index (l, l') and spin (σ, σ'). The spatial integral in the matrix elements is reduced to an integral over the unit cell volume Ω, which is the volume that appears in Eq. (2.3.47). $\hbar\omega_{l'\sigma',l\sigma}(\mathbf{k})$ is the energy difference, $E_{l'\sigma'}(\mathbf{k}) - E_{l\sigma}(\mathbf{k})$, between two eigenstates labeled by band index l and spin σ and band index l' and spin σ'. The first term is the intraband Drude term, and the second term represents the interband optical transition. The matrix elements of the momentum operator $\Pi^{\alpha}_{l\sigma,l'\sigma'}(\mathbf{k})$ are given by

$$\Pi^{\alpha}_{l\sigma,l'\sigma'}(\mathbf{k}) = \int_{\Omega} d\mathbf{r} \psi^{*}_{l\sigma\mathbf{k}}(\mathbf{r}) \left[-i\hbar \frac{\partial}{\partial x_{\alpha}} + \frac{\hbar}{4mc^2} (\boldsymbol{\sigma} \times \boldsymbol{\nabla} V(\mathbf{r}))_{\alpha} \right] \psi_{l'\sigma'\mathbf{k}}(\mathbf{r}) \qquad (2.3.48)$$

The spin-flip term $\hbar/4mc^2(\boldsymbol{\sigma} \times \boldsymbol{\nabla} V(\mathbf{r}))$, given in terms of the Pauli matrices σ and one-electron potential $V(\mathbf{r})$, is usually small and neglected.

In the polar Kerr geometry, the incident light is perpendicular to the surface, we saw that the Kerr angle arose because of the different refractive indexes for left and right circularly polarized light. This can be seen from the Kubo formula by defining $\Pi^{\pm} = \frac{1}{2^{1/2}}(\mp\Pi_x + i\Pi_y)$. The diagonal and off-diagonal elements of the conductivity tensor can be expressed as

$$\sigma_{xx}(\omega) = \frac{e^2}{m^2\hbar} \int d^3k \sum_{l\sigma}' \sum_{l'\sigma'}'' \frac{|\Pi^{+}_{l\sigma,l'\sigma'}|^2 + |\Pi^{-}_{l\sigma,l'\sigma'}|^2}{2\omega_{l'\sigma',l\sigma}}$$

$$\times \left[\frac{1}{(\omega - \omega_{l'\sigma',l\sigma}(\mathbf{k}) + i\delta)} + \frac{1}{(\omega + \omega_{l'\sigma',l\sigma}(\mathbf{k}) + i\delta)} \right] \qquad (2.3.49)$$

$$\sigma_{xy}(\omega) = \frac{e^2}{m^2\hbar} \int d^3k \sum_{l\sigma}' \sum_{l'\sigma'}'' \frac{|\Pi^{+}_{l\sigma,l'\sigma'}|^2 - |\Pi^{-}_{l\sigma,l'\sigma'}|^2}{2\omega_{l'\sigma',l\sigma}}$$

$$\times \left[\frac{1}{(\omega - \omega_{l'\sigma',l\sigma}(\mathbf{k}) + i\delta)} + \frac{1}{(\omega + \omega_{l'\sigma',l\sigma}(\mathbf{k}) + i\delta)} \right] \qquad (2.3.50)$$

This clearly shows the relation to circularly polarized light, since σ_{xy} is seen to be related to the difference and σ_{xx} the average.

The magneto-optic properties are mainly determined by the off-diagonal elements of the conductivity tensor, which are attributed to the combined effects of spin polarization and spin–orbit interaction. Without spin–orbit coupling, the off-diagonal elements vanish. This can be deduced from the time-reversal symmetry of the wave functions ($|l_\mathbf{k}\rangle = |l^*_{-\mathbf{k}}\rangle$). In the case where spin–orbit coupling is present but the magnetization is zero, time reversal symmetry converts the spin-up part of the wave function to spin down and vice versa, and again σ_{xy} vanishes because of the degeneracy between spin up and spin down.

2.4. AB INITIO CALCULATIONS OF THE KERR EFFECT

Over the last few years, there have been a number of first-principle calculations of the Kerr effect. All of these calculations use the wave functions and energy eigenvalues that are the output of a self-consistent LSDA electronic structure calculation and use them to evaluate the conductivity tensor via the Kubo formula. The first studies focused on the elemental ferromagnets Fe, Co, and Ni.[52,56,58,59] In the case of Fe and Co rather close agreement between theory and experiment was obtained. Calculations for Ni, however, showed that the calculated spectra was shifted by about 1 eV when compared with experiment. Shifts between theory and experiment are to be expected since it is well known that the LSDA places excited states at the wrong energy. The best known example of this is the underestimation of the band gap of insulators and semiconductors. The shape of the Kerr spectra is determined by the character of the wave function, and by comparing LSDA calculations to more detailed electronic calculations which correct the band gap error, it is often found that the states in the latter are quite similar to those of the former but merely shifted up in energy. The conclusion of this is that LSDA calculations can be expected to provide a good predictor of the size of the Kerr effect, though the spectra may be shifted in energy when compared to experiment.

There have also been calculations for a number of technologically important materials including FePt and CoPt superlatttices, MnBi and MnSb.[63,64] Al-doped MnBi was studied by Sabiryanov *et al.*[64] after reports of a large Kerr rotation in the alloy.[65,66] Since the purpose of this chapter is to explain the fundamentals of magneto-optic effects, I will simply show two examples of first-principle calculations: Fe and PtMnSb. These illustrate nicely the connection between the underlying electronic structure and the optical properties.

2.4.1. Magneto-Optic Properties of Fe

The optical properties of bcc Fe were calculated by using single-particle wave functions and energies obtained from a self-consistent LSDA electronic structure calculation. The electronic states were obtained from a full potential linearized augmented Slater orbital approach (FLASTO),[58,67] which provides an accurate description of the ground state electronic structure. It is worth pointing out that several different band structure methods which use different types of approximation to the wave functions and electronic potential give rather similar optical properties.[52] In these FLASTO calculations, the exchange-correlation energy and potential was obtained within the LSDA using the Perdew–Zunger parameterization.[68] Spin–orbit coupling term was not included in the self-consistent calculation, rather once the self-consistent electronic potential is obtained, the spin–orbitless wave functions are used as basis functions to construct the Hamiltonian matrix with spin–orbit coupling included. The wave functions and eigenvalues in the presence of the spin–orbit interaction are obtained by diagonalizing this Hamiltonian, and are then used in the Kubo formula to calculate the conductivity tensor elements corresponding to the interband transition. Since the spin–orbit perturbation is small, the resulting solutions are sufficiently accurate for conductivity calculations. The spin–orbit interaction lowers the symmetry of a cubic crystal. With the magnetization **M** aligned along (001), the **k** mesh can only be reduced to 1/16th of the Brillouin zone (as opposed to 1/48th in the absence of spin–orbit coupling).

Lifetime effects are included in conductivity calculation by introducing a non-zero imaginary energy $i\delta$ in the Kubo formula. The relaxation time is estimated from the degree of broadening of the experimental data. We have used $\delta = 0.015$ Hartree (about 0.5 eV) as inverse relaxation time in these calculations. The final results are not too sensitive to the choice of δ. The inclusion of lifetime effects broadens and smooths sharp peaks in the conductivity spectra. The benefits of including a finite broadening in the interband conductivity are the comparison to experiment is facilitated since experimental broadening is always present; the integration over the Brillouin zone is made easier since poles are moved off the real axis; and both real and imaginary components can be extracted at the same time, avoiding the need to perform numerically awkward Kramers Kronig integrals. In addition to interband transitions, intraband transitions, or Drude terms, are also included. The Drude term is given by

$$\sigma_D(\omega) = \frac{\sigma_0}{1 - i\omega\tau_D} \quad (2.4.1)$$

The two Drude parameters σ_0 and τ_D are taken from experiment.[69,70] In the case of Fe they are, in c.g.s. units, 6.4×10^{15} s^{-1}, and 9.12×10^{-15} s, respectively. The real part of σ_{xx} (which includes the parameterized Drude term), and the ω times the imaginary part of σ_{xy} describe absorption are shown in Fig. 2.3. As can be seen, the calculated spectra (solid lines) compare favorably with the results of an independent set of LSDA electronic structure calculations performed by Oppeneer et al.[52] (dashed lines). Experimental values for the real part of σ_{xx} have been reported by several authors.[71-74] Different experiments show somewhat different conductivities, though all have a large peak close to 2.5 eV. Yolken and Kruger's experimental data[74] are in close agreement with the calculated conductivity. Similar arguments apply to the imaginary component of σ_{xy}.

The one large peak in the absorptive part of the conductivity around 2.5 eV rises primarily from transitions between minority spin electrons. The matrix elements provide the usual dipole transition selection rules, namely $\Delta\ell = \pm 1$, and depending on whether left- or right-handed circularly polarized light is being absorbed

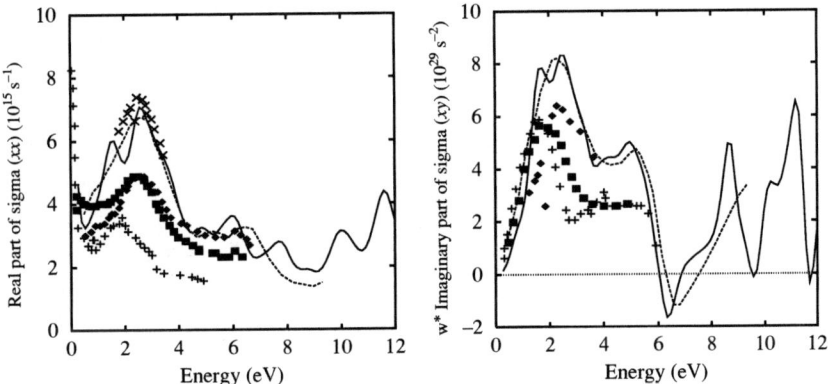

Figure 2.3. $\Re\sigma_{xx}$ and $\omega \times \Im\sigma_{xy}$ for Fe as a function of the photon energy $\hbar\omega$. The theoretical results obtained from the FLASTO calculations are indicated with solid lines, the dotted lines are the results of an independent calculation by Oppeneer et al.[52] Experimental data for $\Re\sigma_{xx}$ shown are as follows: (+) Bolotin et al.[72] (□) Weaver et al.[73] (◊) Johnson and Christy,[71] and (×) Yolken and Kruger.[74] Experimental data for $\omega \times \Im\sigma_{xy}$ are as follows: (+) Krinchik and Artem'ev,[76] (□) van Engen et al.,[132] and (◊) Fergusson and Romagnoli.[133]

$\Delta m = \pm 1$. Neglecting the energy and state dependence of the matrix elements, the peak in the conductivity can be understood simply from an interpretation of the density of states.

As can be seen from Fig. 2.4, the density of states for minority electrons has a large peak above the Fermi energy. The strength of transitions at a particular energy will depend on the joint density of states which measures the product of unoccupied and occupied density of states separated by a particular energy. For energies up to about 4 eV this is large for the minority electrons and small for the majority electrons, since the unoccupied density of states for majority electrons within a few electron volts of the Fermi energy is small. The transition strengths are controlled by the dipole selection rule. An examination of the angular momentum resolved partial density of states shows that while the peaks in the density of states are predominantly from the Fe d states, there is a significant amount of s and p character. The minority transitions are mainly from occupied p to unoccupied d, since there is more occupied than unoccupied p density of states. Three or four electron volts above the Fermi energy, the density of states has a free electron character and transitions are mainly between occupied d to unoccupied p and both spin channels are expected to contribute to the conductivity.

An approximate spin decomposition of the conductivity can be obtained by for each eigenstate in the following manner. The eigenfunction is composed of a sum of contributions of the spin up and spin down eigenfunctions that were obtained from a calculation performed without spin–orbit coupling. Since the spin–orbit coupling is small, the state will be predominantly spin up or spin down. If the eigenstate is predominantly spin up, then the spin-down coefficients are set to zero, and vice versa for a state that is mostly spin down (in which case the spin up coefficients are set to zero). This spin decomposition is shown in Fig. 2.5.

The imaginary part of σ_{xx} and the real part of σ_{xy} are dispersive and are related by Kramers Kronig transformations to the real part of σ_{xx} and the imaginary part

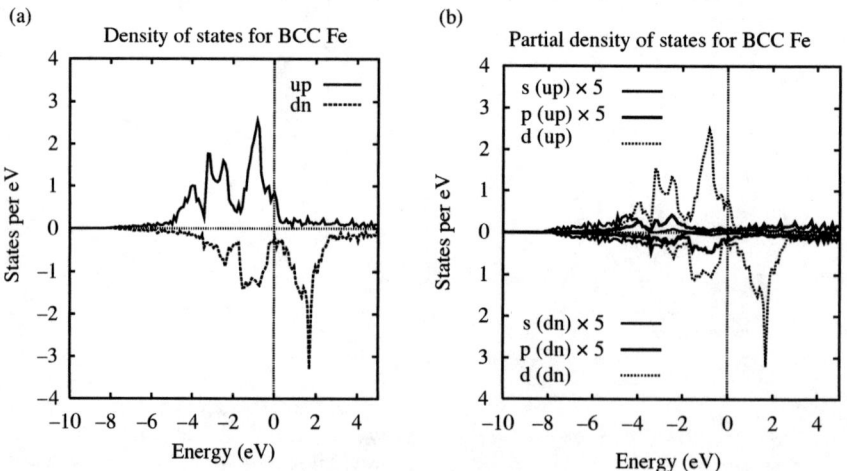

Figure 2.4. Total density of states (a) and partial density of states (b) for Fe for both majority and minority spins.

SECONDARY MAGNETIC PROPERTIES

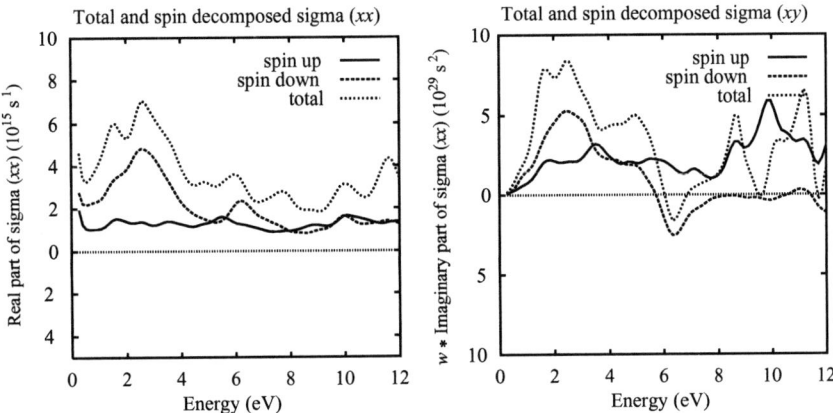

Figure 2.5. Total and spin decomposed contributions to $\Re\sigma_{xx}$ and $\omega \times \Im\sigma_{xy}$ for Fe.

of σ_{xy}. The calculated and measured values of these parts of σ are shown for completeness in Fig. 2.6.

The polar Kerr rotation spectra is calculated from σ_{xx} and σ_{xy} using Eq. (2.2.16) and the results are shown in Fig. 2.7. Oppeneer's calculated results[52] and two different sets of experimental data collected by van Engen[75] and Krinchik and Artem'ev[76] are also presented in the same figure for comparison. All these results presented are in good agreement with each other in the energy range from 0.5 to 5.5 eV, demonstrating the accuracy of *ab initio* electronic structure calculation in this

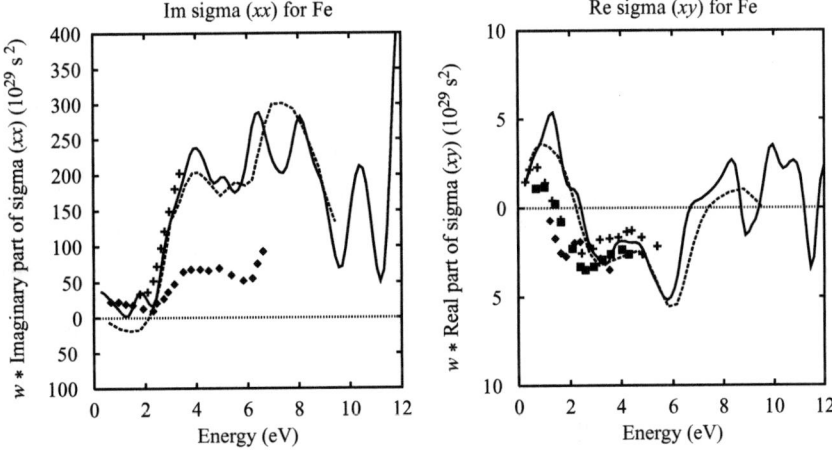

Figure 2.6. $\omega \times \Im\sigma_{xx}$ and $\omega \times \Re\sigma_{xy}$ for Fe. The theoretical results obtained from the FLASTO calculations are indicated with solid lines, the dotted lines are the results of an independent calculation by Oppeneer et.al.[52] Experimental data for $\omega \times \Im\sigma_{xx}$ shown are as follows (+) Yolken and Kruger,[74] (◊) Johnson and Christy.[71] Experimental data for $\omega \times \Re\sigma_{xy}$ are as follows (+) Krinchik and Artem'ev,[76] (□) van Engen,[132] and (◊) Ferguson and Romagnoli.[133]

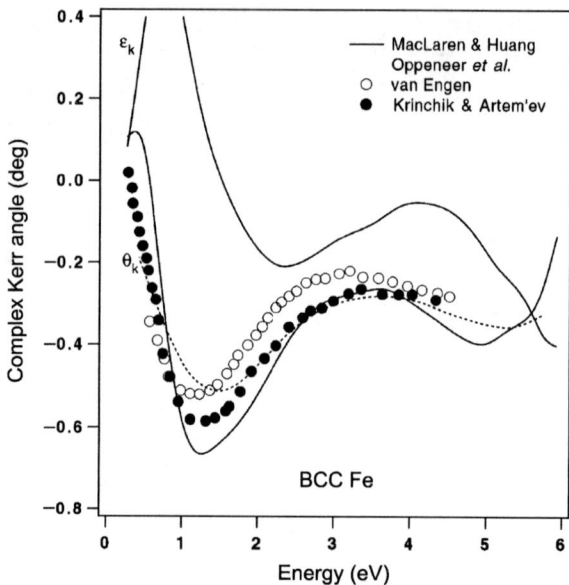

Figure 2.7. The Kerr rotation and ellipticity for bcc Fe.

energy range. The ellipticity calculated by the FLASTO methods are also shown in this figure.

2.4.2. Magneto-Optic Properties of PtMnSb and NiMnSb

PtMnSb and NiMnSb are Heusler alloys that crystallize in the $C1_b$ structure. Both have been predicted on the basis of first-principle calculations to be half-metallic ferromagnets. That is a compound in which one spin channel is semi-conducting and the another is metallic. This leads naturally to an integer spin moment. In the case of PtMnSb, a large polar Kerr effect for red light has also been observed, leading to models that suggested that these two effects were connected. In this physically appealing model in which the effect of $\mathbf{L} \cdot \mathbf{S}$ coupling on valence states in the semiconducting spin channel, around the Γ-point in the Brillouin zone, creates an asymmetry in the optical transitions from those states.[77,78] It is the asymmetry between the absorption of left and right circularly polarized light that produces a Kerr rotation. This mechanism was suggested before detailed electronic structure calculations, which are able to determine the relative strength of optical transitions, had been performed on this material. Similar arguments were recently repeated by Oppeneer et al.[57]

The polar Kerr effect for red light reflected from PtMnSb or NiMnSb crystals is shown by a detailed examination of the underlying electronic structure to be independent of the half-metallic ferromagnetic property of these compounds. The origins of the peak in the Kerr effect at low photon energies can be attributed to interband transitions between parallel sheets of energy bands that project onto the metallic spin channel (Fig. 2.9). Thus, the mechanism in which spin–orbit coupling creates an imbalance in optical transitions between states in the vicinity of the Γ point, in the semiconducting minority spin channel, is incorrect.

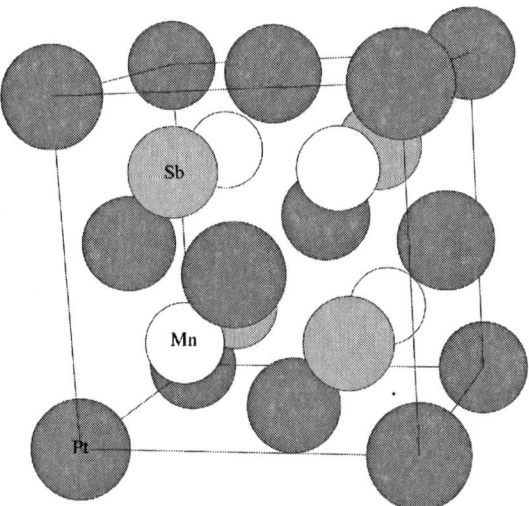

Figure 2.8. Cl$_b$ crystal structure for PtMnSb. The unit cell can be viewed as three interpenetrating fcc lattices.

The relation between the electronic structure of the compounds and the Kerr spectrum is analyzed by decomposing the interband contribution to the Kerr angle for low energy photons (i) according to (projected) spin channel, (ii) along high symmetry directions in reciprocal space, (iii) by selectively switching off the **L · S** interaction, and (iv) by applying hydrostatic pressure to the system.

The optical properties were computed from the underlying electronic structure which was found using two independent theoretical approaches, the FLASTO method (as described previously) and the computationally quicker LMTO approach.[14,15] The LMTO method is one of the most commonly used methods for calculating electronic structure. Its appeal lies in the computational speed and physically transparent basis set. In this case, differences between the results of LMTO calculations and the slower more accurate full potential FLASTO technique are negligible.

Estimates of the Drude parameters σ_0 and τ_D, used in Eq. (2.4.1), were taken from optical conductivity measurements given in Ref. 79. The calculated spin- and orbital magnetic moments agree well with previously published data.[54,80,81] In the absence of **L · S** coupling, both NiMnSb and PtMnSb are found to be half-metallic ferromagnets (LMTO as well as FLASTO). Theoretical LMTO equilibrium lattice constants for NiMnSb and PtMnSb are about 3% and 2% smaller, respectively, than the experimental values.[80] The calculated results for the Kerr spectra are at the experimental lattice constant, unless stated otherwise.

Figure 2.9(a) shows the band structure for the minority semiconducting channel (no **L · S** interaction in the band structure calculation). The doubly degenerate flat bands at 1.4 eV above the Fermi level, ϵ_F, are derived from Mn 3d states (88%), whereas the highest threefold degenerate set of occupied states at the Γ point are of mixed Pt 6p (15%) and Mn 3d (70%) character.

Figure 2.9(b), the band structure for PtMnSb in the presence of **L · S**-coupling, reveals that the manifold at Γ just below ϵ_F indeed does split. A very small fraction of

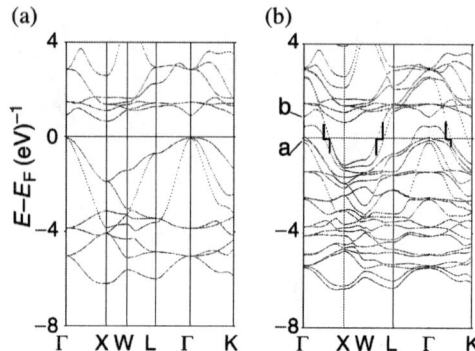

Figure 2.9. PtMnSb band structure for (a) the semiconducting spin channel (no **L · S** coupling) and (b) the band structure in the presence of **L · S** coupling.

the states in the highest of the three bands is depopulated as a result of this splitting. These states, however, do not contribute much to the optical conductivity since the dipole selection rule $\Delta l = \pm 1$ rules out exceptionally strong vertical transitions between the occupied manifold and the Mn 3d states at 1.4 eV above ϵ_F. As it turns out, at low photon energies contributions from the minority spin states in the immediate vicinity of the Γ point to the Kerr spectrum consist of transitions from the three occupied states into the band 1 eV higher in energy, which is a mixture of Pt, Mn, and Sb s states.

Figure 2.10 shows theoretical Kerr rotation spectra for PtMnSb and NiMnSb that were obtained from LMTO and FLASTO calculations, as well as experimental data.[80] First, the agreement between the two computational approaches is good: the effect of a full-potential treatment is rather limited for PtMnSb and NiMnSb. Second, upon inclusion of the semi-empirical Drude term in the conductivity, the experimental trends are clearly reproduced: for PtMnSb, a deep minimum for the Kerr angle at photon energies of about 1.5 eV and after that an increasing ϕ_K with increasing $\hbar\omega$ for PtMnSb. For NiMnSb, the double minimum structure of ϕ_K is reproduced. Additional structure in $\phi_K(\omega)$ at higher $\hbar\omega$ will be affected by energy dependent lifetime effects that are not taken into account in this study.

In PtMnSb, the intraband conductivity affects the extremum in position ($\hbar\omega \simeq 1.5$ eV) and magnitude ($\phi_K = -0.7°$ to $-0.9°$). Usually, the Drude term influences the Kerr spectrum only for energies less than about 1 eV. However, inclusion of the fitted Drude term from Ref. 79 seems to strongly affect the Kerr spectrum over a large part of the visible photon range. Further, it was found that the final Kerr spectrum sensitively depends on the choice for the Drude parameters. Earlier results by Wang et al.[54] closely resemble the Kerr rotation (LMTO+Drude) presented in Fig. 2.10. While the results presented here and in Ref. 54 agree very well, those due to Uspenskii et al.[55] are somewhat different. For both NiMnSb and PtMnSb, these authors compute extremal values for the Kerr effect that are a factor of 2 larger, and that are located at slightly higher photon energies. This may be due to the fact that Uspenskii et al. report values for σ_{xx} that are about a factor of 2 smaller than those found in the present study or in Ref. 54.

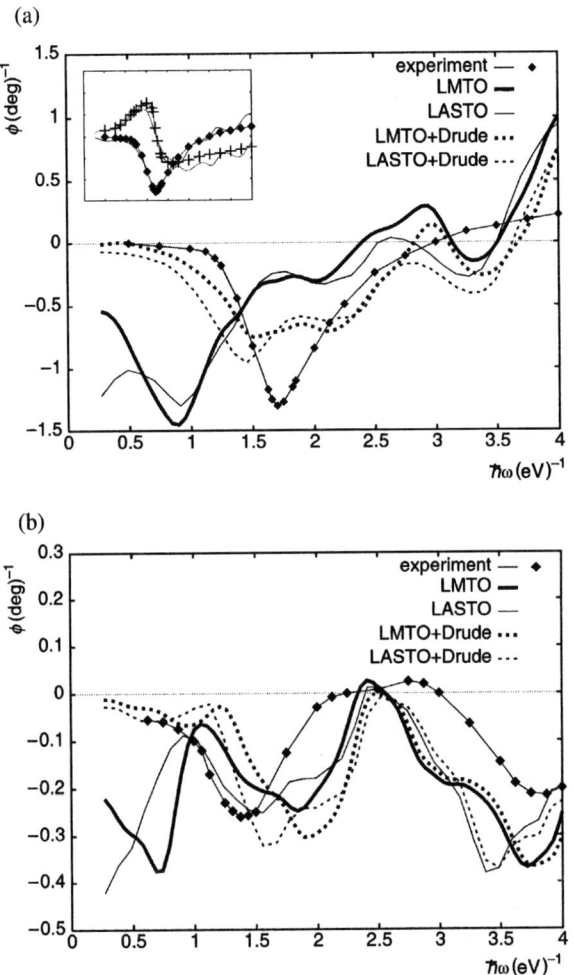

Figure 2.10. Computed Kerr rotation for (a) PtMnSb and (b) NiMnSb with the LMTO method, the FLASTO method, as well as the experimental results from Ref. 80. The inset in (a) compares theoretical and experimental curves for the Kerr rotation and ellipticity at a smaller lattice constant.

At this point, it must be concluded that for PtMnSb reasonable agreement between theory and experiment can only be achieved upon inclusion of the effect of intraband transitions. With model calculations, Feil and Haas[82] illustrated possible enhancement of the Kerr effect in the vicinity of the plasma edge of metals. Feil and Haas only quote the example of PtMnSb, for which there seems to be a big effect, in their paper. However, the diagonal optical conductivity for PtMnSb and NiMnSb behaves very similarly,[79] but *no* strong enhancement due to intraband contributions is observed for NiMnSb.

Results of an approximate spin decomposition as well as the complete curve for the Kerr rotation in PtMnSb are shown in Fig. 2.11. Although this decomposition cannot be rigorously justified, it is clear that for $0.5\,\text{eV} < \hbar\omega < 2\,\text{eV}$, the Kerr

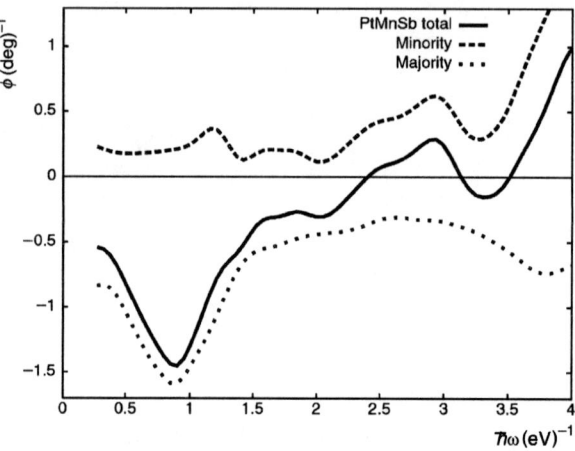

Figure 2.11. Approximate projection of ϕ_K onto minority and majority channels for PtMnSb.

rotation closely follows the contribution from the *metallic* channel and the semiconducting channel contributes very little. At higher $\hbar\omega$, the Kerr rotation seems to follow the semiconducting spin channel. It can therefore be concluded that the half-metallic property of PtMnSb (and NiMnSb) is not essential for the observed large interband Kerr rotation for red light since the metallic spin channel dominates the Kerr effect at low $\hbar\omega$. The results of the electronic structure calculations can be analyzed in more detail, providing a deeper understanding of the connection between the band structure and the Kerr effect.

For example, Fig. 2.12 shows contributions to ϕ_K, at $\hbar\omega = 0.9\,\text{eV}$ along high symmetry directions in reciprocal space, an energy where the interband contribution peaks. The $\vec{k}$-resolved contributions are defined by

$$\phi_K(\vec{k}) = -\frac{1}{8}\sum_{i=1}^{8} \sigma_{xy}(R_i\vec{k}) / \left[\sigma_{xx}\left(1 + \frac{4\pi i}{\omega}\sigma_{xx}\right)^{1/2}\right] \quad (2.4.2)$$

where the sum is over the eight elements of the magnetic point group $D_{2d}(S_4)$ (R_i), so that a properly symmetrized tensor is formed. Along ΓX in Fig. 2.12, the seemingly large contribution to ϕ_K (solid line) at Γ is due to transitions from the three bands labeled 'a' to band 'b' in Fig. 2.9(b), as was found by selectively eliminating transitions in the calculation of $\sigma_{\alpha\beta}$. Contributions from this peak to the integral over the entire Brillouin zone will be modified by a weight roughly proportional to k^2 and, moreover, the sign is opposite to the total computed value for ϕ_K at 0.9 eV. Table 2.1 summarizes characters of bands close to ϵ_F at several k-points in reciprocal space. Bands "a" at Γ are a mixture of Pt-p and Mn-d, while 'b' mainly consists of s on Pt, Mn and Sb. Bands "a" and "b" project onto the semiconducting spin channel as can be seen from Fig. 2.9.

The negative feature in $\phi_K(\vec{k})$ somewhat further along ΓX is associated with a series of transitions between parallel bands that are separated by 0.9 eV, as marked in Fig. 2.9(b). Optical transitions between these sheets of bands are responsible for

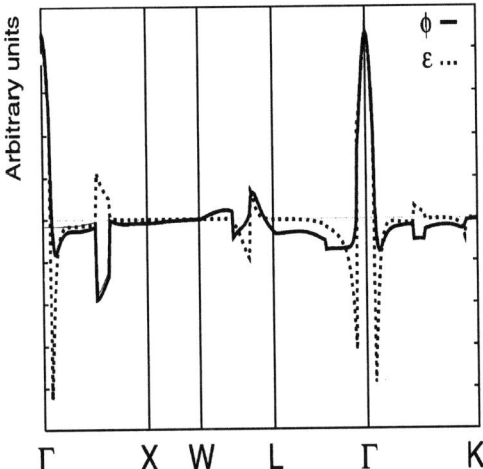

Figure 2.12. Contributions to ϕ_K (thick) and ϵ_K (dashed) at $\hbar\omega = 0.9$ eV for PtMnSb along high symmetry directions in reciprocal space. The thin solid line along ΓX, that mostly coincides with ϕ_K, shows the contribution from states that project onto the metallic spin channel.

the large negative Kerr rotation at 0.9 eV in the theoretical spectrum. Sampling arbitrary radial directions in the three-dimensional Brillouin zone showed that these energy bands form parallel *sheets*, of which the band structure, Fig. 2.9(b), only shows cross-sections. Dominant contributions to the character of the occupied band come from Pt 5d and Mn 3d states, while the two unoccupied states vertically above it consist of Mn 3d and Sb 5p states (see Table 2.1). Marked areas along ΓX, WL and ΓK in Fig. 2.9(b) contribute to the extremal value of the Kerr rotation, and the character of the bands involved is given in Table 2.1. It was explicitly verified that the excitations away from Γ, which dominate the Kerr spectrum, are associated with states that project onto the *metallic* spin channel.

Table 2.1. Dominant Atomic Orbital Character of LMTO Energy Bands below and above E_F in PtMnSb at Selected k-Points as Percentages

k-Point	Pt			Mn			Sb			
	6s	6p	5d	4s	4p	3d	5s	5p	5d	
Γ_{min}		15				70				3 × Occupied
	21			29			33			Unoccupied
$0.6\Gamma X_{maj}$	8		28		6	27	9	8		Occupied
		10	20		8	23		22		2 × Unoccupied
$0.6\,WL_{maj}$		6	16		6	34		13		Occupied
		11	13		10	23		25		Unoccupied
$0.5\Gamma K_{maj}$	6		21			25	7	5		Occupied
			9	19		27		17		Unoccupied

The first column indicates the k-point and whether the band state projects onto the minority or majority spin channel. The last column gives multiplicity and occupation.

Table 2.2. Extremal Kerr Rotation at Low Photon Energies in PtMnSb as a Function of Decreasing Lattice Constant

a (a_0)	$\hbar\omega$ (eV)	ϕ_K
11.72	0.90	$-1.45°$
11.49	0.90	$-1.50°$
11.25	0.69	$-1.54°$
11.02	0.53	$-1.65°$

Selectively switching off the $\mathbf{L} \cdot \mathbf{S}$ interaction for the different atoms in PtMnSb showed that only $\mathbf{L} \cdot \mathbf{S}$ coupling on the Pt site is important for the Kerr spectrum. Hence, spin polarization in the system is provided by the Mn atoms and the effect of $\mathbf{L} \cdot \mathbf{S}$ coupling on the Bloch states entirely originates at the Pt site and is conveyed to the magnetic sites by the wave function. In a simplified picture, $\mathbf{L} \cdot \mathbf{S}$ coupling in the two spin channels gives rise to equal but opposite contributions to the Kerr effect. Due to the exchange splitting of the bands, a net total Kerr effect is observed.

Upon application of hydrostatic pressure to PtMnSb, the magnitude of the minimum in the interband-only Kerr rotation increases and is shifted to lower energies, as given in Table 2.2. The shift towards lower energies is found to correspond to a reduced energy separation between the parallel bands in Fig. 2.9. Interestingly, at $a = 11.02a_0$, PtMnSb no longer has the half-metallic property (no $\mathbf{L} \cdot \mathbf{S}$ coupling).

At the same time ϕ_K is the largest (see Table 2.2), demonstrating once more that the half-metallic property is not relevant to the Kerr effect in $C1_b$ Heusler alloys. It is remarkable that, after the addition of the Drude intraband term, the agreement between theory and experiment is very good at the smallest lattice constant in Table 2.2, see inset in Fig. 2.10. This could point to either surface effects in the experiment or wrong positioning of the bands in the local density approximation (LDA) calculation at the experimental lattice constant.

2.5. MAGNETO-CRYSTALLINE ANISOTROPY

2.5.1. Introduction

As was described in the earlier part of this chapter, it is the spin–orbit interaction that is responsible for the secondary magnetic phenomena such as the magneto-crystalline anisotropy energy. This energy depends on the relative orientation of the spin quantization and crystal axes and manifests itself in easy and hard axes of magnetization. The anisotropy energy is quite small when compared to the electronic total energy. Spin–orbit coupling is also responsible for orbital magnetic moments; however, in transition metals these orbital moments are quenched to a large degree as a result of the crystal field effects. Since the electronic structure depends upon the atomic positions, straining the crystal will also modify the anisotropy, and this can be an important source of magnetic anisotropy in thin films where epitaxial growth on a substrate induces strain naturally. A related phenomenon is magnetostriction, where the crystal changes shape when magnetized. In this case, the crystal minimizes both the magneto-crystalline and magneto-elastic energies (or change in anisotropy energy due to strain) along with changes in the stored elastic energy. In addition, the magneto-elastic coupling also produces changes in the magnetization when a crystal is stressed.

The origin of the magneto-crystalline anisotropy is in the underlying electronic structure of the material, and as previously discussed, in a relativistic treatment of this electronic structure. For most ferromagnets, the valence electrons can be treated by using a two-component Pauli equation which results from taking the limit of the Dirac equation when V/mc^2 is small. In practice, one usually solves the so-called scalar relativistic equations,[19] which also include the mass–velocity and Darwin terms and neglect spin–orbit coupling. Spin–orbit coupling is usually then added back as a perturbing Hamiltonian. The benefits of this approach are primarily computational, since the modifications of the usual numerical integration schemes for the radial Schrödinger equation are minor (see Koelling and Harmon).[19] These approaches work well within the LSDA. There have recently been calculations of magneto-crystalline anisotropy which are based upon the Dirac equation. It appears that the results obtained from the solution of the Dirac equation are similar to those obtained within the scalar relativistic approach.[17,18,83,84] As will be seen, these first-principle electronic structure calculations often agree well with measured anisotropies, quite a tour de force when one considers that the anisotropy energy per atom ranges from tens to hundreds of microelectron volts per atom, and the total energy per atom is often greater than 10,000 eV.

The difficulty associated with treating spin–orbit coupling can be seen by considering perturbation theory. The first-order shift vanishes as a result of time reversal symmetry, and so one is forced to consider second-order perturbation theory. The second-order energy shift involves matrix elements of the spin–orbit coupling in the numerator and an energy difference between states in the denominator. Both numerator and denominator can become small causing numerical problems. For example, this manifests itself in slow convergence in Brillouin zone integrations. Wang et al.[85] proposed a state tracking scheme that focused on those states close to the Fermi energy as a way to improve the numerical stability of the zone integration.

An alternative perspective, which has been successful in understanding various aspects of the magneto-crystalline anisotropy including both the effects of strain and defects, considers a symmetry based expansion of the anisotropy energy. One approach simply expands the energy in terms of direction cosines of the magnetization, keeping only those terms compatible with the underlying crystal symmetry. An excellent review article by Kittel[29] gives a clear exposition of these ideas.

Neel considered a different viewpoint, one in which considered the energy associated with each bond.[34] The expansion Neel started with for the bond energy was written in terms of a series of Legendre polynomials in $\cos\theta$, where the angle between the vector connecting the two atoms and the magnetization was labeled θ. The coefficients depended upon the interatomic distance. The anisotropy energy is invariant under time reversal; however, since the magnetization changes sign, symmetry dictates that only even Legendre polynomials are needed. The energy of the crystal is found from summing over all the atomic pairs. The distance dependence of the expansion coefficients leads naturally to expressions for magnetostriction and magneto-elastic coupling, while the basic interaction can explain many properties of the magneto-crystalline anisotropy. Since the anisotropy is determined primarily by the d electrons, this summation is usually truncated at first nearest neighbors. For bcc based crystals, the proximity of the second shell of neighbors might require the consideration of those atoms too. In the subsequent sections, both first principles and symmetry based models will be discussed.

2.5.2. The Neel Model and Symmetry of the Anisotropy Energy

Before discussing the state of current first-principle electronic structure calculations, it is useful to start with a discussion of the symmetry based Neel model. The basic pairwise bond interaction, which Neel expressed as a series of even Legendre polynomials, can be equivalently written as a series in even powers of $\cos\theta$ as

$$E = E_0 + L(\mathbf{r})\cos^2\theta + Q(\mathbf{r})\cos^4\theta + \cdots \qquad (2.5.1)$$

θ is defined to be the angle between the vector connecting the pair of atoms as shown (Fig. 2.13).

While this expression can include the dipole–dipole interactions between spins, which gives rise to the shape anisotropy, as that interaction has an angular dependence of $\cos^2\theta$, it will be convenient in applying the Neel model to treat that separately. The reason being that the dipole–dipole interaction is long range, since it falls off as $1/r^3$, and hence prohibits truncating the summation over the interatomic bonds at nearest neighbors (nn). Once this magnetostatic interaction has been taken out of the energy, it can be shown from perturbation theory that the coefficients L and Q are of second and fourth order, respectively in the spin–orbit coupling strength.

2.5.3. The Neel Model Applied to Bulk Crystals

Consider first a cubic crystal, and assume that the summation of atomic pairs is restricted to only nn. In this case, it is easy to show that

$$\sum_{nn} \cos^2\theta = \text{constant}$$
$$\sum_{nn} \cos^4\theta = \text{constant} + 2\Lambda(\alpha_1^2\alpha_2^2 + \alpha_2^2\alpha_3^2 + \alpha_1^2\alpha_3^2) \qquad (2.5.2)$$

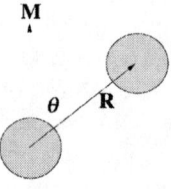

Figure 2.13. Atomic geometry defining the Neel bond energy.

where the coefficients α_i are the direction cosines of the magnetization, and the constant Λ is -2 for a simple cubic crystal, and $16/9$ for a body centered cubic crystal, and 1 for a face centered cubic crystal. The anisotropy per volume can be deduced by multiplying this energy per atom by the number of atoms/m^3 (n) and taking account of double counting by dividing by a factor of 2.

In the case of a hexagonal close-packed lattice, symmetry terms proportional to $\cos^2\theta$ do not vanish when summed over all atomic pairs. In the nn Neel model, however, it can be shown that

$$\sum_{nn} \cos^2\theta = \text{constant} + 4\delta \cos^2\theta + O(\delta^2) \qquad (2.5.3)$$

where $c/a = (1+\delta)(\frac{8}{3})^{1/2}$, and hence only yields an anisotropy when the c/a ratio differs from ideal.

If the crystal is deformed, then in addition to the magneto-crystalline anisotropy energy density of the undeformed crystal, there is the magneto-elastic energy which reflects the change in anisotropy energy resulting from the fact that the interatomic bonds are strained. In Neel's original paper,[34] the change in anisotropy energy with strain is worked out in detail considering the effects of changing both bond lengths and bond angles. Both these effects influence the anisotropy energy given by Eq. (2.5.1). The combined minimization of the elastic and magneto-crystalline anisotropy energy as a function of strain leads to magnetostriction. The expansion of the magneto-elastic coefficients in terms of the Neel interaction parameters is more complex for non-cubic crystals, and has only been developed for a hcp crystal by Bruno.[86]

2.5.4. Application of the Neel Model to Superlattices, Surfaces, and Thin Films

The Neel model is particularly useful for studying the intrinsic anisotropy at a surface or an interface. The latter is responsible for the perpendicular magnetization which is frequently observed in artificially layered magnetic materials. In magnetic multilayer, the shape anisotropy, which favors an in-plane magnetic alignment, can be smaller than the interface magneto-crystalline anisotropy provided the individual magnetic layers are thin enough (typically a few atomic planes), leading to a perpendicular magnetization. The observed anisotropy can often be simply modeled as an interface and bulk term, and is usually an expression for the anisotropy as a function of magnetic layer thickness t as

$$K_u = 2K_s/t + K_v \qquad (2.5.4)$$

In a superlattice because of the breaking of symmetry at the interface, summing terms in $\cos^2\theta$ produces the leading contribution to the anisotropy, and this is the reason for the larger anisotropies measured in magnetic superlattices as compared with bulk crystals. In fact, using the Neel model truncated at nn reproduces the anisotropy versus thickness relation given by Eq. (2.5.4).

The analysis is simplified by first computing effective intralayer and interlayer interactions labeled W_0, W_1, and W_2.... Most superlattices are based on lower Miller index planes, usually the (111) and sometimes (100) planes since close-packed planes are the easiest to grow. In this case, it is sufficient to truncate the interlayer interactions at next to adjacent layers, and then this interaction is only needed for certain superlattices such as an fcc (110) superlattice. To be more specific, we consider a superlattice formed from alternating magnetic and non-magnetic layers. To a good approximation, only two interatomic interactions need to be considered, namely L_f the interaction between two magnetic atoms, and L_m the interaction between a magnetic and non-magnetic atom, since anisotropy requires both magnetization and spin–orbit coupling to be present, the interaction between a pair of non-magnetic atoms can be neglected.

For a superlattice with n magnetic layers and m non-magnetic layers in the repeating unit, and assuming $m > 1$, we have from bond counting

$$E = \begin{cases} nW_0^f + 2W_1^m + 2W_2^m & n = 1 \\ nW_0^f + (n-1)W_1^f + (n-2)W_2^f + 2W_1^m + 2W_2^m & n > 1 \end{cases} \quad (2.5.5)$$

The form of the inter and intralayer interactions depends upon the crystal structure. As an example, consider a fcc-based superlattice with a (100) growth direction. If the in-plane lattice constants is a, then there are four in-plane nn atoms at $(0, \pm\frac{a}{2}, \pm\frac{a}{2})$. W_0, obtained by summing $\cos^2\theta$ over these four atoms is given by

$$W_0 = \frac{L}{2}\sum_{nn}\frac{(\mathbf{M}\cdot\mathbf{R})^2}{R^2} = \frac{L}{2}\left[(M_y - M_z)^2 + (M_y + M_z)^2\right] = L(1 - \cos^2\theta) \quad (2.5.6)$$

where $\mathbf{M}$ is a unit vector along the magnetization direction, and θ is the angle the magnetization makes to surface normal, in this case the x axis. The factor of 2 takes into account doubling counting. The other eight nn are located in the two adjacent layers. Allowing for a possible tetragonal distortion, these other near-neighbor atoms are located at $(\pm\frac{c}{2}, \pm\frac{a}{2}, \pm\frac{a}{2})$. A similar bond counting exercise leads to

$$W_1 = \text{constant} + L(6\alpha^2 - 2)\cos^2\theta \quad (2.5.7)$$

The parameter α characterizes the tetragonal distortion and is given by $\alpha = (c/(c^2 + a^2)^{1/2})$:

$$E = [nL_f(6\alpha_f^2 - 3) + 2L_m(6\alpha_m^2 - 2) - L_f(6\alpha_f^2 - 2)]\cos^2\theta \quad (2.5.8)$$

In the case of an unstrained structure, $c = a$ and $\alpha^2 = \frac{1}{2}$. The final expression is, for $n > 1$,

$$E = (2L_m - L_f)\cos^2\theta \quad (2.5.9)$$

Notice that the volume anisotropy, related to the coefficient in front of n, vanishes for the unstrained superlattice. In this model, only the shape anisotropy will produce a volume contribution in an unstrained superlattice. Table 2.3 summarizes the anisotropy results for several common unstrained superlattices. The results correct some small algebraic errors for the (110) directions that occurred in Neel's original work on the surface anisotropy.[34] Expressions appropriate for a strained superlattice are considerably more complex and have been given by Victora and MacLaren.[87]

2.5.5. Influence of Rough Interfaces

In practice, the growth conditions usually do not produce atomically flat interfaces, but rather structures with rough interdiffused and imperfect interfaces. These imperfect interfaces can be broadly be characterized by two types of roughness, long and short range. For example, an interface may be wavy, and this type of structure would be expected to propagate through the superlattice. This type of imperfection would not be expected to influence the anisotropy dramatically since the short-range order is preserved. On the other hand, an interface could be flat, and yet have have significant interdiffusion, this as will be shown dramatically reduces the interface anisotropy. Both types of roughness may be present to varying degrees in any one sample.

The Neel model can be solved simply under certain simplifying approximations, namely the interface is assumed to be unstrained and the nn are confined to the adjacent layers. A statistical treatment of the interdiffused interface can be developed by defining P_j the probability that a site in layer j is occupied by a ferromagnetic atom). Bond counting then leads to[88]

$$K_i = \frac{1}{2}\sum_{j=1}^{n} P_j^2 W_0^f + 2(1 - P_j)P_j W_0^m + P_j P_{j+1} W_1^f$$
$$+ P_j(1 - P_{j+1})W_1^m + (1 - P_j)P_{j+1} W_1^m$$
$$= \frac{1}{2} K_p \sum_{j=1}^{n}(P_j - P_{j+1})^2 \qquad (2.5.10)$$

where K_p is the anisotropy of the ideal superlattice. In reaching the final result, we have used the fact that for a cubic crystal the anisotropy to this order is zero.

Draaisma et al.[89] have also considered interdiffusion within a Neel model, and arrive at a similar conclusion as to the effects of interdiffusion. Their expression did

Table 2.3. Anisotropy of Unstrained Superlattices Obtained from the Neel Model

System	Anisotropy Energy $\times a^2$
fcc (100)	$(2L_m - L_f)\cos^2\theta$
fcc (111)	$(2L_m - L_f)\cos^2\theta$
fcc (011)	$(2)^{1/2}(2L_m - L_f)\beta_y\beta_z$
bcc (100)	0
bcc (111)	$(2L_m - L_f)(3)^{1/2}/4\cos^2\theta$
bcc (011)	$(2)^{1/2}(2L_m - L_f)\beta_y\beta_z$
sc (100)	$(2L_m - L_f)\cos^2\theta$
sc (111)	0
sc (011)	$(L_f - L_m)\beta_x^2/(2)^{1/2}$

θ is the angle the magnetization makes with the surface normal, and β_i the cosine of the angle the magnetization makes with the vector $\hat{x}_i$. The anisotropy of the corresponding surface can be obtained by setting $L_m = 0$. a is the nearest neighbor distance within the plane.

not explicitly show the connection between the perfect anisotropy, disorder and the imperfect anisotropy, though appeared in the one case considered in detail in their paper to give the same anisotropy reduction.

The connection between interface roughness and loss of anisotropy is often observed experimentally. Several groups have grown 1Co/5Pd superlattices using different growth techniques and different growth conditions. These superlattices show a wide variation in the measured magnetic anisotropy. A 1Co/5Pd superlattice grown by molecular beam epitaxy has the largest anisotropy of 5.9×10^7 ergs/cm^3 (Ref. 90). A 1Co/5Pd superlattice grown epitaxially in UHV also had a large anisotropy (4.9×10^7 ergs/cm^3) (Ref. 91). Sputtered 1Co/5Pd superlattices made by different groups have anisotropies of 2.6×10^7 ergs/cm^3 (Ref. 92) and 2.4×10^7 ergs/cm^3 (Ref. 93), respectively. These can be compared to the theoretical values for a perfect 1Co/5Pd superlattice of 6.6×10^7 ergs/cm^2, (Refs. 94, 95) and 8.7×10^7 ergs/cm^2 (Refs. 96, 97). All of the experimental superlattices will contain defects, with the sputtered samples having the largest concentration. The smallest anisotropies are seen to occur in the sputtered samples which are expected to have the most interdiffused interfaces. The range of values observed is clearly consistent with the theoretical expectations for superlattices with defects.

Another clear example which shows the connection between interface roughness and loss of magneto-crystalline anisotropy can be found in the study of Co/Au superlattices by den Broeder et al.[98] The as-deposited superlattices were subsequently annealed. The anisotropy was observed to increase significantly upon annealing. In addition, the x-ray diffraction pattern showed sharper peaks in the annealed sample suggesting that the annealed sample had more perfect interfaces. Since Co and Au are immiscible, annealing allowed the atoms at the interface to diffuse and form sharper interfaces.

2.5.6. Magneto-Elastic Coupling and Magnetostriction

When a crystal is magnetized its dimensions can change. This important effect is called magnetostriction. As we have already seen that there is an anisotropy energy which depends upon the direction of magnetization. This can be analyzed in terms of bond energies within the Neel model, since these interatomic interactions depend upon the bond length. Thus, the anisotropy energy will change if the crystal is strained since the interatomic distances will also be changed. The coupling between the anisotropy energy and the strain is characterized by coefficients called the magneto-elastic coupling constants, usually labeled B.

Considering a cubic crystal, we have already shown that the magneto-crystalline energy density E_a is given $K(\alpha_1^2 \alpha_2^2 + \alpha_2^2 \alpha_3^2 + \alpha_1^2 \alpha_3^2)$, where α_i is the ith direction cosine of the magnetization. The analysis of magnetostriction presented here follows Kittel's paper.[29] If the cubic crystal is strained, the anisotropy energy takes the form

$$E_a = K(\alpha_1^2 \alpha_2^2 + \alpha_2^2 \alpha_3^2 + \alpha_1^2 \alpha_3^2) + B_1(\alpha_1^2 e_{xx} + \alpha_2^2 e_{yy} + \alpha_3^2 e_{zz})$$
$$+ B_2(\alpha_1 \alpha_2 e_{xy} + \alpha_2 \alpha_3 e_{yz} + \alpha_1 \alpha_3 e_{xz}) \quad (2.5.11)$$

As the strain is assumed to be small, we can write, based upon symmetry, $\partial E_a / \partial e_{ii} = B_1 \alpha_i^2$, and for $i \neq j$, $\partial E_a / \partial e_{ij} = B_2 \alpha_i \alpha_j$. These coefficients can be related to

SECONDARY MAGNETIC PROPERTIES

the Neel bond strength and its first derivative with respect to bond length. An analysis along these lines was presented in Neel's original paper, see Eq. (11) in Ref. 34. If the crystal is strained, then there is a change in elastic energy which is given by

$$E_{el} = \frac{1}{2}c_{11}(e_{xx}^2 + e_{yy}^2 + e_{zz}^2) + \frac{1}{2}c_{44}(e_{xy}^2 + e_{yz}^2 + e_{zx}^2)$$
$$+ c_{12}(e_{yy}e_{zz} + e_{xx}e_{zz} + e_{xx}e_{yy}) \quad (2.5.12)$$

where the c_{ij} are the usual cubic elastic constants. The equilibrium strain is that which minimizes the sum of elastic and anisotropy energies, and results in the following solution:[29]

$$e_{ii} = \frac{B_1[c_{12} - \alpha_i^2(c_{11} + 2c_{12})]}{[(c_{11} - c_{12})(c_{11} + 2c_{12})]} \quad (2.5.13)$$

$$e_{ij} = -B_2\alpha_i\alpha_j/c_{44} \quad i \neq j$$

The conventional first-order expression for magnetostriction used for cubic crystals is

$$\frac{\delta l}{l} = \frac{3}{2}\lambda_{100}\left(\alpha_1^2\beta_1^2 + \alpha_2^2\beta_2^2 + \alpha_3^2\beta_3^2 - \frac{1}{3}\right)$$
$$+ 3\lambda_{111}(\alpha_1\alpha_2\beta_1\beta_2 + \alpha_1\alpha_3\beta_1\beta_3 + \alpha_2\alpha_3\beta_2\beta_3) \quad (2.5.14)$$

The direction cosines β_i define a unit vector which points along the direction $\delta l/l$ is measured. Starting from the usual definition of strains,

$$x_i' = x_i + e_{x_ix_i}x_i + \frac{1}{2}\sum_{i \neq j}e_{x_ix_j}x_j \quad (2.5.15)$$

which characterize the change in the point x, y, z. We can see that

$$\delta(l^2) = 2l \cdot \delta l = 2l^2 \sum_{i \geq j} e_{ij}\beta_i\beta_j \quad (2.5.16)$$

Substituting the equilibrium values of the strain given by Eq. (2.5.13) leads to

$$\frac{\delta l}{l} = -\frac{B_1}{c_{11} - c_{12}}(\alpha_1^2\beta_1^2 + \alpha_2^2\beta_2^2 + \alpha_3^2\beta_3^2)$$
$$- \frac{B_2}{c_{44}}(\alpha_1\alpha_2\beta_1\beta_2 + \alpha_1\alpha_3\beta_1\beta_3 + \alpha_2\alpha_3\beta_2\beta_3) \quad (2.5.17)$$

which, neglecting the term independent of α and β, is the same as the usual expression, Eq. (2.5.14), provided

$$\lambda_{100} = -\frac{2}{3}\frac{B_1}{c_{11} - c_{12}}$$
$$\lambda_{111} = -\frac{1}{3}\frac{B_2}{c_{44}} \quad (2.5.18)$$

A similar symmetry-based analysis of the influence of strain on the anisotropy energy has been given for a hcp crystal by Mason,[99] and as mentioned in the section in bulk crystals within the Neel model, Bruno[86] has connected this energy density expressed in terms of the magneto-elastic coefficients to the Neel parameters. Unlike the cubic case where the two independent magnetostriction coefficients λ_{100} and λ_{111} uniquely determine the Neel parameter L and its first derivative dL/dr, the four independent magnetostriction coefficients for the hcp crystal overdetermine the Neel parameter L and its first derivative. Bruno formed a reasonable determination of L and dL/dr for Co based upon a least squares analysis of this overdetermined set of equations. The surface anisotropy predicted by the value of L obtained from the analysis of bulk magnetostriction data was in satisfactory agreement with the experimental value.

2.6. FIRST-PRINCIPLE CALCULATIONS

Despite the small values of magneto-crystalline energies when compared to electronic total energies, over the past decade first-principle electronic structure calculations have generally been quite successful in calculating magneto-crystalline anisotropies. Theoretical values often match experimental observation quite closely, and theory has reached the stage of being able to predict ahead of measurement. This has been the result of both improved computational algorithms and the increased computational resources.

2.6.1. Isolated Monolayers and Slabs

Systems with symmetry lower than cubic are expected from symmetry arguments to have significantly larger magneto-crystalline anisotropies, and not surprisingly the simplest of these, the isolated monolayer, was the first to be studied using *ab initio* local spin-density calculations. In the pioneering work by Gay and Richter,[100,101] isolated monolayers of Fe, Ni, V, and Co were studied. All of these monolayers were found to be magnetic, with local moments that were larger than those found in the bulk crystal as a result of the reduced dimensionality. The authors found that for Ni and Co, the anisotropy was in-plane, while for Fe and V an out of plane magnetization was found. The over layer calculations were thought to be a reasonable theoretical model of systems such as Fe on Au or Ag, where the coupling between the overlayer and substrate was expected to be small. In fact, this turned out not to be the case. The anisotropy of an isolated Fe layer was found to be 0.38 meV/atom while for an Fe/Ag slab, the anisotropy turned out to be about significantly smaller, only 0.1 meV per atom.[101]

Li et al.[102] arrived at a completely different conclusion for the Fe monolayer which they studied using the more accurate FLAPW approach. In these calculations, the magnetization was found to lie in the plane of the monolayer rather than out of the plane and the anisotropy energy was an order of magnitude smaller at 0.043 meV/atom. This result appears far too small, when compared to other similar monolayer and overlayer systems, and is probably incorrect. In addition, more recent FLAPW calculations by the Northwestern group[85] using the state tracking approach, a technique developed to improve numerical stability, found a perpendicular anisotropy of 0.38 meV/atom for an Fe (100) monolayer whose lattice constant was similar to that used by Gay and Richter. The anisotropy computed is in very close agreement with that earlier work. Wang et al.[85] also noticed considerable sensitivity to the choice of lattice constant. Li et al.[102] also studied the overlayer problem where they found perpendicular magnetizations for (100) Fe/Au, Fe/Ag, and Fe/Pt with anisotropies 0.5, 0.1 and 0.4 meV/atom, respectively. Note that the Fe/Ag result is in accord with the results of Gay and Richter.[101]

The calculations used the force theorem to evaluate the small energy associated with the magneto-crystalline anisotropy. This theorem[103,104] relies on the fact that changing the axis of quantization produces only very small changes in the electronic structure, and the energy difference can be represented by the change in the one electron band energies. This approach has been adopted by most theoretical band structure calculations of the magneto-crystalline anisotropy. It is worth making a few comments about the force theorem at this stage, since it is almost universally used in calculations. The force theorem relies on the variational nature of the total energy functional. This means that any small errors in the density, $\delta\rho$ produce errors that are quadratic in the total energy.

Wang et al.[105] examined the use of the force theorem for the specific case of varying the axis of quantization of the magnetization. They showed that as a result of the stationary nature of the energy functional, and the fact that the spin–orbit coupling produces no first-order shift in the energy eigenvalues (from time reversal symmetry of the unperturbed wave functions), that the errors in energy difference between two magnetization directions obtained by only considering the difference in the band energy term is fourth order in the spin–orbit coupling. Thus, the use of the force theorem is justified for crystals with lower than cubic symmetry, where the anisotropy energy is second order in the spin–orbit coupling. In the case of cubic crystals, the anisotropy is also fourth order in the spin–orbit coupling, and thus the use of the force theorem is not rigorously justified. Despite this, the force theorem has been used in calculations of the magneto-crystalline anisotropy of cubic metals, and reasonable anisotropy energies have been obtained. We shall come back to this point in the section where first-principle calculations for Fe, Ni, and Co are discussed.

In Gay and Richter's calculations, the electronic structure was found self-consistently in absence of spin–orbit coupling. Then two calculations, corresponding to two magnetization directions were performed with the spin–orbit coupling included. The energy difference of the one-electron band sum, taking into account any movement of the Fermi energy, provides an accurate estimate of the full total energy difference between the two calculations. This provides a considerable computational saving since a full relativistic calculation, or a scalar relativistic calculation with spin–orbit coupling present takes typically an order of magnitude longer to complete because the matrix sizes are doubled, and more importantly the absolute value of the total energy does not need to be converged to a level better than the anisotropy energy.

Despite the use of the force theorem, Gay and Richter's calculations were time consuming because the Brillouin zone integration converged rather slowly, requiring over 5000 wave vectors in the Brillouin zone to converge the anisotropy energy. In addition, the symmetry of the Brillouin zone is lower when compared to the same system without spin–orbit coupling, especially for magnetizations in the plane of the overlayer. This requires more wedges of the Brillouin zone to be integrated over.

The slow convergence in k-space suggests that no single k point, or region of k-space dominates the anisotropy energy. In addition, superlattice anisotropies appear, to a good approximation, to be a sum of interface and bulk terms, which suggests a short range real space interaction can describe the anisotropy. This is perhaps to be expected given the success of the near-neighbor real space model proposed by Neel. If only certain regions of k-space were important, such as is the case for understanding interlayer exchange coupling, then a Fourier analysis would be more natural than a real space model.

More recent calculations of Co monolayers by Wang et al.[106] have found similar results to those obtained by Gay and Richter. These calculations used one of the more accurate electronic structure methods, the FLAPW method. An in-plane anisotropy energy of 1.4 meV/atom was obtained for a Co monolayer. The differences in the numerical value are most likely due to the different lattice constants used. In Wang's calculations, a value appropriate for bulk Co was used, while Gay and Richter, whose aim was to understand the overlayer problem chose a lattice constant which was that of the substrate Ag. This is a 10% expansion of the bulk Co lattice constant.

Wang et al.[106] also varied the position of the Fermi energy artificially and observed that the anisotropy energy changed sign as the valence, or Fermi energy, was changed. Assuming that the underlying electronic structure of the transition metal overlayers was essentially similar and that the differences were mainly due to band filling; Wang et al. observed that while the anisotropy of Co (valence 9) is in-plane, Fe (valence 8) is perpendicular. Ni appears to have little anisotropy but is close to a region of in-plane anisotropy. V (valence 5) is also perpendicular. The anisotropy has its origins in the filling of particular d-orbitals, whose energy dispersion is determined by crystal structure and overlap. When either spin band is close to being half-filled (valence 3 or 9) the anisotropy is in-plane, and when the bands are either just filling or almost full the anisotropy is out of plane. A similar picture of the variation of anisotropy energy versus band filling was obtained by Daalderop et al.[107] Band filling arguments also explain why Co/Pd has a perpendicular anisotropy. The strong hybridization between Co and Pd shifts the energy versus band filling curve so that the Fermi energy lies in a region that has perpendicular anisotropy. In contrast, the weaker hybridization between Co and Cu produces a curve intermediate between the Co monolayer and the Co/Pd interface, and results in an anisotropy, i.e., close to zero.[108]

2.6.2. Elemental Ferromagnets Fe, Co, and Ni

There have been several attempts to calculate the magneto-crystalline anisotropy of the elemental ferromagnets Fe, Ni, and Co by using first-principle electronic structure methods. These have been quite challenging because the anisotropy energy is so small. In crystals with cubic symmetry, the second-order contribution to the magneto-crystalline anisotropy vanishes, resulting in an anisotropy that is fourth order in the spin–orbit coupling. All of these calculations, with the exception of the work by Trygg et al.[109] have been based upon muffin–tin or spherical potentials and the force theorem. Trygg et al.[109] used a full-potential LMTO approach and looked at the difference in total energies between easy and hard axes of magnetizations. Some of these calculations have been based upon the a solution of the Dirac

Table 2.4. Calculated Magneto-Crystalline Anisotropy for Three Elemental Ferromagnets

System	Theory (μeV/atom)					Experimental	Easy Axis
bcc Fe	−0.5	−0.4	7.4	−0.5	−1.8	−1.4	(100)
fcc Ni	−0.5	−0.6	10	−0.5	−0.5	2.7	(111)
hcp Co	16	−29		−29	−110	−65	(0001)
fcc Co				0.5	2.2	1.8	(111)

For the cubic metals the anisotropy is defined to be $E_{(001)} - E_{(111)}$, while for hcp Co the anisotropy energy is defined to be $E_{(10\bar{1}0)} - E_{(0001)}$. The first column of data are LMTO calculations by Daalderop et al.[110] using a basis of s, p, and d partial waves. The second column of data are anisotropies LMTO calculations by Daalderop et al.[110] using a basis of s, p, d, and f partial waves. The third column of data are anisotropies obtained by Fritsche et al.[134] using a fully relativistic method. The fourth column of data are anisotropies obtained by Trygg et al.[109] using a full potential LMTO method. The fifth column of data are anisotropies obtained by Trygg et al.[109] using a full potential LMTO method with orbital polarization terms added. The final column of data are the experimental anisotropies. Experimental values are taken from Ref. 135.

equation, while others a scalar relativistic Schrödinger equation. For the early transition metals, the difference between the solutions of the Dirac equation and the scalar relativistic Schrödinger equation are minor. The results of all the calculations are summarized in Table 2.4. As can be seen, the agreement between theory and experiment and between different calculations is on the whole not too bad, with the exception of Ni, where most of the studies predict the incorrect easy axis. The large change in anisotropy seen in the Daalderop et al.[110] calculations for hcp Co when increasing the basis set from spd to spdf is puzzling and may be a result of the hexagonal symmetry. This suggests that a full potential electronic structure calculation is needed.

It would appear that the full potential LMTO calculations by Trygg et al.[109] which do not rely on the force theorem or a shape approximation to the potential or charge density are probably state of the art and any further discrepancy between theory and experiment is probably due to the LSDA. In fact, these authors try to correct one deficiency of the LSDA. A term is added to the Hamiltonian which is called an orbital polarization correction. This correction attempts in an approximate way to account for Hund's second rule. This term is not present in the usual LSDA calculations, since the exchange correlation potential has been derived from the homogeneous electron gas. In LSDA calculations, the orbital moment is usually underestimated because of this deficiency, and one would also expect the anisotropy to also be underestimated. The results of adding this term show modest improvements for the anisotropy energy of bcc Fe overestimate Co, and do not correct the wrong easy axis problem in Ni. The calculated band structure of Ni and its Fermi surface are known to show larger differences with experiment than either Fe or Co. This orbital polarization approach has also had some success in correctly predicting the insulating ground state of some transition metal oxides. The reader is referred to the following references for more details.[111]

There have only been a few first-principle calculations of magnetostriction and magneto-elastic coefficients. Wu and Freeman[112] used the force theorem within the FLAPW method to study the magneto-elastic constants in fcc Co and some simple Co/Cu and Co/Pd (100) thin films. The calculated results appear promising. In the case of fcc Co, calculations that strained the crystal assuming constant volume, gave $\lambda_{100} = 10.2 \times 10^{-5}$ and $B_1 = -1.6 \times 10^8$ ergs/cm^3, which can be compared favorably with the experimental values of $\lambda_{100} = 13 \times 10^{-5}$ and $B_1 = -1.6 \times 10^8$ ergs/cm^3 which were quoted in their manuscript.

2.6.3. Superlattices

Artificially grown superlattices form an important class of magnetic materials with many potential applications. In magnetic recording, layered magnetic materials will be used to read stored data using the giant magnetoresistance (GMR) effect, while future media may take advantage of some of the unique properties of multilayers to achieve high ariel densities. Over the last 10 years, there was a significant effort to produce a media with both a large perpendicular anisotropy and a large polar Kerr effect for use in high density magneto-optic data storage. Co/Pt and Fe/Pt superlattices had the desired perpendicular anisotropy and a reasonable Kerr rotation and there is an extensive literature on these materials. Rather than providing an exhaustive list of references, we refer the reader to recent review articles by Heinrich and Cochran[113] and Johnson et al.,[114] which discusses a wide class of superlattices and their properties. However, with the advent of the GMR as a technology to read high density stored data, interest in magneto-optics has waned over the last few years. Nonetheless, the coercivity of the media is an important material parameter which is influenced by the magneto-crystalline anisotropy. The most common superlattices are composed of magnetic layers separated by non-magnetic spacer layers, examples include Fe/Pt, Co/Pt, Co/Pt, and Co/Cu. The broken symmetry at the interface usually produces a perpendicular magneto-crystalline anisotropy. The shape anisotropy, proportional to the thickness of the magnetic layers, tends to produce an in-plane anisotropy and this term overcomes the interface anisotropy as the thickness of the magnetic layer is increased. The anisotropy thickness relation given by Eq. (2.5.4) is usually observed. Deviations for very thin magnetic layer thicknesses are sometimes seen in materials that are not well lattice matched and results from dislocations and strain in the material.

Rather than reporting on a large number of first-principle calculations, we will focus on a few results which illustrate that, in the case of multilayers where the anisotropy is larger than that of bulk ferromagnets (milli electron volt per cell rather than micro electron volt per cell) that different theoretical approaches are in rather good agreement with each other. In comparing to experiment, it is found that the calculated interface contribution, which results from the intrinsic magneto-crystalline anisotropy, is close to measured values, while the volume term which is smaller per atom can be less accurately found. Thus, while both experiment and theory will show an anisotropy thickness relation given by Eq. (2.5.4), the slopes may disagree. The most probable source of error is thickness dependent strain which leads to a magneto-elastic contribution to the anisotropy, or in the case of (111) based Co superlattices uncertainty as to how much fcc and hcp Co are present, given the low stacking fault energy of Co.

Table 2.5 shows results for several Co-based superlattices. The calculated results were obtained from three independent sets of calculations each using a different technique. As can be seen from the table, consistent results are obtained.

Table 2.5. Calculated Magneto-Crystalline Anisotropies for a Few Superlattices

System	Theory (meV/cell)	
1Co2Pd (111)	0.85[1]	0.83[2]
1Co5Pd (111)	0.65[1]	0.56[2]
1Co1Cu (100)	−0.47[3]	0.08[2]
1Co3Cu (100)	0.48[3]	0.48[2]

The energies are given in milli electron volts per unit cell. [1] Daalderop et al.[96] [2] MacLaren and Victora[95] [3] Freeman et al.[136]

Modern electronic structure methods seem to be able to treat magneto-crystalline anisotropy at least for non-cubic systems. One should keep in mind that even using the force theorem, the energy differences, which are about 1 meV, are quite small compared to the band sum energies which are of the order 10–100 eV.

Several sets of Co/Pd superlattices, with different grown directions, have been fabricated by molecular beam epitaxy by Engel et al.[90] The superlattices all had varying Co layer thickness and a fixed Pd layer thickness. These superlattices were characterized structurally by using x-ray diffraction. On the basis of the diffraction data, all of the samples appeared to be very close to ideal with atomically abrupt interfaces. As such, these samples provided a useful set of experimental data which can be compared to first-principle predictions. The structural information extracted from the x-ray diffraction studies suggested that the samples with a (111) growth direction were unstrained with atoms lying on an ideal fcc lattice. The diffraction data could not, however, distinguish between Co atoms on an fcc or hcp lattice. In the case of (100) and (111) samples, the lattices were tetragonally strained by about 10%. The anisotropies were extracted from measurements of the hysteresis loops of the hard axis.

The comparison between first-principle electronic structure calculations and experiment for the Co/Pd superlattices are shown in Figs. 2.14 and 2.15.

Both clearly show that the anisotropy is a sum of interface and volume contributions, and that theory is able to reproduce quite accurately the intercept (interface anisotropy). The volume and interface anisotropies for the different sets of superlattices are given for convenience in Table 2.6.

Co/Pd (111) and (100) superlattices have also been fabricated and their magnetic properties studied by the Phillips group. The experimental data[92, 115] along with the results of first-principle LMTO calculations have been taken from Refs. 96 and 107 and replotted in Figs. 2.16 and 2.17. Interface and volume anisotropies extracted from anisotropy measurements of a set of vapor deposited (111) Co/Pd and (100) Co/Pd samples are, for K_s, 0.75 ergs/cm^2 and 0.45 ergs/cm^2 for (111) and (100) growth directions, respectively. Values of K_v for (111) and (100) growth directions are -1.3×10^7 and -2.45×10^7 ergs/cm^3. These were extracted from samples grown at 200°C. Smaller values of anisotropy were found for samples grown at 50°C: K_s, K_v for the (111) superlattices were 0.26 ergs/cm^2 and -0.72×10^7 ergs/cm^3; and 0.32 ergs/cm^2 and -2.19×10^7 ergs/cm^3 for the (100) superlattices. These authors reported that the superlattices grown at 200°C had smoother interfaces.

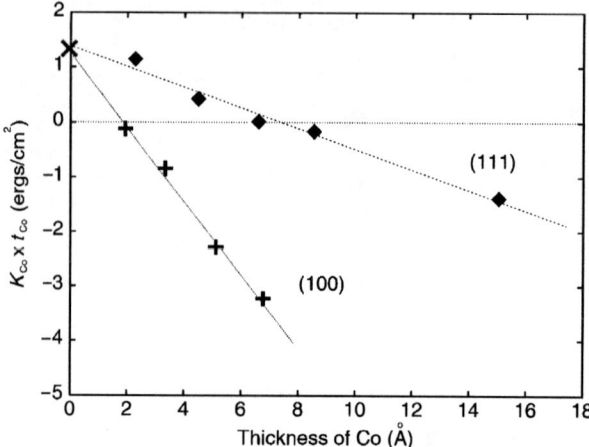

Figure 2.14. The theoretical uniaxial anisotropy energy times the Co thickness versus Co thickness. Lines indicate best fit to each orientation. Experimental interface anisotropy is marked by the X and taken from Ref. 90.

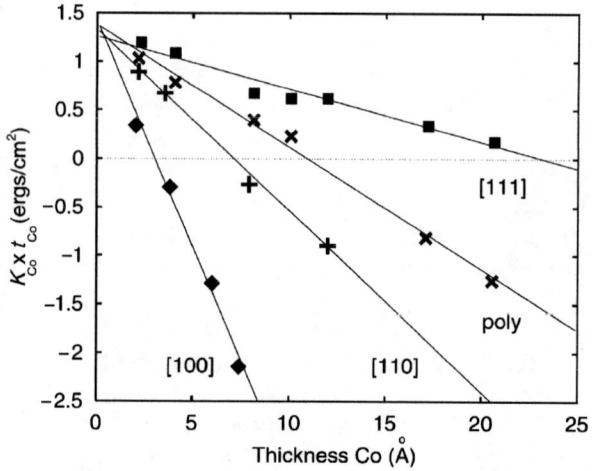

Figure 2.15. The experimental uniaxial anisotropy energy times the Co thickness versus Co thickness, data replotted from Ref. 90.

Table 2.6. Theoretical[87] and Experimental[90] Interface and Volume Anisotropies for Various Co/Pd Superlattices

Co/Pd	$2K_s$ (ergs/cm^2)		K_v (ergs/cm^3)	
	Theory	Experiment	Theory	Experiment
(111)	1.32	1.3	-1.8×10^7	-0.6×10^7
(100)	1.14	1.2	-6.4×10^7	-4.5×10^7
(011)		1.3		-1.9×10^7

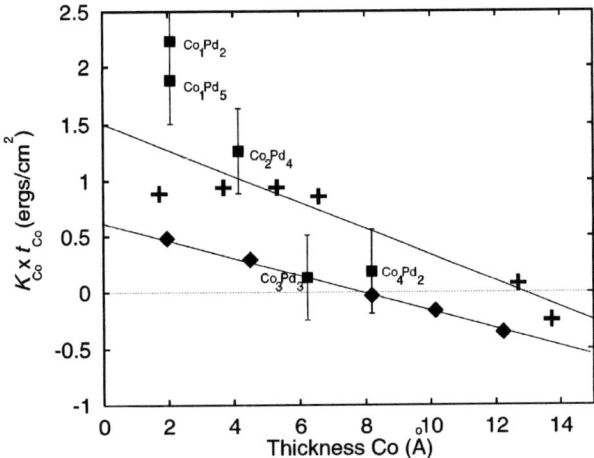

Figure 2.16. Measured and calculated anisotropies for (111) and Co/Pd multilayers grown by the Phillips group. The plusses (diamonds) represent superlattices grown at 200°C (50°C).[92,115] The theoretical calculations, by Daalderop et al.,[96] are marked as filled boxes.

The measured anisotropies are somewhat different than those obtained by Engel[90] which presumably reflects differences in structure and strain associated with different growths. The largest discrepancies are seen for the (100) direction where the effects of strain and a correspondingly larger magneto-elastic contribution to the anisotropy are to be expected. In fact, the anisotropy versus thickness for Phillips samples do not seem to simply be the sum of volume and interface contributions over the range of Co thicknesses studied. This suggests that there are other factors present which are thickness dependent which influence the anisotropy.

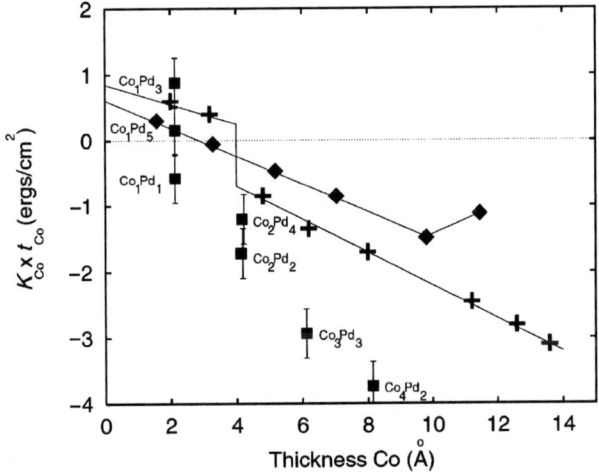

Figure 2.17. Measured and calculated anisotropies for (100) and Co/Pd multilayers grown by the Phillips group. The plusses (diamonds) represent superlattices grown at 200°C (50°C).[92,115] The theoretical calculations, by Daalderop et al.,[96] are marked as filled boxes.

The electronic structure calculations can be used to fit the two parameters (L_f and L_m) used in the Neel model description. This was done for both Co/Pd and Co/Pt, allowing the anisotropy of (110) Co/Pd and the Co/Pt superlattices to be predicted. Details on the fitting can be found in Ref. 94. The anisotropy versus thickness for all six superlattices is summarized below. The close level of agreement found validates the approach. The theoretical anisotropy thickness relations derived from the Neel model[87] are

$$K_u \times t_{Co} = 1.3 \text{ ergs/cm}^2 - t_{Co}(1.3 \times 10^7 \text{ ergs/cm}^3) \quad (111)$$
$$K_u \times t_{Co} = 1.1 \text{ ergs/cm}^2 - t_{Co}(6.4 \times 10^7 \text{ ergs/cm}^3) \quad (001)$$
$$K_u \times t_{Co} = 1.2 \text{ ergs/cm}^2 - t_{Co}(4.2 \times 10^7 \text{ ergs/cm}^3) \quad (011)$$

which should be compared to Engel et al.'s data[90]

$$K_u \times t_{Co} = 1.2 \text{ ergs/cm}^2 - t_{Co}(0.6 \times 10^7 \text{ ergs/cm}^3) \quad (111)$$
$$K_u \times t_{Co} = 1.3 \text{ ergs/cm}^2 - t_{Co}(4.5 \times 10^7 \text{ ergs/cm}^3) \quad (001)$$
$$K_u \times t_{Co} = 1.3 \text{ ergs/cm}^2 - t_{Co}(1.9 \times 10^7 \text{ ergs/cm}^3) \quad (011)$$

A similar comparison has been made for Co/Pt superlattices studied by Farrow et al.[116] In modeling Co/Pt, the value of the Co–Co bond strength was assumed to be the same as that obtained from fitting the Co/Pd superlattices, since the superlattices are structurally similar. The theoretical anisotropy thickness relations are

$$K_u \times t_{Co} = 1.7 \text{ ergs/cm}^2 - t_{Co}(1.3 \times 10^7 \text{ ergs/cm}^3) \quad (111)$$
$$K_u \times t_{Co} = 1.2 \text{ ergs/cm}^2 - t_{Co}(6.4 \times 10^7 \text{ ergs/cm}^3) \quad (001)$$
$$K_u \times t_{Co} = 1.4 \text{ ergs/cm}^2 - t_{Co}(4.2 \times 10^7 \text{ ergs/cm}^3) \quad (011)$$

which again compared favorably with the following measured anisotropies for the three different growth directions:

$$K_u \times t_{Co} = 1.9 \text{ ergs/cm}^2 - t_{Co}(0.7 \times 10^7 \text{ ergs/cm}^3) \quad (111)$$
$$K_u \times t_{Co} = 1.2 \text{ ergs/cm}^2 - t_{Co}(5.9 \times 10^7 \text{ ergs/cm}^3) \quad (001)$$
$$K_u \times t_{Co} = 0.8 \text{ ergs/cm}^2 - t_{Co}(2.0 \times 10^7 \text{ ergs/cm}^3) \quad (011)$$

2.6.4. Superlattices with Diffuse Interfaces

In the case of an unstrained (111) superlattice, the Neel model provided a simple expression which shows how the interface anisotropy was reduced significantly when there was interdiffusion of atoms at the interface. This loss of anisotropy with increasing roughness is consistent with experimental observations as mentioned previously.

Some first-principle calculations of rough interfaces have been performed. The interdiffused interface was treated in the coherent potential approximation. Several Co_1Pt_3 and Co_1Pd_3 with various degrees of interdiffusion were studied. The central Pt or Pd layer was assumed to be ordered and a simple trapezium distribution

Table 2.7. Magneto-Crystalline Anisotropy of Some Co_1Pd_3 and Co_1Pt_3 Superlattices Calculated from First-Principle Electronic Structure Calculations and Estimated from the Neel Model

Center Layer	Interface Layer	Buffer Layer	Energy (µRy)		Difference	
			Defect	Perfect	LKKR	Neel
$Co_{0.8}Pt_{.2}$	$Co_{0.1}Pt_{.9}$	Pt	16.8	64.6	$-1.5K_p$	$-1.0K_p$
$Co_{0.8}Pd_{.2}$	$Co_{0.1}Pd_{.9}$	Pd	28.8	56.6	$-1.0K_p$	$-1.0K_p$
$Co_{0.6}Pt_{.4}$	$Co_{0.2}Pt_{.8}$	Pt	8.0	64.6	$-1.8K_p$	$-1.6K_p$
$Co_{0.6}Pd_{.4}$	$Co_{0.2}Pd_{.8}$	Pd	14.6	56.6	$-1.5K_p$	$-1.6K_p$

representing the interdiffusion profile was assumed. As can be seen from Table 2.7 the first-principle calculations and Neel theory are in close agreement, and that modest amounts of interdiffusion have a dramatic influence on the size of the anisotropy. More comparisons can be found in Ref. 88.

It is interesting to note that the reduction in interface anisotropy due to alloying may find an application in a patterned recording media. In some recent experimental work from IBM, a Co/Pt superlattice which has sufficient interface anisotropy to produce a perpendicular magnetization is covered by a mask with an array of holes in it. The sample is irradiated with N ions, which cause roughening at the interfaces in the exposed regions. The reduced interface anisotropy is now insufficient to overcome the shape anisotropy and an array of regions with in-plane anisotropy are formed.

ACKNOWLEDGMENTS

I would like to acknowledge the support of DARPA under grant MDA972-97-1-003, and NSF award numbers DMR # 9971573 and NSF/LEQSF (2001-04)-RII-03. In addition, I have benefited from many valuable discussions with colleagues, students, and post-docs, including Randall Victora, Jan van Ek, Weidong Huang, and Stephan Kurth.

APPENDIX A: A BRIEF OVERVIEW OF DENSITY FUNCTIONAL THEORY

Electronic structure calculations by using density functional theory have been successful in providing a better understanding of the chemical and physical properties of solids. These calculations are capable of predicting many ground-state properties, including crystal structures and equilibrium lattice constants. In order to calculate the electronic structure and hence physical properties of solids, several approximations need to be made. Since the electrons are much lighter than the nuclei, these nuclei can be viewed as fixed point charges, providing an electrostatic field in which the electrons move. This is known as the Born–Oppenheimer approximation. One can first solve for the electronic states with the nuclei treated as frozen.

The ground state energy obtained, as a function of nuclear positions, can be used as a potential energy to solve for the motion of the nuclei. The resulting interacting electron problem is still intractable for a solid which contains many interacting electrons, and further approximation is needed. The most successful approach is that of density functional theory, and this is the most commonly used approach in modern electronic structure calculations. Electronic structure calculation which are based upon density functional theory are founded in work performed in the 1960s and described in a paper by Hohenberg and Kohn,[5] and a later paper by Kohn and Sham.[6]

In the paper by Hohenberg and Kohn, two important theorems were proved. First, it was shown that the total energy of the ground state of an inhomogeneous interacting electron gas in an external potential, $V_{ext}(\mathbf{r})$, was a functional of the density, and that the total energy E of such a system can be written as

$$E = \int V_{ext}(\mathbf{r})n(\mathbf{r})d\mathbf{r} + \frac{1}{2} \int \int \frac{n(\mathbf{r})n(\mathbf{r}')}{|\mathbf{r}-\mathbf{r}'|} d\mathbf{r}d\mathbf{r}' + G[n] \qquad (2A.1)$$

where $n(\mathbf{r})$ is the electron density, and $G[n]$ is a universal functional of the electron density $n(\mathbf{r})$. Second, the total energy of the system is minimized for the true ground-state electron density. A variational principle for the energy was derived. This result was shown to follow from the variational property of the energy as computed via the Schrödinger equation. This variational principle, subject to conserving the total number of electrons, results in the following Euler equation:

$$\frac{\delta E}{\delta n[r]} = \mu \qquad (2A.2)$$

μ is a Lagrange multiplier associated with conserving the total number of electrons. Kohn and Sham observed that these results would also apply to a non-interacting system. Then, the universal functional G is divided into a kinetic energy term and the remainder which is called the exchange and correlation energy:

$$G[n] = T_s[n] + E_{xc}[n] \qquad (2A.3)$$

From the stationary property of Eq. (2A.1), subject to the condition of conserving the total number of electrons,

$$\int \delta n(\mathbf{r})d\mathbf{r} = 0 \qquad (2A.4)$$

we have

$$\int \delta n(\mathbf{r}) \left[\frac{\delta T_s[n]}{\delta n(\mathbf{r})} + V_{ext}(\mathbf{r}) + \int \frac{n(\mathbf{r}')}{|\mathbf{r}-\mathbf{r}'|} d\mathbf{r}' + V_{xc}[n(\mathbf{r})] \right] d\mathbf{r} = 0 \qquad (2A.5)$$

Kohn and Sham made the important observation that the solution to this equation was equivalent to solving a set of non-interacting single-particle Schrödinger equations:

$$\left[-\frac{1}{2}\nabla^2 + V_{ext}(\mathbf{r}) + \int \frac{n(\mathbf{r}')}{|\mathbf{r}-\mathbf{r}'|} d\mathbf{r}' + V_{xc}[n(\mathbf{r})] \right] \psi_i(\mathbf{r}) = \epsilon_i \psi_i(\mathbf{r}) \qquad (2A.6)$$

equations for a particular effective potential.* The effective potential consists of the sum of the external potential, the Hartree potential, and the exchange correlation potential which is given by the functional derivative with respect to the density of the exchange correlation energy:

$$V_{xc}[n(\mathbf{r})] = \frac{\delta E_{xc}[n]}{\delta n} \quad (2A.7)$$

The density is found by occupying the Kohn–Sham orbitals according to the Aufbau principle, i.e., by just populating the lowest lying orbitals:

$$n(\mathbf{r}) = \sum_{j}^{occ.} |\psi_j|^2 \quad (2A.8)$$

The Kohn–Sham formalism has an appealing physical picture, since each electron is considered as an independent particle moving in the mean field of the other electrons and the field of nuclei. While the Kohn–Sham orbitals are those of a fictitious system which has the same density and total energy as the real many-body system, and as such do not have a physical meaning; in practice, the Kohn–Sham orbitals and the single-particle density of states resulting from them can provide a framework for interpretation and calculation of physical properties. The Kohn–Sham wave functions and eigenvalues are often used in an evaluation of the Kubo formula for the optical conductivity. Results are usually in good agreement with measured optical properties. Other spectroscopies including photoemission, electron energy loss spectroscopy, and low energy electron diffraction, e.g., are also described quite accurately within a Kohn–Sham framework.

The exact functional form of the exchange correlation energy is not known, though there are several approximate forms which are quite accurate, these approximate forms are mostly founded on one further simplification, the LDA. The LDA assumes that at each point in space, the energy can be written in terms of the exchange correlation energy per electron of a uniform electron gas at that density, i.e.,

$$E_{xc}[n(\mathbf{r})] = \int n(\mathbf{r}) \epsilon_{xc}[n(\mathbf{r})] \quad (2A.9)$$

Simple parameterized estimates for ϵ_{xc} have been obtained for the uniform electron gas by several researchers.[6,68,117–125] Most are based upon random phase approximation calculations for the electron gas, though one, the most sophisticated, is based upon Monte Carlo simulations by Ceperley and Alder.[126,127] Most of these functionals give somewhat similar properties.

Efforts have been made to improve the LDA by including information about the gradient of electron density into the exchange-correlation energy functional. The most successful of these is the generalized gradient approximation (GGA), and a parameterization suitable for use in electronic structure calculations can be found in Refs. 128–130.

---

*Hartree atomic units have been used in this Schrödinger equation, in which $\hbar = m = e = 1$, and 1 Hartree ≈ 27.2 eV.

APPENDIX B: MATRIX ELEMENTS OF THE SPIN–ORBIT COUPLING AND GRADIENT OPERATORS

The matrix elements in an ℓ, m, σ basis of the spin–orbit coupling operator are most easily obtained by noting that

$$\mathbf{L}\cdot\mathbf{S} = L_x S_x + L_y S_y + L_z S_z = \frac{1}{2}(L_+ S_- + L_- S_+) + L_z S_z \qquad (2\text{B}.1)$$

where $L_\pm = L_x \pm iL_y$ and $S_\pm = S_x \pm iS_y$ are the usual orbital and spin angular momentum raising and lowering operators. Using the usual representation of the spin operators, we can write in spin space

$$2\mathbf{L}\cdot\mathbf{S} = \begin{pmatrix} L_z & L_- \\ L_+ & -L_z \end{pmatrix} \qquad (2\text{B}.2)$$

For a magnetization quantization direction which is described by a polar angle θ and an azimuthal angle ϕ, the two spin functions with eigenvalues $\pm\hbar/2$ diagonalize $\mathbf{S}\cdot\hat{n}$, where $\hat{n}$ is a unit vector along the quantization axis. These eigenfunctions are[1,95]

$$|\uparrow\rangle = \begin{bmatrix} \exp(-i\phi/2)\,\cos\theta/2 \\ \exp(i\phi/2)\,\sin\theta/2 \end{bmatrix}$$

$$|\downarrow\rangle = \begin{bmatrix} -\exp(-i\phi/2)\,\sin\theta/2 \\ \exp(i\phi/2)\,\cos\theta/2 \end{bmatrix} \qquad (2\text{B}.3)$$

Using these spin-up and spin-down functions and the properties of the raising and lowering operators

$$J_+|j,m\rangle = [j(j+1) - m(m+1)]^{1/2}|j,m+1\rangle$$
$$J_-|j,m\rangle = [j(j+1) - m(m-1)]^{1/2}|j,m-1\rangle \qquad (2\text{B}.4)$$

where $J_\pm$ can be either $L_\pm$ or $S_\pm$, the four spin-blocks are

$$\langle \ell m \uparrow |2\mathbf{L}\cdot\mathbf{S}|\ell'm'\uparrow\rangle = m\cos\theta\,\delta_{m,m'}\delta_{\ell,\ell'}$$

$$+ \frac{1}{2}[\ell(\ell+1) - m(m-1)]^{1/2}\sin\theta\,\exp(-i\phi)\delta_{m,m'+1}\delta_{\ell,\ell'}$$

$$+ \frac{1}{2}[\ell(\ell+1) - m(m+1)]^{1/2}\sin\theta\,\exp(i\phi)\delta_{m,m'-1}\delta_{\ell,\ell'}$$

$$\langle \ell m \uparrow |2\mathbf{L}\cdot\mathbf{S}|\ell'm'\downarrow\rangle = m\sin\theta\,\delta_{m,m'}\delta_{\ell,\ell'}$$

$$+ \frac{1}{2}[\ell(\ell+1) - m(m-1)]^{1/2}\sin^2\theta/2\,\exp(-i\phi)\delta_{m,m'+1}\delta_{\ell,\ell'}$$

$$- \frac{1}{2}[\ell(\ell+1) - m(m+1)]^{1/2}\cos^2\theta/2\,\exp(i\phi)\delta_{m,m'-1}\delta_{\ell,\ell'}$$

$$\langle \ell m \downarrow |2\mathbf{L}\cdot\mathbf{S}|\ell'm'\uparrow\rangle = m\sin\theta\,\delta_{m,m'}\delta_{\ell,\ell'}$$
$$-\frac{1}{2}[\ell(\ell+1)-m(m-1)]^{1/2}\cos^2\theta/2\,\exp(-i\phi)\delta_{m,m'+1}\delta_{\ell,\ell'}$$
$$+\frac{1}{2}[\ell(\ell+1)-m(m+1)]^{1/2}\sin^2\theta/2\,\exp(i\phi)\delta_{m,m'-1}\delta_{\ell,\ell'} \quad (2\text{B}.5)$$
$$\langle \ell m \downarrow |2\mathbf{L}\cdot\mathbf{S}|\ell'm'\downarrow\rangle = -m\cos\theta\,\delta_{m,m'}\delta_{\ell,\ell'}$$
$$-\frac{1}{2}[\ell(\ell+1)-m(m-1)]^{1/2}\sin\theta\,\exp(-i\phi)\delta_{m,m'+1}\delta_{\ell,\ell'}$$
$$-\frac{1}{2}[\ell(\ell+1)-m(m+1)]^{1/2}\sin\theta\,\exp(i\phi)\delta_{m,m'-1}\delta_{\ell,\ell'}$$

In computing the optical conductivity, the matrix elements of the gradient operator are required. These are somewhat tedious to derive for a function expressed as a product of a radial function times a spherical harmonic.[131] These are

$$\nabla_0 F(r) Y_{\ell m}(\hat{r}) = \left[\frac{(\ell+m+1)(\ell-m+1)}{(2\ell+1)(2\ell+3)}\right]^{1/2}\left[\frac{dF}{dr}-\frac{\ell}{r}F\right]Y_{\ell+1m}(\hat{r})$$
$$+ \left[\frac{(\ell+m)(\ell-m)}{(2\ell-1)(2\ell+1)}\right]^{1/2}\left[\frac{dF}{dr}+\frac{\ell+1}{r}F\right]Y_{\ell-1m}(\hat{r})$$

$$\nabla_+ F(r) Y_{\ell m}(\hat{r}) = \left[\frac{(\ell+m+1)(\ell+m+2)}{2(2\ell+1)(2\ell+3)}\right]^{1/2}\left[\frac{dF}{dr}-\frac{\ell}{r}F\right]Y_{\ell+1m+1}(\hat{r}) \quad (2\text{B}.6)$$
$$- \left[\frac{(\ell-m)(\ell-m-1)}{2(2\ell-1)(2\ell+1)}\right]^{1/2}\left[\frac{dF}{dr}+\frac{\ell+1}{r}F\right]Y_{\ell-1m+1}(\hat{r})$$

$$\nabla_- F(r) Y_{\ell m}(\hat{r}) = \left[\frac{(\ell-m+1)(\ell-m+2)}{2(2\ell+1)(2\ell+3)}\right]^{1/2}\left[\frac{dF}{dr}-\frac{\ell}{r}F\right]Y_{\ell+1m-1}(\hat{r})$$
$$- \left[\frac{(\ell+m)(\ell+m-1)}{2(2\ell-1)(2\ell+1)}\right]^{1/2}\left[\frac{dF}{dr}+\frac{\ell+1}{r}F\right]Y_{\ell-1m-1}(\hat{r})$$

where

$$\nabla_0 = \frac{\partial}{\partial z}$$
$$\nabla_+ = -\frac{1}{2^{1/2}}\left(\frac{\partial}{\partial x}+i\frac{\partial}{\partial y}\right) \quad (2\text{B}.7)$$
$$\nabla_- = \frac{1}{2^{1/2}}\left(\frac{\partial}{\partial x}-i\frac{\partial}{\partial y}\right)$$

REFERENCES

1. M. Rose, *Relativistic Electron Theory* (Wiley, New York, 1961).
2. J. D. Jackson, *Classical Electrodynamics*, 2nd ed. (Wiley, New York, 1975).
3. J. Slater, *Quantum Theory of Atomic Structure*, vol. II (McGraw-Hill, New York, 1960).
4. J. Gay, in *Ultrathin Magnetic Films*, ed. J. A. C. Bland and B. Heinrich (Springer Verlag, Berlin, 1994).

5. H. Hohenberg and W. Kohn, *Phys. Rev.* **136**, B864 (1964).
6. W. Kohn and L. Sham, *Phys. Rev.* **140**, A1133 (1965).
7. A. Rajagopal, *J. Phys. C* **11**, L943 (1978).
8. A. MacDonald and S. Vosko, *J. Phys. C* **12**, 2977 (1979).
9. D. Singh, *Planewaves, Pseudopotentials and the LAPW Method* (Kluwer Academic Publishers, Boston, 1994).
10. R. Feder and F. Rosicky, *Z. Phys. B* **52**, 52 (1983).
11. P. Strange, J. Staunton, and B. Gyorffy, *J. Phys. C* **17**, 3355 (1984).
12. E. Tamura, *Phys. Rev. B*, 45 (1992).AQ
13. L. Szunyogh, B. Újfalussy, and P. Weinberger, *Phys. Rev. B* **51**, 9552 (1995).
14. H. L. Skriver, *The LMTO Method, Muffin–Tin Orbitals and Electronic Structure* (Springer-Verlag, Berlin, 1984).
15. O. Andersen, *Phys. Rev. B* **12**, 3060 (1975).
16. H. Ebert, *Phys. Rev. B* **38**, 9381 (1988).
17. G. Guo, W. Temmerman, and H. Ebert, *J. Phys.: Condens. Matter* **3**, 8205 (1991).
18. G. Y. Guo and H. Ebert, *Phys. Rev. B* **50**, 10377 (1994).
19. D. Koelling and B. Harmon, *J. Phys. C* **10**, 3107 (1977).
20. M. Faraday, *Trans. R. Soc. (London)* **5**, 592 (1846).
21. J. Kerr, *Philos. Mag.* **3**, 339 (1877).
22. J. Kerr, *Philos. Mag.* **5**, 161 (1878).
23. G. Metzger, P. Pluvinage, and R. Torguet, *Ann. Phys.* **10**, 5 (1965).
24. M. Freiser, *IEEE Trans. Magn.* **MAG-4**, 152 (1967).
25. A. V. Sokolov, *Optical Properties of Metals* (Elsevier, New York, 1968).
26. H. Bennett and E. Stern, *Phys. Rev.* **137**, A448 (1965).
27. J. Erskine and E. Stern, *Phys. Rev. B* **8**, 1239 (1973).
28. J. Erskine and E. Stern, *Phys. Rev. B* **12**, 5016 (1975).
29. C. Kittel, *Rev. Mod. Phys.* **21**, 541 (1949).
30. F. Bloch and G. Gentile, *Z. Phys.* **70**, 395 (1931).
31. J. van Vleck, *Phys. Rev.* **52**, 1178 (1937).
32. H. Brooks, *Phys. Rev.* **58**, 909 (1940).
33. G. Fletcher, *Proc. Phys. Soc. (London)* **A67**, 505 (1954).
34. L. Neel, *J. de Physique et le Radium* **15**, 225 (1954).
35. A. Bennett and B. Cooper, *Phys. Rev. B* **3**, 1642 (1971).
36. H. Takayama, K.-P. Bohnen, and P. Fulde, *Phys. Rev. B* **14**, 2287 (1976).
37. P. Bruno, *Phys. Rev. B* **39**, 865 (1989).
38. M. Cinal, D. Edwards, and J. Mathan, *Phys. Rev. B* **50**, 3754 (1994).
39. C.-W. Chen, *Magnetism and Metallurgy of Soft Magnetic Materials* (Dover, Minealo, New York, 1986).
40. R. Bozorth, *Ferromagnetism* (IEEE Press, New York, 1993).
41. S. Chikazumi, *Physics of Magnetism* (Wiley, New York, 1964).
42. J. Zak, E. Moog, C. Liu, and S. Bader, *J. Appl. Phys.* **68**, 4203 (1990a).
43. J. Zak, E. Moog, C. Liu, and S. Bader, *Phys. Rev. B* **45**, 6423 (1990b).
44. M. Mansuripur, *J. Appl. Phys.* **67**, 6466 (1990).
45. L. Landau and E. Lifshitz, *Electrodynamics of Continuous Media* (Pergamon Press, London, 1960).
46. N. McGee, M. Johnson, J. de Vries, and J. aan de Stegge, *J. Appl. Phys.* **73**, 3418 (1993).
47. S. Purcell, M. Johnson, N. McGee, J. de Vries, W. Zeper, and W. Hoving, *J. Appl. Phys.* **73**, 1360 (1993).
48. J. Florczak and E. D. Dahlberg, *Phys. Rev B* **44**, 9338 (1913).
49. C. Wang and J. Callaway, *Phys. Rev. B* **9**, 4897 (1974).
50. J. Callaway, *Quantum Theory of the Solid State*, 2nd ed. (Academic Press, New York, 1991).
51. H. Ebert, P. Strange, and B. Gyorffy, *J. Appl. Phys.* **63**, 3055 (1988).
52. P. Oppeneer, T. Maurer, J. Sticht, and J. Kübler, *Phys. Rev. B* **45**, 10924 (1992a).
53. P. M. Oppeneer, J. Sticht, T. Maurer, and J. Kübler, *Z. Phys. B* **88**, 309 (1992b).
54. X. Wang, V. Antropov, and B. Harmon, *IEEE Trans. Magn.* **30**, 4458 (1994).
55. Y. A. Uspenskii, E. Kulatov, and S. Khalilov, *JETP* **80**, 952 (1995).
56. G. Y. Guo and H. Ebert, *Phys. Rev. B* **51**, 12633 (1995).

57. P. Oppeneer, V. Antonov, T. Kraft, H. Eschrig, A. Yaresko, and A. Perlov, *Solid State Commun.* **94**, 255 (1995).
58. J. MacLaren and W. Huang, *J. Appl. Phys.* **79**, 6196 (1996).
59. N. Mainkar, D. Browne, and J. Callaway, *Phys. Rev. B* **53**, 3692 (1996).
60. J. van Ek and J. MacLaren, *Phys. Rev. B* **56**, R2924 (1997).
61. J. van Ek, W. Huang, and J. MacLaren, , *J. Appl. Phys.* **81**, 5429 (1997).
62. G. D. Mahan, *Many Particles Physics* (Plenum, New York, 1981).
63. P. Oppeneer, V. Antonov, A. Yaresko, A. Perlov, T. Kraft, and H. Eschrig, in Proceedings of the Magneto-Optical Recording International Symposium '96, *J. Mag. Soc. Japan* **20**, S1, p. 41 (1996).
64. R. F. Sabiryanov and S. S. Jaswal, *Phys. Rev. B* **53**, 313 (1996).
65. C. Shang, *J. Phys. D* **29**, 277 (1996).
66. G. Q. Di, S. Iwata and S. Uchiyama, *J. Magn. Magn. Mater.* **131**, 242 (1994).
67. G. Fernando, J. Davenport, R. Watson, and M. Weinert, *Phys. Rev. B* **40**, 2757 (1989).
68. J. Perdew and A. Zunger, *Phys. Rev. B* **23**, 5048 (1981).
69. A. P. Lenham and D. M. Treherne, *J. Opt. Soc. Am.* **56**, 1137 (1966a). [This paper contains the corrected optical data appearing in the previous reference.]
70. A. P. Lenham and D. M. Treherne, *Optical Properties and Electronic Structure of Metals and Alloys*. p. 196 (North Holland, Amsterdam, 1966b).
71. P. B. Johnson and R. W. Christy, *Phys. Rev. B* **9**, 5056 (1974).
72. G. A. Bolotin, M. M. Kirillova, and V. M. Mayevskiy, *Fiz. Met. Metalloved* **27**, 224 (1969).
73. J. H. Weaver, E. Colavita, D. W. Lynch, and R. Rosei, *Phys. Rev. B* **19**, 3850 (1979).
74. H. T. Yolken and J. Kruger, *J. Opt. Soc. Am.* **55**, 892 (1965).
75. K. Buschow, *Ferromagnetic Materials*, vol. 4. (Elsevier, Amsterdam, 1988).
76. G. S. Krinchik and V. A. Artem'ev, *Sov. Phys. JETP* **26**, 1080 (1968).
77. R. de Groot, F. Mueller, P. van Engen, and K. Buschow, *Phys. Rev. Lett.* **50**, 2024 (1983).
78. R. de Groot, F. Mueller, P. van Engen, and K. Buschow, *J. Appl. Phys.* **55**, 2151 (1984).
79. P. van der Heide, W. Baelde, R. de Groot, A. de Vroomen, P. van Engen, and K. Buschow, *J. Phys. F: Met. Phys.* **15**, L75 (1985).
80. P. van Engen, K. Buschow, R. Jongebreur, and M. Erman, *Appl. Phys. Lett.* **42**, 202 (1983a).
81. S. Youn and B. Min, *Phys. Rev. B* **51**, 10436 (1995).
82. H. Feil and C. Haas, *Phys. Rev. Lett.* **58**, 65 (1987).
83. G. Guo, W. Temmerman, and H. Ebert, *J. Magn. Magn. Mater.* **104–107**, 1772 (1992).
84. B. Újfalussy, L. Szunyogh, and P. Weinberger, *Phys. Rev. B* **54**, 9833 (1996).
85. D.-S. Wang, R. Wu, and A. Freeman, *Phys. Rev. Lett.* **70**, 869 (1993a).
86. P. Bruno, *J. Phys. F: Met. Phys.* **18**, 1291 (1988).
87. R. Victora and J. MacLaren, *J. Appl. Phys.* **73**, 6415 (1993a).
88. J. MacLaren and R. Victora, *J. Appl. Phys.* **76**, 6069 (1994).
89. H. Draaisma, F. den Broeder, and W. de Jonge, *J. Appl. Phys.* **63**, 3479 (1988).
90. B. Engel, C. England, R. V. Leeuwen, M. Wiedmann, and C. Falco, *Phys. Rev. Lett.* **67**, 1910 (1991).
91. F. den Broeder, W. Hoving, and P. Bloemen, *J. Magn. Magn. Mater.* **93**, 562 (1991).
92. H. Draaisma, W. de Jonge, and F. den Broeder, *J. Magn. Magn. Mater.* **66**, 351 (1987).
93. D. Stinson and S.-C. Shin, *J. Appl. Phys.* **67**, 4459 (1990).
94. R. Victora and J. MacLaren, *Phys. Rev. B* **47**, 11919 (1993b).
95. J. MacLaren and R. Victora, *IEEE Trans. Magn.* **29**, 3034 (1993).
96. G. Daalderop, P. Kelly, and M. Schuurmans, *Phys. Rev. B* **42**, 7270 (1990a).
97. G. Daalderop, P. Kelly, and M. Schuurmans, *Science and Technology of Nanostructured Magnetic Materials*, p. 185 (Plenum, New York and London, 1990c).
98. F. den Broeder, D. Kuiper, A. van de Mosselaar, and W. Hoving, *Phys. Rev. Lett.* **60**, 2769 (1988).
99. W. Mason, *Phys. Rev.* **96**, 302 (1954).
100. J. Gay and R. Richter, *Phys. Rev. Lett.* **56**, 2728 (1986).
101. J. Gay and R. Richter, *J. Appl. Phys.* **61**, 3362 (1987).
102. C. Li, A. J. Freeman, H. Jansen, and C. Fu, *Phys. Rev. B* **42**, 5433 (1990).
103. A. Mackintosh and O. Andersen, *Electrons at the Fermi Surface*. ed. M. Springford (Cambridge University Press, Cambridge, 1980).
104. M. Weinert, R. Watson, and J. Davenport, *Phys. Rev. B* **32**, 2115 (1985).
105. X. Wang, D.-S. Wang, R. Wu, and A. Freeman, *J. Magn. Magn. Mater.* **159**, 337 (1996).

106. D.-S. Wang, R. Wu, and A. Freeman, *J. Appl. Phys.* **73**, 6745 (1993b).
107. G. Daalderop, P. Kelly, and M. Schuurmans, in *Ultrathin Magnetic Films*, ed. J. A. C. Bland and B. Heinrich (Springer Verlag, Berlin, 1994).
108. D.-S. Wang, R. Wu, and A. Freeman, *Phys. Rev. B* **48**, 15886 (1993c).
109. J. Trygg, B. Johansson, O. Eriksson, and J. Wills, *Phys. Rev. Lett.* **75**, 2871 (1995).
110. G. Daalderop, P. Kelly, and M. Schuurmans, *Phys. Rev. B* **41**, 11919 (1990b).
111. O. Eriksson, M. Brooks, and B. Johansson, *Phys. Rev. B* **41**, 7311 (1990).
112. R. Wu and A. Freeman, *J. Appl. Phys.* **79**, 6209 (1996).
113. B. Heinrich and J. F. Cochran *Adv. Phys.* **42**, 523 (1993).
114. M. Johnson, P. Bloemen, F. den Broeder, and J. de Vries, *Rep. Prog. Phys.* **59**, 1409 (1996).
115. F. den Broeder, D. Kuiper, H. Donkerslot, and W. Hoving, *Appl. Phys. A.* **49**, 507 (1989).
116. R. Farrow, C. Lee, R. Marks, P. Harp, M. Toney, T. Rabedeau, D. Weller, and H. Brandle, in *Proceedings of the NATO Advanced Study Institute Series B*, Vol. 309, ed. R. F. C. Farrow, B. Dieny, M. H. Donath, A. Fert, B. D. Hermsmeier (Plenum, New York and London, 1993).
117. J. Slater, *Phys. Rev.* **81**, 385 (1951).
118. R. Gaspár, *Acta. Phys. Hungaria* **3**, 263 (1954).
119. L. Hedin and B. Lundqvist, *J. Phys. C* **4**, 2064 (1971).
120. O. Gunnarsson, B. Lundqvist, and S. Lundqvist, *Solid State Commun.* **11**, 149 (1972).
121. O. Gunnarsson, B. Lundqvist, and J. Wilkins, *Phys. Rev. B* **10**, 1319 (1974).
122. O. Gunnarsson and B. Lundqvist, *Phys. Rev. B* **13**, 4274 (1976).
123. U. von Barth and L. Hedin, *J. Phys. C* **5**, 1629 (1972).
124. J. Janak, V. Moruzzi, and A. Williams, *Phys. Rev. B* **12**, 1257 (1975).
125. S. Vosko, L. Wilk, and M. Nusair, *Can. J. Phys.* **58**, 1200 (1980).
126. D. Ceperley and B. Alder, *Phys. Rev. B* **18**, 3126 (1978).
127. D. Ceperley and B. Alder, *Phys. Rev. Lett.* **45**, 566 (1980).
128. J. Perdew, J. Chevary, S. Vosko, K. Jackson, M. Pederson, D. Singh, and C. Fiolhais, *Phys. Rev. B* **46**, 6671 (1992).
129. J. Perdew, J. Chevary, S. Vosko, K. Jackson, M. Pederson, D. Singh, and C. Fiolhais, *Phys. Rev. B* **48**, 4978 (1993).
130. S. Kurth, J. Perdew, and P. Blaha, *Int. J. Quantum Chem. Symp.* **75**, 889 (1999).
131. A. Edmonds, *Angular Momentum*, 2nd ed. (Princeton University Press, Princeton, New Jersey, 1960).
132. P. G. van Engen, K. H. Buschow, and M. Erman, *J. Magn. Magn. Mater.* **30**, 374 (1983b).
133. P. E. Ferguson and R. J. Romagnoli, *J. Appl. Phys.* **40**, 1236 (1969).
134. L. Fritsche, J. Noffke, and H. Eckardt, *J. Phys. F* **17**, 943 (1987).
135. M. Stearns, "3d, 4d and 5d Elements, Alloys and Compounds," in *Landolt Börnstein*, Vol. 19a of New Series III, p. 41 (Springer Verlag, Berlin).
136. A. Freeman, O. Myrasov, D.-S. Wang, and R. Wu, *Mater. Sci. Eng. B* **31**, 225 (1995).

3

Spin-Dependent Transport in Magnetic Multilayers

William Butler*

3.1. INTRODUCTION

It has long been known that electrons in metals have two spin states and that when an electric field is applied to a metal, two approximately independent currents flow. In non-magnetic metals such as copper, these two spin channels are equivalent, in the sense that they have the same Fermi energy density of states (DOS) and the same electron velocities, but in the ferromagnetic transition metals they are usually quite different. Until recently, however, it was only the total conductivity resulting from these two parallel currents or spin channels that was important for most physically observable phenomena. This situation changed dramatically with the discovery of the giant magnetoresistance effect[1,2] and the advent of reliable and reproducible observations of spin-dependent tunneling.[3-5] These technologically important effects exist only because of the differences in electron transport between the two spin channels.

In order to understand spin-dependent transport, we must first understand how applied electric fields lead to currents in metals. In the transition and noble metals that we will be concerned with here, it is often a good approximation to consider the currents carried by the two spin channels to be independent. This approximation is usually called the "two-current model" and is valid as long as we agree to neglect the (usually) small coupling between the spin and orbital motions of the electrons, and we assume that the moments on the atoms are collinear. In this case, spin-dependent transport is conceptually no more complicated than ordinary transport. Simply, we must calculate the conductances for each of the two spin channels and add them to get the total conductance.

In Section 3.2, we shall discuss the conductivity as the quantum mechanical linear response of a system to an applied electric field; in Section 3.3, we shall obtain some closed form solutions for the simplest case we can think of, free electrons with random point scatterers; and in Section 3.4, we shall develop the semi-classical model for electron transport for realistic systems.

---

*University of Alabama, Tuscaloosa, Alabama and Metals and Ceramics Division, Oak Ridge National Laboratory, Oak Ridge, Tennessee.

3.2. KUBO–GREENWOOD FORMULA FOR CONDUCTIVITY

Kubo[6] and Greenwood[7] have shown that the conductivity (defined to be the *linear* response of the current to an applied electric field) may be obtained from the wave functions of the system evaluated in the *absence* of the applied field. Thus, the zero-temperature dc conductivity is given by

$$\sigma_{\mu\nu} = \frac{\pi\hbar}{V}\left\langle \sum_{\alpha,\alpha'}\langle\alpha|j_\mu|\alpha'\rangle\langle\alpha'|j_\nu|\alpha\rangle\right\rangle\delta(E_F - E_{\alpha'})\delta(E_F - E_\alpha) \quad (3.2.1)$$

where j_μ is the current operator $(-ie\hbar/m)\partial/\partial r_\mu$, V is the volume, and the quantum states $|\alpha\rangle$ are the exact eigenfunctions of a particular configuration of the atoms. The large angle brackets indicate an average over the atomic configurations, i.e., over the types, positions, etc. of the atoms. Equation (3.2.1) can be written in the form*

$$\sigma_{\mu\nu} = \frac{\hbar}{\pi V}\text{Tr}\langle j_\mu \text{Im } G(E_F)j'_\nu \text{Im } G(E_F)\rangle \quad (3.2.2)$$

by using the Green function, defined as $G = [E - H]^{-1}$ which is related to the sum over states in Eq. (3.2.1) through

$$\sum_\alpha |\alpha\rangle\langle\alpha|\delta(E - E_\alpha) = -\frac{1}{\pi}\lim_{\eta\to 0}\text{Im } G(E + i\eta) \quad (3.2.3)$$

Note that the Kubo–Greenwood formula requires an average over the product of two Green functions, $\langle GG\rangle$, rather than the product of the average of Green functions, $\langle G\rangle\langle G\rangle$. The error made when the former is approximated by the latter is known as "the neglect of vertex corrections" and is in general, quite serious. In the semi-classical limit, this approximation is equivalent to the "life-time" approximation or the "neglect of the scattering-in term of the Boltzmann equation."

Based on the Kubo–Greenwood formula, it is possible to write the conductivity as a non-local kernel:

$$\sigma_{\mu\nu}(\mathbf{r},\mathbf{r}') = \frac{\hbar}{\pi}\langle j_\mu\text{Im } G(\mathbf{r},\mathbf{r}';E_F)j'_\nu\text{Im } G(\mathbf{r}',\mathbf{r};E_F)\rangle \quad (3.2.4)$$

which emphasizes that the current at a point $\mathbf{r}$ depends not only on the electric field at that point but also on the field at points $\mathbf{r}'$ within the vicinity (approximately the electronic mean-free path) of $\mathbf{r}$:

$$J_\mu(\mathbf{r}) = \int d^3r' \sigma_{\mu\nu}(\mathbf{r},\mathbf{r}')\mathcal{E}_\nu(\mathbf{r}') \quad (3.2.5)$$

Thus, $\sigma_{\mu\nu}(\mathbf{r},\mathbf{r}')$ is the current in the direction μ at point $\mathbf{r}$ induced by an electric field in direction ν that exists at point $\mathbf{r}'$. The realization that the conductivity is non-local, i.e., that the current at one point depends on fields applied at other points is a key to understanding giant magnetoresistance for the technologically important case in which the current flows parallel to the planes of the multilayer.

---

*In the following, it will be assumed that the energy variable of the Green function has an infinitesimal positive imaginary part.

3.3. FREE ELECTRONS WITH RANDOM POINT SCATTERERS

In order to get a better understanding of this non-local conductivity, let us evaluate it for free electrons. The free electron Green function, $G_0(\mathbf{r},\mathbf{r}')$ is defined by

$$\left[\frac{\hbar^2}{2m}\nabla^2 + E\right]G_0(\mathbf{r},\mathbf{r}') = \delta(\mathbf{r}-\mathbf{r}') \qquad (3.3.1)$$

and can be written as

$$G_0(\mathbf{r},\mathbf{r}') = \frac{1}{(2\pi)^3}\int d^3k\,\frac{\exp[i\mathbf{k}\cdot(\mathbf{r}-\mathbf{r}')]}{E - \frac{\hbar^2 k^2}{2m}} = \frac{2m}{\hbar^2}\frac{\exp(i\kappa\cdot|\mathbf{r}-\mathbf{r}'|)}{4\pi|\mathbf{r}-\mathbf{r}'|} \qquad (3.3.2)$$

where $\kappa = (2mE)^{1/2}/\hbar$.

Without something to scatter the electrons, the electrical conductivity would be infinite.* In our example, we will choose the simplest scattering that we can think of, namely randomly distributed point scatterers. We shall call this model the FERPS model for "Free Electrons with Random Point Scatterers." The Green function in the presence of these scatterers located at points, R_i, satisfies

$$\left[\frac{\hbar^2}{2m}\nabla^2 + E - \sum_i v(\mathbf{r}-\mathbf{R}_i)\right]G(\mathbf{r},\mathbf{r}') = \delta(\mathbf{r}-\mathbf{r}') \qquad (3.3.3)$$

and can be expanded as follows:

$$G(\mathbf{r},\mathbf{r}') = G_0(\mathbf{r},\mathbf{r}') + \int d\mathbf{r}_1 G_0(\mathbf{r},\mathbf{r}_1) \sum_i v(\mathbf{r}_1 - \mathbf{R}_i) G(\mathbf{r}_1,\mathbf{r}') \qquad (3.3.4)$$

This integral equation for the Green function is known as the Lippmann–Schwinger equation and can be verified by substituting Eq. (3.3.4) into Eq. (3.3.3).

Let us write the Lippmann–Schwinger expression for the Green function including the random point scatterers by using the simplified notation, $G = G_0 + \sum_i G_0 v_i G$. Then, we can expand by substituting the entire expression for the G on the right-hand side:

$$G = G_0 + \sum_i G_0 v_i G_0 + \sum_{i,j} G_0 v_i G_0 v_j G_0 + \sum_{i,j,k} G_0 v_i G_0 v_j G_0 v_k G_0 + \cdots \qquad (3.3.5)$$

If we use angle brackets to denote an average over configurations, i.e., over the possible positions of the scatterers then we can assume that $\langle v_i \rangle = 0$, since a shift in the average potential can be accommodated as a shift in the energy zero; then, $\Delta v_i = v_i - \langle v_i \rangle$ and we can write

$$\langle G \rangle = G_0 + \sum_i G_0 \langle \Delta v_i G_0 \Delta v_i \rangle G_0 + \sum_i G_0 \langle \Delta v_i G_0 \Delta v_i G_0 \Delta v_i \rangle G_0$$
$$+ \sum_{ij} G_0 \langle \Delta v_i G_0 \Delta v_i \rangle G_0 \langle \Delta v_j G_0 \Delta v_j \rangle G_0 + \cdots$$

---

*More precisely, we would have ballistic rather than diffusive transport, and the resistance would not be proportional to the length of the sample in the direction of the field.

Then, we can write

$$\langle G \rangle = G_0 + G_0 \Sigma G_0 + G_0 \Sigma G_0 \Sigma G_0 + \cdots \qquad (3.3.6)$$

where

$$\Sigma = \langle \Delta v_i G_0 \Delta v_i \rangle + \langle \Delta v_i G_0 \Delta v_i G_0 \Delta v_i \rangle + \cdots \qquad (3.3.7)$$

If the potential differences, Δv_i, are very short ranged, relatively weak, and randomly distributed in space, Σ will be approximately independent of the electron momentum and position. It will, however, generally be a function of energy.

Finally, we have made enough approximations to be able to write down the average Green Function in the FERPS approximation:

$$\left[\frac{\hbar^2}{2m} \nabla^2 + E - \Sigma \right] \langle G(\mathbf{r}, \mathbf{r}') \rangle = \delta(\mathbf{r} - \mathbf{r}') \qquad (3.3.8)$$

$$\langle G(\mathbf{r}, \mathbf{r}') \rangle = \frac{1}{(2\pi)^3} \int d^3 \kappa \frac{\exp(i\mathbf{k} \cdot (\mathbf{r} - \mathbf{r}'))}{E - \Sigma - \frac{\hbar^2 k^2}{2m}} = \frac{2m}{\hbar^2} \frac{\exp(i\kappa |\mathbf{r} - \mathbf{r}'|)}{4\pi |\mathbf{r} - \mathbf{r}'|} \qquad (3.3.9)$$

where $\kappa = 2m(E - \Sigma)^{1/2}/\hbar$. In the following, we omit the angle brackets to simplify the notation, but it should be remembered that we are concerned with the average of the Green function over the atomic configurations.

To remind you of why we want the averaged Green function, here is the expression for the non-local conductivity again:

$$\sigma_{\mu\nu}(\mathbf{r}, \mathbf{r}') = \frac{\hbar}{\pi} j_\mu \text{Im } G(\mathbf{r}, \mathbf{r}') j'_\nu \text{Im } G(\mathbf{r}', \mathbf{r}) \qquad (3.3.10)$$

Note that here we have neglected the difference between $\langle GG \rangle$ and $\langle G \rangle \langle G \rangle$. This is called the neglect of "vertex corrections". It is an approximation that can be made in both the quantum and in the semi-classical approaches to transport. In the latter case, this approximation is called "neglect of the scattering-in terms." We shall show that whether or not these terms can be neglected for a layered systems depends on the geometry. For the case in which the current is perpendicular to the layers (CPP), we shall show in Section 3.4.6 that at least an approximate treatment of the vertex corrections is necessary for a consistent theory. However, for the case in which the current is in the plane of the layers (CIP) or for a homogeneous system, these terms do not contribute to the current if the scattering is isotropic. Since we will be primarily concerned with the latter case, we will neglect them for the time being.

We can now evaluate the conductivity for a homogeneous system by integrating over $\mathbf{r}'$ and averaging over $\mathbf{r}$ and directions (μ). Thus, $\mathbf{J} = \sigma_0 \mathcal{E}$, where

$$\sigma_0 = \frac{1}{3V} \sum_\mu \int d^3\mathbf{r} \int d^3\mathbf{r}' \sigma_{\mu\mu}(\mathbf{r}, \mathbf{r}')$$

$$= -\frac{1}{3} \frac{e^2 \hbar^3}{\pi m^2 V} \int d^3\mathbf{r} \int d^3\mathbf{r}' \nabla \text{Im } G(\mathbf{r}, \mathbf{r}') \cdot \nabla' \text{Im } G(\mathbf{r}', \mathbf{r})$$

Let $\mathbf{R} = \mathbf{r} - \mathbf{r}'$ and $|\mathbf{R}| = R$, we have

$$\sigma_0 = \frac{e^2\hbar^3}{3\pi m^2} \int d^3\mathbf{R} [\nabla \text{Im } G(R)]^2 \tag{3.3.11}$$

This integration can be performed exactly by using elementary techniques and yields

$$\sigma_0 = \frac{e^2}{12\pi^2\hbar} \frac{\kappa_R^2}{\kappa_I} \tag{3.3.12}$$

where $\kappa_R = \{\text{Re}[2m(E - \Sigma)]^{1/2}/\hbar\}$ and $\kappa_I = \text{Im}\{[2m(E - \Sigma)]^{1/2}/\hbar\}$. This is the same as the usual expression for free electrons if we identify $\kappa_R = k_F$ and $2\kappa_I = 1/\lambda$, where λ is the mean-free path:

$$\sigma_0 = \frac{e^2 k_F^2 \lambda}{6\pi^2\hbar} = \frac{k_F^3}{6\pi^2} \frac{e^2\tau}{m} = \frac{Ne^2\tau}{m} \tag{3.3.13}$$

Here, we used $\lambda = v_F \tau = \hbar k_F \tau/m$, where v_F is the Fermi velocity, τ, the electron lifetime, and N, the number of free electrons per unit volume.

The above exercise shows that the Kubo–Greenwood quantum mechanical linear response formalism applied to the FERPS model gives familiar results for a homogeneous system. In the preparation for dealing with layered systems, let us treat a homogeneous system as an artificial layered system by calculating the non-local conductivity that would arise if we could apply an electric field in a plane of vanishing thickness. Thus, the current density $J_\mu(z)$ induced in the direction μ in the plane z due to an electric field, $\mathcal{E}_\nu(z')$ applied in the direction ν to the plane z' (see Fig. 3.1) are related through

$$J_\mu(z) = \int dz' \sum_\nu \sigma_{\mu\nu}(z, z') \mathcal{E}_\nu(z') \tag{3.3.14}$$

This non-local conductivity is given by

$$\sigma_{\mu\nu}(z, z') = -\frac{e^2\hbar^3}{\pi m^2 A} \int dx\, dy \int dx'\, dy' \nabla_\mu \text{Im } G(\mathbf{r}, \mathbf{r}') \nabla'_\nu \text{Im } G(\mathbf{r}', \mathbf{r}) \tag{3.3.15}$$

which also can be evaluated exactly. The details of the evaluation are given in the appendix and in Ref. 8. The results for the current and field perpendicular to the planes z and z' can be written (using $Z = |z - z'|$) as

$$\sigma_{zz}(Z) = \sigma_0 \frac{3\kappa_I}{2\kappa_R^2}$$

$$\times [\kappa^2 E_3(-2i\kappa Z) + \kappa^{*2} E_3(2i\kappa^* Z) + 2|\kappa|^2 E_3(2\kappa_I Z)$$

$$+ \frac{2i\kappa}{Z} E_4(-2i\kappa Z) - \frac{2i\kappa^*}{Z} E_4(2i\kappa^* Z) + \frac{4\kappa_I}{Z} E_4(2\kappa_I Z)$$

$$- \frac{1}{Z^2} E_5(-2i\kappa Z) - \frac{1}{Z^2} E_5(2i\kappa^* Z) + \frac{2}{Z^2} E_5(2\kappa_I Z)] \tag{3.3.16}$$

and for the current and field parallel to the planes z and z',

$$\sigma_{xx}(Z) = \sigma_0 \frac{3\kappa_I}{4\kappa_R^2}$$
$$\times [\kappa^2 E_1(-2i\kappa Z) + \kappa^{*2} E_1(2i\kappa^* Z) + 2|\kappa|^2 E_1(2\kappa_I Z)$$
$$+ \frac{2i\kappa}{Z} E_2(-2i\kappa Z) - \frac{2i\kappa^*}{Z} E_2(2i\kappa^* Z) + \frac{4\kappa_I}{Z} E_2(2\kappa_I Z)$$
$$- \frac{1}{Z^2} E_3(-2i\kappa Z) - \frac{1}{Z^2} E_3(2i\kappa^* Z) + \frac{2}{Z^2} E_3(2\kappa_I Z)] - \frac{1}{2}\sigma_{zz}(Z) \quad (3.3.17)$$

Here, the functions, $E_n(x)$, are the exponential integrals which are defined by

$$E_n(x) = \int_1^\infty \frac{\exp(-xt)}{t^n} dt \quad (3.3.18)$$

The functions $\sigma_{zz}(Z)$ and $\sigma_{xx}(Z)$ give the current induced in the z and x directions, respectively, for electric fields applied at a distance z away in the z direction from

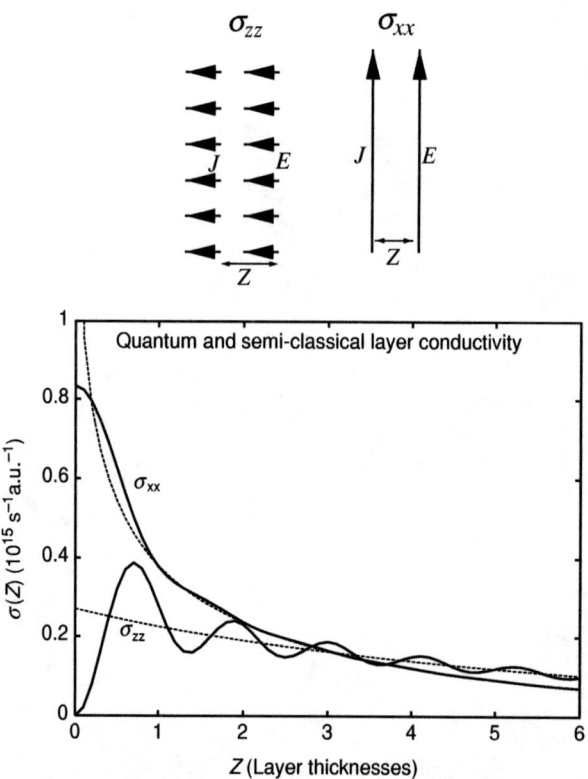

Figure 3.1. Top panel: geometry for σ_{zz} and σ_{xx}. Bottom panel: quantum and semi-classical non-local layer conductivity as a function of layer separation, Z, for a homogeneous free electron system. Solid lines are the quantum conductivity. Dashed lines are the semi-classical approximation. In this example, the lattice constant is that of copper (0.3615 nm); the Fermi momentum corresponds to 0.5 electrons per spin channel; and Z is measured in terms of the thickness of (111) layers of copper (0.209 nm). $\sigma(Z)$ is measured in units of 10^{15} s^{-1} a.u.$^{-1}$ where 1 a.u. = 0.0529 nm. Z is the distance between the plane at which the field is applied and the plane at which the current is induced.

SPIN-DEPENDENT TRANSPORT

the point where the current is induced. The currents are in the same direction as the applied fields. The geometry for σ_{zz} and σ_{xx} is indicated in Fig. 3.1. σ_{xx} is instructive concerning electron transport for the CIP geometry, while σ_{zz} is instructive concerning transport in the CPP geometry. The non-local conductivities, $\sigma_{xx}(Z)$ and $\sigma_{zz}(Z)$ are shown in Fig. 3.1 for an electron density approximately equal to that of copper. The CIP non-local conductivity decreases monotonically as a function of the distance between applied field and induced current, whereas the CPP non-local conductivity oscillates. Interestingly, for CPP, the applied field induces zero current at the plane where it is applied.

3.3.1. Semi-Classical Limit

The expressions for the non-local conductivity contain quantum interference effects. It is usually easier to understand the transport in the semi-classical approximation in which the electrons are assumed to obey classical mechanics but the Fermi statistics. We can obtain the semi-classical analogs of Eqs. (3.3.16) and (3.3.17) by replacing the oscillatory exponential integrals with decaying ones which have the same decay length and volume integral. Thus, we obtain for the semi-classical limit of the non-local conductivity,

$$\sigma_{zz}^{sc}(Z) = 3\sigma_0 \kappa_I \, E_3(2\kappa_I Z) \qquad (3.3.19)$$

$$\sigma_{xx}^{sc}(Z) = \frac{3}{2}\sigma_0 \kappa_I [E_1(2\kappa_I Z) - E_3(2\kappa_I Z)] \qquad (3.3.20)$$

From Fig. 3.1, we can see that the quantum and semi-classical results for the non-local conductivity are rather similar. They differ less when the current and electric field are parallel to the layers except near $z = z'$, where the quantum conductivity is finite and the semi-classical conductivity has an integrable singularity. When the current and field are perpendicular to the planes, there are perceptible oscillations in the quantum non-local conductivity, and we observe the rather surprising result that $\lim_{Z \to 0} \sigma_{zz}(Z) = 0$, i.e., that a field applied in plane z does not induce a current in that plane.

3.3.2. Conductivity for Layered Systems

Within the FERPS model, it is relatively easy to calculate the exact conductivity for a layered system. If the self-energy, Σ is a function of z, only then we can obtain the Green function by solving a one-dimensional Schrödinger equation. The Green function is given by

$$\left[\frac{\hbar^2}{2m}\nabla^2 + E - \Sigma(z)\right]G(\mathbf{r},\mathbf{r}') = \delta(\mathbf{r}-\mathbf{r}') \qquad (3.3.21)$$

Its x and y dependence can be expressed by using a Fourier transform

$$G(\mathbf{r},\mathbf{r}') = \int \frac{dk_x dk_y}{(2\pi)^2} \exp\{i[k_x(x-x') + k_y(y-y')]\} G(k_x, k_y; z, z') \qquad (3.3.22)$$

This leaves us with a one-dimensional inhomogeneous Schrödinger equation to solve

$$\left[-\frac{\hbar^2}{2m}\left(k_x^2 + k_y^2 - \frac{\partial^2}{\partial z^2}\right) + E - \Sigma(z)\right] G(k_x, k_y; z, z') = \delta(z - z') \quad (3.3.23)$$

The solution of this one-dimensional Schrödinger equation can be written down quite simply in terms of wave functions $\psi_L(z)$ which satisfy a boundary condition on the left and $\psi_R(z)$ which satisfy a boundary condition on the right. As an example; for an infinite homogenous system, we might have $\psi_L(z) = \exp(-i\kappa z)$ and $\psi_R(z) = \exp(i\kappa z)$ where $\kappa = (E - \Sigma - k_x^2 - k_y^2)^{1/2}$, corresponding to outgoing wave boundary conditions. The Green function defined in terms of these wave functions is

$$G(k_x, k_y; z, z') = \left(\frac{2m}{\hbar^2}\right) \frac{\psi_L(z_<)\psi_R(z_>)}{W} \quad (3.3.24)$$

where W is the wronskian of the wave functions,

$$W = \psi_L(z)\psi_R'(z) - \psi_L'(z)\psi_R(z) \quad (3.3.25)$$

which can be shown to be independent of z.

Once the Green function is known, the non-local in-plane conductivity can be calculated by using Eq. (3.3.15) which yields

$$\sigma_{xx}(z, z') = \frac{e^2\hbar^3}{\pi m^2} \int \frac{dk_x}{2\pi} \int \frac{dk_y}{2\pi} k_x^2 [\text{Im } G(k_x, k_y; z, z')]^2$$

$$= \frac{e^2\hbar^3}{8\pi^3 m^2} \int dk_\parallel^2 k_\parallel^2 [\text{Im } G(k_\parallel; z, z')]^2 \quad (3.3.26)$$

where $k_\parallel^2 = k_x^2 + k_y^2$.

3.3.3. Semi-Classical Conductivity for Layered Systems

It is also possible to calculate the semi-classical conductivity for a layered system[9] by starting from the Kubo–Greenwood formula. Consider Eq. (3.3.23) for the one-dimensional Green function. If Σ were independent of z, then the Green function would be

$$G(k_x, k_y; z, z') = \frac{2m}{\hbar^2} \frac{1}{2ik} \exp(ik|z - z'|) \quad (3.3.27)$$

where $k = [2m/\hbar^2(E - \Sigma) - k_x^2 - k_y^2]^{1/2}$. If we assume that $\Sigma(z)$ is not exactly constant, but changes very slowly then we might hope that this expression might still be a good approximation. One situation where this can be achieved is the case in which the real part of Σ is constant and its imaginary part is small compared to $E - \text{Real}\Sigma$. In this case, we can write the Green function as

$$G_{sc}(k_x, k_y; z, z') = \frac{m}{i\hbar^2(k_F^2 - k_x^2 - k_y^2)^{1/2}} \exp\left[i\int_{z_<}^{z_>} dz\, k(z)\right] \quad (3.3.28)$$

where $k(z) = [k_F^2 - k_x^2 - k_y^2 + ik_F/\lambda(z)]^{1/2}$ with $E - \Sigma = \hbar^2 k_F^2/2m$. Here, we have identified $\lambda(z) = 1/(2 \operatorname{Im} k_F)$ with the local semi-classical mean-free path. The factor of 2 comes from the fact that the electron probability decays twice as fast as the electron amplitude.

From Eq. (3.3.26) we need $[\operatorname{Im} G(k_\parallel; z, z')]^2$,

$$[\operatorname{Im} G(k_\parallel; z, z')]^2 = \frac{m^2}{2\hbar^4 k_z^2} \cos^2 k_z(z - z') \exp[-k_F \phi(z, z')/k_z] \quad (3.3.29)$$

where $k_z = (k_F^2 - k_\parallel^2)^{1/2}$ and the phase function $\phi(z, z')$ is

$$\phi(z, z') = \int_{z_<}^{z_>} dz'' \frac{1}{\lambda(z'')} \quad (3.3.30)$$

In the spirit of the semi-classical approximation, the oscillations of Eq. (3.3.29) are "averaged" yielding

$$[\operatorname{Im} G(k_\parallel; z, z')]^2 \approx \frac{m^2}{4\hbar^4 k_z^2} \exp[-k_F \phi(z, z')/k_z] \quad (3.3.31)$$

Finally, we obtain the semi-classical limit for the FERPS model with a slowly varying mean-free path, $\lambda(z)$,

$$\sigma_{xx}(z, z') = \frac{e^2 k_F^2}{8\pi^2 \hbar} \int_1^\infty dt \left(\frac{1}{t} - \frac{1}{t^3} \right) \exp[-\phi(z, z')t] \quad (3.3.32)$$

where $t = k_F/(k_F^2 - k_x^2 - k_y^2)^{1/2}$. For a homogeneous system, Eq. (3.3.32) is the result we obtained by taking the semi-classical limit of the conductivity evaluated using the Kubo–Greenwood formula (3.3.20).

Figures 3.2 and 3.3 show a comparison between quantum and semi-classical calculations of the CIP z-dependent conductivity (the current density per unit applied

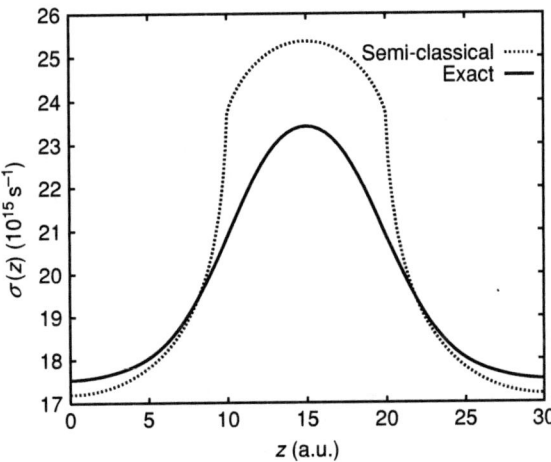

Figure 3.2. CIP conductivity for a trilayer consisting of a central clean layer between 10 and 20 a.u. surrounded by two dirty layers. The mean-free path for the clean layer is $\lambda = 360$ a.u. and for the dirty layer is $\lambda = 36$ a.u.

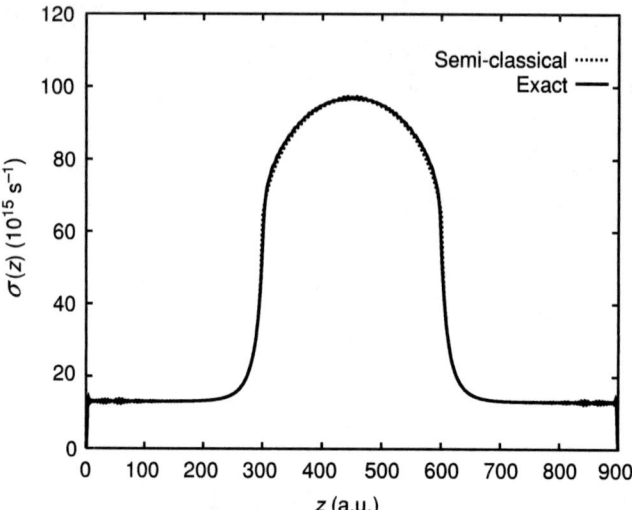

Figure 3.3. CIP conductivity for a trilayer consisting of a central clean layer (300 a.u. thick, $\lambda = 36$ a.u.) surrounded by dirty layers (600 a.u. thick, $\lambda = 360$ a.u.).

field) for a trilayer system in which the real part of the self-energy is fixed and the imaginary part (scattering rate) changes between the layers. There is a significant difference between the quantum and semi-classical conductivities when the layers are thin compared to the mean-free paths, but the difference is very small when the layers are thick compared to the mean-free path.

The Kubo–Greenwood approach has also been applied to electronic transport for layered systems within the context of first-principle electronic structure theory.[10–14] In this case, we can define a non-local intersite conductivity, $\sigma_{\mu\nu}(i,j)$ by integrating $\mathbf{r}'$ in Eq. (3.2.4) over atomic cell j and averaging $\mathbf{r}$ over atomic cell i.[10] By summing $\sigma_{\mu\nu}(i,j)$ over the sites, i, in atomic layer I, and the sites, j in atomic layer J, we obtain a layer-dependent non-local conductivity $\sigma_{\mu\nu}(I,J)$ analogous to the non-local conductivity $\sigma_{\mu\nu}(z,z')$ for free electrons described in this section. The advantage of the first-principle approach is that it allows a much more realistic treatment of transport in films and multilayers because it can include a more realistic representation of their electronic structure.

3.4. BOLTZMANN EQUATION

Overall, the semi-classical theory agrees with the quantum-mechanical linear response theory rather well in many cases. Since the semi-classical theory is much easier to deal with computationally and conceptually, we shall study it in more detail. Semi-classical transport theory is usually derived from the Boltzmann equation. This approach has the advantage compared to the semi-classical approach of Section 3.4.6 that not only the imaginary part of the electron self-energy, but also the real part can be allowed to vary from one layer to the next. In the semi-classical Boltzmann approach, the electrons are assumed to behave like classical particles.

SPIN-DEPENDENT TRANSPORT

The only concession we make to their quantum nature is the use of Fermi statistics (which implies that it is the Fermi energy electrons that are important for transport) and the use of quantum mechanics to calculate the relation between electron energy and momentum.

Consider the number of electrons with given values of momentum **k** and spin s, at position **r** at time t. This is called the electron distribution function, f^s (**k**, **r**, t). If there were no applied fields, the electrons would presumably be at equilibrium and the distribution function would be the equilibrium distribution function $f_0(E_k^s - \mu_0) = [1 + \exp(E_k^s - \mu_0/k_B T)]^{-1}$. Now, we imagine that a field has been applied but that we have waited long enough that the system is in a steady state, i.e., the distribution function is no longer changing so that $df/dt = 0$. The Boltzmann equation is obtained by balancing the changes caused by the applied field against processes that act to bring it back towards equilibrium.

Thus at steady state, the time rate of change of the distribution function is given by

$$\frac{df}{dt} = \frac{\partial f}{\partial t}\bigg|_{\text{drift}} + \frac{\partial f}{\partial t}\bigg|_{\text{field}} + \frac{\partial f}{\partial t}\bigg|_{\text{scatt}} = 0 \quad (3.4.1)$$

The drift term can be evaluated from the fact that the electrons entering a volume near point **r** at time t were previously at position $\mathbf{r} - \mathbf{v}\, dt$ at time $t - dt$. Thus,

$$\frac{\partial f(\mathbf{r}, \mathbf{k}, t)}{\partial t}\bigg|_{\text{drift}} = -\mathbf{v}(\mathbf{k}) \cdot \nabla_{\mathbf{r}} f(\mathbf{r}, \mathbf{k}, t) \quad (3.4.2)$$

Similarly, the field term can be evaluated from the fact that the electrons entering a volume of momentum space near point **k** at time t were previously located in momentum space at $\mathbf{k} - (d\mathbf{k}/dt)dt$ at time $t - dt$. Then, using Newton's second law, $-e\mathbf{E} = \hbar d\mathbf{k}/dt$ we have

$$\frac{\partial f(\mathbf{r}, \mathbf{k}, t)}{\partial t}\bigg|_{\text{field}} = -\frac{\partial f(\mathbf{r}, \mathbf{k}, t)}{\partial \mathbf{k}} \frac{\partial \mathbf{k}}{\partial t} = \frac{e}{\hbar} \nabla_{\mathbf{k}} f(\mathbf{r}, \mathbf{k}, t) \cdot \mathbf{E} \quad (3.4.3)$$

where the symbol e represents the *magnitude* of the electronic charge.

The scattering term can be written in terms of the probability, $P_{\mathbf{kk}'}$, for an electron to scatter between momentum states **k** and **k**'. It will be the sum of the probabilities for an electron to scatter into state **k** from some other momentum state minus the probability for an electron to scatter out of state **k**:

$$\frac{\partial f(\mathbf{r}, \mathbf{k}, t)}{\partial t}\bigg|_{\text{scatt}} = \sum_{\mathbf{k}'} P_{\mathbf{kk}'} \{f(\mathbf{r}, \mathbf{k}', t)[1 - f(\mathbf{r}, \mathbf{k}, t)] - f(\mathbf{r}, \mathbf{k}, t)[1 - f(\mathbf{r}, \mathbf{k}', t)]\}$$

$$= \sum_{\mathbf{k}'} P_{\mathbf{kk}'} [f(\mathbf{r}, \mathbf{k}', t) - f(\mathbf{r}, \mathbf{k}, t)] \quad (3.4.4)$$

In principle, the scattering probability should also include processes that scatter electrons from one spin channel to another, but we neglect those here because we are concentrating on the two-current model.

Assembling the three terms, and assuming steady state, we obtain the Boltzmann equation

$$-\mathbf{v}(\mathbf{k}) \cdot \nabla_r f(\mathbf{r},\mathbf{k}) + \frac{e}{\hbar}\nabla_\mathbf{k} f(\mathbf{r},\mathbf{k}) \cdot \mathbf{E} + \sum_{\mathbf{k}'} P_{\mathbf{k}\mathbf{k}'}\{f(\mathbf{r},\mathbf{k}') - f(\mathbf{r},\mathbf{k})\} = 0 \quad (3.4.5)$$

We attempt to calculate a linear response, i.e., proportional to the field so we write the distribution function as the equilibrium distribution function plus a correction term called the "deviation" function that describes the deviation from equilibrium, $f(\mathbf{r},\mathbf{k},t) = f_0(E_\mathbf{k} - \mu_0) + g(\mathbf{r},\mathbf{k})$. Substituting this form into the Boltzmann equation, we obtain

$$-\mathbf{v}(\mathbf{k}) \cdot \nabla_r g(\mathbf{r},\mathbf{k}) + \frac{e}{\hbar}\nabla_\mathbf{k} f_0(E_\mathbf{k} - \mu_0) \cdot \mathbf{E} + \sum_{\mathbf{k}'} P_{\mathbf{k}\mathbf{k}'}\{g(\mathbf{r},\mathbf{k}') - g(\mathbf{r},\mathbf{k})\} = 0 \quad (3.4.6)$$

where we have only retained the lowest order contribution to the field term because the field, E is assumed to be small. The field term can be further simplified by using

$$\nabla_\mathbf{k} f_0(E_\mathbf{k} - \mu_0) = \frac{\partial f_0(E_\mathbf{k} - \mu_0)}{\partial E_\mathbf{k}}\nabla_\mathbf{k} E_\mathbf{k} = \frac{\partial f_0(E_\mathbf{k} - \mu_0)}{\partial E_\mathbf{k}}\hbar\mathbf{v}(\mathbf{k}) \quad (3.4.7)$$

and the "scattering-out" term can be simplified by using

$$\sum_{\mathbf{k}'} P_{\mathbf{k}\mathbf{k}'} = 1/\tau_\mathbf{k} \quad (3.4.8)$$

which defines the lifetime as the inverse of the total scattering rate for electrons to scatter out of momentum state $\mathbf{k}$. Thus, the Boltzmann equation becomes

$$-\mathbf{v}(\mathbf{k}) \cdot \nabla_r g(\mathbf{r},\mathbf{k}) - \frac{g(\mathbf{r},\mathbf{k})}{\tau_\mathbf{k}} + \sum_{\mathbf{k}'} P_{\mathbf{k}\mathbf{k}'} g(\mathbf{r},\mathbf{k}') = -e\frac{\partial f_0(E_\mathbf{k} - \mu_0)}{\partial E_\mathbf{k}}\mathbf{v}(\mathbf{k}) \cdot \mathbf{E} \quad (3.4.9)$$

and the current density is given by

$$\mathbf{J}(\mathbf{r}) = -\frac{e}{V}\sum_\mathbf{k} \mathbf{v}(\mathbf{k}) g(\mathbf{r},\mathbf{k}) \quad (3.4.10)$$

Very often, we do not know very much about the details of the scattering probability $P_{\mathbf{k}\mathbf{k}'}$. In these cases, it is very popular to make the "lifetime approximation" which consists of dropping the scattering-in term, $\sum_{\mathbf{k}'} P_{\mathbf{k}\mathbf{k}'} f(\mathbf{k}')$ The $\mathbf{k}$ dependence of the lifetime is also often neglected. If the scattering is isotropic in the sense that $P_{\mathbf{k}\mathbf{k}'}$ does not depend on the angle between $\mathbf{k}$ and $\mathbf{k}'$, then one can argue that the scattering-in term vanishes because of symmetry because $g(\mathbf{k}')$ usually vanishes when summed over $\mathbf{k}'$. In general, however, the scattering is not isotropic and the neglect of the scattering-in term is an important, non-trivial approximation.

For a bulk system in the lifetime approximation, the deviation function is

$$g(\mathbf{k}) = e\tau_\mathbf{k} \frac{\partial f_0(E_\mathbf{k} - \mu_0)}{\partial E_\mathbf{k}}\mathbf{v}(\mathbf{k}) \cdot \mathbf{E} \quad (3.4.11)$$

and the current is

$$\mathbf{J} = -\frac{e^2}{V}\sum_\mathbf{k} \frac{\partial f_0(E_\mathbf{k} - \mu_0)}{\partial E_\mathbf{k}}\mathbf{v}(\mathbf{k})\mathbf{v}(\mathbf{k}) \cdot \mathbf{E}\tau_\mathbf{k} \quad (3.4.12)$$

For an isotropic or cubic system, we will have $\mathbf{J} = \sigma\mathbf{E}$ with the conductivity given by

$$\sigma = \frac{e^2}{3V} \sum_{\mathbf{k}} v^2(\mathbf{k}) \tau_{\mathbf{k}} \delta(E_{\mathbf{k}} - E_F) \tag{3.4.13}$$

3.4.1. "Scattering-in" Term

In order to obtain some insight into the effects of the scattering-in term which was neglected in Eqs. (3.4.11)–(3.4.13), let us rewrite the Boltzmann Eq. (3.4.9) for an infinite homogeneous system retaining the scattering-in terms as

$$g(\mathbf{k}) = e\tau_{\mathbf{k}} \frac{\partial f_0(E_{\mathbf{k}} - \mu_0)}{\partial E_{\mathbf{k}}} \mathbf{v}(\mathbf{k}) \cdot \mathbf{E} + \tau_{\mathbf{k}} \sum_{\mathbf{k}'} P_{\mathbf{kk}'} g(\mathbf{k}') \tag{3.4.14}$$

This can be written as

$$g_{\mathbf{k}} = \tau_{\mathbf{k}} x_{\mathbf{k}} + \tau_{\mathbf{k}} \sum_{\mathbf{k}'} P_{\mathbf{kk}'} g_{\mathbf{k}'} \tag{3.4.15}$$

where $x_{\mathbf{k}} = e[\partial f_0(E_{\mathbf{k}} - \mu_0)/\partial E_{\mathbf{k}}]\mathbf{v}(\mathbf{k}) \cdot \mathbf{E}$. We see that in general, the "scattering-in" term presents us with a complicated linear equation. The analytical or computational difficulties associated with solving this equation are not the only issue. As a matter of practice, we occasionally know enough about the details of the scattering probability $P_{\mathbf{kk}'}$, to make it a sensible proposition to attempt a detailed solution. Although calculations have been performed for particular types of scattering,[15,16] we usually have several types of scattering operating simultaneously, and, especially in films, it is difficult to treat them all simultaneously.

The procedure, that is, usually followed is to neglect the "scattering-in" term and to retain the scattering out term as a single parameter $1/\tau$ that describes the overall scattering rate. We can gain some insight into how good or bad an approximation this is by considering a special form of the scattering probability. Let us assume that the scattering probability can be approximated by an isotropic term and a term proportional to $\mathbf{v}_{\mathbf{k}} \cdot \mathbf{v}_{\mathbf{k}'}$:

$$P_{\mathbf{kk}'} = \frac{\delta(E_{\mathbf{k}} - E_{\mathbf{k}'})}{n\tau_0} + \frac{\mathbf{v}_{\mathbf{k}} \cdot \mathbf{v}_{\mathbf{k}'} \delta(E_{\mathbf{k}} - E_{\mathbf{k}'})}{\tau_1 n \langle v_x^2 \rangle} \tag{3.4.16}$$

where n is the Fermi energy DOS, $1/\tau_0$, the isotropic scattering rate, and $1/\tau_1$, the anisotropic "p-wave" scattering rate. The angle brackets indicate an average over the Fermi surface:

$$\langle x \rangle \equiv \frac{\sum_{\mathbf{k}} x_{\mathbf{k}} \delta(E_{\mathbf{k}} - E_F)}{\sum_{\mathbf{k}} \delta(E_{\mathbf{k}} - E_F)} \tag{3.4.17}$$

The parameter τ_1 is displayed as a lifetime even though the anisotropic term could, in principle, have either sign. Usually, however, one expects additional *forward* scattering especially if tightly bound d-states are not involved in the scattering. Actually, the form Eq. (3.4.16) is not too bad as a first approximation and it will help us to qualitatively understand the effect of the "scattering-in" term.

Equation (3.4.15) can be expanded in the form

$$g_{\mathbf{k}} = \tau_{\mathbf{k}} x_{\mathbf{k}} + \tau_{\mathbf{k}} \sum_{\mathbf{k}'} P_{\mathbf{k}\mathbf{k}'} \left[\tau_{\mathbf{k}'} x_{\mathbf{k}'} + \tau_{\mathbf{k}'} \sum_{\mathbf{k}''} P_{\mathbf{k}'\mathbf{k}''} g_{\mathbf{k}''} \right] \qquad (3.4.18)$$

If $1/\tau_1$ is sufficiently small, iteration of Eq. (3.4.18) should converge. Consider the term, $\sum_{\mathbf{k}'} P_{\mathbf{k}\mathbf{k}'} \tau_0 x_{\mathbf{k}'}$. This can be evaluated as

$$\sum_{\mathbf{k}'} P_{\mathbf{k}\mathbf{k}'} \tau_0 x_{\mathbf{k}'} = x_{\mathbf{k}} \frac{\tau_0}{\tau_1} \qquad (3.4.19)$$

Thus, for this particular form of the scattering probability, the distribution function can be evaluated as

$$g_{\mathbf{k}} = x_{\mathbf{k}} \frac{\tau_0}{1 - \frac{\tau_0}{\tau_1}} \qquad (3.4.20)$$

In this case, the effect of the "scattering-in" term is simply to renormalize the lifetime by a factor $\tau_1/(\tau_1 - \tau_0)$. The current and thus the conductivity will be increased by this factor as well. Thus, one effect of the scattering-in term will be to change the effective lifetimes from what they would be in their absence. By itself this would not be a major complication since the lifetime is often treated as a parameter anyway, but the effect can be more subtle in multilayers as we shall see later.

3.4.2. Layered Systems

If the system is composed of layers of different material stacked in the z direction, it is often a good approximation in magnetic multilayers, especially those optimized for GMR to assume that we have two-dimensional periodicity within each layer. If the layers are not too thin, we may also imagine that within each layer we can use the dispersion relation appropriate to that material in bulk. We would, of course, have to worry about obtaining the correct relative placement of the energy bands because, in general, when two materials are brought together a dipole layer forms at the interface to balance the electrochemical potentials and allow the materials to have their correct Fermi energies far from the interfaces. Figure 3.4 shows the self-consistent electronic structure of interfaces between permalloy and cobalt and also between cobalt and copper. For these metallic systems of similar atomic size and electronegativity, there are only small perturbations in the number of electrons per atom and spin channel on the layers near the interface. In layered systems such as these, it may be a good approximation to apply the Boltzmann Eq. (3.4.9), within each layer.

Because we have boundaries and interfaces, the distribution function will vary with z and will satisfy Eq. (3.4.9) specialized to our layered geometry:

$$\left[v_z(\mathbf{k}) \frac{\partial}{\partial z} + \frac{1}{\tau_{\mathbf{k}}} \right] g(z, \mathbf{k}) = x_{\mathbf{k}} + \sum_{\mathbf{k}'} P_{\mathbf{k}\mathbf{k}'} g(z, \mathbf{k}') \qquad (3.4.21)$$

where $x_{\mathbf{k}} = e[\partial f_0(E_{\mathbf{k}} - \mu_0)/\partial E_{\mathbf{k}}] \mathbf{v}(\mathbf{k}) \cdot \mathbf{E}$. The solution to a differential equation, of course, is not unique until a proper set of boundary conditions is specified. The key to applying the boundary conditions for layered systems is to realize that electrons

SPIN-DEPENDENT TRANSPORT

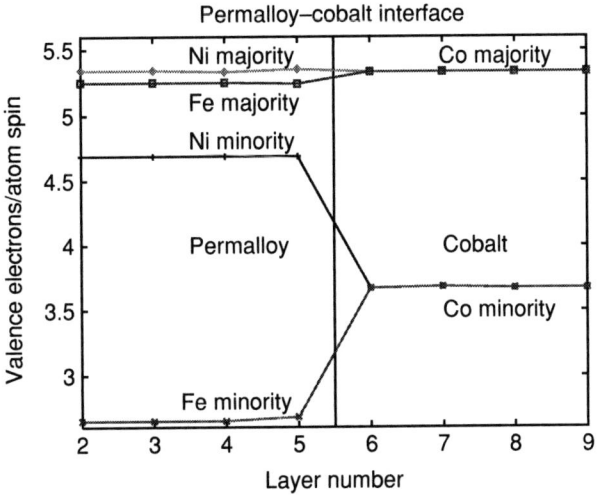

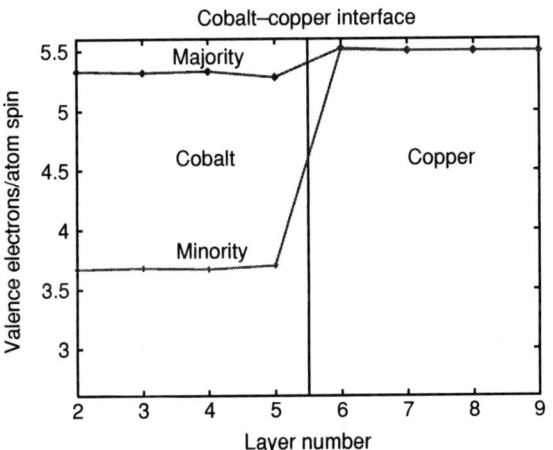

Figure 3.4. Permalloy–coblat and cobalt–copper interfaces.

traveling in the $+z$ direction satisfy a different boundary condition from those traveling in the $-z$ direction. This was first worked out for single layer films by Fuchs[17] and the generalization to multilayers[18–21] is relatively straightforward.

3.4.3. Boundary Conditions for Layered Systems

Before deriving the boundary conditions satisfied by the distribution function on the interfaces, it is helpful to explicitly acknowledge that transport in metals occurs at the Fermi energy. Because of the factor $\partial f_0(E_\mathbf{k} - \mu_0)/\partial E_k$ in the inhomogeneous term in the Boltzmann equation, the deviation function will contain a similar factor [see Eq. (3.4.11)]. This factor, which in metals at room temperature is effectively a delta function, allows us to simplify the boundary conditions for layered systems. Thus defining, $g^{\pm}(z,\mathbf{k}) = h^{\pm}(z,\mathbf{k})\delta(E_\mathbf{k} - E_F)$, we can write the Boltzmann equation as

$$\left[v_z(\mathbf{k})\frac{\partial}{\partial z} + \frac{1}{\tau_\mathbf{k}}\right]h(z,\mathbf{k}) = -e\mathbf{v}(\mathbf{k}) \cdot \mathbf{E} + \sum_{\mathbf{k}'} P_{\mathbf{k}\mathbf{k}'}h(z,\mathbf{k}') \qquad (3.4.22)$$

where we have assumed that $P_{\mathbf{k}\mathbf{k}'}$ contains an energy conserving delta function.

The current perpendicular to the layers is now given by

$$\begin{aligned}
J_z(z) &= -\frac{e}{V}\sum_\mathbf{k} v_z(\mathbf{k})h(z,\mathbf{k})\delta(E_\mathbf{k} - E_F) \\
&= -\frac{e}{2\pi A}\sum_{\mathbf{k}_\parallel,j}\left[\frac{h^{+,j}(z,\mathbf{k}_\parallel)v_z^+(\mathbf{k}_\parallel,j)}{|v_z^+(\mathbf{k},j)|} + \frac{h^{-,j}(z,\mathbf{k}_\parallel)v_z^-(\mathbf{k}_\parallel,j)}{|v_z^-(\mathbf{k},j)|}\right] \\
&= -\frac{e}{2\pi A}\sum_{\mathbf{k}_\parallel,j}[h^{+,j}(z,\mathbf{k}_\parallel) - h^{-,j}(z,\mathbf{k}_\parallel)] \qquad (3.4.23)
\end{aligned}$$

To obtain this result, we took explicit advantage of the fact that the delta function confines the summation to states at the Fermi energy so that the Bloch states can be described in terms of their transverse momentum, $\mathbf{k}_\parallel$ and a band index j. The band index is needed because for each value of $\mathbf{k}_\parallel$ there may be more than one Bloch state. In addition, it is important to note that for every value of $\mathbf{k}_\parallel$ and j, there will be two states one for which $v_z > 0$ and another with $v_z < 0$. This is true even if the Fermi surface does not have mirror symmetry around the plane $k_z = 0$, as occurs, e.g., in fcc (111). The boundary conditions on $h^{\pm}(z,j,\mathbf{k}_\parallel)$ are obtained by requiring particle conservation at each of the interfaces. Since $h_i^+(z,j,\mathbf{k}_\parallel)$, and $h_i^-(z,j\mathbf{k}_\parallel)$ represent the distribution functions in layer i for electrons traveling in the $+z$ and $-z$ directions, respectively, we can express the relationships between the distribution

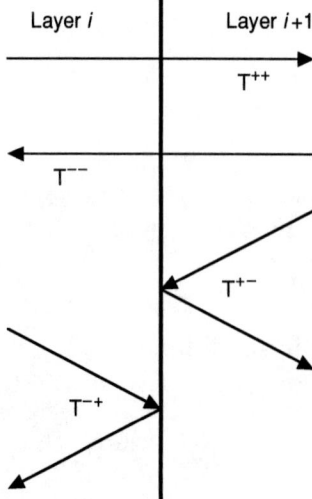

Figure 3.5. Convention for the transmission and reflection probabilities.

SPIN-DEPENDENT TRANSPORT

functions in layers i and $i+1$ (with interface at z_i) in terms of the transmission (T_i^{++}, T_i^{--}) and reflection (T_i^{+-}, T_i^{-+}) probabilities of the interfaces. We use a convention illustrated in Fig. 3.5 in which, e.g., $T_i^{+-}(\mathbf{k},\mathbf{k}')$ is the probability for a $-z$ going electron in Bloch state $\mathbf{k}'$ incident on interface i to leave the interface going in the $+z$ direction in Bloch state $\mathbf{k}$. Consider the flux of electrons leaving this interface traveling in the $+z$ direction (in layer $i+1$), $\sum_{j,\mathbf{k}_\|} h_{i+1}^+(z,j\mathbf{k}_\|)$. This flux is the sum of the transmitted flux of $+z$ going electrons from layer i and the reflected flux from those electrons originally traveling in the $-z$ direction in layer $i+1$. A similar flux conservation argument relates the $-z$ going electron flux leaving the interface to the incoming fluxes in the two layers; thus,

$$h_{i+1}^{+,j}(z_i^+,\mathbf{k}_\|) = \sum_{j',\mathbf{k}_\|'}^{N_R} T_i^{+-}(j\mathbf{k}_\|,j'\mathbf{k}_\|')h_{i+1}^{-,j'}(z_i^+,\mathbf{k}_\|')$$
$$+ \sum_{j',\mathbf{k}_\|'}^{N_L} T_i^{++}(j\mathbf{k}_\|,j'\mathbf{k}_\|')h_i^{+,j'}(z_i^-,\mathbf{k}_\|')$$

$$h_i^{-,j}(z_i^-,\mathbf{k}_\|) = \sum_{j',\mathbf{k}_\|'}^{N_L} T_i^{-+}(j\mathbf{k}_\|,j'\mathbf{k}_\|')h_i^{+,j'}(z_i^-,\mathbf{k}_\|')$$
$$+ \sum_{j',\mathbf{k}_\|'}^{N_R} T_i^{--}(j\mathbf{k}_\|,j'\mathbf{k}_\|')h_{i+1}^{-,j'}(z_i^+,\mathbf{k}_\|')$$

(3.4.24)

Here, N_L and N_R denote the number of states on the left or right of the interface, respectively, for a given value of $k_\|'$. If we assume that the layers have two-dimensional periodicity, so that the momentum parallel to the interface is conserved on transmission or reflection, the boundary conditions become

$$h_{i+1}^{+,j}(z_i^+,\mathbf{k}_\|) = \sum_{j'}^{N_R} T_i^{+-}(j,j')h_{i+1}^{-,j'}(z_i^+,\mathbf{k}_\|) + \sum_{j'}^{N_L} T_i^{++}(j,j')h_i^{+,j'}(z_i^-,\mathbf{k}_\|)$$
$$h_i^{-,j}(z_i^-,\mathbf{k}_\|) = \sum_{j'}^{N_L} T_i^{-+}(j,j')h_i^{+,j'}(z_i^-,\mathbf{k}_\|) + \sum_{j'}^{N_R} T_i^{--}(j,j')h_{i+1}^{-,j'}(z_i^+,\mathbf{k}_\|')$$

(3.4.25)

The first of these relations is shown in Fig. 3.6.

The transmission and reflection matrices can be calculated from the underlying electronic structure of the layers and their interface.[22] Figures 3.7 and 3.8 show the transmission and reflection probabilities for the Bloch waves in copper incident on cobalt. The transmission and reflection probabilities conserve electron flux. Thus, considering incident left- and right-going waves of unit flux, respectively, we can derive the following conservation rules:

$$\sum_j^{N_R} T^{++}(j,j') + \sum_j^{N_L} T^{-+}(j,j') = 1$$
$$\sum_j^{N_L} T^{--}(j,j') + \sum_j^{N_R} T^{+-}(j,j') = 1$$

(3.4.26)

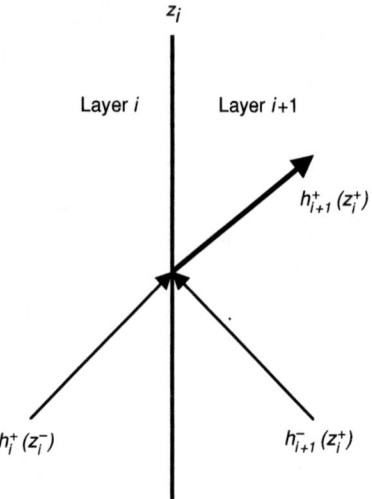

Figure 3.6. The right-going beam in the layer $i+1$, h_{i+1}^+, is the sum of the reflected part of the left-going beam in that layer, h_{i+1}^- and the transmitted part of the right-going beam in the layer i, h_i^+.

and considering unit left- and right-going fluxes leaving the interface, we obtain

$$\sum_{j'}^{N_L} T^{++}(j,j') + \sum_{j'}^{N_R} T^{+-}(j,j') = 1$$
$$\sum_{j'}^{N_R} T^{--}(j,j') + \sum_{j'}^{N_L} T^{-+}(j,j') = 1$$
(3.4.27)

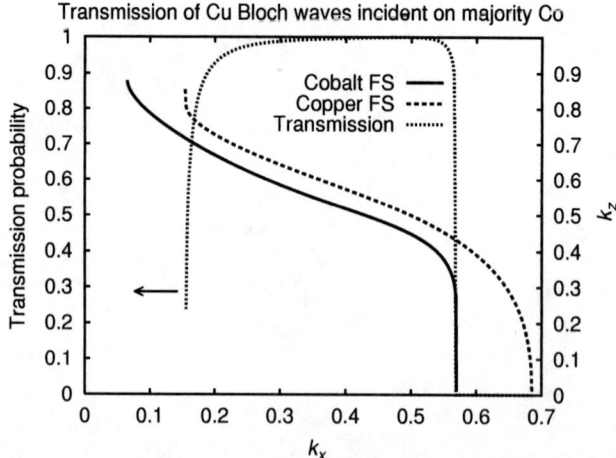

Figure 3.7. Transmission probabilities of copper electrons incident on majority cobalt for a cut through the Fermi surface with $k_y = 0$. The Fermi surfaces of cobalt and copper are shown as k_z as a function of k_x. The transmission probabilities are also given as a function of k_x for $k_y = 0$. Note that there is a perfect reflection for values of k_x for which there are no cobalt states to receive the electrons.

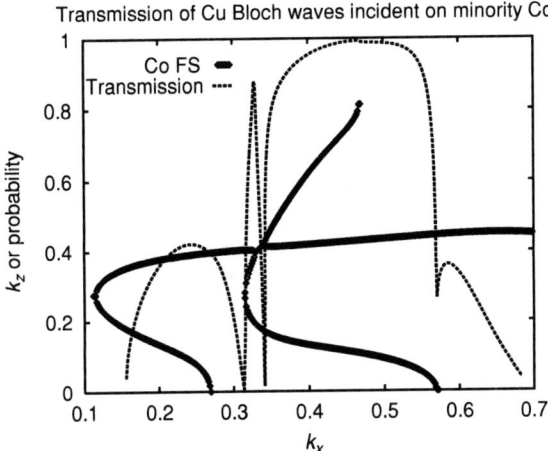

Figure 3.8. Transmission probabilities of copper electrons incident on minority cobalt for a cut through the Fermi surface with $k_y = 0$. The Fermi surface of minority cobalt is also shown. Note that the transmission probability goes to zero if the velocity of the state receiving the electrons goes to zero. This is a consequence of flux conservation.

If the interface is disordered, flux will still be conserved, but the conservation rules must be extended to include the scattering of electrons between different values of $\mathbf{k}_\parallel$. For CIP, it is possible to include such a diffuse scattering by the interfaces phenomenologically by including a specularity parameter for each interface so that Eq. (3.4.25) becomes

$$h_{i+1}^{+,j}(z, \mathbf{k}_\parallel) = S_i \left[\sum_{j'}^{N_R} T_i^{+-}(j,j') h_{i+1}^{-,j'}(z_i^+, \mathbf{k}_\parallel) + \sum_{j'}^{N_L} T_i^{++}(j,j') h_i^{+,j'}(z_i^-, \mathbf{k}_\parallel) \right]$$

$$h_i^{-,j}(z_i^-, \mathbf{k}_\parallel) = S_i \left[\sum_{j'}^{N_L} T_i^{-+}(j,j') h_i^{+,j'}(z_i^-, \mathbf{k}_\parallel) + \sum_{j'}^{N_R} T_i^{--}(j,j') h_{i+1}^{-,j'}(z_i^+, \mathbf{k}_\parallel') \right]$$

(3.4.28)

Here, $S_i = 1$ for purely specular scattering and $S_i = 0$ for purely diffuse scattering. The assumption made here is that diffusely scattered electrons completely lose their memory of their momentum before scattering and therefore no longer contribute to the CIP current. The interfacial specularity parameters, S_i were introduced by Hood and Falicov[20] as a slight generalization of the Fuch[17] use of a similar specularity parameter which he denoted by p to describe the effects of scattering from the boundary of a film.

Diffuse interfacial scattering, at least as modeled in the manner described, rapidly reduces the GMR. A more sophisticated treatment of scattering from disordered interfaces is needed, however, and could be modeled on the widely used theory of diffuse scattering of x-rays.[23,24]

3.4.4. Solving the Boltzmann Equation for Layered Systems

Neglecting the scattering-in term for the time being, we can write the Boltzmann Eq. (3.4.21) as

$$\frac{\partial h_{i\sigma}^{\pm,j}(z,\mathbf{k}_{\|})}{\partial z} v_{zij\sigma}^{\pm}(\mathbf{k}_{\|}) + \frac{h_{i\sigma}^{\pm,j}(z,\mathbf{k}_{\|})}{\tau_{i\sigma}} = -e\mathbf{v}_{ij\sigma}^{\pm} \cdot \mathbf{E} \qquad (3.4.29)$$

Here, $\pm$ refers to whether the z-component of the electron velocity is positive or negative. It can be shown that there will be equal numbers of states with positive and negative v_z for a given value of $\mathbf{k}_{\|}$. Since we will, in general, have more than one band for a given value of $\mathbf{k}_{\|}$ we need an index, j, to label them. The layer number is denoted by i and we have added an index, σ, to indicate the spin channel. Within layer i, this has solution

$$h_{i\sigma}^{\pm,j}(z) = -e\tau_{i\sigma}\mathbf{v}_{ij\sigma}^{\pm} \cdot \mathbf{E}\{1 + F_{ij\sigma}^{\pm} \exp[\mp z/(\tau_{i\sigma}|v_{zij\sigma}^{\pm}|)]\} \qquad (3.4.30)$$

where the coefficients $F_{ij\sigma}^{\pm}$ are determined by the boundary conditions, Eq. (3.4.27).

Let us specialize, for the present, to the case in which the field and current are parallel to the layers (CIP). The current density can be obtained from Eq. (3.4.10) and is given by

$$J_{i\sigma}(z) = e^2 E_x \sum_{j,\pm} \int \frac{d\mathbf{k}_{\|}}{\hbar|v_{zij\sigma}^{\pm}(\mathbf{k}_{\|})|} (v_{xij\sigma}^{\pm}(\mathbf{k}_{\|}))^2 \tau_{i\sigma}\{1 + F_{ij\sigma}^{\pm} \exp[\mp z/(\tau_{i\sigma}|v_{zij\sigma}^{\pm}|)]\} \qquad (3.4.31)$$

Figure 3.9 shows current densities calculated by using this approach for (111) copper films. The lifetime is the same for the seven films and is set to give a bulk resistivity of 3 $\mu\Omega$ cm. The label "$p=0$" indicates that perfectly diffuse scattering has been assumed at the boundaries of the films. This means that $h^+(z,\mathbf{k}_{\|})$ has been set to zero at the left-hand boundary and $h^-(z,\mathbf{k}_{\|})$ has been set to zero at the right-hand boundary of the film for all values of $\mathbf{k}_{\|}$. Note that the boundary scattering significantly depresses the conductivity as the films are made thinner. Qualitatively,

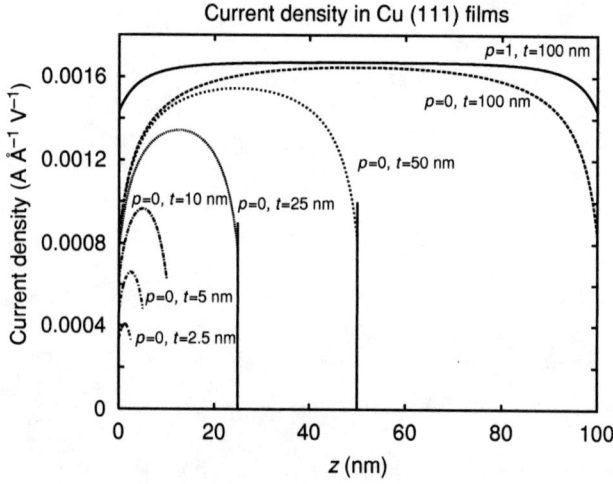

Figure 3.9. In-plane current density, $J(z)$, for Cu (111) films of different thicknesses. Boundary scattering depresses the current density for thin films.

we can say that in the thinner films, the electrons are likely to strike a boundary before they have traveled a distance equal to their mean-free path in the bulk.

Note that the current is also depressed near the boundaries for a film with "$p=1$", i.e., for a film with boundaries that reflect all electrons specularly. It is sometimes stated that a film of any thickness with perfectly reflecting boundaries should have the bulk resistivity and current density, but this is only true if $v_z^+(\mathbf{k}_\|) = v_z^-(\mathbf{k}_\|)$, a condition that is not satisfied for fcc (111) layers. A similar calculation for (100) or (110) layers does yield the expected result of a uniform current density equal to the bulk value because when z is in these directions, the condition, $v_z^+(\mathbf{k}_\|) = v_z^-(\mathbf{k}_\|)$, is satisfied.

Figure 3.10 shows the calculated resistivity as a function of Cu (111) film thickness for diffuse ($p = 0$) boundary scattering. These films correspond to those of Fig. 3.9 and have a scattering rate that corresponds to a bulk resistivity of 3 $\mu\Omega$ cm. The solid curve shows the free electron result for comparison. For free electrons, with $p = 0$ boundary conditions, it is easy to show that the resistivity of a film should vary with thickness as

$$\rho = \frac{\rho_{\text{bulk}}}{1 - (4l/t)[(1/3) - E_4(l/t)] + (4l/t)[(1/5) - E_6(l/t)]} \tag{3.4.32}$$

It has been pointed out that the parameter, p, which describes the diffuseness of the boundary scattering should depend on $\mathbf{k}_\|$.[8,25]

We earlier mentioned the importance of vertex corrections. For a bulk system, their effect, at least in the approximation that we used in Section 3.4.1 is simply to renormalize the lifetime and the conductivity by the factor $\tau_1/(\tau_1 - \tau_0)$. Even with the approximate form for $P_{\mathbf{kk}'}$ assumed in Section 3.4.1, it is somewhat complicated to solve the Boltzmann equation for a film or multilayer in the presence of vertex corrections. It can be done,[26] however, and results of calculations which include

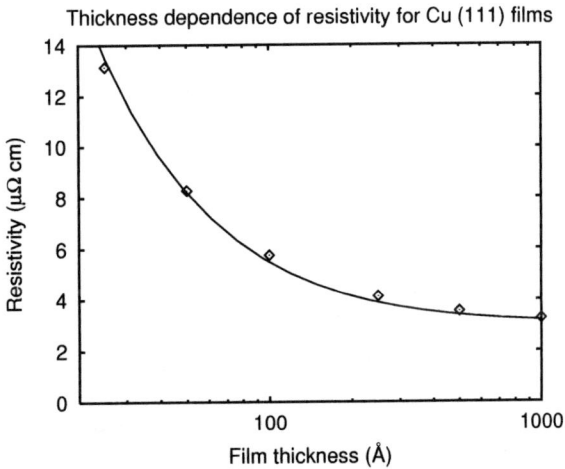

Figure 3.10. Calculated resistivity of Cu (111) films with diffuse boundary scattering. The solid line is the free electron model. Points are calculated by using the Fermi surface of copper.

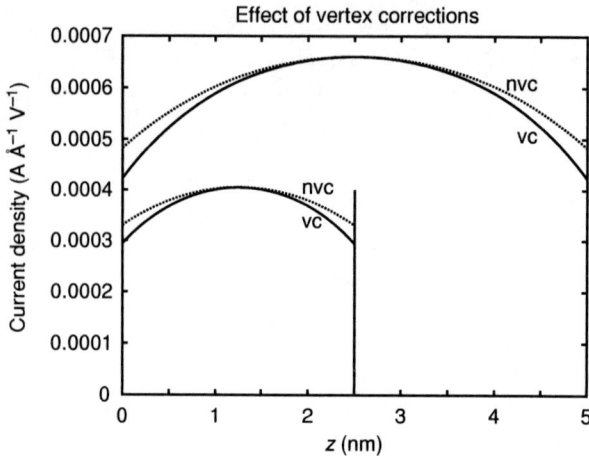

Figure 3.11. Effect of vertex corrections on calculated resistivity of Cu (111) films with diffuse boundary scattering. "nvc" indicates neglect of vertex corrections. "vc" indicates "vertex" corrections are included with $\tau_0/\tau_1 = 0.5$. τ_0 is also reduced by a factor of 0.5 for the "vc" case compared to the "nvc" case.

vertex corrections are shown in Fig. 3.11. The calculations are for Cu (111) films of thickness 25 and 50 Å. It is assumed that $\tau_0/\tau_1 = 0.5$ so that the scattering-in term would double the conductivity of a bulk sample if τ_0, the lifetime for isotropic scattering were held fixed. Here, however, we have assumed that the bulk resistivity is 3 µΩ cm as in Fig. 3.9. Thus, for the curves labeled "vc" in Fig. 3.11, the lifetime for isotropic scattering for Fig. 3.11 is only one-half its value for the curves labeled "nvc" or "no vertex-corrections". It can be seen that the effect of the "scattering-in" term (for fixed bulk resistivity) is an additional decrease in the current densities near the boundaries.

3.4.5. Giant Magnetoresistance

One contribution to the giant magnetoresistance effect can be thought of as an effect similar to the effect of a boundary decreasing the conductivity. Consider a three-layer system, e.g., a layer of copper sandwiched between two layers of cobalt. Suppose that copper and cobalt matched perfectly in the majority channel. Then, the majority electrons, when the moments of the two cobalt layers are aligned, would effectively see a film thickness equal to the sum of the thicknesses of the three layers while the minority electrons would tend to be confined within the individual layers because of the changes in electronic structure at the interfaces. When the moments are antialigned, however, both of the spin channels would see effectively two layers. This situation is shown in Fig. 3.12. It is also assumed in this example that the mean-free path is much longer for the majority cobalt channel than for the minority.

The sum of the "uu" and "dd" currents gives the total current for parallel alignment and the sum of the "ud" and "du" currents give the total current for anti-parallel alignment. These are shown in Fig. 3.13 together with the difference which yields the giant magnetoconcuctance.

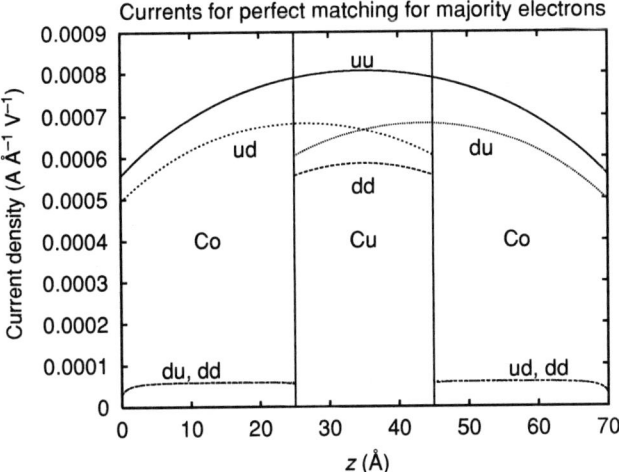

Figure 3.12. Currents for spin channels for an idealized Co–Cu spin valve with perfect matching in the majority channel. uu and dd refer to the majority and minority channels, respectively, for aligned Co moments. ud and du refer to the channels that are locally majority on the left and on the right, respectively, for the case of anti-parallel alignment of the Co moments.

In fact, of course, as indicated in Fig. 3.4, there is a difference between cobalt and copper in the majority channel. The copper majority Fermi surface is larger than that of cobalt. It holds 0.5 electrons while that of cobalt holds only 0.3. A cut through the Fermi surfaces of copper and majority cobalt is shown in Fig. 3.7. The z direction (perpendicular to the layers) is towards the top of the figure. The directions perpendicular to this direction are in the plane of the layers. If the interfaces are smooth on an atomic scale then the component of the momentum parallel to

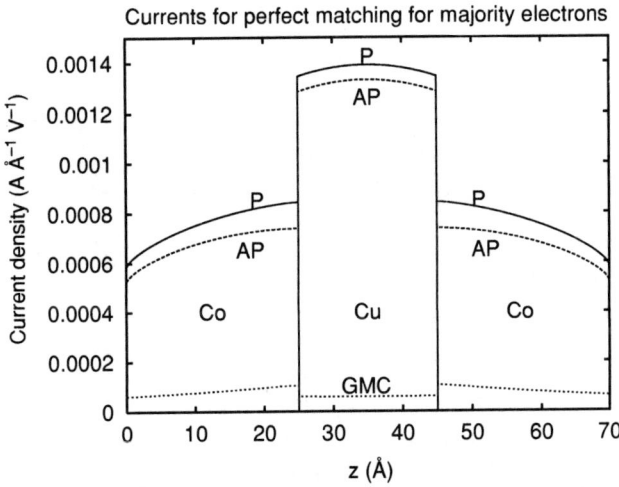

Figure 3.13. Currents for parallel (P) and anti-parallel (AP) alignments of the cobalt moments. The difference is the giant magnetoconductance (GMC).

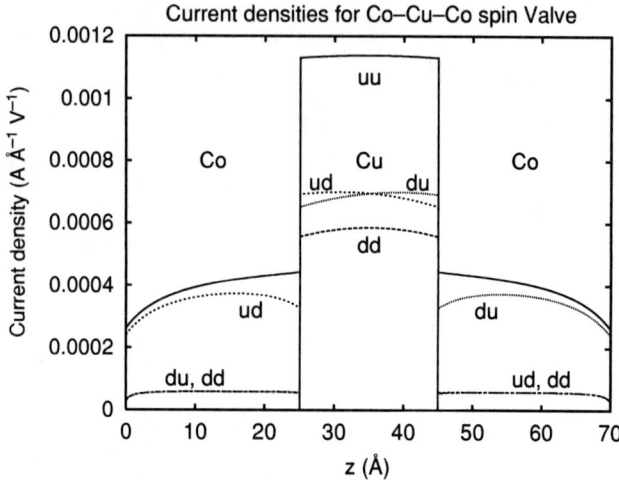

Figure 3.14. Current Densities for Co–Cu–Co spin valves. The scattering rates correspond to bulk resistivities of 3 μΩ cm for copper and 15 μΩ cm for cobalt with the majority lifetime 10 times longer than the minority lifetime for cobalt.

the interface ($k_\parallel$) does not change on reflection or refraction at an interface. Thus, from Fig. 3.7, it is clear that there are values of $k_\parallel$ for which states exist in the copper but not in the cobalt. This means that these states cannot refract into the cobalt, they must reflect back into the copper. This can lead to a significant contribution to the GMR if the interface is sufficiently smooth because some of the majority electrons can be "trapped" inside the copper where the resistance is significantly lower for both spin channels than for cobalt. This "trapping" of the electrons inside the copper layer is analogous to the trapping of light waves within a waveguide.[27]

A calculation for the current density in a cobalt–copper–cobalt spin valve using realistic electronic structures is shown in Figs. 3.14 and 3.15. Figure 3.14 shows the majority and minority currents for both parallel (uu, dd) and anti-parallel (ud, du) alignment. In this example, the scattering rate in the copper is chosen to give the copper a resistivity of 3 μΩ cm, a typical value for sputter deposited copper films at room temperature. The scattering rates for cobalt were chosen to give it a resistance of 15 μΩ cm which is also typical of sputtered films. A much higher scattering rate was chosen for the minority than for the majority cobalt. It can be seen that the current density is significantly higher in the copper than in the cobalt. It can be seen from Fig. 3.15 that the largest contributions to the giant magnetoconductance arise from the copper spacer layer.

3.4.6. Current Perpendicular to the Planes

When the system is inhomogeneous in the direction in which the field is applied, it is necessary to deal with the accumulation of spin and charges. It is useful to think first about a simple system with a local but spatially varying conductivity. Let us assume that the system is homogeneous in the x and y directions and has

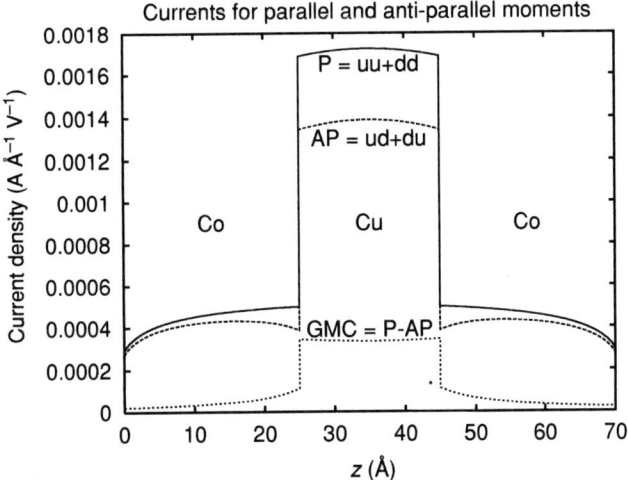

Figure 3.15. Current densities for parallel alignment, anti-parallel alignment, and difference or giant magnetoconductance.

a *local* conductivity that depends on z, the direction which sustains a current, J and a potential difference, ΔV. In this case, the current is related to the local electric field through the conductivity

$$J(z) = \sigma(z)E(z) \tag{3.4.33}$$

The integral of this local field gives the potential difference

$$\int dz E(z) = -\Delta V \tag{3.4.34}$$

In steady state, $J(z)$ must be independent of z if charge is to be conserved. Thus, the local field, $E(z)$, must vary as $1/\sigma(z)$. We can think of this local field as arising from an applied field together with the fields due to the inhomogeneous distribution of electrons, i.e., set up by the current passing through the sample. We could ignore these charge accumulation effects in the previous subsections because there we assumed that the system was homogeneous in the direction of the applied field.

For the more general case in which the spatial inhomogeneities have a scale comparable to or smaller than the electron mean-free path, we cannot assume a local conductivity. We can, however, use the Boltzmann Eq. (3.4.9), specialized to our geometry

$$v_z(\mathbf{k}) \frac{\partial f(z,\mathbf{k})}{\partial z} - \sum_{\mathbf{k}'} P_{\mathbf{k}\mathbf{k}'}[f(z,\mathbf{k}') - f(z,\mathbf{k})] = ev_z(\mathbf{k})\delta(E_\mathbf{k} - E_F)\frac{\partial V}{\partial z} \tag{3.4.35}$$

but we must still deal with charge accumulation effects and variable local electric fields. For simplicity, let us assume that the scattering is isotropic:

$$P_{\mathbf{k}\mathbf{k}'} = \frac{\delta(E_\mathbf{k} - E_{\mathbf{k}'})}{n\tau_0} \tag{3.4.36}$$

Taking, $f(z,\mathbf{k}) = f_0(E_\mathbf{k} - \mu_0) + \delta(E_\mathbf{k} - E_F)h(z,\mathbf{k})$, and taking into account the fact that the distribution function, $f(z,\mathbf{k})$, may contain spatial charge inhomogeneities, we can write

$$\sum_\mathbf{k} f(z,\mathbf{k}) = N + n\mu(z) \tag{3.4.37}$$

where N is the number of electrons and n is the Fermi energy DOS. Thus, the scattering terms of the Boltzmann equation are given by

$$\sum_{\mathbf{k}'} P_{\mathbf{k}\mathbf{k}'} h(z,\mathbf{k})\delta(E_\mathbf{k} - E_F) = h(z,\mathbf{k})\delta(E_\mathbf{k} - E_F)/\tau_0 \tag{3.4.38}$$

and

$$\sum_{\mathbf{k}'} P_{\mathbf{k}\mathbf{k}'} h(z,\mathbf{k}')\delta(E_{\mathbf{k}'} - E_F) = \mu(z)\delta(E_\mathbf{k} - E_F)/\tau_0 \tag{3.4.39}$$

Note that for CPP the scattering-in term cannot be neglected as it usually is for CIP. Thus, the Boltzmann equation can be written as

$$v_z(\mathbf{k})\frac{\partial h(z,\mathbf{k})}{\partial z} + \frac{h(z,\mathbf{k}) - \mu(z)}{\tau_0} = ev_z(\mathbf{k})\frac{\partial V}{\partial z} \tag{3.4.40}$$

If we define the anisotropic part of the deviation function as, $h^A(z,\mathbf{k}) = h(z,\mathbf{k}) - \mu(z)$, we can write the Boltzmann equation in the form

$$v_z(\mathbf{k})\frac{\partial h^A(z,\mathbf{k})}{\partial z} + \frac{h^A(z,\mathbf{k})}{\tau_0} = -v_z(\mathbf{k})\frac{\partial \bar{\mu}}{\partial z} \tag{3.4.41}$$

where $\bar{\mu} = \mu - eV$. At a first glance, this appears to be no more complicated than Eq. (3.4.29) which we used for the CIP case, but it is really much more complicated because $h^A(z,k)$ must be anisotropic for all z. In other words, it must satisfy

$$\sum_\mathbf{k} h^A(z,\mathbf{k})\delta(E_\mathbf{k} - E_F) = \sum_\mathbf{k} [h(z,\mathbf{k}) - \mu(z)]\delta(E_\mathbf{k} - E_F) = 0 \tag{3.4.42}$$

We know of only one case that has been solved to date, the almost trivial case in which the electronic structure does not change from layer to layer.* If the only property that changes between the layers is the scattering rate, then Eq. (3.4.41) is solved by $h^A(z,\mathbf{k}) = -v_z(\mathbf{k})\tau_0(\partial\bar{\mu}/\partial z)$ and by $(\partial\bar{\mu}/\partial z)$ proportional to the the scattering rate, $1/\tau_0$. Note that Eq. (3.4.42) is also satisfied and that it is not possible to separate $\bar{\mu}$ into μ and eV. This solution was provided by Camblong et al.[28] (CLZ). Valet and Fert[29] later extended their solution to the case in which there is scattering between the spin channels.

One important feature of the Camblong–Levy–Zhang solution is that the film resistance depends only on the integral over the scattering rate, $\int dz\, 1/\tau_0(z)$. This means that if one considers a multilayer system made up of with two types of layers with different scattering rates, these layers could be arranged in any order and any set of thicknesses could be used. As long as the total thicknesses of the two types of

---

*During the three years since this was written, significant progress has been made. The reader is referred to W. H. Butler, X.-G. Zhang, and J. M. MacLaren, J. Appl. Phys. **87**, 5173 (2000) and references therein.

SPIN-DEPENDENT TRANSPORT

films remained the same, the resistivity would remain the same. The term "self-averaging" was applied to these systems by CLZ. This feature seems to be peculiar to the assumption made by CLZ that the electronic structure does not change from layer to layer. It is not clear how badly the self-averaging property is violated by more realistic models.

3.5. LANDAUER FORMULA FOR CONDUCTANCE

There is another approach to CPP conduction that can be applied in the limit in which the transport in ballistic rather than diffusive. Usually transport in metallic systems is diffusive in nature; between the time that an electron enters the sample from one lead and an electron leaves the sample through another lead many scattering events occur. Devices (usually on the nanometer scale) can be constructed, however, in which very few scattering events take place between the two leads. In this limit, sometimes called the ballistic limit, there is a very simple expression due to Landauer[30] which relates the conductance to the probability of an electron being transmitted through the sample.

To understand the Landauer conductance formula, it is helpful to consider two reservoirs of electrons connected by a sample as shown in Fig. 3.16. If we imagine the left reservoir, with chemical potential μ_1, to be an emitter of right-going electrons, we can write the current density of those electrons that leave the reservoir on the left and enter the reservoir on the right as

$$J^+ = \frac{e}{(2\pi)^3} \int d^3k v_z^+(\mathbf{k}) f_0^+(\mu_1) T^+(\mathbf{k}) \tag{3.5.1}$$

where

$$T^+(\mathbf{k}) \equiv \sum_{\mathbf{k}'} T^{++}(\mathbf{k}, \mathbf{k}') \tag{3.5.2}$$

and z is the direction from reservoir 1 to reservoir 2. We can perform the integral over k_z as

$$J^+ = \frac{e}{A} \sum_{\mathbf{k}_\parallel, j} \frac{1}{2\pi} \int dk_z \frac{1}{\hbar} \frac{\partial \varepsilon}{\partial k_z} f_0(\mu_1) T^+(\mathbf{k}) \tag{3.5.3}$$

which yields an expression for the current

$$I^+ = \frac{e}{h} \int^{\mu_1} d\varepsilon \sum_{\mathbf{k}_\parallel, j} T^+(\mathbf{k}_\parallel, j) \tag{3.5.4}$$

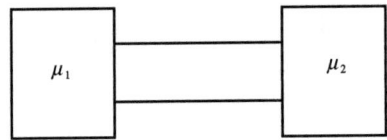

Figure 3.16. Two electron reservoirs connected by a sample.

A similar line of reasoning leads to an expression for the current of electrons emitted in the $-z$ direction by the reservoir on the right which enter the reservoir on the left:

$$I^- = \frac{e}{h} \int^{\mu_2} d\varepsilon \sum_{k_\parallel,j} T^-(k_\parallel,j) \tag{3.5.5}$$

Assuming time reversal invariance, we can equate T^+ and T^-. This allows us to write the net current as

$$I = I^+ - I^- = \frac{e^2}{h} \sum_{k_\parallel,j} T^+(k_\parallel,j) \frac{\mu_1 - \mu_2}{e} \tag{3.5.6}$$

which yields the Landauer conductance formula

$$G = \frac{e^2}{h} \sum_{k_\parallel,j} T^+(k_\parallel,j) \tag{3.5.7}$$

The original Landauer formula has the ratio of transmission probability divided by reflection probability (T/R), where we have only the transmission probability in Eq. (3.5.7). It is argued that this additional factor of $1/R$ arises from the reflected electrons changing the chemical potential of the reservoirs. Now, it is usually accepted that this additional factor of $1/R$ is present or not depending on exactly how the measurement is performed, i.e., on whether or not one measures current and voltage using the same leads, as is assumed in the derivation here, or whether a separate set of probes is used to determine the voltage across the sample. In the application in Section 3.6, the difference between the two formulas will usually be negligible because tunneling transmission probabilities are very small and the reflection probabilities are near unity.

3.6. SPIN-DEPENDENT TUNNELING

One important application of the Landauer conductance formula is to the calculation of tunneling conductance. It is assumed that the two electrodes act as electron reservoirs and that the electrons traverse the barrier region in a single tunneling event. Recently, it has been observed that the tunneling conductance between two ferromagnetic electrodes depends on the relative orientation of the magnetic moments in the two electrodes.[4,5] Generally, the tunneling rate is greater when the moments of the two electrodes are parallel.

The first observation of spin-dependent tunneling as well as the first theory was due to Julliere.[3] Julliere's theory was based on the reasonable assumption that the tunneling rate was proportional to the products of the Fermi energy densities of states of the two electrodes. Thus, the tunneling current for parallel and anti-parallel alignment of the moments in the ferromagnetic electrodes would be

$$\begin{aligned} I_P &\propto n_L^\uparrow n_R^\uparrow + n_L^\downarrow n_R^\downarrow \\ I_A &\propto n_L^\uparrow n_R^\downarrow + n_L^\downarrow n_R^\uparrow \end{aligned} \tag{3.6.1}$$

where $n_{L,R}^{\uparrow,\downarrow}$ are the majority and minority Fermi energy densities of states for the left and right electrodes, respectively. Defining the polarization of the Fermi energy DOS for the two electrodes as

$$P_L = \frac{n_L^\uparrow - n_L^\downarrow}{n_L^\uparrow + n_L^\downarrow} \quad P_R = \frac{n_R^\uparrow - n_R^\downarrow}{n_R^\uparrow + n_R^\downarrow} \tag{3.6.2}$$

we can write the magnetoresistance as

$$\frac{I_P - I_A}{I_A} = \frac{2P_L P_R}{1 - P_L P_R} \tag{3.6.3}$$

Unfortunately, the assumption that the tunneling rate is proportional to the DOS is not supported by experimental observations of tunneling from ferromagnets into superconductors.[32] These observations indicate that it is almost always the majority channel conductance which dominates the tunneling from ferromagnets, despite the fact that the minority Fermi energy DOS is often many times larger than the minority, e.g., in cobalt and nickel. Despite this rather serious problem, this approach is still widely used to analyze and rationalize experimental data. The argument is typically made that the densities of states and polarizations refer to "those electrons that participate in the tunneling" and that these "effective" polarizations can be obtained from superconducting tunneling experiments.

Recently, the Landauer approach was used to evaluate the the spin-dependent tunneling conductance for some epitaxial systems of the form Fe-S-Fe where S represents an insulator or semiconductor. The same techniques that were used to calculate the transmission and reflection probabilities in metals can be used to calculate the transmission probabilities for spin-dependent tunneling. The semiconductors Ge, GaAs, and ZnSe which have lattices that match almost perfectly to Fe by using (100) interfaces for both materials[33,34] were studied, and it was found that these systems show a remarkably large magnetoresistance the microscopic origins of which can be analyzed in detail.

As the semiconducting layer is made thicker, the tunneling conductance decreases exponentially for all values of $k_\|$, but this decrease is the slowest for $k_\| = 0$, i.e., for those electrons whose momentum in the iron is perpendicular to the interface with the semiconductor. Figure 3.17 shows the DOS for each of the bands in Fe at the Fermi energy for $k_\| = 0$. It also shows how the DOS decays within the semiconductor which in this case is ZnSe. This DOS is calculated for a particular incident Bloch state on the left, all possible reflected Bloch states on the left, and all possible transmitted Bloch states on the right. It can be seen that the incident Bloch states differ in how well they are *injected* into the semiconductor, their *rate of decay* within the semiconductor and how well they are *extracted* from the semiconductor to form the transmitted wave on the right.

For both majority and minority Fe for $k_\| = (0,0)$ at E_F, there are four bands. A doubly degenerate Δ_5 band (compatible with p and d DOS), a $\Delta_{2'}$ band (compatible with d DOS) are found for both majority and minority. In addition, there is a majority Δ_1 band (compatible with s, p, and d DOS) and a minority Δ_2 band (compatible with d DOS).

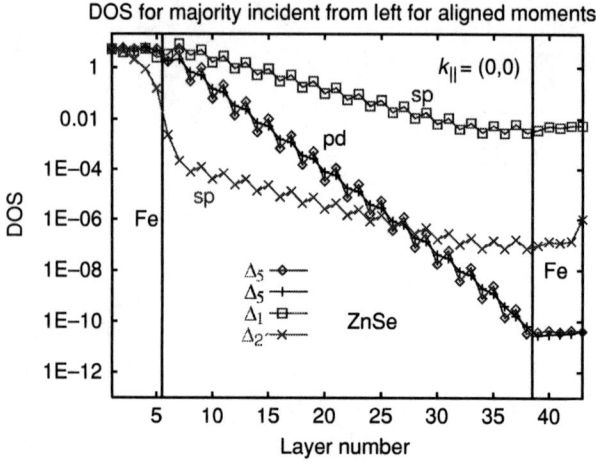

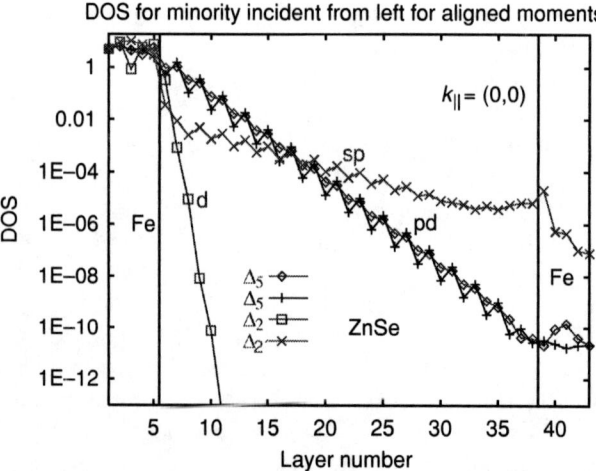

Figure 3.17. Majority (upper panel) and minority (lower panel) DOS for each of the Fe Fermi energy Bloch states for $k_\parallel = 0$. The magnetic moments on the two Fe layers are aligned parallel.

Three decay rates can be discerned within the semiconductor, one associated with the Δ_1 and $\Delta_{2'}$ bands, one connected with the Δ_5 bands, and a very rapid decay associated with the Δ_2 band. The angular momentum composition of the DOS within the semiconductor is noted on the figure for each of the bands. It can be seen that the Δ_1 and Δ_5 bands are efficiently injected into the semiconductor on the left and extracted from it on the right. However, the Δ_5 bands couple into evanescent states in the semiconductor that decay relatively rapidly within the semiconductor. The $\Delta_{2'}$ bands, decay relatively slowly within the ZnSe, but are injected and extracted with very low efficiency. Only the majority Δ_1 band is both injected efficiently and decays slowly. It should be clear, because of the logarithmic scale, that for reasonably thick samples, the parallel alignment conductance is dominated by the Δ_1 band and the majority spin channel.

SPIN-DEPENDENT TRANSPORT

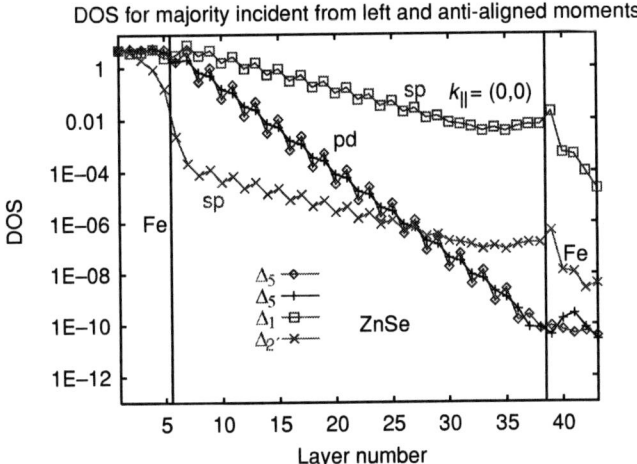

Figure 3.18. DOS for each of the Fe Fermi energy Bloch states for $k_\parallel = 0$. The magnetic moments on the two Fe layers are aligned parallel. The Fe majority Bloch states are incident from the left.

Figure 3.18 shows the DOS for majority Bloch states incident from the left when the Fe moments are aligned anti-parallel. The DOS is similar to the upper panel of Fig. 3.17 with the major difference that the Δ_1 band does not couple efficiently from the ZnSe into the minority Fe bands. Because of this poor extraction efficiency, the conductance for anti-parallel alignment is much lower than for parallel alignment.

These results give insight into a very deep mystery concerning spin-dependent tunneling. Everytime that it has been possible to determine which of the spin channels has the highest tunneling rate, it has been the majority channel.[32] This is true even for systems such as nickel for which the minority DOS at E_F exceeds the majority by an order of magnitude. One possible reason for this is that not all electrons tunnel equally well. For the particular states shown here, it is those bands which have some s character that seem to be most efficient at tunneling. Many of the magnetic systems, e.g., nickel and cobalt have filled majority d-bands. These "strong" magnets will clearly have more states with s-character in the majority channel than in the minority channel. Fe does not have a filled majority d-band, but, at least in the 100 direction, the band with s-character dominates.*

ACKNOWLEDGMENTS

Work sponsored by the Defense Advanced Research Projects Agency and Office of Naval Research (N00014–02–1–0590) and by the Division of Materials Sciences Office of Basic Energy Sciences of the US DOE under contract DE-AC05-96OR22464 with UT-Batelle LLC.

---

*Recent advances in understanding spin-dependent tunneling can be found in "Spin-Dependent Tunneling Conductance of Fe/MgO/Fe Sandwiches," W. H. Butler, X.-G. Zhang, and T. C. Shulthess, *Phys. Rev. B* **63**, 054416 (2001).

APPENDIX: LAYER CONDUCTIVITY

Here, we derive Eqs. (3.3.16) and (3.3.17) starting from Eq. (3.3.15). Since the system is homogeneous, we can write

$$\sigma_{zz}(Z) = \frac{e^2 Z^2}{4\pi^3 \hbar} \int dX \int dY \left\{ \text{Im}\left[\exp(i\kappa R)\left(i\kappa - \frac{1}{R} \right) \right] \right\}^2 \frac{1}{R^4} \quad (A3.1)$$

where $X = x - x'$, $Y = y - y'$, $Z = z - z'$, and $R^2 = X^2 + Y^2 + Z^2$. Using $\rho^2 = X^2 + Y^2$ and

$$\int dX \int dY = 2\pi \int_0^\infty \rho \, d\rho = 2\pi \int_Z^\infty R \, dR \quad (A3.2)$$

we obtain

$$\sigma_{zz}(Z) = \frac{e^2 Z^2}{8\pi^2 \hbar} \int_Z^\infty \frac{dR}{R^3} \left[\exp(i\kappa R)\left(i\kappa - \frac{1}{R} \right) - \exp(-i\kappa^* R)\left(-i\kappa^* - \frac{1}{R} \right) \right]^2 \quad (A3.3)$$

Making a change of variables $R \to Zt$ and by using the definition of the exponential integral, Eq. (3.3.18) we obtain Eq. (3.3.16).

Similarly, $\sigma_{xx}(Z)$ can be derived by using

$$\sigma_{xx}(Z) = \frac{1}{2}(\sigma_{xx}(Z) + \sigma_{yy}(Z)) \quad (A3.4)$$

$$\sigma_{xx}(Z) = \frac{e^2}{16\pi^2 \hbar} \int_Z^\infty \frac{dR}{R^3} (R^2 - Z^2) \left[\exp(i\kappa R)\left(i\kappa - \frac{1}{R} \right) \right.$$
$$\left. - \exp(-i\kappa^* R)\left(-i\kappa^* - \frac{1}{R} \right) \right]^2 \quad (A3.5)$$

With the same change of variable used above, we obtain Eq. (3.3.17).

REFERENCES

1. M. N. Baibich, J. M. Broto, A. Fert, F. Nguyen Van Dau, F. Petroff, P. Eitenne, G. Creuzet, A. Friederich, and J. Chazelas, *Phys. Rev. Lett.* **61**, 2472 (1988).
2. G. Binach, P. Grünberg, F. Saurenbach, and W. Zinn, *Phys. Rev. B* **39**, 4828 (1989).
3. M. Julliere, *Phys. Lett.* **54A**, 225 (1975).
4. J. S. Moodera, L. R. Kinder, T. M. Wong, and R. Meservey, *Phys. Rev. Lett.* **74**, 3273 (1995).
5. T. Miyazaki and N. Tezuka, *J. Magn. Magn. Mater.* **151**, 403 (1995).
6. R. Kubo, *J. Phys. Soc. Jpn.* **12**, 570 (1957).
7. D. A. Greenwood, *Proc. Phys. Soc. London* **71**, 585 (1958).
8. X.-G. Zhang and W. H. Butler, *Phys. Rev. B* **51**, 10085 (1995).
9. H. E. Camblong and P. A. Levy, *J. Appl. Phys.* **73**, 5533 (1993).
10. W. H. Butler, X.-G. Zhang, D. M. C. Nicholson, and J. M. MacLaren, *J. Appl. Phys.* **76**, 6808 (1994).
11. W. H. Butler, X.-G. Zhang, D. M. C. Nicholson, T. C. Schulthess, and J. M. MacLaren, *J. Appl. Phys.* **79**, 5282 (1996).
12. W. H. Butler, X.-G. Zhang, D. M. C. Nicholson, and J. M. MacLaren, *Phys. Rev. B* **52**, 13399 (1995).
13. W. H. Butler, X.-G. Zhang, T. C. Schulthess, D. M. C. Nicholson, J. M. MacLaren, V. S. Speriosu, and B. A. Gurney, *Phys. Rev. B* **56**, 14574 (1997).

14. P. Weinberger, P. M. Levy, J. Barnhart, L. Szunyogh, and B. Ujafalussy, *J. Phys: Condens. Matter* **8**, 7677 (1996).
15. I. Mertig, R. Zeller, and P. H. Dederichs, *Mater Res. Soc. Symp. Proc.* **253**, (W. H. Butler, P. H. Dederichs, A. Gonis, and R. L. Weaver, eds.) Materials Research Society, Pittsburgh (1992).
16. W. H. Butler and G. M. Stocks, *Phys. Rev. B* **29**, 4217 (1984); J. C. Swihart, W. H. Butler, G. M. Stocks, D. M. Nicholson, and R. C. Ward, *Phys. Rev. Lett.* **57**, 1181 (1986).
17. K. Fuchs, *Proc. Philos. Camb. Soc.* **34**, 100 (1938); E. H. Sondheimer, *Adv. Phys.* **1**, 1 (1952).
18. P. F. Carcia and A. Suna, *J. Appl. Phys.* **54**, 2000 (1983).
19. R. E. Camely and J. Barnás, *Phys. Rev. Lett.* **63**, 664 (1989).
20. R. Q. Hood and L. M. Falicov, *Phys. Rev. B* **46**, 8287 (1992); L. M. Falicov and R. Q. Hood, *J. Appl. Phys.* **76**, 6595 (1994).
21. W. H. Butler, X.-G. Zhang, and J. M. MacLaren, *IEEE Trans. Magn.* **34**, 927 (1998).
22. J. M. MacLaren, X.-G. Zhang, and W. H. Butler, *Phys. Rev. B* **59**, 5470 (1999).
23. G. H. Vinyard, *Phys. Rev. B* **26**, 4146 (1982).
24. S. K. Sinha, E. B. Sirota, S. Garoff, and H. B. Stanley, *Phys. Rev. B* **38**, 2297 (1988).
25. S. B. Soffer, *J. Appl. Phys.* **38**, 1710 (1967).
26. W. H. Butler and X.-G. Zhang (unpublished).
27. W. H. Butler, X.-G. Zhang, D. M. C. Nicholson, T. C. Schulthess, and J. M. MacLaren, *J. Appl. Phys.* **79**, 5282 (1996).
28. H. E. E. Camblong, S. Zhang, and P. M. Levy, *Phys. Rev. B* **47**, 4735 (1993); H. E. E. Camblong, P. M. Levy, and S. Zhang, *Phys. Rev. B* **51**, 16052 (1995).
29. T. Valet and A. Fert, *Phys. Rev. B* **48**, 7099 (1993); T. Valet and A. Fert, *Phys. Rev. B* **52**, 13399 (1995).
30. R. Landauer, *IBM J. Res. Develop.* **1**, 223 (1957).
31. S. S. P. Parkin, Paper GA-03 (this symposium).
32. R. Meservey and P. M. Tedrow, *Phys. Rep.* **238**, 173 (1994).
33. W. H. Butler, X.-G. Zhang, X.-D. Wang, J. van Ek, and J. M. MacLaren *J. Appl. Phys.* **81**, 5518–5520 (1997).
34. J. M. MacLaren, W. H. Butler, and X. G. Zhang, *J. Appl. Phys.* **83**, 6521 (1998).

4

MAGNETOTRANSPORT (EXPERIMENTAL)

Jack Bass*

4.1. INTRODUCTION AND OVERVIEW

This chapter is intended to provide an overview of experimental information about magnetotransport in systems involving ferromagnetic (F) metals. The simplest such systems are the three F-elements, Fe, Co, and Ni, and their F-alloys. Magnetotransport in these metals and alloys has been studied for decades and much is reasonably well understood.[1] Of most current interest is magnetotransport in three different systems: (1) Alternating thin layers (multilayers), or granular composites, of F-metals and non-magnetic (N) metals. These are called giant magnetoresistance (GMR) systems.[2-6] (2) F/I/F sandwiches, or granular composites of F-metals plus insulators (I). Here, we have tunneling MR (TMR).[7] (3) Materials that undergo phase transitions with decreasing temperature or increasing magnetic field H from a paramagnetic insulator (PI) to a ferromagnetic metal (FM) state. These are called colossal MR (CMR) materials.[8,9] All of these phenomena, plus others to be described, are now collected under the rubric spin-polarized transport (SPT), because they involve differences in scattering of electrons with spins along or opposite to the local sample magnetization M.

To allow coverage of a wide range of topics within a reasonable length, none is treated in the depth appropriate for a separate review, and discussion of theory is limited to the minimum necessary to comprehend the significance of the data. An attempt has been made to provide representative examples of the most significant phenomena, along with a judicious selection of references, including reviews where available. Topics not covered include spin-glasses[10] and structures involving ferromagnetic metals plus superconductors.[11]

This review focuses on magnetoresistance (MR) for three reasons: (1) it is the transport property most often measured; (2) it usually allows the most direct insight into underlying physics; (3) it usually has the most direct technological application. But three other transport properties, magneto-thermal conductivity, the Hall effect, and magneto-thermopower, have sometimes been measured, and examples of their behaviors will be included.

In a homogeneous metal or alloy, MR is traditionally defined as the percentage change in resistance upon application of a magnetic field H:

---

*Department of Physics and Astronomy, Michigan State University, East Lansing, Michigan.

$$\mathrm{MR} = 100\%[R(H) - R(0)]/R(0) = 100\%\Delta R/R(0) \quad (4.1.1a)$$

MR is, thus, a dimensionless quantity. For the usual positive MRs $[R(H) > R(0)]$ of N-metals,[12] this MR is unbounded. But for the negative MRs $[R(H) < R(0)]$ that predominate in the systems covered in the present review, the MR defined in this way is limited to a maximum value of 100%. More often, one now sees the alternative definition:

$$\mathrm{MR} = 100\%[R(H) - R(H_{\max})]/R(H_{\max}) \quad (4.1.1b)$$

This definition was initially introduced in GMR for samples where $R(H_{\max}) = R_s$ is the saturation resistance at high H, is well defined, but $R(H=0)$ is history dependent[13] (Fig. 4.1).[14] It is now used both for this reason, and because for the usual "negative" MR it is unbounded.

Interestingly, in a completely free electron N-metal, $\mathrm{MR} = 0$. The Lorentz force due to H acting on the conduction electrons pushes these electrons toward one side of the sample. This process leads to a buildup of charge on that side of the sample that repels the remaining electrons in just the way needed to leave the current exactly the same as it was before H was applied. Instead of changing R, the magnetic field gives rise to a transverse voltage, called the Hall voltage. Real metals are not completely free-electron-like. Electrons on different parts of the Fermi surface are deflected differently, making it impossible to "balance" all of them at once. Elemental N-metals exhibit non-zero Lorentz MRs that are invariably positive—i.e., $R(H)$ increases with increasing H. At small fields, these MRs increase as H^2, becoming large only when $\omega_c \tau > 1$, where $\omega_c = eH/mc$ is the cyclotron frequency and τ is the characteristic electron relaxation time for the metal.[12] At liquid helium temperature,

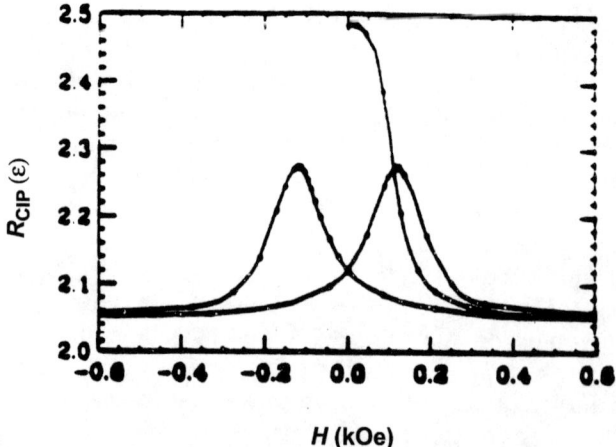

Figure 4.1. Example of a history-dependent MR in a magnetic multilayer; the sample resistance $R = 2.48\,\Omega$ is larger in the as-prepared sample [designated $R(0)$] than it is ever again after the sample is taken to saturation. When similar samples were demagnetized by reducing the magnitude of the field from 0.5 kOe while oscillating its sign, the maximum value of R was found to lie between $R(0)$ and $R(\mathrm{Pk})$ (after Schroeder,[14] p. 6611).

4.2 K, high-purity metals have large τ (long mean-free-paths, $\lambda = v_F\tau$, where v_F is the Fermi velocity), and positive MRs which, using Eq. (4.1.1a), can exceed 100% or even 1,000,000% at high fields.[15] Especially large MRs are seen in compensated metals, where the numbers of electrons, n_0, and holes, n_h, are the same, for then the MR continues to increase indefinitely as H^2.[12,15,16] Non-magnetic alloys, in contrast, or high-purity non-magnetic metals at much higher temperatures, such as room temperature (~295 K), have much smaller τ, and their MRs are usually small, typically <0.1% at $H = 1$ T.[17]

Ferromagnetic elements and alloys display an MR of a different type, called anisotropic MR (AMR) that is briefly described in Section 4.2.2. The term AMR comes from the observation that the MR is positive when the field is applied along the direction of current flow, but negative when the field is transverse to the current. The largest AMR at room temperature is only a few percent in permalloy (Py = $Ni_{1-x}Fe_x$ with $x \sim 0.2$). Nonetheless, the reproducible and stable achievement in thin permalloy films of a 2% AMR at 300 K in a field of only ~10 Oe was until recently the basis of MR read-heads.[18]

As noted above, the systems of main interest in this chapter are composite structures, or materials displaying PI to FM transitions. Several such systems are of potential technological interest because they manifest MRs much larger than the 2% AMR of thin film Py. Larger MRs at 300 K and low enough fields would allow room temperature readout of smaller bits of magnetic information, and denser magnetic memory storage. In fact, GMR F/N multilayers have now taken over as MR read-heads.[18,19] Scientific interest lies in understanding the physics underlying these large MRs.

The rest of this chapter is organized as follows. Section 4.2 provides background on transport in F-metals and F-based alloys. Section 4.2.1 describes the experimental geometry used for standard transport measurements and reviews the equations that define the transport quantities of interest. Section 4.2.2 describes the AMR and extraordinary Hall effect seen in F-metals and alloys. Section 4.2.3 introduces the concepts of spin-polarized transport and of spin-direction dependent scattering anisotropy. Section 4.3 describes spin-dependent tunneling and the concept of "Polarization" of an F-metal. Section 4.4 describes spin-injection studies. Section 4.5 covers GMR in F/N multilayers and granular alloys. Section 4.6 covers TMR. Section 4.7 covers CMR. In conclusion, Section 4.8 briefly describes some miscellaneous topics not already covered.

4.2. ELECTRONIC TRANSPORT IN F-METALS AND F-BASED ALLOYS

The structures of interest in this chapter are assumed to contain enough scattering centers (impurities, structural defects, phonons—quantized lattice vibrations, or magnons—quantized spin-waves of magnetism) so that they have $\omega_c\tau << 1$. Changes in their transport properties with applied magnetic field H are then dominated primarily by effects of changing magnetic order within the structure. More complex behaviors of conductors when $\omega_c\tau \geq 1$ are described elsewhere.[12,16]

4.2.1. Current-in-Plane (CIP) Measuring Geometries

Figure 4.2(a) shows the standard geometry for measuring the electrical resistivity, ρ_{xx}, the Hall resistivity, ρ_{xy}, the thermal conductivity, κ_{xx}, and the thermopower S_{xx}, of a ferromagnetic metal. From $\rho_{xx}(H)$, one can, of course, determine the MR through Eqs. (4.1.1a) and (4.1.1b). Current is sent through a long, narrow foil or film of width W, thickness t, and distance L between longitudinal potential leads.

The linear transport equations used to define the quantities of interest experimentally are[16]

$$E = \rho(H) \cdot j + S(H) \cdot \nabla T \qquad (4.2.1a)$$

$$\dot{Q} = \Pi(H) \cdot j - \kappa(H) \cdot \nabla T \qquad (4.2.1b)$$

Here, ρ, κ, Π, and S are tensors relating two vectors (the electric field E and the heat current density Q to two others (the electrical current density j and the temperature gradient ∇T), and H is the magnitude of the applied magnetic field. Π and S are further related by the Kelvin–Onsager relation, $\Pi(-H) = TS(H)$. In this chapter, we

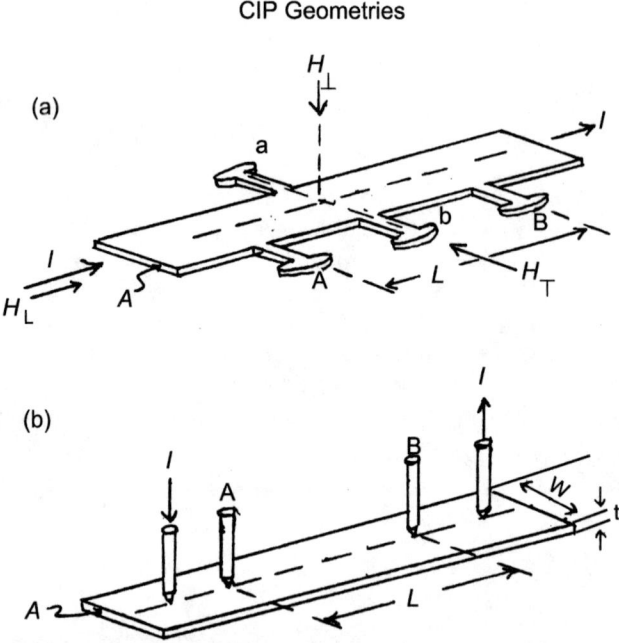

Figure 4.2. Standard geometries for measuring transport properties. (a) Geometry for measuring ρ_{xx}, R_H, κ, and S. For ρ_{xx}, current I_x is injected through area A (width W times thickness t) and V_x is measured over length L. For R_H, I_x is injected through A and V_y is measured over width W. For κ, heat current Q_x is injected and temperature difference ΔT_x is measured across L. For S, heat current Q_x is injected and V_x is measured across L. H can be applied tangential to the film (H_T) or perpendicular to it ($H_\perp$). (b) Geometry for measuring MR. Current I_x is injected and removed as shown and V_x is measured across L. Because I_x is not distributed uniformly within the sample, determination of ρ_{xx} requires calculation of a correction factor. But, so long as the spatial distribution of I_x is independent of H, $MR = [R(H) - R(0)]/R(0)$ will be reliable. $R - V_{AB}/I$; $R_H = V_{ab}/I$; $\rho_{xx} = RA/L$; $\rho_{xy} = R_H t$

focus upon three diagonal elements, ρ_{xx}, κ_{xx}, and S_{xx}, where E_x, j_x, $\dot{Q}_x$, and $\nabla_x T$ are all along the x-axis, and one off-diagonal element, ρ_{xy}, where j_x is in the x-direction, but E_y is measured in the y-direction. For measurements of ρ_{xx}, κ_{xx}, and S_{xx}, the external field, H, is usually applied in the x–y plane. ρ_{xy}, in contrast, is defined for H along the z-axis [see Fig. 4.2(a)].

For N-metals, the direction of H is usually of little consequence for determining the diagonal transport coefficients, so long as $\omega_c \tau \ll 1$, and the sample dimensions are large enough so that finite size effects remain negligible. For F-metals, in contrast, the situation is more complex, since increasing H increases the sample magnetization M, until M reaches its saturation value M_s. When H is applied in the plane of the film, H inside (H_I) the F-metal is the same as H outside ($H_A = H_{applied}$), because Maxwell's equations require H tangential to be continuous.[20] As the sample becomes magnetized, only the applied H orients the local magnetic moments to increase M. When, however, H is applied perpendicular to the film plane, partial alignment of the internal moments to produce $M \neq 0$, causes $H_I = H_A - KM$ to be reduced from H_A by KM, where K is a "demagnetizing factor." K ranges from 4π for H perpendicular to a film, to 2π for H perpendicular to a wire, to 0 for H parallel to a film. For the topics in the present section, a non-zero K mostly just increases the field H needed to reach M_s.

To determine ρ_{xx} and ρ_{xy}, an electrical current must be sent through the foil or film of Fig. 4.2(a). For H in the x–y plane of the film, H along y is called transverse, and H along x longitudinal. If the total current $I_x = A j_x$ flows uniformly through the sample, the experimentally measured resistances R and R_H are related to the intrinsic quantities ρ_{xx} and ρ_{xy} by the voltages $V_x = L E_x$ and $V_y = W E_y$ through

$$R = V_x/I_x = E_x L / A j_x = \rho_{xx} L / A \tag{4.2.2a}$$

and

$$R_H = V_y/I_x = E_y W / A j_x = \rho_{xy}/t = \rho_H/t \tag{4.2.2b}$$

For these measurements, the sample is usually immersed in a thermally conducting (but electrically insulating) fluid to keep it at constant temperature.

Similarly, κ_{xx} and S_{xx} are determined by passing a uniform heat current density $\dot{Q}_x$ through the foil and measuring the resulting temperature difference $\Delta_x T$ or the voltage difference $V_x = L E_x =$ over length L, respectively. The relations of interest are

$$K_{xx} = \dot{Q}_x/\nabla_x T = \kappa_{xx}(L/A) \tag{4.2.3}$$

and

$$S_{xx} = V_x/\Delta_x T = E_x/\nabla_x T \tag{4.2.4}$$

where K_{xx} is the thermal conductance and $\nabla_x T = \Delta_x T / L$.

For these measurements, the sample is usually in vacuum to minimize heat loss by conduction through any surrounding gas. For precise determination of κ, it is also necessary that heat loss by conduction through any substrate holding the sample, and by radiation, be small.

The alternative geometry of Fig. 4.2(b) allows the determination of just the dimensionless MR(H), so long as the non-uniform current distribution does not change with H. Then, determining ρ_{xx} then requires calculations of a correction factor for the non-uniform current.

Other steps needed to ensure reliable experimental data are covered in Refs. 21 and 22.

4.2.2. Anisotropic Magnetoresistance (AMR); Extraordinary Hall Effect

4.2.2.1. AMR. The MR in an F-metal usually shows hysteresis, and depends on the angle between the current I and the magnetization M. Figure 4.3 shows schematically the differences in behavior observed when H is applied in the plane of a flat sample [Fig. 4.2(a)] but either (a) parallel (longitudinal geometry) or (b) perpendicular (transverse geometry) to I.[23] This difference is the AMR, which is attributed to anisotropy in the scattering rates of s-electrons off of spin ↑ or spin ↓ d-electrons, due to spin–orbit coupling.[24] When the current-in-plane (CIP) MRs for more complex structures are small, AMR corrections can be significant. As noted above, the 2% room-temperature AMR at 10 Oe in thin films of Py was until recently the basis of read-heads.

4.2.2.2. Extraordinary Hall Effect. The Hall effect in a ferromagnet always involves both demagnetization [because H is applied perpendicular to the plane of the sample film—Fig. 4.2(a)], as well as hysteresis. Figure 4.4 shows schematically the behavior of ρ_H in the simplest circumstance, $\omega_c \tau \ll 1$. The straight line at high fields is the ordinary Hall effect. The rise from $H = 0$ to meet this line is due to the extraordinary Hall effect, which is determined experimentally by subtracting the high-field linear term from the total Hall effect. The extraordinary Hall effect is proportional to the magnetization M, and is a consequence of asymmetric scattering of electrons, sources of which (primarily the spin–orbit interaction) are described in Refs. 21 and 25.

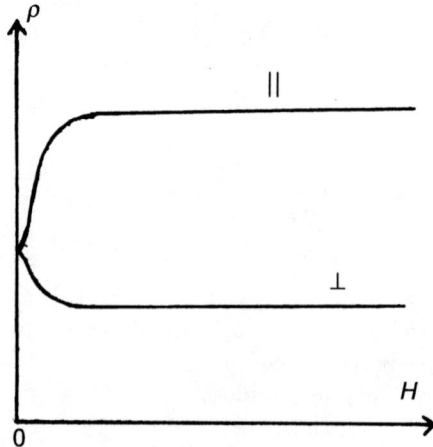

Figure 4.3. Schematic drawing of AMR for a ferromagnetic metal.

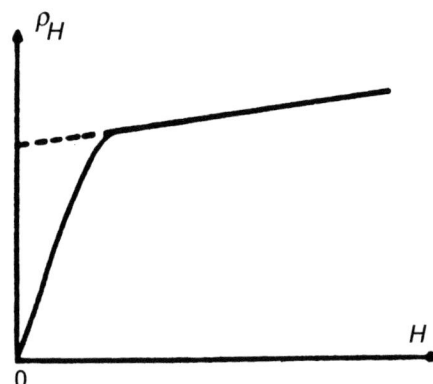

Figure 4.4. Schematic drawing of the Hall resistivity for a ferromagnetic metal. The linear variation of $\rho_{xy} = \rho_H$ with H at high fields indicates the simplest behavior for the normal Hall resistivity. The more rapid increase at low fields is the extraordinary Hall resistivity.

4.2.2.3. Spin-Direction Scattering Anisotropy; Slater–Pauling Plot.

For transport in an N-metal or alloy, the fact that electrons have spin can usually be neglected, since interactions normally do not depend upon the direction in which this spin points. To first approximation, one often takes the resistivity, $\rho_0(c)$, due to scattering from a concentration "c" of impurities or other defects, to be temperature independent and simply additive to the temperature-dependent resistivity, $\rho_p(T)$, due to phonon scattering in the pure host metal (Matthiessen's rule):[12]

$$\rho_T(T) = \rho_0(c) + \rho_p(T) \tag{4.2.5}$$

In practice, even N-alloys always display deviations from Matthiessen's rule, which in different circumstances can be comparable to either term on the right-hand side of Eq. (4.2.5).[26]

In the F-metals on which we focus in this review, it is necessary to allow scattering of electrons with spins along (↑) the local magnetization M_i to differ from that for electrons with spins opposite (↓) to M_i. The simplest model of such scattering is the extreme s–d model, in which spin (↑) and (↓) band electrons are subdivided into unhybridized s- and d-electrons, with only the s-electrons carrying current. Figure 4.5 illustrates the case where the d (↑) band is completely filled. At low temperatures, where no magnons are excited, spin-flip scattering is negligible, and the resistivity is determined by independent parallel conduction of current by spin (↑) and (↓) s-electrons, with stronger scattering of the spin (↓) s-electrons because they can be scattering into non-conducting spin (↓) d-states. Adding conductivities ($\sigma = 1/\rho$) and inverting gives[1]

$$\rho = \frac{\rho\uparrow \rho\downarrow}{\rho\uparrow + \rho\downarrow} \tag{4.2.6}$$

At high temperatures, electron–magnon scattering will transfer momentum between the two currents—spin-mixing, and one must add a spin-mixing resistivity $\rho\uparrow\downarrow$,

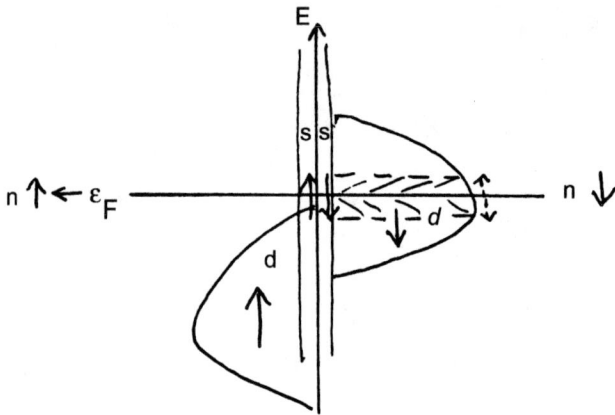

Figure 4.5. Schematic drawing of the spin-up (↑) and spin-down (↓) branches of the s- and d-bands of a ferromagnetic (F-) metal in the extreme s–d model (no mixing) limit, and for the case where the majority d-band is below the Fermi level, ε_F. The dashed regions in the minority d-band indicate the changes in filling of that band, in the rigid band limit, upon adding impurities to the left of the host F-metal in the periodic table (hatched area decreasing from ε_F) or to the right (hatched area increasing from ε_F). Adding impurities to the left decreases the filling, thereby increasing μ, the magnetic moment/atom. Adding impurities to the right increases the filling, thereby decreasing μ. Such "normal" behavior is shown as the "backbone" line in Fig. 4.8.

giving a more complex expression[1] not needed here. When $\rho\uparrow$ and $\rho\downarrow$ are each due to more than one scattering process [e.g., impurities and phonons as in Eq. (4.2.5), or two different impurities], the simplest assumption is that Matthiessen's rule (additive resistivities) applies within each spin-channel. Separate conductance in each channel then leads to a more complex form of Eq. (4.2.6), involving the contributions from the different spin-dependent components of $\rho\uparrow$ and $\rho\downarrow$.[1]

Mott[27] introduced a two-current s–d model to analyze $\rho(T)$ of F-metals near the Curie temperature, T_C. Nowadays, this model is used mainly at much lower temperatures. The fundamental quantity in such a two-current model is the ratio $\rho\downarrow/\rho\uparrow$, for which we define alternative variables $\alpha_F = (1+\beta)/(1-\beta) = \rho\downarrow/\rho\uparrow$.[1,28] In this review, we will be interested mainly in α_F (or its related β) for F-alloys consisting of a given impurity in one of the host F-metals Ni, Fe, and Co. By analogy, we will assume that there is similar anisotropy, and similar quantities $\alpha_{F/N} = (1+\gamma)/(1-\gamma) = R_{F/N}\downarrow/R_{F/N}\uparrow$, at F/N interfaces ($\alpha_{F/N}$ and γ might depend upon details of the interface structure).

A question of interest in this review will be whether the parameters β estimated for dilute F-based alloys, as we describe next, are similar to those derived from measurements of GMR in magnetic multilayers described in later sections.

Experimentally, values of α_F (or β) for dilute F-based alloys have been estimated from measurements of deviations from Matthiessen's rule (additivity of resistivities) in either the low temperature (residual) resistivities of ternary alloys, or in the temperature-dependent resistivities of binary alloys.[1] If the values of α_B are similar for two impurities, a and b, then the residual resistivity of a ternary alloy should vary linearly between the values for the individual impurities [Fig. 4.6(a)]. If they differ substantially, the resistivity of a ternary alloy should deviate from a linear variation

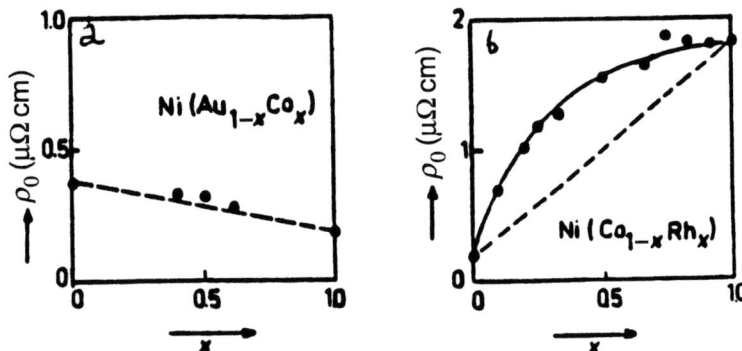

Figure 4.6. Residual resistivities versus x for Ni(Au$_{1-x}$Co$_x$) and Ni(Co$_{1-x}$Rh$_x$) alloys. The straight line behavior for the Ni(Au$_{1-x}$Co$_x$) occurs because the values of α_{Co} (~20) and α_{Au} (~6) are "similar" (i.e., both large). The large deviations from linearity (deviations from Matthiessen's rule) for the Ni(Co$_{1-x}$Rh$_x$) alloys result from very different values of α_{Co} (~20) and α_{Rh} (~0.25) (from Campbell and Fert,[1] p. 767 © 1982, Elsevier Science).

[Fig. 4.6(b)]. From data such as those in Fig. 4.6, values of α_F (or β) have been estimated for a wide range of Ni-, Fe-, and Co-based alloys.[1] The variation of α_F from one impurity to the next is illustrated in Fig. 4.7, which shows that theory [Fig. 4.7(a)][29] and experiment [Fig. 4.7(b)][1] can yield similar patterns. Care must be exercised, however, about the absolute values in either case. Values inferred from experiments by different groups often scatter widely,[1] and the calculations are still incomplete—e.g., they do not properly include phenomena such as spin–orbit coupling.[29]

As illustrated by the cross-hatched areas in Fig. 4.6, if the band structure of the host metal remains "rigid" as a dilute fraction of impurities is added, then adding an impurity to the right of the host in the periodic table should increase the number of electrons in the (↓) d-band, reducing the difference between majority (↑) and minority (↓) d-electron concentrations, thereby reducing the magnetic moment, μ. Conversely, adding an impurity to the left should increase μ. As shown in Fig. 4.8,[30] such a behavior is observed for data falling along the linear "backbone" of a Slater–Pauling plot of μ versus electron concentration. Impurities can, however, sometimes behave differently. For example, when Cr is added to Ni, Co, or Fe, its local magnetic moment orients itself opposite to that of the host F-metal, thereby decreasing μ and forming a "virtual bound state" that strongly scatters majority (↑) s-electrons. Such a scattering greatly increases $\rho\uparrow$ (Fig. 4.7), leading to $\alpha_F < 1$ ($\beta < 0$), and to data that fall away from the "linear backbone" as the electron concentration decreases.

4.3. SPIN-DEPENDENT TUNNELING AND "POLARIZATION"

We now turn to early experiments on a different phenomenon, spin-dependent tunneling from an F-metal through an oxide. There, we are no longer dealing with diffusive transport, but with quantum mechanical tunneling. Provided that

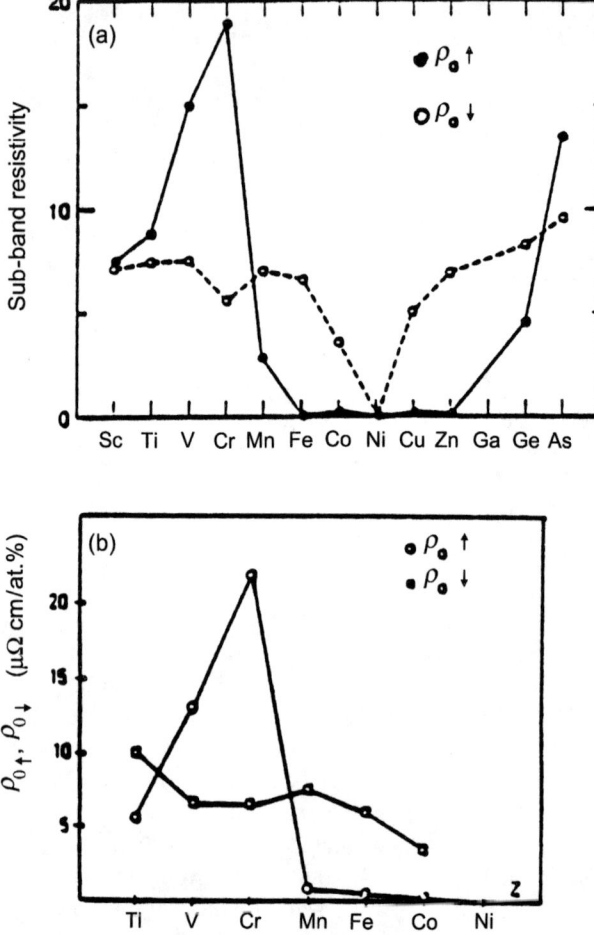

Figure 4.7. Sub-band residual resistivities $\rho_{0\uparrow}$ and $\rho_{0\downarrow}$ for 3d impurities in Ni. (a) Calculations (from Mertig et al.,[29] p. 16182). (b) Experiment (from Campbell and Fert,[1] p. 769 © 1982, Elsevier Science).

the current density is not so large as to cause significant Joule heating, diffusive transport obeys Ohm's law for a constant resistance—the measured voltage V is linearly related to the applied current I. In contrast, for a large enough applied voltage V, tunneling involves a non-linear relation between V and the resulting current I.

The first tunneling measurements using F-metals involved tunneling into a superconductor (S), and focused upon trying to determine the "spin polarization" of the tunneling current. To explain what is measured, we start with tunneling from an N-metal into an S-metal. Figure 4.9 shows[31] (a) the band structure; (b) the broadened tunneling kernel at finite temperature; and (c) the expected non-linear tunneling spectrum dI/dV versus eV (e is the electron charge), for tunneling from an N-metal into an S-metal in the presence of a field H large enough to Zeeman split the spin ↑ and spin ↓ levels in the S-metal as shown in (a). An applied voltage $\pm V$

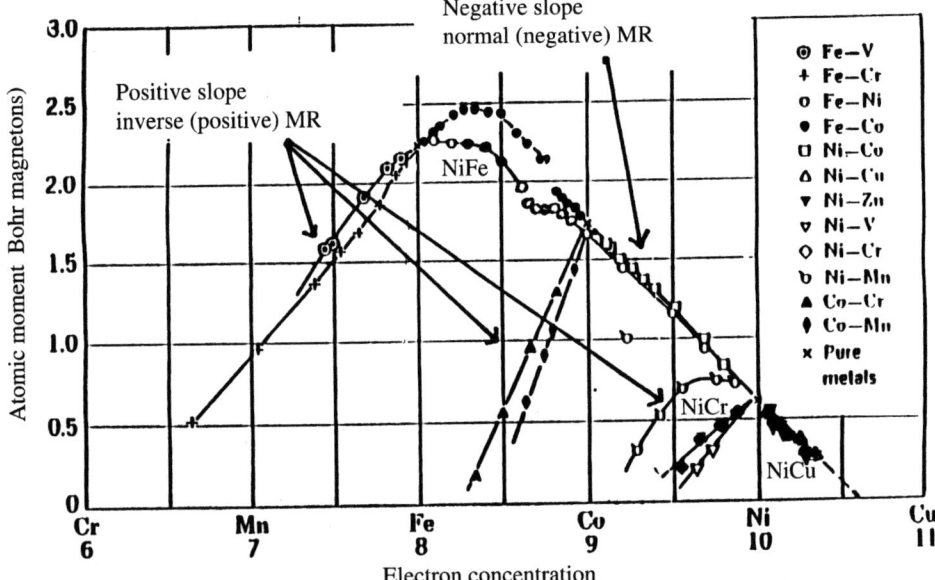

Figure 4.8. Slater–Pauling plot for F-metal based alloys. The straight line rising to the middle of the figure from the lower right indicates the "normal" backbone behavior described in Fig. 4.5. The lines angling down to the left deviate from this "normal" curve (after Kittel[30]).

(energy = $\pm eV$) shifts the location of the Fermi level of the N-metal (not shown) up ($+eV$) or down ($-eV$) relative to that for the S-metal. If, first, we neglect the Zeeman splitting, then application of a voltage V giving an energy eV between 0 and $\pm \Delta/2$ (where Δ is the energy gap) would leave the Fermi level of the N-metal below the value needed to start tunneling from the filled states in the N-metal into the empty excited states in the S-metal ($+eV$), or from the filled states in the S-metal to the empty states in the N-metal ($-eV$). If the tunneling kernel in (b) was just a delta-function ($T = 0$ K), no tunneling would occur. Any observed tunneling in this region, thus requires a finite width of the tunneling kernel. As eV grows further in magnitude, dV/dI should map out the shape of the electron band in the superconductor. Figure 4.9 shows additional structure, because the electron band in the S-metal is Zeeman-split into spin-↑ and spin-↓ sub-bands.

If, now, the N-metal is replaced by an F-metal that has different numbers of ↑ and ↓ electrons at its Fermi level (see, e.g., the d-density of states in Fig. 4.5), and/or different tunneling probabilities for ↑ and ↓ electrons, then moving the Fermi level of this F-metal up or down will lead to an asymmetry in the tunneling spectrum as shown in Fig. 4.10.[31] This figure illustrates the behavior expected if ↑ electrons are either more prevalent and/or tunnel more easily. Such measurements led to the inferred "tunneling" spin polarizations listed in Table 4.1,[31,32] where + means tunneling dominated by majority (↑) electrons. Here, the polarization is defined by[33]

$$P = (N_\uparrow - N_\downarrow)/(N_\uparrow + N_\downarrow) \qquad (4.3.1)$$

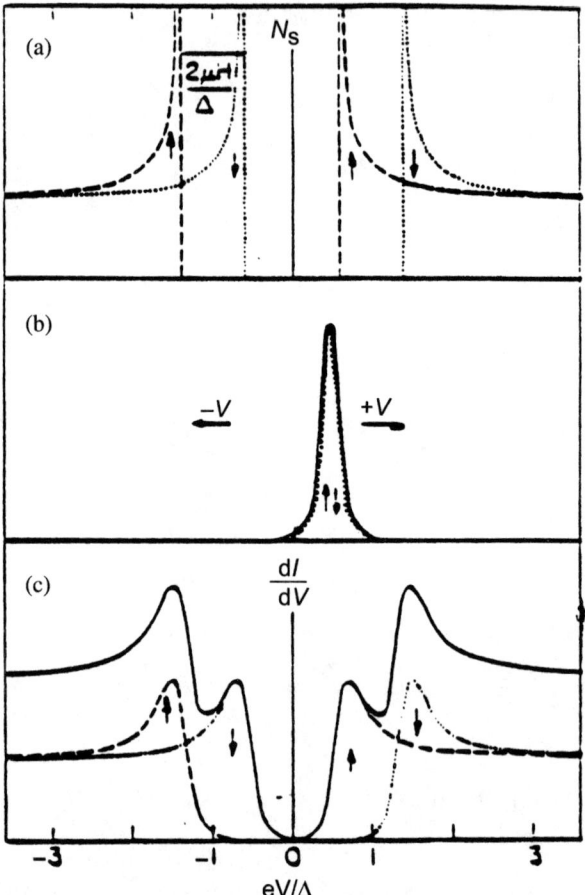

Figure 4.9. Diagrams for tunneling spectrum between a non-magnetic (N) metal and a superconductor (S) in a magnetic field large enough to produce significant Zeeman splitting between the spin-up (↑) and spin-down (↓) branches of the S-metal electrons. (a) Field splitting of the quasiparticle states. (b) The spin and temperature dependent kernel in the tunneling current integral. Tunneling only occurs over the energy range allowed by this kernal. (c) Spin-up conductance (dashed), spin-down conductance (dotted), and total conductance (solid curve). The bottom curves are mapped out by moving the Fermi level of the N-metal up ($+eV$) or down ($-eV$) relative to the S-metal structure in part (a) (from Meservey and Tedrow,[31] p. 189, © 1994, Elsevier Science).

where $N_{\uparrow,\downarrow} = n_{\uparrow,\downarrow} |T_{\uparrow,\downarrow}|^2$, with $n_{\uparrow,\downarrow}$ the ↑ or ↓ electron density of states, and $T_{\uparrow,\downarrow}$ the matrix element for tunneling (transmission) of ↑ or ↓ electrons.

The + values listed in Table 4.1 are opposite in sign to what would be expected from Fig. 4.5 if the tunneling was dominated by the density of states of the minority (↓) d-electrons at the Fermi surface. Rather, these values indicate that the majority (↑) electrons dominate the tunneling, just as ↑ electrons usually carry current more easily in diffusive transport (Section 4.2.3).

Recently, point contact measurements of the Andreev reflection at the interfaces between F- and S-metals have also been used to infer "spin polarizations"

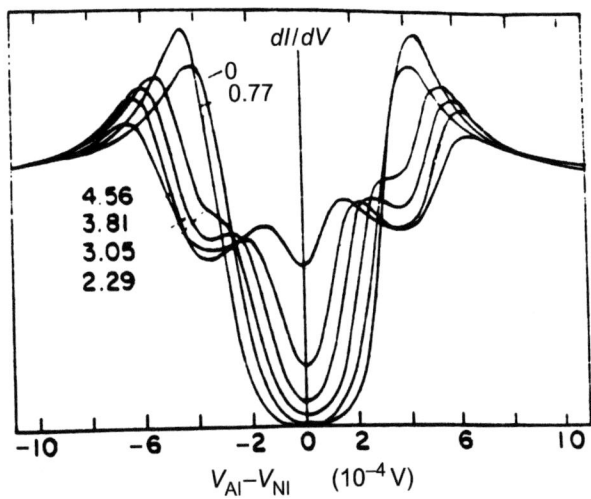

Figure 4.10. Measured conductance versus voltage V for a typical Al/Al$_2$O$_3$/Fe junction at several values of the magnetic field (T). The "polarization" of the Fe can be derived from the asymmetry of the conductance (from Meservey and Tedrow,[31] p. 201, © 1994, Elsevier Science).

Table 4.1. Spin Polarizations (%) Derived from Different Measurements

Material	S-point Contacts	Tunneling (A)[31]	Tunneling (B)[32]	Point Contact (A)[34]	Point Contact (B)[3]
Co	Nb, Pb	35 ± 3		37 ± 2	42 ± 2
Ni	Nb, Ta, Pb	23 ± 3		32 ± 2	44 ± 3
Ni$_{0.8}$Fe$_{0.2}$ = Py	Nb	25 ± 2	45		37 ± 5
Fe	Ta, Nb, V	25 ± 2			45 ± 3
NiMnSb	Nb				58 ± 2
La$_{0.7}$Sr$_{0.3}$MnO$_3$	Nb				78 ± 4
CrO$_2$	Nb				90 ± 4

Specified uncertainties have been rounded up or down to the nearest integer.

of electrons in F-metals.[34,35] Because of the difference in physics between Andreev reflection and tunneling, these "spin polarizations" might have been expected to have no simple relation to each other. However, Table 4.1 shows that the Andreev and tunneling "spin polarizations" for Co, Fe, and the F-alloy Py are rather similar. Of interest for use later in this chapter are the inferred values for the three compounds NiMnSb, CrO$_2$, and La$_{0.7}$Sr$_{0.3}$MnO$_3$. NiMnSb[36] and CrO$_2$[37] were predicted to be 100% polarized at the Fermi energy (designated "half-metallic" ferromagnets). But the predictions were not confirmed by spin-resolved photoemission, where NiMnSb was found to be only ~ 50% polarized;[38] and CrO$_2$ to be 100% polarized ~2eV below the Fermi energy, but to have very few metallic carriers at the Fermi energy.[39] La$_{0.7}$Sr$_{0.3}$MnO$_3$, in contrast, was predicted to be only ~36%

polarized.[40] But here spin-polarized tunneling studies in F/I/F trilayers suggest polarizations of 54%[41] or 81%,[42] and a spin-polarized photoemission study suggests 100% polarization at the Fermi energy.[43] Some caution must be exercised concerning spin-polarized photoemission studies, since they are extremely surface sensitive.

4.4. SPIN-INJECTION STUDIES

Injection of spin-polarized current from an F-metal into an N-metal at one location, and detection of the polarization remaining a distance L away, provides a way to measure the loss of spin-direction memory (spin relaxation), governed by the spin diffusion length, l_{sf}, in the N-metal. The first such measurements were made by Johnson and Silsbee[44] and followed up by Johnson.[45] Figure 4.11 shows the "linear" geometry first used. Polarized current was injected into a thin film evaporated onto a cold-rolled Al foil, and detected by measuring the voltage difference, V_d, between a similar film, with its magnetization oriented parallel to that of the first film, and a peripheral point on the sample. The relaxation time of electrons in Al was measured using the Hanle effect, where application of a small magnetic field in the plane of the film causes the magnetization of the injected current to precess, thereby changing its orientation relative to that of the detector. Figure 4.12 shows the reduction in V_d with H for two different values of L and Fig. 4.13 shows that the inferred relaxation times, T_2, are comparable with those estimated from transmission electron spin-resonance (TESR). The resulting spin-diffusion length in Al, $l_{sf}^{Al} = (2DT_2)^{1/2}$ was ~170 µm.

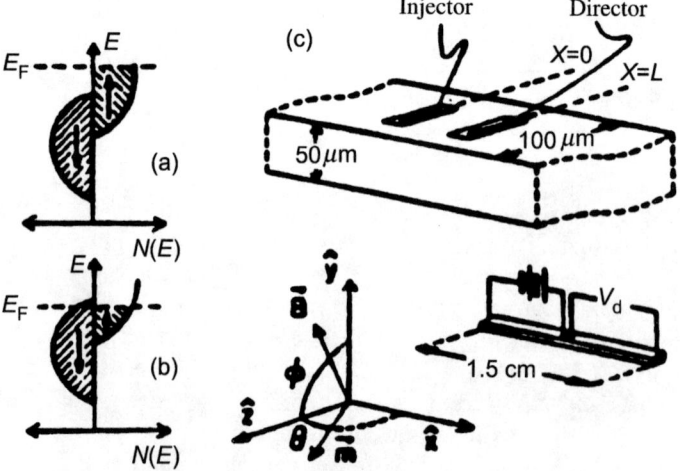

Figure 4.11. (c) Linear geometry for spin-injection studies. Overidealized Stoner ferromagnet with (a) one full sub-band, or (b) inequivalent sub-bands (from Johnson and Silsbee,[44] p. 1790, © 1985 APS).

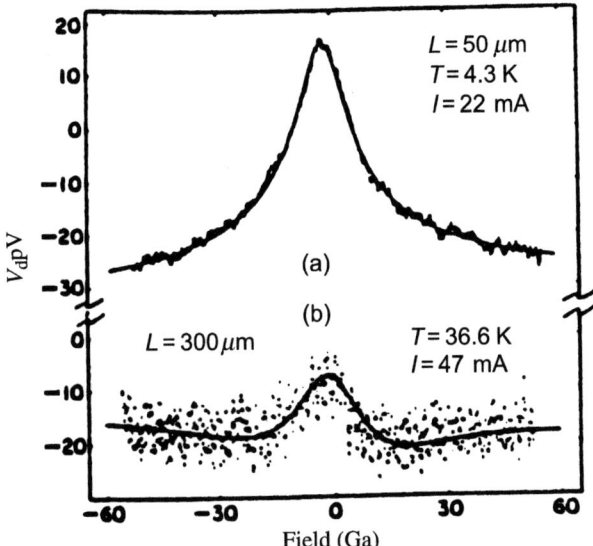

Figure 4.12. Sample data for the Hanle technique. The voltage detected in the geometry of Fig. 4.11 is plotted versus a small applied magnetic field (from Johnson and Silsbee,[44] p. 1793, © 1985 APS).

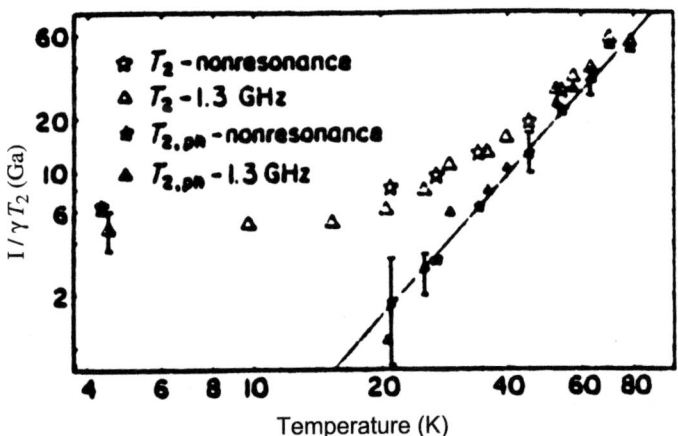

Figure 4.13. Comparison of $1/T_2$ inferred from data such as in Fig. 4.12 with TESR results at 1.3 GHz (from Johnson and Silsbee[44] p. 1793, © 1985 APS).

Johnson[45] subsequently extended spin-polarization studies to the "perpendicular" geometry of Fig. 4.14(a), where the injecting and detecting films face each other. The measured quantity is now the difference in V_d between the parallel (P) and anti-parallel (AP) states of the magnetizations of the injector and detector, where shape differences cause one to "flip" before the other. The expected signal is

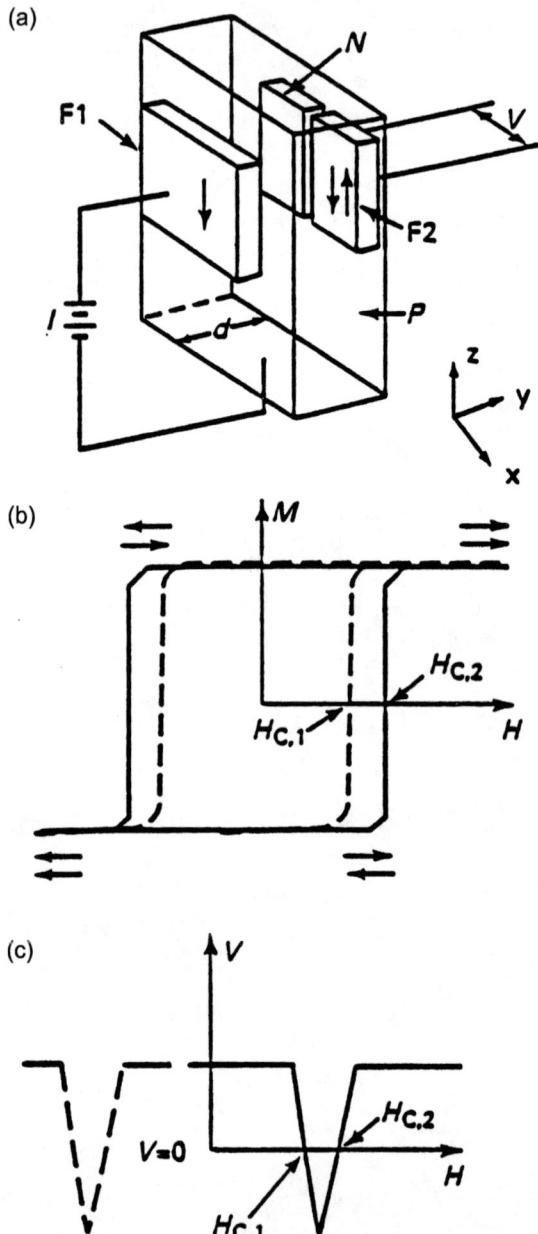

Figure 4.14. (a) Perpendicular geometry for spin-injection studies. (b) Schematic of the expected hysteresis loop. (c) Schematic of the expected voltage pattern versus field (after Johnson,[45] p. 6716).

illustrated in Fig. 4.14(c) and a real signal is shown in Fig. 4.15. A value of $l_{sf}^{Au} \sim 1.5 \pm 0.4$ μm was inferred for an evaporated Au film from the data of Fig. 4.16.

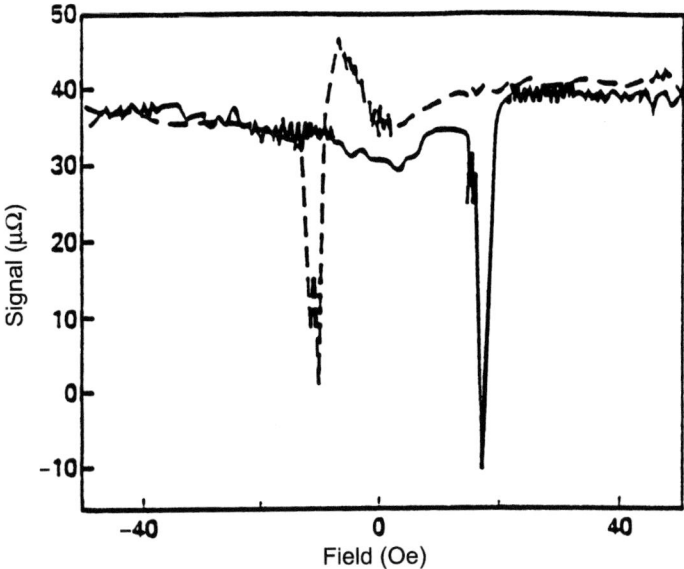

Figure 4.15. Experimental data for Au between permalloy layers in the geometry of Fig. 4.14 (from Johnson,[45] p. 6716).

This geometry, in which the current and voltage are measured across different contacts, has the potential to produce "amplification."[46] Its disadvantages for deriving the spin-diffusion length of interest are that: the current is non-uniform; the magnetic states are not well controlled; and a question has been raised about the analysis,[47] which assumed, as illustrated in Fig. 4.11(a), that the ferromagnet is half-metallic—i.e., contains only a single fully polarized conduction band. Alternative ways to access the same parameters are described in Section 4.5.3.11.

4.5. GIANT MAGNETORESISTANCE IN F/N MULTILAYERS AND GRANULAR ALLOYS

4.5.1. Overview and Background

The term GMR designates the observed decrease in the resistance of an F/N multilayer (or an F–N granular alloy) when a large enough magnetic field H reorients the magnetizations M_i of the F-layers (or F-inclusions) from a state with total magnetization $M \approx 0$ into the state of maximum $M = M_s$ (saturation magnetization). In this latter state, the M_i are aligned parallel (P) to each other. The GMR is largest when at low fields the M_i of adjacent F-layers are aligned antiparallel (AP). Calculations usually focus upon the MR between these two states, $MR = [R(AP) - R(P)]/R(P)$. References 4–6 review facets of GMR.

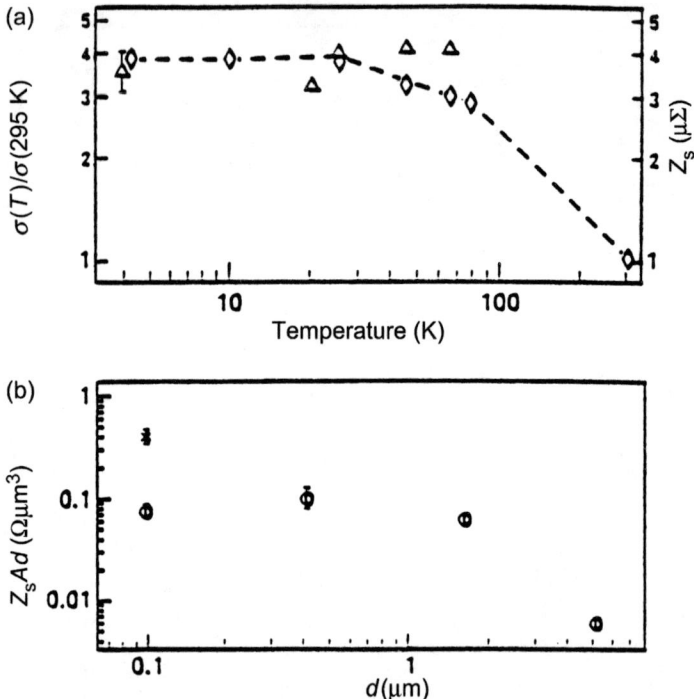

Figure 4.16. (a) Comparison of the inferred ratio $\sigma(T)/\sigma(295 \text{ K})$ (left-hand axis, ◊) with $Z_s(T) = V/I$ (right-hand axis, △) for Au thickness $d = 1.6$ μm. (b) $Z_s A d$ versus d. Z_s is referred to the current through a single window. Most data are for F1 and F2 both Py; the △ data are for F1 = Py and F2 = Co (from Johnson,[45] p. 2144, © 1993 APS).

In the standard picture, GMR results from spin-dependent scattering within the F-metal or at the F/N interfaces,[1] characterized by the experimental parameters, α_B (or β) and $\alpha_{F/N}$ (or γ) defined in Section 4.2.2.3. The upper-half of Fig. 4.17 illustrates the physical source of GMR for spin-dependent scattering inside the F-layers in the usual case where $\alpha_B > 1$ ($\beta > 0$).[3] Spin-dependent scattering at the F/N interfaces can be treated similarly. The injected electrons are divided into two "channels", spin up or down with respect to a fixed direction (e.g., the applied field $+H$), and electron spin-direction is taken to be conserved as the electrons propagate through the multilayer ($l_{sf} = \infty$ in both the F- and N-metals). In the low-field AP state, electrons of each spin are scattered strongly in alternate layers. In the high-field P state, in contrast, electrons of one spin direction are scattered strongly in every F-layer, but electrons of the other spin are not scattered strongly in any. These latter electrons "short out" the sample, reducing its resistance. For identical F-layers, exactly the same result occurs in the opposite case where $\alpha_B < 1$ ($\beta < 0$). The "normal" behavior is thus a decrease in R, as shown in Fig. 4.1.

If, however, a multilayer is made of two different F-metals, one with $\alpha_B > 1$ ($\beta > 0$), and the another with $\alpha_B < 1$ ($\beta < 0$) then, as shown in the lower-half of Fig. 4.17, the MR "inverts"—$R(P)$ is larger than $R(AP)$. Such a behavior can be used to test for negative β or negative γ. Assume that $\gamma 1$, $\gamma 2$, and $\beta 1$ are all positive,

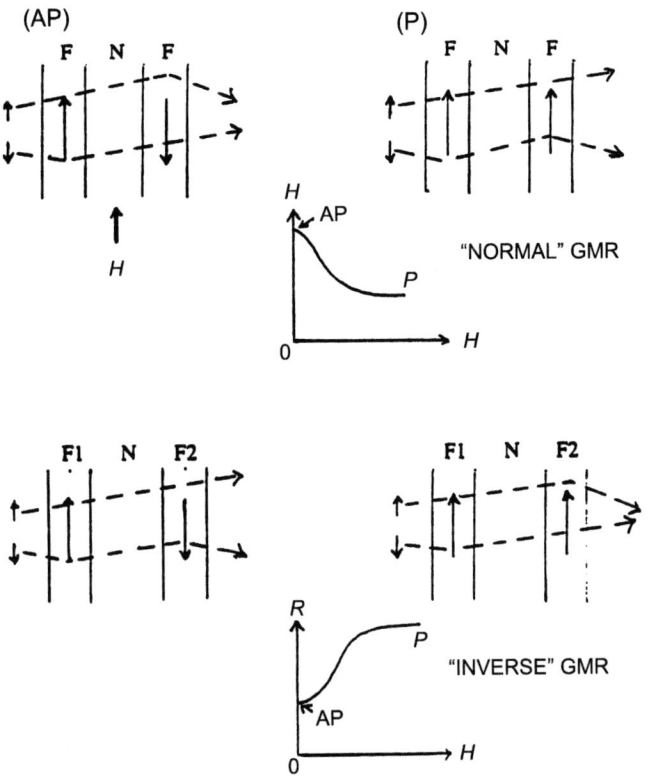

Figure 4.17. Schematic drawings of the physics underlying GMR assuming anistropic scattering only in the bulk of the F-metal. Upper-half: Normal GMR, showing the AP and P state behaviors when scattering in both F-layers is identical. Lower-half: Inverse GMR, showing the AP and P state behaviors when scattering in one F-layer is normal (i.e., weaker for electron moment parallel to the layer magnetization), but that in the other layer is "inverse" (i.e., stronger for electron moment parallel to the layer magnetization). GMR—underlying physics Consider bulk scattering only. Also smilar contribution from interface scattering. Case 1: Single F-metal —$\rho\downarrow/\rho\uparrow > 1 (\beta > 0)$ [Note: Same result for— $\rho\downarrow/\rho\uparrow < 1$ ($\beta < 0$)]. Case 2: F1, F2— $\rho 1\downarrow/\rho 1\uparrow > 1(\beta_1 > 0)$; $\rho 2\downarrow/\rho 2\uparrow < 1$ ($\beta_2 < 0$) [Note: Same result for —($\beta_1 < 0$, $\beta_2 > 0$)].

but $\beta 2$ is negative. Then, multilayers with very thin t_2 should yield a normal MR, because $\gamma 2$ dominates over $\beta 2$. However when t_2 becomes large enough to dominate over $\gamma 2$, the MR should change sign and become inverse. We will see such a behavior in Sections 4.5.2.8 and 4.5.3.9.

There are two basic geometries for studying GMR, the CIP geometry of Figs. 4.2 and 4.18(a), and the current-perpendicular to the planes (CPP) geometry of Fig. 4.18(b).

The CIP resistance for a typical "macroscopic" multilayer with length = $L = 1$ cm and area = $Wt = 10^{-5}$ cm^2 is $R \sim 0.01$–1 Ω, easily measured with a digital voltmeter or lock-in amplifier, and large enough for technology.[18,19] However, it is difficult to separate the contributions from within the F-layer and the F/N interfaces in this geometry; the required equations contain many variables,[48] and

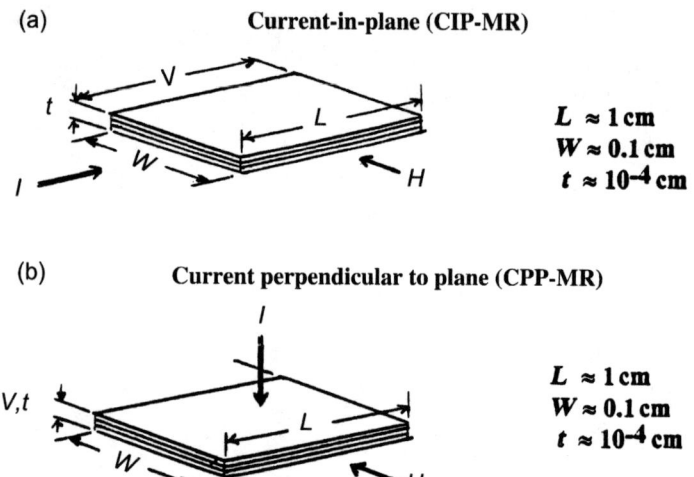

Figure 4.18. (a) CIP-MR and (b) CPP-MR measuring geometries and characteristics. (a) $R = V/I = \rho L/Wt \approx 1\,\Omega$; Easy to measure; Other advantage: large enough for technology. Problem: hard to separate bulk (F) from interface (F/N) scattering (must model data, and momentum changing mean-free-path is fundamental length). (b) $R = V/I = \rho t/WL \approx 10^{-7}\,\Omega$; Need SQUID to measure; (b) Lithography, $\rightarrow \geq 10^{-2}\,\Omega$ nanowires; easy to measure; might have technological uses.

the momentum changing mean-free-path is a fundamental length, as we will explain further.

The CPP resistance for a typical "macroscopic" multilayer, with area now $= LW = 10^{-2}$ cm² and "length" $= t = 10^{-4}$ cm, is $R \sim 10^{-7}$–$10^{-8}\,\Omega$. To measure such a resistance in the presence of magnetic fields requires both care and more sophisticated techniques, such as superconducting quantum interference devices (SQUIDS)[49] or low temperature ac amplifiers.[50] Resistances large enough to be measured with standard digital voltmeters or lock-in amplifiers can be obtained by lithography[51] or by producing nanowires,[52–54] as we will see in Section 4.5.3.1. However it is measured, the CPP-MR has the advantages that it is usually larger than the CIP-MR, and is usually described by simpler equations that allow more straightforward separation of bulk and interface contributions.[28,55,56]

The qualitative difference between the CIP- and CPP-MRs is illustrated in Fig. 4.19.

In the CIP-MR [Fig. 4.19(a)], the drift current flows parallel to the layers. But GMR requires communication perpendicular to the layers. If the mean-free-path, λ_N, in the N-metal is much shorter than t_N, the electrons cannot communicate across N, and thus cannot distinguish between the P and AP states (more precisely, λ_N delimits the spread of the electron's wave packet).[4] There is, then, no GMR. More generally, the CIP-MR depends exponentially on the ratios λ_N/t_N, $\lambda_F^\uparrow/t_F$, and $\lambda_F^\downarrow/t_F$, where $\lambda_F^\uparrow$ and $\lambda_F^\downarrow$ are the momentum changing mean-free-paths for majority and minority electrons.[4]

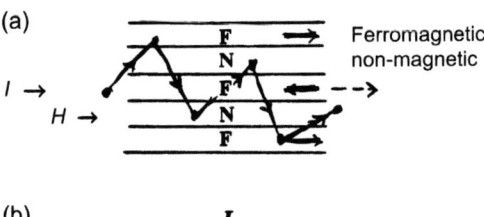

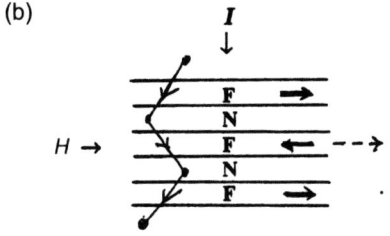

Figure 4.19. Schematic drawings of the different behaviors associated with the electron mean-free-paths in: (a) the CIP-MR, and (b) the CPP-MR. Mean-free-paths in CIP- and CPP-MR (a) CPP-MR varies mainly with λ_{el}. Add impurities → Changes in λ_{el} dominate. (b) But when $\ell_{sf} \gg \lambda_{el}, t_f, t_N$, spin direction stays fixed, and CPP-MR is independent of λ_{el} [except through ρ_F and ρ_N]. "Every electron" crosses all layers, independent of λ_{el}. AR(AP) for $[F/N]_N$ multilayer is simply the series sum of contributions from each layer and interface: $AR(AP) = N(\rho_N t_N + \rho^*_F t_F + ZAR^*_{F/N})$ where $N = \#$ of bilayers, $\rho^*_F = (\rho \uparrow_F + \rho \downarrow_F)/2$, $AR^*_{F/N} = (AR \uparrow_{F/N} + AR \downarrow_{F/N})/2$, where ↑ and ↓ mean electron spin along or opposite to the local magetization.

For the CPP-MR, in contrast [Fig. 4.19(b)], each electron that passes completely through the multilayer, and thus contributes to the current, passes through every F-layer. AP and P states are, thus, clearly distinct. In the limit of long l_{sf} in the F- and N-layers, the CPP-MR is "self-averaging"[4]—the total resistance R for each separate spin direction is just the sum of the contributions from the individual layers and interfaces.[28,55,56] The λs are not characteristic lengths and drop out of the expressions for the CPP-MR, which become quite simple, as we will see below.

The trigger for the discovery of GMR was the discovery by Grunberg's group[57] that thin layers of Fe exchange couple antiferromagnetically across a particular thickness of Cr ($t_{Cr} \approx 0.9$ nm), leading to a spontaneous AP state at $H = 0$ for this thickness. Two groups[2,3] showed that application of an H large enough to convert this AP state into a P state generated a large decrease in resistance—GMR. It soon became clear, however, that the antiferromagnetic coupling was not crucial to GMR, only the change in magnetic order from AP to P. Below, we will describe ways to reproducibly control this change.

The major discoveries in GMR, to be covered in more detail below, are the following. First, the presence of a large (Giant) CIP-MR in Fe/Cr sandwiches[2] and multilayers[3] (soon followed by similar results in other F/N pairs). Second, that the coupling between F-layers across N-layers oscillates with increasing t_N,[58] thereby complicating systematic studies of GMR as a function of t_N. Third, that the CPP-MR is usually larger than the CIP-MR,[13,55] and usually gives simpler access to the physical parameters underlying GMR.[28,55,56] Fourth, that GMR appears also in granular F-alloys.[59,60]

4.5.2. Current-in-Plane Magnetoresistance

4.5.2.1. Discovery. Binash et al.[2] showed that the room-temperature CIP-MR of a simple Fe/Cr/Fe sandwich with the t_{Cr} that gives antiferromagnetic alignment at $H=0$: (a) was fairly large, ~1.5%; (b) changed in correlation with the magnetization M of the sandwich [compare Fig. 4.20(a) with (c) and (b) with (d)]; (c) had a maximum value independent of demagnetizing effects [compare Fig. 4.20(c) and (d)]; and (d) was much larger than the AMR of a 25-nm-thick Fe film. At 4.2 K, the MR increased to above 10%, i.e., more than a factor of 5 larger than at room temperature. The appellation "giant magnetoresistance" was coined to describe the much larger MRs found almost simultaneously in Fe/Cr multilayers at 4.2 K by Baibich et al.[3] (Figs. 4.21 and 4.22). Of comparable importance, Baibich et al. also proposed what has become the standard model used to describe GMR (Fig. 4.17).

Together, these two papers stimulated an explosive growth of GMR studies that, within a decade, have led to technological applications.[19] It was quickly

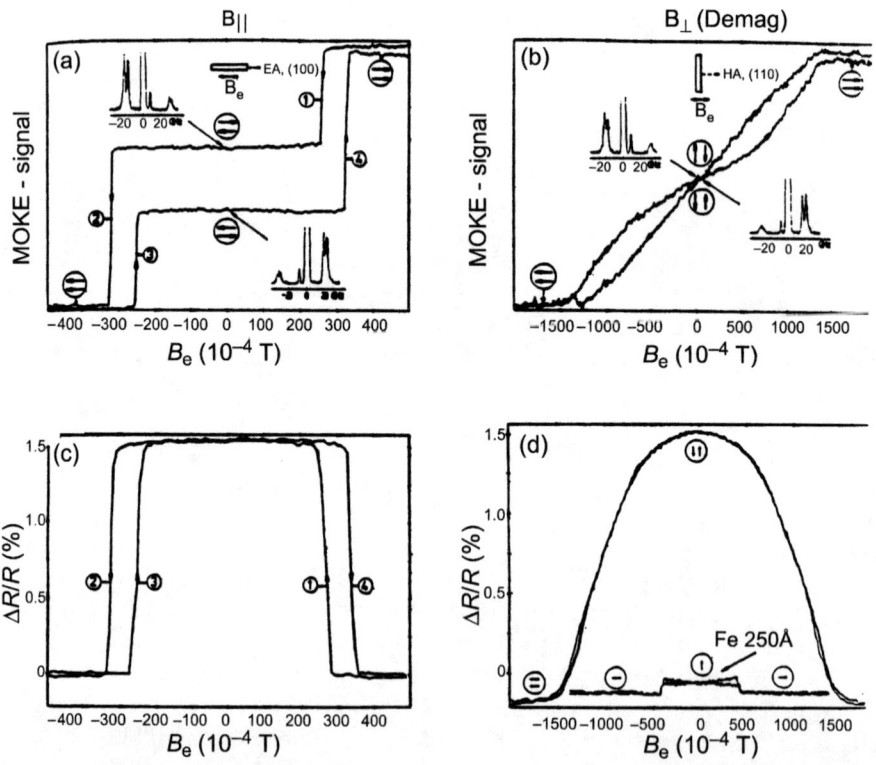

Figure 4.20. (a) and (b) MagnetOptic Kerr Effect (MOKE) hysteresis curves, and (c) and (d) magnetoresistance. $\Delta R/R = [R(H) - R(P)]/R(P)$ for Fe double layers separated by the Cr thickness needed to give antiferromagnetic coupling. (a) and (c) H in the plane of the layers. (b) and (d) H perpendicular to the layers. For comparison, also shown in (d) is the anisotropic MR of a 25 nm thick Fe film. All measurements are at room temperature (From Binash et al.,[2] p. 4829, © 1989 APS).

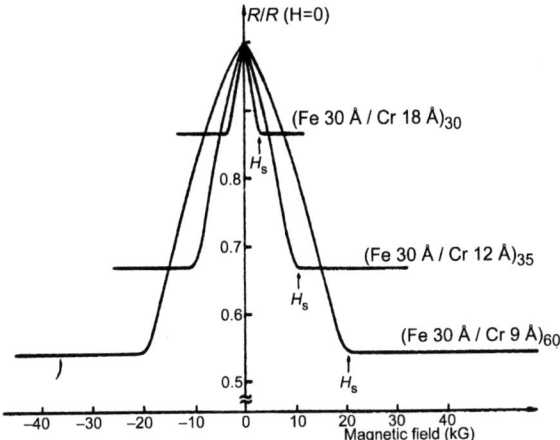

Figure 4.21. "Giant" MR of three Fe/Cr superlattices (layer thicknesses and number of layers indicated) at 4.2 K. The current and applied field are along the same [1 1 0] axis in the plane of the layers (from Baibich et al.,[3] p. 2472, © 1988 APS).

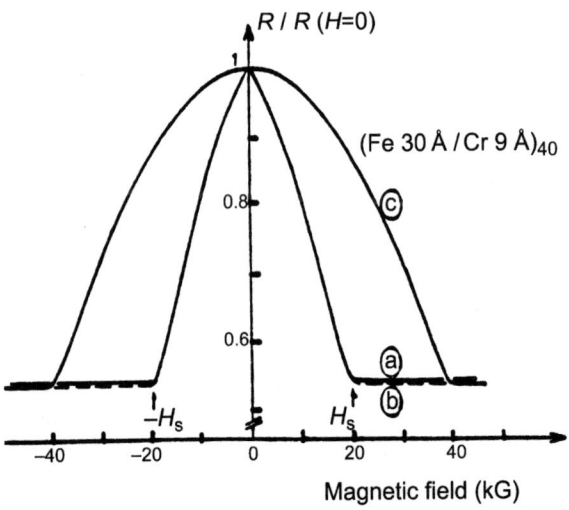

Figure 4.22. $R(H)/R(H=0)$ versus H for a [Fe(3 nm)/Cr(0.9 nm)]$_{40}$ multilayer with H: (a) in the layer plane parallel to the current; (b) in the layer plane perpendicular to the layer current; and (c) perpendicular to the layer plane. A small hysteresis is not shown (from Baibich et al.,[3] p. 2472, © 1988 APS).

discovered that a hybrid multilayer of the form [Co/Cu/Py/Cu]$_N$ could give a much larger CIP-MR (~9%) at room temperature than the 2% of the Py AMR, with the change occurring over a field range (<100 Oe) low enough for device potential.[61]

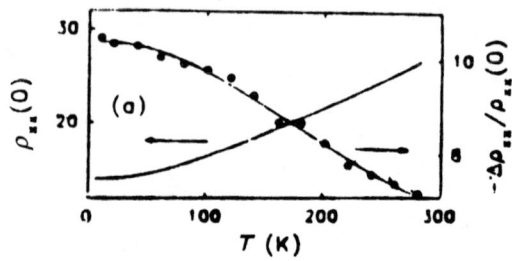

Figure 4.23. Temperature dependences of zero-field resistivity (solid rising curve) and MR (P state assumed at 5.7 T) for a [Co(1.5 nm)/Cu(0.8 nm)]₃₀ multilayer prepared by MBE. (from Tsui et al.,[62] p. 741, © 1994 APS).

4.5.2.2. Temperature Dependence.

We noted earlier that the MR of an Fe/Cr sandwich decreased by ~85% from 4.2 K (MR > 10%) to 300 K (MR ~ 1.5%). In contrast, Fig. 4.23[62] shows that the MR for a Co/Cu multilayer decreased by only ~30% from 4.2 to 300 K, while its resistivity increased by more than a factor of 5. Here, the change in resistance between AP and P states must have substantially increased from 4.2 to 300 K.

4.5.2.3. Oscillation with N-Layer Thickness.

The next fundamental discovery, by Parkin et al.,[58,63] was that the exchange coupling between F-layers does not decrease monotonically as t_N increases, but rather oscillates with t_N, between ferromagnetic and antiferromagnetic states, with gradually decreasing coupling strength. This observation spawned an independent subfield of experimental and theoretical studies of exchange coupling between F-layers across intervening N-layers and (more recently) I-layers, which is outside the scope of the present review. Figure 4.24[63] shows that this oscillatory coupling gives rise to an oscillatory MR, because antiferromagnetic coupling allows a large MR, whereas ferromagnetic coupling causes the moments of the different layers to want to remain aligned parallel to each other, inhibiting any MR. Only at the "first peak" in Fig. 4.24 ($t_N \sim 0.9$ nm) is the antiferromagnetic exchange coupling so strong that it produces an essentially unique AP state with little hysteresis. And only for $t_N > 6$ nm is the exchange coupling weak enough for the CIP-MR to decrease monotonically with increasing t_N. Between these limits, GMR oscillates with t_N in "phase" with the coupling, with the field required to saturate the magnetization (and thus also the MR) decreasing as the exchange coupling weakens. For $t_N > 1$ nm, the MR generally displays hysteresis and non-unique behavior at $H = 0$ (see Fig. 4.1), indicating that a complete AP state is rarely achieved in simple F/N multilayers for such N-layer thicknesses.

4.5.2.4. Relation of MR to Magnetization, M.

While Fig. 4.20 shows a CIP-MR closely correlated with M, such correlation is not universal. For example, in Fig. 4.25,[64] M saturates to a constant value above ~1 T, whereas the MR continues to decrease to above 2.5 T. This observation is not yet understood, but

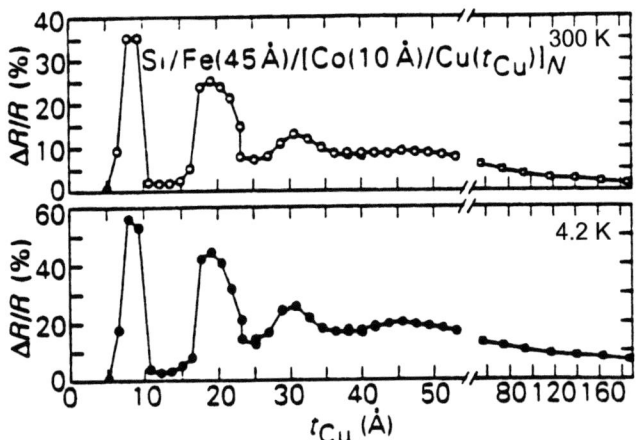

Figure 4.24. $\Delta R/R = [R(\text{Pk})-R(\text{P})]/R(\text{P})$ versus t_{Cu} for $[\text{Co}(1\text{ nm})/\text{Cu}(t_{Cu})]_N$ multilayers where $N = 16$ for $t_{Cu} < 5.5$ nm (circles—open for 300 K and filled for 4.2 K) and $N = 8$ for $t_{Cu} > 5.5$ nm (squares—open for 300 K and filled for 4.2 K). Here, $R(\text{Pk})$ is the maximum value achieved after the sample has been taken to saturation field (from Parkin et al.,[63] p. 2153, © 1991 APS).

qualitatively similar behavior has been seen in some granular alloys (see Section 4.6.4).

4.5.2.5. Control of the AP State.

As noted earlier, the AP orientation desired for analysis does not occur straightfowardly in simple two-component F/N multilayers over a wide range of N- and F-layer thicknesses. More flexible control has been achieved by using two techniques that we call hybrid spin-valves (hybrid-SVs)[61] and exchange-biased SVs (EBSV).[65] A hybrid-SV has the form [F1/N/F2/N]$_N$. If the saturation field of F1 is much less than that of F2, i.e., $H_{s1} \ll H_{s2}$, then the magnetizations of the F1-layers will fully rotate by H_{s1}, while those of the F2-layers will not fully rotate until H_{s2}. A closely AP state can then occur between H_{s1} and H_{s2}, as shown in Fig. 4.26. In an EBSV (AF/F/N/F), an AF metal is used to exchange/pin the adjacent F-layer to a much higher saturation field, H_p, than that of the "free" layer, H_f. To allow maximum flexibility, the two F-layers are usually separated by a thickness t_N large enough to make the exchange coupling between them weak. Figure 4.27 shows the hysteresis curves of M and MR for such an EBSV. Here, the AP state can persist over a wide field range.

4.5.2.6. Intrinsic Quantity—Sheet Conductance.

The intrinsic electronic transport quantity in a bulk metal is the resistivity ρ. To determine the equivalent intrinsic quantity in the CIP-MR, measurements were made of three nominally identical spin-valves, on top of which were deposited different shunting layers.[66] The task was to determine if any quantity was independent of the shunting layers. Figure 4.28 shows that ΔR and MR gave different results for the three samples, but that the change in sheet conductance, $\Delta G = \Delta R/R^2$, gave essentially the same

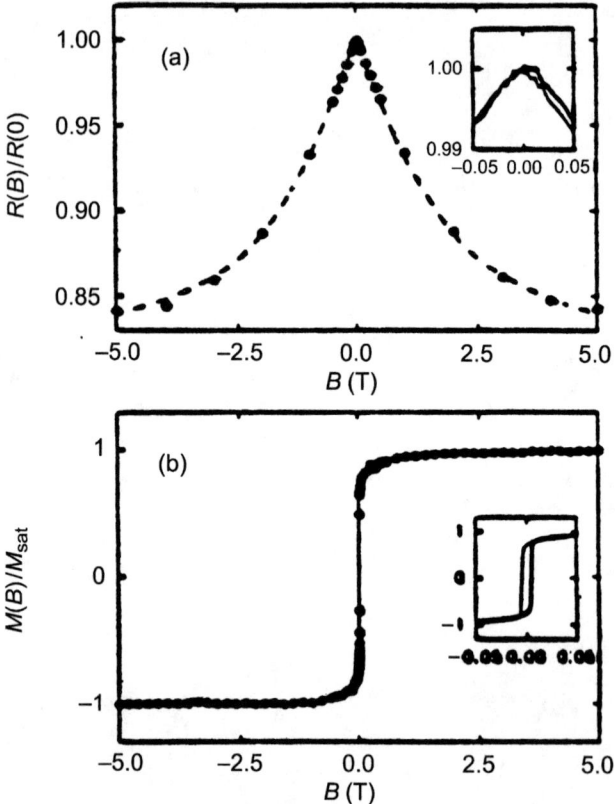

Figure 4.25. (a) Ambient temperature MR, $R(B)/R(0)$ for a [Co(7.5 ML)/Cu(5.5 ML)]$_{26}$ superlattice. The dashed line is a theoretical fit to a Langevin function. Inset: low-field MR versus B. (b) Magnetization versus B for the same sample. $M_{sat} = 1400$ emu/cm^3. Inset: low-field magnetization curve. $T = 300$ K (from Bartlett et al.,[64] p. 1522, © 1994 APS).

results for all three. Since it is independent of the shunting layer, the sheet conductance must be the intrinsic quantity in the CIP-MR.

4.5.2.7. Bulk versus Interface Contributions.

The combination of the difficulty in reliably producing an AP state in a simple F/N multilayer (except at the first peak of Fig. 4.24), with the complexity of the equations needed to analyze the CIP-MR, make it difficult to uniquely and reliably derive parameters such as α_B and $\alpha_{F/N}$ from the CIP-MR in such multilayers. Also, analyses of different samples by different investigators can yield rather different values. As examples, estimated values of α_F (or β) for Py range from ≥ 9 ($\beta \geq 0.8$)[67] to 2.1 ($\beta \approx 0.35$),[68] and for Co from ≥ 5.5 ($\beta \approx 0.7$)[67] to 2.3 ($\beta \approx 0.4$)[69] to 1 ($\beta \approx 0$).[70]

Semiquantitative studies have been somewhat more successful, as illustrated in Fig. 4.29.[71] This figure shows the variation of the CIP-MR of Co/Cu multilayers over a wide range of t_N. The expected oscillation for $t_N < 5$ nm is followed by

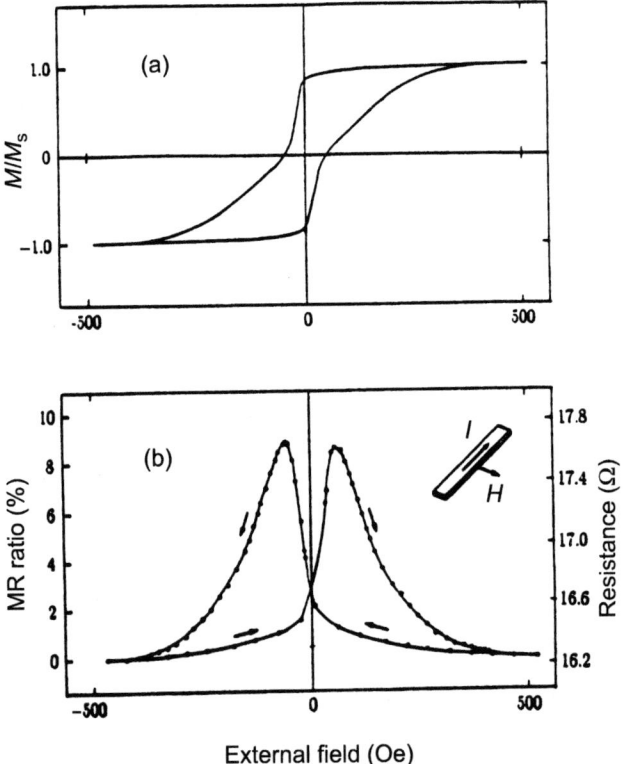

Figure 4.26. (a) Magnetization and (b) room-temperature CIP-MR curves for a Co/Cu/Py/Cu "hybrid spin-valve" multilayer (from Shinjo and Yamamoto,[61] p. 3062).

a monotonic decrease for larger t_N. The data at 300 K were fit approximately to the form $(1/t_N) \exp(-At_{Cu})$ and those at 4.2 K simply to $(1/t_N)$ (see the figure caption for subtleties). The data were interpreted as showing that the change in the CIP-MR with t_N for these samples is dominated by a simple "dilution" effect, involving increased shunting of current through the Cu layers as they become thicker, superimposed (at 300 K) upon an exponential mean-free-path decay. This interpretation is probably basically correct. But we call the analysis only semiquantitative because the MR for large t_N was taken to be that at the local maximum after saturation (see Fig. 4.1), and we will see in Section 4.5.3.4 that this state is not necessarily a good approximation to the required AP state.

In the CIP-MR, the electrons drift parallel to the interfaces, causing the (usually) lower resistivity N-layers to carry higher current density than the (usually) higher resistivity F-layers. Such a preferential shunting of the electrons into the N-layers leads one to expect the interfaces to play an important role in determining the CIP-MR. In fact, for the thin F- and N-metal layers often used in GMR studies, the CIP-MR is generally dominated by interface effects. Direct evidence of the importance of interfaces has been obtained by examining the effects of depositing small amounts of a second F-metal (F2) at an F1/N interface. The top

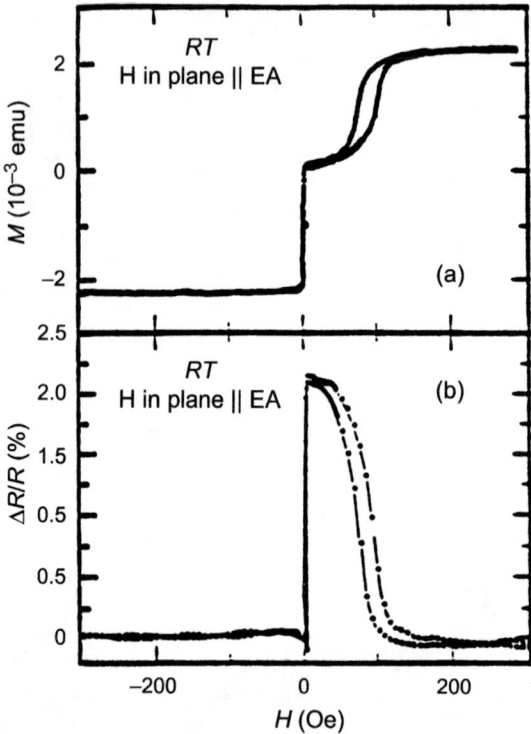

Figure 4.27. $M(H)$ (a) and CIP-MR(H) (b) for a Si/Py(15 nm)/Cu(2.6 nm)/Py(15 nm)/FeMn(10 nm) EBSV with a 2 nm Ag protective cover. H is parallel to the exchange anisotropy (EA) field created by the FeMn, and the current is perpendicular to H (from Dieny et al.,[65] p. 1298, © 1991 APS).

portion of Fig. 4.30 (from Ref. 72) and part (a) of the bottom portion, show a significant MR enhancement upon depositing at the interfaces of a Py/Cu spin-valve, an impurity (here Co) that produces a larger MR with the Cu than does the Py. Conversely, parts (b) and (c) of the bottom portion of Fig. 4.30 show that the MR of Co/Cu spin-valves can be reduced both by similarly depositing Py at the Co/Cu interfaces, or by depositing Co inserts into the middle of the Py layers.

4.5.2.8. Inverse CIP-MR. As explained in Section 4.5.1, by constructing a multilayer involving two different F-metals with opposite signs of β, it should be possible to produce an inverse MR, i.e., to obtain the largest resistance when the sample is in the P-state. The first evidence of such a behavior was reported in Ref. 73, which used an expected negative γ for Fe/Cr interfaces to construct an artificial "negative β" within the F-layers of an Fe/Cr/Fe trilayer. The Cr thickness was made so thin that the two Fe-layers coupled together ferromagnetically, so that they initially rotated together in a field. The anisotropic scattering within the sandwich was dominated by the two Fe/Cr interfaces, making the sandwich behave at low fields as if it was a single F-metal layer with a negative β. Subsequently, an inverse CIP-MR due directly to a negative β in FeV (see Fig. 4.8) was reported.[74]

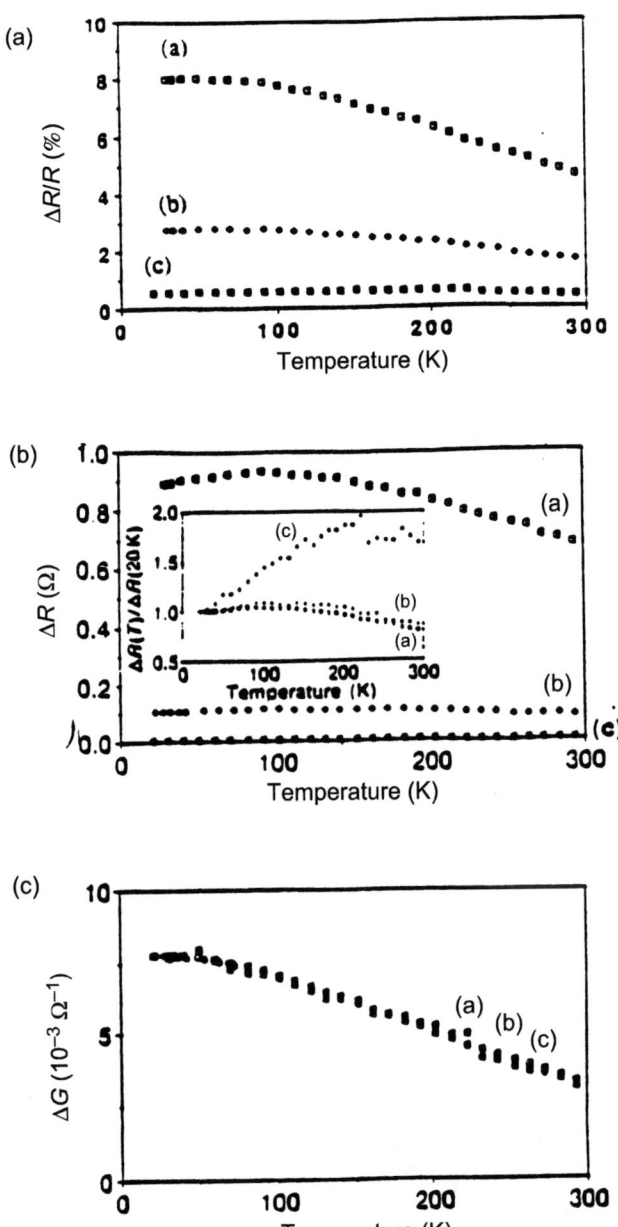

Figure 4.28. Comparison, for three Co/Cu/Py EBSVs with different "capping layers" of Cu and Ta, of: (a) MR = $\Delta R/R$(%); (b) ΔR (B); and (c) ΔG; versus T for (a) no capping layer; (b) capping layer = 5 nm of Ta; (c) capping layer = 30 nm of Cu and 5 nm of Ta (from Dieny et al.,[66] pp. 2112 and 2113).

4.5.2.9. Effects of Interfacial Roughness.
Because interface scattering is usually important in the CIP-MR, numerous attempts have been made to try to establish the effect on the CIP-MR of "interfacial roughness."

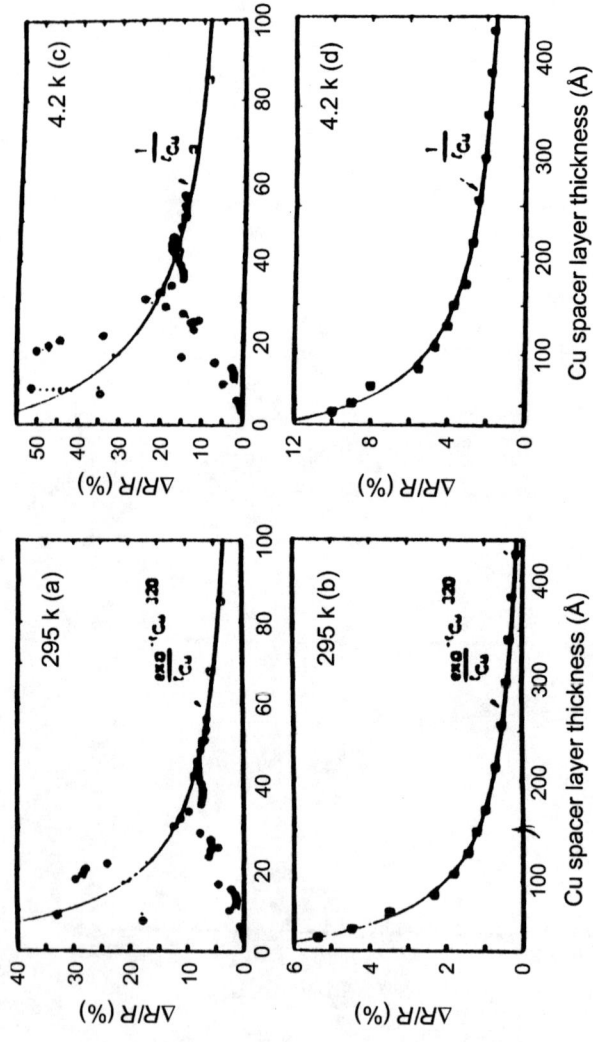

Figure 4.29. Variation at 295 K (a and b) and 4.2 K (c and d) of CIP-MR versus t_{Cu} for Si(111)/Ru(5 nm)[Co(1.1 nm)/Cu(t_{Cu})/Ru(1.5 nm)] multilayers. $N = 20$ (circles) or 6 (closed squares); the $N = 6$ data have been increased by empirical factor of 1.6 to make comparison with the $N = 20$ data easier. The curves drawn through the data are of the form $\Delta R/R = 289/(4.3 + t_{Cu}) \exp(-(t_{Cu}/318))$ at 295 K and $\Delta R/R = 0.28 + [554(1/13 + t_{Cu})]$ at 4.2 K. The curves have been scaled by an empirical factor of 1.6 in (a) and (c) (from Parkin *et al.*,[71] p. 9138, © 1993 APS).

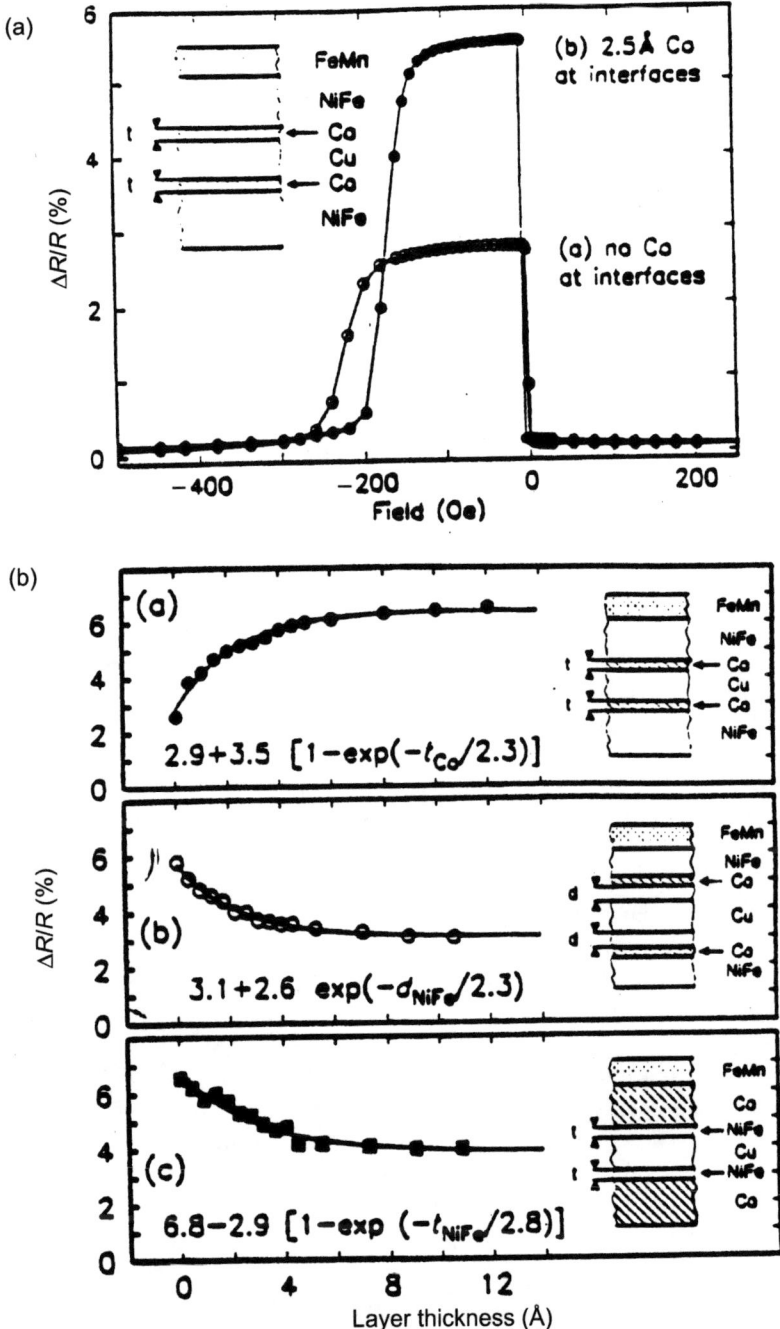

Figure 4.30. Enhancing CIP-MR by interfacial doping. (a) Comparison of MR for Py-based EBSVs with no Co and 0.25 nm of Co at the interfaces. (b) MR versus appropriate layer thickness (Angstroms) for: (a) Py-based EBSVs with Co doping at the interfaces; (b) Py-based EBSVs with Co doping inside the Py layers; and (c) Co-based EBSVs with Py doping at the interfaces (from Parkin,[72] p. 1642, © 1993 APS).

The problems involved in reaching reliable conclusions are illustrated by a sequence of studies on FeCr. Fullerton et al.[75] inferred that increasing roughness increases the MR. Two groups[76] inferred that increasing roughness decreases the MR. Rensing et al.[77] concluded that the main effect of increasing roughness was just to increase R. Belien et al.[78] concluded that increasing the number of steps at interfaces increases the MR, while increasing interdiffusion, or roughness that is more random, suppresses it. The difficulties in isolating interface from bulk effects were discussed by Schad et al.[79] It seems likely that different kinds of roughness (e.g., chemical mixing, physical roughness involving simply steps, or more heavily disordered interfaces), as well as different length scales of roughness, can produce different (even opposite) effects, and that bulk and interface contributions interact. In experiments where antiferromagnetic coupling between the F-layers is initially strong, one must also distinguish between direct effects of increased interface roughening on the MR, and the indirect effect of weakened coupling simply reducing the AP order needed for maximum MR.

4.5.2.10. Angular Variation. Theorists have argued that spin-dependent potential steps at interfaces may be important sources of GMR.[80] Unfortunately, few experiments distinguish between effects of spin-dependent potential steps and those due to spin-dependent scattering within interfacial alloys. Studies of the angular dependence of the CIP-MR have been proposed to distinguish between them. Spin-dependent potential steps have been predicted to cause the CIP-MR to deviate from a simple linear dependence on $\cos(\theta_1-\theta_2)$, the angle between the magnetizations of adjacent F-layers. Figure 4.31[81] shows experimental data for an (FeMn/NiFe/Cu/NiFe) EBSV. No deviations from linearity (to within experimental error) are seen, after the data are corrected for a small AMR contribution. Such a simple linearity was also seen later.[82]

4.5.2.11. Noise. Experiments on resistance noise in GMR are still in their infancy.[83] $1/f$ noise has been observed, and found to be maximal when the GMR is changing most rapidly with H. Resistive Barkhausen noise in uncoupled multilayers has also been seen, and found to manifest asymmetric hysteresis, attributed to hysteresis in the domain structure.

4.5.2.12. Thermal Conductivity, Hall Effect, and Thermopower. Measurements of transport properties other than MR, such as the thermal conductivity, κ, the Hall resistivity, ρ_{xy}, and the thermopower, S, have been made with hopes of gaining additional insight into the physics underlying GMR. For such relatively "dirty" (low-residual resistance ratio) materials as sputtered or MBE prepared multilayers, κ would be expected to be related to ρ rather well by the Wiedemann–Franz law,[12] $\kappa\rho/T = L_0$. Figure 4.32 from Ref. 84 shows a case where $\Delta\kappa = \kappa(H)-\kappa(0)$ is, indeed, so related to $\Delta\rho = \rho(H)-\rho(0)$. But the relation in multilayers need not always be perfect.[85]

The behaviors of the anomalous Hall coefficient,[62,85] and S[62,85] are more complex. But Fig. 4.33–4.35[62] show that they can scale with the resistivity. Interpretations of these results are given in Refs. 62 and 85, the latter arguing that the S data can be understood without including effects of potential steps.

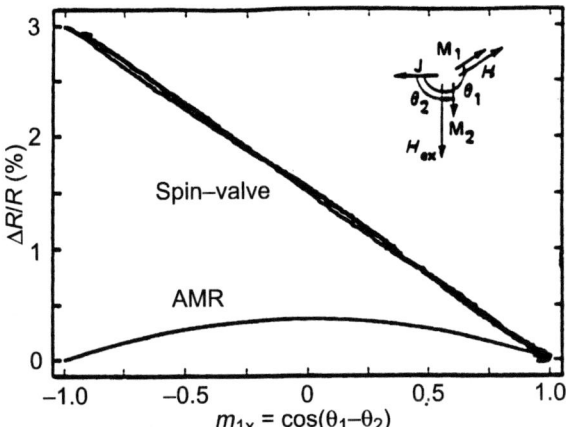

Figure 4.31. $\Delta R/R(\%)$ (after correction for the indicated AMR) versus the cosine of the angle between the magnetizations of the NiFe layers in a Si/(Py(6 nm)/Cu(2.6 nm)/Py(3 nm)/FeMn(6 nm)/Ag(2 nm) CIP-EBSV. The inset shows the orientation of the current I, exchange field, H_{ex}, applied field H, and magnetizations M_1 and M_2 (from Dieny et al.,[81] p. 1298, © 1991 APS).

4.5.3. Current Perpendicular to Plane Magnetoresistance (Including CAP-MR)

4.5.3.1. Measurement Techniques—Advantages and Disadvantages.
Several different techniques have been developed to measure the CPP-MR. Figure 4.36 shows the oldest,[13,49] using crossed superconducting (S) strips, one above

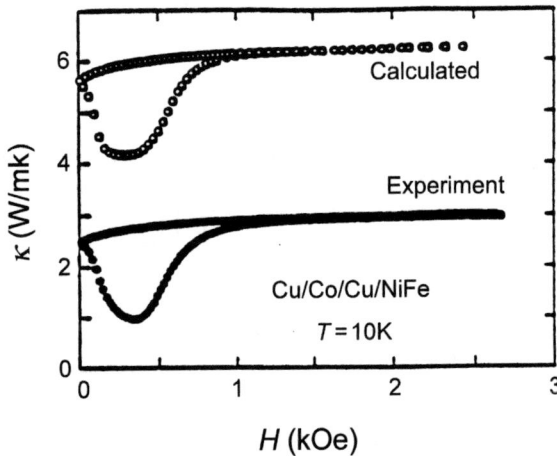

Figure 4.32. Measured hysteretic behavior of $\kappa(H)$ (lower curve) and values calculated (upper curve) by applying the Wiedemann–Franz law to the measured $\rho(H)$ for a Cu/Co/Cu/Py multilayer. The experimental curves are arbitrarily placed in the vertical direction (from Sato,[84] p. 101, © 1995, Elsevier Science).

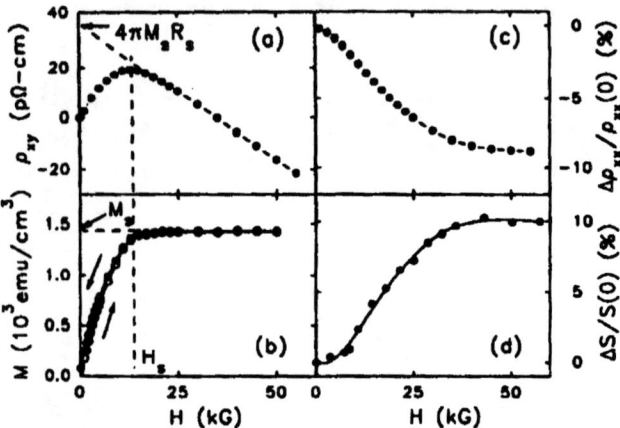

Figure 4.33. Field dependences of (a) Hall resistivity, (b) magnetization, (c) magnetoresistivity, and (d) magnetothermopower for [Co(1.5 nm)/Cu(0.8 nm)]$_{30}$ multilayers at $T = 161$ K, for H along the [111] growth direction. The vertical dashed line indicates the perpendicular saturation field for the Co layers. The linear part of the Hall resistivity at high fields corresponds to the ordinary Hall effect and the low-field magnetization-dependent behavior to the extraordinary Hall effect. The extraordinary Hall coefficient is obtained by extrapolating the high-field linear behavior to zero field, as indicated by the dashed line in (a) (from Tsui et al.,[62] p. 741, © 1994 APS).

and one below the multilayer. Such strips are equipotentials, even when current flows. In a short ($t \sim 1$ μm), wide ($W \sim 1$ mm) sample, edge effects are small, just as in a thin parallel plate capacitor, giving a uniform current through the overlap area A of the strips.[49,86] The second technique (Fig. 4.37) involves lithography with non-S leads,[51,87,88] and the third (Figs. 4.38 and 4.39) electrodeposition of nanowire multilayers into nanotubes produced by etching after ion-bombardment.[52–54,89] Lastly, the MR has been measured[90,91] on samples deposited onto a stepped surface, giving current at an angle to the planes (CAP-MR) (Fig. 4.40). Combining CIP- and

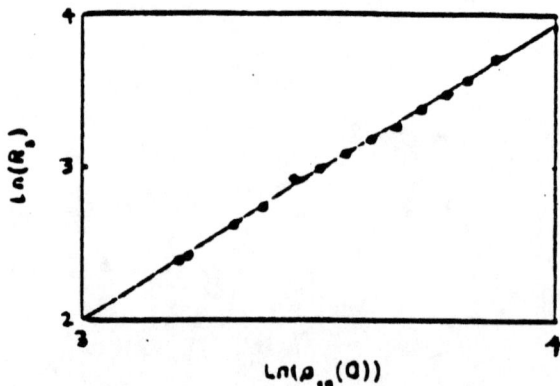

Figure 4.34. Scaling behavior between the extraordinary Hall coefficient and the resistivity for the sample of Fig. 4.33. The slope of the log–log plot is 2, indicating that $R_H \propto \rho_{xx}(0)^2$ (from Tsui et al.,[62] p. 742, © 1994 APS).

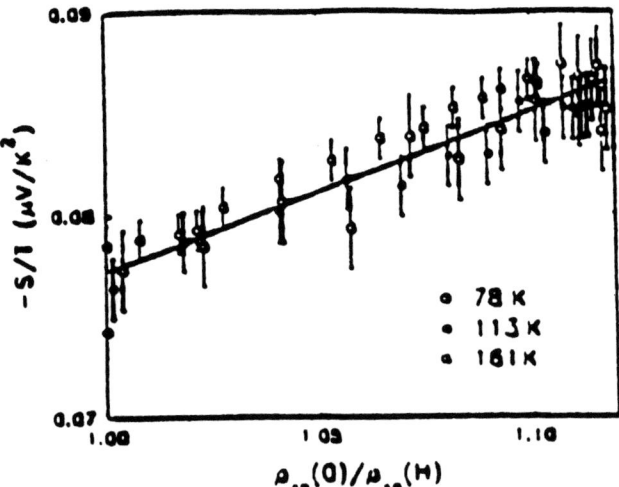

Figure 4.35. Scaling behavior of magnetothermopower and MR for a [Co(1.5 nm)/Cu(1 nm)]$_{30}$ multilayer at various temperatures and fields (from Tsui et al.,[62] p. 742, © 1994 APS).

CAP-MRs can give the CPP-MR (Fig. 4.41); Levy et al.[92] found reasonable agreement between CPP-MRs deduced for [Co/Cu/Py/Cu]$_N$ multilayers and directly measured ones.[93] All of the techniques except the first allow measurements at room temperature, and thus might have device potential.

(1) The crossed S-lead technique is recognized as the "method of choice for understanding fundamental physics,"[5] because (a) it provides uniform current

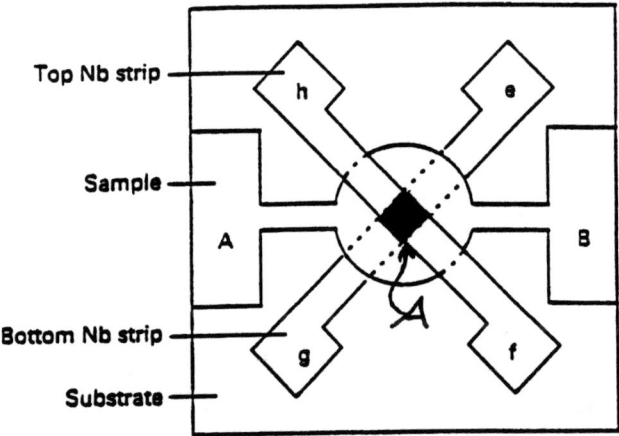

Figure 4.36. Crossed superconducting strip geometry for CPP-MR and CIP-MR measurements. The CPP-MR voltage is measured between f and g with current injected at e and removed at h. The current is uniform through the indicated area A. The CIP-MR voltage is measured between A and B with current flowing from A to B (after Slaughter et al.[49]).

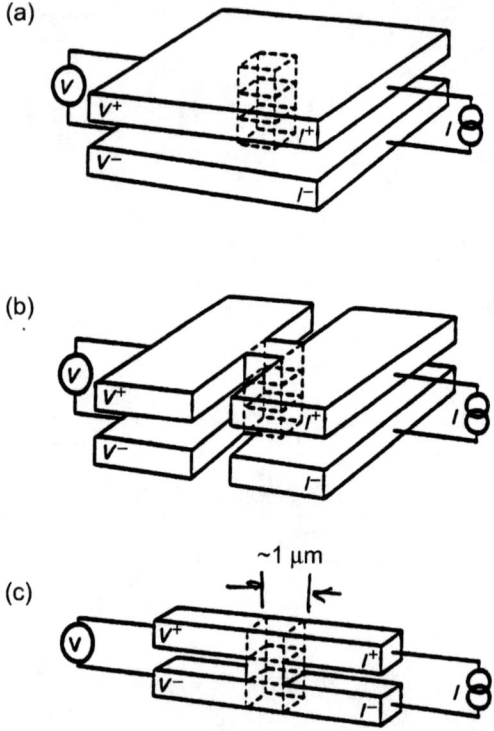

Figure 4.37. Schematic drawings of alternate geometries for measuring the CPP-MR of lithographically prepared samples (a) and (b) from Gijs et al.,[87] p. 6711; (c) from Vavra et al.,[88] p. 2580.

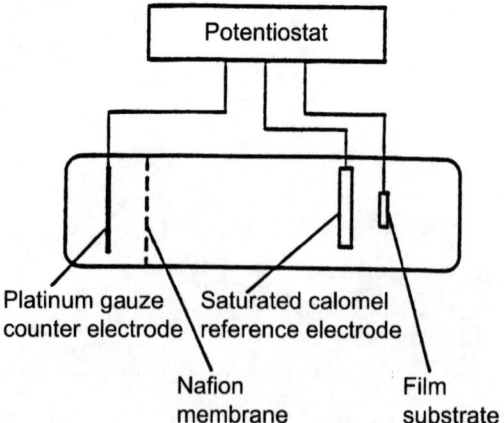

Figure 4.38. Schematic drawing of an electrochemical cell for growing metallic multilayers (from Alper et al.,[89] p. 2144).

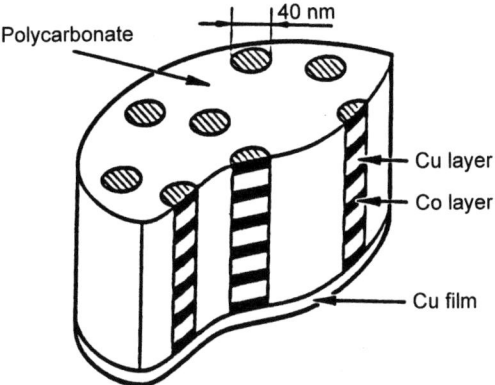

Figure 4.39. Schematic drawing of an array of electrodeposited nanowires in an insulating polymer matrix (from Piraux et al.,[52] p. 2484).

through the sample, allowing direct measurements of $AR(P)$, $AR(AP)$, and $A\Delta R$; (b) the S/F contact resistance is reproducible and can be separately measured; (c) the CPP- and CIP-MRs can be measured on the same sample; and (d) using either

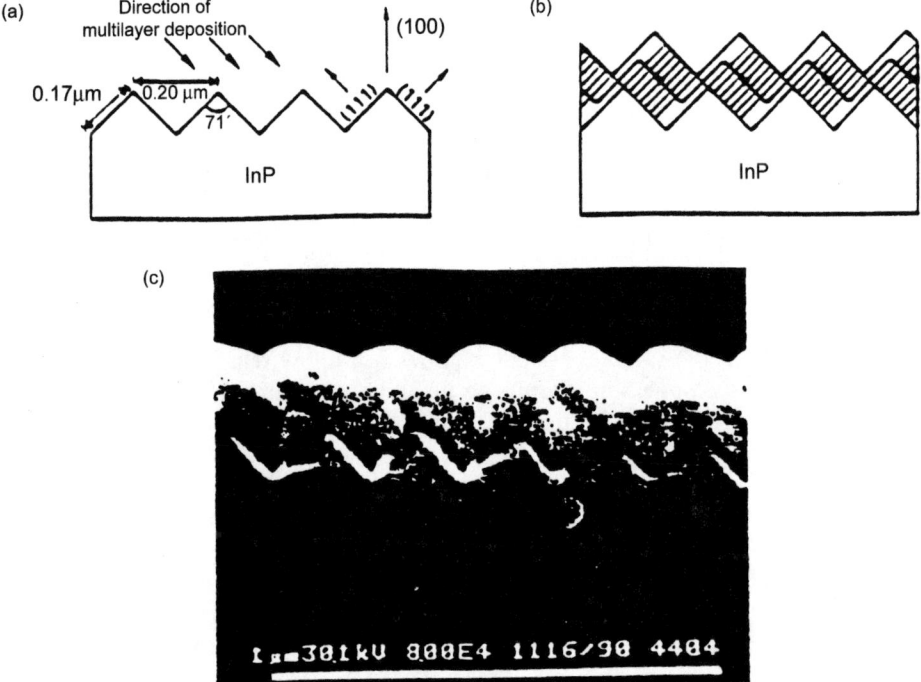

Figure 4.40. Schematic diagrams of (a) a grooved substrate surface and (b) a multilayer deposited at an angle to the surface (curves with arrows indicate idealized current flow). (c) Scanning electron micrograph of the cross-section of a real structure (from Gijs et al.,[90] p. 337, © 1995, Elsevier Science).

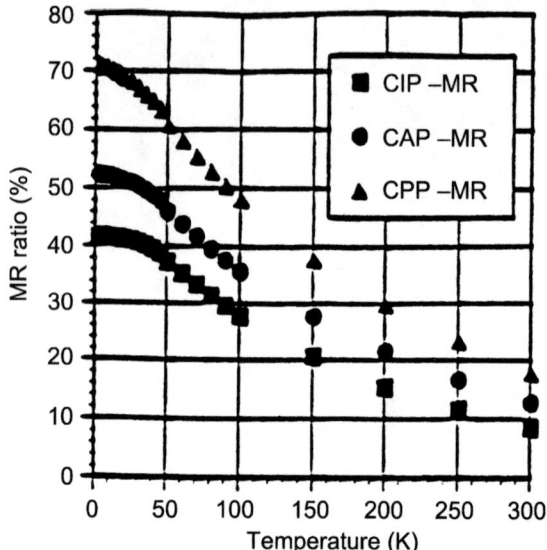

Figure 4.41. The measured CIP- and CAP-MRs for corrugated multilayers of the form [Co(1.2 nm)/Cu(5.8 nm)/Py(1.2 nm)/Cu(5.8 nm)]$_{167}$ and the CPP-MR predicted as described in Ref. 92 (from Levy et al.,[92] p. 16052, © 1995 APS).

sputtering[13] or molecular beam epitaxy (MBE),[94] the number of F- and N-materials in a sample is limited only by the number of sputtering targets (or evaporation or e-beam sources), so the AP state can be controlled using both hybrid-SVs and EBSVs. The only disadvantage is its limitation to low temperatures. But low temperatures are actually best for elucidating most of the physics underlying GMR, since they eliminate complications of scattering and current mixing by phonons and magnons.

(2) Lithography has most of the advantages of S-leads noted just above, but two problems: (a) non-uniform current[95] (because the lead and contact resistances are comparable to the sample resistance), does not allow reliable measurements of AR for the sample alone; and (b) strong magnetostatic interactions between small domains complicate low-field studies.[96] Figure 4.42(a) shows an example of a non-uniform current distribution giving an unphysical result, an inferred room-temperature resistance lower than that at 4.2 K. On the other hand, if the current distribution is independent of H, a non-uniform current can have little effect on the MR, because the MR is a ratio. Figure 4.42(b) shows a case where the MR was nearly independent of contact diameter, even when current non-uniformity gave an incorrect R. Figure 4.43 shows that reducing the F-layer widths down to micrometers produces interactions between the layers due to edge fields that broaden the MR curve, by an amount that increases with increasing F-layer thickness.

(3) Measuring the CPP-MR of nanowires electrodeposited into tiny holes (diameter $\geq$ 30 nm) in an insulator gives uniform current. Initially, contact

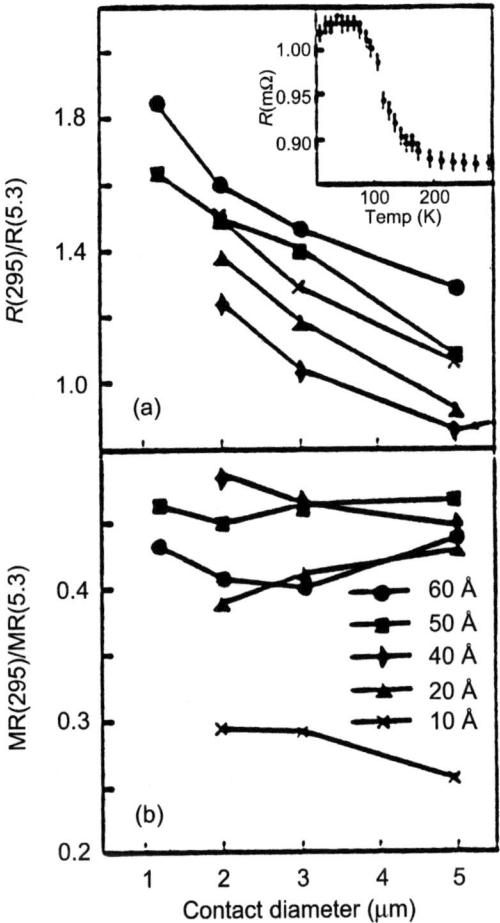

Figure 4.42. (a) $R(295\text{ K})/R(5.3\text{ K})$ versus contact diameter d for a series of lithographically prepared Co/Cu/NiFeCo/Cu multilayers. (b) MR(295 K)/MR(5.3 K) for the same samples. Note especially the points in the lower right corner of (a), where $R(295\text{ K})/R(5.3\text{ K}) < 1$ (from Vavra et al.,[88] p. 2580).

resistances were irreproducible and unknown. Samples contained an unknown number of wires and AR had to be estimated indirectly by assuming Matthiessen's rule.[97] Now, however, connections to single wires are becoming feasible.[98] Depositing the F- and N-metals from a single bath tends to contaminate each with the other and limits samples to just a single F-metal (or F-alloy) and a single N-metal (or N-alloy). Attempts have been made to electrodeposit samples with layers of more than one F-metal, or more than one N-metal, by changing the bath composition between depositions, or using multiple baths. But satisfactory multilayers have not yet been achieved. Also, the magnetostatic interactions between the F-layers in the tiny wires can be very strong; for "long" layers they can cause the M_i to orient perpendicular to the layers—i.e., along the wire.

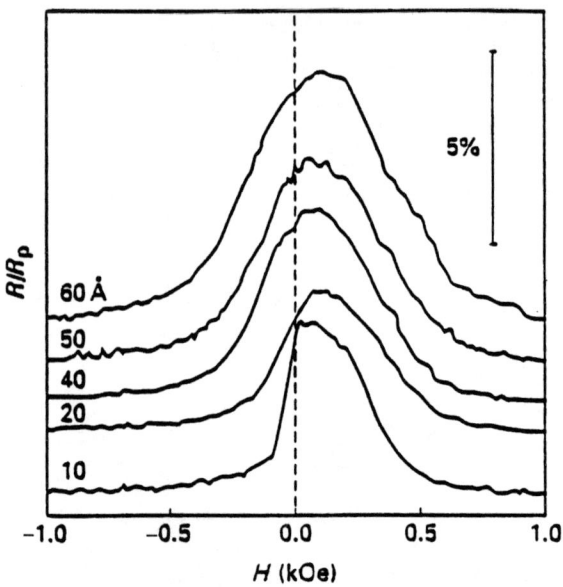

Figure 4.43. Effect of edge dipole fields on the room temperature MR of 2.4 × 2.4 µm² multilayer stacks with fixed total thickness. Thicker F-layers produce broader MR curves (from Vavra et al.,[88] p. 2581).

4.5.3.2. Comparison of CPP-MR and CIP-MR. Figures 4.44–4.47 show that the CPP-MR is usually several times larger than the CIP-MR of a given multilayer, both at 4.2 K (Figs. 4.44 and 4.45) and at 300 K (Figs. 4.46 and 4.47). Figure 4.44[99] shows that the CPP-MR and CIP-MR for [Co(1.5 nm)/Cu(d)] multilayers with fixed total thickness $t_T = 360$ nm display similar oscillations with increasing Cu thickness. Fig. 4.45 compares the CPP-MR(H) and CIP-MR(H) with M(H) for a sputtered [Co(6 nm)/Ag(6 nm)]$_{60}$ multilayer, and with the CPP-MR(H) for a single Co film.[13] The shapes of the CPP- and CIP-MRs are very similar and correlate closely with M(H), but the CPP-MR is several times larger than the CIP-MR. The bottom panel in Figure 4.45 proves that the CPP-MR derives from the multilayer, since the MR for a single Co layer is less than 1%.

4.5.3.3. Temperature Dependence. Figure 4.46[100] shows that the CPP-MR of a Co/Cu multilayer decreases by less than 50% from 4.2 and 300 K, whereas Fig. 4.47[51] shows that the CPP-MRs of Fe/Cr multilayers drop to much smaller fractions of their initial values. This much smaller temperature dependence in Co/Cu mirrors the behavior seen in the CIP-MR (see Section 4.5.2.2). Figure 4.48[96] shows that the decrease in CPP-MR from 4.2 to 300 K for a [Co/Cu/NiFeCo/Cu] hybrid spin-valve is due mostly to the increase in resistance of the multilayer, rather than to the decrease in $\Delta R = R(AP) - R(P)$. In general, the decrease of the CPP-MR with increasing temperature involves a complex combination of: increases in the resistivities of the components of the multilayer; changes in the anisotropy parameters; an increase in spin-mixing due to scattering of electrons by increasing

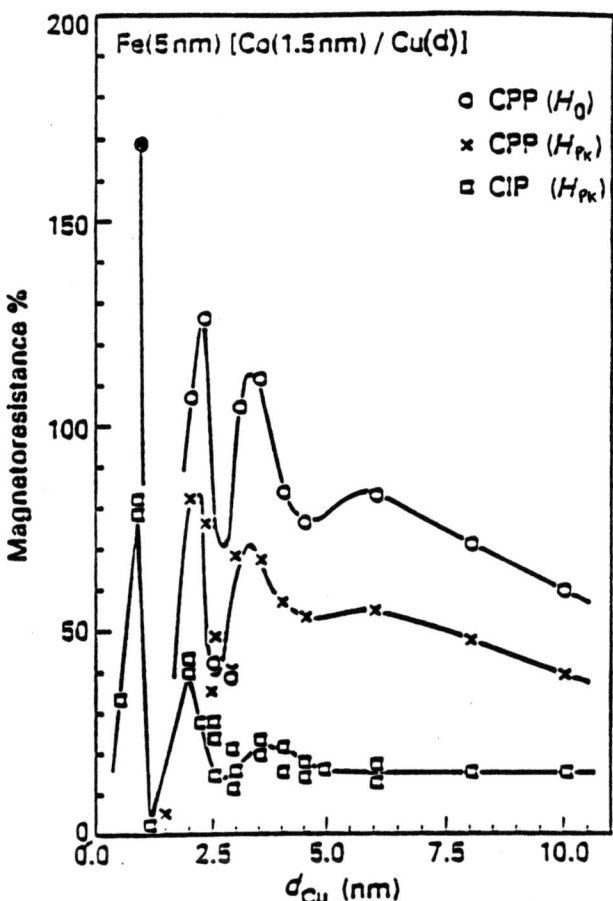

Figure 4.44. CPP-MR at the fields H_0 and H_{Pk}, and CIP-MR at H_{Pk}, versus Cu layer thickness for Co(1.5 nm)/Cu(d_{Cu}) multilayers (from Schroeder et al.,[99] p. 47).

numbers of magnons; and decreases in spin-diffusion lengths (especially in the F-metal).[101,102]

4.5.3.4. Estimating or Achieving the AP State.

In addition to the similarities in shape, and differences in magnitude, of the CPP- and CIP-MRs, the data of Figs. 4.44 and 4.45 display another feature often seen in samples with t_N large enough so that exchange coupling is small; namely, both MRs are larger in the "as-prepared" state, $R(0)$, than in the "peak" state, $R(Pk)$, that occurs near the coercive field, H_c, after the sample has been taken to saturation (see also Fig. 4.1). Because $R(Pk)$ reproduces after saturation, whereas $R(0)$ does not, most investigators (e.g., Figs. 4.24 and 4.29) have used MR(Pk) to estimate the desired MR(AP). However, this choice is problematic, since demagnetization by reducing the magnitude of H while oscillating its sign often produces an MR(D) larger than

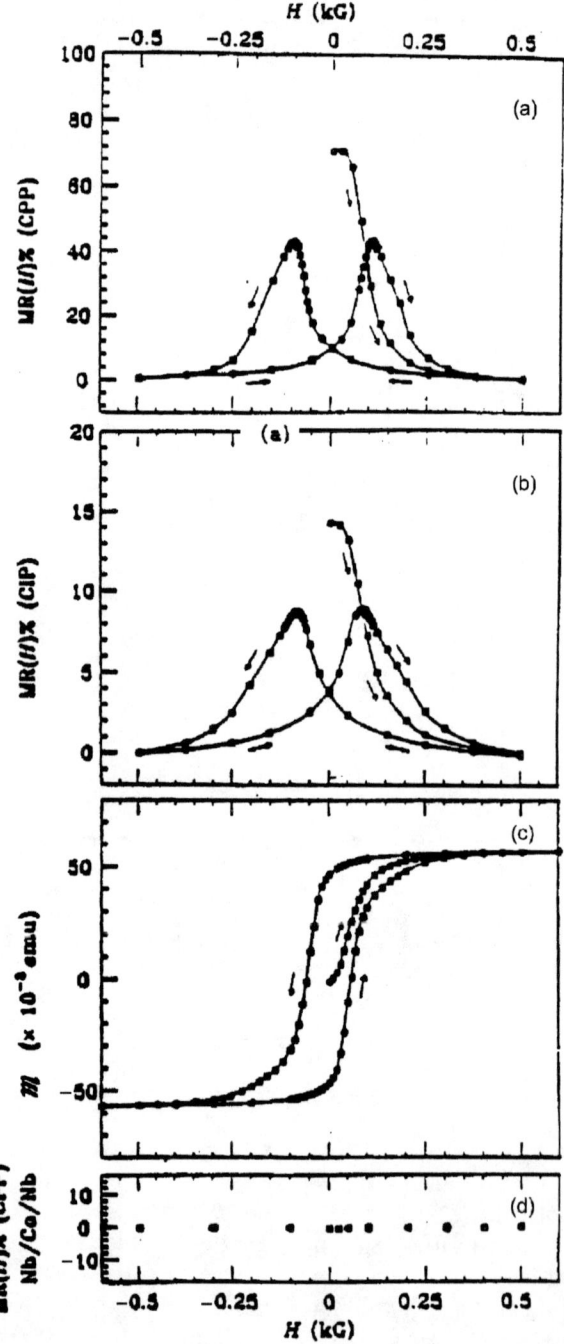

Figure 4.45. (a)–(c) Data versus H for a [Ag(6 nm)/Co(6 nm)]$_{60}$ multilayer. (a) CPP-MR(H) measured at 4.2 K using crossed Nb strips. (b) CIP-MR(H) at 4.2 K. (c) Magnetization $M(H)$ at 12 K to minimize the contribution of the Nb. (d) CPP-MR(H) for a simple Nb/Co(9 nm)/Nb sandwich (from Pratt et al.,[13] p. 3061, © 1991 APS).

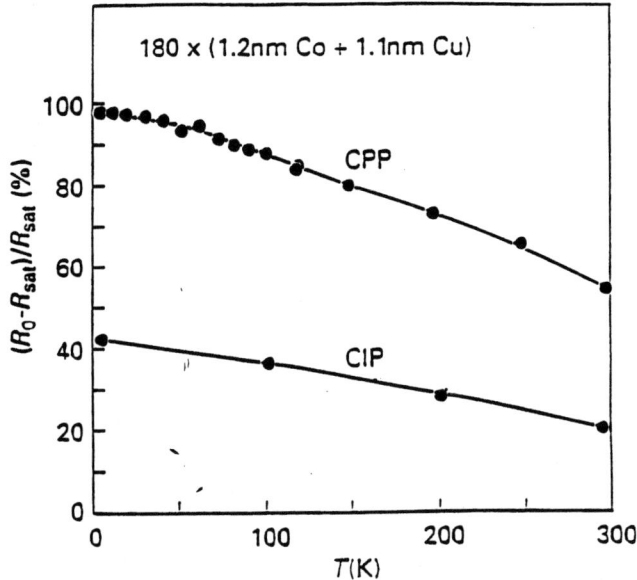

Figure 4.46. CPP-MR(T) versus CIP-MR(T) for a [Co(1.2 nm)/Cu(1.1 nm)]$_{180}$ multilayer. The CPP samples were prepared lithographically (from Gijs,[87] p. 6713).

MR(Pk) (but usually ≤MR(0)[103]). Recent neutron scattering and SEMPA studies revealed that the MR(0) state in [Co(6 nm)/Cu(6 nm)]$_{10,20}$ multilayers is approximately AP aligned, whereas the MR(Pk) state is magnetically disordered.[104] Thus, caution must be exercised in estimating MR(AP) for such multilayers. If reliable values of R(P) and R(AP) are required for ranges of thicknesses of F- and/or N-metals, the AP state should be controlled by using hybrid- or exchange-biased-SVs (see Section 4.5.2.5). The importance of such a control has recently been emphasized.[105]

4.5.3.5. Intrinsic Quantity—Specific Resistance. The intrinsic quantity in the short-wide CPP-geometry is the conductance per unit area. For CPP analysis, it is usually more convenient to use its inverse, the specific resistance AR.[28]

4.5.3.6. Bulk versus Interface Contributions. The theory underlying the CPP-MR is reviewed in Refs. 4 and 5. Here, we examine the data in terms of the models most widely used by the experimentalists, the two-current series-resistor (2CSR) model, or its generalization to finite l_{sf} by Valet and Fert.[56] In the limit of long l_{sf} in both the F- and N-layers, the 2CSR model gives AR for each fixed direction of incident electron spin (+ or −) as simply the sum of the effective resistivities times thicknesses ($\rho_i t_i$) of the individual layers, and the areas times effective resistances (AR_i) of the various interfaces. Here, the terms "effective resistivity" and "effective resistance" are needed because the resistivities within the F-metal, and the resistances at the F/N interfaces, are different for electron spin

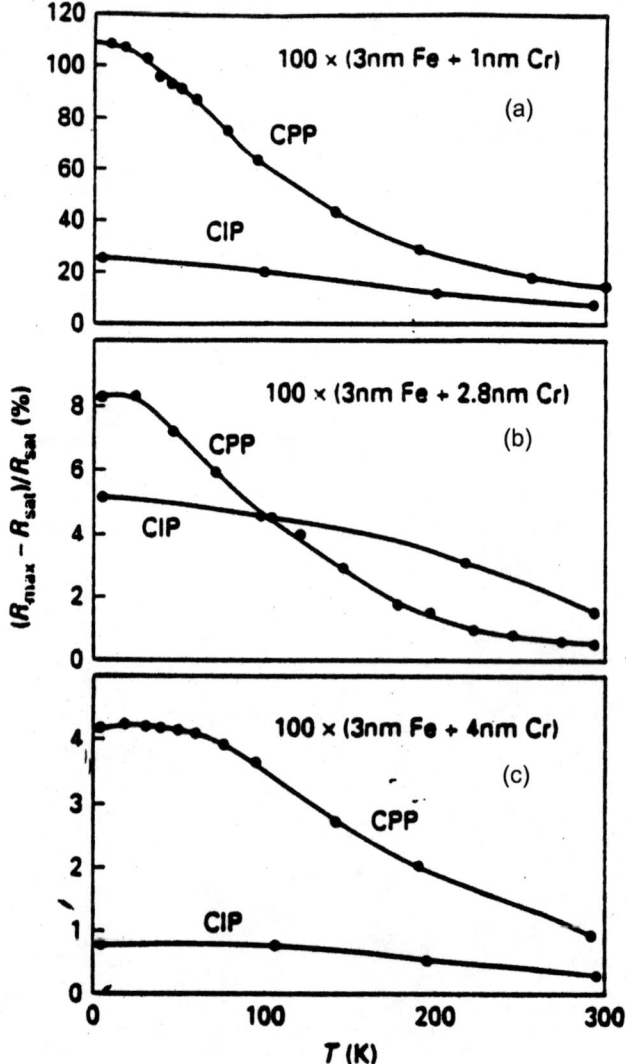

Figure 4.47. CPP-MR and CIP-MR versus T for [Fe(3 nm)/Cr (t_{Cr})]$_{100}$ multilayers. (a) $t_{Cr} = 1$ nm; (b) $t_{Cr} = 2.8$ nm; (c) $t_{Cr} = 4$ nm (from Gijs et al.,[51] p. 3345, © 1993 APS).

along (↑) or opposite to (↓) the local M_i (see Figs. 4.5 and 4.17). The simplest case is the AP state, where the resistances of $+$ and $-$ electrons are the same, since they see the same structure, just reversed in order. The total resistance is just half of that for either, and the "effective" resistivities and resistances are just the averages of those for (↑) and (↓) states. If we define $\rho_F^{\uparrow,\downarrow} = 2\rho_F/(1 \pm \beta)$ as the F-layer resistivity for electron spin ↑ or ↓, the "effective" resistivity is $\rho_F^* = \rho_F/(1-\beta^2)$, and the contribution to AR from the N F-layers is simply $N\rho_F^* t_F$. We similarly define an "effective" F/N interface specific resistance $AR_{F/N}^*$, and we assume that the S/F interface specific resistance, $AR_{S/F}$, is independent of the orientation of the F-layer

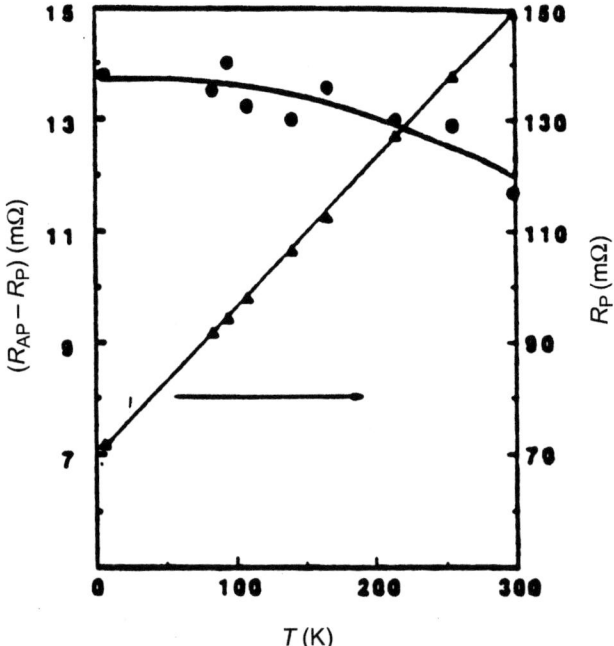

Figure 4.48. ($R_{AP} - R_P$) (left scale) and R_P (right scale) versus T for a [Co/Cu/NiFeCo/Cu] multilayer (from Krebs et al.,[96] p. 6086).

magnetization. If we neglect the difference between N and $N-1$, $AR(AP)$ for an F/N multilayer with N-bilayers should then be,[28,56]

$$AR(AP) = 2AR_{S/F} + N\rho_N t_N, + N\rho_F^* t_F + 2NAR_{F/N}^* \quad (4.5.1a)$$

For a series of samples with fixed total thickness $t_T = Nt_N + Nt_F$, t_N can be eliminated giving

$$AR(AP) = [2AR_{S/F} + \rho_N t_T] + N[(\rho_F^* - \rho_N)t_F + 2AR_{F/N}^*] \quad (4.5.1b)$$

Equation (4.5.1b) has the nice features that it is linear in N with an ordinate intercept that is independent of the magnetic quantities in the multilayer. $2AR_{S/F}$ can be independently determined from measurements of AR versus t_F for S/F/S sandwiches, and ρ_N can be independently measured on thick (≥ 300 nm) films sputtered in the same way as the multilayer. Perhaps a bit surprisingly, it has so far been possible to analyze most CPP-MR data using such independently measured values for both ρ_N and ρ_F. Both may thus be taken as predetermined parameters in Eq. (4.5.1), rather than as unknowns. If ρ_N is controllably increased by adding impurities to the N-metal, the top graph in Fig. 4.49 shows that Eq. (4.5.1b) predicts that the intercept should increase in a well-defined way and that the slope should become smaller (since ρ_N is subtracted from ρ^*_F). Figure 4.50 shows $AR(0)$ data for [Co(6 nm)/Ag(t)]$_N$ and [Co(6 nm)/AgSn(t)]$_N$ multilayers plotted in the form of

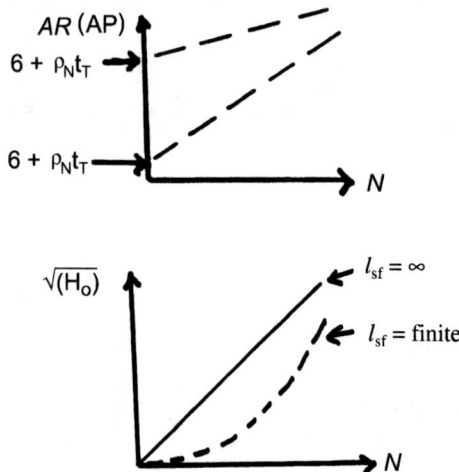

Figure 4.49. (1) Expected behavior of $AR(AP)$ versus N for samples described by Eq. (4.5.1b) with two different N-layer resistivities. (2) Expected behavior of $\sqrt{H_o} = \sqrt{[AR(AP) - AR(P)]AR(AP)}$ versus N (Eq. 4.5.2) for the same samples as in (1). If l_{sf}^N is long, and t_F is fixed, both sets of data should fall on the same straight line passing through the origin. If, however, l_{sf}^N becomes less than the N-layer thickness, t_N, the electron should lose spin-memory as it passes through N, forget the orientation of the magnetization of the layer from which it came, and fall below (dashed curve) the straight line, by a larger amount the larger t_N (i.e., the smaller N).

Eq. (4.5.1b). The arrows on the ordinate axis indicate the independently predicted values of the intercepts. The Co/Ag data are nicely consistent with the predictions of Eq. (4.5.1b). The Co/AgSn data, while not in as complete agreement, display the expected behaviors.

A more stringent test can be provided by examining the difference in ARs for the AP and P states in the form of Eq. (4.5.2),[28]

$$\sqrt{[AR(AP) - AR(P)]AR(AP)} = N(\beta \rho_F^* t_F + 2\gamma AR_{F/N}^*) \qquad (4.5.2)$$

As was true for Eq. (4.5.1b), Eq. (4.5.2) is linear in N, but now with zero intercept, and also with the right-hand side independent of ρ_N. As illustrated in the lower graph in Fig. 4.49, a plot of the experimental left-hand side square-root versus N should then give a straight line passing through the origin, and the slope of the line should be independent of ρ_N. That is, the two very different sets of data in Fig. 4.4.9(1) should both fall on exactly the same line in the graph in Fig. 4.4.9(2). Choosing $AR(0)$ for $AR(AP)$, Fig. 4.51 shows that this prediction is obeyed for the data of Fig. 4.50.

4.5.3.7. Test of Predictability of CPP-MR.

If data such as those in Figs. 4.50 and 4.51 closely represent the behavior for $AR(AP)$ and $AR(P)$, then measurements of sample sets with different thicknesses of t_F and t_N can be used to derive the parameters β, γ, and $AR_{F/N}^*$ for given F–N pairs. If F-metals F1 and F2 have very different saturation fields, $H_1 \ll H_2$, then combining these metals into an $(F1/N/F2/N)_N$ hybrid SV should allow both $AR(AP)$ and $AR(P)$ to be reliably produced. If the parameters for F1/N and F2/N are known, if the 2CSR model is valid, and if the layers in the hybrid SV behave the same as those in the F1/N and F2/N multilayers, then it is straightforward to predict $AR(AP)$, $AR(P)$, and $A\Delta R$ for the hybrid SV. Reference 93 shows that the M and $[AR(H)-AR(P)]$ hysteresis curves for a $[Py/Cu/Co/Cu]_N$ hybrid SV behave as expected for properly ordered magnetic states—i.e., M and AR are constant in the high field P-state, the Py-layer magnetization reverses rapidly at low field, and the Co-layer magnetization reverses more slowly at higher fields.

Figure 4.52 compares the experimental data for two (Py/Cu/Co/Cu) hybrid SVs with predictions made with no adjustable parameters. The solid lines were predicted based on the analysis of the Py/Cu data assuming that l_{sf}^{Py} is long.[93] The agreement is reasonable—within 15–20%. However, as we will describe in Section 4.5.3.11, it now looks as if l_{sf}^{Py} is short, only ~5.5 nm.[106] Reanalysis of the Py/Cu data in terms of such a short l_{sf}^{Py} using the theory of Valet and Fert,[56] slightly modified the other parameters, leading to the dashed curves in Fig. 4.52.[107] These curves fall within the uncertainties (~10%) of the data.

4.5.3.8. Spin-Direction Scattering Anisotropy Parameters.

We noted in Section 4.5.2.7 that the bulk parameters α_F (or β) derived by different investigators for a given F-metal or F-alloy from CIP-MR data could be quite different. Because of the simpler equations for the CPP-MR, one might hope to find closer agreement for such parameters. Table 4.3 compares values of both β, and the interfacial anisotropy parameter γ, estimated for Co/Cu multilayers by investigators using not only different measuring techniques (crossed S-strips, lithography, or nanowires), but very different sample preparation techniques (sputtering, MBE, or electrodeposition) that gave different values of ρ_{Cu} and ρ_{Co}^*. Recognizing, also, the differences expected between analyses that approximate $AR(AP)$ by $AR(0)$ or by $AR(Pk)$, the inferred values are mostly quite similar.

4.5.3.9. Inverse CPP-MR and Unified Picture.

As noted in Section 4.5.1, if one combines into a hybrid SV of the form $(F1/N/F2/N)_N$ metals F1 and F2 that have the same signs of γ (say $\gamma > 0$), but opposite signs of β, say $\beta_1 > 0$ but $\beta_2 < 0$, then the MR should be "normal" for thicknesses t_{F2} small enough so that scattering at the F2/N interface is dominant, but should change to "inverse" as t_{F2} increases to where scattering in the bulk of F2 becomes dominant. From the older bulk F-metal measurements,[1] and more recent calculations,[29] discussed in Section 4.2.3, positive β is expected for Co containing just structural defects, and for Py = NiFe, but negative β are expected for alloys of Co, Ni, or Fe with Cr or V. Figures 4.53 and 4.54[111] show that, indeed, the CPP-MR changed sign with increasing t_{NiCr} for [g-Co/Cu/NiCr/Cu]$_{20}$ multilayers. Here, g-Co indicates granular-Co, obtained by making the Co layers so thin (0.4 nm) that they mixed with the surrounding Cu to give a granular Co layer. This procedure was chosen because g-Co has a high H_s, which was needed to achieve an AP state (Fig. 4.53). Similar "inversion" of the CPP-MR was also seen when g-Co was combined with FeCr and FeV.[111,112] Interestingly, in none of these cases did the CIP-MR become inverse. Such a different behavior shows that the

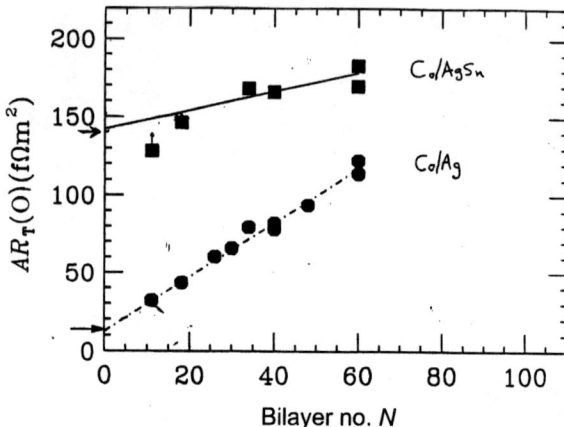

Figure 4.50. Test of Eq. (4.5.1b) with [Co(6 nm)/Ag(t_{Ag})]$_N$ (circles) and [Co(6 nm)/AgSn(t_{AgSn})]$_N$ (squares) multilayers, all with fixed $t_T = 720$ nm. $AR(AP)$ is approximated by $AR(0)$. The arrows on the ordinate axes indicate the independently predicted intercepts. The solid line for Co/Ag is a least-squares fit to the data. The solid line for Co/AgSn is drawn to go through the data yet intersect the arrow (after Lee et al.,[28] p. L3, © 1993, Elsevier science).

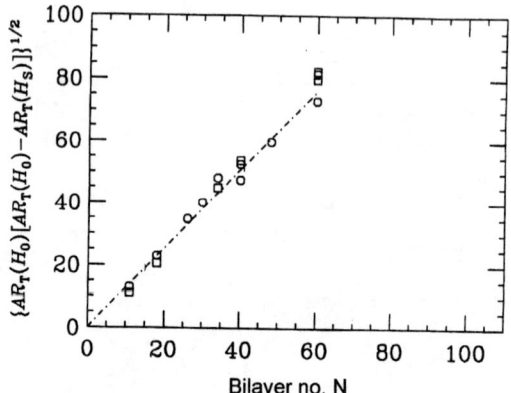

Figure 4.51. Test of Eq. (4.5.2) using the same multilayers as in Fig. 4.50. The two sets of data are consistent within their uncertainties with the same straight line passing through the origin (from Lee et al.,[28] p. L5, © 1993, Elsevier Science).

relative importances of bulk and interface scattering can be very different in the CIP- and CPP-MRs. Lastly, an opposite transition from "inverse" to "normal" CPP-MR was seen in Fe/Cr multilayers.[113] The interpretation was that the γ of the Fe/Cr interface is negative, as assumed in the analysis in Section 4.5.2.8. In all of these cases, the signs of β and γ are consistent with those for dilute bulk F-alloys[1] and with calculations.[29] In this sense, the data support a "unified picture" of spin-dependent anisotropy, where the anisotropies for CPP-MR, dilute F-alloys, and calculations, are semi-quantitatively similar. Further studies are needed to clarify the generality of this apparent relationship.

Table 4.2. Comparison of CPP-MR Parameters for Co/Cu Multilayers Prepared in Different Ways

	MSU[93] Sup. Strips Sputtering 4.2 K $AR(0)$	MSU[108] Sup. Strips Sputtering 4.2 K $AR(Pk)$	Leeds[109] Sup. Strips MBE 4.2 K $AR(0)$	Doudin et al.[110] Nanowires Electrodep. 20 K $AR(Pk)$	Piraux et al.[97] Nanowires Electrodep. 77 K $AR(Pk)$
ρ_{Cu} (nΩ m)	4.5 ± 1	6 ± 1	13 ± 3	$13 - 33$	~ 31
ρ^*_{Co} (nΩ m)	76 ± 5	66 ± 5	30 ± 6	$510 - 570$	180 ± 20
β	0.46 ± 0.08	0.38 ± 0.06	0.48 ± 0.04	0.5 ± 0.08	0.36 ± 0.04
γ	0.75 ± 0.05	0.71 ± 0.05	0.71 ± 0.02	$0.3 - 0.6$	0.85 ± 0.1
$AR^*_{Co/Cu}$ (fΩ m^2)	0.52 ± 0.02	0.38 ± 0.03	0.43 ± 0.04	$0.3 - 1.1$	0.3 ± 0.05

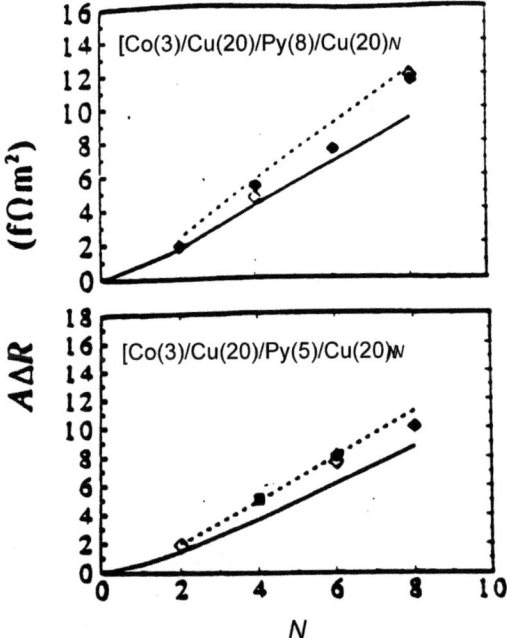

Figure 4.52. Measured (symbols) and predicted (curves) values of AΔR versus N for [Co(3 nm)/Cu(20 nm)/Py(8 nm)/Cu(20 nm)]$_N$ and [Co(3 nm)/Cu(20 nm)/Py(5 nm)/Cu(20 nm)]$_N$ hybrid-SVs. The solid curves are predictions with earlier parameters based upon the assumption that $l^{Py}_{sf} = \infty$. The dashed curves are predictions taking parameters consistent with $l^{Py}_{sf} = 5.5$ nm. Open and filled symbols indicate samples sputtered in different runs (from Pratt et al.,[107] p. 3509, © 1997, IEEE).

4.5.3.10. Test of NiMnSb as a "Half-Metallic" Ferromagnet.

As discussed in Section 4.3, band structure calculations predict that NiMnSb should be a half-metallic ferromagnet (100% polarized), but the predication is not supported by electron-emission studies or estimates of polarization. Half-metallic behavior can

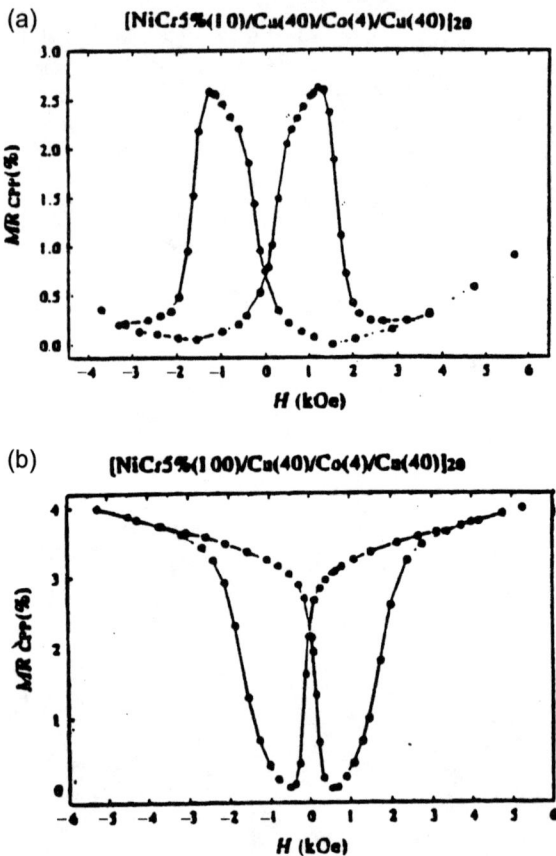

Figure 4.53. CPP-MRs for sputtered [NiCr5%(t_{NiCr})/Cu(4 nm)/Co(0.4 nm)/Cu(4 nm)]$_{20}$ multilayers with (a) $t_{NiCr} = 1$ nm ("Normal" MR) and (b) $t_{NiCr} = 10$ nm (After C. Vouille et al.,[111] p. 6720, © 1999 APS).

be directly tested by measuring perpendicular transport, since the perpendicular resistance for either transport or tunneling between two half-metallic ferromagnets with their polarizations oriented opposite to each other (AP state) should be infinite. As illustrated in Figs. 4.55 and 4.56, so far neither CIP-MR nor CPP-MR measurements involving NiMnSb give results compatible with half-metallic behavior.[114] We will see in Section 4.6.4 that CPP-tunneling experiments do not support "half-metallicity" of NiMnSb either. But further studies are still needed to rule out problems with the quality of the NiMnSb layers.

4.5.3.11. Finite Spin-Diffusion Lengths (Spin-Relaxation) in N- and F-Metals and -Alloys.

The analysis so far (see especially Section 4.5.3.6) has assumed very long spin-diffusion lengths, l_{sf} in both the N- and F-metals. But what if l_{sf}^N or l_{sf}^F are reduced, leading to perceptible spin-relaxation?

l_{sf}^N can be reduced by adding impurities to the N-metal that cause spins to flip, either by spin–orbit or spin–spin interactions. The experimental data for

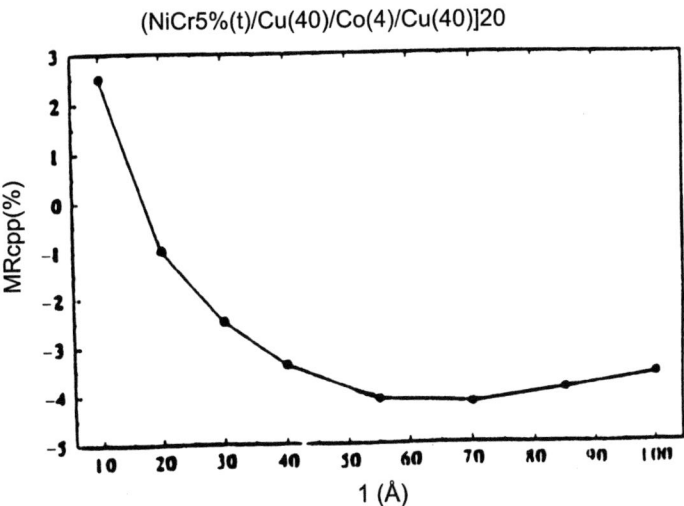

Figure 4.54. CPP-MR versus t_{NiCr} for the samples of Fig. 4.53 plus additional ones, showing that the CPP-MR first changes sign with increasing t_{NiCr}, and then approximately saturates in value (After Vouille et al.,[111] p. 4574).

the square-root-functions (left-hand side) of Eq. (4.5.2) should then fall below the predicted straight line. Qualitatively, if the electron spins flip many times (relax) while the electrons traverse the N-metal, the orientation of the magnetization of the F-layer from which the electrons emerged is forgotten, and the orientation of the magnetization of the F-layer at which they arrive no longer matters. $A\Delta R$ then falls toward zero as the N-layer thickness becomes larger than l_{sf}. Quantitatively, such "spin-relaxation" can be analyzed using the more general Valet–Fert (VF) theory for finite l_{sf}.[56] For Co/Ag and Co/AgSn multilayers, l_{sf} is expected to be relatively long.[115,116] Figure 4.57[115] shows that the square-root-functions for these multilayers fall closely along a single straight line passing through the origin. When, however, Cu or Ag are alloyed with Pt (a heavy impurity → strong spin–orbit interaction) or Mn (an impurity with a magnetic moment → strong spin–spin interaction) l_{sf}^{N} should become shorter. The solid lines are fits to the data for CuPt and CuMn using the VF theory. Table 4.3 compares the values of l_{sf} derived from such fits for Cu- and Ag-based alloys with values predicted from independent data on the spin–orbit interaction[115,116] or with values estimated from spin–spin effects.[116] The agreements are suprisingly good.

l_{sf}^{F} in F-metals can be estimated either by measuring how the MR of simple F/N multilayers decreases with increasing F-metal thickness, t_F, or how $A\Delta R$ in EBSVs increases with increasing t_F. Measurements applying the first technique to Co/Cu multilayer nanowires gave estimates of $l_{sf}^{Co} \sim 59$ nm at 77 K and ~ 38 nm at 300 K.[102] The analysis was sufficiently complex that we refer the reader to the paper for details. $l_{sf}^{Py} \sim 5.5$ nm[105,106] and $l_{sf}^{CoFe} \sim 12$ nm[117] have been estimated using the second technique at 4.2 K. Figure 4.58 shows that $A\Delta R$ for a symmetric [FeMn/Py/Cu/Py] EBSV (the same Py thickness, t_{Py}, for both Py layers) does not

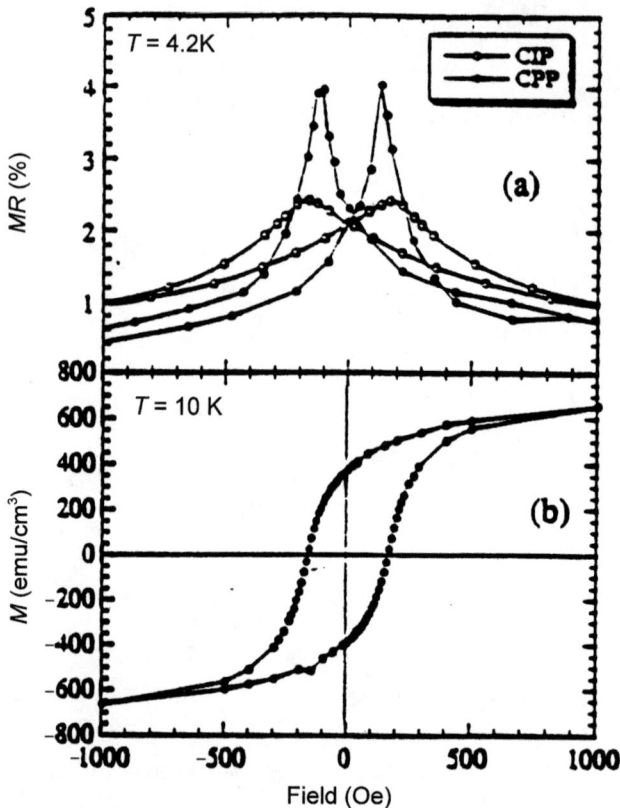

Figure 4.55. (a) CIP-MR (open circles, $T = 10$ K) and CPP-MR (closed circles, $T = 4.2$ K) for a [NiMnSb(3 nm)/Cu(4.5 nm)]$_{10}$ multilayer on a glass substrate. (b) Magnetic hysteresis loop of the same sample at 10 K (from Caballero et al.,[114] p. 1803).

continue to increase monotonically with increasing t_{Py}, as expected for long l_{sf}^F. Rather, it "saturates" to a constant value above about $t_{Py} = 20$ nm. The solid curve shows the VF fit to these data, attributing the saturation to a finite l_{sf}^{Py}. The resulting value of $l_{sf}^{Py} = 5.5 \pm 1$ nm was much shorter than had been expected. The validity of the parameters and model was tested by predicting the MR $= A\Delta R/R(P)$ of new data for "double" spin-valves with one FeMn layer in the middle or with two FeMn layers on the outside. In both cases the agreement between the data and the non-adjustable predictions was quite good.[107] A short l_{sf}^{Py} (~4.3 $\pm$ 1 nm at 77 K) was subsequently also derived from measurements on multilayered Py/Cu nanowires.[118] These latter authors gave a plausibility argument why l_{sf}^{Py} might be so short.

Finally, l_{sf}^N in N-metals and alloys, and spin-relaxation at N1/N2 interfaces, can be measured using a "spin-relaxation detector" such as the one shown in Fig. 4.59(a). Here, an insert I of the material of interest—a metal, an alloy, or a multilayer—is deposited in the middle of an EBSV, separated from the two F-layers by thick enough layers of the usual N-metal (e.g., Cu or Ag) so that exchange coupling between the two F-metals is small. If this insert causes no spin-relaxation, it

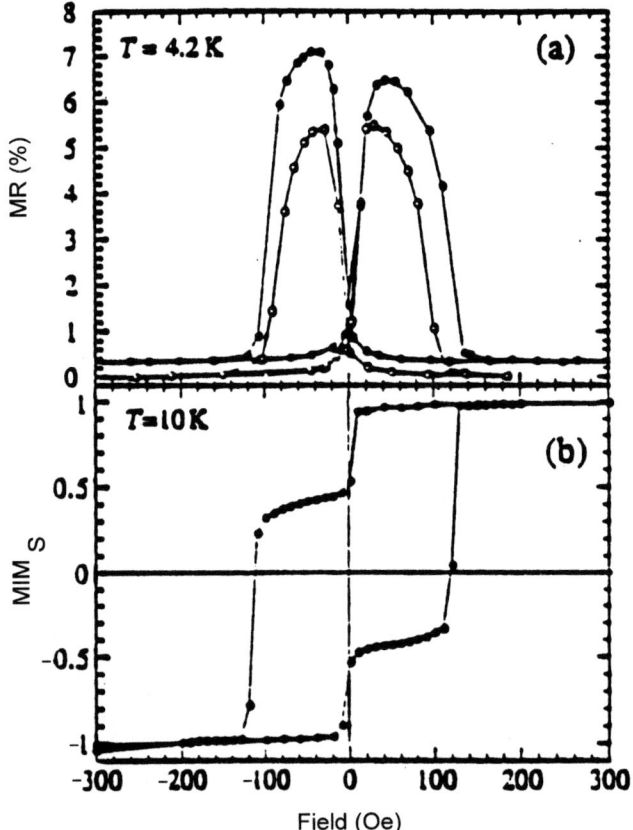

Figure 4.56. (a) CPP-MR(H) at $T = 4.2$ K for [NiMnSb(20 nm)/Cu(t_{Cu})/NiFe(6 nm)] hybrid SVs with $t_{Cu} = 10$ nm (filled circles) and 12 nm (open circles). The data have been corrected for contributions due to the FeMn "pinning layer". (b) Magnetic hysteresis loop at $T = 10$ K for a sample with $t_{Cu} = 10$ nm (from Caballero et al.,[114] p. 1803).

will reduce $A\Delta R$ only because of the additional specific resistance that it contributes, and this reduction will be only an algebraic function of the thicknesses of the inserted layers or the number of inserted interfaces. If, however, either the material of the insert, or the N1/N2 interfaces, causes spin-relaxation, then $A\Delta R$ will decrease exponentially with the layer thickness or the number of interfaces. Figure 4.59(b)[119] illustrates such behaviors for a multilayer insert of the form $I = [\text{Nb/Cu}]_N$, in a [FeMn/Py/Cu/I/Cu/Py] EBSV. The dashed curve shows the reduction in $A\Delta R$ expected for no spin-relaxation at the Nb/Cu interfaces. The data fall well below this curve, suggesting a surprisingly large (~15%/interface) spin-relaxation. This large value was tentatively attributed to strong spin–orbit scattering in a highly resistive Nb(Cu) alloy ($\rho_{NbCu} \sim 150\ \mu\Omega$ cm) formed at the Nb/Cu interfaces.[119]

4.5.3.12. Angular Variation.
A major contentious issue of GMR is whether anisotropic scattering at interfaces arises mainly from interfacial alloying or

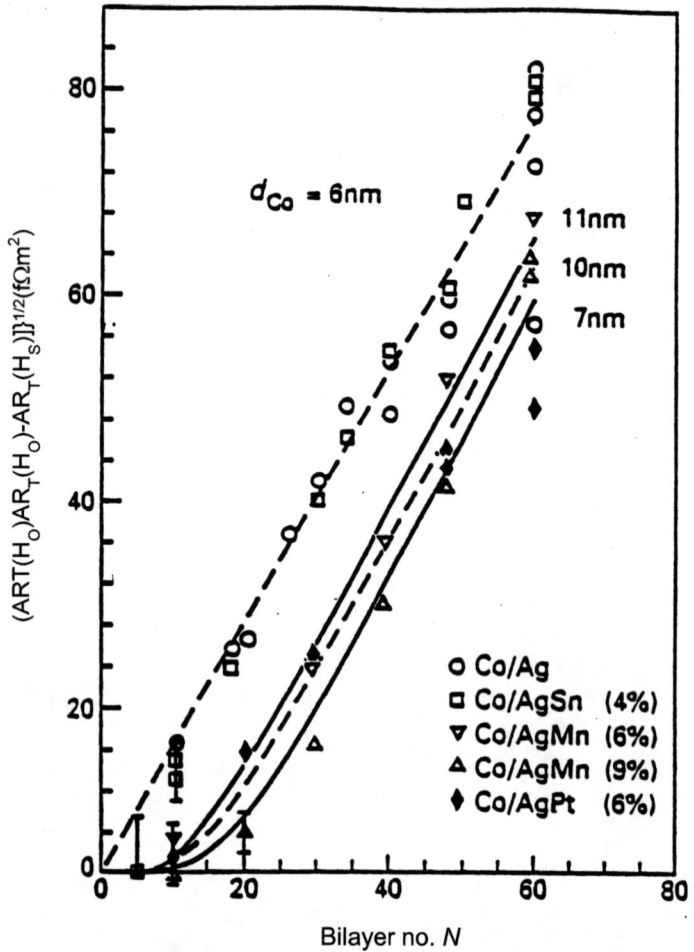

Figure 4.57. $\{([AR(0) - AR(H_s)]AR(0)\}^{1/2}$ versus N for Co/Ag, Co/AgSn(4 at.%), Co/AgMn(6 at.%), Co/AgMn(9 at.%), and Co/AgPt (6 at.%) multilayers with fixed $t_{Co} = 6$ nm. The dashed line through the origin is for $l_{sf} = \infty$. The other dashed and solid curves are fits with finite l_{sf} to the Valet–Fert model; the best fit values of l_{sf} are indicated (from Yang et al.,[115] p. 3276, © 1994 APS).

from anisotropic potential steps. As noted above, most measurements do not distinguish between the two. However, measurements of the angular variation of the CPP-MR might. Strong contributions from potential steps have been predicted[120] to induce deviations from a simple $\cos^2(\theta/2)$ variation, where θ is the angle between the magnetizations of neighboring F-layers. Such a deviation in the CPP-MR was first observed in $[Py/Ag/Co/Ag]_N$ hybrid SVs as shown in Fig. 4.60.[82,121] Unfortunately, the model used to analyze the data was appropriate for the case of only a single F-metal, so deviations from linearity could be attributed to non-validity of the model. Deviations were later found in $[Py(t_{Py})/Ag(1.1 \text{ nm})]_{50}$ multilayers where the Ag thickness was chosen to give antiferromagnetic coupling.[122] Here, the angular dependence was found to vary substantially with the thickness of the Py

Table 4.3. Measured versus Calculated Spin-Diffusion Lengths for N-Metal Alloys

Metal or Alloy	l_{sf}^N (Measured)	l_{sf}^N (Calculated)
Ag	Long	~500 nm
AgSn(4 at.%)	≥25 nm	~26 nm
AgMn(6 at.%)	~10 nm	~12 nm
AgMn(9 at.%)	~7 nm	~9 nm
AgPt(6 at.%)	~10 nm	≥9 nm
Cu	Long	~450 nm
CuMn(7 at.%)	~3 nm	~3 nm
CuPt(6 at.%)	~9 nm	~7 nm
CuNi(7 at.%)	~23 nm	~22 nm
CuNi(23 at.%)	~7.5 nm	~7 nm.

layer. A model to describe these variations has been proposed.[120] But further studies are still needed to sort out the physics involved.

4.5.3.13. TESTS OF NOVEL PREDICTIONS.
We conclude our discussion of the CPP-MR with brief mention of some studies done in response to novel predictions.

Schep et al.[123] predicted that ballistic multilayers could give rise to a very large MR (~120%) due to band-structure effects, even in the absence of scattering. An

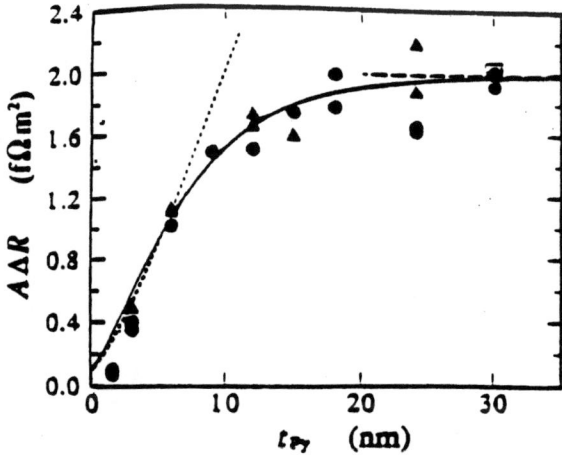

Figure 4.58. $A\Delta R$ versus t_{Py} for FeMn(8 nm)/Py(t_{Py})/Cu(2 nm)/Py(t_{Py}) EBSVs (filled symbols). The solid curve is a numerical fit to the VF model with $l_{sf}^{Py} = 5.5$ nm. The short dashed curve is the expected behavior for $l_{sf}^{Py} = \infty$ with all of the other parameters the same. The horizontal dashed line is the limit $t_{Py} \gg 5$ nm for the same parameters. The open square is $A\Delta R$ for an asymmetric EBSV, with $t_{py} = 15$ nm for the "pinned" layer and 45 nm for the "free" layer. VF theory predicts that this $A\Delta R$ should be indistinguishable from that for $t_{Py} = 3$ nm in both layers (from Pratt et al.,[107] p. 3506, © 1997 IEEE).

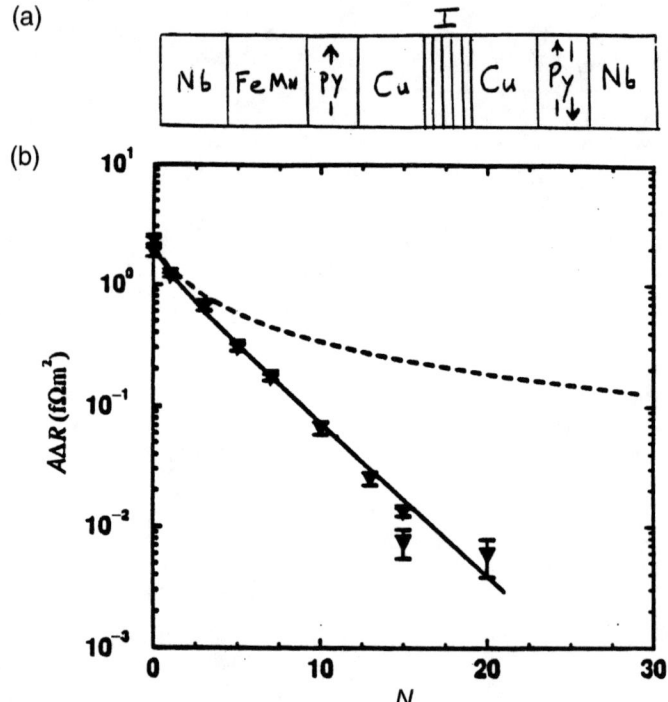

Figure 4.59. (a) Schematic of the structure for measuring spin-relaxation in a multilayer insert I. (b) $A\Delta R$ versus number of bilayers N in the insert for [Cu(10 nm)/Nb(1.5 nm)]$_N$ inserts. The dashed curve is the reduction in $A\Delta R$ expected for no spin-relaxation in the insert. The solid curve is a fit to the theory of Valet–Fert with ~15% spin-relaxation at each interface (after Baxter et al.,[119] p. 4548).

attempt to see such large effects in "real" multilayers was made using point contacts.[124] Any MRs associated with the contacts were ≤7%, much less than predicted. However, the multilayers almost surely did not satisfy the ballistic propagation condition assumed in the analysis, mainly because of strong scattering at the interfaces. A fully rigorous test could require samples with atomically flat and unmixed interfaces.

Mathon[125] predicted an unusually large GMR in samples with pseudorandomly varying F-layer thicknesses. In the AP state, such a variation was predicted to induce localization, and thus a very high resistance. Here too, a search[126] failed to uncover the predicted large MR but, again, the samples may not have satisfied the conditions required for the large MR to occur.

Berger and Slonczewski[127] independently predicted two novel effects in the CPP-geometry of multilayers, that a sufficiently large current density would: (a) generate spin-waves and (b) reverse layer magnetizations. There in now clear evidence for both phenomena[128]. There is still debate about the physics that dominates these phenomena[129].

Lastly, both theoretical and experimental challenges have recently been raised to the 2CSR and VF models. Ref. 130 contains a selection of papers for the interested reader. The main questions seems to be whether scattering form adjacent interfaces in present day multilayers is partially coherent or completely incoherent.

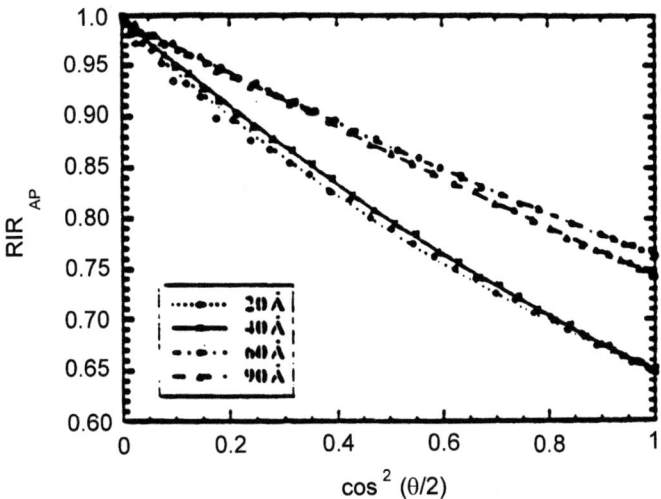

Figure 4.60. Angular dependence of the CPP-MR at 4.2 K of [(Ag(4 nm)/Co(0.4 nm)/Ag(4 nm)/Py(t)]₂₀ multilayers with different Py thicknesses. The symbols are the data. The curves are fits to the form $R(\vartheta)/R_{AP} = 1 - a\cos^2(\vartheta/2) + b\cos^4(\vartheta/2)$ (from Dauguet et al.,[121] p. 1085, © 1996 APS).

4.6. GRANULAR MAGNETORESISTANCE

4.6.1. Discovery

In 1992, two groups[59,60] independently discovered that metallic wires containing small F-metal inclusions (F-metal clusters), called granular magnetic alloys, could display large MRs comparable to those seen in magnetic multilayers. We designate this phenomenon g-MR, for granular-MR. The first g-MR samples were prepared by sputtering from a single alloy target,[59] or by co-sputtering from independent targets.[60] Samples have since been successfully prepared by other techniques, such as rapid quenching from the melt–melt-spinning.[131] Figures 4.61 and 4.62 illustrate the behaviors obtained at cryogenic temperatures for Cu(Co) samples, and how they are affected by annealing the sample, thereby changing the cluster size. It was hoped that such alloys might be advantageous for technology, since they are usually simpler to make than multilayers. Several problems have so far forstalled these hopes, but the physics of g-MR is still quite interesting.

4.6.2. Dependence on F-Cluster Shape

It has been argued that the small spherical inclusions that give large MRs require large fields to align their moments.[132] The required fields are seen in Figs. 4.61 and 4.62. Oriented disk-like inclusions that can realign in smaller fields have been fabricated,[133] but the process is not technologically advantageous, since it requires starting with multilayers.

4.6.3. Dependence on F-Cluster Size and Measuring Temperature

Both g-MR and its temperature dependence vary strongly with F-cluster size, which in turn depends upon annealing history.

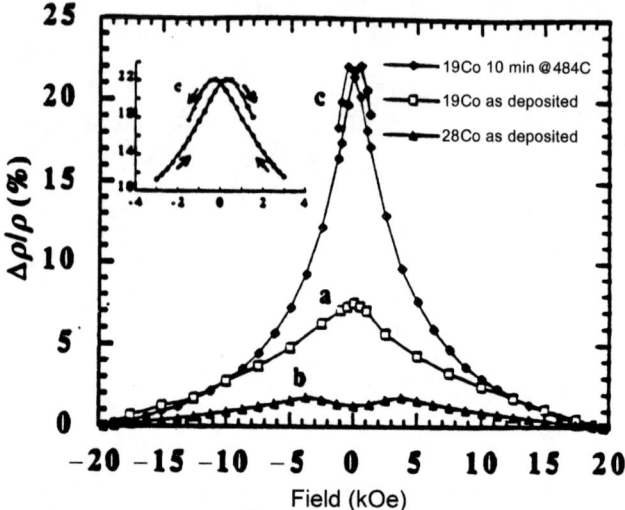

Figure 4.61. Field dependence of $\Delta\rho/\rho = [\rho(H)-\rho(H=20 \text{ kOe})]/\rho(H=20 \text{ kOe})$ for two $Cu_{100-x}Co_x$ alloys with $x=19$ or 28 as-deposited and one alloy with $x=19$ annealed at 484°C. Curves a and b were measured at $T=100$ K; curve c was measured at 10 K. The field is parallel to the current (from Berkowitz et al.,[59] p. 3746, © 1992 APS).

Generally, annealing increases the cluster size, leading to a decrease in $\Delta\rho$, but not necessarily in the MR = $\Delta\rho/\rho$ (Fig. 4.63[134]). Figure 4.64[135] shows how the g-MR varies with Co content for both as-prepared Ag(Co) samples, and the same samples annealed to produce maximum g-MR. Above the percolation concentration of 55%, there is no g-MR, only the residual AMR of the F-metal.[135] Figure 4.65 shows a series of Ag(Fe) samples for which $\Delta\rho_{xx}$ and ρ_{xx} are both proportional to $1/r_0$, the inverse of the average Fe particle radius.[136] Such results suggest that the g-MR is

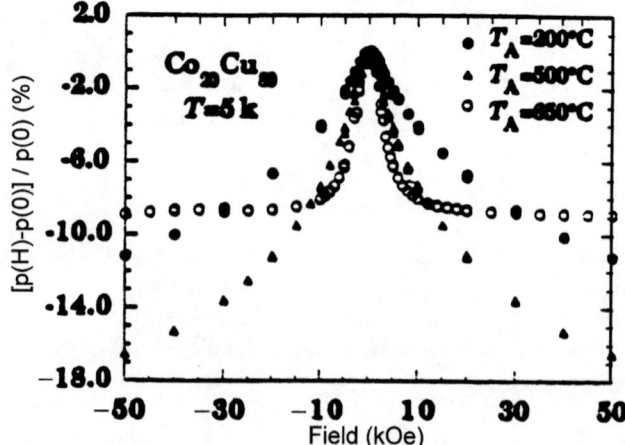

Figure 4.62. MR versus H at 5 K for a phase-separated $Cu_{80}Co_{20}$ sample annealed for 10 min each at $T_A = 200$, 500, and 650°C (from Xiao et al.,[60] p. 3750, © 1992 APS).

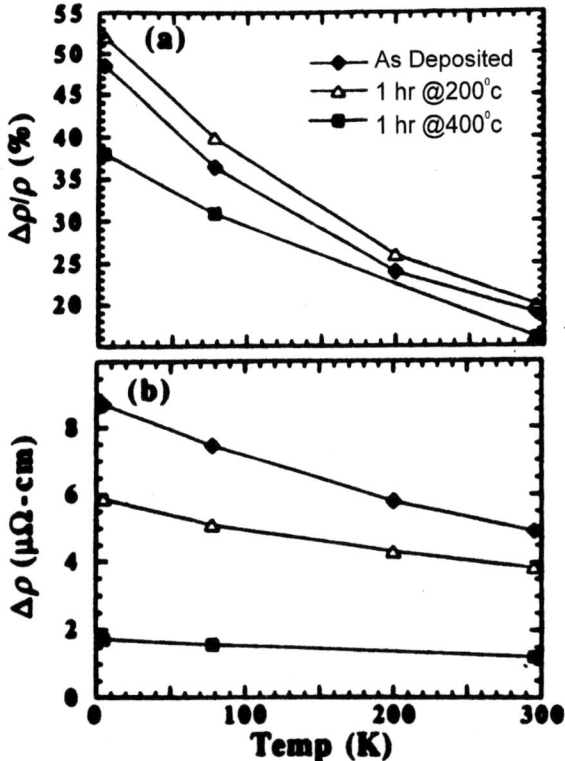

Figure 4.63. Temperature dependences of $\Delta\rho/\rho$ and $\Delta\rho$ for a $Ag_{74}Co_{26}$ sample under the conditions indicated (from Carey et al.,[134] p. 2937).

approximately proportional to the surface/volume ratio of the inclusions; i.e., that g-MR is dominated by scattering at the inclusion surface.

Figure 4.66[137] shows that the granular alloys manifest differences in dc susceptibility below a "blocking temperature" when one compares measurements on samples cooled in zero field (ZFC) or in a finite field (FC). Below the blocking temperature, the magnetic moments of the predominant clusters are frozen in direction, but above the blocking temperature the sample behaves like a "superparamagnet." Figure 4.66 shows that the clusters in samples of different concentrations can reorient at temperatures from a few kelvin to above room temperature. Interestingly, there is little correlation between the blocking temperature and the magnitude of the g-MR, presumably because the electron scattering time is much shorter than the superparamagnetic relaxation time. However, the saturation field becomes larger above the blocking temperature.

4.6.4. Relation of g-MR to M

Figures 4.67 and 4.68 show that the g-MR is sometimes a simple function of M. But Fig. 4.69[138] shows that it is not always simply related to M.[138] This latter situation

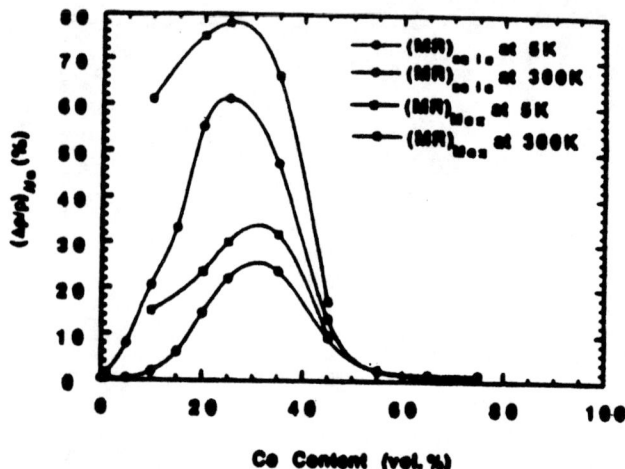

Figure 4.64. MR at 5 K (filled symbols) and 300 K (open symbols) of granular Co–Ag samples as a function of Co volume fraction. Circles indicate as-prepared samples deposited at 300 K and squares indicate samples annealed to give maximum g-MR for the given Co content (from Chien et al.,[135] p. 5315).

is probably a consequence of the presence of a range of cluster diameters. If M is proportional to the volume of a cluster, but the g-MR is inversely proportional to its radius, large clusters will contribute more to M than to the g-MR, while smaller ones will contribute more to the g-MR than to M. Reorientation of different cluster sizes at different fields will then give different field dependences of M and the g-MR. Different cluster sizes can also have different relative importances at different temperatures.[139] At higher F-metal concentrations, the clusters also interact magnetically with each other, generally complicating the magnetic behavior of the sample.[139]

4.6.5. Thermal Conductivity, Hall Effect, and Thermopower

Complementary measurements have been made of κ, the extraordinary Hall effect, R_H^M, and the thermopower of g-MR materials. As expected for such "dirty alloys", $\kappa(H)$ correlates well with $\rho(H)$ through the Wiedemann–Franz law (Fig. 4.70).[85,140] The extraordinary Hall effect has been reported to display an unusually simple correlation with M (Fig. 4.71),[141] but how fundamental and widespread is this simplicity is a topic of debate.[142]

Simple correlations of S with $1/\rho$ have also been observed, as shown in Fig. 4.72.[85,141,143]

4.7. TUNNELING MAGNETORESISTANCE

4.7.1. Discovery, I–V Characteristics, and the Julliere Model

As noted in Section 4.3, whereas the diffusive phenomena we have discussed up to now involve a linear relation between an applied current I and a resulting voltage V (except for the point contact studies of section 4.5.3.13, the tunneling studies that we cover in this section involve a non-linear relation between high enough applied voltage

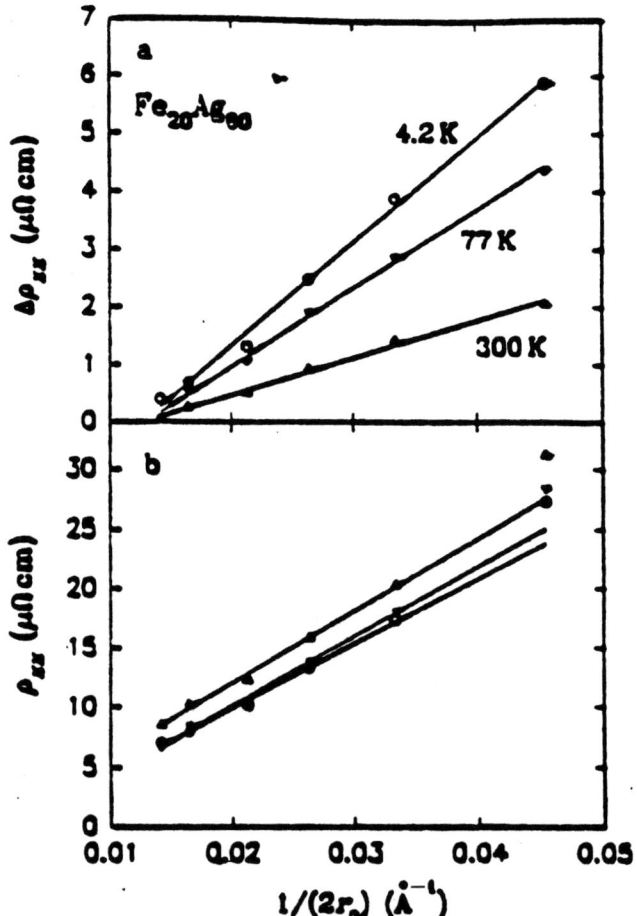

Figure 4.65. (a) Change in resistivity, $\Delta\rho_{xx}$, and (b) zero field resistivity, $\rho_{xx}(0)$, versus inverse Fe particle diameter, $1/(2r_0)$, for different $Ag_{80}Fe_{20}$ granular alloys measured at 4.2 K (circles), 77 K (inverted triangles), and 300 K (erect triangles) (from Wang and Xiao,[136] p. 3992, © 1994 APS).

V and the resulting current I. Figure 4.73 shows examples of both linear behavior for very low voltages and non-linear behavior for higher ones.[7] Demonstrating non-linearity is taken as a requirement for asserting that one is observing tunneling.

The first measurements of tunneling between two F-metals through an insulator (I) were reported in 1975 by Julliere,[144] who demonstrated both a non-linear I–V curve, indicating tunneling, and a TMR $= [R(AP) - R(P)]/R(AP)$ of 14% at 4.2 K. He used the semiconductor Ge as the tunneling barrier, and achieved an approximate AP state by using two different F-metals, Fe and Co. The basic TMR (F/I/F) geometry is shown in Fig. 4.74,[145] and Julliere's data for TMR as a function of bias voltage V are shown in Fig. 4.75.

Julliere proposed a simple model for this TMR. Assuming that a and a' are the fractions of tunneling electrons in Fe and Co that have their magnetic moments

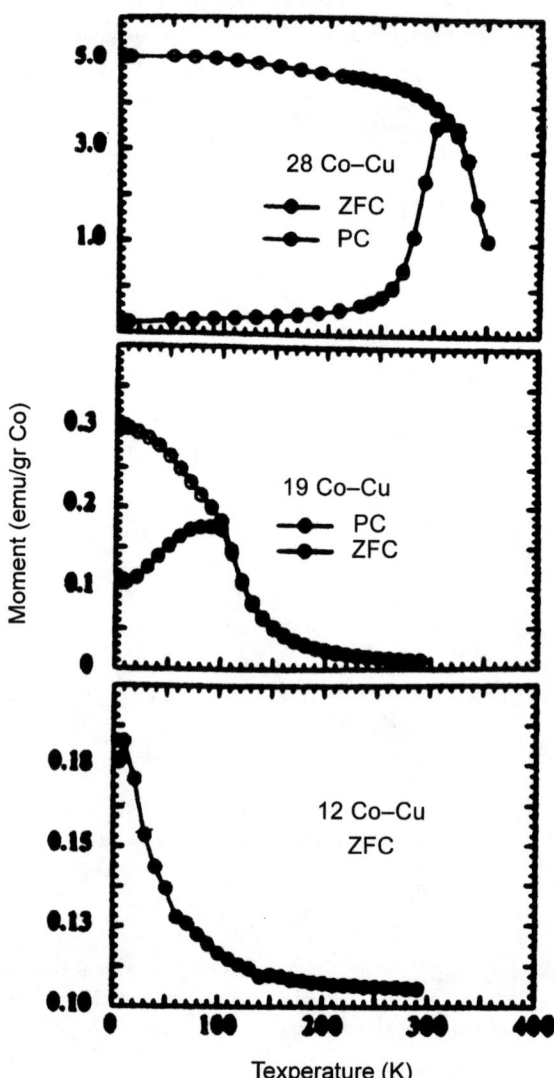

Figure 4.66. Zero-field-cooled (ZFC) and field-cooled (FC) dc susceptibilities versus temperature of $Co_{100-x}Cu_x$ samples. The applied field was 10 Oe for $x = 19$ and 28 and 50 Oe for $x = 12$ (from Berkowitz et al.,[137] p. 5321).

parallel to the magnetization, he defined $P_{Fe} = 2a-1$ and $P_{Co} = 2a'-1$ as the conduction electron spin polarizations. From equations given by Tedrow and Meservey,[146] and assuming spin-direction conservation during tunneling, he derived a result that can be written in two forms:

$$\text{TMR} = [R(\text{AP}) - R(\text{P})]/R(\text{AP}) = 2P_{F1}P_{F2}/[1 + P_{F1}P_{F2}] \quad (4.7.1a)$$

or

$$\text{TMR}' = [R(\text{AP}) - R(\text{P})]/R(\text{P}) = 2P_{F1}P_{F2}/[1 - P_{F1}P_{F2}] \quad (4.7.1b)$$

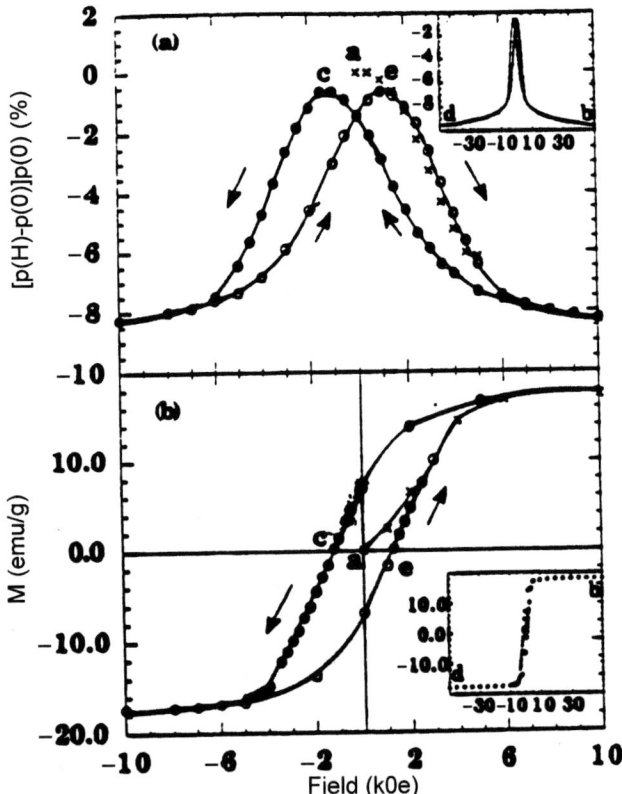

Figure 4.67. Hysteresis loops for (a) MR $= (\rho(H)-\rho(0))/\rho(0)$ and (b) Magnetization M, at 5 K for a phase-separated $Cu_{84}Co_{16}$ film deposited at 350°C. Crosses, filled, and open circles, denote the initial curve (a→b), and branches of decreasing (b→c) and increasing (d→e→b) fields (from Xiao et al.,[60] p. 3750, © 1992 APS).

depending on whether the denominator is chosen to be $R(AP)$ or $R(P)$. Inserting Tedrow and Meservey's values of $P_{Fe} = 44\%$ and $P_{Co} = 34\%$, he predicted TMR $= 26\%$. He attributed the difference between this value and his observed 14% to a combination of magnetic coupling between the F-films plus some spin-flip scattering. It has since been suggested[31] that his large MR might have been an effect of a zero-bias anomaly, due to magnetic moments in the barrier.

Magnetoresistances of F/I/F sandwiches of several percent at 4.2 K, and up to ~3% at room temperature, were reported in the early 1990s using various insulators.[147] But the larger room-temperature MRs usually involved rather low sample resistances, a circumstance that must elicit caution, as will be explained in Section 4.6.1. A review of much of this work is given in Ref. 31. Major interest in the subject of TMR was stimulated when Moodera et al.[7] reported achieving a large (12%) room-temperature TMR, using Al_2O_3 insulating barriers that gave junction resistances high enough to produce unequivocal tunneling and to ensure uniform current flow. Miyazaki and Tezuka[148] reported an even larger (~18%)

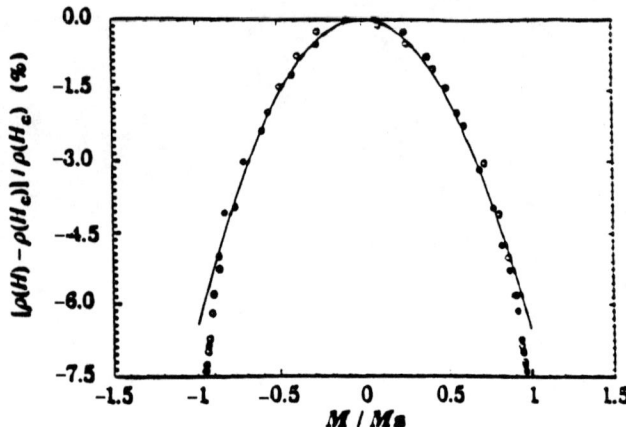

Figure 4.68. MR versus normalized magnetization, M/M_s, for the data of Fig. 4.67. Open and filled circles are for increasing and decreasing fields. The solid curve is $\Delta\rho/\rho = -0.065(M/M_s)^2$ (from Xiao et al.,[60] p. 3750, © 1992 APS).

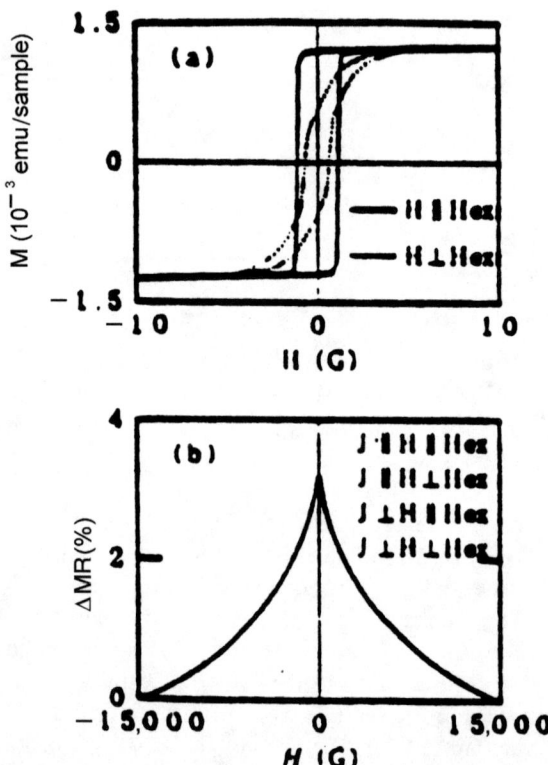

Figure 4.69. (a) $M(H)$ and (b) $MR(H)$ at room temperature of a 100 nm thick $Ag_{52}Fe_{48}$ film prepared in a magnetic field. Here, $MR(H)$ does not closely follow $M(H)$ (from Maeda et al.,[138] p. 3751).

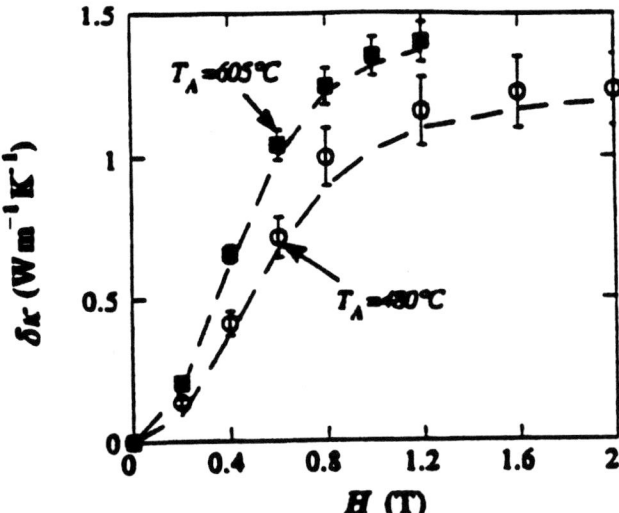

Figure 4.70. Test of the Wiedemann–Franz law by comparing giant electrical and thermal magnetoconductivities at $T = 6$ K for $Ag_{80}Co_{20}$ samples annealed at $T_A = 480$ and $605°C$. The symbols are $\delta\kappa$ (the change in κ between low and high field), and the dashed lines indicate $L_0 T \delta\sigma$ (from Piraux et al.,[140] p. 224, © 1994, Elsevier Science).

room-temperature TMR, but with low junction resistance. A variety of alternative barriers to Al_2O_3 have been examined, but none have yet produced as large TMRs at room temperature.[149] There is great interest in the possibility of using TMR for devices. To compete for magnetic memory, the junction specific resistance has to be reduced to ~ 1–10 k$\Omega(\mu m)^2$ with high TMR.[150]. There has been recent success in approaching this resistance with room temperature MRs ~ 10–15%.[151] Reviews of theoretical and experimental progress in TMR are given in Ref. 152.

4.7.2. TMR Amplitude versus Julliere Model

Figure 4.76[7] compares the TMR at 295 K for a $CoFe/Al_2O_3/Co$ "hybrid-SV" tunneling junction with the AMRs for single Co and CoFe films. The TMR is 10^2 times larger than the AMRs. Figure 4.73 showed that such junctions displayed the I–V characteristics expected for tunneling.

As noted in Section 4.5.3.3, in addition to hybrid SVs, an AP state can also be reliably achieved via EBSVs, and TMR has been observed in [AF/F/I/F] EBSVs.[153] Because TMR involves tunneling across an insulating layer, it is exponentially sensitive to the I-layer thickness, t_I. Sensitivity of the TMR to a particular thin region of the I-layer is potentially a problem. However, Fig. 4.77[153] shows that the total resistance, R, of successful EBSV junctions of Co and Py increases closely linearly with decreasing junction area, while the TMR is (within a factor of 2) relatively insensitive to the junction area.

This study and others[32] indicate that, so far, the maximum observed TMRs often seem to be closely related to the Julliere expression, using for the P_Fs essentially the values derived from tunneling into superconductors (see Table 4.1).

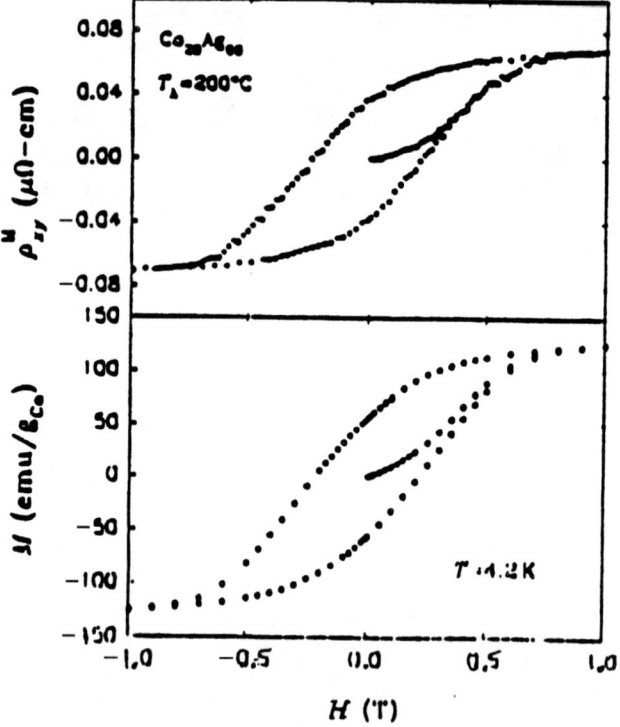

Figure 4.71. Hysteresis loops at $T = 4.2$ K of the extraordinary Hall resistivity, ρ_{xy}^M (inverted in sign), and the magnetization, M, for a Co–Ag sample annealed at 200°C. Note the close correspondence between the two (from Xiong et al.,[141] p. 3221, © 1992 APS).

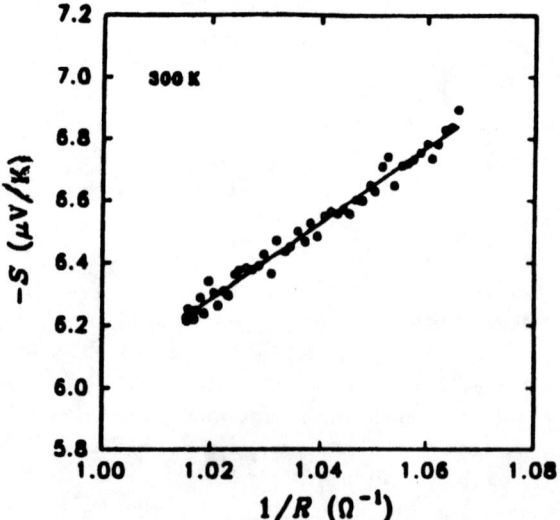

Figure 4.72. $-S$ versus $1/R$ for a $Co_{20}Ag_{80}$ granular film (from Shi et al.,[143] p. 16120, © 1993 APS).

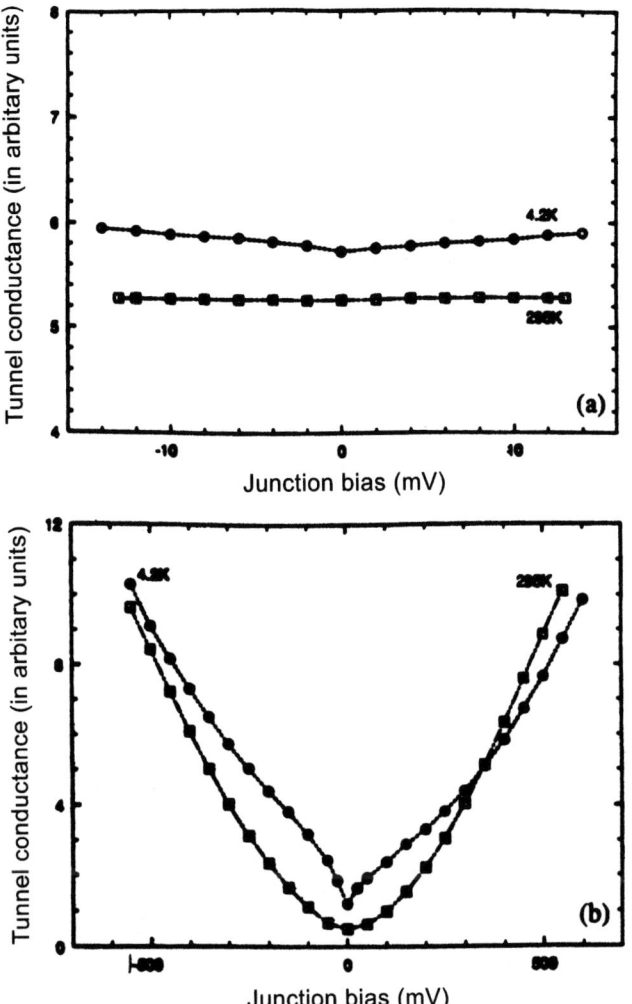

Figure 4.73. Zero field dynamic conductance (dI/dV) versus dc bias at 4.2 and 295 K for a CoFe/Al_2O_3/Py tunnel junction. (a) Very low bias and (b) higher bias (from Moodera et al.,[7] p. 3274, © 1995 APS).

4.7.3. Temperature Dependence

Figures 4.78 and 4.79 show that the tunneling MR for standard F-metals and alloys decreases with increasing temperature, apparently by a larger fraction when the interface is "dirtier".[32] One source of this decrease is temperature-induced magnetic disorder in the F-metal,[33] most likely especially at the surface of the F-metal in contact with the I-layer.[32] Additional sources have also been predicted.[154]

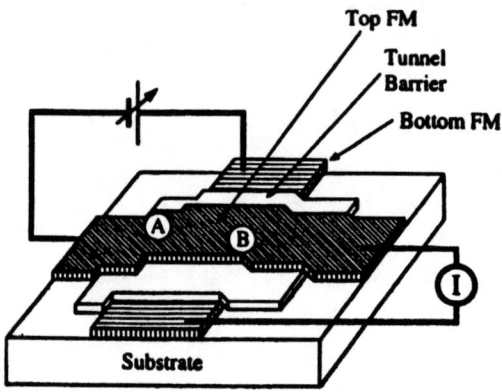

Figure 4.74. Schematic of a crossed-electrode tunnel junction showing the current and voltage connections for measuring the tunneling resistance R_T. If the lead resistance $R_L \ll R_T$, the current density is uniform. But if $R_L \geq R_T$, the current density becomes non-uniform (higher in region A than in region B) (from Moodera et al.,[145] p. 708).

4.7.4. Junction Bias Dependence

Figures 4.78 and 4.79 also illustrate that the observed TMR is usually very sensitive to junction bias, retaining its maximum value only for biases below about 0.1 V.

4.7.5. TMR with "Half-Metallic" F-Metals

As with the CPP-MR, use of a half-metallic F-metal ($P = 100\%$) in a tunnel junction would be expected to give an unusually large TMR—in fact, an EBSV with two half-metallic layers should give TMR $= \infty$. Figure 4.80[155] shows an example of TMR for the putative half-metallic F-metal NiMnSb against Py. The results are much

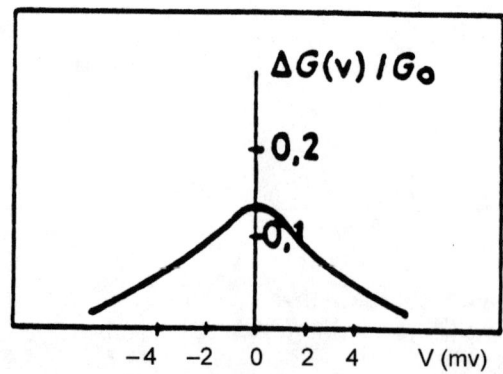

Figure 4.75. $\Delta G/G_0$ at 4.2 K versus bias voltage V (mV) for an Fe/Ge/Co sandwich. ΔG is the difference between the conductances corresponding to parallel and antiparallel magnetizations of the two F-layers. (The field needed for this transition was not specified) (from Julliere[144] p. 225, © 1975 Elsevier Science).

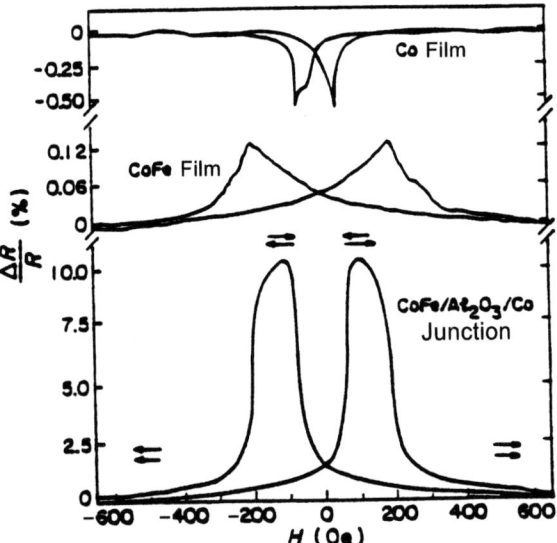

Figure 4.76. TMR at 295 K of a CoFe/Al$_2$O$_3$/Co junction versus H in the film plane. Also shown are the variation of the resistances of CoFe and Co films. The arrows indicate the direction of M in the films (from Moodera et al.,[7] p. 3275, © 1995 APS).

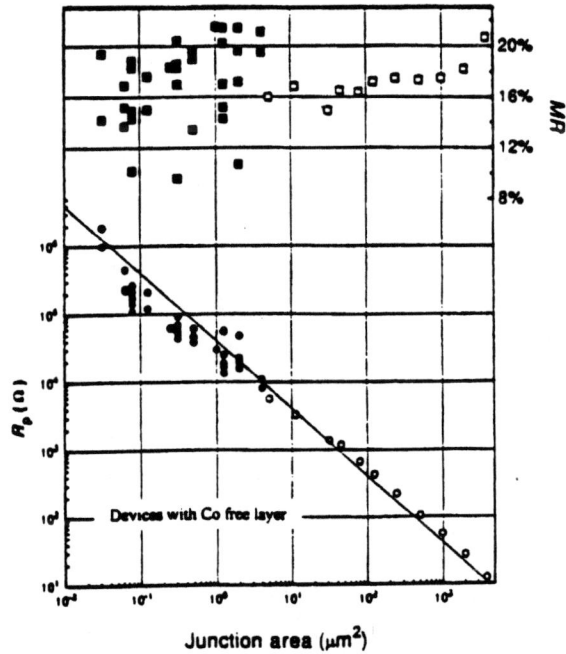

Figure 4.77. Room-temperature junction resistance, R_J (left scale) and TMR (right scale) versus nominal junction area for MnFe/Py/Al$_2$O$_3$/Co EBSVs. The closed circles indicate samples patterned using electron-beam lithography, and the open circles ones using optical lithography (from Gallagher et al.,[153] p. 3744).

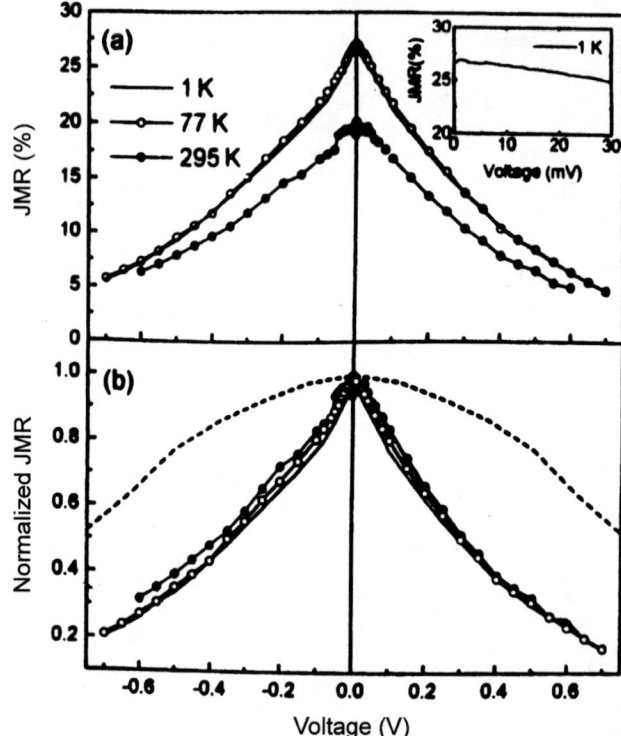

Figure 4.78. TMR versus dc bias at 1, 77, and 295 K for a $Co/Al_2O_3/(Py = Ni_{80}Fe_{20})$ junction. (a) actual percentages, (b) normalized at zero bias. The inset shows that the TMR is almost constant in the very low bias region. The dashed line in (b) is a predicted variation for a $Fe/Al_2O_3/Fe$ junction with an assumed 3 eV barrier height (from Moodera et al.,[32] p. 2942, © 1998 APS).

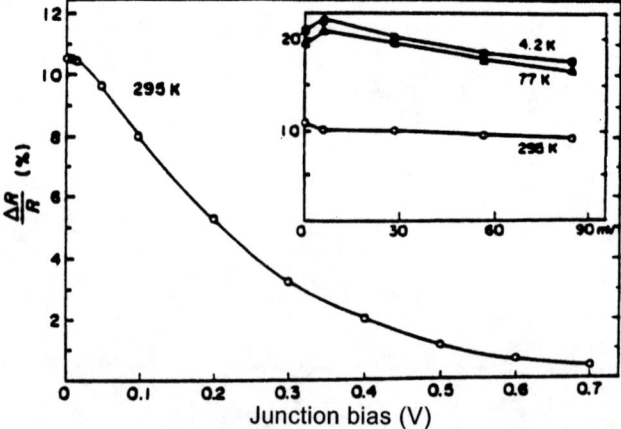

Figure 4.79. TMR versus dc bias of a $CoFe/Al_2O_3/Co$ junction at 295 K. The inset shows the low bias region at three different temperatures. The abscissa in the inset is for 4.2 K, which are twice the values at 295 K (from Moodera et al.,[7] p. 3275, © 1995 APS).

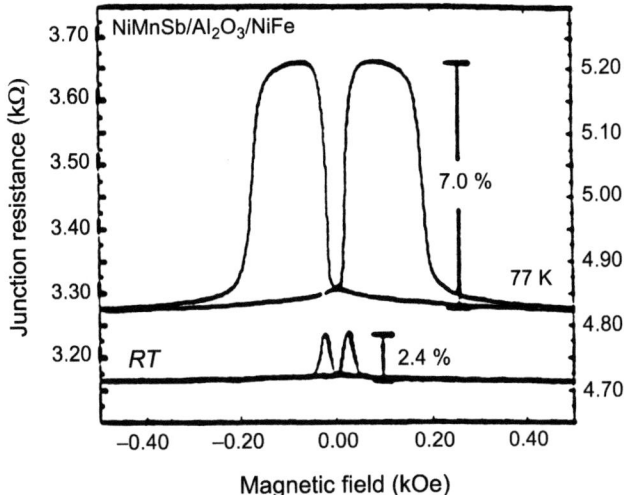

Figure 4.80. TMR versus H of a NiMnSb/Al$_2$O$_3$/Py junction at room temperature (RT) (left scale) and 77 K (right scale). The broad peak at 77 K indicates well separated coercive fields of the F-metals and AP orientation at the peak. The sharp peak at RT indicates closely spaced coercieve forces and incomplete AP ordering at the peak (From Tanaka et al.,[155] p. 5517).

less than expected for true half-metallic behavior. But, as noted in Table 4.1, NiMnSb might not be "half-metallic", and as considered also in Section 4.5.3.10, sample quality may also be a problem. As indicated in Table 4.1, other possibilities for high polarizability are La$_{0.7}$Sr$_{0.3}$MnO$_3$ and related manganites, as well as other oxides such as CrO$_2$. Figures. 4.81[156] and 4.82[41,42] illustrate that large TMRs have been seen

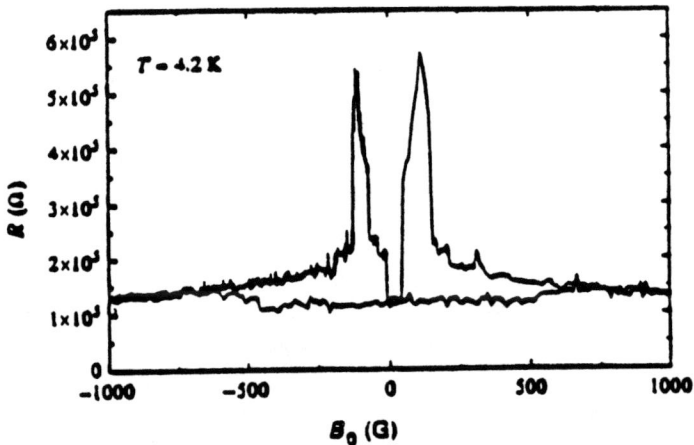

Figure 4.81. $R(B)$ versus B at 4.2 K for a La$_{0.7}$Sr$_{0.3}$MnO$_3$/SrTiO$_3$/La$_{0.7}$Sr$_{0.3}$MnO$_3$ TMR sandwich. The resistance increases by a factor of 5 in the AP state (from Viret et al.,[156] p. 547).

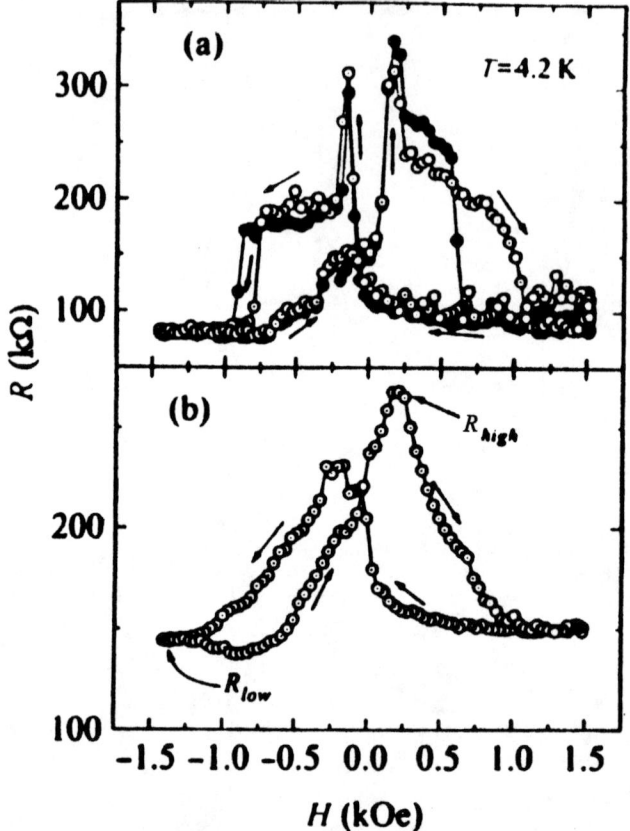

Figure 4.82. $R(H)$ versus H curves for a $La_{0.67}Sr_{0.33}MnO_3/SrTiO_3/La_{0.67}Sr_{0.33}MnO_3$ junction at a measuring frequency of 1.3 Hz. (a) Two single-trace $R(H)$ loops taken within a few seconds of each other. (b) an $R(H)$ loop representing a 25-trace average. Arrows indicate the direction of field sweep. The lower switching field,~100 Oe, correponds to the coercivity of LSMO films measured separately. The upper switching field, up to ~1 kOe, might relate to shape anisotropy (from Sun et al.[42] p. 1770).

with manganites at low temperatures,[41,42,156] but these materials seem to lose their surface magnetization above a temperature that is still well below room temperature.[157]

4.7.6. Enhanced TMR in Low-Resistance Junctions

If the resistance of the "insulating" layer becomes so low that it is comparable to the resistances of the contacts to the F-metal, the current flow in the junction will be non-uniform—i.e., more current will flow through the portion of the "insulating" layer nearest to where it is injected. Figure 4.83[145] shows that the apparent TMR can then become extremely large. In fact, Fig. 4.83(c) shows that R can even become negative. An ability to reproducibly achieve such behavior might be useful for technology.[145] But a non-uniform current limits utility for deriving fundamental physics.

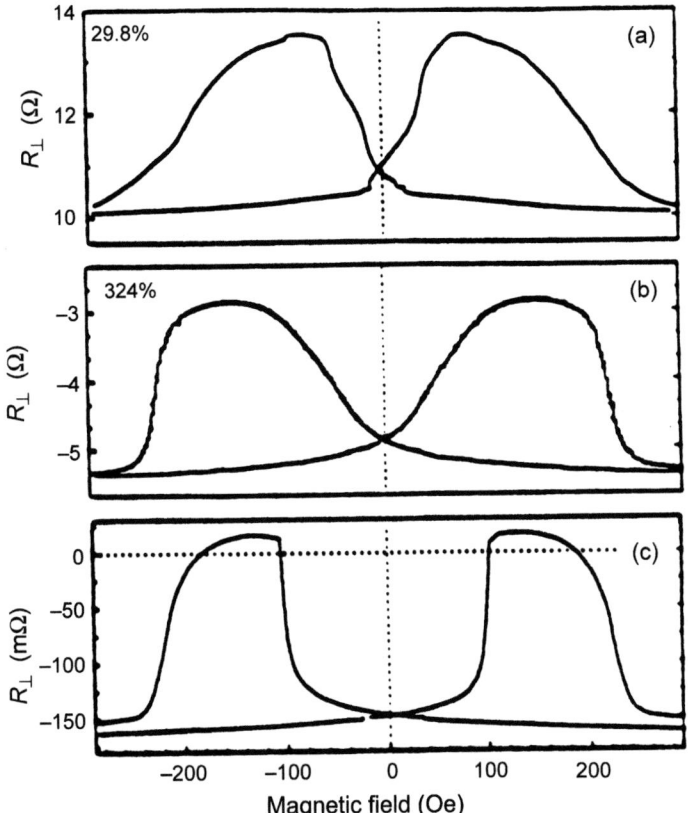

Figure 4.83. Magnetoresistances of low resistance CoFe/Al$_2$O$_3$/Py junctions at room temperature, showing apparent TMRs. In (c) the barrier is AlN, and the signal changes sign (from Moodera et al.,[145] p. 709).

4.7.7. TMR with Granular Samples

A number of investigators have seen TMR behavior in granular samples involving F-metal clusters embedded in an insulating matrix. If the insulating barriers are thin enough, and the clusters numerous enough, a combination of tunneling plus percolation has been found to produce large TMRs at low temperatures.[157,158] Large low temperature TMRs are also found in CrO$_2$ powder compacts.[159]

4.7.8. TMR with Coulomb Blockade

Finally, TMR has been investigated in the regime of Coulomb blockade ($k_BT \ll e^2/2C_I$, where C_I is the junction capacitance), where transfer of one electron across the junction stops other electrons from following until the first is taken away. Three groups, using tunnel junctions,[160–162] reported a significantly enhanced MR in this regime (see, e.g., Fig. 4.84). In contrast, one,[163] studying

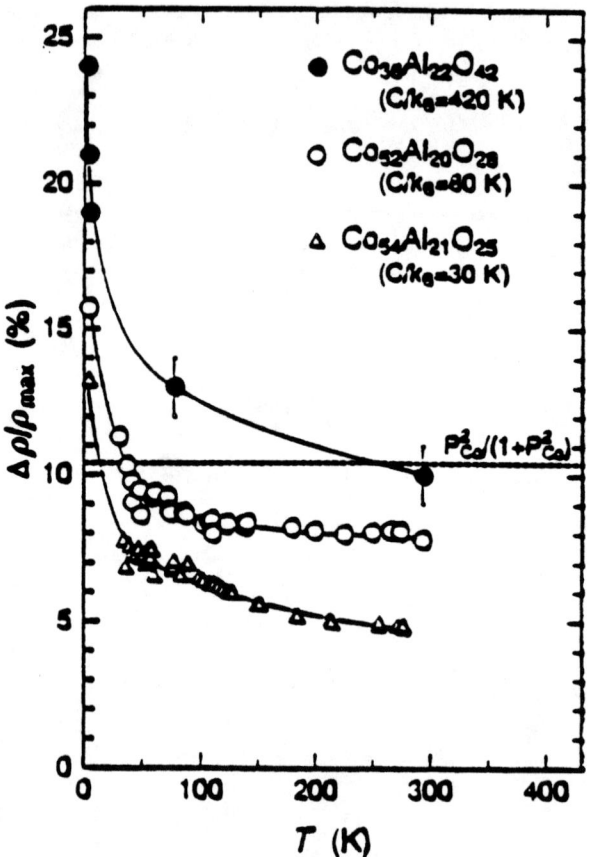

Figure 4.84. TMR versus tenoeratyre T for Co–Al–O insulating granular films. The horizontal line indicates the Julliere prediction for TMR between Co metal particles. Estimated values of C/k_B are listed (from Mitani et al.[161] p. 6525).

a granular tunneling system reported little or no change. Localized tunneling to a single Co cluster has been probed using a point contact.[164]

4.8. COLOSSAL MR

The discovery of GMR, and its applications to sensors, has stimulated new interest in alloys and compounds that undergo magnetic phase transitions that can be driven by magnetic fields. Major experimental and theoretical effort is now being expended on manganese-oxide-based (manganite) perovskites, where a large enough magnetic field can drive a transition from a paramagnetic insulator (PI) to a ferromagnetic metal (FM). The consequent decrease in resistance can be so large that the MR in such materials has been given the sobriquet "Colossal MR" (CMR).[8] CMR has been recently reviewed [see Ref. 9]. So far, the intrinsic CMR requires fields of several Tesla and is usually very temperature sensitive. Fairly large low-field MRs have been observed for intergrain transport, but not yet at room temperature. As described in Section 4.6.5, nearly "half-metallic" CMR compounds (see Table 4.1) can produce large TMRs at low temperatures.

Field-induced changes from antiferromagnetic to ferromagnetic metallic states have also been found to give MRs ranging from a few percent to 85%[165–167] at large fields, low temperatures, or both. Other families of compounds are also being explored, such as pyrochlores[9,168] and spinels.[9,169]

4.8.1. Discovery and Background

The existence of a PI to FM transition in certain perovskites,[170] associated with a large decrease in resistance,[171] and even a significant MR,[172] has been

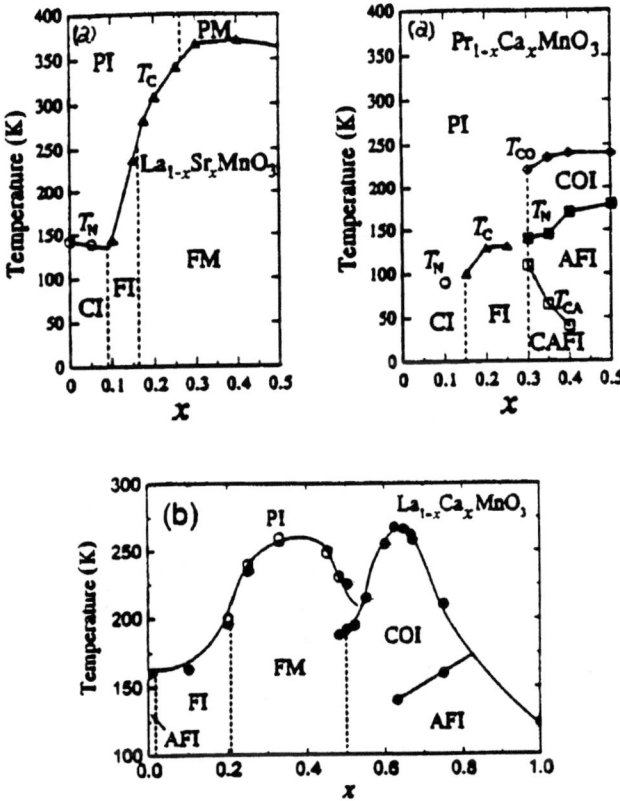

Figure 4.85. (a) Magnetic and electronic phase diagrams of $La_{1-x}Sr_xMnO_3$ and $Pr_{1-x}Ca_xMnO_3$. The states are paramagnetic insulator (PI), paramagnetic metal (PM), canted insulator (CI), ferromagnetic insulator(FI), ferromagnetic metal (FM), canted antiferromagnetic insulator (CFI), and charge ordered insulator (COI). T_C, T_N, and T_{CO} are Curie, Neel, and charge-ordering temperatures respectively (b) Phase diagram for $La_{1-x}Ca_xMnO_3$. Phase labeling is as in (a) (from Ramirex,[9] after Tomioka et al.,[173] (a) and Schiffer et al.,[174] (b)).

known for five decades. The prototype CMR compound is derived from the perovskite $LaMnO_3$ by "hole doping", substituting divalent Ca or Sr for trivalent La. In $La_{1-x}Ca_xMnO_3$, the PI to FM transition occurs for $x \approx$ 0.2–0.4. Figure 4.85 shows phase diagrams for representative $A_{1-x}B_xMnO_3$ CMR oxides.[173,174] The PI to FM transition underlying CMR behavior usually occurs over fairly narrow ranges of x. Examples of how $\rho_x(T)$ varies with x are shown in Fig. 4.86 for $La_{1-x}Sr_xMnO_3$.[175]

The source of CMR is that application of a large enough H leads to magnetic ordering that favors the FM state, shifting the PI to FM transition to higher temperature. The large shift in R that can be produced, and the sensitivity of the MR to temperature, are illustrated in Fig. 4.87, which compares for $La_{0.75}Ca_{0.25}MnO_3$ the temperature variations of the magnetization M, resistivity ρ, and MR [defined using the bounded Eq. (4.1.1a) to have a maximum value of 100%]. The first modern report of a large MR in a perovskite was for polycrystalline $Nd_{0.5}Pb_{0.5}MnO_3$ (Fig. 4.88), where a 20 T CMR of over 10,000% [using the unbounded definition of Eq. (4.1.1b)] was found at $\sim$ 180 K.[176] A drop in resistance of 60% at room

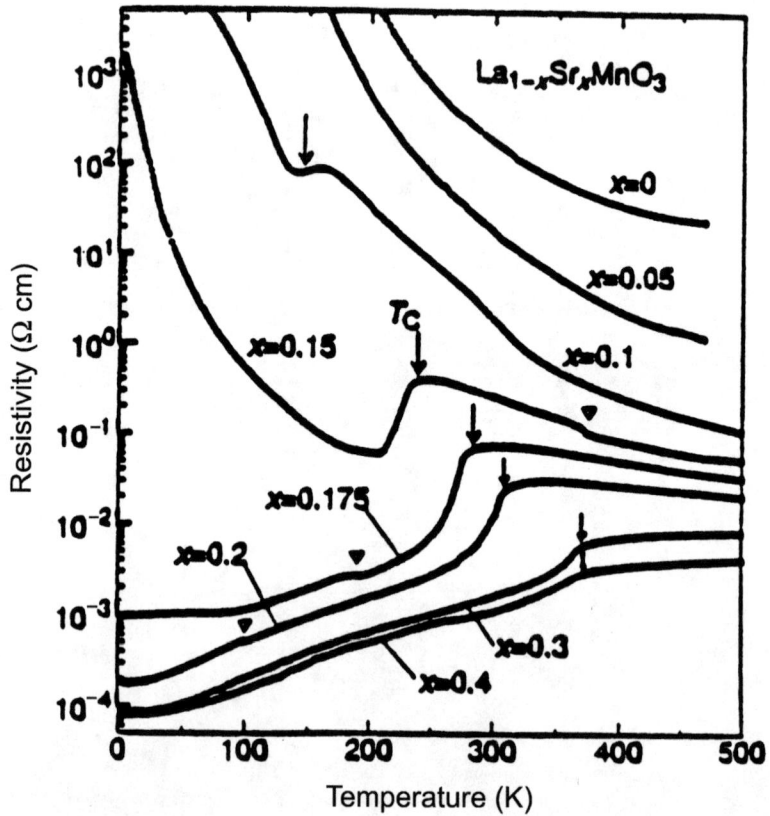

Figure 4.86. Resistivity versus T for $La_{1-x}Sr_xMnO_3$ for various values of x. The arrows denote the transition as determined from magnetization measurements (From Urushibara et al.,[175] p. 14105, © 1995 APS).

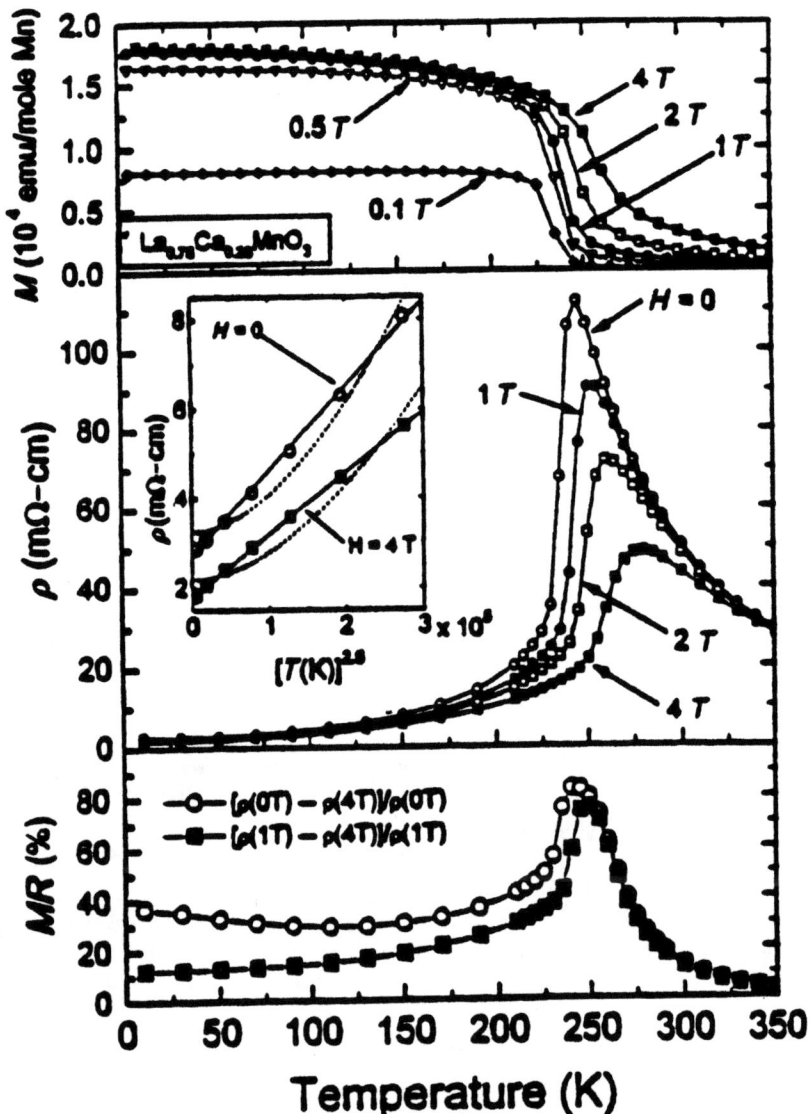

Figure 4.87. Top frame: M versus T for $La_{0.75}Ca_{0.25}MnO_3$ for various values of H. Middle frame: resistivity versus T. The inset shows the low-temperature resistivity compared to $T^{2.5}$ (solid line) and $T^{4.5}$ (dashed line). Bottom frame: MR versus T. Open symbols reflect low-field behavior and solid symbols high-field behavior (from Schiffer et al. [174] p. 3337, © 1995 APS).

temperature was subsequently found in $La_{2/3}Ba_{1/3}MnO_3$ for a field of 7 T.[177] The observation of an MR in thin films of $La_{2/3}Ca_{1/3}MnO_3$ of over 100,000% at 77 K led to coining of the name CMR.[8] An order of magnitude larger MRs were soon seen in sintered bulk $La_{0.6}Y_{0.07}Ca_{0.33}$ samples,[178] and even larger in $Nd_{0.7}Sr_{0.3}MnO_{3-\delta}$.[179]

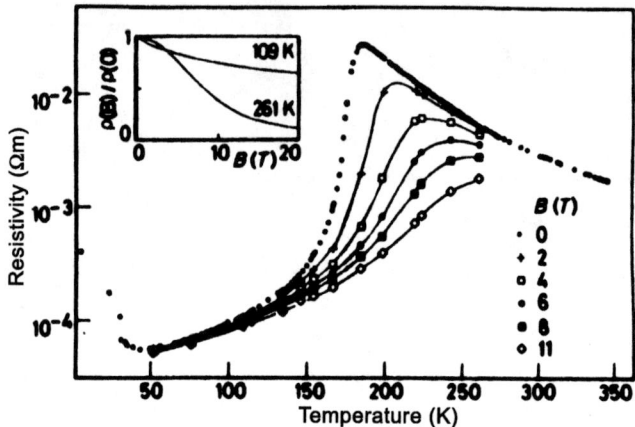

Figure 4.88. Resistivity versus temperature and magnetic field B for a [110] bar of $Nd_{0.5}Pb_{0.5}MnO_3$. The data are shown as points; the curves are simply guides to the eye. The inset illustrates the behavior of the MR at temperatures (261 K) above and below (109 K) the magnetic ordering temperature (from Kusters et al,[176] p. 363, © 1989, Elsevier Science).

Major theoretical and experimental effort is now being expended to understand the microscopic physics underlying CMR and to produce materials and geometries with device potential.

The theory of CMR is still under development.[9] In the 1950s, a double-exchange model was proposed to explain how CMR materials could be so highly resistive in the PI state yet metallic in the FM state. More recently, it has been argued that double exchange alone is not enough,[180] but that the behavior of the perovskites requires a combination of double exchange plus strong electron phonon coupling[181,182]—"...for $T > T_c$, the strong electron–phonon coupling localizes the conduction band electrons as polarons, but the polaron effect is "turned off" as T decreases through T_c, permitting the formation of a metallic state."[181] Electron–electron interactions have also been proposed as important.[183,184] Several different phenomena have been taken as evidence for the presence of polarons in the PI state.[9] Whether such polarons are intrinsic to the compounds, or arise from defects, remains a topic of debate. It has been noted that this model based on double-exchange "does not carry over easily"[9] to two other materials that exhibit CMR behavior, the pychlores and spinels (described earlier).

4.8.2. Modifying CMR Behavior

Under investigation are techniques to increase the CMR, reduce its temperature sensitivity, reduce the fields needed to produce it, or to provide them in unusual ways.[185] In $La_{0.67}Ba_{0.33}MnO_z$, decreasing oxygen content was found to shift the magnetic transition to lower temperature, increasing the CMR (Fig. 4.89).[186] Studies that have led to lower fields and/or broadened transitions, but also to not such large MRs, include: grain boundary tunneling in ceramic samples,[174,187–189] interplanar tunneling in layered manganite structures;[190] control

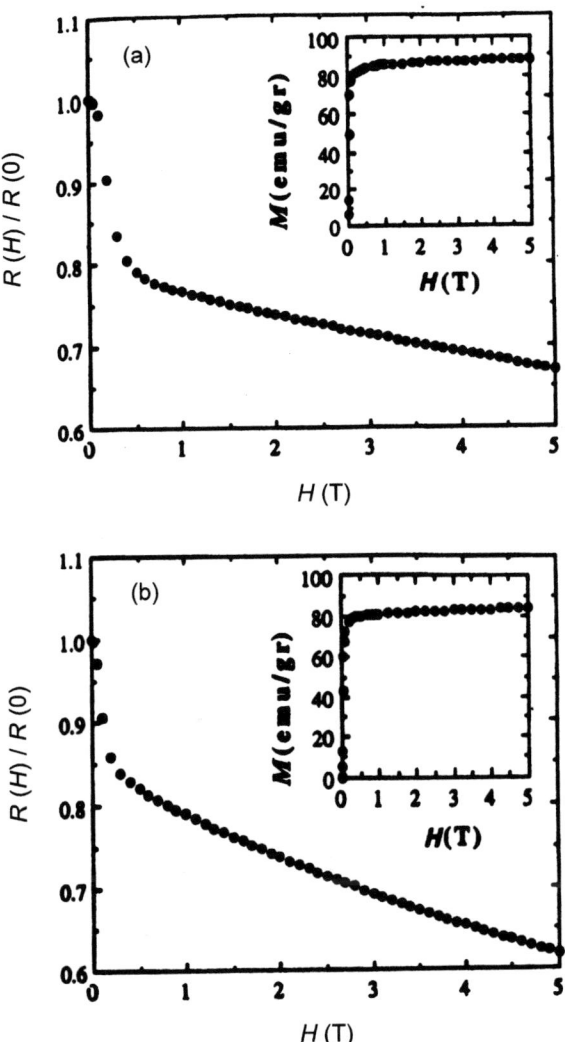

Figure 4.89. Normalized resistance versus H at 5 K for (a) a $La_{0.67}Ba_{0.33}MnO_{0.299}$ sample, and (b) a $La_{0.67}Ba_{0.33}Mn_{0.290}$ sample. Insets show the magnetizations versus H (from Ju et al.,[186] p. 6145, © 1995 APS).

by doping of one electron bandwidth;[191] and $La_{0.7}C_{0.3}MnO_3/SrTiO_3$ superlattice structures with layer thicknesses of a few nanometers.[192] Figure 4.90[187] illustrates low-field effects of grain boundary tunneling by comparing the ratio $\rho(H)/\rho_0$ versus H from 5 to 280 K for a single-crystal sample and two polycrystalline samples annealed at different temperatures. Replacing Mn atoms with another magnetic atom, such as Fe or Cr, seems to move the peak in total resistance to lower temperature,[193,194] and generally the maximum MR too (Fig. 4.91).

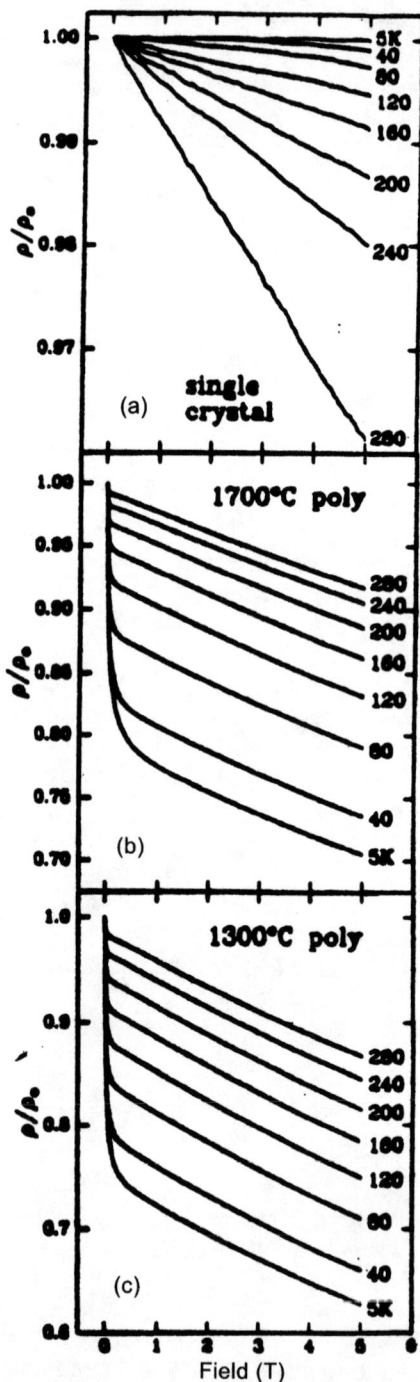

Figure 4.90. Normalized resistivity versus field H at various temperatures from 5 to 280 K for samples of $La_{2/3}Sr_{1/3}MnO_3$. (a) single-crystal sample; (b) polycrystalline sample annealed at 1700°C; (c) polycrystalline sample annealed at 1300°C. The larger low-field MRs for the polycrystalline samples are ascribed to intergrain tunneling (from Hwang et al.,[187] p. 2042, © 1996 APS).

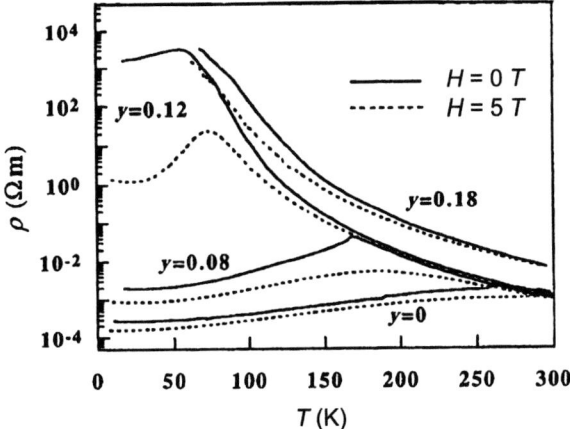

Figure 4.91. Resistivity versus T at $H=0$ (solid curves) and $H=5$ T (dashed curves) for $La_{0.63}Ca_{0.37}Mn_{1-y}Fe_yO_3$ with $y = 0$, 0.08, 0.12, and 0.18 (from Ahn et al.,[193] p. 15300, © 1996 APS).

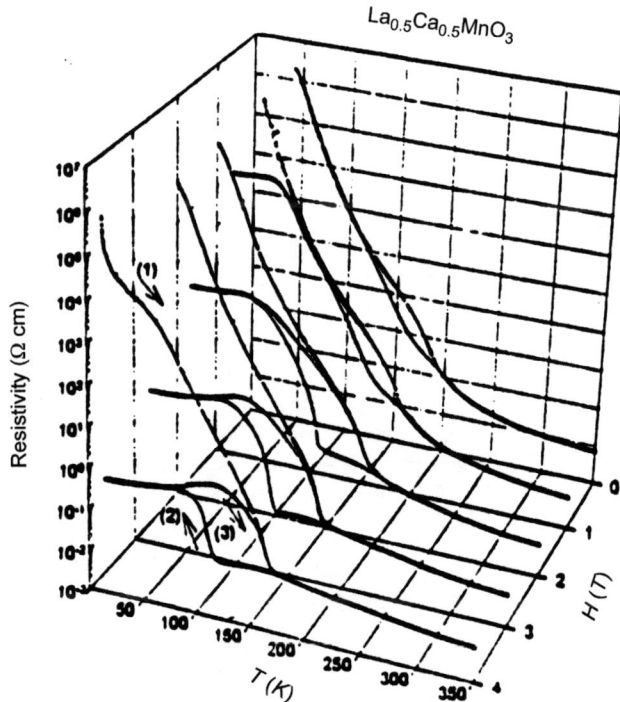

Figure 4.92. Resistivity versus T at various fields for a $La_{0.5}Ca_{0.5}MnO_3$ sample. For each measurement, the sample was first cooled to 15 K in zero field. Then, a field was applied and the resistivity measured as (1) T was increased to 350 K, (2) decreased to 15 K, and (3) again increased to 350 K (from Xiao et al.,[195] p. 5328).

Also, under study is hysteresis associated with CMR, examples of which are shown in Fig. 4.92.[195]

4.8.3. Noise

Noise related to magnetic domain fluctuations was examined in $La_{0.6}Y_{0.07}Ca_{0.33}MnO_3$ thin films around the FM to PI transition.[196] The noise was found to be closely $1/f$, much larger than typical for metal films, and consistent with magnetization fluctuations in the FM-phase. At low frequencies (~1–25 Hz), the noise was large enough to probably preclude competitive use of such CMR compounds, but by 10 MHz, it was compatible with device needs.

4.8.4. Thermal Conductivity, Hall Effect, and Thermopower

Measurements of κ, R_H, and S in both the PI and FM states have been made to help elucidate the fundamental physics underlying CMR. But because the insulating

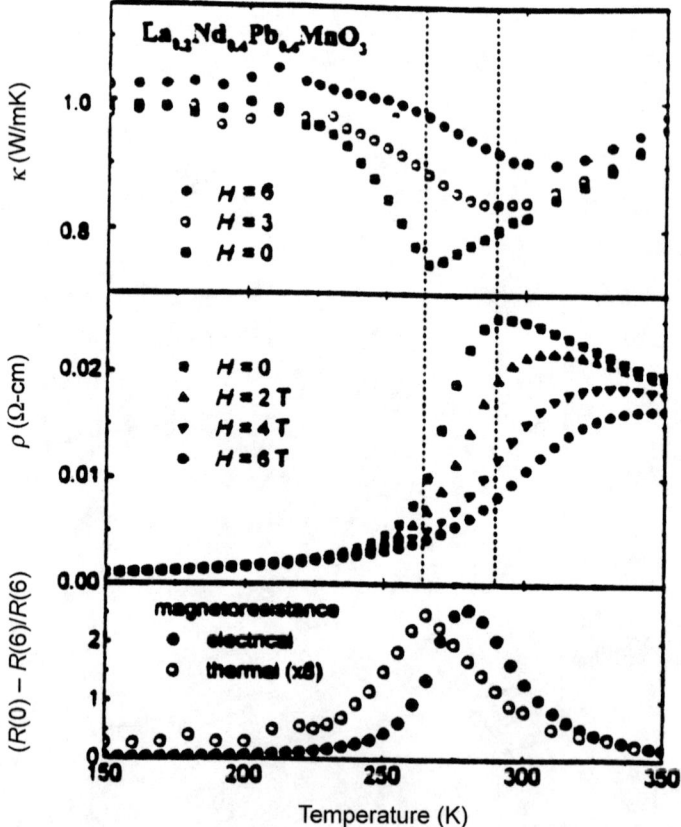

Figure 4.93. Transport Properties of $La_{0.2}Nb_{0.4}Pb_{0.4}MnO_3$ versus temperature at different applied magnetic fields. Top frame: Thermal conductivity κ at three fields. Middle frame: Resistivity, ρ, at three fields. The vertical dashed lines span a temperature range where $\kappa(T)$ and $\rho(T)$ display behaviors opposite to that expected from the Wiedemann–Franz law. Bottom frame: Electrical and thermal MRs at $H = 6\,T$, normalized to the high field value (from Visser et al.,[197] p. 3949, © 1997 APS).

state behavior is harder to understand, most studies have focused on it. The data shown in Figs. 4.93–4.96[197–199] have been taken as support for a polaronic model of transport in the PI state. Exponential variation of R with T—$R \propto \exp(E/k_b T)$—in the PI state shows activated behavior. S varying as Δ_s/T is also consistent with semiconducting behavior. The simplest polaron models predict $\Delta_s < < E$, which is as observed.

4.9. MISCELLANEOUS PHENOMENA

We conclude with some other interesting magnetotransport phenomena now being studied.

4.9.1. Giant Magneto-Impedance

In 1992, Mohri et al.[200] discovered a giant magneto-impedance effect at small fields in thin amorphous magnetic wires and ribbons. Figure 4.97 shows examples of fractional changes over 100% in the amplitude E_w of the ac voltage measured across FeCoSiB wires. For some other studies see Ref. 201.

4.9.2. Domain Wall Scattering

Domain walls have been reported to increase the resistance of thin Co and Ni films,[202] but to decrease the resistivities of Ni wires with thickness ~ width ~ 20 nm[203] and Fe wires with thickness ~ 100 nm and widths ~ 0.65–20 nm.[204] Published calculations of both Boltzmann[205] and quantum transport[206] do not yet seem to be able to explain all of the observed behaviors.

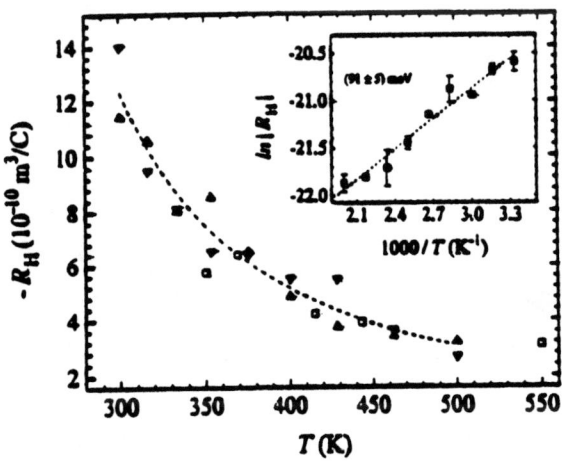

Figure 4.94. Magnitude of the Hall coefficient, R_H, versus temperature, for $(La_{1-x}Gd_x)_{0.67}Ca_{0.33}MnO_3$ films. Open squares are for $x = 0$ and have large error bars. Other samples are for $x = 0.25$ when the field was ramped up (solid triangles) and down (solid nablas). The dashed line is an Arrhenius fit. Inset: The natural logarithmn of the Hall coefficient (average of up and down field values) for $x = 0.25$ versus $1000/T$. Here, the dashed line is a linear fit giving an activation energy of 91 meV (from Jaime et al.,[198] p. 953, © 1997, APS).

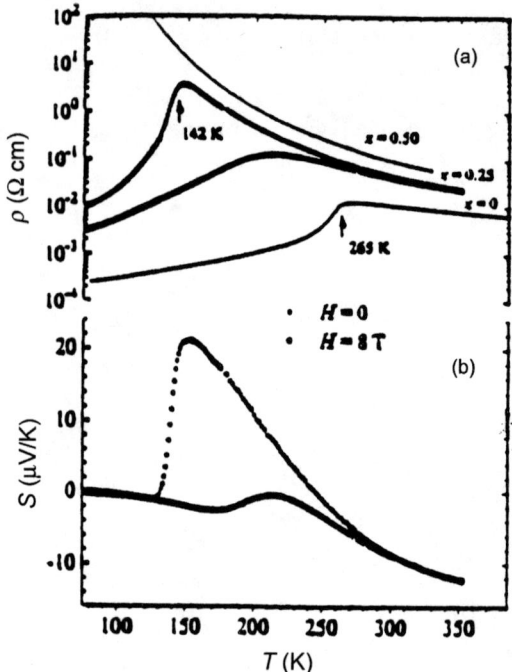

Figure 4.95. (a) Resistivity ρ versus T, and (b) thermopower S versus T, for a $(La_{1-x}Gd_x)_{0.67}Ca_{0.33}MnO_3$ film with $x = 0.25$ at fields $H = 0$ (filled symbols) and $H = 8\ T$ (open symbols). In (a), ρ data for $x = 0$ and $x = 0.5$ are shown for comparison. In (b) note that $S(H)$ saturates for $T < T_c = 142$ K (from Jaime et al.,[198] p. 952, © 1997 APS).

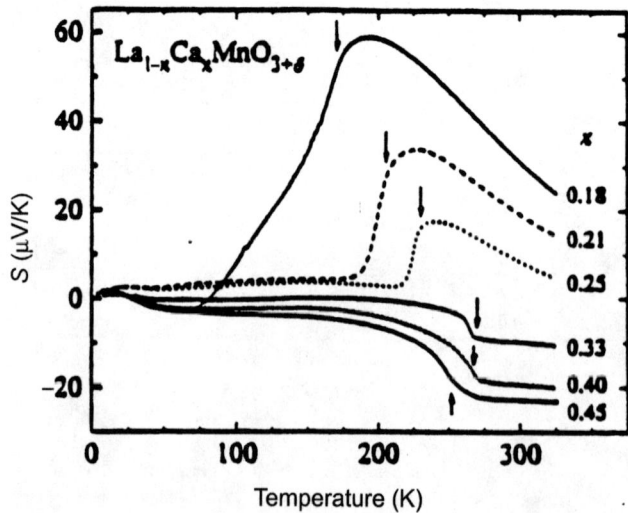

Figure 4.96. S versus T for $La_{1-x}Ca_xMnO_3$ with varying Ca^{2+} concentration. Arrows indicate magnetic ordering temperatures (from Hundley and Neumeier,[199] p. 11512, © 1997 APS).

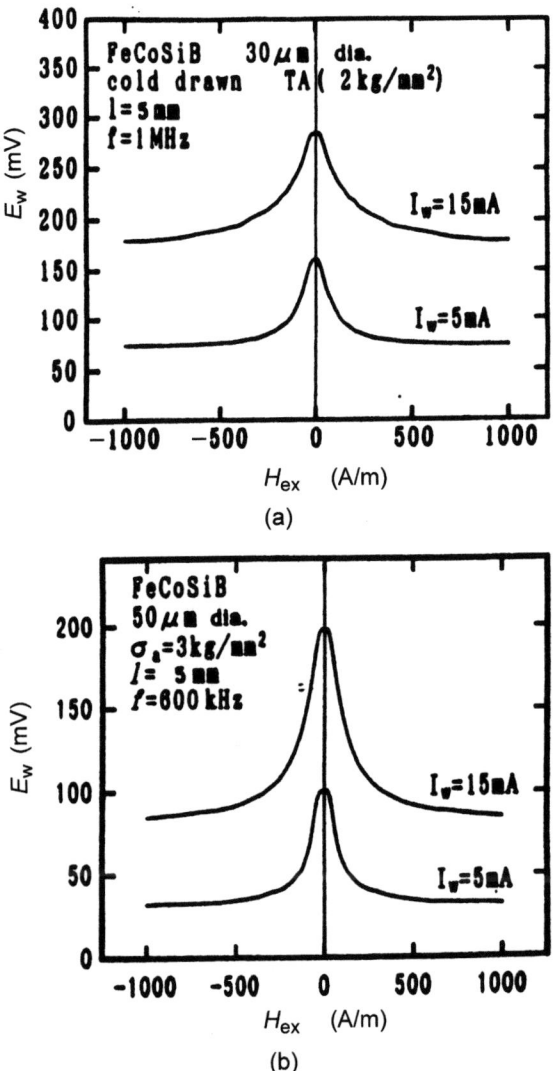

Figure 4.97. Fractional changes in the amplitude E_w of the measured ac voltage across very thin FeCoSiB wires for two different current amplitudes, I_w. The measuring frequencies and some sample characteristics are indicated. 1000 A/m = 4π Oe (from Mohri et al.,[200] p. 2457, © 1995 IEEE).

4.9.3. Spin-Valve Transistor

Monsma et al.[207,208] have explored spin-dependent scattering above the Fermi level, using a solid-state spin-valve transistor, in which electrons are injected from one Si(100) semiconductor (emitter), through a Co/Cu/Co/Pt magnetic multilayer, into a second Si(100) semiconductor (collector). The most recent study showed significant MR at room temperature. (Figure 4.98 shows schematically the device

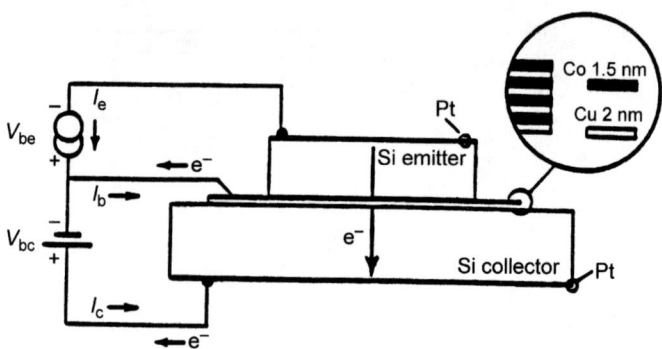

Figure 4.98. Cross-section of a spin-valve transistor. The emitter is directly bonded onto the multilayer in the base region. Bias voltages and transport currents are schematically indicated (from Monsma et al.,[207] p. 5261, © 1995 APS).

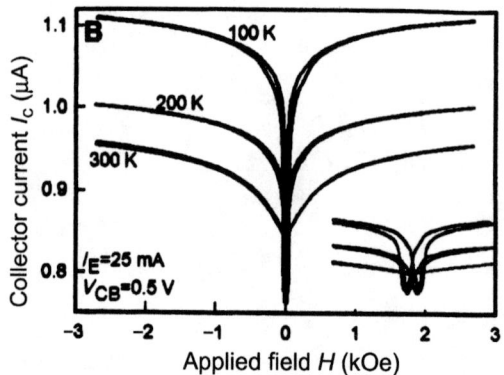

Figure 4.99. Collector current versus applied field in the spin-valve transistor versus applied magnetic field H for temperatures $T = 100$, 200, and 300 K. The relative changes in collector current are 42, 32, and 15%, respectively. The inset accentuates small-field behavior; the field range is from -399 to 300 Oe (from Monsma et al.,[208] p. 408, © 1998 AAAS).

and an energy band diagram, and Fig. 4.99 examples of data.) Kinno et al.[209] reported qualitatively similar data for an Fe/Au/Fe multilayer measured at 77 K, as well as an extension to localized injection from a tunneling tip at temperatures of 250–260 K.

4.9.4. MR Steps and Additional Structure

Doudin et al.[210] recently reported finding MR steps of 4–50% (Fig. 4.100) in ultra-small ($A < 0.01$ μm^2) electrodeposited Ni/NiO/Co junctions. They attributed these steps mostly to two-level fluctuations involving spin-dependent trapping and untrapping of electrons in the oxide barrier, evidence for which included telegraph noise.

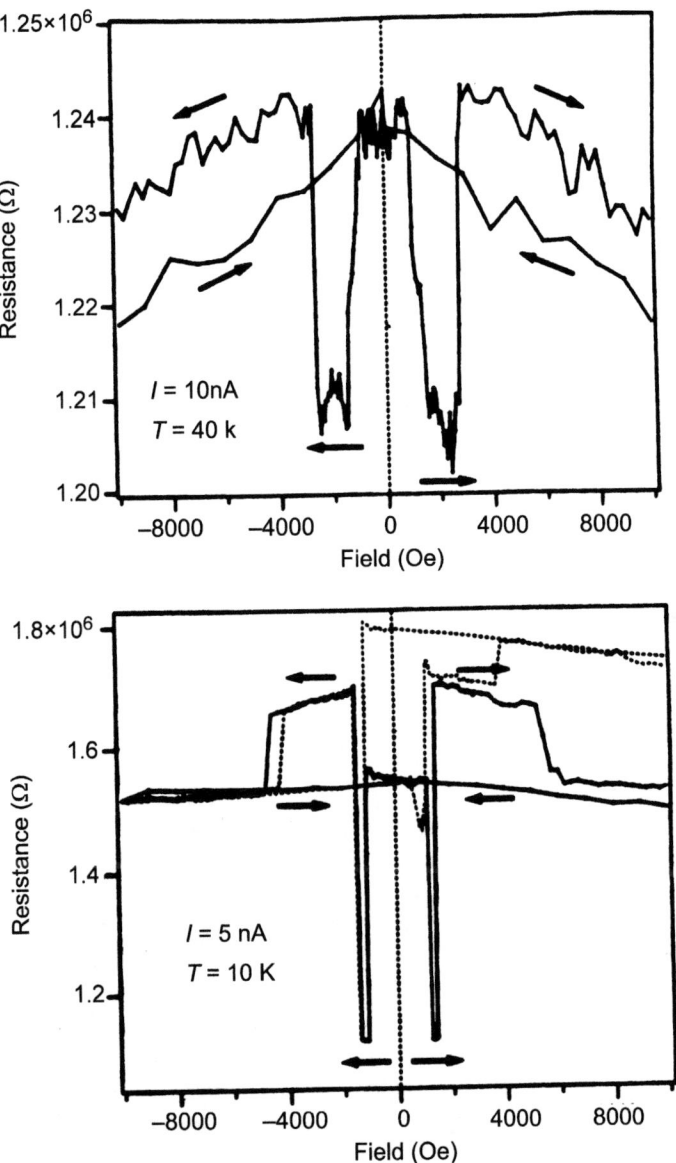

Figure 4.100. Magnetoresistance curves for Ni/NiO/Co junctions with area <0.01 µm^2, showing positive MR and the influence of magnetic history. Dotted lines show the behavior when the sample was saturated at ± 20 kOe, solid lines at ± 70 kOe, and arrows indicate the direction of field change (from Doudin et al.,[210] p. 934, © 1997 APS).

Rogers et al.[211] found that large bias currents produced additional structure in the CIP-MR of Py/Ag multilayers (Fig. 4.101), and that the peaks in the CIP-MR correlated with regions of high $1/f$ noise (Fig. 4.102).

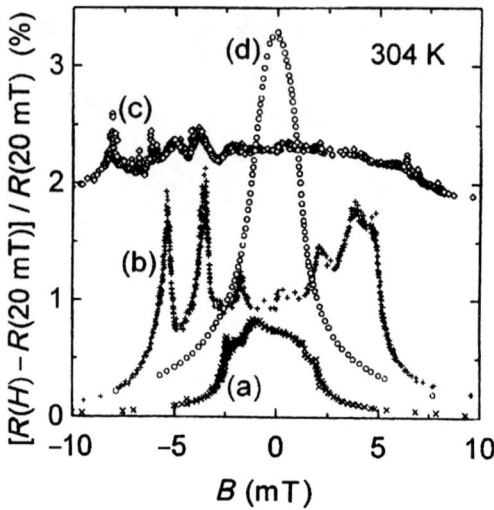

Figure 4.101. MR versus magnetic field B for three different Py/Ag multilayer devices at 304 K, showing a correlation of the number of MR structures with the number of Py layers. All devices are nominally 8-μm wide. (a) A 5-layer, 32-μm long, unannealed resistor at $I = 40$ mA (1.5×10^7 A/cm^2); $R(20$ mT$) \approx 41.1$ Ω. (b) A 7-layer, 160 μm long, annealed resistor at 65 mA (1.8×10^7 A/cm^2); $R(20$ mT$) \approx 64.3$ Ω. (c) A 9-layer, 32 μm long, annealed resistor at 50 mA (0.8×10^7 A/cm^2); $R(20$ mT$) \approx 20.7$ Ω. (d) The 7-layer annealed resistor of (b) at zero bias current, R (20 mT)≈ 59.2 Ω. The increase in R under bias is due largely to Joule heating. Curve (c) is offset by 1.8% for clarity (from Rogers et al.,[211] p. R8504, © 1997 APS).

Ono et al.[212] observed GMR steps in a 0.5-μm wide, 20-μm long, trilayer wire of NiFe(20 nm)/Cu(10 nm)/NiFe(5 nm), which contained a 0.35-μm wide "neck" about 1/3 of the way from one end. They attributed the steps to reversal of the

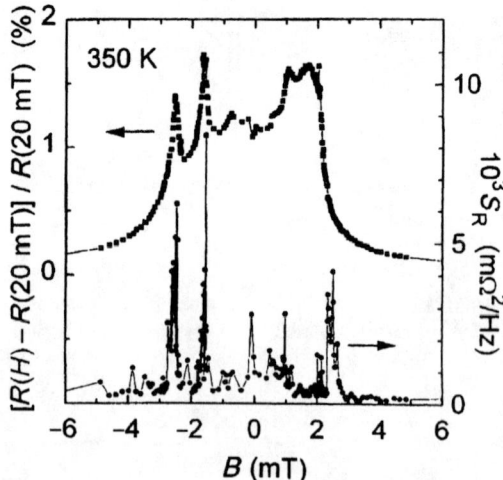

Figure 4.102. MR (left scale) and resistance noise power spectral density, S_R (right scale), at 1 Hz for a seven-layer, 16-μm wide, 160-μm long device at 0.8×10^7 A/cm^2 and 350 K; R (20 mT) ≈ 30.9 Ω. Note the correlation between high noise and peaks in MR (from Rogers et al.,[211] p. R8504, © 1997 APS).

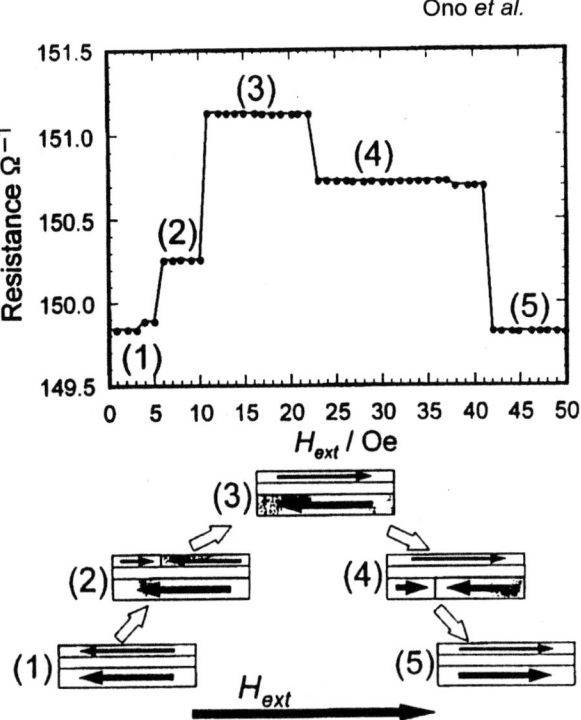

Figure 4.103. Resistance versus external field H_{ext} at 300 K for a 0.5-μm wide, 20-μm long, Py(20 nm)/Cu(10 nm)/Py(5 nm) trilayer wire with a 0.35-μm thick neck about one-third of the way from one end. The schematic drawings show the magnetic domain structures inferred from the resistance steps (from Ono et al., [212] p. 1117).

magnetizations of individual magnetic domains in each of the two NiFe layers, with the domains constrained to one side or the other of the neck as shown schematically at the bottom of Fig. 4.103.

ACKNOWLEDGMENTS

This work was supported in part by the MSU-CFMR, and by the MSU MRSEC Center for Sensor Materials supported by NSF grants DMR-94-00417 and DMR-98-09688. The author thanks colleagues P.A. Schroeder, W.P. Pratt Jr., and N. Birge for helpful suggestions about this manuscript, and the first two plus his students, postdocs, and other colleagues, for their collaboration in the research on GMR in which he has been involved. He also thanks P.M. Levy and S. Zhang for a preprint of their review of TMR.[152]

REFERENCES

1. I. A. Campbell and A. Fert, Ferromagnetic materials, in: *Transport Properties of Ferromagnets*, edited by E. P. Wohlfarth, vol 3 (North Holland, Amsterdam, 1982), p. 748.

2. G. Binash et al., *Phys. Rev. B* **39**, 4828 (1989).
3. M. N. Baibich et al., *Phys. Rev. Lett.* **61**, 2472 (1988).
4. P. M. Levy, *Solid State Phys. Ser.* **47**, 367 (1994).
5. M. A. M. Gijs and G. E. W. Bauer, *Adv. Phys.* **46**, 285 (1997).
6. J. Bass and W. P. Pratt Jr., *J. Magn. Magn. Mater.* **200**, 274 (1999); J. Bass et al., *Commun. Condens. Matter. Phys.* **18**, 223 (1998); A. Fert and L. Piraux, *J. Magn. Magn. Mater.* **200**, 338 (1999).
7. J. S. Moodera et al., *Phys. Rev. Lett.* **74**, 3273 (1995), © 1989 by the American Physical Society.
8. S. Jin et al., *Science* **264**, 413 (1994).
9. A. P. Ramirez, *J. Phys. Condens. Mater.* **9**, 8171 (1997); Y. Tokura and Y. Tomioka, *J. Magn. Magn. Mater.* **200**, 1 (1999); A. Gupta and J. Z. Sun, *J. Magn. Magn. Mater.* **200**, 24 (1999).
10. J. A. Mydosh, *Spin Glasses: An Experimental Introduction* (Taylor and Francis, London, 1993).
11. See, e.g., C. Fierz et al., *J. Phys. Condens. Mater.* **2**, 9701 (1990); Proc. LT-19, *Physica B* **165–166**, 453 (1990); S. Oh et al., *Appl. Phys. Lett.* **71**, 2376 (1997).
12. N. W. Ashcroft and N. D. Mermin, *Solid State Physics* (W. B. Saunders Co., Philadelphia, 1976).
13. W. P. Pratt Jr. et al., *Phys. Rev. Lett.* **66**, 3060 (1991), © 1989 by the American Physical Society.
14. P. A. Schroeder et al., *J. Appl. Phys.* **76**, 6610 (1994).
15. E. Fawcett, *Adv. Phys.* **13**, 139 (1964); C. Kittel, *Quantum Theory of Solids* (John Wiley and Sons, New York, 1963).
16. A. A. Abrikosov, *Fundamentals of the Theory of Metals* (North Holland, Amsterdam, 1988).
17. J. L. Olsen, *Electronic Transport in Metals* (John Wiley and Sons, New York, 1962).
18. J. A. Brug et al., *J. Appl. Phys.* **79**, 449 (1996).
19. See, e.g., M. H. Kryder et al., *J. Appl. Phys.* **79**, 4485 (1996), and the web pages of IBM Almaden, Seagate Technology, etc.
20. J. D. Jackson, *Classical Electrodynamics* (John Wiley and Sons, New York, 1962).
21. C. M. Hurd, *The Hall Effect in Metals and Alloys* (Plenum Press, New York, 1972).
22. J. P. Jan, *Solid State Phys. Ser.* **5**, 1 (1957).
23. R. M. Bozorth, *Ferromagnetism* (D. Van Nostrand Co., New York, 1951).
24. I. A. Campbell et al., *J. Phys. F. Metal Phys. Suppl.* **1**, S95 (1970).
25. L. Berger and G. Bergmann, in: *The Hall Effect and its Applications*, edited by C. L. Chien and C. R. Westgate (Plenum Press, New York, 1980), p. 55.
26. J. Bass, *Adv. Phys.* **21**, 431 (1972).
27. N. F. Mott, *Proc. Roy. Soc.* **153**, 699 (1936); **156**, 368 (1936); *Adv. Phys.* **13**, 325 (1964).
28. S. F. Lee et al., *J. Magn. Magn. Mater.* **118**, L1 (1993).
29. I. Mertig et al., *Phys. Rev.* **47**, 16178 (1993); *Phys. Rev. B* **49**, 11767 (1993); *J. Magn. Magn. Mater.* **151**, 363 (1995).
30. C. Kittel, *Introduction to Solid State Physics*, 6th edn. (John Wiley and Sons, New York, 1986).
31. R. Meservey and P. M. Tedrow, *Phys. Rep.* **238**, 173 (1994).
32. J. S. Moodera et al., *Phys. Rev. Lett.* **80**, 2941 (1998).
33. A. H. MacDonald et al., *Phys. Rev. Lett.* **81**, 705 (1998).
34. S. K. Upadhyay et al., *Phys. Rev. Lett.* **81**, 3247 (1998).
35. R. J. Soulen Jr. et al., *Science*, **282**, 85 (1998).
36. R. A. deGroot et al., *Phys. Rev. Lett.* **50**, 2024 (1983).
37. K. Schwartz, *J. Phys. F*, **16**, L211 (1986).
38. G. L. Bona et al., *Solid. State Commun.* **56**, 391 (1985).
39. K. P. Kamper et al., *Phys. Rev. Lett.* **59**, 2788 (1987).
40. W. E. Pickett and D. J. Singh, *Phys. Rev. B* **53**, 1146 (1996).
41. Y. Lu et al., *Phys. Rev. B* **54**, R8357(1996).
42. J. Z. Sun et al., *Appl. Phys. Lett.* **70**, 1769 (1997).
43. J.-H. Park et al., *Nature* **392**, 794 (1998).
44. M. Johnson and R. H. Silsbee, *Phys. Rev. Lett.* **55**, 1790 (1985); **60**, 377 (1988).
45. M. Johnson, *Phys. Rev. Lett.* **70**, 2142 (1993); *J. Appl. Phys.* **75**, 6714 (1994).
46. M. Johnson, *Appl. Phys. Lett.* **65**, 1460 (1994).
47. A. Fert and S. F. Lee, *Phys. Rev. B* **53**, 6554 (1993).
48. See, e.g., R. Q. Hood and L. M. Falicov, *Phys. Rev. B* **46**, 8287 (1992); B. Dieny, *Europhys. Lett.* **17**, 261 (1992).
49. J. M. Slaughter et al., *Rev. Sci. Instrum.* **60**, 127 (1989).

50. P. Dauguet et al., J. Appl. Phys. **79**, 5823 (1996).
51. M. J. M. Gijs et al., Phys. Rev. Lett. **70**, 3343 (1993).
52. L. Piraux et al., Appl. Phys. Lett. **65**, 2484 (1994).
53. A. Blondel et al., Appl. Phys. Lett. **65**, 3019 (1994).
54. K. Liu et al., Phys. Rev. B **51**, 7381 (1995).
55. S. Zhang and P. M. Levy, J. Appl. Phys. **69**, 4786 (1991).
56. T. Valet and A. Fert, Phys. Rev. B **48**, 7099 (1993).
57. P. Grunberg et al., Phys. Rev. Lett. **57**, 2442 (1986).
58. S. S. P. Parkin et al., Phys. Rev. Lett. **64**, 2304 (1990).
59. A. Berkowitz et al., Phys. Rev. Lett. **68**, 3745 (1992).
60. J. Q. Xiao et al., Phys. Rev. Lett. **68**, 3749 (1992).
61. T. Shinjo and H. Yamamoto, J. Phys. Soc. Jpn. **59**, 3061 (1990).
62. F. Tsui et al., Phys. Rev. Lett. **72**, 740 (1994).
63. S. S. P. Parkin et al., Phys. Rev. Lett. **66**, 2152 (1991); D. H. Mosca et al., J. Magn. Magn. Mater. **94**, L1 (1991).
64. D. Bartlett et al., Phys. Rev. B **49**, 1521 (1994).
65. B. Dieny et al., Phys. Rev. B **43**, 1297 (1991).
66. B. Dieny et al., Appl. Phys. Lett. **61**, 2111 (1992).
67. B. A. Gurney et al., Phys. Rev. Lett. **71**, 4023 (1993).
68. S. K. J. Lenczowski et al., Mater. Res. Soc. Symp. Proc. **384**, 341 (1995).
69. S. K. J. Lenczowski et al., Phys. Rev. B **50**, 9982 (1994).
70. G. J. Strijkers et al., Phys. Rev. B **54**, 9365 (1996).
71. S. S. P. Parkin et al., Phys. Rev. B **47**, 9136 (1993).
72. S. S. P. Parkin, Phys. Rev. Lett. **71**, 1641 (1993).
73. J. M. George et al., Phys. Rev. Lett. **72**, 408 (1994).
74. J. P. Renard et al., Phys. Rev. B **51**, 12821 (1995).
75. E. E. Fullerton et al., Phys. Rev. Lett. **68**, 859 (1992).
76. K. Takanashi et al., J. Phys. Soc. Jpn. **61**, 1169 (1992); A. Kamijo and H. Igarashi, J. Appl. Phys. **72**, 3497 (1992).
77. N. M. Rensing, et al., J. Magn. Magn. Mater. **121**, 436 (1993).
78. R. Belien et al., Phys. Rev. B **50**, 9957 (1994).
79. R. Shad et al., Phys. Rev. B **57**, 13692 (1998).
80. See, e.g., J. Barnas and A. Fert, Phys. Rev. B **49**, 12835 (1994); W. H. Butler et al., J. Magn. Magn. Mater. **151**, 354 (1995); M. D. Stiles, J. Appl. Phys. **79**, 5605 (1996).
81. B. Dieny et al., Phys. Rev. B **43**, 1297 (1991).
82. L. Steren et al., Phys. Rev. B **51**, 292 (1995) and references therein.
83. See, e.g., H. T. Hardner et al., Phys. Rev. B **48**, 16156 (1993); J. Appl. Phys. **79**, 7751 (1996).
84. H. Sato, Mater. Sci. Eng. B **31**, 101 (1995).
85. J. Shi et al., J. Magn. Magn. Mater. **125**, L251 (1993); Phys. Rev. **54**, 15273 (1996).
86. S. F. Lee et al., Phys. Rev. B **52**, 15426 (1995).
87. M. A. M. Gijs et al., J. Appl. Phys. **75**, 6709 (1994).
88. W. Vavra et al., Appl. Phys. Lett. **66**, 2579 (1995).
89. M. Alper et al., Appl. Phys. Lett. **63**, 2144 (1993).
90. M. A. M. Gijs et al., J. Magn. Magn. Mater. **151**, 333 (1995).
91. T. Ono and T. Shinjo, J. Phys. Soc. Jpn. **64**, 363 (1995); T. Shinjo and T. Ono, J. Magn. Magn. Mater. **177–181**, 31 (1998).
92. P. M. Levy et al., Phys. Rev. B **52**, 16049 (1995).
93. Q. Yang et al., Phys. Rev. B **51**, 3226 (1995); W. P. Pratt et al., J. Appl. Phys. **79**, 5811 (1996).
94. For example, N. J. List et al., Mater. Res. Soc. Symp. Proc. **384**, 329 (1995).
95. M. A. M. Gijs et al., Appl. Phys. Lett. **63**, 111 (1993); S. K. J. Lenczowski et al., J. Appl. Phys. **75**, 5154 (1994).
96. J. J. Krebs et al., J. Appl. Phys. **79**, 6084 (1996).
97. For example, L. Piraux et al., J. Magn. Magn. Mater. **156**, 317 (1996).
98. J.-E. Wegrowe et al., Helv. Phys. Acta **70**, S5 (1997), and private communications from: L. Piraux, B. Doudin, and J.-Ph. Ansermet.
99. P. A. Schroeder et al., Mater. Res. Soc. Symp. Proc. **313**, 47 (1993).

100. M. A. M. Gijs et al., *Phys. Rev. B* **50**, 16733 (1994).
101. W. Oepts et al., *Phys. Rev. B* **50**, 14024 (1996).
102. L. Piraux et al., *Eur. Phys. J. B* **4**, 413 (1998).
103. P. A. Schroeder et al., *J. Appl. Phys.* **76**, 6610 (1994).
104. J. A. Borchers et al., *Phys. Rev. Lett.* **82**, 2796 (1999).
105. J.-Ph. Ansermet, *J. Phys. Condens. Mater.* **10**, 6027 (1998).
106. S. Steenwyk et al., *J. Magn. Magn. Mater.* **L1** (1997).
107. W. P. Pratt Jr. et al., *IEEE Trans Magn.* **33**, 3505 (1997).
108. P. Holody, Ph.D. Thesis, *Michigan State University*, 1996 (unpublished).
109. M. Howson et al., unpublished.
110. B. Doudin et al., *J. Appl. Phys.* **79**, 6090 (1996).
111. C. Vouille et al., *Phys. Rev. B* **60**, 6710, (1999); *J. Appl. Phys.* **81**, 4573 (1997).
112. S. Y. Hsu et al., *Phys. Rev. Lett.* **78**, 2652 (1997).
113. P. A. Schroeder et al., *J. Magn. Magn. Mater.* **177–181**, 1464 (1998).
114. J. A. Caballero et al., *J. Vac. Sci. Technol.* **16A**, 1801 (1998); *J. Magn. Magn. Mater.* **198–199**, 55 (1999).
115. Q. Yang et al., *Phys. Rev. Lett.* **72**, 3274 (1994); S. Y. Hsu et al., *Phys. Rev. B* **54**, 9027 (1996).
116. A. Fert, J.-L. Duvail, and T. Valet, *Phys. Rev. B* **52**, 6513 (1995).
117. A. Reilly et al., *J. Magn. Magn. Mater.* **195**, L269 (1999).
118. S. Dubois et al., *Phys. Rev. B* **60**, 477 (1999).
119. D. Baxter et al., *J. Appl. Phys.* **85**, 4545 (1999).
120. A. Vedyayev et al., *Phys. Rev. B* **55**, 3728 (1997).
121. P. Dauguet et al., *Phys. Rev. B* **54**, 1083 (1996).
122. B. Dieny et al., *J. Appl. Phys.* **79**, 6370 (1996).
123. K. M. Schep et al., *Phys. Rev. Lett.* **74**, 586 (1995).
124. M. V. Tsoi et al., *J. Appl. Phys.* **81**, 5530 (1997).
125. J. Mathon, *Phys. Rev. B* **54**, 55 (1996); **55**, 960 (1997).
126. W.-C. Chiang et al., *Proc. Mater. Res. Soc.* **475**, 451 (1997).
127. L. Berger, *Phys. Rev. B* **54**, 9353 (1996); *J. Appl. Phys.* **81**, 4880 (1997); *IEEE Trans. Magn.* **34**, 3837 (1998); *Phys. Rev. B* **59**, 11465 (1999); J. C. Slonczewski, *J. Magn. Magn. Mater.* **159**, L1 (1996); *J. Magn. Magn. Mater.* **195**, L261 (1999).
128. M. V. Tsoi et al., *Phys. Rev. Lett.* **80**, 4281 (1998); *Nature* **406**, 46 (2000); E. B. Myers et al., *Science* **285**, 867 (1999); J. A. Katine et al., *Phys. Rev. Lett.* **84**, 3149 (2000).
129. S. Zhang, P.M. Levy, and A. Fert, *Phys. Rev. Lett.* **88**, 236601(2002).
130. D. Bozec, M. A. Howson, and B. J. Hickey, *Phys. Rev. Lett.* **85**, 1314 (2000). E. Y. Tsymbal, *Phys. Rev. B* **62**, R3608 (2000); A. Shpiro and P. M. Levy, *Phys. Rev. B* **63**, 014419 (2001); K. Eid et al., *J. Magn. Magn. Mater.* **224** L205 (2001); K. Eid et al., *Phys. Rev. B* **65**, 054424 (2002); K. Eid et al. *J. Magn. Magn. Mater.* (in press).
131. B. J. Hickey et al., *Phys. Rev. B* **51**, 667 (1995).
132. T. L. Hylton, *Appl. Phys. Lett.* **62**, 2431 (1993).
133. T. L. Hylton et al., *Science*, **261**, 1021 (1993); *J. Appl. Phys.* **75**, 7058 (1994).
134. M. J. Carey et al., *Appl. Phys. Lett.* **61**, 2935 (1992).
135. C. L. Chien et al., *J. Appl. Phys.* **73**, 5309 (1993).
136. J. Q. Wang and G. Xiao, *Phys. Rev. B* **49**, 3982 (1994).
137. A. E. Berkowitz et al., *J. Appl. Phys.* **73**, 5320 (1993).
138. A. Maeda et al., *J. Appl. Phys.* **76**, 6793 (1994).
139. J. F. Gregg et al., *Phys. Rev. B* **49**, 1064 (1994); K. Ono et al., *J. Phys. Soc. Jpn.* **65**, 3449 (1996).
140. L. Piraux et al., *J. Magn. Magn. Mater.* **136**, 221 (1994).
141. P. Xiong et al., *Phys. Rev. Lett.* **69**, 3220 (1992).
142. H. Sato, *Mater. Sci. Eng. B* **31**, 101 (1995).
143. J. Shi et al., *Phys. Rev. B* **48**, 16119 (1993).
144. M. Julliere, *Phys. Lett.* **54A**, 225 (1975).
145. J. S. Moodera et al., *Appl. Phys. Lett.* **69**, 708 (1996).
146. P. M. Tedrow and R. Meservey, *Phys. Rev. Lett.* **26**, 192 (1971).
147. See, e.g., T. Miyazaki et al., *J. Magn. Magn. Mater.* **98**, L7 (1991); J. Nowak and J. Rauluszkiewicz, *J. Magn. Magn. Mater.* **109**, 79 (1992); T. Yaoi et al., *J. Magn. Magn. Mater.* **126**, 430 (1993) as well as further references in Refs. 7, 31.

148. T. Miyazaki and N. Tezuka, *J. Magn. Magn. Mater.* **139**, L231 (1995); **151**, 403 (1995).
149. See, e.g., C. L. Platt *et al.*, *J. Appl. Phys.* **81**, 5523 (1997); K. Ono *et al.*, *J. Phys. Soc. Jpn.* **65**, 3449 (1996).
150. J. M. Daughton, *J. Appl. Phys.* **81**, 3758 (1997).
151. H. Tsuge and T. Mitsuzuka, *Appl. Phys. Lett.* **71**, 3296 (1997); J. J. Sun *et al.*, *Appl. Phys. Lett.* **74**, 448 (1999).
152. P. M. Levy and S. Zhang, Curr. Opin. Solid State Mater. Sci. **4**, 231 (1999); J. S. Moodera and G. Mathon, *J. Magn. Magn. Mater.* **200**, 248 (1999).
153. W. J. Gallagher *et al.*, *J. Appl. Phys.* **81**, 3741 (1997).
154. See, e.g., S. Zhang *et al.*, *Phys. Rev. Lett.* **79**, 3744 (1997).
155. C. T. Tanaka *et al.*, *J. Appl. Phys.* **81**, 5515 (1997).
156. M. Viret *et al.*, *Europhys. Lett.* **39**, 545 (1997); Z. W. Li *et al.*, *J. Appl. Phys.* **81**, 5509 (1997); J. Z. Sun *et al.*, *Appl. Phys. Lett.* **73**, 1008 (1998).
157. J. H. Park *et al.*, *Phys. Rev. Lett.* **81**, 1953 (1998).
158. H. Fujimori *et al.*, *Mater. Sci. Eng.* B **31**, 219 (1995); A. Milner *et al.*, *Phys. Rev. Lett.* **76**, 475 (1996). A. Gerber *et al.*, *Phys. Rev. B* **55**, 6446 (1997); S. Sankar *et al.*, *J. Appl. Phys.* **81**, 5513 (1997); S. Honda *et al.*, *J. Magn. Magn. Mater.* **165**, 153 (1997); S. Honda *et al. Phys. Rev. B* **56**, 14,566 (1997); K. Inomata and Y. Saito, *Appl. Phys. Lett.* **73**, 1143 (1998); B. Dieny *et al.*, *J. Magn. Magn. Mater.* **185**, 283 (1998).
159. J. M. D. Coey *et al.*, *Phys. Rev. Lett.* **80**, 3815 (1998).
160. K. Ono *et al.*, *J. Phys. Soc. Jpn.* **66**, 1261 (1997).
161. S. Mitani *et al.*, *J Appl. Phys.* **83**, 6524 (1998).
162. H. Bruckl *et al.*, *Phys. Rev. B* **58**, R8893 (1998).
163. F. Schelp *et al.*, *Phys. Rev. B* **56**, R5747 (1997).
164. R. Desmicht *et al.*, *Appl. Phys. Lett.* **72**, 386 (1998).
165. R. B. van Dover *et al.*, *Phys. Rev. B* **47**, 6134 (1993).
166. P. A. Algarabel *et al.*, *Appl. Phys. Lett.* **66**, 3062 (1995).
167. C. Mazumdar *et al.*, *J. Appl. Phys.* **81**, 5781 (1997).
168. Y. Shimakawa, Y. Kubo, and T. Manako, *Nature*, **379**, 53 (1996).
169. A. P. Ramirez, R. J. Cava, and J. Krajewski, *Nature*, 387, 268 (1997).
170. G. H. Jonker and J. H. van Santen, *Physica* **16**, 337 (1950).
171. J. H. van Santen and G. H. Jonker, *Physica* **16**, 599 (1950).
172. J. Volger, *Physica* **20**, 49 (1954).
173. Y. Tomioka *et al.*, *Phys. Rev. B* **53**, R1689 (1996).
174. P. Schiffer *et al.*, *Phys. Rev. Lett.* **75**, 3336 (1995).
175. A. Urushibara *et al.*, *Phys. Rev. B* **51**, 14103 (1995).
176. R. M. Kusters *et al.*, *Physica B* **155**, 362 (1989).
177. R. von Helmolt *et al.*, *Phys. Rev. Lett.* **71**, 2331 (1993).
178. S. Jin *et al.*, *Appl. Phys. Lett.* **66**, 382 (1995).
179. G. C. Xiong *et al.*, *Appl. Phys. Lett.* **66**, 1427 (1995).
180. A. J. Millis *et al.*, *Phys. Rev. Lett.* **74**, 5144 (1995).
181. A. J. Millis *et al.*, *Phys. Rev. Lett.* **77**, 175 (1996); A. J. Millis, *J. Appl. Phys.* **81**, 5502 (1997).
182. H. Roder *et al.*, *Phys. Rev. Lett.* **76**, 1356 (1996).
183. C. M. Varma, *Phys. Rev. B* **54**, 7328 (1996).
184. S. Ishihara *et al.*, *Physica C* **263**, 130 (1996).
185. M. Johnson *et al.*, *Appl. Phys. Lett.* **71**, 974 (1997).
186. H. L. Ju *et al.*, *Phys. Rev. B* **51**, 6143 (1995); G. C. Xiong *et al.*, *Appl. Phys. Lett.* **66**, 1689 (1996).
187. H. W. Hwang *et al.*, *Phys. Rev. Lett.* **77**, 2041 (1996).
188. N. D. Mathur *et al.*, *Nature* **387**, 266 (1997).
189. A. Gupta *et al.*, *Phys. Rev. B* **54**, R15629 (1996).
190. T. Kimura *et al.*, *Science* **274**, 1698 (1996).
191. H. Kuwahara *et al.*, *J. Appl. Phys.* **81**, 4954 (1997).
192. C. Kwon *et al.*, *J. Appl. Phys.* **81**, 4951 (1997).
193. K. H. Ahn *et al.*, *Phys. Rev. B* **54**, 15299 (1996).
194. A. Barnabe *et al.*, *Appl. Phys. Lett.* **71**, 3907 (1997).
195. G. Xiao *et al.*, *J. Appl. Phys.* **81**, 5324 (1997).

196. G. B. Alers *et al.*, *Appl. Phys. Lett.* **68**, 3644 (1996).
197. D. W. Visser *et al.*, *Phys. Rev. Lett.* **78**, 3947 (1997); A. P. Ramirez *et al.*, *J. Appl. Phys.* **81**, 5337 (1997).
198. M. Jaime *et al.*, *Phys. Rev. Lett.* **78**, 951 (1997).
199. M. F. Hundley and J. J. Neumeier, *Phys. Rev. B* **55**, 11511 (1997).
200. K. Mohri *et al.*, *IEEE Trans. Magn.* **28**, 3150 (1992); **29**, 1245 (1993); **31**, 2455 (1995).
201. R. S. Beach and A. E. Berkowitz, *Appl. Phys. Lett.* **64**, 3652 (1994); J. Valazquez *et al.*, *Phys. Rev. B* **50**, 16737 (1994); F. L. A. Machado *et al.*, *Phys. Rev. B* **51**, 3926 (1995); R. L. Sommer and C. L. Chien, *Appl. Phys. Lett.* **67**, 857, 33465 (1995), *Phys. Rev. B* **53**, R5982 (1996).
202. M. Viret *et al.*, *Phys. Rev. B* 53, 8464 (1996); J. F. Gregg *et al.*, *Phys. Rev. Lett.* **77**, 1580 (1996); W. Allen *et al.*, *J. Magn. Magn. Mater.* **165**, 121 (1997).
203. K. Hong and N. Giordano, *J. Phys. Condens. Mater.* **10**, L401 (1998) and earlier references therein.
204. U. Ruediger *et al.*, *Phys. Rev. Lett.* **80**, 5639 (1998).
205. P. M. Levy and S. Zhang, *Phys. Rev. Lett.* **79**, 5110 (1997).
206. G. Tatara and H. Fukuyama, *Phys. Rev. Lett.* **78**, 3773 (1997); Y. Lyanda-Geller *et al.*, *Phys. Rev. Lett.* **81**, 3215 (1998).
207. D. J. Monsma *et al.*, *Phys. Rev. Lett.* **74**, 5260 (1995)
208. D. J. Monsma *et al.*, *Science* **281**, 407 (1998).
209. T. Kinno *et al.*, *Phys. Rev. B* **56**, R4391 (1997).
210. B. Doudin *et al.*, *Phys. Rev. Lett.* **79**, 933 (1997).
211. C. T. Rogers *et al.*, *Phys. Rev. B* **56**, R8503 (1997).
212. T. Ono *et al.*, *Appl. Phys. Lett.* **72**, 1116 (1998).

5

Magnetic Characterization of Materials

Chia-Ling Chien*

5.1. INTRODUCTION

Magnetic materials have been of importance to mankind for several thousand years. Lodestone is believed to have been instrumental in winning a decisive battle for the first emperor of China fought in a dense fog against the barbarians *ca.* 2600 BC. Many branches of science, and indeed our daily lives, depend on magnetic materials and magnetic characterization of one form or another, from high-density magnetic recording media and read-heads in computers to magnetic resonance imaging. Condensed matter physicists and materials researchers also rely heavily on magnetic characterization for the understanding and the exploration of new materials.

All materials, magnetic or otherwise, respond more or less to a magnetic field. A discussion of magnetic characterization should include the magnetic behavior of all materials, and not just the instrumentation that measures the magnetic properties. Furthermore, magnetic characterization, like other forms of measurements, is quantitative. A practitioner of magnetic characterization must understand and account for the measured results. Although some instruments, such as a magnetometer, measure the magnetic properties directly, other techniques, using phenomena such as the Mössbauer effect (ME), indirectly provide information concerning the magnetic properties. Both types of measurements are included in this discussion.

Finally, there is probably no field other than magnetism, where the system of units is so confusing in the literature. We therefore begin our discussion of magnetic characterization with the systems of units.

5.2. UNITS

The cgs units are preferred by most practitioners in magnetic measurements, in apparent defiance towards the mandate of exclusively using the SI units. On the instrumentation side, the cgs units also persist; most magnetometers are calibrated to display magnetic moment in emu, and most gaussmeters measure the magnetic field in oersteds. At the risk of being accused of perpetrating the illegitimate, we will express all quantities in cgs units. This is more a reflection of the status quo in the research literature rather than trying to be a countercurrent.

In cgs units, the relationship among magnetic induction or magnetic flux density (**B**), magnetic field intensity (**H**), and magnetization (**M**) is

---

*Department of Physics and Astronomy, Johns Hopkins University, Baltimore, Maryland.

Table 5.1. Conversion Between cgs Units and SI Units for Selected Magnetic Quantities

Quantity	Symbol	cgs Units $\mathbf{B} = \mathbf{H} + 4\pi\mathbf{M}$	Conversion Factor C (cgs units) = $C \times$ (SI units)	SI Units $\mathbf{B} = \mu_0(\mathbf{H}+\mathbf{M})$ $\mu_0/4\pi = 10^{-7}$ H/m
Magnetic induction	**B**	Gauss (G)	10^{-4}	Tesla (T) Wb/m²
Magnetic flux	Φ	G cm², Maxwell (Mx)	10^{-8}	Weber (Wb)
Magnetic field intensity	**H**	Oersted (Oe)	$10^3/4\pi$	A/m
Magnetic moment	μ	emu, erg/Oe	10^{-3}	(A m²), J/T
(Volume) magnetization	**M**	emu/cm³	10^3	A/m
	4π**M**	G	$10^3/4\pi$	A/m
Mass magnetization	M_g	emu/g	1	(A m²)/kg
			$4\pi \times 10^{-7}$	(Wb m)/kg
Volume susceptibility	χ	Dimensionless	4π	Dimensionless
Mass susceptibility	χ_g	cm³/g	$4\pi \times 10^{-3}$	m³/kg
Permeability	μ	Dimensionless	$4\pi \times 10^{-7}$	Wb/(A m)

$$\mathbf{B} = \mathbf{H} + 4\pi\mathbf{M} \tag{5.2.1}$$

universally used without ambiguity, where the units for **B** (gauss = G), **H** (oersted = Oe), and **M** (emu /cm³) have the *same* dimension. In SI units, the corresponding relation is

$$\mathbf{B} = \mu_0(\mathbf{H} + \mathbf{M}) \tag{5.2.2}$$

and the units for **H** and **M** are in A/m, while **B** is in tesla, or Wb/m². Most unfortunately, other forms, such as $\mathbf{B} = \mu_0\mathbf{H} + \mathbf{M}$, have *also* been used by many, creating endless confusion. In Table 5.1, various common quantities of magnetic properties are shown in cgs and SI units.

5.3. MAGNETIC MOMENT, MAGNETIZATION, AND SUSCEPTIBILITY

In the absence of compelling evidence of the existence of magnetic monopoles, the basic entity of magnetism remains to be the *magnetic dipole moment* (μ). A macroscopic dipole moment may be a bar magnet for which μ is the product of the pole strength and the separation of the two poles with a direction from the "south" pole to the "north" pole, or a current loop of area A and of current I with a magnetic moment $\mu = IA$ and a direction normal to the loop. On a microscopic level, magnetic moments are the results of angular momenta, both orbital and spin, of the particles or ions. Susceptibility is a measure of the response of a magnetic material to the external field.

5.3.1. Magnetic Moment and Magnetization

The magnetic moment in cgs units is expressed in erg/Oe, or in *the electromagnetic units for magnetic moment* (emu), where emu = erg/Oe. With a collection of magnetic moments, the *magnetization* (**M**) is defined as magnetic moments per volume:

$$\mathbf{M} = \frac{\sum \mu_i}{V} \quad [\text{erg}/(\text{Oe cm}^3) \text{ or emu/cm}^3] \tag{5.3.1}$$

where V is the volume in which the vectorial sum $\sum \mu_i$ is performed. We can also define *mass magnetization* as

$$\mathbf{M}_g = \frac{\sum \mu_i}{W} = \frac{\sum \mu_i}{\rho V} = \frac{\mathbf{M}}{\rho} \quad (\text{emu/g}) \tag{5.3.2}$$

where W and ρ are the mass and density of the specimen, respectively.

5.3.2. Atomic Magnetic Moment

The atomic magnetic moments are measured in units of *atomic Bohr magneton*,

$$\mu_B = \frac{eh}{4\pi m_e c} = 0.927 \times 10^{-20} \text{ erg/Oe} \quad (\text{emu}) \tag{5.3.3}$$

where h is Planck's constant, m_e, the electron rest mass, e, the electronic charge, and c, the speed of light.

The magnetic moments in some magnetic materials (e.g., insulators) are *localized* to the atomic sites, while those in other materials (e.g., metals) may be *itinerant*, originated from band effects. The sources of localized atomic magnetic moments are total angular momentum $\mathbf{J} = \mathbf{L} + \mathbf{S}$ of the electrons in the *partially filled* electronic shell, where $\mathbf{L}$ and $\mathbf{S}$ are the total orbital angular momentum and the total spin angular momentum, respectively. The *atomic magnetic dipole moment*, also known as the *ordered moment*, is given by

$$\mu = -g\,\mu_B\,\mathbf{J} = -\gamma\,\hbar\,\mathbf{J} \tag{5.3.4}$$

where γ is the *atomic gyromagnetic ratio* and g is the *Lande g factor*,

$$g = 1 + \frac{J(J+1) + S(S+1) - L(L+1)}{2J(J+1)} \tag{5.3.5}$$

The atomic ground state configuration, often the most relevant one, can be obtained from the Hund's rules. As shown in Tables 5.2 and 5.3, Hund's rules can be used to determine the ground state configurations and the magnetic moments of all the transition metal (3d) ions and the rare-earth (4f) ions, which are important for magnetism. This single-ion description works usually well for the rare-earth systems, but not for the transition-metal solids, for which the J values are often not good quantum numbers. Note that the g factor is of order unity, as shown in Table 5.2.

Experimentally, the magnetic moment of a material can be measured by a number of methods. The *ordered moment* can be measured at low temperatures, when magnetic ordering occurs, using magnetometry, neutron diffraction, or inferred through hyperfine interactions. For magnetic materials with itinerant magnetic moments, one can still define a value for the magnetic moment (expressed in μ_B) per atom. The *paramagnetic moment* (or *effective moment*), described later, can be measured through its response to a magnetic field at high temperatures.

Table 5.2. The Number of 4f Electrons, the Ground State Configuration According to Hund's Rules, the g Factor, the Ordered Moment, the Theoretical and the Experimental Values of Effective Moments of Rare-Earth Ions

Rare-Earth Ion	Number of 4f Electrons	Ground State	g	Ordered Moment gJ (μ_B)	Theory $p_{eff} = g[J(J+1)]^{1/2}$	Experiment $(p_{eff})_{exp}$
La^{3+}	0	$^1S_0$	0	0	0.0	0.0
Ce^{3+}, Pr^{4+}	1	$^2S_{5/2}$	6/7	15/7	2.54	2.4
Pr^{3+}	2	$^3H_4$	4/5	16/5	3.58	3.6
Nd^{3+}	3	$^4I_{9/2}$	8/11	36/11	3.62	3.6
Pm^{3+}	4	$^5I_4$	3/5	12/5	2.68	—
Sm^{3+}	5	$^6H_{5/2}$	2/7	5/7	0.84	1.5
Eu^{3+}, Sm^{2+}	6	$^7F_0$	0	0	0.0	3.6
Gd^{3+}, Eu^{2+}	7	$^8S_{7/2}$	2	7	7.94	8.0
Tb^{3+}	8	$^7F_6$	3/2	9	9.72	9.6
Dy^{3+}	9	$^6H_{15/2}$	4/3	10	10.63	10.6
Ho^{3+}	10	$^5I_8$	5/4	10	10.6	10.4
Er^{3+}	11	$^4I_{15/2}$	6/5	9	9.59	9.4
Tm^{3+}	12	$^3H_6$	7/6	7	7.57	7.3
Yb^{3+}	13	$^2F_{7/2}$	8/7	4	4.54	4.5
Lu^{3+}, Yb^{2+}	14	$^1S_0$	0	0	0.0	0.0

5.3.3. Nuclear Magnetic Moment

All particles (neutron, proton, nuclei, muons, etc.) with non-zero spin also have magnetic moments. For nuclei, the magnetic moments are expressed in units of nuclear Bohr magneton

$$\mu_n = \frac{eh}{4\pi M_p c} = 5 \times 10^{-24} \text{ erg/Oe} \quad \text{(or emu)} \qquad (5.3.6)$$

Table 5.3. The Number of 3d Electrons, the Ground State Configuration According to Hund's Rules, the Calculated Values Using J and S and the Experimental Values of Effective Moments of Transition-Metal Ions

Transition-Metal Ion	Number of 3d Electrons	Ground State	Theory $p_{eff} = g[J(J+1)]^{1/2}$	Theory $p_{eff} = g[S(S+1)]^{1/2}$	Experiment $(p_{eff})_{exp}$
K^+, Ca^{3+}	0	$^1S_0$	0.0	0.0	0.0
Ti^{3+}, V^{4+}	1	$^2D_{3/2}$	1.55	1.73	1.7
V^{3+}	2	$^3F_2$	1.63	2.83	2.8
Cr^{3+}, V^{2+}, Mn^{4+}	3	$^4F_{3/2}$	0.77	3.87	3.8
Mn^{3+}, Cr^{2+}	4	$^5D_0$	0.0	4.9	4.9
Fe^{3+}, Mn^{2+}	5	$^6S_{5/2}$	5.92	5.92	5.9
Fe^{2+}	6	$^5D_4$	6.7	4.9	5.4
Co^{2+}	7	$^4F_{9/2}$	6.64	3.87	4.8
Ni^{2+}	8	$^3F_4$	5.6	2.83	3.2
Cu^{2+}	9	$^2D_{5/2}$	3.55	1.73	1.9
Zn^{2+}, Cu^+	10	$^2S_0$	0.0	0.0	0.0

MAGNETIC CHARACTERIZATION OF MATERIALS

where $\mathbf{M}_p$ is the proton rest mass. The nuclear Bohr magneton is 1835 times smaller than its atomic counterpart. A nucleus in a state with a spin I has a nuclear magnetic moment of

$$\mu = g_n \, \mu_n \, I = \gamma_n \hbar \, I \tag{5.3.7}$$

where g_n is the *nuclear g factor* and γ_n is the *nuclear gyromagnetic ratio*, both may be either positive or negative. For example, proton has a spin $I = 1/2$ and a magnetic moment of $\mu = 2.793 \, \mu_n$, with $g_n = 5.586$ and $\gamma_n = 2.68 \times 10^4$ s G.

Here, we consider a simple example to illustrate the different ways of expressing the magnetic moment and magnetization of a ferromagnet. Metallic iron is ferromagnetic, and at low temperatures its moments can be aligned parallel to each other by a modest field (e.g., 50 Oe). Magnetometry measurement of a piece of Fe with a mass of 2.3 mg gives a total moment of 0.5 emu. We can neglect the contributions from the nuclear moments. From the measured result, a mass magnetization of $M_g = 0.5/0.0023 = 217$ emu/g can be determined. The Fe sample of 2.3 mg contains 2.4×10^{19} Fe atoms, each with a magnetic moment of 0.21×10^{-19} emu/atom, or $0.21 \times 10^{-19}/0.927 \times 10^{-20} = 2.2 \, \mu_B$. With a density of $\rho = 7.87$ g/cm^3, the magnetization of Fe is $M = 217 \times 7.87 = 1711$ emu/cm^3 = 1711 G. Sometimes, the magnetization is expressed as saturation flux density $B_S = 4\pi M = 21.7$ kG. Hence, the magnetization of Fe metal can be expressed as

$$\text{mass magnetization} = M_g = 217 \text{ emu/g},$$
$$\text{magnetization} = M = 1711 \text{ emu/cm}^3 = 1.711 \text{ kG},$$
$$\text{magnetic moment per Fe} = 2.2 \, \mu_B,$$
$$\text{saturation flux density } B_S = 4\pi M = 21.7 \text{ kG}.$$

It should be noted that we have obtained these values for Fe from the total emu *and* the mass of the specimen. Useful magnetic information requires the knowledge of the sample, such as its mass, volume, and composition. It should be noted that we have been quite sloppy in units, a luxury of using the cgs units. Similar information about Fe, Co, Ni, and Gd, the common ferromagnetic metals, are tabulated in Table 5.4.

Table 5.4. Magnetic Moment (μ), Curie Temperature (T_C), Curie–Weiss Constant (θ_C), Density (ρ), and Magnetization (**M**) of Ferromagnetic Metals of Fe, Co, Ni, and Gd

	Fe (bcc)	Co (hcp)	Ni (fcc)	Gd (hcp)
μ (μ_B) at 0 K	2.22	1.72	0.615	7.56
T_C (K)	1044	1388*	627	293
θ_C (K)	1100	1415*	650	317
ρ (g/cm^3)	7.875	8.804	8.912	7.898
M (0 K) in emu/g	221.7	163	58.5	268
M (0 K) in emu/cm^3 (G)	1746	1435	520	2116
$4\pi M$ (0) in kG	21.9	18.1	6.55	26.6
M (300 K) in emu/g	217	161	55	0
M (300 K) in emu/cm^3 (G)	1709	1417	490	0

*fcc state of Co.

5.3.4. Demagnetizing Field

The field $\mathbf{H}' = \mathbf{H} - D\mathbf{M}$ inside a magnetic specimen can be *different* from the applied $\mathbf{H}$ field due to the magnetization of the specimen, where D is the *demagnetization factor*. Only for samples with an *ellipsoidal* shape, with uniform $\mathbf{B}$, $\mathbf{H}'$, and $\mathbf{M}$ inside, closed-form analytical results for D can be obtained. The value of D is between 0 and 4π depending on the *shape* of the specimen. In the limit of a flattened ellipsoid, we have a thin disk, as would be encountered in a thin film sample. In the limit of an elongated ellipsoid, we have a long rod. The ellipsoid becomes a sphere when all the axes are the same. For these special shapes, the demagnetization factors are

For a thin disk: $\quad D \approx 4\pi$ (field applied $\perp$ disk plane) $\quad$ (5.3.8)

$\quad\quad\quad\quad\quad\quad\quad\quad D \approx 0$ (field $\parallel$ to disk plane) $\quad$ (5.3.9)

For a long rod: $\quad D \approx 2\pi$ (field applied $\perp$ long rod) $\quad$ (5.3.10)

$\quad\quad\quad\quad\quad\quad\quad\quad D \approx 0$ (field applied $\parallel$ long rod) $\quad$ (5.3.11)

For a sphere: $\quad D = \dfrac{4\pi}{3}$ (any direction) $\quad$ (5.3.12)

Because of the demagnetization factor, due to the sample shape, the measured results can be dramatically different when the magnetic field is applied along different directions. For example, for a soft Fe thin film, the magnetization can be saturated with a field of only 50 Oe when applied in the film plane. If the field is applied perpendicular to the film plane, saturation does not occur until an external field of about $4\pi M \approx 21$ kOe. The demagnetization factor also accounts for the *shape anisotropy*. The magnetic moments of a "soft" ferromagnetic film are often found to be lying in the sample plane.

5.3.5. Energy and Motion of a Magnetic Moment in a Uniform Magnetic Field

The energy of the magnetic moment $\boldsymbol{\mu}$ in an external $\mathbf{H}$ field is

$$E = -\boldsymbol{\mu} \cdot \mathbf{H} \quad (5.3.13)$$

The lowest energy state is for $\boldsymbol{\mu}$ to align with $\mathbf{H}$. If an angular momentum $\mathbf{J}$ is associated with the dipole moment, then Eq. (5.3.12) leads to a Zeeman splitting of $2\mathbf{J} + 1$ levels.

For an atomic moment $\boldsymbol{\mu} = -\gamma\hbar\mathbf{J} = -g\mu_B\mathbf{J}$ in a field $\mathbf{H}$, the energies are

$$E_j = g\mu_B m_j H = \gamma\hbar m_j H \quad (5.3.14)$$

with an energy level splitting of

$$\Delta E = g\mu_B H = \gamma\hbar H \quad (5.3.15)$$

where m_j is the *magnetic quantum number* with the values of $m_j = -\mathbf{J}, -\mathbf{J}+1, \ldots, \mathbf{J}-1, \mathbf{J}$. Semi-classically, there is a specific orientation of $\boldsymbol{\mu}$ associated with each of these energy levels. For a nuclear moment $\boldsymbol{\mu} = g_n\mu_n\mathbf{I} = \gamma_n\hbar\mathbf{I}$ in a field $\mathbf{H}$, the energies are

$$E_j = g_n\mu_n m_j H = \gamma_n \hbar m_j H \tag{5.3.16}$$

with an energy level splitting of

$$\Delta E = |g_n|\mu_n H = |\gamma_n|\hbar H \tag{5.3.17}$$

In a uniform magnetic field **H**, the magnetic moment **μ** receives a torque of **N** = **μ** × **H**. A magnetic moment with an angular momentum **μ** = $\gamma\hbar$**J**, atomic or nuclear, precesses around the **H** field according to the equation of motion of

$$\frac{d\boldsymbol{\mu}}{dt} = \boldsymbol{\mu} \times \gamma\mathbf{H} \tag{5.3.18}$$

with an angular precession frequency, the *Larmor angular frequency*, of

$$\omega_0 = \Delta E/\hbar = \gamma H \tag{5.3.19}$$

For an atomic moment $\gamma = g\mu_B/\hbar$, the Larmor frequency $\nu = \omega/2\pi$:

$$\nu \text{ (GHz)} = 1.4\, gH \text{ (kOe)} \tag{5.3.20}$$

is in the gigaHertz range, as encountered in ferromagnetic resonance (FMR). For nuclear moments, $\gamma_n = g_n\mu_n/\hbar$, the frequency

$$\nu \text{ (MHz)} = 0.7\, |g_n|\, H \text{ (kOe)} \tag{5.3.21}$$

is in the megaHertz range, as encountered in nuclear magnetic resonance (NMR).

5.3.6. Magnetic Susceptibility

When **B**, **H**, and **M** are parallel, it is useful to define the *permeability* μ by **B** = μ**H** and the *susceptibility* χ by **M** = χ**H**, where $\mu = 1 + 4\pi\chi$. In cgs units, μ and χ are dimensionless, and their values can be used to classify magnetic materials. The small induced magnetization (hence χ is small) of diamagnetism and paramagnetism disappear when the applied field is turned off. However, for both ferromagnetic and ferrimagnetic materials, part of the magnetization is retained even after the field has been turned off. Furthermore, for ferromagnetic materials, the values of both μ and χ are large and dependent on **H** as well as the history of the specimen.

The *volume susceptibility* (χ), or susceptibility per volume, is defined as

$$\chi = \frac{M}{H} = \frac{\text{emu}}{VH} \text{ (emu/Oe cm}^3 = \text{G/Oe} = \text{dimensionless)} \tag{5.3.22}$$

where V is the sample volume. For sufficiently small **H**, χ is independent of **H**. We can also define *mass susceptibility* as $\chi_g = \text{emu}/W\,H = \chi/\rho$, and *atomic susceptibility* as $\chi_A = \chi_g A$, where W is the mass, ρ, the density of the sample, and A, the atomic mass of the element.

5.4. DIAMAGNETISM AND PARAMAGNETISM

A magnetic material responds strongly to an external magnetic field when the material is magnetically ordered. However, at sufficiently high temperatures where there is no magnetic ordering, the material responds weakly to a magnetic field mainly due to Curie–Weiss paramagnetism. In addition, all materials, magnetic or otherwise, exhibit diamagnetism.

5.4.1. Diamagnetism

An external magnetic field interacts with the electron orbital motion, which could be viewed as a "current loop." Due to Lenz's law, a magnetic field will induce a change in the "current loop" resulting in an induced magnetic moment. The diamagnetic susceptibility per volume of an elemental specimen is

$$\chi_{\text{diam}} = -N \frac{e^2}{6m_e c^2} \sum_{i=1}^{Z} \langle r_i^2 \rangle \qquad (5.4.1)$$

where N is the number of atoms per cubic centimeter, Z, the atomic number of the element, and $\langle r_i^2 \rangle$, the mean square radius of an electron "current loop," and the summation includes all the electrons of the element.

It should be noted that diamagnetism is a property of *all* materials, magnetic or otherwise. The quantity χ_{diam} is always *negative*, small (10^{-6}–10^{-5}), independent of T, and appears in all measurements of χ as a temperature-independent term. These features are shown in Fig. 5.1. Although diamagnetism is weak, it may be the dominant contribution to χ in weakly magnetic systems. For a ferromagnetic thin film on a thick substrate, the diamagnetic signal from the substrate can also be comparable to that of the ferromagnetic thin film.

5.4.2. Diamagnetism in Superconductors

Diamagnetism is small in all materials except superconductors, where the diamagnetism is large. Below the transition temperature, a superconductor exhibits the Meissner effect with no magnetic field ($\mathbf{B} = 0$) inside the superconductor, i.e., perfect diamagnetism. The magnetization curve for a superconductor is a straight line of $-\mathbf{M} = \mathbf{H}/4\pi$, up to the critical field $\mathbf{H}_c$ for type I superconductors. Above $\mathbf{H}_c$, both perfect diamagnetism and superconductivity disappear. For type II superconductors, a complete Meissner state of $-\mathbf{M} = \mathbf{H}/4\pi$ is maintained up to the lower critical field $\mathbf{H}_{c1}$. Above $\mathbf{H}_{c1}$, B is no longer zero inside the superconductor, which is in the vortex state. Above the upper critical field $\mathbf{H}_{c2}$ the superconductor becomes normal.

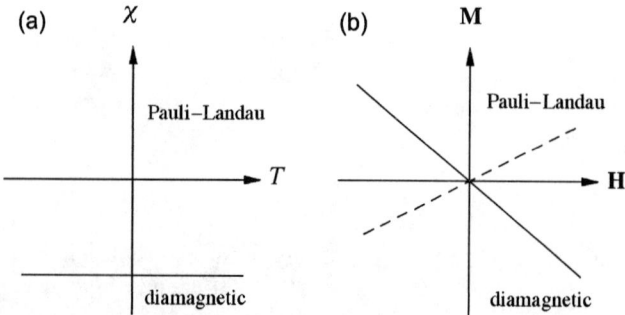

Figure 5.1. Temperature (a) and magnetic filed (b) dependence of Pauli–Landau paramagnetic and diamagnetic susceptibilities (courtesy of F. Y. Yang).

5.4.3. Pauli–Landau Paramagnetism

In a metal there are free electrons, each of which carries a magnetic moment of μ_B. Under an external field **H**, the spin-up and the spin-down electronic bands shift relative to each other by an amount of $2\mu_B H$. The induced magnetization is the product of μ_B and the population difference in the two bands. The Pauli–Landau paramagnetic susceptibility is $n_e \mu_B^2 / k_B T_F$, where n_e is the electron density and T_F the Fermi temperature. The Pauli–Landau susceptibility is *positive*, *small* (10^{-6}–10^{-5}) and *independent of T*, as shown in Fig. 5.1. As an estimate, by taking $n_e = 10^{23}$ cm^{-3} and $T_F = 10^4$ K, the Pauli susceptibility is about 5×10^{-6}, a magnitude similar to that of the diamagnetic susceptibility.

5.4.4. Curie–Weiss Paramagnetism

The most important paramagnetic effect is that of the individual moments. The interaction among the magnetic moments causes magnetic ordering to occur below the magnetic ordering temperature T_O, which may be the Curie temperature T_C of a ferromagnet, or the Néel temperature T_N of an anti-ferromagnet, or other types of magnetic ordering. At $T > T_O$, there is no magnetic ordering, and the material is said to be in the *paramagnetic state* with a collection of interacting paramagnetic moments. The magnetic susceptibility at temperatures sufficiently higher than T_O has the form:

$$\chi_{\text{para}} = \frac{ng^2 \mu_B^2 J(J+1)}{3k_B(T-\theta)} = \frac{n(p_{\text{eff}}\mu_B)^2}{3k_B(T-\theta)} = \frac{C}{T-\theta} \quad (5.4.2)$$

where n is the number of paramagnetic moments per volume, $p_{\text{eff}}\mu_B = g[J(J+1)]^{1/2}\mu_B$ is the *effective moment*, or *paramagnetic moment*, which is *different* from the ordered moment $gJ\mu_B$, $C = ng^2\mu_B^2 J(J+1)/3k_B$ is the Curie–Weiss constant, and θ is known as the *Curie–Weiss temperature*.

From the susceptibility (χ) data of a material in the paramagnetic state, one can determine the value of $(p_{\text{eff}})_{\text{exp}} = [3\chi_{\text{para}}k_B(T-\theta)/n\mu_B^2]^{1/2}$ experimentally and compare it with the theoretical value. In Table 5.2, we show the calculated and the experimental values at *room temperature* of p_{eff} of the rare earth ions. The agreement is excellent for most of the rare earth ions using the ground state configuration. However, the agreements are poor for Sm^{3+}, Eu^{3+}, and Sm^{2+}, because in these cases, the excited states are not sufficiently above the ground state. The element Pm is radioactive with few experimental results. In general, the rare earth ions in both insulators and metals have *localized* moments, whose values are largely determined by the single-ion values and are insensitive to chemistry, because the partially filled $4f$ shell is an *inner* shell shielded by the $5s^2 5p^6$ shells.

In Table 5.3, we show the calculated and the experimental values at room temperature of p_{eff} of the transition metal ions. It is immediately clear that agreement between the experimental and the calculated value using the appropriate **J** value is generally poor. On the other hand, if we had used the spin value of the ground state configuration and proceeded to calculate p_{eff}, the agreement is quite good in most cases. This is because the unfilled 3d shell is the outermost shell, where effects due to chemical bonding and crystal fields are so strong that the orbital angular momenta may be *quenched*.

The Curie–Weiss temperature θ gives information about the magnetic interactions among the magnetic moments. It is well known from the mean-field theory that

$$\theta = \frac{2\mathbf{J}(\mathbf{J}+1)}{3k_B} \times (\text{sum of magnetic interactions on a moment}) \qquad (5.4.3)$$

whose value can be used to classify magnetic materials. For example, for a ferromagnet with a bcc lattice structure, there are eight nearest-neighbor moments each with a quantum number $\mathbf{J}$. Suppose the nearest-neighbor interaction strength is $\mathcal{J}_1$. Then $\theta = |2\mathbf{J}(\mathbf{J}+1)/3k_B|(8\mathcal{J}_1)$ in this case, if we include only the nearest neighbors. If the second-neighbor interaction strength is $\mathcal{J}_2$, then $\theta = |2\mathbf{J}(\mathbf{J}+1)/3k_B| (8\mathcal{J}_1 + 6\mathcal{J}_2)$.

Combining diamagnetism and paramagnetism, the magnetic susceptibility of a system with magnetic moments at sufficiently high temperatures can be expected to be

$$\chi = \chi_0 + \frac{n\mu_{\text{eff}}^2}{3k_B(T-\theta)} \qquad (5.4.4)$$

or

$$\chi_M = \chi - \chi_0 = \frac{C}{T-\theta} \qquad (5.4.5)$$

where χ_0 is the temperature-independent part, whose value may be positive or negative, depending on the size of the diamagnetic contribution, and χ_M is the Curie–Weiss term.

For paramagnets with negligible interactions among the magnetic moments, and spin glasses with competing interactions, $\theta \approx 0$. In both cases, $1/\chi_M$ will be linearly dependent on T and extending close to $T = 0$ K, sometimes known as the Curie law as shown in Fig. 5.2(a). In real specimens, there are often paramagnetic impurities in the sample or the substrate. The contributions from the paramagnetic impurities are negligible at high temperatures, but large at low temperatures because of the $1/T$ dependence.

For ferromagnets, it is generally true that $\theta = \theta_C > 0$, and θ_C is close to, although often slightly higher than, T_C, as shown in Table 5.4, for the common ferromagnetic metals. The value of χ_M becomes large when T_C is approached from higher temperatures, hence easily measurable. The value of $1/\chi_M$ is linearly dependent upon T and extrapolates to θ_C, but perhaps bending slightly as T approaches T_C from high temperatures, as shown in Fig. 5.2(b).

For anti-ferromagnets, usually $\theta = \theta_N < 0$, but θ_N has no simple relationship with T_N. In most cases, θ_N/T_N is between -5 and -1. Here, θ_N is not experimentally accessible, but through extrapolation only. Furthermore, even though χ_M shows a cusp at T_N, χ_M remains small even at T close to T_N, as shown in Fig. 5.2(d). The value of $1/\chi_M$ is linearly dependent on T for $T \geqslant T_N$ and extrapolates to θ_N, which is on the negative T side, as shown in Fig. 5.2(c).

From an experimentalist's standpoint, it should be emphasized that not only the experimental results of χ qualitatively should be described in Eq. (5.4.4), but also all the values (χ, χ_0, n, μ_{eff}, and θ) must *quantitatively* be reasonable. Furthermore,

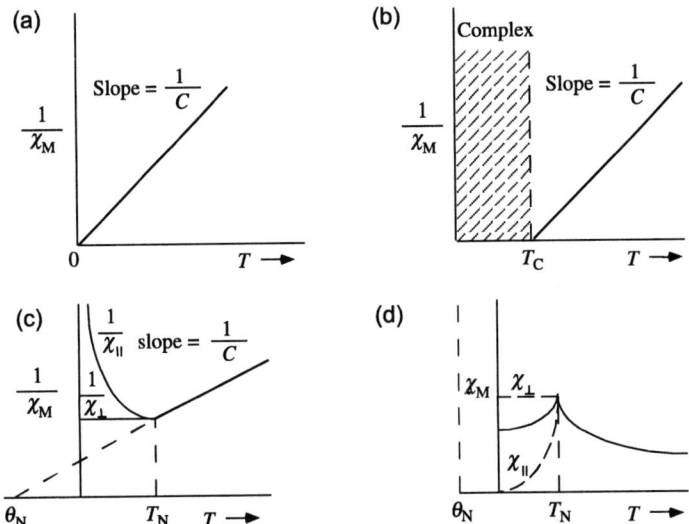

Figure 5.2. Temperature dependence of paramagnetic susceptibility of: (a) a paramagnet with non-interacting moments; (b) a ferromagnet at $T > T_C$, and (c, d) a uniaxial antiferromagnet at $T > T_N$, with the magnetic field parallel and perpendicular to the uniaxial axis (courtesy of Tom Ambrose).

the size of the magnetic field during measurement should be suitably chosen. All the formalisms of $\chi = M/H$ described in this section assume that the magnetic field **H** is sufficiently small, such that **M/H** is independent of **H**. Within this realm, a larger **H** would give a larger **M**, thus facilitating the measurement. However, when **H** is too large, the measured value of **M/H** deviates from the $\chi = M/H$ value defined for small **H**. A useful guideline for selecting the strength of the magnetic field **H** is through the condition of $\mu_{\text{eff}} H/k_B T \ll 1$, where μ_{eff} is the effective moment and T the temperature. For example, for moments of 1 μ_B at $H = 10$ kOe, $\mu_B H$ is about 10^{-16} erg or about 1 K. At $T > 30$ K and under a 10 kOe field, the condition of $\mu_{\text{eff}} H/k_B T \ll 1$ is safely satisfied.

5.5. MAGNETIC ORDERING

The magnetic moments in a solid interact with each other. The strongest interaction is between the nearest neighbors. For simplicity, let us assume all the magnetic moments are the same and that they occupy crystallographically equivalent sites. The interaction among the moments can be described, e.g., by the Heisenberg-type interaction with a Hamiltonian of

$$\mathbf{H} = -\sum_{ij} \mathcal{J}_{ij} \boldsymbol{\mu}_i \cdot \boldsymbol{\mu}_j \tag{5.5.1}$$

where $\boldsymbol{\mu}_i$ and $\boldsymbol{\mu}_j$ are two such magnetic moments on lattice sites i and j, and $\mathcal{J}_{ij}$ is the strength of the interaction between them. It is remarkable that such a two-body interaction can lead to magnetic ordering for the entire solid that undergoes a phase transition at the ordering temperature, and that the sign of $\mathcal{J}_{ij}$ leads to entirely different ferromagnetic (if $\mathcal{J}_{ij}$ is largely positive) ordering and anti-ferromagnetic (if $\mathcal{J}_{ij}$ is largely negative) ordering.

We shall neglect other more complicated magnetic orderings, such as ferrimagnetic ordering, where there are two or more kinds of magnetic moments, and spin glass ordering involving competing magnetic interactions, even though some of the technologically important magnetic materials are ferrimagnets (e.g., ferrites and garnets).

5.5.1. Ferromagnetism

Ferromagnetism is a phenomenon in which the interactions ($\mathcal{J}_{ij} > 0$) among the magnetic moments favor parallel alignment. Neglecting anisotropy and magnetic domains for the time being, all the magnetic moments are aligned at $T = 0$ K, with a spontaneous magnetization of

$$\mathbf{M}_{\text{spon}}(0) = n\mu(0) = n_g \mu_B J \tag{5.5.2}$$

which is the largest value attainable. At a finite temperature, the *spontaneous magnetization* (with $\mathbf{H} = 0$), which is an intrinsic property of a ferromagnet, is

$$\mathbf{M}_{\text{spon}}(T) = n\mu(T) = n_g \mu_B \langle J \rangle \tag{5.5.3}$$

where $\langle J \rangle$ is the expectation value of $\mathbf{J}$. The value of $\mathbf{M}_{\text{spon}}(T)$ decreases with increasing T and vanishes at $T \geq T_C$, where T_C is the *Curie temperature*, above which the ferromagnet is in the paramagnetic state.

The mean-field approximation describes the ferromagnets remarkably well qualitatively, but not quantitatively. For example, at low temperatures, the spontaneous magnetization of most ferromagnets has a $T^{3/2}$ dependence, the *Bloch's law*,

$$\mathbf{M}(T) = \mathbf{M}(0)[1 - BT^{3/2} \cdots] \tag{5.5.4}$$

due to the excitations of *spin-wave* (*magnons*), which cannot be accounted for in the mean-field description. The coefficient B is related to the spin-wave stiffness constant, which can be measured by neutron diffraction from the dispersion relation of the spin waves. Another well-known discrepancy is in the critical behavior of ferromagnets as T approaches T_C,

$$\mathbf{M} \propto \left(1 - \frac{T}{T_C}\right)^\beta \tag{5.5.5}$$

for which the mean-field model gives $\beta = 1/2$, whereas most experiments and rigorous theories show $\beta \approx 1/3$.

The above descriptions of the intrinsic properties of a ferromagnet exclude anisotropy and domains. However, all real ferromagnetic materials have magnetic anisotropy, which prefers the alignment of the moments to be along certain directions. Furthermore, small magnetic domains within the ferromagnet are formed to reduce the magnetostatic energy. Thus most ferromagnetic materials, in the absence of a magnetic field, are found in the partially magnetized or unmagnetized state ($\mathbf{M} = 0$), where the ferromagnet is partitioned into magnetic domains. Ferromagnetic alignment of magnetic moments occurs only within the small magnetic domains, and the magnetization directions between domains are not aligned.

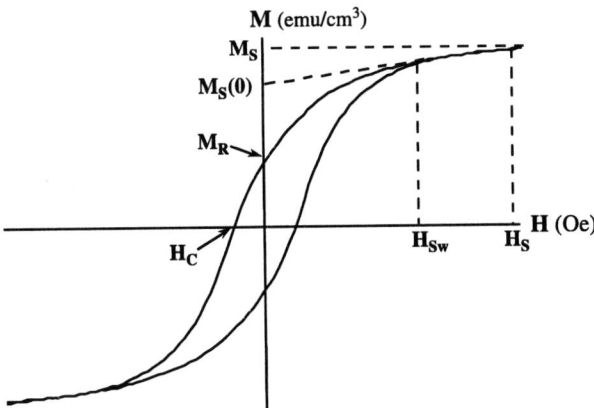

Figure 5.3. Hysteresis loop of a typical ferromagnet showing saturation magnetization [M_S], spontaneous magnetization [$M_S(0)$], remnant magnetization [M_R], and coercivity [H_C] (courtesy of Tom Ambrose).

Under an increasing external field, the magnetization of the sample increases as domains are being swept. When the external field **H** overcomes all sources of magnetic anisotropy and aligns all the moments, one obtains the saturation magnetization $M_S(H, T)$. The actual manner with which **M** increases with **H** depends on various sources of magnetic anisotropy. One of the characteristics of ferromagnetic materials is the *hysteresis loop*, which may be plotted as either **M** versus **H** or **B** versus **H**. Consider an example of the **M** versus **H** hysteresis loop shown in Fig. 5.3. Starting from the origin, when the sample is in the original unmagnetized state, **M** follows the initial magnetization curve during increasing **H**. The saturation magnetization is eventually reached at sufficiently large fields. Upon decreasing the field, the magnetization does *not* follow the initial curve, but instead traces out a hysteresis loop, which intercepts the **M**-axis at M_R, the *remnant magnetization* and the **H**-axis at H_C, the *coercivity*. Sometimes the *squareness* ($SQ = M_R/M_S$) is used instead of M_R. Crudely speaking, a "soft" (small H_C and large SQ) and a "hard" (large H_C) ferromagnet are distinguished by the values of H_C and squareness. Metallic Fe, permalloy ($Ni_{80}Fe_{20}$), and Metglass (e.g., a-$Fe_{80}B_{20}$) are examples of soft ferromagnets. They can be attracted by a magnet, but they themselves are incapable of attracting other ferromagnets.

A permanent magnet is a "hard" ferromagnet with a large magnetic anisotropy and large coercivity H_C. With proper heat treatment in an external field, a large magnetization can be retained when the external field is removed, hence, a permanent magnet. The **B** versus **H** plot, or the **B–H** hysteresis loop, which can be directly measured using a **B–H** loop tracer or obtained from a **M** versus **H** plot, is more often used in describing permanent magnets. For a good permanent magnet, both B_R (*remnant flux density*) and $(H_C)_B$ are large. The prowess of a permanent magnet is measured by the magnetic energy (BH), given by a point in the second quadrant of a **B–H** loop at which (BH) is maximum, known as the *energy product* [$(BH)_{max}$]. In the order of increasing energy product, commercially available permanent magnets include ceramic magnets (mainly ferrites), transition-metal alloy magnets (e.g.,

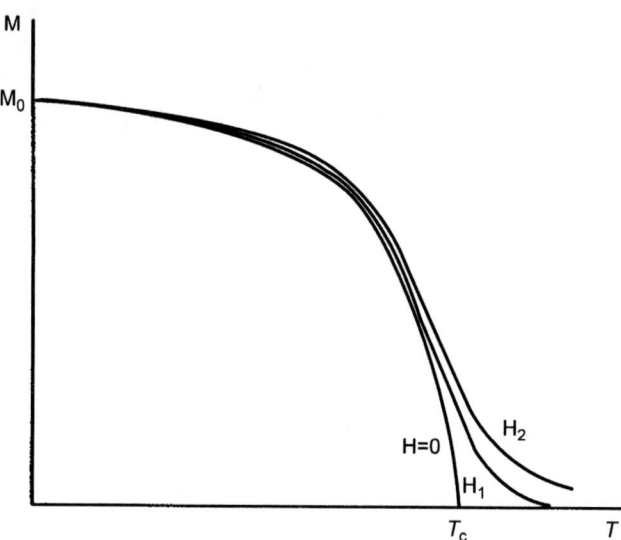

Figure 5.4. Temperature dependence of saturation magnetization $M_S(H, T)$ at different applied fields **H**, where $H_2 > H_1 > 0$, and the curve with $H = 0$ is the spontaneous magnetization (courtesy of F. Y. Yang).

ALNICO consists of Al, Fe, Ni and Co), rare-earth transition-metal magnets (e.g., $SmCo_5$), and the neodymium–iron magnets (Nd–Fe–B), which, with an energy product of 40 MG Oe, is the reigning champion.

The temperature dependence of the saturation magnetization $M_S(H, T)$ of a ferromagnet at various field values is schematically shown in Fig. 5.4. It should be noted that $M_S(H, T)$ is *not* the spontaneous magnetization $M(T)$, and in fact, $M_S(H, T) \geq M(T)$. Furthermore, $M_S(H, T)$ *increases* with **H**, but not to exceed $M(0) = ng\mu_B J$. One notes that at a given temperature a higher **H** field produces a larger $M_S(H, T)$. Of special interest is $M_S(0, T)$, the one with zero applied field, i.e., the spontaneous magnetization $M(T)$. Experimentally, spontaneous magnetization may be obtained using magnetometry by extrapolating to $H = 0$, a non-trivial method more successful in the case of soft magnets. Alternatively, the spontaneous magnetization and its temperature dependence can be measured by microscopic techniques (neutron diffraction, Mössbauer spectroscopy, NMR, muon spin resonance, etc.) where an external field is *not* present.

The Curie temperature (T_C) is a crucial characteristic of a ferromagnet. As long as an external field is used as in magnetometry, as shown in Fig. 5.5, T_C cannot be accurately determined. One could resort to the method of *Arrott plot* in which one plots $M(H, T)^{1/\beta}$ versus $(H/M)^{1/\gamma}$ for a number of temperatures near T_C, where $\beta \approx 1/3$ and $\gamma \approx 4/3$ are the critical exponents for spontaneous magnetization and critical isotherm respectively. The value of T_C can be determined by noting the temperature for which the curve $M(H, T)^{1/\beta}$ versus $(H/M)^{1/\gamma}$ passes through the origin. On the other hand, if a microscopic technique is used, the location of T_C can be cleanly determined by the temperature at which the spontaneous magnetization vanishes.

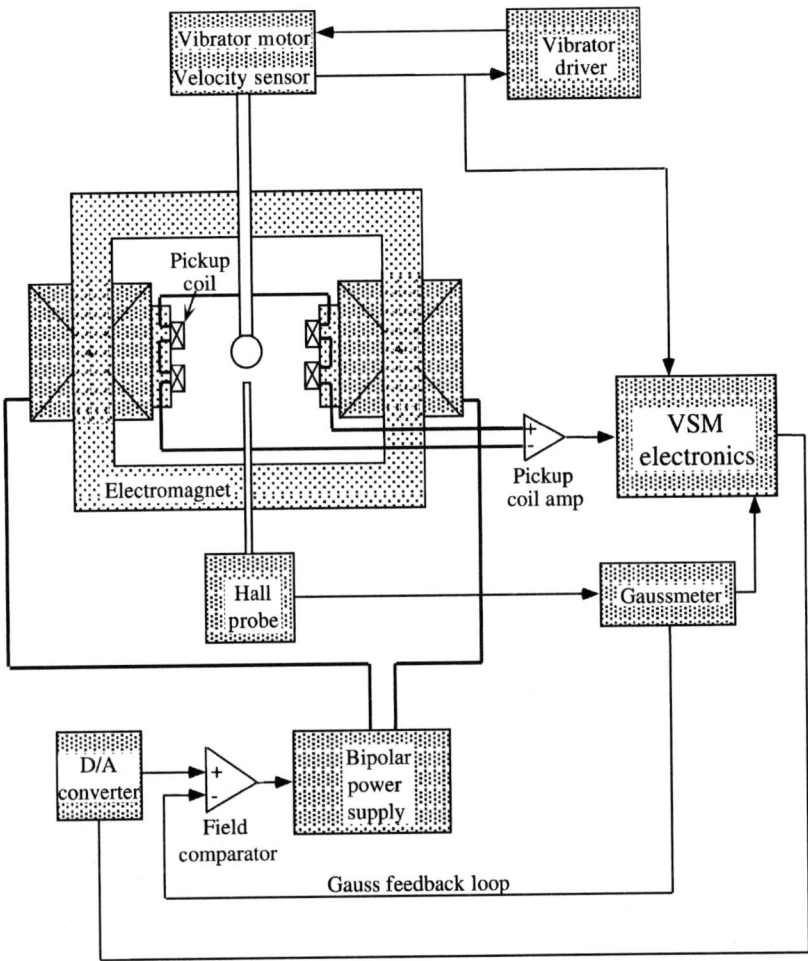

Figure 5.5. Block diagram of a VSM (courtesy of Tom Ambrose).

5.5.2. Anti-Ferromagnetism

In Eq. (5.5.1) if $\mathcal{J}_{ij}$ is <0, then an anti-ferromagnetic ordering is formed at low temperatures in which the magnetic atoms can be divided into two or more sublattices. In the ordered state at $T < T_N$, where T_N is the *Néel temperature*, all the moments on each sublattice are ferromagnetically aligned, but their directions are different from those of the other sublattices, such that the total magnetization of the specimen remains zero. The simplest anti-ferromagnet consists of two sublattices with anti-parallel moments. In the absence of an external field, there is no net magnetization even though the sublattice magnetization is large. The sublattice magnetization of anti-ferromagnets *cannot* be measured by magnetometry, except by microscopic techniques and, most importantly, neutron diffraction. Indeed, anti-ferro magnetism was mainly a conjecture until the confirmation by neutron diffraction.

In comparison to ferromagnetic ordering, the anti-ferromagnetic ordering is more fascinating, such as the large variety of interesting spin structures and a rich phase diagram. However, anti-ferromagnetic materials are not technologically useful, in part because the response of an anti-ferromagnet to an external field is weak even at $T < T_N$ due to the nearly zero total magnetization. The paramagnetic behavior at $T > T_N$ has been discussed previously. The location of T_N can be revealed by a small cusp at T_N in the temperature dependence of induced magnetization, as shown in Fig. 5.2(d). For a uniaxial anti-ferromagnet, the susceptibility perpendicular ($\chi_\perp$) and parallel ($\chi_\parallel$) to the uniaxial anisotropy axis are different, as shown in Fig. 5.2(c) and (d). Furthermore, by increasing the **H** field along the uniaxial anisotropy axis, under suitable conditions, a spin-flop transition may occur where the sublattice magnetizations turn nearly 90° to be perpendicular to the applied field.

5.6. MAGNETOMETRY

A number of magnetometers have been developed over the years for the purpose of measuring the magnetization **M**, and in some cases, the magnetic induction **B**, of magnetic materials. First of all, it is important to recognize that these magnetometers make *integral* measurements of **M** along a certain direction, most commonly the external field direction. The magnetometer does *not* distinguish a magnitude change in |**M**| from a directional change of **M**. By the judicious use of the magnetic field, one may be able to distinguish the difference. Secondly, not only the contribution from the sample of interest, but those of the substrate, the sample holder, and other materials present in the magnetometer are also measured. It is contingent upon the operator to extract the contribution from the sample of interest. Thirdly, most magnetometers measure only the total moment in emu, and not the magnetization of the sample. As we have alluded to earlier, other characteristics of the sample (dimensions, volume, mass, etc.) are also needed in order to obtain useful information about the magnetization of the sample. Finally, since an external magnetic field is usually applied, the applied magnetic field often alters the quantity that we wish to measure. The inability to measure the spontaneous magnetization is one such example.

There are many kinds of magnetometers, and we will limit our attention to only a few common ones. Most magnetometers utilize either a closed or an open magnetic circuit. The **B-H** loop tracer is an example of a closed magnetic circuit, where the sample is in a toroidal shape with coils wrapped around the circumference. The magnetic flux remains entirely in the sample, useful for measuring very "soft" magnetic materials. It is of course quite a nuisance to shape the sample into a toroidal shape. In the open magnetic circuits, the magnetic flux enters the sample from one region and exits the other. Most of the commercial magnetometers including the Faraday balance, the alternating gradient magnetometer (AGM), the vibrating sample magnetometer (VSM), and the superconducting quantum interference device (SQUID) magnetometer employ open magnetic circuits.

There are two kinds of common magnetometers with open magnetic circuits. Some measure **M** via a change of the *magnetic force* on a sample in a *non-uniform* field. The Faraday balance method and the AGM are notable examples. Other magnetometers, such as the VSM and the SQUID magnetometer, which are the most widely used magnetometers, measure **M** via a change of *magnetic flux* due to the sample in a *uniform* magnetic field.

5.6.1. Vibrating Sample Magnetometer

In a VSM, a sample mounted on a sample holder is placed in the gap of a magnet, which supplies a uniform and static magnetic field of a certain value. A pair of opposite-wound pick-up coils are placed in the vicinity of the sample, most commonly on the pole faces, as shown in Fig. 5.5. The sample is set to vibrate at a certain frequency (e.g., 77 Hz) with a small amplitude (e.g., 1 mm) to eliminate all static contributions. The magnetic flux change due to the vibrating sample induces a signal in the pick-up coils, which is then measured by a lock-in amplifier. One often takes repeated measurements at a given settings of field and temperature to improve accuracy. A hysteresis loop, e.g., can be obtained by successively measuring **M** at various field values. Most VSMs utilize electromagnets, which have limitation in the maximum magnetic fields of no more than about 20 kOe. The magnetic field value is feed-back controlled and monitored by a gaussmeter. More than one pair of pick-up coils can also be placed in several orientations to measure other components of **M** in the so-called vector VSM.

5.6.2. SQUID Magnetometer

A SQUID magnetometer utilizes SQUID technology to detect the magnetization of a sample. At the heart of a SQUID detection system are coupled Josephson junctions with a weak link. Depending upon the number of junctions on the superconducting ring, the detection is via an RF SQUID (single junction) or a DC SQUID (double junction). A schematic circuit diagram of an RF SQUID is shown in Fig. 5.6. The current in the superconducting ring is modulated by the amount of flux passing through the ring in units of flux quantum, $\Phi_0 = hc/2e = 2 \times 10^{-7}$ G cm^2. When a sample is moved through the sensing coils, a current proportional to the magnetic flux is induced in the ring. The sensing coils are a single superconducting wire wound into three coils as a second-order gradiometer to reduce noise in the detection circuit, as shown in Fig. 5.7. The amplitude of the sample movement is large (several centimeters) and the frequency is low compared with those of a VSM. Most SQUID

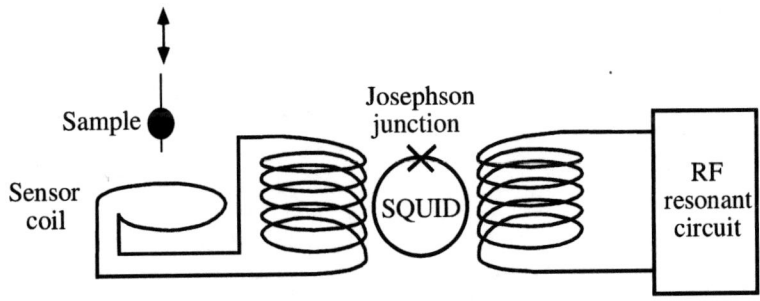

Figure 5.6. Schematic diagram of the RF SQUID detection system (courtesy of Tom Ambrose).

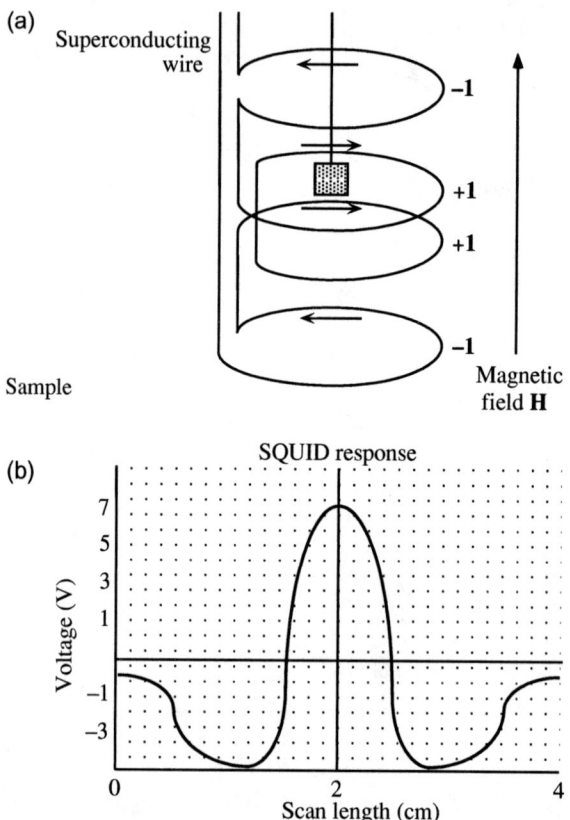

Figure 5.7. (a) Schematic view of superconducting gradiometer detection coil, through which the sample is moved. (b) The output of the SQUID as a magnetic dipole is moved through the gradiometer pickup coil (courtesy of Tom Ambrose).

magnetometers are provided with a superconducting solenoid, which requires liquid helium at all times, as does the SQUID detection system.

The selection of a magnetometer is based on various considerations, such as the cost of the instrument, ease of operation, running expense, magnetic field range, and sensitivity. It is perhaps useful to compare the key differences between VSM and SQUID magnetometers. The costs of commercial VSM and SQUID magnetometers are similar. A SQUID magnetometer has a higher sensitivity of about 10^{-7} emu than the value of 10^{-5} emu common for a VSM. In proper perspectives, the values of 10^{-7} and 10^{-5} emu are moment values of Fe thin films 1 cm × 0.5 cm × 1 Å monolayer and 1 cm × 0.5 cm × 1 monolayer, respectively. The time for one measurement is much longer in the SQUID because of the slow movement of the sample. Furthermore, the higher sensitivity of the SQUID is accomplished in part by a superconducting shield, which must be warmed up and cooled down at every field value. The consumption of liquid helium is a major expense of operating a SQUID magnetometer at any temperature, whereas electrical power is the only running expense for operating a VSM at room temperature. Because the superconducting magnet in the SQUID magnetometer is usually operating in the persistent current mode, the actual magnetic field is often not monitored by an independent gaussmeter. Finally, due to trapped flux in the superconducting magnet, it is not a trivial task to set a small magnetic field reliably in the superconducting magnet. In short, the main advantages of SQUID magnetometers are high sensitivity and higher magnetic fields, which are offset by the ease of operation and the lower running expenses enjoyed by VSMs.

5.7. MAGNETIC CHARACTERIZATIONS THROUGH HYPERFINE INTERACTIONS

Hyperfine interactions are those between mainly the s electrons of an atom and its nucleus. The energy scale of hyperfine interaction, as the name implies, is very small, of the order of 10^{-9}–10^{-6} eV. However, there are nuclear effects that can readily measure such small energies with high accuracy. Ironically, although these are nuclear effects, they provide valuable information about the condensed matter and not the nuclei. The basic reasons are that hyperfine interaction energies are products of two factors: a nuclear factor and an extra-nuclear factor. For example, the hyperfine magnetic interaction energy is a product of the nuclear magnetic moment and the hyperfine magnetic field, and the hyperfine quadrupole interaction energy is a product of the nuclear quadruple moment and the electrical field gradient at the nuclear site. The nuclear magnetic moment and quadrupole moment of a given nuclear state do *not* vary for various materials, whereas the extra-nuclear factors (magnetic hyperfine field and electric field gradient) are *different* for various materials. For example, the value of the magnetic moment of Fe nucleus remains the same whether the Fe atom is in steel, a Nd–Fe–B permanent magnet, or blood. On the other hand, the hyperfine magnetic fields in the three materials are totally different.

Perturbed angular correlation (PAC), and especially ME and NMR are three well-known nuclear techniques that are important for condensed matter physics, materials research, and indeed our health, as in the case of magnetic resonance imaging (MRI), which is based on NMR.

Every nuclear isotope has specific natural abundance and energy levels, each of which has a definite nuclear spin and parity. Only the lowest few nuclear levels are important for NMR, ME, and PAC. Consider the Fe^{57} nucleus, the 14.4 and the 136-keV states are the first two excited states above the ground state. The technique of PAC involves all three states, ME involves the 14.4 keV and the ground state, and NMR involves only the ground state. We will limit our discussions to ME and NMR only, because PAC is rarely used nowadays.

5.7.1. Mössbauer Effect

ME is also known as the *recoilless nuclear gamma-ray resonance*. For a nucleus whose atom is part of a solid, there is a finite probability (the Debye–Waller factor) for the nuclear gamma-ray resonance to occur without any recoil from either the emitting or the absorbing nucleus. The ME has been observed in many nuclear isotopes, of those, Fe^{57}, Sn^{119}, Dy^{161}, Eu^{151}, Tm^{169}, and Sb^{121} are some of the

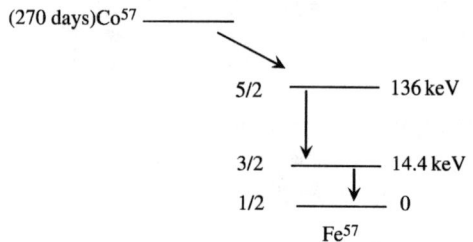

Figure 5.8. Decay scheme of radioactive Co^{57} to the 136 keV, the 14.4 keV, and the ground states of Fe^{57}. The spin and parity values are also shown.

"good" ME isotopes for which the effect can be observed with relative ease and with good resolution. Without any question, the Fe^{57} ME is the most important, commanding more than 95% of the ME literature. It is particularly useful for studying magnetic materials for which Fe is often a major constituent. We will use the 14.4-keV Fe^{57} ME for illustration here.

Consider the 14.4-keV ME of Fe^{57}. The decay of radioactive Co^{57} (half life = 270 days), populates the 136-keV state, which in turn decays to the 14.4-keV state with the emission of a 122-keV photon as shown in Fig. 5.8. The transition from the 14.4-keV state ($I = 3/2$, 98 ns half-life) to the ground state ($I = 1/2$) results in the emission of a photon of 14.4 keV 10% of the time, and through the internal conversion process 90% of the time. In the latter, internally converted electrons and x-rays of about 6 keV are emitted.

In the usual transmission Fe^{57} ME, one counts the absorption of the $E_0 = 14.4$-keV photons by the sample (absorber) as schematically shown in Fig. 5.9. The experimental setup involves a radioactive source of Co^{57}, whose 14.4-keV photons transmit through a Fe^{57}-containing sample (absorber) of interest. A proportional counter is placed behind the absorber to measure the transmitted count rate of all low energy photons. The single-channel analyzer selects out the signals of the 14.4-keV events from the pulse height spectrum, and sends them to the multi-channel analyzer. Resonance is accomplished by changing the velocity of the source relative to the stationary absorber so that the photon energy is minutely altered by the amount of $\Delta E = (\Delta v/c)E_0$ through the Doppler effect. At the resonance condition, there will be a decrease of count rate when the source is moving at a certain Doppler velocity. The multi-channel analyzer consists of a number of channels (e.g., 512) of which only one particular channel is opened when the radioactive source is being driven at a specific Doppler velocity. An ME spectrum is a record of counts versus Δv in which the latter is usually expressed in millimeter per second.

For magnetic materials, the hyperfine magnetic interaction is the most important. The basic interaction is the Zeeman interaction of the nuclear magnetic moment in an effective magnetic field $\mathbf{H}_{eff}$:

$$g_n \mu_n \, \mathbf{I} \cdot \mathbf{H}_{eff} \tag{5.7.1}$$

For Fe^{57} ME, the magnetic hyperfine spectrum consists of six peaks because of the selection rules of the 3/2 to 1/2 transition, as shown in Fig. 5.10. The separation of the six lines in the spectrum is proportional to $|\mathbf{H}_{eff}|$ and their intensities have a ratio

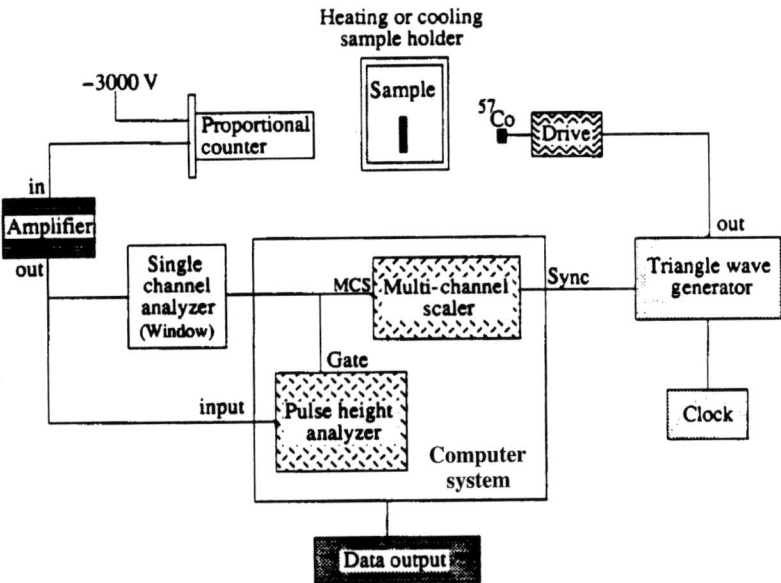

Figure 5.9. Schematic of a Mössbauer spectrometer (courtesy of J. Q. Xiao).

of 3 : b : 1 : 1 : b : 3, where $b = 4\sin^2\theta/(1+\cos^2\theta)$ and θ is the angle between the gamma-ray direction and that of H_{eff}. The values of the hyperfine magnetic field H_{eff} for Fe^{57} ME are large, up to about 350 kOe for metallic systems and more than 500 kOe for insulating systems.

There are several unique features about the magnetic hyperfine interaction. First of all, the effective magnetic field H_{eff} is along the direction of the atomic moment, hence the nuclear moment $\mu = g_n\mu_n I$ will be parallel or anti-parallel to the atomic moment. Thus, the direction of the nuclear moment gives away the information of the direction of the atomic moment. For example, if the moments (atomic and nuclear) are in the sample plane (e.g., in a soft magnetic thin film), b would be about 4. If the moments are perpendicular to the sample plane, b would be 0. For all other cases, b is between 0 and 4.

Secondly, the magnetic hyperfine interaction can be obtained without the use of an external magnetic field. Thirdly, the magnitude of $|H_{\text{eff}}(T)|$ is sensitive to the value of the atomic moment and its local environment. Hence, ME can determine several different $H_{\text{eff}}(T)$ values for materials with multiple-sites and a distribution of H_{eff} such as those in amorphous ferromagnets. Fourthly, the temperature dependence of the $|H_{\text{eff}}(T)|$, to a good approximation, is the same as that of the spontaneous magnetization of a ferromagnet or the sublattice magnetization of an anti-ferromagnet. Finally, the information obtained by ME does not depend on the amount of sample, which only affects the ease of the measurement.

In short, the directional information and the magnitude information of magnetization M can be *separately* obtained even *without* an external field. One can now measure spontaneous magnetization or sublattice magnetization by using ME in zero external field. Because ME is a microscopic technique, it can measure a sample

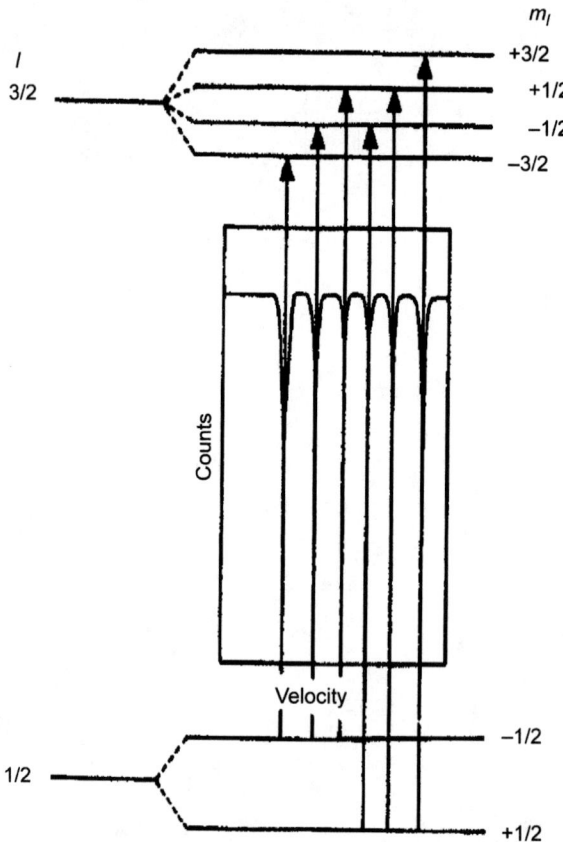

Figure 5.10. The six-line Fe^{57} Mössbauer absorption spectrum due to the transitions between the 3/2 and 1/2 sublevels of the 14.4 keV state and the ground state with an intensity ratio of $3:b:1:1:b:3$ (courtesy of F. Y. Yang).

that contains more than one phase, each with more than one site. Thus, the ME can be used to do phase identification as well as to separately follow the temperature dependence of several $H_{eff}(T)$ in the sample. These are the unique capabilities of ME that magnetometry cannot match.

As mentioned earlier, 90% of the time the decay of the 14.4-keV state is through internal conversion with the emission of internally converted electrons and photons, which have very short ranges in a solid. If we now count these low-energy electrons and photons as in *conversion electron Mössbauer spectroscopy* (CEMS), only those emitted near the top surface of the sample can be detected. Hence, CEMS is a surface-sensitive technique, suitable for studying surfaces and very thin films that contain Fe^{57}. Another difference of the CEMS is that, at resonance, the count rate increases instead of decreasing as in the transmission ME. Because the natural abundance of Fe^{57} is only about 2%, it is often necessary to use enriched Fe^{57} to fabricate Fe containing thin films, while in bulk samples natural Fe is usually adequate.

5.7.2. Nuclear Magnetic Resonance

NMR occurs in the ground state of a nucleus having a nuclear magnetic moment of $g_n\mu_n I = \gamma_n \hbar I$, for which the smallest spin value for NMR is $I = 1/2$. The case of $I = 1/2$ is also the simplest because there are only two spin orientations, and there will be no complication due to the nuclear quadrupole interaction. We shall confine the discussions of NMR to $I = 1/2$, and without loss of generality, assume g_n and γ_n to be positive.

It is useful to discuss NMR of a bare nucleus before moving on to the case of a collection of nuclei in a condensed matter. Under a magnetic field H_{eff} (for the moment, ignore the source of this magnetic field), the nuclear state splits into two states (spin up and spin down) separated by an energy difference of $\Delta E = g_n\mu_n H_{eff} = \gamma_n \hbar H_{eff}$ (Eq. 5.3.17), which is equivalent to an angular Larmor frequency of $\omega_0 = g_n\mu_n H_{eff}/\hbar = \gamma_n H_{eff}$ (Eq. 5.3.19). If one now inserts the expression for the nuclear Bohr magneton μ_n (Eq. 5.3.6), one finds that $\omega_0 = g_n c H_{eff}/2M_p c$, in which the quantity $\hbar$ has dropped out, suggesting that a semi-classical description of NMR would be useful and informative.

The nucleus, in either the spin-up or the spin-down state, will precess around the magnetic field H_{eff} with the Larmor frequency. A bare nucleus without energy exchange will remain in the spin-up or the spin-down state indefinitely. For a nucleus to make a transition from one spin state to the another, an energy of precisely $\Delta E = \hbar\omega_0 = \gamma_n \hbar H_{eff}$ must be provided. When energy of this amount is exchanged with the nucleus, its spin orientation changes from one to the another. As mentioned earlier in Eq. (5.3.21), the Larmor frequency for the nuclear moments is in the megahertz range, the radio frequency (RF) range.

The basic NMR experimental setup consists of a static magnetic field H_{eff}, which splits the nuclear state into two (in general $2I + 1$) states Eq. (5.7.1). One also applies an RF magnetic field H_{RF} of frequency ω perpendicular to the static field H_{eff}. NMR occurs when $\omega = \omega_0 = \gamma_n H_{eff}$. To sweep through the resonance condition, one can either use a fixed H_{eff} and sweep ω, or more practically use a fixed ω and sweep H_{eff}. The above description is for NMR operating in the CW mode in which the RF power is on.

There are two equivalent descriptions of NMR. The first description is RF-induced transitions of nuclear levels as we discussed earlier. The second is the motion of nuclear magnetization in an RF field as described later. Consider a system of spin 1/2 nuclei under a magnetic field H_{eff}, which splits the nuclear state into spin-up and spin-down states, separated by an energy difference of $\Delta E = \gamma_n \hbar H_{eff}$. NMR causes changes in the spin orientation, thus affecting the population of the two spin states, hence the magnetization. Let us assume we start out with a population of N_1 and N_2 for the two spin states. The magnetization of the nuclear magnetic moments is due to the population difference $n = N_1 - N_2$ of the two states. For spin 1/2 nuclei, the nuclear magnetization $\mathbf{M} = (\Sigma\mu_i)/V$ has a value of

$$\mathbf{M} = \gamma_n \hbar (N_1 - N_2)/2 = \gamma_n/2 \tag{5.7.2}$$

with a total Zeeman energy of $-\mathbf{M} \cdot \mathbf{H}_{eff} = -M_z H_{eff}$. Since each of the moment precesses around $\mathbf{H}_{eff}$ with the Larmor frequency, so would $\mathbf{M}$, which has the equation of motion of

$$\frac{d\mathbf{M}}{dt} = \mathbf{M} \times \gamma \mathbf{H}_{\text{eff}} \qquad (5.7.3)$$

If the magnetic moments are not interacting with the surrounding nor with each other, $\mathbf{M}$ would precess indefinitely around $\mathbf{H}_{\text{eff}}$ according to Eq. (5.7.3). The vector $\mathbf{M}$ will be moving in a conic motion in which $\mathbf{M}_z$ will be a constant, and the tip of $\mathbf{M}$ will be tracing out a circle.

If however, as in condensed matter, the magnetic moments interact with the surrounding (called the spin–lattice interaction) and with other moments (called the spin–spin interaction), then energy will be exchanged. In time, $\mathbf{M}$ will gradually evolve to $\mathbf{M}_0$, the equilibrium magnetization, which is

$$\mathbf{M}_0 = \gamma_n \hbar (N_{10} - N_{20})/2 = \gamma_n \hbar n_0/2 \qquad (5.7.4)$$

where the population numbers N_{10} and N_{20} are determined by the Boltzmann factor $N_{20}/N_{10} = \exp(-\Delta E/k_B T)$. If we were to follow the motion of the vector $\mathbf{M}$ visually during its evolution, it would precess but with a changing $\mathbf{M}_z$. The tip of $\mathbf{M}$ would not be tracing out a circle but a spiral.

It is useful to decompose $\mathbf{M}$ in to $\mathbf{M} = \mathbf{M}_z + \mathbf{M}_\perp$, which are the magnetization components parallel and perpendicular to $\mathbf{H}_{\text{eff}}$. During the evolution from $\mathbf{M}$ to $\mathbf{M}_0$, both $\mathbf{M}_z$ and $\mathbf{M}_\perp$ change. However, the total energy depends only on $\mathbf{M}_z$, which varies due to interaction with the lattice (the spin–lattice interaction), whereas the spin–spin interaction involves $\mathbf{M}_\perp$ with no change in energy. Another important characteristic is that T_1, the spin–lattice relaxation time, is much longer than T_2, the spin–spin relaxation time, by a few orders of magnitude. Furthermore, T_1 varies as $1/T$, whereas T_2 is roughly temperature independent. Because $T_1 \gg T_2$, if one performs the NMR experiment within a short time, say of the order of T_2, then the spin–lattice relaxation never has a chance to participate. For all practical purposes, $\mathbf{M}_z$ will remain unchanged and only $\mathbf{M}_\perp$ varies. This advantage is utilized in pulsed NMR, of which spin-echo NMR is one of the most useful for magnetic materials.

Before the description of the spin-echo NMR, we need to discuss $\mathbf{H}_{\text{eff}}$, which is the effective magnetic field that the nucleus experiences. There are several sources of $\mathbf{H}_{\text{eff}}$. The core polarization and contributions from the rest of the condensed matter, particularly those from the nearest neighbors, are the most important. The value of $\mathbf{H}_{\text{eff}}$ is typically in the hundreds of kilo-oersted range, much larger than the actual applied magnetic field. Because the value of $\mathbf{H}_{\text{eff}}$ is sensitive to the surroundings of the nucleus, nuclei will experience different values of $\mathbf{H}_{\text{eff}}$ (hence different Larmor frequency) when they are situated in different environment. NMR is therefore capable of detecting various values of $\mathbf{H}_{\text{eff}}$ and a distribution of $\mathbf{H}_{\text{eff}}$—a capability similar to the ME.

In the spin-echo NMR, one applies RF pulses with well-defined durations and well-orchestrated separations. We will assume that $T_1 \gg T_2$, and that all the events occur during a shorter time period compared to T_1. The distribution of $\mathbf{H}_{\text{eff}}$ are represented by six magnetizations $\mathbf{M}_1$ through $\mathbf{M}_6$, in the order of their Larmor frequencies. A succession of events are schematically shown in Fig. 5.11 and described below:

(a) At $t < 0$, $\mathbf{M}$ is in the z-direction.
(b) At $t = 0$, a 90° RF pulse with the proper duration is applied along the x^*-direction. This 90° pulse rotates $\mathbf{M}$ into the x^*y^*-plane and points in

MAGNETIC CHARACTERIZATION OF MATERIALS

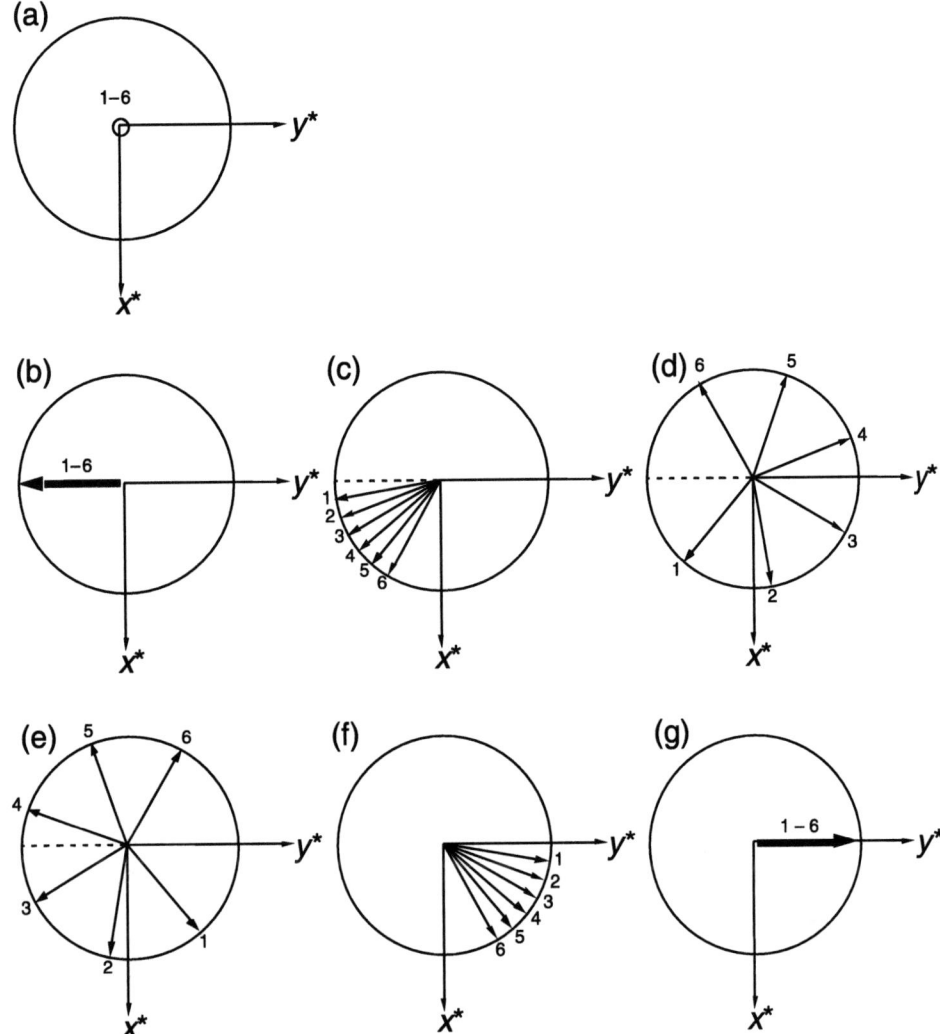

Figure 5.11. Illustration of the spin-echo NMR phenomena (see the text). A distribution of magnetizations is represented by six components labeled 1–6, all of which precess in a counter-clockwise manner. The situation in (b) is right after a 90° pulse has been applied. After a 180° pulse has been applied, the situation changes from (d) to (e) (courtesy of F. Y. Yang).

the $-y^*$ direction. At the instance when the 90° pulse is removed, there is a net magnetization of **M** pointing in the $-y^*$ direction. The NMR spectrometer will duly record a strong signal.

(c) Because there is a distribution of H_{eff}, a faster precessor advances more than the slower precessor, causing **M** to spread, or fan out, while precessing. Of the six representative M_i, M_6 precesses faster than M_5, M_5 precesses faster than M_4, and so on. This is like racing around a circular race track in a counter-clockwise manner, where M_6 is the fastest and M_1 is the slowest. As time goes on, the spread between M_1 and M_6 becomes wider and the net magnetization decreases. The signal in the NMR spectrometer becomes smaller, i.e., the NMR signal is decaying with time.

(d) Since it is a circular track, M_6 approaches M_1 from behind and will eventually overtake M_1. The spread becomes so wide that there is no net magnetization left. The NMR signal has decayed to zero. The total time interval from the instance that the 90° pulse has been applied is t_0.

(e) At $t = t_0$, one applies a 180° RF pulse with a longer duration along the x^*-direction, which causes all the moments to flip 180° about the x^*-axis and still end up in the x^*y^*-plane. Now, there is a complete reversal of fortune. The slowest (M_1) is now leading the pack, while the fastest (M_6) ends up to be the last.

(f) As time goes on, the spread becomes narrower, the net magnetization is now increasing, and so is the NMR signal.

(g) If one examines the spread at around $t = 2t_0$, one can determine the distribution of H_{eff} from the spin-echo NMR spectrum, which is the spin-echo intensity as a function of H_{eff} in T, or RF frequency in MHz, because of Eq. (5.3.21). Note that in the measurement of the spin-echo spectrum, there is no RF field present.

An example of the Co^{59} spin-echo NMR spectrum is shown in Fig. 5.12. Notice that Co in hexagonal close-packed (hcp) and face-centered cubic (fcc) structures, and those in the interfacial regions appear at different locations in the spectrum. In other words, the NMR spectrum provides microscopic information about the Co atoms in the specimen.

There are a large number of nuclei for which NMR is possible but not necessarily simple and easy, hence useful for characterizing magnetic materials.

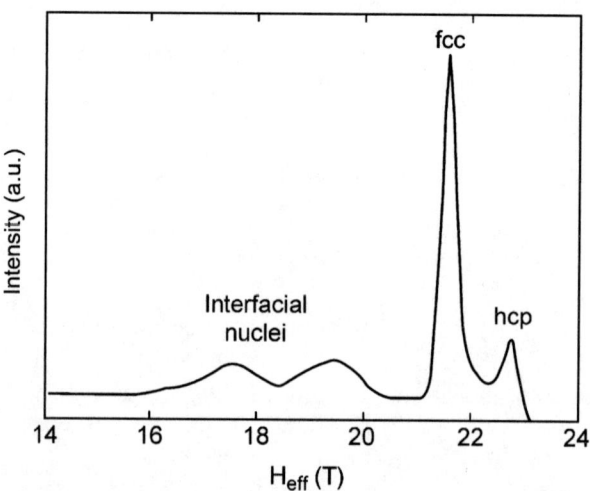

Figure 5.12. A schematic Co^{59} spin-echo NMR spectrum.

Two major considerations are natural abundance of the nucleus and its spin-echo sensitivity. It is customary to compare these values with those of Co^{59}, which has a natural abundance of 100% and a sensitivity of $S = 1$. The sensitivity of $S = 1$ translates to the capability of measuring a sample containing about 10^{18} Co atoms. The other "good" spin-echo NMR nuclei are V^{51} (99.8%, $S = 1.4$), Cu^{63} (69%, $S = 0.3$), Cu^{65} (31%, $S = 0.4$), Mn^{55} (100%, $S = 0.6$), and Al^{27} (100%, $S = 0.7$).

5.8. CONCLUDING REMARKS

In this chapter, we have briefly described the essential information necessary for magnetic characterization. Because of the limitation of space, we have not been able to include other relevant information, other techniques, nor sufficient examples. This chapter only intended to introduce the subject matters and entice the motivated to consult other chapters in this book and more voluminous treatises listed below in the references.

ACKNOWLEDGMENTS

I am grateful to various former and present associates, whose graphs I have used here, and the support from NSF and ONR that sustains our research.

BIBLIOGRAPHY

Magnetism and Magnetic Materials

C. Kittel, *Introduction to Solid State Physics*, 7th ed. (Wiley, New York, 1996).
N. W. Ashcroft and N. D. Mermin, *Solid State Physics* (Saunders, Philadelphia, PA, 1976).
A. H. Morrish, *The Physical Principles of Magnetism* (Wiley, New York, 1965).
B. D. Cullity, *Introduction to Magnetic Materials* (Addison-Wesley, Reading, MA, 1972).

Mössbauer Spectroscopy

T. C. Gibbs, *Principles of Mössbauer Spectroscopy* (Chapman and Hall, London, 1971).
D. P. E. Dickson and F. J. Berry, *Mössbauer Spectroscopy* (Cambridge University Press, Cambridge, 1986).

Nuclear Magnetic Resonance

C. P. Slichter, *Principles of Magnetic Resonance*, 3d Edition (Springer, New York, 1990).
P. Panissod, in *Microscopic Methods in Metals*, Chapter 12 (Springer, New York, 1986).

6

Magnetic Domain Imaging of Spintronic Devices

Robert J. Celotta,* John Unguris,* and Daniel T. Pierce*

6.1. INTRODUCTION

In the analysis of magnetoelectronic structures, the primary goal of a magnetic imaging technique is to provide a spatially resolved picture of the magnetization vector, $\mathbf{M}(x,y,z)$, throughout the sample or device. Knowing this magnetic structure is at the heart of understanding how magnetic devices work. Properties such as the magnetoresistance, the magnetic stray field, and the response to an applied magnetic field all depend critically on the magnetic structure. In this chapter, we will review various methods of imaging magnetic structure. In particular, we will focus on methods that may be applied to spintronic devices and materials.

The magnetization in a magnetic sample can be non-uniform over a large range of length scales for many reasons. For dimensions greater than a micrometer, the most common structures are related to magnetic domains. These are regions where the magnitude of the magnetization is constant, but the magnetization direction varies in order to reduce the stray field emanating from the edges of a magnetized structure. This reduction in energy associated with the long-range magnetostatic interactions is balanced by the cost in the short-range exchange energy of forming domain walls, which typically have dimensions on the order of a tenth of a micrometer. In addition, domain walls may also have internal structure such as chirality and singularities that depend on specific properties of the magnetic system such as film thickness. As the size of the magnetic system becomes smaller, concepts such as domains and domain walls may no longer apply, but the magnetization may still have structure. For example, magnetic particles or lithographically patterned thin film elements that are smaller than a domain wall width might be expected to be single domain, yet these structures quite often exhibit some partial rotation of the magnetization near their edges. Finally, for dimensions less than a nanometer, a continuum picture of uniform magnetization no longer applies and one must consider the magnetization at the atomic scale. For example, different elements in an alloy may have different magnetization, or the atomic moments near an interface may be different than in the bulk.

Although magnetic domain structures can be quite complicated, the fundamental physics governing their formation is well known. The domain structure is

---

*National Institute of Standards and Technology, Gaithersburg, Maryland.

determined by minimizing the total energy, which consists primarily of contributions from the exchange, anisotropy, self-field, applied field, and magnetostriction. In fact one might reasonably ask why we need magnetic imaging techniques at all; why cannot we simply compute the magnetic structure? The answer, for systems larger than a few micrometers lies in the complexity and variety of possible domain structures. In such cases, magnetic domain theory can be used to *explain* the observed domain structures, but theory usually cannot *predict* the domain structure. For smaller, simpler structures, e.g., single domain patterned thin film elements, micromagnetic modeling has become much more successful at predicting magnetic structures.[1] Real magnetic devices, however, are usually more complicated than the simple model structures, so that magnetic imaging tools are still necessary in order to understand why a real magnetic device does not behave in the same way as its model counterpart.

We can ask what sort of information an ideal magnetic imaging technique would provide. It should be able to image the magnetic structure with nearly atomic resolution both laterally and as a function of depth. It should be able to image the magnetization while applying an arbitrary magnetic field. The imaging technique should be fast enough to follow the magnetization dynamics on a timescale comparable to that of the spin precession. To be useful to device manufacturers, the technique should be able to image the magnetization in working devices that may be buried under non-magnetic overlayers or deposited on top of complex structures. Finally, all of this imaging must be done without disturbing the magnetic structure of the device. No single, current imaging technique can satisfy all of these demands, but some are better at particular aspects. In order to get a more complete picture of the magnetic structure, therefore, multiple complementary imaging methods should be used whenever possible.

There are several general comments that apply to every magnetic imaging technique we discuss. First, the contrast is related either to a sample's magnetization or to the magnetic field produced by that magnetization. While the magnetic field can always be calculated from the complete magnetization distribution, determining a unique magnetization distribution from the magnetic field is not possible. However, in the case of magnetic recording media, for example, the stray field image may be the desired information. Second, probing depths limit all of the methods to, at most, the top few micrometers of a sample. This surface sensitivity is usually not a problem for thin films, but it means that bulk domain structures cannot be directly determined. Third, all of the techniques require smooth, damage-free surfaces. In general, surfaces need to have a mirror finish, so that topography or surface stress does not affect the measurement. Fourth, correlations may exist between parameters describing a technique. For example, it may be impossible to achieve a technique's ultimate magnetic sensitivity at the highest possible resolution. Finally, most of the techniques use digital signal acquisition and rely heavily on modern image processing tools.

In the following sections, we discuss several techniques that are particularly well suited for imaging the magnetic microstructure in the types of structures one might encounter in dealing with spintronic devices. Specifically, we limit our discussion to scanning electron microscopy with polarization analysis (SEMPA), transmission electron microscopy (TEM), magneto-optic, magnetic force microscopy (MFM), and magnetic circular dichroism (MCD). More complete reviews of these and other domain imaging techniques are available.[2,3]

6.2. SCANNING ELECTRON MICROSCOPY WITH POLARIZATION ANALYSIS

Scanning electron microscopy with polarization analysis directly provides an image of the surface magnetization of a sample. SEMPA measures the spin polarization of the secondary electrons that exit from a magnetic sample as the finely focused (unpolarized) beam of the scanning electron microscope rasters over the sample, as shown schematically in Fig. 6.1. SEMPA depends on the fact that the polarization of the secondary electrons reflects the net spin density of the material.

For the purposes of SEMPA, it is sufficient to treat each component of the vector polarization separately. The polarization along the *x*-direction is

$$P_x = (N_\uparrow - N_\downarrow)/(N_\uparrow + N_\downarrow) \quad (6.2.1)$$

where $N_\uparrow$ ($N_\downarrow$) are the number of electrons with spins parallel (antiparallel) to the *x*-direction. Measurements of the energy distribution of spin polarized electrons from a ferromagnet showed a significant polarization of the secondary electron peak suggesting[4] the possibility of using this effect for magnetic imaging.[5,6] There is even an enhancement of the polarization at low secondary kinetic energy. In a ferromagnetic material, there are more unfilled minority (down spin) states for electrons to scatter into during the secondary cascade process, thereby preferentially filtering out minority spin electrons and increasing the polarization.[7] At higher kinetic energies, between 10 and 20 eV, there is reasonably good agreement between the measured polarization for Fe, Co,[8] and Ni[9] and the expected polarization, 28, 19, and 5% for Fe, Co, and Ni, respectively. Here, we assume the cascade electrons represent a uniform excitation of the valence band. In this simple model,

$$P = n_B/n_v \quad (6.2.2)$$

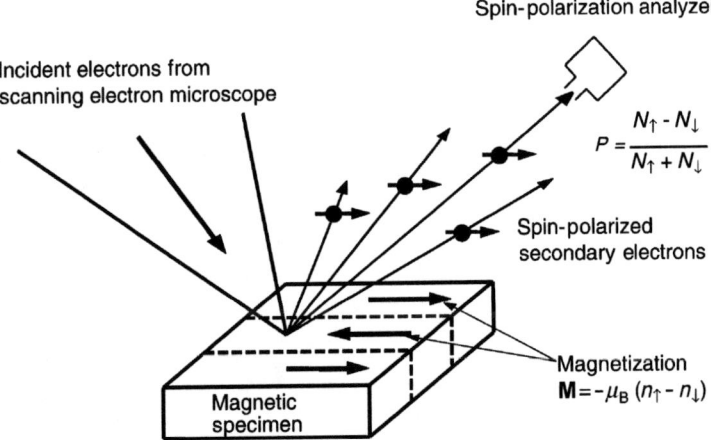

Figure 6.1. A schematic depiction of a SEMPA apparatus. Spin-polarized secondary electrons, emitted when a finely focused incident electron beam hits the sample, are spin analyzed to determine the magnetization direction in the region under the incident beam.

where n_v is the total number of valence electrons per atom and n_B is the number of Bohr magnetons per atom. Since the Bohr magneton number is the net spin density per atom $(n_\uparrow - n_\downarrow)$, the measured polarization is directly related to the spin part of the magnetization, which is

$$\mathbf{M} = -\mu_B(n_\uparrow - n_\downarrow) \tag{6.2.3}$$

The magnetization and polarization are of opposite sign because the electron spin magnetic moment (units of the Bohr magneton, μ_B) and the electron spin are in opposite directions. The spin part of the magnetization is a close approximation to the total magnetization in a transition metal ferromagnet where the orbital moment is quenched.

In this chapter, our aim is to illustrate the features of SEMPA, both strengths and weaknesses, which should be considered for applications to domain imaging of spintronic devices. Further discussion of the detailed implementation of SEMPA can be found elsewhere.[10-13]

The spatial resolution is determined largely by the electron beam diameter of the SEM. Because the beam current decreases rapidly with decreasing beam diameter, there are practical resolution limits. These restrictions are determined by the current required to obtain a polarization image in a reasonable time (limited by sample drift, deterioration of the sample surface, and operator patience). As a rule of thumb, a beam current of ~1 nA is required to obtain a SEMPA image in about 1 h; this leads to resolution limits of approximately 50 nm for LaB_6 and 10 nm for field emission SEM electron gun cathodes. Better spatial resolution is obtained for SEM intensity images that do not suffer from the inefficiency of electron spin polarization analyzers.[10-13]

SEMPA is a surface sensitive technique with a probing depth of about 1 nm because of the short escape depth of secondary electrons. Therefore, sample surfaces must be clean; contaminants would dilute the polarization or, in the case of a thick overlayer, obscure it completely. Conventional surface science preparation techniques are used to prepare SEMPA samples *in situ*. Samples fabricated elsewhere and inserted into the SEM will have a layer of surface contamination, such as a surface oxide, that can be removed by ion bombardment. Depending on the sample, the ion bombardment may be accompanied by annealing to relieve any induced strain. An ultrahigh vacuum environment is required in the SEM chamber. Commercial scanning electron microscopes with ultrahigh vacuum capability are usually sold as scanning Auger microprobes. Compositional mapping is then also available to correlate with the magnetic images. In certain cases, ion milling and the Auger analysis may be combined with SEMPA to depth profile the magnetic structure of multilayer devices, e.g., a Cu/Co GMR structure.[14] SEMPA enjoys other advantages that are typical of an SEM. These include a large depth of field, easily variable magnification to look at regions of the surface ranging from a few millimeters to a few hundred nanometers, and a large working distance. Image acquisition time runs from 1 to 100 min, depending on resolution, magnetization, and image size. SEMPA has a high sensitivity and can detect a couple of tenths of an atomic layer of Fe, which at high resolution corresponds to about 10^3 atoms or 10^{-17} emu. On the negative side, stray magnetic fields (≥ 10 Oe) must be avoided. In addition, to avoid charging effects, conductive samples are required.

An example of a SEMPA magnetization image of the surface of a Fe–3%Si single crystal is shown in Fig. 6.2. The intensity and two components of the magnetization are imaged simultaneously in a SEMPA measurement. The intensity image of Fig. 6.2(a) is the familiar topography image of the SEM. This image is obtained from the sum $(N_\uparrow + N_\downarrow)$, which appears in the denominator of Eq. (6.2.1). Spin analyzers are capable of measuring two components of the magnetization which in Fig. 6.2(b) and (c) are the x and y components in the plane of the sample. The magnetization is expected to lie in plane to minimize the magnetostatic energy. Out-of-plane magnetization, which is present in special circumstances, can be measured by electrostatically diverting the secondary electrons to a spin analyzer at 90° with respect to the one that measures the in-plane components[15] or by using a spin rotator.[16] In the M_x image, white corresponds to magnetization to the right and black to the left. The intermediate gray regions correspond to magnetization in the y-direction which are seen as white and black in the M_y image for up and down magnetization, respectively. The magnetization is expected to be uniform in magnitude which can be tested by processing the signal to obtain the magnitude, $(M_x^2 + M_y^2)^{1/2}$, shown in Fig. 6.2(d). The reduced magnetization at the domain walls is an artifact and does not represent an out of plane M_z component of a Bloch wall. Close inspection at high resolution reveals that the walls at the surface are in-plane Néel caps on the interior Bloch walls. The domain wall artifacts in image Fig. 6.2(d) occur for a lower-resolution image such as this when the beam is wider than the wall.

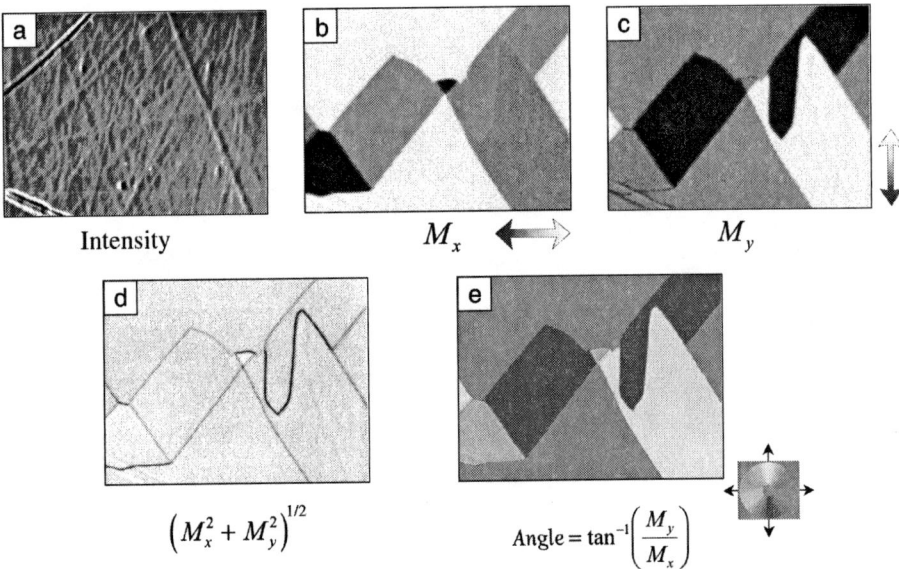

Figure 6.2. SEMPA images of an Fe–3%Si single crystal. (a) An intensity image showing the topography as would normally be seen with an SEM. (b) Polarization image where the gray scale intensity is proportional to the component of the magnetization in the x-direction. (c) Polarization image for y component of magnetization. (d) Image of the magnitude of the observed polarization computed from the measured M_x and M_y components. (e) Image giving the direction of the magnetization as computed from the measured M_x and M_y components.

Then, electrons excited from each side of the wall add to a zero polarization for a 180° wall and a reduced polarization for a 90° wall. In Fig. 6.2(e), the angle of magnetization, which equals $\tan^{-1}(M_x/M_y)$, is calculated and plotted in color with the directions corresponding to the accompanying color wheel.

The images of Fig. 6.2 illustrate some characteristics of SEMPA. First, the magnetization vector **M** is imaged, not the stray magnetic field **H**. Any two components of **M** can be simultaneously imaged with the topography. The images are formed with the same electrons from exactly the same sample area so features present in the magnetic microstructure can be directly correlated with the surface topography. The magnetization image should be independent of the topography as can be seen from Eq. (6.2.1). Nevertheless, sometimes "topographic feedthrough" can be seen as is evident from scrutiny of the lower left region of Fig. 6.2(a) and (c). A procedure involving a second measurement exists to minimize the topographic feedthrough further.[13] Topographic feedthrough is a particular problem at edges, such as would be encountered in a thin film magnetic element, and special care is necessary to measure such structures at high resolution.

As an illustration of the SEMPA technique, we show some results of SEMPA applied to the study of the coupling of ferromagnetic layers separated by nonmagnetic layers. Artificially layered magnetic structures allow one to tailor transport and magnetic properties to fit special requirements, such as for giant magnetoresistance or spin valve devices. As a first example in Fig. 6.3, we show three stages of creating an Fe/Cr/Fe(001) trilayer. The top panels, a–c, display M_y, and the bottom panels, d–f, display M_x. Figure 6.3(a) and (d) at the left show magnetization images

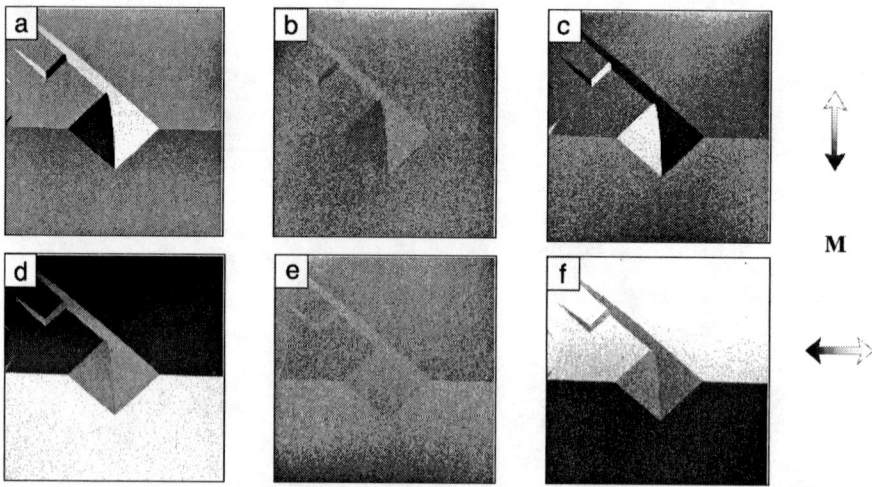

Figure 6.3. SEMPA images displaying the magnetization observed in different layers of an Fe/Cr/Fe trilayer. (a) Shows the M_y component of the magnetization of an interesting domain structure on an Fe whisker substrate. (b) Shows the same region after deposition of 1 nm of Cr. (c) Shows the same region following the addition of 2 nm of Fe to make a Fe/Cr/Fe trilayer. Note the reversal of magnetization directions between (a) and (c). (d–f) The M_x component of magnetization for the same conditions as in (a–c).

of the clean Fe(001) single crystal whisker. There is a domain wall running horizontally, with M_x in the bottom of the whisker to the right and the top half to the left. There is a diamond-shaped domain in the middle with magnetization exhibited in the M_y image. The middle panels show this same region of the whisker after deposition of 1 nm of Cr. The domains of the whisker are largely obscured but can still be discerned through the 1 nm of Cr, giving a vivid visual demonstration of the probing depth of SEMPA. The images at the right show the magnetization at the surface of a 2 nm Fe layer deposited on top of the Cr. Note that each component of the magnetization in this trilayer sandwich is antiparallel to that on the clean Fe whisker. This antiparallel coupling of two Fe layers separated by a particular thickness of Cr was first observed by Grünberg et al.[17] Parkin et al.[18] subsequently found that the coupling of the layers oscillated between antiferromagnetic and ferromagnetic depending on the thickness of the spacer layer.

In order to study the dependence of the interlayer coupling on the thickness of the spacer layer, SEMPA was applied to measure the magnetization of the top layer of a structure like that shown in Fig. 6.4(a).[19] In this structure, the thickness of the Cr spacer is varied continuously in a very shallow wedge that increases approximately 10 nm in thickness over approximately 1 mm in length. In contrast to the alternative method of creating trilayer structures as in Fig. 6.3, where fluctuations in preparation conditions could occur from one film to the next, the wedge structure provides all spacer layer thicknesses in a single deposition. In the case of layer by layer growth of the Cr, it is possible to use the SEM, equipped with a phosphor screen below the sample, to do reflection high energy electron diffraction (RHEED) along the wedge. In this way, one can obtain a very accurate measure of the wedge thickness from the RHEED intensity oscillations.[19]

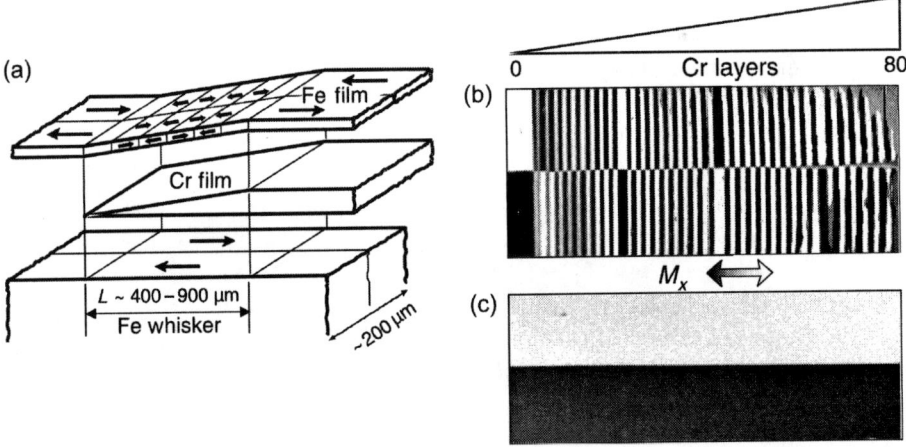

Figure 6.4. (a) A schematic depiction of the Fe/Cr/Fe wedge geometry, as used in SEMPA. It consists of a single crystal, Fe whisker substrate with two oppositely directed domains, an epitaxial Cr layer with a linearly varying average thickness, and an epitaxial Fe overlayer. (b) The M_x component showing the alternation of magnetization direction of the Fe overlayer with each single layer increase in the Cr layer thickness. (c) A SEMPA measurement of the magnetization of the bare Fe substrate. Modified from Ref. 19.

Magnetization images from this wedge structure are shown in Fig. 6.4(b) and (c). The M_x image of the clean Fe whisker with two domains is shown in Fig. 6.4(c). The magnetization of a 2 nm Fe layer grown on top of the Cr wedge is shown in Fig. 6.4(b). The Cr was grown at an Fe substrate temperature of approximately 300°C which produces layer by layer growth. The magnetization of the Fe overlayer is coupled ferromagnetically to the whisker up to a thickness of four Cr layers and then reverses the direction of the coupling with each additional Cr layer thickness. This reversal continues until at a thickness of 24 layers, the Cr thickness increases by two layers before the Fe magnetization reverses. This phase slip in the reversal process that occurs again at 44 and 64 layers results because the period of the oscillatory exchange coupling is measured to be 2.105 ± 0.005 layers, which is slightly incommensurate with the lattice spacing. Such a precise determination of the period by SEMPA was possible because of the many oscillations observed and the accurate thickness measurement from RHEED intensity oscillations. SEMPA, which is uniquely suited for such studies, provided precise measurements of the periods of oscillatory coupling which could be tied to Fermi surface properties of the spacer layer. On the other hand, to measure the strength of the coupling requires the application of magnetic field that would disturb the secondary electrons. For coupling strength measurements, samples prepared and checked in the SEMPA system had to be coated with an Au protective layer for measurement in a Kerr microscope[20] as described in Section 6.4.

As a final example of SEMPA applications, we show a magnetization image of an array of Fe nanowires in Fig. 6.5. The array was fabricated by oblique deposition of Fe onto a template of Cr lines made by laser-focused atom deposition.[21] The Fe wires are approximately 100 nm wide and 0.15 mm long, spaced by 213 nm. The low magnification image of Fig. 6.5(a) shows the Fe nanowires as white lines with magnetization M_y up, and black lines with magnetization down. The gray regions between the white and black domains correspond to the nonmagnetic Cr underlayer exposed between the magnetic Fe lines. Where there is a magnetization reversal (change from white to black in the image), there must be a domain wall.

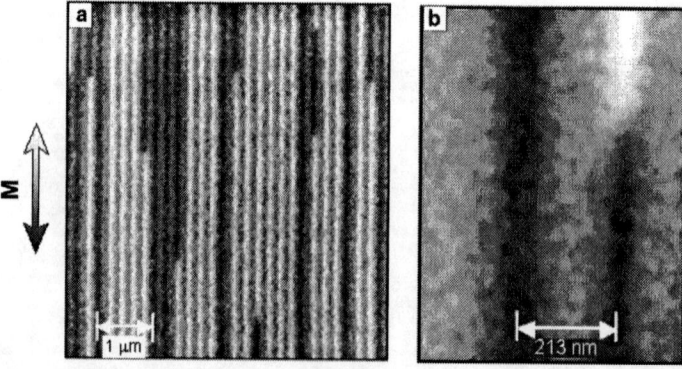

Figure 6.5. (a) SEMPA image of an array of Fe nanowires showing their in-plane magnetization, M_y, where white (black) indicates up (down). (b) A high-resolution SEMPA image showing a domain wall within a wire. From Ref. 21.

A high-resolution image of such a region is shown in Fig. 6.5(b). It can be seen that the domain wall forms at an angle of 45° with respect to the Fe lines.

In summary, SEMPA has a number of features useful for measurements on spintronics devices. It provides images of two components of the magnetization that are independent of the topography, which is imaged simultaneously. It is a non-perturbative measurement that offers high spatial resolution. At the same time, SEMPA maintains the depth of field capability of an SEM and the ability to look at large areas and then magnify regions of interest. It is a surface sensitive technique, which can be advantageous for investigating thin film devices.

6.3. MAGNETIC FORCE MICROSCOPY

Magnetic force microscopy[22-24] developed as an extension to atomic force microscopy (AFM),[25] is the most widely used technique for domain imaging on the nanometer scale. In AFM, a sharp tip, i.e., 10–20 nm in radius, is scanned along the surface of the sample in a non-destructive manner. This is accomplished by mounting the sharp probe at the end of a highly compliant cantilever and sensing the interaction between tip and surface in a variety of ways. In the contact mode, the deflection of the very soft cantilever is detected and used to generate a topograph. In the non-contacting mode, the attraction of the long-range van der Waals force is sensed through its effect on the natural resonant frequency of the cantilever. Finally, in the intermittent contact mode the tip probes the repulsive part of the van der Walls potential, i.e., taps the surface, once for each oscillation.

Magnetic sensitivity is obtained by coating the silicon tip of the cantilever with a ferromagnetic coating to add a magnetic interaction between the tip and the surface. Although a smooth surface is needed, it is possible to sense the magnetic field through non-magnetic layers so surface cleanliness is not of great importance. This allows for in-air imaging and greatly simplifies imaging structures that, in practice, use coatings. Resolutions in the range of 40–90 nm have been demonstrated[26] and non-conductive samples can be used. It is also convenient to be able to apply an external magnetic field of up to about 800 kA/m in strength. As a relatively inexpensive add-on to commercial AFMs, this technique has found a wide audience, particularly within the magnetic information storage industry.

While it is invitingly simple to view this as the interaction of a "magnetic monopole" tip experiencing a force dependent on the magnetic field above an unperturbed ferromagnetic sample, the real situation is far more complex. In actuality, the tip and the sample each consist of a distribution of both surface and bulk magnetic charges that may interact with each other. Interpreting an MFM image means understanding the relationship between the two distributions of magnetic multipoles, the exact nature of which may vary to minimize the total magnetostatic energy of the system.

In domain imaging, the objective is generally to visualize the magnetization of a sample. In the ideal case, an MFM measurement would produce an image related to the stray field above the sample and from this measured field distribution the sample magnetization would be deduced. Unfortunately, it is not possible to uniquely determine the sample magnetization from the field distribution. Further, the imaging mechanism for the MFM is complex and the instrument response is

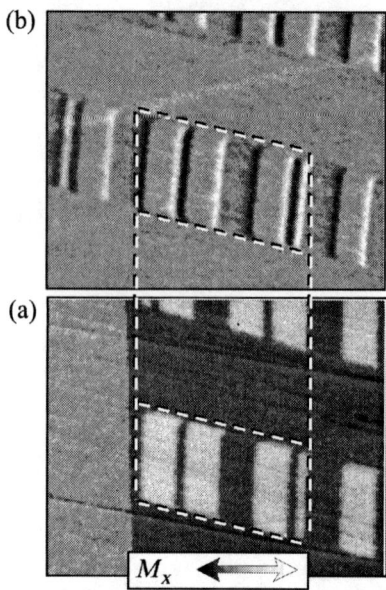

Figure 6.6. (a) A SEMPA image showing the M_x component of the magnetization of a high density bit pattern written on a hard magnetic disk. (b) Same structure shown in (a) as imaged using MFM. An Au mask, which can be seen in (a), covers the left-hand side of the sample. From Ref. 27.

difficult to quantify. Generally, the MFM does not respond to the magnetic field, but is usually much more sensitive to the higher spatial derivatives of the field components. This means the MFM generally produces images that denote the positions of magnetic charge, e.g., the magnetic charge present inside of domain walls. Such a charge can be thought to arise from the divergence of the magnetization, i.e., $-\nabla \cdot \mathbf{M}$, internally and at surfaces.

In order to discuss how magnetization information might be inferred from MFM data, and better understand the MFM imaging process, we will compare measurements of the same domain structure as seen by SEMPA and MFM. In Fig. 6.6(a), we see a SEMPA[27] image of the magnetization of a magnetic bit pattern written on a thin film recording disk[28] and in Fig. 6.6(b) an MFM image[27] of exactly the same region. Note how the Au mask reduces the magnetization measured by SEMPA to zero while the long range of the magnetic field allows the MFM to still image the region through the Au.

Ideally, given the magnetization determined by the SEMPA measurement, it should be possible to predict the MFM response. The line traces shown in Fig. 6.7 were calculated based on a bit pattern seen in SEMPA results similar to those of Fig. 6.6(a). The top trace reflects the x component of the magnetization, M_x, as seen in SEMPA. White (black) areas are magnetized to the right (left). The next lower trace depicts the x component of the field, B_x, as calculated from M_x, at a distance of 100 nm above the sample surface. Next, we calculate and plot B_z, the z component perpendicular to the surface, and the first and second spatial derivatives of B_z. Assume the tip to be a monopole, i.e., it is long, slender, and magnetized along its length (in the z-direction) so opposite poles are located either near the surface or far from the sample's fields. The interaction between tip and B_x would then produce a force in the x-direction that would not be sensed by the MFM, and an interaction with B_z that would. Note from the B_z trace that the domain walls appear as either

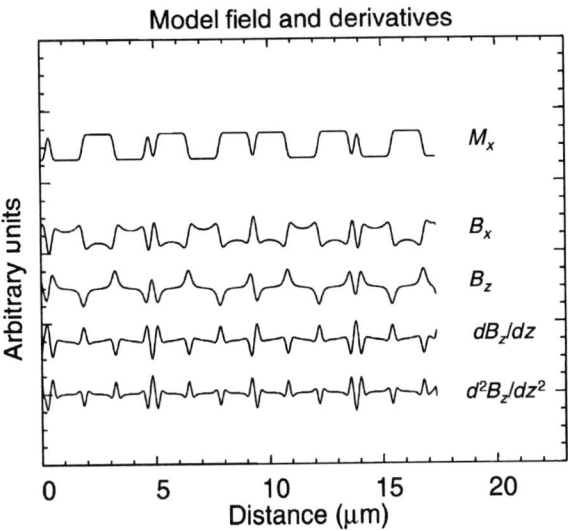

Figure 6.7. A magnetization distribution modeled after SEMPA measurements similar to those shown in Fig. 6.6(a) and several fields and field gradients derived from that distribution. From Ref. 27.

maxima or minima while within the domains the value of B_z is roughly the same independent of the sign of M_x. This argument would predict that MFM domain images of recorded bits would tend to have alternating bright and dark domain boundaries with the bits themselves appearing gray, independent of their magnetization direction, exactly the appearance seen in Fig. 6.6(b). Notice how the alternating light and dark domain edges give the illusion of height to some bits and appears to distinguish between those magnetized to the right and to the left. A line trace through the MFM image of these bits shows little or no difference in intensity between domains magnetized to the right or left.

An alternative imaging model involves sensing the force gradient, either by measuring the effect of the field gradient on a monopole tip oscillating above the surface or the force on a static dipole tip. Either of these cases corresponds to the derivative curve in Fig. 6.7; were a dipole tip to be used in the oscillating mode, the second derivative curve of Fig. 6.7 would best reflect the interaction. With the exception of the B_x curve, it can be seen that all of the other curves are similar. None display domain contrast; all have structure at domain walls, i.e., at the location of magnetic charge.

Understanding MFM imaging in terms of magnetic charge contrast[29] is dependent on the tip–sample interaction being sufficiently weak as to preclude changes in the magnetic structure of either. Ideal tips would be both hard, e.g., unchanging in their magnetic structure in the sample's field, and simultaneously magnetically weak, e.g., producing a field adequate enough to allow a measurable interaction yet too small to affect the magnetization distribution in the sample. Changes in the magnetic state of the sample or tip that occur for overly strong interactions may be either reversible or hysteretic. The latter are demonstrated when MFM tips are used to set the state of a magnetic bit,[30] for example. Such changes may limit the applicability of MFM imaging in the study of low coercivity thin films used in spintronic devices.

Changes in height during scanning introduce another complication in MFM image interpretation. In trying to predict the MFM instrument response to the fields calculated in Fig. 6.7, it is important to remember that each trace is calculated for a fixed tip–sample distance. The MFM may vary this distance during each scan to maintain a constant force. More significant is the fact that surface topography variations will feed through and appear in the MFM magnetic image. Techniques have been developed to minimize this effect.[31],* The most popular method, called the interleave or "liftmode,"* makes two scans. First, a scan of the surface is made in the intermittent contact mode to determine the topography. Second, a scan is made in which magnetic forces are sensed while the tip is programmed to follow a constant height contour determined from the topography. In this way, separate topographic and magnetic images can be obtained. Of course, tapping the surface with a magnetic tip in the initial scan may significantly modify the magnetic structure of the sample or the tip.

The MFM offers several important advantages when applied to spintronic devices. It is readily available as a commercial instrument that will provide high-resolution images of domain wall locations with resolution in the tens of nanometers range. External magnetic fields can be applied to observe device response and the MFM can image through non-magnetic overlayers. The sample need not be thinned, so spintronic devices on thick substrates present no difficulty.

Sample topography can present a problem that needs to be considered. Imaging is over a limited area and is slower than most of the other methods so it may be necessary to use MFM in conjunction with a different survey method or have other means of locating the area of study. Image formation is complex and images may be difficult to interpret in terms of a unique distribution of the underlying magnetization. This will present a particular problem in small spintronic devices where there are no domain walls, but it is the subtle changes in magnetization direction that are important to image and understand. Perhaps most difficult is obtaining a probe that will neither modify the magnetic configuration of the device under study[32] nor change its own magnetic configuration.

6.4. MAGNETO-OPTIC IMAGING

The weak interaction between polarized light and a material's magnetization leads to a large variety of very useful domain imaging methods. The primary magneto-optic interactions depend directly on the magnetization, so that most magneto-optic techniques directly image the magnetic structure, or at least the part of the magnetization that is optically active. For optically transparent samples and transmitted light, the magneto-optic interaction is usually referred to as the Faraday

---

*Digital Instruments, Santa Barbara, California. Certain commercial equipment, instruments, or materials are identified in this paper in order to specify the experimental procedure adequately. Such an identification is not intended to imply recommendation or endorsement by the National Institute of Standards and Technology, nor is it intended to imply that the materials or equipment identified are necessarily the best available for the purpose.

effect, while for reflected light the interaction is commonly called the Kerr effect. The basic magneto-optic interactions, however, are the same for reflected and transmitted lights. Several excellent comprehensive reviews of magneto-optical domain imaging are available.[3,33]

The physics of the complete magneto-optic interaction is well understood, but somewhat complex.[2,34] The primary magneto-optic interactions involve the frequency-dependent interaction of the electric field vector of the polarized light and the dielectric permittivity tensor of the magnetic material. For the purposes of this discussion, however, it is sufficient to treat the magneto-optic interaction simply as a rotation of the polarization plane of linearly polarized light upon reflection from, or transmission through, a magnetic material. Note, however, that this rotation may also be accompanied by a change in phase leading to elliptical polarization of the light.

The relative geometry between the magnetization, **M**, and the polarization, **E**, vectors determines which component of the magnetization will be visible in a particular magneto-optic image. These geometric sensitivities can be determined using a classical Lorentz force picture of the magneto-optic interaction. Polarized light causes electrons in the sample to oscillate with velocity along **E**. The sample magnetization interacts through the Lorentz force with the electrons generating a small polarization component in the direction of $-\mathbf{M} \times \mathbf{E}$. However, one must be cautious applying this classical picture. Although it is useful in determining the direction of the magneto-optic signal, the interaction is not due to a classical Lorentz force but to a relativistic spin–orbit interaction with the solid.

Figure 6.8(a) and (b) shows geometries for observing magnetization perpendicular to and in the sample surface plane, respectively. The polar Faraday or Kerr effects are used to image perpendicular magnetization. From Fig. 6.8(a) and applying the simple Lorentz force picture, one can immediately see that the contrast for perpendicular magnetization is maximized for normally incident, $\theta = 0$, light and is roughly independent of the incident polarization direction. For in-plane magnetization, several scattering geometries are possible. Figure 6.8(b) shows the most common arrangement known as the longitudinal effect in which the magnetization lies in the scattering plane of the light. The longitudinal effect vanishes for normal incidence light and is a maximum for $\theta \sim 60°$. The longitudinal effect can also be observed for polarization perpendicular to the optical scattering plane. The trans-

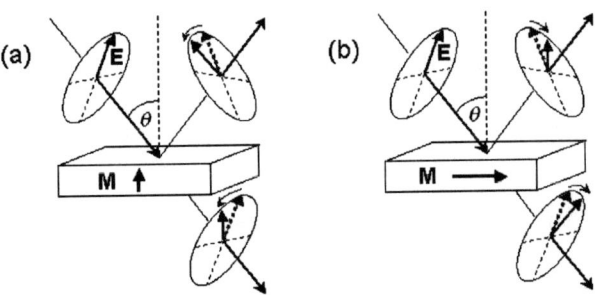

Figure 6.8. Light and magnetization geometries for observing the polar (a), and longitudinal (b) magneto-optic effects for either reflected or transmitted light. Modified from Ref. 2.

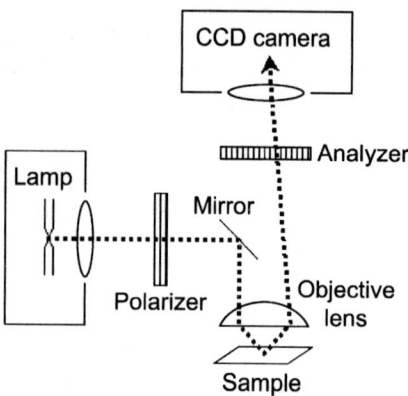

Figure 6.9. A schematic of a magneto-optical microscope for high-resolution imaging of magnetic microstructure in the longitudinal Kerr mode. Modified from Ref. 2.

verse effect allows the observation of magnetization, that is in the sample plane, but perpendicular to the scattering plane. The transverse mode is used much less, however, since it does not induce a change in polarization, but only a change in intensity for reflected light and no magneto-optic effect for transmitted light.

Domain imaging instrumentation based on the magneto-optic effects is conceptually straightforward. One simply needs to shine polarized light on a sample and look at the reflected or transmitted light through a crossed polarizer. In practice, however, high-quality optics and significant image processing are required to separate the small magneto-optic signal from the large non-magnetic background.[35–37] A schematic of a high resolution magneto-optic microscope for imaging in-plane magnetization with the longitudinal Kerr effect is shown in Fig. 6.9. These microscopes are usually based on high quality polarized light microscopes that have been specifically modified for magneto-optic imaging. In this case of longitudinal Kerr imaging, the incident and reflected light both pass through the objective lens, and the required oblique illumination and separation of incident and reflected light is obtained by suitable placement of mirrors and apertures. One interesting feature of this design is that the magnetic contrast increases as the resolution increases, since the numerical aperture increases with increasing magnification resulting in illumination that is more oblique. Typical resolution with this type of microscope is about 1 μm, but using a high numerical aperture oil immersion objective lens and blue light illumination such a microscope can achieve a diffraction limited resolution of 0.3 μm.

Other variations of the basic magneto-optic microscope are possible, each with its own specific advantages.[33] For example, scanned images can also be generated by using a focused laser beam for illumination and rastering either the laser beam or the specimen.[38] The scanned images take longer to acquire than conventional microscopy, but the laser provides very intense illumination and hence a large signal. For larger fields of view and lower magnifications, the illumination and imaging optics for longitudinal imaging are usually separated; however, this leads to imaging a tilted sample and the associated depth of field problems.

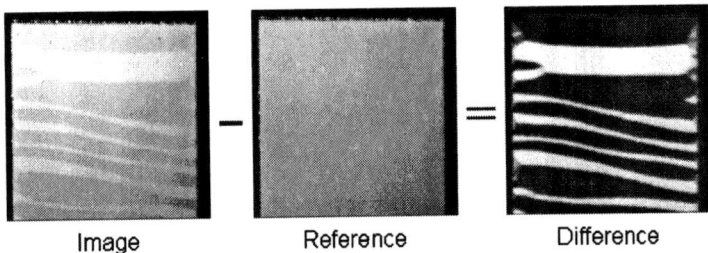

Figure 6.10. An example of digital image subtraction to reduce the non-magnetic background signal in a longitudinal Kerr image. A reference image of the Permalloy rectangle, acquired in a magnetic field large enough to saturate the magnetization, was subtracted from an image acquired at zero applied field.

The small size of the magneto-optic signal, especially in the case of the longitudinal Kerr effect, means that some sort of signal processing is usually required to extract a satisfactory magnetic image. The most common method is to remove the large non-magnetic background signal by taking the difference between the current image and a stored reference image usually taken in a field large enough to saturate the magnetization. In practical terms, this means that magneto-optic imaging is best applied to samples where the magnetization can be changed by applying a field. For static magnetic domain patterns that one cannot or does not want to alter, such as written bits in recording media, high quality, magneto-optic images are difficult to acquire.

Image subtraction is most easily done digitally, so that digital image acquisition with large dynamic range (at least 12 bits) is essential. An example of this method is shown in Fig. 6.10. In this case, the image acquisition and processing with a digital CCD camera and standard software took about 10 s. Near video rate processing is also possible by using a video rate CCD and dedicated image processing electronics.[35,36]

Perhaps the greatest advantage of magneto-optic imaging is the speed with which magnetic images can be acquired. Magneto-optic imaging can yield a great deal of information about magnetization dynamics in a magnetic material or device, since arbitrarily large magnetic fields may be applied to the sample while imaging. While video rate imaging of domain dynamics is routinely achieved using standard arc lamp illumination, pulsed laser illumination can reveal domain wall motion that occurs over timescales as short as a few nanoseconds. Transient changes can occasionally be captured in a single pulse, while reproducible domain motion can be imaged stroboscopically.[39] Some of the best examples of high-speed magneto-optic imaging can be found in studies of domain dynamics in thin film recording heads.[40,41]

The information depth in the Kerr imaging mode is determined by the penetration depth of the light and is about 20 nm in a metal. The technique is, therefore, moderately surface sensitive and can be used to image magnetic domains that are only a few monolayers thick, as well as domains that are coated with thin non-magnetic coatings. Figure 6.11 compares the probing depth of Kerr with that of SEMPA with domain images of a multilayer structure similar to the one described in Fig. 6.4.[20] While SEMPA only sees the magnetization of the top Fe film, the Kerr image contains magnetic contrast from both the top Fe film and the Fe whisker

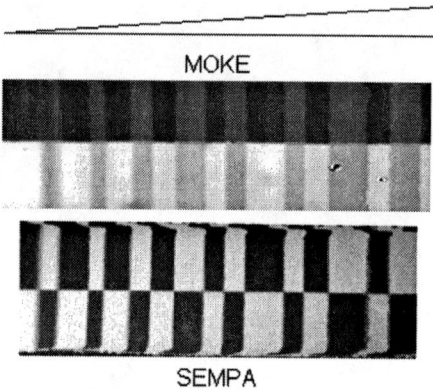

Figure 6.11. Magneto-optic and SEMPA images of the same multilayer structure. The difference in the appearance between the two images is due to the different probing depths of the two methods. The sample consists of an Fe whisker substrate, 0–23 layer Au wedge, 12 layers of Fe, and six Au layer overcoat. Modified from Ref. 20.

substrate. Hence, the reduced contrast for Au spacer layer thicknesses where the Fe film magnetization is opposite to that of the Fe whisker.

The magneto-optic signal can also provide additional information about the layer-dependent magnetization in a magnetic multilayer, because the phase of the reflected Kerr amplitude depends on the depth and the interfaces.[42] This difference in phase can be exploited to selectively cancel the magnetic contrast from a particular layer of the sample, and obtain depth-dependent magnetization information. Figure 6.12 shows an example of this layer-dependent cancellation in an Fe/Cr/Fe wedge multilayer similar to the one described in Fig. 6.4. As the analyzing polarizer angle is rotated, the images either show magnetic contrast from both the top Fe film and the Fe whisker substrate, or from just the Fe film or Fe substrate alone.

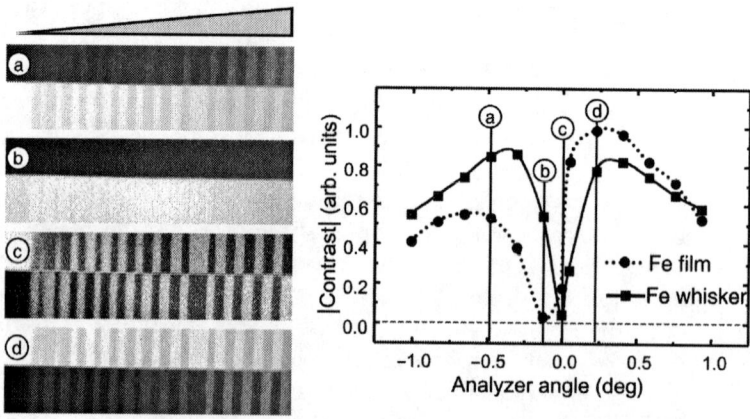

Figure 6.12. (a–d) Magneto-optic images of an (Fe whisker)/(Cr wedge)/(Fe film) structure as the analyzer is rotated. Magnetic contrast from the top film is extinguished in image (b), whereas the Fe whisker contrast is extinguished in (c). The absolute value of the contrast is shown in the adjoining plot.

Unfortunately, since this method only cancels the magnetic contrast from one layer at a time, it is difficult to sort out the magnetic structure if more than two magnetic layers are present.

The small sampling depth and topographic sensitivity of the Kerr imaging mode require preparing samples that have optically flat damage free surfaces. Bulk samples can be prepared by mechanical polishing, followed by chemical polishing or annealing to remove the remaining damage. High-quality surfaces for imaging can also be generated by evaporation or electrodeposition of thin films on flat, polished substrates. Samples may also be coated by thin, nonmagnetic films, without significantly affecting the magneto-optic images. In fact, appropriate antireflective coatings can be applied to samples in order to increase the magneto-optical contrast.[43]

There are also several new developments in magneto-optical imaging that are worth noting because of their potential future impact. First, near-field optical techniques are being used to overcome the resolution limits of conventional diffraction limited optics. By using scanned apertures or tips much closer than an optical wavelength from a surface, resolutions on the order of $\lambda/10$ have been achieved.[44,45] So far, near-field techniques have worked best for transparent samples and transmitted light. Second, magneto-optic indicator films have been developed as an alternative to the Bitter pattern imaging.[46] In this method, a thin, free-standing garnet film is placed against a magnetic sample and the resulting domain pattern, induced by the sample's stray field, is imaged using a conventional polarized light microscope. Compared with the Bitter imaging, this relatively simple and inexpensive technique has the advantage of faster response to applied magnetic fields and no sample contamination. Finally, intense laser illumination has made imaging by using second harmonic Kerr effects possible.[47] The second harmonic mode can image structures such as domains in antiferromagnets that are not visible with conventional Kerr imaging.

In conclusion, magneto-optic imaging is a relatively straightforward technique that can directly image the magnetic structure of a wide range of materials. Magneto-optic imaging is fast, can be used with arbitrarily large applied magnetic fields, and can be used in air. The major drawback, especially for spintronics device applications, is its limited resolution of, at best, a few tenths of a micrometer, although near-field techniques may improve this in the future.

6.5. TRANSMISSION ELECTRON MICROSCOPY

Transmission electron microscopy (TEM), as realized on both the conventional transmission electron microscope (CTEM) and scanning transmission electron microscope (STEM),[48] are important tools for domain imaging and understanding spintronic devices. Utilizing highly sophisticated electron lens design and high-energy electrons, a CTEM or STEM can deliver spatial resolution of a few tenths of a nanometer. It is not surprising, therefore, that such instruments would be applied to high resolution domain imaging.

There are a wide variety of imaging modes[48] used in transmission microscopy. In general, an electron optically bright, high energy (200 keV) electron beam is passed through a thinned (<150 nm) sample and an image is formed. This image will reflect both the physical structure, e.g., thickness variations, crystal grain boundaries, etc., and the magnetic structure, e.g., magnetization, domain wall location, etc., of the sample. Transmission methods designed to elucidate magnetic structure are generally referred to as either Lorentz techniques or holographic techniques. In the Lorentz methods, electrons in the beam are viewed as particles deflected by the Lorentz force produced by the magnetic field resulting from nearby magnetic material. The holographic methods are understood by viewing the microscope's field emitter as a highly coherent source of electron waves that exhibit interference when they take alternate paths to the same point in the imaging plane. The magnetic flux enclosed by these alternate paths affects this interference pattern and thereby permits the magnetic structure to be determined. We will first describe several variations of the Lorentz technique and then provide a description of one of the holographic methods.

There are three variations of Lorentz microscopy,[49–52] which are frequently referred to as Fresnel, Foucault, and differential phase contrast (DPC) microscopy. All sense the deflection of the electron beam as it travels through a magnetic field. In the Fresnel mode (sometimes referred to as the "defocused mode"), illustrated in Fig. 6.13(a), a defocused electron beam is transmitted through a thinned sample which has the domain structure shown schematically in Fig. 6.14(a). As depicted in Fig. 6.13(a), the Lorentz force, $\mathbf{F} = e(\mathbf{v} \times \mathbf{B})$, deflects the transmitted beam either toward or away from domain walls. The accompanying intensity curve shows the consequential reduction and increase in electron signal strength that signals

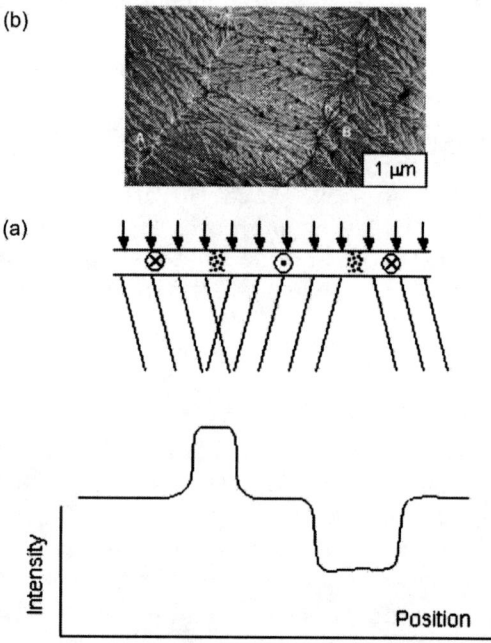

Figure 6.13. (a) Schematic depiction of deflection of defocused incident electron beam upon transmission through a thin magnetic sample illustrating the formation of bright and dark, domain wall structures. (b) Example of two domain walls, white (A) and black (B), in a thin film $Fe_{81}B_{13.5}Si_{3.5}C_2$ sample.[57]

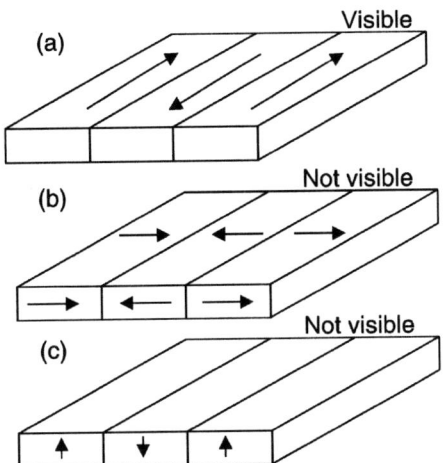

Figure 6.14. Three possible domain geometries, which will lead to observable (a) or non-observable (b,c) domain walls by using the Fresnel mode. Modified from Ref. 2.

the existence of a domain wall. Figure 6.13(b) shows an image[53] with two domain walls, one intensified (white) and the another reduced in intensity (black).

The Fresnel mode is operationally straightforward. Domain walls are imaged with high contrast, but it is very difficult to image the internal structure of the wall itself. Also, the magnetization directions of the domains bounded by the walls must be inferred, e.g., by making use of the fact that the magnetization direction is perpendicular to the direction of "ripple" seen in the magnetization of polycrystalline samples.[54] The domain magnetization direction can also be determined by noting the reaction of the walls to the application of external magnetic fields. One example of the Fresnel mode applied to imaging spin valves can be found in the study of the dependence of magnetization reversal on coupling strength and direction of applied field in permalloy spin valve elements.[55] A second example is the study of the magnetization reversal in CoCu multilayers as a function of the number of bilayers.[56]

If the domain configuration to be measured is as shown in Fig. 6.14(b) or (c), the domain wall contrast mechanism is not so straightforward. In the case shown in Fig. 6.14(b), the beam deflection would be along the wall and cancellation from stray fields would occur. In the case of Fig. 6.14(c), v and $\mathbf{B}$ are parallel in the sample so no deflection would occur unless the sample was tilted to generate a magnetization component similar to Fig. 6.14(a).

The Foucault or in-focus Lorentz mode of domain imaging is illustrated in Fig. 6.15(a). In this mode, an in-focus image is formed but an edge inserted in the objective plane is maneuvered to discriminate against electrons that have been deflected to one side. This corresponds to increasing the image intensity for domains magnetized in one of the directions parallel to the edge and reducing the intensity for those domains oppositely magnetized. By moving the objective aperture, it is possible to intensify any desired magnetization direction. Figure 6.15(b) shows a Foucault image of the same region imaged by the Fresnel method in Fig. 6.13(b). The Fresnel method provides wall contrast while the Foucault method gives domain

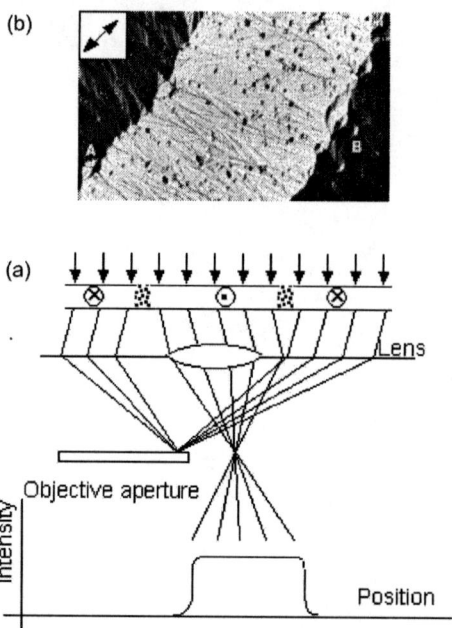

Figure 6.15. (a) Schematic of the Foucault or in-focus mode of domain imaging showing the use of an objective aperture to select out rays deflected in a particular direction to provide domain contrast. (b) An example a domain [also shown in Fig. 6.13(b)] as seen in the Foucault mode.[50,53,57]

contrast. When used together, these modes of TEM give a good overall description of the domain structure. However, switching between modes to image the same region with both techniques requires changes in the electron optics.

The Foucault technique has been used recently to image an active spin valve element and show the domain structure during magnetization reversal for a different spin valve structures, compositions and applied currents.[58] Figure 6.16(a) shows the magnetization reversal for a NiFe/Cu/Co/NiFe/MnNi spin valve structure, measuring 10 μm × 30 μm that has been deposited on a 40 nm alumina membrane. Currents of three different values were applied to the element to measure the GMR *in situ*. In Fig. 6.16(b) and (c), we see the domain patterns for points marked on the curves of Fig. 6.16(a) for currents of 0.3 and 3.5 mA, respectively. The difference between the domain structures for the high and low current cases is attributed to heating effects.

The DPC mode, with two exceptions,[60,61] makes use of a STEM. As depicted in Fig. 6.17, the focused beam is scanned across the domain structure and the Lorentz force deflects the transmitted beam. It is detected predominantly by the right or left detector elements depending on the magnetization direction in the domain. De-scan coils have been used to center the beam on the detectors in the absence of a magnetic sample. The magnetic contrast is derived from the difference in signal level at opposite detectors, which may be either half circles or quadrants. Differential phase contrast microscopy offers a more straightforward interpretation of the magnetic image than the other Lorentz methods, at the expense of requiring a more complex instrument. It shares a difficulty with the other Lorentz methods; crystallographic

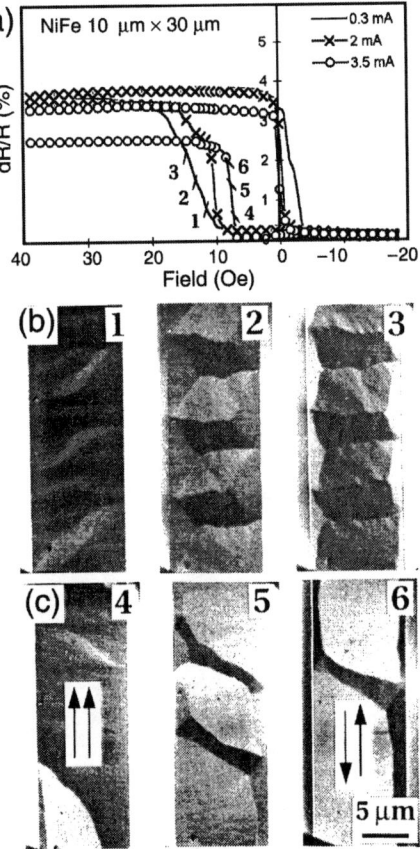

Figure 6.16. (a) GMR curves of a NiFe spin valve measured *in situ* during Foucault mode domain observations shown in (b) and (c). Three different currents were applied to the device to measure the GMR. (b), (c) Domain images of the spin valve corresponding to the measurement points indicated in (a) for applied currents of 0.3 and 3.5 mA, respectively.[58,59]

structure can be difficult to distinguish from magnetic structure.[50] Modification of the DPC detection system has proven useful in minimizing this effect.[50,62] Differential phase contrast has been used to study the magnetization reversal in micrometer sized, permalloy spin valve elements to see directly the effect of size on the magnetization reversal mechanism.[63] In a slightly earlier study using both the Fresnel and DPC methods, NiFe/Cu/NiFe/FeMn spin valve structures were imaged and studied in detail.[64]

Electron holography,[51,66–68] suggested by Gabor in 1948[69] as a way to reduce the effect of aberrations at high magnification, has been applied to the imaging of magnetic domains. In this technique, a field-emission electron gun is typically used to form a very bright, highly coherent source of electrons. The small source size produces the necessary lateral coherence and the relatively narrow energy distribution ensures that the electrons have a significant temporal coherence in the beam. Using such a source, it is possible to demonstrate interference between electrons that traverse paths of different length before detection, as illustrated in Fig. 6.18. Here, electrons

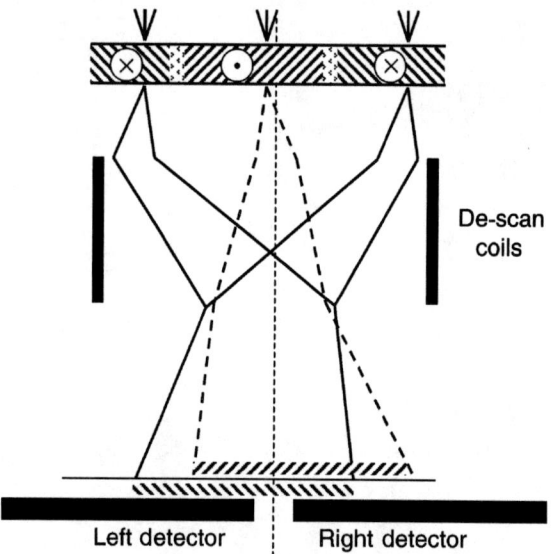

Figure 6.17. A schematic depiction of the DPC mode of imaging in which split detectors are used to monitor the angular deflections that electrons incur on transmission through ferromagnetic domains.[52,65]

can pass either to the right or left of a positively charged filament between two negatively charged plates; this configuration of elements is known as an "electron biprism."[70] The interference fringes depicted result from the difference in phase accumulated between the two beams. The shift in phase caused by a magnetic material is visualized in Fig. 6.19 where the initial plane wavefront is distorted by the enclosed magnetic flux according to the equation,

$$\Delta\phi/\hbar = -2\pi(e/\hbar) \int \mathbf{B} \cdot \mathbf{dS} \qquad (6.5.1)$$

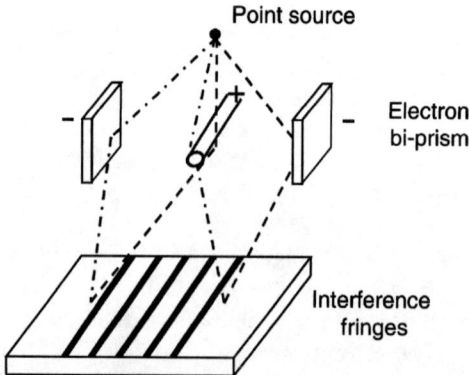

Figure 6.18. A schematic showing the interference resulting from electrons following paths of differing lengths from a high coherence source through the electron optical equivalent of a bi-prism. Modified from Ref. 66.

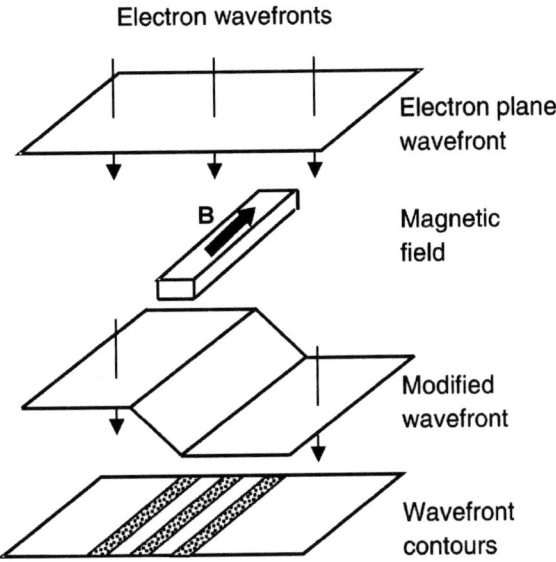

Figure 6.19. A schematic showing the evolution of a plane wave wavefront as it passes a magnetic material. Modified from Ref. 66.

Here, $\Delta\phi$ is the difference in phase between electrons that arrive at the detector over two different paths that form a closed loop. The magnetic field, **B**, passes through the surface **S** enclosed by these paths; the integral gives the total magnetic flux passing through **S**. Since the phase difference is proportional to the total enclosed flux, the wavefront contours can be interpreted directly as magnetic lines of force. An experiment that allows the measurement of the wavefront phase can therefore directly measure the enclosed flux. Indeed, the flux measurement is absolute in the sense that the surface element that corresponds to the difference between adjacent contour lines has a flux of $h/e \approx 4.1 \times 10^{-15}$ Wb flowing through it.

The absolute mode of holographic domain imaging is shown schematically in Fig. 6.20. Here, a field emission TEM incorporating an electron bi-prism is used. The object is off center so the incident beam is both transmitted through the magnetic object and passes by it in (ideally) a field free region. The electron bi-prism causes the transmitted beam and the external (reference) beam to interfere. The resulting magnified interference pattern or hologram is composed not only of interference due to path length differences, but due to phase changes resulting from the different paths enclosing magnetic flux. The resulting hologram must be processed by optical means or, more recently, through digital processing, to display the magnetic flux pattern.

An example of a holographic image[71] of the magnetic flux both in and around a thin magnetic tape is shown in Fig. 6.21. The top 75% depicts the flux within the tape, while the bottom 25% of the interferogram represents the fringing field. The magnetic material consisted of a 45 nm thick cobalt film. It is important to remember that the reference beam may not always be in a field free region and the effect of stray flux may be included in images taken in this way.

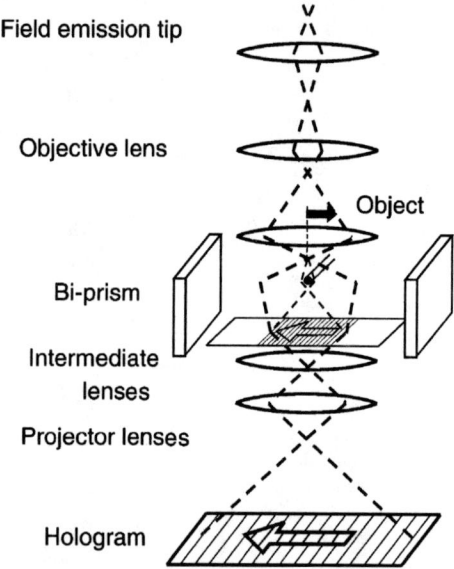

Figure 6.20. A schematic showing a setup for absolute holography. Here, a part of the wave passes through the object (right) and another part of it does not (left). A bi-prism is used to recombine the waves and produce an interferogram. Modified from Ref. 66.

In Fig. 6.22, we see another example[72] of absolute mode holographic imaging. Here, two 30 nm cobalt rectangles have been deposited on a 55 nm thick silicon nitride membrane. The rectangle on the left is 220 nm wide by 275 nm high, the one on the right is 300 nm wide by 275 nm high, and they are separated by 170 nm. The field is applied from right to left. The lines shown follow paths of constant magnetization. With no applied field, the magnetization lies parallel to the edges of

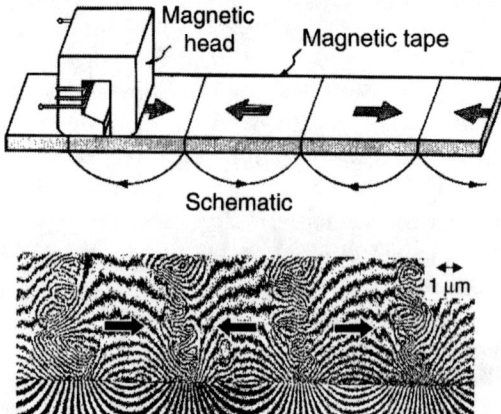

Figure 6.21. Top: Schematic depiction of recording method and recorded pattern. Bottom: Interference pattern revealing magnetic lines of force. The 45 nm thick Co film occupies the upper 3/4 of the image and the lower 1/4 is free space. The magnetic lines of force are seen to lie along the magnetization direction within the Co and as a fringe field extending past the edge of the film. Modified from Ref. 67.

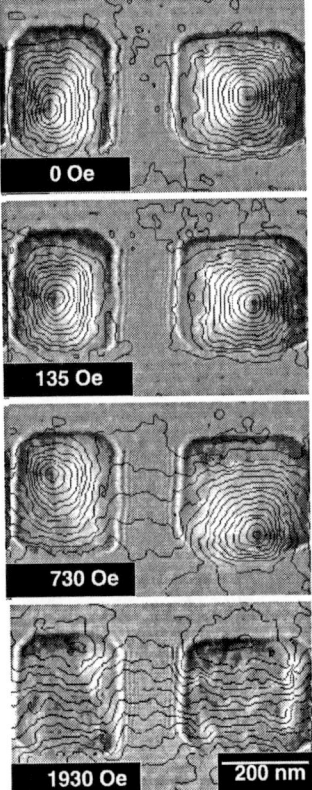

Figure 6.22. Holographic determination of magnetic lines of flux in Co rectangles of different shape for four values of an external magnetic field. Note how the two field distributions make a transition from a solenoidal configuration at zero applied field to a near-saturated configuration at 1930 Oe.[72,73]

the films; the lack of surface poles at the edges help to minimize the magnetostatic energy of the system. The magnetization of the two films circulates in opposite directions, as partially indicated by the shaded background in the figure. As the applied field increases the magnetic vortices of the film on the left and right move up and down, respectively, eventually disappearing as the films approach saturation at 1930 Oe of applied field.

Holographic domain imaging can also be done in what is referred to as the STEM differential mode.[74] Here again, an electron bi-prism is used and a hologram is recorded that reflects the phase variation due to the enclosed magnetic flux. However, in this case, there is no reference beam external to the sample. The interference is between two almost parallel beams that pass through the sample displaced by a distance of as little as 10 nm. For a uniformly magnetized sample, the enclosed flux remains constant as the beams are scanned, and the phase difference remains constant as well. This method is a particularly useful way to image small structures or domain wall profiles.[74]

All of the transmission methods described above have several common features. They all require thinned samples, i.e., having a thickness ≤150 nm. Sample preparation for TEM examination requires significant skill and it is important that a uniform thickness be achieved. Generally, the thinning requirement precludes being able to examine an actual spintronic device. However, if an element of the spintronic device involves a magnetic film or films that are already thin and can be prepared on a suitable, thin substrate, then, these methods can be very useful.[58]

If the magnetic element must be thinned, then the thinning may have an important consequence; it may change the micromagnetics. Domain formation depends on energy minimization and as the device thickness decreases the proximity of the surface becomes increasingly more important. For example, Bloch walls in the interior of a magnetic material will contain spins that are perpendicular to the surface of the material. As these walls approach the surface, the energy cost of spins oriented perpendicular to the surface gradually forces the spins to lie in the surface plane and the wall takes on the character of a Néel wall. Such a structure is known as a Néel cap.[75] As the sample thickness is reduced, the Néel caps present where the Bloch wall intersects each surface approach each other and eventually a vortex is formed. Transmission techniques would then respond to the average field distribution of the vortex, instead of the field distribution in the sample before preparation for observation.

As mentioned above, one great advantage of TEM methods is the high resolution available in spatial imaging. In magnetic imaging, the Fresnel, Foucault, DPC, and holographic modes have demonstrated "best" resolutions of ~10 nm and below.

The high spatial resolution available in TEM comes about, in part, from having the sample in the magnetic field of the objective lens element. When imaging a magnetic sample, this external magnetic field can seriously perturb the magnetic state of the device under study and must be considered. Possible remedies, which may reduce the resolution, include using a special low field lens, switching off the objective lens, or moving the sample outside the lens field. A useful aspect of the field is the ability to change the component of the field in the sample plane by tilting the sample in the lens field.[76]

Electron holography has the advantage of being able to produce a direct display of magnetic lines of force and the absolute mode is capable of measuring, in an absolute sense, the magnetic induction. However, it has been pointed out that in domain imaging, the direction of the magnetization is generally a much more important quantity than the size of the induction.[77] The absolute method relies on a path for the electrons that passes outside of the sample and this can place restrictions on the fabrication and accessibility of a device under study. This has been avoided in one recent study[72] where the "free space" reference path was routed through the SiN substrate material adjacent to a patterned magnetic structure.

6.6. MAGNETIC IMAGING WITH X-RAY DICHROISM

Magnetic imaging with x-ray dichroism takes advantage of the existence of powerful new sources of synchrotron radiation. The absorption of circularly polarized x-rays in a magnetic material depends on the relative orientation of the photon helicity, σ, and the sample magnetization, **M**. The difference in the absorption of light, as a function of polarization, is known as dichroism, giving rise to the name x-ray magnetic circular dichroism (XMCD) for this effect. This effect can be relatively large near atomic absorption edges. For transition metals, the spin–orbit split L_3 and L_2 edges, corresponding to transitions from core 2p to valence 3d states, are usually used in XMCD measurements. Herein lies one of the main strengths of the technique; by tuning the x-rays to a particular absorption edge, elemental specificity can be achieved. The dichroism arises because the circularly polarized photons create partially polarized photoelectrons owing to the coupling between the x-ray helicity and the orbital angular momentum of the excited electron, which, in turn, is coupled to the spin of the electron by the spin–orbit interaction. The polarized electrons are excited into empty states above the Fermi level that are polarized because of the exchange interaction in the ferromagnet.

The secondary electron yield is proportional to the x-ray absorption and therefore, is sensitive to the dichroism. The dichroism, which is proportional to $\sigma \cdot \mathbf{M}$, will cause a spatial variation in the secondary electron intensity as the direction of magnetization $\mathbf{M}$ changes from one domain to another. A magnetic image is obtained by imaging these secondary electrons as is done, for example, in a photoemission electron microscope (PEEM). XMCD domain imaging was first demonstrated[78] using an immersion lens photoelectron microscope shown in Fig. 6.23. The two stage electrostatic lens system magnifies the image of the sample and projects it on to the double microchannel plate where it is amplified before forming an image on the phosphor screen.[79] A digital camera and associated data acquisition electronics outside the vacuum system are not shown. Because the photons are incident on the sample at fairly grazing incidence, and it is the projection of the magnetization on the direction of photon spin that is measured, this geometry is most sensitive to magnetization in the plane of the sample. XMCD domain imaging is thus accomplished by coupling circularly polarized x-rays from a synchrotron radiation source with a PEEM system.

This domain imaging technique is nicely illustrated by the first results, which were obtained from a CoPtCr hard disk with a test pattern of alternating in-plane magnetic domains written at different recording densities.[78] Three XMCD images of the same 200 μm diameter region of the disk but taken at different photon energies are shown in Fig. 6.24(a). The XMCD spectrum in Fig. 6.24(b), around the Co L edges, shows the relation between the images and the photon energies. At photon energies below the L edge, there is no magnetic contrast due to XMCD and the image at the left of Fig. 6.24(a) displays the topography of the sample surface. The middle image at the L_3 resonance energy shows the domain image of the written bits. The squares are 10 μm × 10 μm and the rectangles in the row below are 10 μm high by 2 μm wide. The dashed (solid) line in the spectrum is for magnetization parallel (antiparallel) to the photon spin. The two different spectra can be obtained by either reversing the photon helicity or the magnetization depending on the experimental situation. Thus, at the L_3 energy where the dashed line is higher in the spectrum, the bright regions in the image correspond to magnetization in the direction of the photon spin. At the L_2 energy, the contrast in the image is just reversed. The image is sensitive only to one component of the magnetization; domains with $\mathbf{M}$ perpendicular to σ exhibit no magnetic contrast. The XMCD effect is strong enough that domains can be seen in the raw image. Clearer images, as shown in Fig. 6.24(a), were obtained by dividing the raw image by an image taken at 810 eV photon energy in order to remove non-uniformity in the response of the optical system.

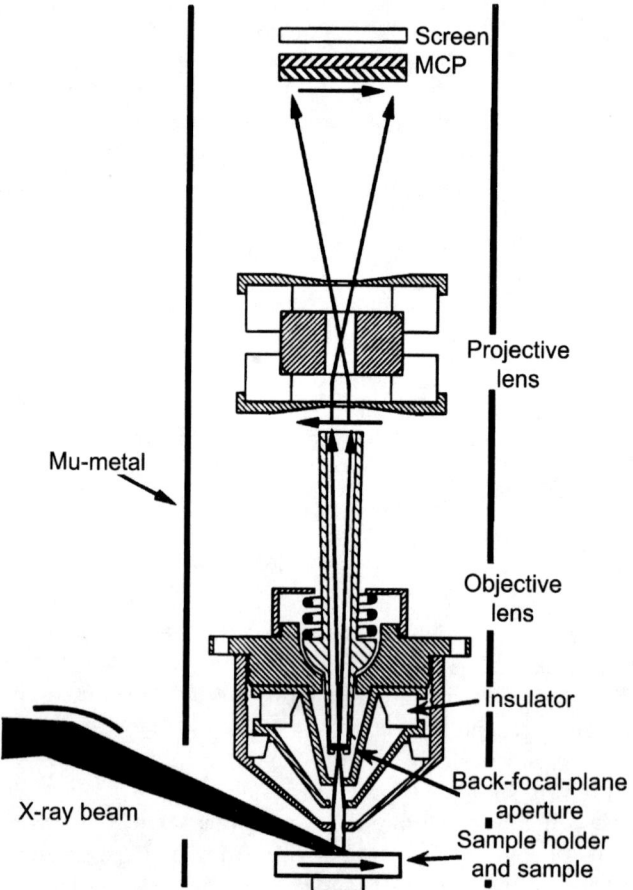

Figure 6.23. Schematic depiction of an apparatus that uses circularly polarized x-rays to produce secondary electrons that are then focused electrostatically to form a highly magnified image on a microchannel-plate (MCP) intensifier.[79,80]

At higher magnifications where the signal is lower, the contrast can be improved by looking at the difference between appropriately scaled images at the Co L_2 and L_3 energies. A row of bits 1 μm wide (the second row below the squares) was then clearly resolved. The resolution in this first experiment was about 1 μm. Improvements in the electron optics and higher x-ray fluxes from third-generation synchrotron radiation sources are predicted to lead to a 10 nm resolution.[78]

There are a number of strengths of XMCD domain imaging.[81] Its elemental specificity makes it a very powerful technique for particular problems. The magnetic measurement can also, in principle, be correlated with other core level measurements that give information on the local site, symmetry and chemical state. Because the spin and orbital moments can be determined in XMCD, the magnitude of the magnetization can be quantitatively determined. XMCD imaging has been shown sensitive enough to image 0.1 atomic layer of magnetic material.[82] The information

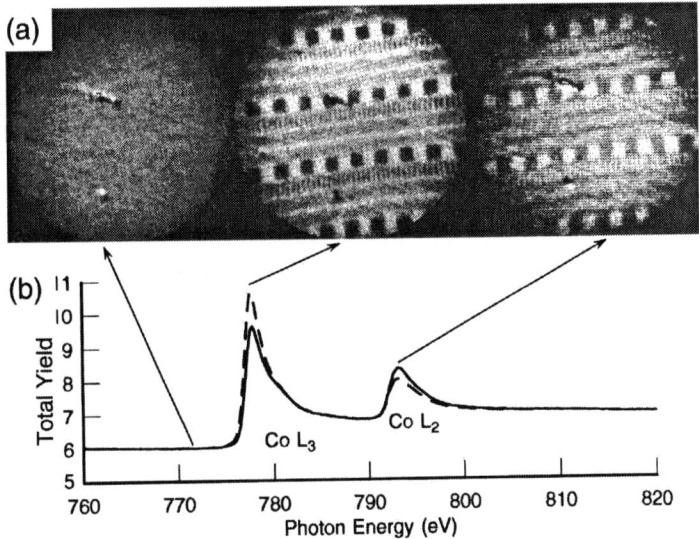

Figure 6.24. (a) Intensity image (left) and magnetic domain images (right) from a recording disk as seen by MCXD using circularly polarized x-rays of different energies. The domain rows have dimensions of 10 μm × 10 μm, 10 μm × 2 μm, 10 μm × 1 μm, and 10 μm × 0.5 μm, respectively. The domain magnetization lies along the direction of the rows. (b) The spectra of the L-edge dichroism and specifies the energies at which the images were taken.[78,83]

depth varies with the material. It is possible to "see through" contamination or coatings. The images in Fig. 6.24 are from a disk covered with 13 nm of carbon and 4 nm of an organic fluorocarbon lubricant. The domain image depends on the relative number of secondary electrons from one domain to another, which is partially preserved in the electron cascade process that takes place as the electrons pass through overcoats. Domains can be observed through 2–4 nm of a transition metal.[82] This is unlike the surface sensitivity of SEMPA where the spin must be preserved and not diluted by scattering or generation of unpolarized electrons.

The XMCD can also be detected by directly measuring the helicity-dependent x-ray absorption in a transmission x-ray microscope. A condenser zone-plate images the x-ray source on the sample. A magnified image of the transmitted x-rays is formed by a second zone-plate. The spatial resolution of the first experiments on GdFe multilayer was about 60 nm.[84] Information is provided about the magnetic properties integrated over the path of the transmitted x-rays through the sample. The component of magnetization along the photon spin, perpendicular to the sample surface, is measured. The sample thickness should be such that the transmission is approximately 10%, which is typically of order 100 nm for transition metals. Magnetic microstructure is thickness dependent and will change on thinning a sample; this technique is particularly well suited when the sample to be measured is of the appropriate thickness. For XMCD imaging in transmission, magnetic fields can be applied, the sample need not be conducting, and the measurement is insensitive to surface contamination or moderate roughness. Scanning x-ray microscopy is a related implementation that focuses the x-ray spot on the sample and either scans the sample or the x-ray spot while monitoring the transmitted intensity or the fluorescent x-rays from the sample.[82]

Linearly polarized x-rays coupled with a PEEM system to image the total electron yield have also been used to image magnetic domains.[85] This measurement, which is the x-ray analog of the transverse magneto-optic Kerr effect, is sensitive to the magnetization component perpendicular to the plane of incidence of the x-rays and to the electric vector of the obliquely incident p-polarized radiation. However, there is no equivalent to changing the helicity as is done in XMCD. Therefore, contrast in a domain image is obtained by taking the difference between two images measured at photon energies that have the largest difference in the total electron yield on reversal of the magnetization. The sensitivity of the domain image to **M** perpendicular to the plane of incidence is complementary to XMCD which determines **M** along the photon spin, and hence, in the plane of incidence. When the component perpendicular to the plane of incidence is required, or when linear polarized x-rays are most available, domain imaging using linear polarization can be useful, even though the image contrast signal is about an order of magnitude less than with circularly polarized light.

Yet, a different effect allows the imaging of domains where there is an alignment of the magnetic moments, as in an antiferromagnet, but not a net magnetization as in a ferromagnet. Near the x-ray absorption threshold, the absorption depends on whether the magnetic alignment is parallel or perpendicular to the x-ray linear polarization. This dichroism signal, sometimes referred to as x-ray magnetic linear dichroism (XMLD), is proportional to $|\mathbf{M}|^2$ rather than **M** as in XMCD.[86] Whereas XMCD requires spin-polarized d final states and spin–orbit split core levels such as the $2p_{1/2}$ and $2p_{3/2}$ states for an effect, XMLD additionally requires that the multiplet structure within the L_2 and L_3 edges can be observed. Selection rules cause a change in the spectrum depending on whether the x-ray polarization is parallel to the magnetization ($\Delta m = 0$) or perpendicular to it ($\Delta m = \pm 1$). XMLD was observed[86] in the antiferromagnet Fe_2O_3; in that report, the imaging of antiferromagnetic domains using an x-ray microscope was also suggested. Preliminary evidence of antiferromagnetic domains was seen in images of NiO acquired by scanning the sample under the focused linearly polarized x-ray beam.[87] These first measurements point the way to measuring antiferromagnetic domains, but significant developments and refinements are necessary before this becomes a routine antiferromagnetic domain imaging technique. Such techniques may prove valuable in understanding the exchange biasing of a ferromagnet by an antiferromagnet, which is important for spintronics device implementation.

Finally, we mention that if instead of measuring x-ray absorption, the angular distribution of emitted electrons is measured to determine the electron wave vector **k**, a rich variety of phenomena occur on reversing the magnetization which are loosely termed "dichroism."[88] For example, using an imaging x-ray spectrometer, XMCD images obtained with angle resolved Auger electrons illustrate that a "dichroism" in the emitted electrons can be observed even when σ is perpendicular to **M**.[89] There are also magnetic effects in photoemission angular distributions obtained with linearly polarized light.[90] While these chiral effects may be employed in special cases for magnetic imaging, we expect that most imaging will use dichroism in absorption and exploit the large total electron yield signal.

6.7. CONCLUSIONS

Some of the more relevant characteristics of the magnetic imaging methods discussed in this chapter are summarized in Table 6.1. This table serves as a useful starting point for selecting a particular imaging method, but the reader should be cautious about judging a technique based on the numbers in this table alone. For example, some important characteristics have been left out because they are difficult to quantify, such as a technique's ease of use and cost, or the difficulty in interpreting the measurements. In addition, the various parameters listed in the table are usually not independent. Optimizing one characteristic may have to be done at the expense of degrading several others. For example, an image with high spatial resolution will usually require a long acquisition time.

The row of Table 6.1 labeled resolution requires further comment. We list the best resolution demonstrated for each method as well as a value that is more typical of routine practice. However, there are several problems encountered in comparing the best resolution available for different techniques. First, it is difficult to find samples with magnetic structure known on the length scale of interest to serve as calibration samples. Second, high resolution measurements can be very demanding and consequently, are not frequently performed. Third, authors generally do not quote the resolution of their measurements. Fourth, and finally, different definitions of resolution may be used. In Table 6.1, we give our estimates of the best demonstrated resolution as obtained from the literature, and provide the references used. Several methods have the potential for improving on the quoted values.

Table 6.1. Characteristics of Magnetic Imaging Methods

	SEMPA	MFM	Magneto-optic	Fresnel, Foucault	DPC	Holography	XMCD
Contrast origin	M	VB	M	$\nabla \times B$, B	B	B, Φ_B	M
Resolution Best [Ref] Typical (nm)	20 [91] 200	40 [26] 100	300 [36] 1000	~10 [52] 50	~2 [92] 20	~5 [93] 20	300 [82] 500
Information depth (nm)	2	20–500	20	Sample thickness ($\leq$150 nm)			2–20
Acquisition time	1–100 min	5–30 min	10 ns–1 s	0.04–30 s	5–50 s	0.03–10 s	0.03 s–10 min
External Field (kA/m)	<1	<800	No limit	<500 (vert.) <100 (horiz.)		<100	<1
Insulators	No	Yes	Yes	No	No	No	No
Vacuum	UHV	None	None	HV	HV	HV	UHV
Topographic/ crystallographic sensitivity	Low	High	High	Moderate	Moderate	Moderate	Low
Special sample requirements	Clean surface	Flat surface	Flat surface	Thin ($\leq$150 nm) sample + substrate			

It should be clear from Table 6.1 that no single imaging technique can solve all of the imaging problems one might encounter with spintronic devices. Rather, each technique works best for specific types of samples and each provides different magnetic information. SEMPA works well for clean ultrathin magnetic films on opaque substrates. The MFM can image the magnetic fields of buried magnetic structures. Magneto-optic techniques allow high speed imaging of magnetization dynamics in devices that are larger than a micrometer. The TEM techniques can all produce very high resolution images of patterned magnetic elements grown on transparent membrane substrates. XMCD provides element specific magnetic imaging in magnetic alloys or multilayers. In addition, the best approach to understanding some particular magnetic structure may quite often involve using several complementary techniques.

This purpose of this chapter has been to provide the reader with a limited introduction to techniques that can image the magnetic microstructure of spintronic devices. More information can be obtained by reading the larger reviews of magnetic imaging, as listed in the references, or by speaking with some of the expert practitioners of the various techniques. Ultimately, however, trying the technique is likely to be the best approach to learning which technique provides the information you need to solve your problem. You may well find that only by using several methods can you develop a complete understanding of the complex magnetic structures and interactions that can occur in nanoscale spintronic devices.

ACKNOWLEDGMENTS

We would like to acknowledge very helpful conversations with Michael Kelley, John Chapman, Stephen Lorentz, Michael Scheinfein and John Moreland during the preparation of this manuscript. This work was supported in part by the Office of Naval Research.

REFERENCES

1. J. Zhu, *Magnetic Interactions and Spin Transport*.
2. A. Hubert and R. Schaefer, *Magnetic Domains* (Springer, Berlin, 1998).
3. D. J. Craik, in: *Methods of Experimental Physics*, Vol. 11, edited by R. V. Coleman (Academic Press, New York, 1974), pp. 675–743.
4. J. Unguris, D. T. Pierce, A. Galejs, and R. J. Celotta, *Phys. Rev. Lett.* **49**, 72–76 (1982).
5. K. Koike and K. Hayakawa, *Jpn. J. Appl. Phys.* **23**, L187–L188 (1984).
6. J. Unguris, G. G. Hembree, R. J. Celotta, and D. T. Pierce, *J. Microsc.* **139**, RP1–RP2 (1985).
7. D. R. Penn, S. P. Apell, and S. M. Girvin, *Phys. Rev. Lett.* **55**, 518–521 (1985); D. R. Penn, S. P. Apell, and S. M. Girvin, *Phys. Rev. B* **32**, 7753–7768 (1985).
8. E. Kisker, W. Gudat, and K. Schröder, *Solid State Commun.* **44**, 591–595 (1982).
9. H. Hopster, R. Raue, E. Kisker, G. Güntherodt, and M. Campagna, *Phys. Rev. Lett.* **50**, 70–73 (1983).
10. G. G. Hembree, J. Unguris, R. J. Celotta, and D. T. Pierce, *Scanning Microsc. Suppl.* **1**, 229–240 (1987).
11. K. Koike, H. Matsuyama, and K. Hayakawa, *Scanning Microsc. Suppl.* **1**, 241–253 (1987).
12. H. P. Oepen and J. Kirschner, *Scanning Microsc.* **5**, 1–16 (1991).
13. M. R. Scheinfein, J. Unguris, M. H. Kelley, D. T. Pierce, and R. J. Celotta, *Rev. Sci. Instrum.* **61**, 2501–2526 (1990).
14. J. A. Borchers, J. A. Dura, C. F. Majkrzak, J. Unguris, D. Tulchinsky, M. H. Kelley, S. Y. Hsu, R. Loloee, W. P. Pratt Jr and J. Bass, *Phys. Rev. Lett.* **82**, 2796–2799 (1999).

15. J. Unguris, M. R. Scheinfein, D. T. Pierce, and R. J. Celotta, *Appl. Phys. Lett.* **55**, 2553–2555 (1989).
16. J. Barnes, L. Mei, B. M. Lairson, and F. B. Dunning, *Rev. Sci. Instrum.* **70**, 246–247 (1999).
17. P. Grünberg, R. Schreiber, Y. Pang, M. B. Brodsky, and H. Sowers, *Phys. Rev. Lett.* **57**, 2442–2445 (1986).
18. S. S. P. Parkin, N. More, and K. P. Roche, *Phys. Rev. Lett.* **64**, 2304–2307 (1990).
19. J. Unguris, R. J. Celotta, and D. T. Pierce, *Phys. Rev. Lett.* **67**, 140–143 (1991); J. Unguris, R. J. Celotta, and D. T. Pierce, **69**, 1125–1128 (1992).
20. J. Unguris, R. J. Celotta, and D. T. Pierce, *Phys. Rev. Lett.* **79**, 2734–2737 (1997).
21. D. A. Tulchinsky, M. H. Kelley, J. J. McClelland, R. Gupta, and R. J. Celotta, *J. Vac. Sci. Technol.* A **16**, 1817–1819 (1998).
22. P. Grütter, H. J. Mamin, and D. Rugar, in: *Scanning Tunneling Microscopy*, Vol. II, edited by H. J. Güntherodt and R. Wiesendanger (Springer, Berlin, 1992) pp. 151–207.
23. P. Grütter, *MSA Bull.* **24**, 416–425 (1994).
24. E. D. Dahlberg and J.-G. Zhu, *Phys. Today* **48**, 34–40 (1995).
25. D. Sarid, *Scanning Force Microscopy* (Oxford University Press, Oxford, 1991).
26. L. Abelmann, S. Porthun, M. Haast, C. Lodder, A. Moser, M. E. Best, P. J. A. van Schendel, B. Steifel, H. J. Hug, G. P. Heydon, A. Farley, S. R. Hoon, T. Pfaaffelhuber, R. Proksch, and K. Babcock, *J. Magn. Magn. Mater.* **190**, 135–147 (1998).
27. M. Kelley, private communication.
28. P. Rice, S. E. Russek, and B. Hines, *IEEE Trans. Magn.* **32**, 4133–4137 (1996).
29. A. Hubert and R. Schaefer, *Magnetic Domains* (Springer, Berlin, 1998). Section 2.6.1.
30. M. Kleiber, F. Kümmerlen, M. Löhndorf, A. Wadas, D. Weiss, and R. Wiesendanger, *Phys. Rev.* B **58**, 5563–5567 (1998).
31. C. Schönenberger, S. F. Alvarado, S. E. Lambert, and I. L. Sanders, *J. Appl. Phys.* **67**, 7278–7280 (1990).
32. M. R. Scheinfein, J. Unguris, D. T. Pierce, and R. J. Celotta, *J. Appl. Phys.* **67**, 5932–5937 (1990).
33. A. Hubert and R. Schaefer, *Magnetic Domains* (Springer, Berlin, 1998). Section 2.3.
34. S. D. Bader and J. L. Erskine, Magneto-optical effects in ultrathin magnetic structures, in: *Ultrathin Magnetic Structures* II, edited by B. Heinrich and J. A. C. Bland, (Springer, Berlin, 1994), p. 297.
35. B. E. Argyle, in: *Proc. Electr. Chem. Soc. Symposium on Magnetic Materials and Devices*, edited by L. T. Romankiw and D. A. Herman, Vol. 90–98, 85 (1990).
36. F. Schmidt and A. Hubert, *J. Magn. Magn. Mater.* **61**, 307–320 (1986).
37. A. Green and M. Prutton, *J. Sci. Instrum.* **39**, 244–245 (1962).
38. G. L. Ping, C. W. See, M. G. Somekh, M. B. Suddendorf, J. H. Vincent, and P. K. Footner, *Scanning* **18**, 8–12 (1996).
39. M. Du, S. S. Xue, W. Epler, and M. H. Kryder, *IEEE Trans. Magn.* **31**, 3250–3252 (1995).
40. B. Petek, P. L. Trouilloud, and B. E. Argyle, *IEEE Trans. Magn.* **26**, 1328–1330 (1990).
41. F. H. Liu, M. D. Schultz, and M. H. Kryder, *IEEE Trans. Magn.* **26**, 1340–1342 (1990).
42. R. Schafer, *J. Magn. Magn. Mater.* **148**, 226–231 (1995).
43. A. Hubert and R. Schaefer, *Magnetic Domains* (Springer, Berlin, 1998). Section 2.3.4.
44. E. Betzig, J. K. Trautman, R. Wolfe, E. M. Gyorgy, P. L. Finn, M. H. Kryder, and C. H. Chang, *Appl. Phys. Lett.* **61**, 142–144 (1992).
45. T. J. Silva and S. Schultz, *Rev. Sci. Instrum.* **67**, 715–725 (1996).
46. V. I. Nikitenko, V. S. Gornakov, L. M. Dedukh, Yu. P. Kabanov, A. F. Khapikov, L. H. Bennett, P. J. Chen, R. D. McMichael, M. J. Donahue, L. J. Swartzendruber, A. J. Shapiro, H. J. Brown, and W. F. Egelhoff, *IEEE Trans. Magn.* **32**, 4639–4641 (1996).
47. V. Kirilyuk, A. Kirilyuk, and Th. Rasing, *Appl. Phys. Lett.* **70**, 2306–2308 (1997).
48. L. Reimer, Transmission electron microscopy: physics of image formation and microanalysis, *Springer Series in Optical Sciences*, vol. 36 (Springer, New York, 1997).
49. A. Hubert and R. Schaefer, *Magnetic Domains* (Springer, Berlin, 1998). Section 2.4.
50. I. R. McFadyen and J. N. Chapman, *EMSA Bull.* **22**, 64–75 (1992).
51. M. Mankos, M. R. Scheinfein, and J. M. Cowley, *Adv. Imaging Electron Phys.* **98**, 323–426 (1996).
52. J. N. Chapman, *J. Phys.* D **17**, 623–647 (1984).
53. L. J. Heyderman, J. N. Chapman, M. R. J. Gibbs, and C. Shearwood, *J. Magn. Magn. Mater.* **148**, 433–445 (1997).
54. A. Hubert and R. Schaefer, *Magnetic Domains* (Springer, Berlin, 1998). Section 5.5.2.
55. J. P. King, J. N. Chapman, and J. C. S. Kools, *J. Magn. Magn. Mater.* **177–181**, 896–897 (1998).

56. P. R. Aitchison, J. N. Chapman, D. B. Jardine, and J. E. Evetts, *J. Appl. Phys.* **81**, 3775–3777 (1997).
57. Modified from Ref. 54, Copyright 1997, with permission of Elsevier Science.
58. X. Portier, A. K. Petford-Long, T. C. Anthony, and J. A. Burg, *J. Appl. Phys.* **83**, 6840–6842 (1998).
59. Reprinted with permission from Ref. 58, Copyright 1998, American Institute of Physics.
60. M. R. McCartney, P. Kruit, A. H. Buist, and M. R. Scheinfein, *Ultramicroscopy* **65**, 179–186 (1996).
61. A. C. Daykin and J. P. Jakubovics, *J. Appl. Phys.* **80**, 3408–3411 (1996).
62. J. N. Chapman, I. R. McFadyen, and S. McVitie, *IEEE Trans. Magn.* **26**, 1506–1511 (1990).
63. J. N. Chapman, P. R. Aitchison, K. J. Kirk, S. McVitie, J. C. S. Kools, and M. F. Gillies, *J. Appl. Phys.* **83**, 5321–5325 (1998).
64. M. F. Gillies, J. N. Chapman, and J. C. Kools, *J. Appl. Phys.* **78**, 5554–5562 (1995).
65. Modified from Ref. 52, Copyright 1984, The Institute of Physics.
66. A. Tonomura, *Electron Holography* (Springer, Berlin, 1994).
67. A. Tonomura, *Adv. Phys.* **41**, 59–103 (1992).
68. M. Mankos, J. M. Cowley, and M. Scheinfein, *Phys. Stat. Sol. (a)* **154**, 469–504 (1996).
69. D. Gabor, *Nature* **161**, 777–778 (1948).
70. G. Möllenstedt and H. Düker, *Z. Phys.* **145**, 377–397 (1956).
71. N. Osakabe, K. Yoshida, Y. Horiuchi, T. Matsuda, H. Tanabe, T. Okuwaki, J. Endo, H. Fujiwara, and A. Tonomura, *Appl. Phys. Lett.* **42**, 746–748 (1983).
72. R. E. Dunin-Borkowski, M. R. McCartney, B. Kardynal, and D. J. Smith, *J. Appl. Phys.* **84**, 374–378 (1998).
73. Reprinted, in part, with permission from Ref. 72, Copyright 1998, American Institute of Physics.
74. M. Mankos, M. R. Scheinfein, J. M. Cowley, *J. Appl. Phys.* **75**, 7418–7424 (1994).
75. M.R. Scheinfein, J. Unguris, J. L. Blue, K.J. Coakley, D. T. Pierce, R. J. Celotta, and P.J. Ryan, *Phys. Rev. B* **43**, 3395–3422 (1991).
76. S. McVitie, J. N. Chapman, L. Zhou, L. J. Heyderman, and W. A. P. Nicholson, *J. Magn. Magn. Mater.* **148**, 232–236 (1995).
77. A. Hubert and R. Schaefer, *Magnetic Domains* (Springer, Berlin, 1998). Section 2.4.4.
78. J. Stöhr, Y. Wu, B. D. Hermsmeier, M. G. Samant, G. R. Harp, S. Koranda, D. Dunham, and B. P. Tonner, *Science* **259**, 658–661 (1993).
79. B. P. Tonner, D. Dunham, J. Zhang, W. L. O'Brien, M. Samant, D. Weller, B. D. Hermsmeier, and J. Stöhr, *Nucl. Instrum. Methods A* **347**, 142–147 (1994).
80. Modified from Ref. 79, Copyright 1994, with permission of Elsevier Science.
81. J. Stöhr, H. A. Padmore, S. Anders, T. Stammler, and M. R. Scheinfein, *Surf. Rev. Lett.* **5**, 1297–1308 (1998).
82. C. M. Schneider, *J. Magn. Magn. Mater.* **175**, 160–176 (1997).
83. J. Stöhr, private communication.
84. P. Fischer, G. Schütz, G. Schmahl, P. Gutmann, and D. Raasch, *Z. Phys. B* **101**, 313–316 (1966).
85. F. U. Hillebrecht, T. Kinoshita, D. Spanke, J. Dresselhaus, C. Roth, H. B. Rose, and E. Kisker, *Phys. Rev. Lett.* **75**, 2224–2227 (1995).
86. P. Kuiper, B. G. Searle, P. Rudolf, L. H. Tjeng, and C. T. Chen, *Phys. Rev. Lett.* **70**, 1549–1552 (1993).
87. D. Spanke, V. Solinus, D. Knabben, F. U. Hillebrecht, F. Ciccacci, L. Gregoratti, and M. Marsi, *Phys. Rev. B* **58**, 5201–5204 (1998).
88. D. Venus, *Phys. Rev. B* **48**, 6144–6151 (1993); D. Venus, *Phys. Rev. B* **49**, 8821–8829 (1994).
89. C. M. Schneider, Z. Celinski, M. Neuber, C. Wilde, M. Grunze, K. Meinel, and J. Kirschner, *J. Phys.: Condens. Matter* **6**, 1177–1182 (1994).
90. C. Roth, F. U. Hillebrecht, H. B. Rose, and E. Kisker, *Phys. Rev. Lett.* **70**, 3479–3482 (1993).
91. D. A. Tulchinsky, M. H. Kelley, J. J. McClelland, R. Gupta, and R. J. Celotta, *J. Vac. Sci. Technol. A* **16**(3), 1817–1819 (1998), Fig. 6.5 (b), this paper; H. Matsuyama and K. Koike, *J. Electron Microsc.* **43**, 157–163 (1994).
92. G. R. Morrison, H. Gong, J. N. Chapman, and V. Hrnciar, *J. Appl. Phys.* **64**, 1338–1342 (1988).
93. M. R. McCartney and Y. Zhu, *J. Appl. Phys.* **83**, 6414–6416 (1998); M. R. McCartney and Y. Zhu, *Appl. Phys. Lett.* **72**, 1380–1382 (1998).

7

Domain Dynamics and Magnetic Noise

Heidi Hardner*

7.1. INTRODUCTION

In this chapter, we will discuss the sources of electrical noise encountered when employing magnetic materials in device applications. This will include magnetic contributions to the $1/f$ noise floor in magnetic sensors, magnetic noise due to domain switching and other magnetic instabilities, and also recording media noise. In addition to examples of efforts to reduce all of these types of noise in practical applications, this chapter will describe the use of noise measurements as a tool to better understand magnetic materials and devices.

We will start out with a brief overview of the different ways that the magnetic noise can enter into device applications and a comparison with non-magnetic noise sources. Magnetic noise has its fundamental origin in the fluctuation of magnetic moments in the device. Fluctuations of large numbers of moments together can occur when domains or domain walls fluctuate, and this has the potential to cause greater noise than the fluctuation of a single moment. We will consider the basics of domain formation and magnetization reversal. This will demonstrate various ways in which magnetization fluctuations can take place. Section 7.4 will give some general background on electrical noise. We will see how noise is measured and characterized. Then, we will move on to look at how magnetization fluctuations create electrical noise in magnetoresistive (MR) devices. Several examples will illustrate different types of MR noise. We will see that the biggest magnetic noise problem in sensors is magnetic instability, and Section 7.6 considers various attempts to achieve magnetic stability in sensors. Section 7.7 gives a brief overview of recent work on noise reduction in thin film longitudinal recording media. We will see that both the desired magnetic properties and the issues involved in noise reduction are different for media than for sensors. Media noise tends to slightly dominate head noise in magnetic recording, and the reduction of media noise is a very active area of inquiry. Several examples of recent research will be considered.

7.2. OVERVIEW OF MAGNETIC NOISE IN DEVICE APPLICATIONS

Many different magnetic thin film devices produce an electrical output from magnetic fields or changes in magnetic fields. Sensors based on the Hall effect,

---

*Seagate Technology, Bloomington, Minnesota.

MR effects, spin-dependent tunneling, and even magnetostriction are some examples. This chapter will focus on devices based on MR materials, though much of the discussion could be applied to other magnetic systems as well. Both anisotropic magnetoresistive (AMR) and giant magnetoresistive (GMR) materials have been described in detail in previous chapters. They are currently the prime focus of three major industrial applications: magnetic recording heads, sensors, and memory. In these applications, current flows in a MR element and the voltage changes along with the magnetization due to the MR effect. The purpose of the device might be an absolute measurement of an external magnetic field or the detection of changes in field associated with bits of recorded information passing by the sensor or, in the case of magnetoresistive random access memory (MRAM), the detection of the magnetic state the device was last left in. These applications will be discussed in detail in later chapters.

As illustrated in Fig. 7.1, there are broadly three areas where noise enters into these systems: the sense element, external fields, and the detection electronics. Figure 7.1 is a very schematic view of magnetic recording but the same general noise sources apply to the other applications. Sensor noise can be broken down into three categories: Johnson noise, $1/f$ noise, and magnetic instability. Only the last two

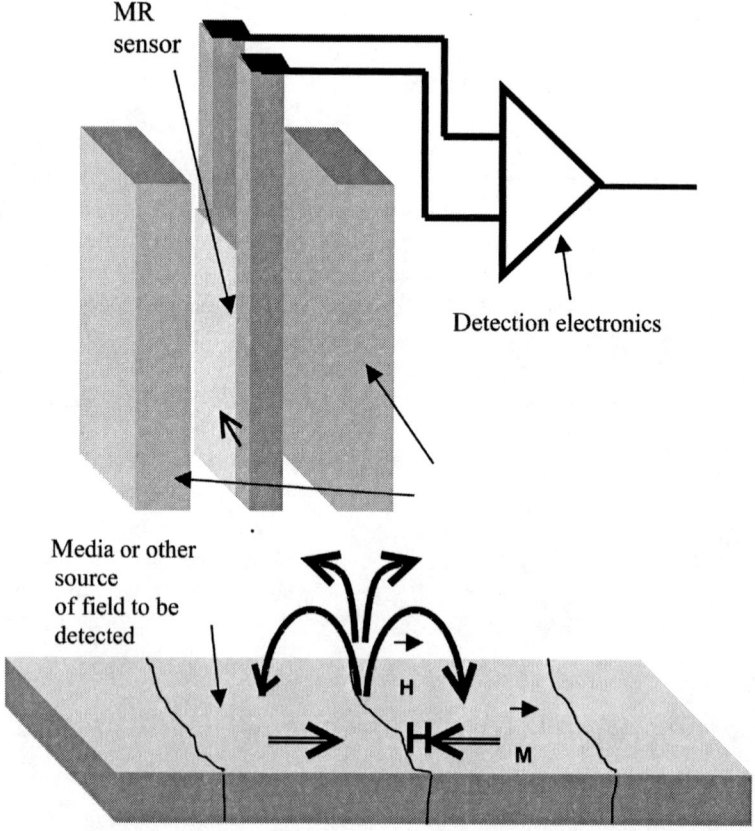

Figure 7.1. Sources of noise in magnetic device applications illustrated by magnetic recording.

are magnetic but, as we will see, the first is the limiting background noise in magnetic recording heads. Magnetic instability can be a catastrophic problem in all kinds of sensors while magnetic $1/f$ noise can be a limiting background in low-frequency applications. If the signal itself is noisy, the sensor will necessarily pick that up. Sometimes that is a good thing. You might want a very sensitive field detector that will reveal small fluctuations in field. However, in some devices, the field to be detected is setup as part of the overall system; it provides a specific signal that is to be detected above extraneous noise. Then, noise in the signal must be reduced at the same time the sense element is made more sensitive. Media noise in magnetic recording is a prime example of this. Other extraneous external field noise can come from domain instability in other magnetic materials in the device such as shield layers or magnetic thin films used to bias or stabilize a sensor. Electronics noise is not magnetic in origin and will not be discussed further in this chapter. Nevertheless, it is an important contributor to the noise background in many device applications, and a significant effort is put into electronics design in order to keep pace with the increasing sensitivity of magnetic sensors.

7.3. MAGNETIC DOMAINS AND MAGNETIZATION REVERSAL

A magnetic domain is a volume of magnetic material within which the magnetization is aligned in a single direction. The exchange interaction between magnetic moments favors their relative alignment while various types of anisotropy set a preferred direction. If the material is a single crystal, spin–orbit coupling will give rise to a preferred direction of magnetization relative to the crystallographic axes. This is referred to as crystalline anisotropy and the preferred direction is called an easy axis. Directions along which the energy is maximized are called hard axes. In polycrystalline thin films, an easy axis is often produced by depositing or annealing the film in a magnetic field. This is called an induced anisotropy. The anisotropy is called uniaxial if there is a single easy axis.

A patterned rectangular element in a single-domain state is essentially a bar magnet with poles at either end. These poles generate a magnetic field called the demagnetizing field, H_d, which, for a given material, depends on the magnetization of the element and its geometry. The geometrical factor is called the demagnetization factor, N_d which gives us $\mathbf{H}_d = -N_d \mathbf{M}$. The demagnetization factor is specific to a particular direction relative to the shape of the element. For our rectangular element, e.g., N_d for the direction of the long axis will be smaller than for the short axis since the same magnetization will produce fewer free poles at the ends. The magnetostatic energy will have a dependence on the direction of $\mathbf{M}$ relative to these geometrical axes that is referred to as shape anisotropy. The magnetization will prefer to align along the axis with lowest N_d just as it prefers to align along the easy axis in a film with induced anisotropy or crystalline anisotropy.

So, in the absence of an external magnetic field a magnetized element is in fact subject to an internal magnetic field H_d in the direction opposite to the magnetization and when an external magnetic field is applied the actual field experienced by the element will be $\mathbf{H}_{\text{applied}} + \mathbf{H}_d$. The resulting magnetostatic energy of a single-domain element with no applied field (demagnetizing field only) is equal to $(1/2)\, N_d\, \mathbf{M}^2$. For a large single-domain specimen of ferromagnetic material this can become quite significant, and the magnetization will break up into a domain structure that reduces $\mathbf{M}$ in order to reduce the magnetostatic energy as long as the energy cost of doing so is not too high.

The cost is the energy required to form domain walls, the boundaries between magnetic domains. Domain walls can be of various types and are often described by the angle through which the magnetization rotates in the width of the wall. This angle is often 180° because uniaxial anisotropy is common. In crystalline cubic materials, there are two perpendicular equivalent easy directions and 90° walls are typical. Why does the domain wall cost energy? The exchange interaction favors alignment between magnetic moments. A boundary between domains which are internally magnetized in a single direction will involve some angle between the moments in the boundary and that will cause an increase in exchange energy. This angle can be made very small by making the transition very wide—a gradual rotation of the magnetization over many individual moments as depicted in Fig. 7.2(a). On the other hand, an abrupt transition [Fig. 7.2(b)] minimizes the anisotropy energy associated with having many moments at some finite angle with respect to the easy directions. The competition between these two energy terms determines the width of a domain wall. The anisotropy term favors narrow walls while the exchange term favors wide ones. The situation can be more complicated in heterogeneous magnetic structures. For example, the antiferromagnetic coupling between layers in a GMR multilayer acts as an effective anisotropy field and causes a reduction in wall width.[1] Typical wall widths in thin films are hundreds of Angstroms but the width also depends on the type of wall. The wall type depends in turn on another energy minimization.

In Fig. 7.2(a), the magnetization is depicted as rotating in the plane of the magnetization. This would be the typical case for a magnetic thin film—both the

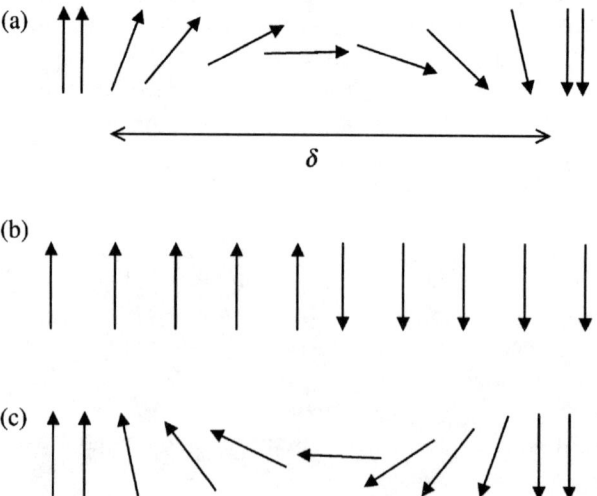

Figure 7.2. (a) Moments in a domain wall of width δ. (b) An abrupt transition. (c) The same transition as in (a) but with opposite chirality.

magnetization and the rotation that takes place in the domain wall lie in the film plane. This type of wall is a Neél wall. A Bloch wall is a domain wall in which the magnetization rotates out of the plane of the magnetization. This is the typical case in thick films or bulk material. In a thin film, the out-of-plane magnetization of a Bloch wall causes magnetic poles at the film surface which can be responsible for a large magnetostatic energy which increases as film thickness decreases. At the same time, the magnetostatic energy of a Neél wall decreases as the film gets thinner. The Neél wall has magnetostatic energy from its in-plane magnetization. The total moment in the wall decreases as the magnet is thinned, reducing the magnetostatic energy and making the Neél wall more favorable in thin films.

Various more complicated wall types are possible. Figure 7.2(a) shows the magnetization rotating through the wall with an increasing angle in the clockwise direction from left to right. For this 180° wall, the rotation could equally well have gone in a counter-clockwise direction as shown in Fig. 7.2(c). The walls in Fig. 2(a) and (c) are said to have opposite chirality. The chirality of a wall can change along its length at a discontinuity called a Bloch line, opening up many possibilities for complex wall structures. A cross-tie wall is a complex wall type that can be stable in an intermediate thickness range between Neél and Bloch walls. A schematic is shown in Fig. 7.3. The main length of the wall is a Neél wall that has broken up into segments of opposite chirality in order to reduce the magnetostatic energy. The segments are separated by Bloch lines where the wall chirality changes. The cross-ties are short segments of the Neél wall which allow the flux lines from the magnetic poles in the wall to run parallel to the magnetization in the bulk of the domain. Without the cross-ties some flux lines would run opposite to the magnetization as shown by the dotted line in Fig. 7.3. That would more than offset the decrease in magnetostatic energy that comes from reducing the average magnetization in the wall. Cross-tie walls are stable in permalloy thin films in a thickness range between 300 and 1000 Å. We will see an example of a cross-tie wall structure in permalloy in a later section.

A schematic of a closed loop 360° wall structure is shown in Fig. 7.4.[2] As we will see shortly this complex structure can form during the magnetization reversal process when domain walls come together. This structure does have intrinsically higher energy than the uniformly magnetized state but because the magnetization is almost all in one direction this structure can be stable to very high fields once it has formed. In a device that normally operates over a limited field range such a structure might essentially be permanent. These 360° structures tend to form and stabilize near defects[3] or other places where they can minimize the magnetostatic energy associated with the magnetic charges shown in Fig. 7.4. For example, in a multilayered structure a 360° wall might position itself over a region of localized magnetic charge associated with the roughness of the layers. Imagine an island of magnetic material sticking up above an otherwise flat magnetic layer of infinite extent. In the flat part of the film, the magnetization lies completely in the plane but the island will have magnetic poles associated with it. This is the magnetic charge that gives rise to a magnetostatic coupling between the layers called "orange-peel" coupling or Neél coupling.

Multilayered structures provide other opportunities for complex wall structure. In coupled multilayers, the magnetization within domain walls can also be coupled between layers and the wall type can depend on the relationship between neighboring layers. A theoretical study has shown that with high enough antiferromagnetic

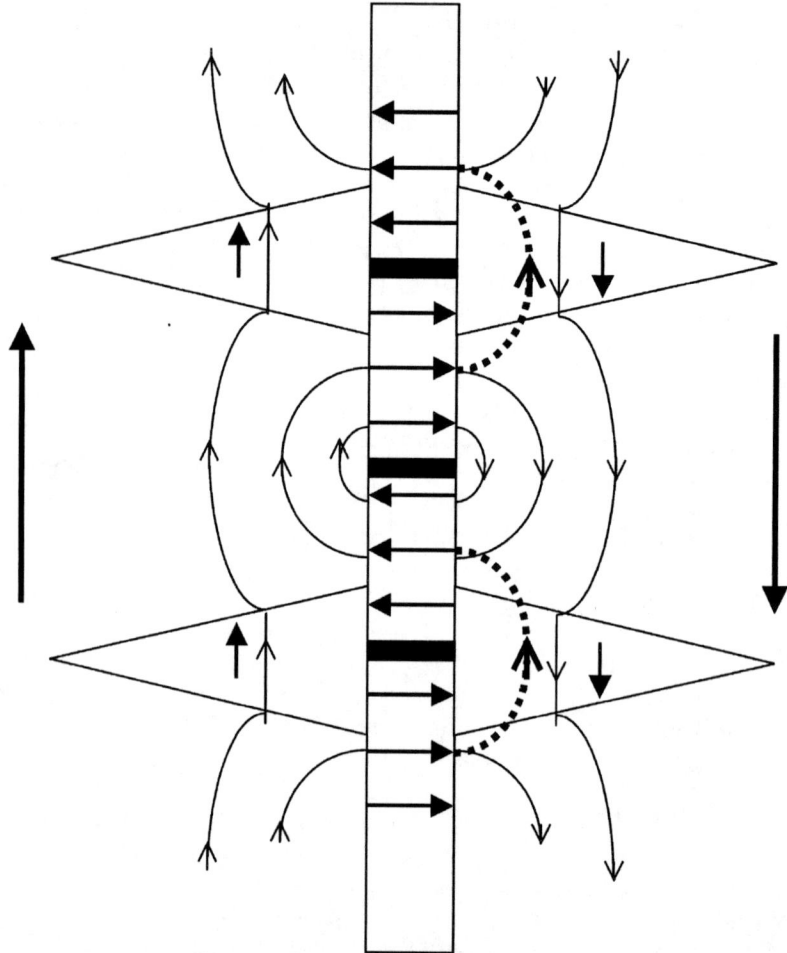

Figure 7.3. Schematic of a cross-tie wall. The dotted field lines on one side show the fields antiparallel to the domain magnetization which would arise without the cross-ties.

coupling an antiparallel Bloch wall structure can be stable.[1] This is a Bloch wall with opposite chirality in succeeding layers throughout a multilayer stack.

Now that we have looked at the basic wall types as well as some more complex examples, let us move on to consider the overall domain patterns we expect to see in thin films patterned into simple rectangular device elements. Some kind of closure domain pattern is commonly observed. A simple example is shown in Fig. 7.5. This structure has zero net magnetization and the 90° walls on the end of the element allow for flux closure, eliminating poles. Small edge domains acting as closure domains can form any time there is a component of the magnetization normal to the edge. Sometimes, when the energy benefit is great enough, closure domains with 90° walls are even stable in films with uniaxial anisotropy. In very small patterned elements, a vortex structure can form eliminating the domain wall energy. This is shown schematically in Fig. 7.6.

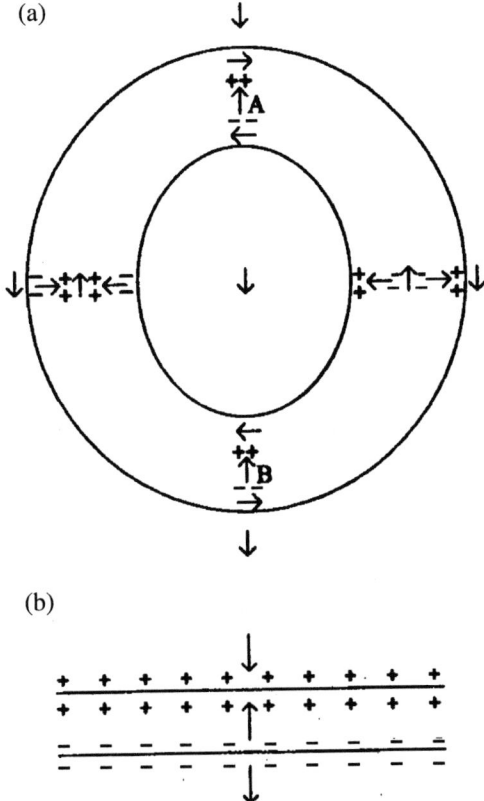

Figure 7.4. (a) Schematic of a 360° wall structure. (b) Magnetic charge distribution for a straight segment of the loop in the vicinity of points A and B. Reprinted with permission from M. F. Gillies, J. N. Chapman, and J. C. S. Kools, *J. Appl. Phys.* **78**, 5554 (1995). © 1995 American Institute of Physics.

Finally, we need to consider the magnetization reversal process. How do the domain configurations we have considered form and change as the material is magnetized? When an external magnetic field is applied to a ferromagnetic material the magnetization can change by essentially two mechanisms. The first is the growth of favorably oriented domains by the nucleation and translation or bowing of domain walls and the second is the rotation of the magnetization direction within

Figure 7.5. A simple closure domain structure.

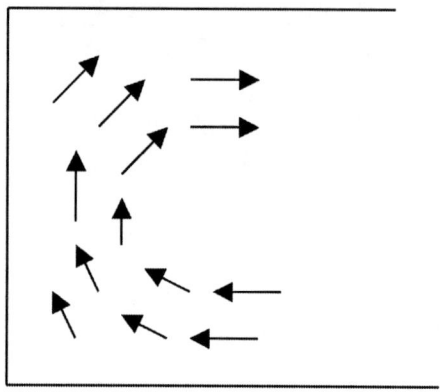

Figure 7.6. Vortex magnetization pattern.

individual domains. Domain wall motion tends to happen first because the magnetic moments within a wall are freer to rotate than those in the middle of the domain. The moments deep within the domain are strongly exchange coupled to their neighbors and are also probably pointing in a direction that gives an energy minimum dictated by the local anisotropy. The moments in the wall are already at various angles with respect to the easy axes and their precise direction is the result of the precarious energy balance described earlier. Under the influence of an applied field, a wall propagates via rotations of the moments in and near the wall like a wave pulse propagating through water. Domain rotation tends to commence at higher fields where the anisotropy energy of unfavorably oriented domains is overcome and the magnetization rotates to a different easy direction. Finally, the relatively favorably oriented domains that have filled up the entire volume slowly rotate into more perfect alignment with the applied field.

Both magnetization rotation and wall motion can be either reversible or irreversible. Imagine a magnetization reversal occurring as a 180° domain wall moves through a magnetic thin film while an applied magnetic field is increased. If the position of the wall as a function of field is repeatable when the field is decreased then the process is reversible. This can occur for domain wall motion but generally only in a very pure single crystal specimen. More commonly wall motion is irreversible. The wall pins and de-pins on impurities, grain boundaries or regions of localized strain. The magnetization remains constant as field is increased until a pinning barrier is overcome. Then, there is a discontinuous increase in magnetization as the wall breaks free. To reverse the motion of the wall, a similar barrier must be overcome so the same magnetization value is not reached until the field has been decreased by some amount. In this way, irreversible magnetization processes produce hysteresis, or history dependence, in the **M–H** loop. The discrete changes in magnetization associated with irreversible domain activity as an external field is varied create an important type of magnetic noise called the Barkhausen noise. Small thermal fluctuations in either wall position or domain orientation at fixed magnetic field also create magnetization noise that can translate into electrical noise in magnetic devices, as we will see later in this chapter.

Having considered the basic reversal mechanisms, it is worth turning to a few examples of how the magnetization process actually takes place in some technologically relevant materials and structures. This is an area of much current research activity where there are many unanswered questions. The main point is that the reversal process can be quite complex, especially in the heterogeneous systems that are of so much recent interest. As should be expected, the details of the reversal process can depend on the direction of applied field with respect to the easy axes in the film. We will first consider the case of a thin permalloy sheet film. The magnetization process has been studied by the Lorentz microscopy.[2] When the applied field is along the easy axis of the film, the reversal process proceeds as in the simple example given above. A 180° domain wall moves across the sample as shown in Fig. 7.7(a). When the applied field is along the hard axis of the film the process is different. The field is perpendicular to the easy axis so the magnetization can rotate either clockwise or counter-clockwise towards an easy axis as the field is reduced from saturation. As areas of the sample choose different directions of rotation low-angle domain walls are formed. The magnetization within the domains rotates until all the domains are aligned in the reverse direction. A snapshot of these domains as the wall angle increased is shown in Fig. 7.7(b).

The same study also considered the cases of permalloy exchange biased to FeMn (as if it were the pinned layer in a spin valve) and of permalloy free-layer reversal in a permalloy/Cu/permalloy/FeMn spin valve.[2] Both of these cases involved more complex reversal processes than those observed in the single permalloy film. The reversal involved the nucleation of domain walls and the growth of more favorable domains by domain wall motion as described earlier. However, wall mobility was lower than in the case of a single permalloy film and as unfavorably oriented domains shrank the walls on either side did not always annihilate. Instead, they sometimes formed 360° wall structures. These structures appeared to be stabilized in certain positions and were often stable until very high fields were applied. During the spin valve reversal, thermal fluctuations could be observed in the domain wall positions and additional domain wall nucleation continued creating a very intricate domain structure.

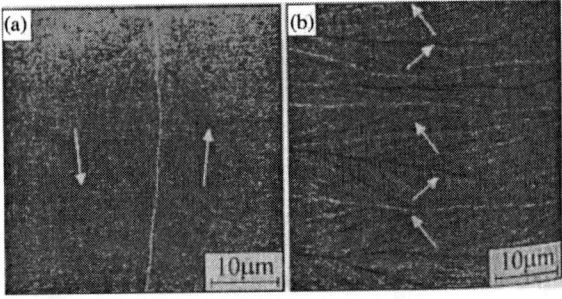

Figure 7.7. Lorentz microscopy images of magnetization reversal in permalloy with the applied field along the easy axis (a) and along the hard axis (b). Approximate magnetization directions are shown by arrows. Reprinted with permission from M. F. Gillies, J. N. Chapman, and J. C. S. Kools, *J. Appl. Phys.* **78**, 5554 (1995). © 1995 American Institute of Physics.

Much of the rest of this chapter will be devoted to understanding how fluctuations in domain wall positions and domain orientations, both reversible and irreversible, cause electrical noise in MR devices and what efforts are being made to control this noise. For the most part, the general idea is to try to avoid complex domain structures, and we will discuss various ways of dealing with the demagnetizationfields that cause a device to break up into a multidomain state. At the beginning of this section, we noted that a magnetic thin film element will break up into domains as long as the energy cost associated with forming the domain walls is not too high. Are there circumstances where it is energetically unfavorable to form domain walls? It turns out that the domain wall energy scales with the wall area while the magnetostatic energy scales with domain volume. Below a critical size, a magnetic particle will not support domain walls. The actual wall energy per unit area depends on the wall type and width as well as the film thickness and anisotropy. The magnetostatic energy of the domain depends on shape as well as volume as we have already seen. Example calculations of domain wall energies and single-domain critical sizes can be found in the bibliography. The typical critical radius for single-domain particles is hundreds of Angstroms which is unfortunately much smaller than a magnetic sensor element.

However, when we discuss media noise in Section 7.7, we will be concerned with thin film media where the magnetic particles are small single-domain magnetic grains. These particles have no domain walls so the magnetization reversal takes place purely through rotation. Thermal fluctuations of the magnetization direction of these small particles create noise. We will see that making the grains smaller reduces the noise but another size limit prevents the unlimited reduction of particle size. When the magnetic particles get too small thermal energy overwhelms the anisotropy energy and a collection of these particles can spontaneously demagnetize. This behavior is referred to as superparamagnetism,[4] and it is analogous to the Langevin paramagnetism of atoms except with larger moments. Behaving as a paramagnet, the system no longer has any coercivity and is thus no longer useful as a storage medium. This lower size limit for particles for magnetic recording media is referred to as the superparamagnetic limit. Even above the critical diameter for superparamagnetism the coercivity decreases with decreasing particle size due to thermal fluctuations. The thermal stability of single-domain magnetic grains is often characterized by the ratio of the anisotropy energy given by KV, the product of the anisotropy constant and the grain volume, and the thermal energy kT, where k is Boltzmann's constant and T is the temperature. This ratio needs to be above 50, at least, for acceptable thermal stability of information stored on thin film media.

7.4. ELECTRICAL NOISE

This section describes an experimental setup to measure noise in MR thin films and discusses electrical noise in general. Electrical noise in a resistive device consists of unwanted voltage fluctuations that obscure the intended signal and thus degrade device operation. However, in some cases, the study of these fluctuations can help one to understand the behavior of the materials in the device. Whether the goal is the study or the elimination of noise, the first step is to measure and characterize the fluctuations.

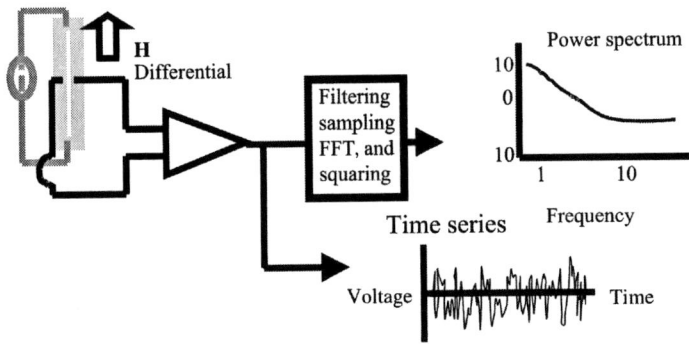

Figure 7.8. Schematic of resistance noise measurments on a patterned thin film sample. Both time and frequency domain measurements are shown. Notice both a $1/f$ component and a Johnson noise component to the example power spectrum are shown. The sample is thin film photolithographically patterned in to a bridge. Constant current is applied to the bridge and the fluctuating voltage across the bridge is measured. A magnetic field can be applied in the film plane.

Figure 7.8 outlines the collection of noise data from a MR thin film sample. The sample depicted has been patterned into a bridge to reduce the pickup of noise common to the two segments and to minimize the dc signal. Constant current is applied to the bridge. A battery with a large, quiet bias resistor can be used for a very low noise current source. The resulting voltage fluctuations across the bridge are passed through low noise amplification and anti-alias filtering and are then digitally sampled. The anti-alias filter removes everything above half of the sampling frequency. The Nyquist theorem asserts that components of the signal above this frequency will result in alias signals at lower frequencies in the sampled data. The basic idea of aliasing is simple. Imagine a 100 Hz sine wave sampled at 1 kHz. Ten points are collected for every cycle giving a reasonable reproduction of the sine wave. If the same signal was sampled at exactly 100 Hz each data point would sample the same place on the sine wave and the resulting signal would be dc. If the sampling frequency were not exactly 100 Hz but close to it, the output would be a very low frequency signal as each data point sampled a slightly different point on the sine wave.

The fluctuating voltage can be recorded directly as a function of time (time series) or it can be Fourier transformed and squared to produce a power spectrum in frequency space. In Section 7.5, we will look at both power spectra and time series collected at various dc applied magnetic fields as well as time series collected both at constant and continuously varied magnetic fields.

Let us stop and consider how this simple test setup for a patterned thin film element compares to actual device operation and product testing. For many applications, actual operation of a magnetic device will involve dynamically varying fields. Magnetic sensors might operate at very low frequencies but recording heads and memory elements need to respond at very high frequencies. During fabrication recording heads undergo a test called "transfer curve" in which a constant current and an ac magnetic field are applied to the head and the voltage signal is measured as a function of field. This is a little bit more realistic than a completely dc test but the field frequency range is very low (only hundreds of Hertz) compared to the actual drive operation. Fast dynamic measurements of magnetic response at the wafer level are only in the research stage. A group at NIST has fabricated high-speed test structures for pseudo-spin-valve MRAM elements.[5] The sense and write lines are microwave transmission lines. A fast pulse is sent on the write line and the shape of the resulting voltage pulse on the sense line can reveal the details of the high-speed switching process. Switching from pulses less than 0.5 ns wide has been observed. In addition to this frequency issue, actual device operation often involves very non-uniform, localized fields. It can be very difficult to replicate this in experimental testing of the device especially during fabrication. Typically, in such a testing, a uniform field is applied to the entire device or indeed to an entire wafer full of devices. Finally, few devices are fabricated in a bridge configuration as shown in Fig. 7.8 though a differential measurement across a bridge helps to eliminate effects common to both sides of the bridge like thermal drift and electrical pickup. It can also be difficult to sort out effects of the contacts if there are only two leads to the device as is often the case in actual sensors. Despite these drawbacks noise measurements can be done across any resistive element from specially designed test structures to finished packaged devices and that can sometimes be a great advantage in comparison to other techniques for studying magnetic structure and behavior.

Returning to the simple test setup shown in Fig. 7.8, we need to consider how the noise is quantified from the time series data that is collected. Typically, the voltage time series is sampled, Fourier transformed, and squared. Many such individual power spectra collected over a period of time are averaged resulting in a power spectral density, $S_V(f)$, with units of V^2/Hz. Both the frequency dependence of $S(f)$ and the total noise power in some bandwidth can be of interest. The square root of $S(f)$ is also often reported (voltage spectrum rather than power spectrum). In device applications, a figure of merit comparing the size of the noise to the size of the signal is often used. Most commonly this is the signal-to-noise ratio (SNR)—the size of the voltage signal divided by the size of the noise in volts at the operating frequency of the device.

The counterpart to the power spectral density is the correlation function,

$$R(t_1, t_2) = \langle \delta V(t_1) \delta V(t_2) \rangle \tag{7.4.1}$$

which is the cosine transform of the power spectral density. The correlation function is perhaps more intuitive as a characteristic of a fluctuating signal. δV is the variation of the voltage from its average value at a given time. The correlation function takes an average of the product of δV at the two times of interest over a large ensemble of identically prepared systems. Experiments are generally not done by using a large ensemble but rather by averaging over a long time. When the correlation functions obtained by these two methods are the same, the system is said to be ergodic. Physically, this means that the measurement time was long enough for the single system under study to visit a large enough sampling of the available states to be representative of all the states that contribute to the mean behavior. This is not always the case. The system under study could contain many states that are global energy minima separated by energy barriers too large to allow transitions on a reasonable experimental time scale. A relevant example could be a set of many possible domain configurations that a given device will not sample at room temperature without resetting the system by changing the magnetic field. Data collected over a long time at fixed field would then only probe the behavior of one particular state of the system while collecting data after many resets of the system might approximate a large ensemble of systems and probe a wider range of states.

A stationary process is one for which the correlation function depends only on the difference in times, $t_1 - t_2$, not the particular time at which the system is studied. Media noise is an example of non-stationary noise since there are unique times where the behavior is different—the times associated with bit transitions as compared to times when the signal comes from the center of a bit. The correlation function is used extensively in the study of media noise, a noise correlation length being related to the transition parameter in magnetic recording.[6] The transition parameter characterizes the width of the transition between bits on media (which would of course ideally be very sharp). Intuitively, it is the correlation of the signals from magnetic grains which are on opposite sides of the transition center which gives width to the transition.

Either the correlation function or the power spectrum can completely describe the Gaussian noise—noise comprised of the fluctuations of a large normal distribution of independent fluctuators. If the voltage noise is non-Gaussian (either due to details of the conduction process in the material or to a small sample size limiting the number of fluctuators) further statistical study of voltage time series can be interesting. The size, duration, and frequency of discrete voltage steps in time series can illuminate the nature of the fluctuators. In magnetic materials, such steps can be related to irreversible domain wall motion as we will see in Section 7.5. The extreme case is the isolation of a single fluctuator. We will see examples of this also in Section 7.5. When individual discrete steps are not visible but the noise is not Gaussian, more sophisticated techniques for statistical study of the distribution of fluctuators such as the second spectrum can be utilized as described in detail by Weissman.[7] Such an investigation of the distribution of fluctuators is mostly used to compare with various theories of the detailed physics of the transport properties. The magnetic example of noise in rare-earth manganites will come up in Section 7.5.

Equilibrium fluctuations can be mathematically related to the dissipation that arises when a linear, dissipative system is driven by an external force. This is described by the fluctuation–dissipation (FD) theorem.[8] This relationship arises because the random thermal motion that causes a measured quantity such as voltage to fluctuate in equilibrium is also responsible for the friction or damping that occurs when the system is driven. By the FD theorem, the power spectral density of equilibrium fluctuations in a quantity is proportional to the out-of-phase part of the ac response of that quantity to a driving force.

Consider the case of a resistor. Voltage fluctuations in equilibrium are related to the response of the voltage to an applied current. By Ohm's law, the response is just the resistance. These voltage fluctuations are known as Johnson's noise or Nyquist's noise. The power spectral density (at room temperature) is equal to $4kTR$, where k is Boltzmann's constant, T, the temperature, and R, the resistance. The Johnson noise is present in all resistors. It depends only on the resistance and temperature and is independent of frequency. At the high operating frequencies of magnetic recording heads, this non-magnetic Johnson noise is the background noise from the head and minimizing the sensor resistance is the way to reduce this background noise.

Noise with $S_V(f)$ proportional to $1/f$ is ubiquitous in resistors which are driven by a current. The power spectral density is proportional to the square of the current. This means that for a fixed current, the frequency above which the Johnson noise is larger than $1/f$ noise decreases with increased resistance. Significant at low frequencies and large current densities, MR $1/f$ noise can be an important background noise in some MR sensors. Generally, the fluctuating forces that could be associated with $1/f$ resistance noise via a FD relation are not easily applied experimentally. Localized strain forces are an example. The strong coupling between resistance and magnetization in GMR materials has been exploited to demonstrate an FD relation for $1/f$ noise in a GMR element.[9] The experiment will be described in Section 7.5 as we move on to look at how domain dynamics and magnetic fluctuations cause electrical noise.

7.5. FROM MAGNETIC FLUCTUATIONS TO ELECTRICAL NOISE

In MR materials, resistance is coupled to magnetization and local fluctuations in magnetization give rise to resistance noise. For example, in antiferromagnetically coupled Co/Cu GMR multilayers the resistance depends on the magnetic alignment of the layers nearly following the simple relationship:[10,11]

$$R = R_{AP} - (\mathbf{M}/\mathbf{M}_s)^2(R_{AP}-R_P) \qquad (7.5.1)$$

where R_{AP} is the resistance with the magnetizations of the layers parallel, R_P, the resistance with layers parallel, $\mathbf{M}$, the magnetization and $\mathbf{M}_s$, the saturation magnetization. We can write down an FD relation for magnetization noise relating the magnetization fluctuations to the response of the magnetization to an applied ac magnetic field:

$$f\, S_M(f) = (2/\pi)\chi''(f)kT/V \qquad (7.5.2)$$

where $S_M(f)$ is the power spectral density of magnetization fluctuations, $\chi''(f)$, the out-of-phase magnetic susceptibility, T, the temperature, k, Boltzmann's constant and V, the sample volume. Then, we can use the derivative of Eq. (7.5.1) with respect to magnetization to convert Eq. (7.5.2) from magnetization to resistance. This will convert the out-of-phase ac magnetic susceptibility into the out-of-phase response of the resistance to an ac magnetic field, χ''_R. The result looks like this:

$$\alpha(f,\mathbf{H}) = (4/\pi)(kT/\mu_s)|\chi''_R(f,\mathbf{H})|(\Delta R)^{1/2}[R_{AP}-R(\mathbf{H})]^{1/2}/R^2 \qquad (7.5.3)$$

where $\mu_s = M_s V/N$, the saturation moment per atom and where

$$\alpha(f) \equiv f\, S_R(f)N/R^2 \qquad (7.5.4)$$

is a dimensionless noise parameter proportional to the power spectral density and the total number of atoms in the sample, N.[12] We now have a prediction for the resistance noise that we expect to arise from thermal fluctuations in the overall parallel/antiparallel ordering of the layers in a GMR multilayer. The comparison of the actual noise and the ac resistance response to magnetic field from this material will let us test the FD theorem for $1/f$ noise in a resistor for the first time. We will also see what fraction of the magnetic fluctuations in our sample come from the fluctuations in ordering we have used for our prediction.

This experiment was conducted on a sputtered antiferromagnetically coupled Co/Cu multilayer. The noise measurement was performed as described in Section 7.4. Power spectra were collected at a series of fixed dc magnetic fields in the plane of the sample so that at each data point was collected over a period of time with a constant field. Then, the field was changed, the sample allowed to equilibrate and more data were collected. The dimensionless noise parameter, α, and the resistance are both shown in Fig. 7.9 as a function of field. The increase in noise near 500 Oe

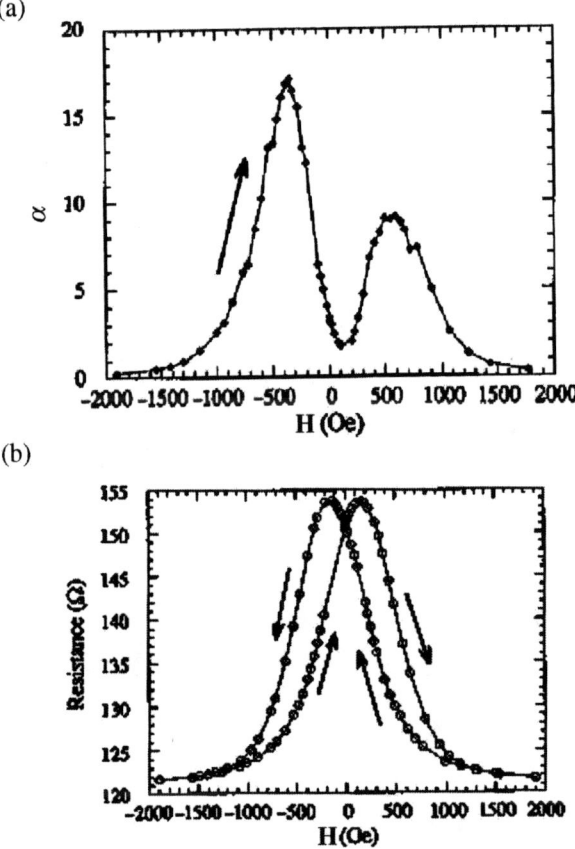

Figure 7.9. Noise (a) and resistance (b) as a function of applied dc magnetic field for an antiferromagnetically coupled GMR multilayer. Data were collected with field constant at each point. After each change in field, the system was allowed to equilibrate and data were collected at the new dc field. The data were taken in the order indicated by the arrow. A noise peak occurs in the region where the dR/dH is the greatest. Reprinted with permission from H. T. Hardner, M. B. Weissman, M. B. Salamon, and S. S. P. Parkin, *Phys. Rev. B* **48**, 16156 (1993). © 1993 The American Physical Society.

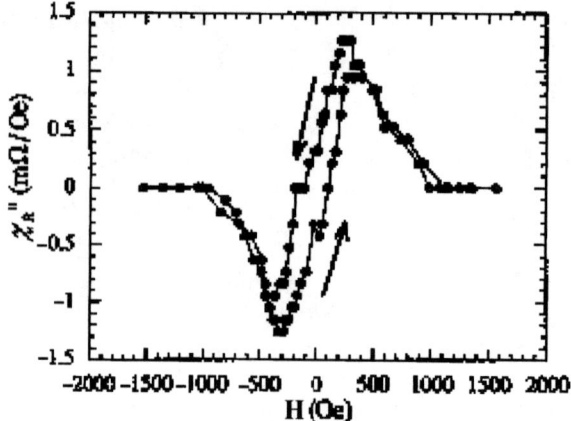

Figure 7.10. The out-of-phase response of the resistance to a small ac applied field as a function of dc field. Again the data were collected at each fixed dc field after a period of equilibration. The ac field was superimposed on the fixed dc field. The arrows indicate the order in which the points were taken demonstrating a hysteresis qualitatively like that seen in the power spectral density. Reprinted with permission from H. T. Hardner, M. B. Weissman, M. B. Salamon, and S. S. P. Parkin, *Phys. Rev.* B **48**, 16156 (1993). © 1993 The American Physical Society.

corresponds to the field range where the slope of the resistance with magnetic field is the greatest for this sample. The out-of-phase resistance response was measured with a lock-in amplifier. The dc field was stepped as in the noise measurement but a small ac field was superimposed on it. These data are shown in Fig. 7.10.

Figure 7.11 compares the measured noise to that predicted by the FD relation by using the measured resistance response. There are no adjustable parameters in this comparison. The good agreement between the two curves shows that most of the noise in this system does come from fluctuations in the parallel/antiparallel ordering

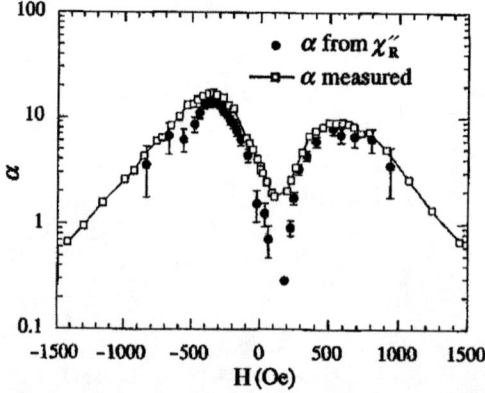

Figure 7.11. Comparison of the measured noise (●) and the noise predicted by the FD relation using the measured out-of-phase resistance response (○) as a function of applied dc field. Reprinted with permission from H. T. Hardner, M. B. Weissman, M. B. Salamon, and S. S. P. Parkin, *Phys. Rev.* B **48**, 16156 (1993). © 1993 The American Physical Society.

of the layers via the GMR relationship in Eq. (7.5.1). However, there is also a significant magnetic noise near zero field that is not described by this mechanism. It was speculated that this noise might result from fluctuations in the direction of antiparallel-ordered domains since at zero field there is no preferred direction for the antiparallel domains.

This experiment primarily illustrates the use of noise measurements to understand magnetic fluctuations and the mechanism by which they create resistance noise. However, the observation of a peak in the power spectral density of $1/f$ noise in the most useful field range also provides a basis for optimization and comparisons of $1/f$ noise among various materials and sensor geometries for low-frequency applications.[13,14] A noise peak around the field where dR/dH is greatest has now been observed in many MR materials.[15,16] A further example from a Co/Cu spin valve is shown in Fig. 7.12.[17] In this case, at low field the resistance as a function of field is an odd function and there is only one noisy region since the top Co layer is pinned by an antiferromagnetic cobalt oxide layer.

Even at high current densities in the noisiest of these materials, this magnetic $1/f$ noise falls below the Johnson noise at frequencies well below the operating frequencies for magnetic recording. However, this does not mean that magnetic noise is not a problem for sensors that operate at high frequencies. The background noise from thermal fluctuations is non-magnetic at high frequency but catastrophic noise problems can arise from irreversible changes in the magnetization state of the device. As we saw earlier, the same coupling between resistance and magnetization that makes the material useful as a sensor provides the mechanism for jumps in the output voltage if there are discrete changes in the local magnetization.

The first example of this that we will consider is referred to as the Barkhausen noise. It arises from discrete changes in magnetization as applied magnetic field is varied. This is shown schematically in Fig. 7.13. The jumps in magnetization can come from the irreversible motion of domain walls or from sudden changes in

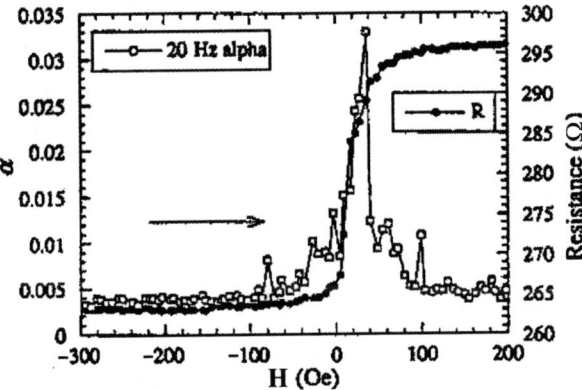

Figure 7.12. Resistance noise and resistance as a function of applied dc field in a Co/Cu spin valve with a CoO layer pinning the top Co layer. Reprinted with permission from H. T. Hardner, M. B. Weissman, B. Miller, R. Loloee, and S. S. P. Parkin, *J. Appl. Phys.* **79**, 7751 (1996). © 1996 American Institute of Physics.

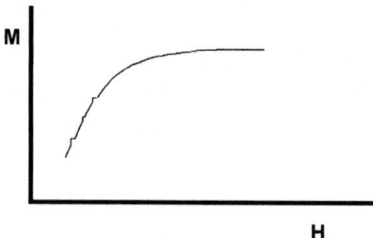

Figure 7.13. Illustration of the Barkhausen noise. As the applied field **H** is smoothly varied, the magnetization makes discrete jumps due to irreversible domain rotation or domain wall motion.

domain orientation. This effect was originally detected with a pickup coil wound around iron[18] and is still used to characterize the microstructural quality of steels. Domain walls sweeping through the material as the field is varied are pinned at defect sites and create the Barkhausen noise as they break free revealing the density of defects in the material. In MR materials, discrete changes in magnetization couple to the resistance causing jumps in the output signal from a sensor. The Barkhausen noise from the Co/Cu multilayer specimen used for the FD experiment is shown in Fig. 7.14. This is time series data collected while the applied magnetic field was slowly ramped. Large noise is visible as the field is swept over the same range where the large power spectral density was observed at fixed field. A time series trace taken at the fixed field that produced a peak in power spectral density is shown to illustrate that the Barkhausen noise is much larger than the noise at fixed magnetic field. The Barkhausen noise from the spin valve sample is shown in Fig. 7.15. This was a physically smaller sample and individual steps in the voltage are clearly visible.

From the size of these voltage steps, an estimate of domain size can be made and used to compare various materials. Imagine that a step in the voltage

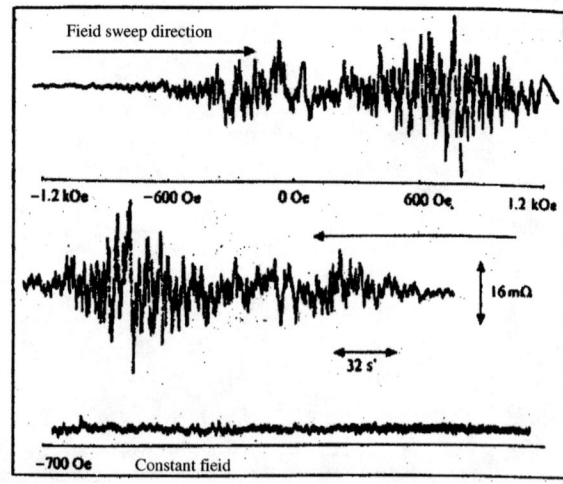

Figure 7.14. The Barkhausen noise in the Co/Cu multilayer discussed earlier. The magnetic field was varied linearly with time as indicated by the arrows. The bottom time series was taken at fixed field where one of the peaks in the power spectral density was observed. Reprinted with permission from H. T. Hardner, S. S. P. Parkin, M. B. Weissman, M. B. Salamon, and E. Kita, *J. Appl. Phys.* **75**, 6531 (1994). © 1994 American Institute of Physics.

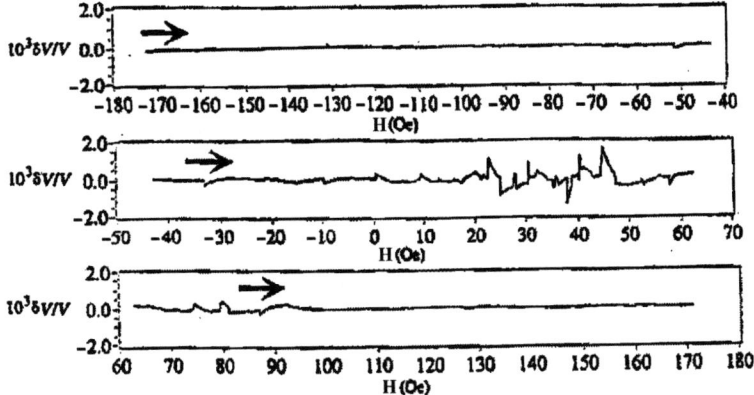

Figure 7.15. The Barkhausen noise in voltage time series from the Co/Cu spin valve device. Magnetic field was ramped linearly over time in the direction of the arrows. This sample was patterned smaller than the multilayer and discrete steps in the time series are visible. Reprinted with permission from H. T. Hardner, M. B. Weissman, B. Miller, R. Loloee, and S. S. P. Parkin, *J. Appl. Phys.* 79, 7751 (1996). © 1996 American Institute of Physics.

corresponded to a domain in a single magnetic layer rotating 180° so that it was parallel to the adjoining layer instead of antiparallel. Knowing the total change in resistance of the whole sample in a complete field sweep we could use the change in resistance in this single step to find the size of the domain relative to the size of the device. That simplistic calculation looks like this:

$$V_D = (V_s * (\delta V/V))/(\text{GMR}) \tag{7.5.5}$$

where V_D is the domain volume, V_s, the sample volume, GMR, the total change in R/R average, and $\delta V/V$, the voltage step size divided by the average voltage. Since a given step might correspond to a smaller rotation of a larger domain this estimate gives a minimum domain size. On the other hand, the step might correspond to the motion of a domain wall. In that case, the size corresponds to the increase in domain size caused by that movement rather than the size of the domain itself but again there must be a domain at least that much large.

Many different GMR samples have been compared by using this domain size estimate. Figure 7.16(a) gives an example comparing Co/Cu multilayers with different Cu layer thicknesses. Figure 7.16(b) shows how the coupling depends on spacer thickness in these structures.[19] More weakly coupled multilayers had larger Barkhausen noise steps corresponding to larger domains. Figure 7.17 shows a comparison of GMR sandwiches patterned into different areas.* The domain size appears to be limited by the sample size and the implied domain sizes are quite large.

An example of Barkhausen noise in a spin valve GMR recording head is shown in Fig. 7.18(a).[20] The inset is data from the transfer curve measurement described earlier, the voltage from the head while an applied field is swept back and forth.

---

*GMR sandwiches grown, patterned and packaged by Nonvolatile Electronics, Inc., Eden Prairie, MN.

(a)

Multilayer	Coupling	Estimated Domain Area in μm²
Co/Cu (10 Å /10 Å) 39 layers	Strong AF	0.25
Co/Cu (10 Å /21 Å) 39 layers	Weaker AF	5
Co/Cu (10 Å /11 Å) 39 layers	Nearly uncoupled	9

(b)

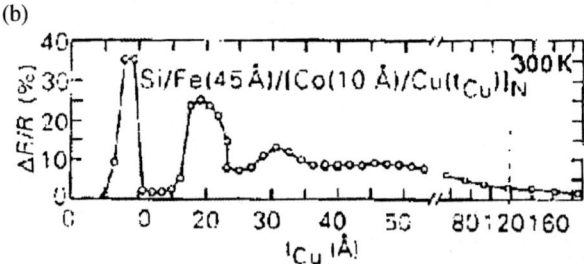

Figure 7.16. (a) GMR as a function of Cu thickness showing the variation in the interlayer coupling with Cu thickness. (b) Comparison of estimated domain size from Barkhausen noise for Co/Cu multilayers with different Cu thicknesses provided different amounts of interlayer coupling. (b) is reprinted with permission from S. S. P. Parkin, R. Bhadra, and K. P. Roche, *Phys. Rev. Lett.* **66**, 2152 (1991). © 1991 American Physical Society.

Jumps in the voltage creating hysteresis and openness in the curve are Barkhausen noise indicating magnetic instability in the device. A head with magnetic instability will not give repeatable results and will be unacceptable for use in a disk drive. The experimenter also observed discrete steps in voltage time series without applied

Width (μm)	Strip Area (μm²)	Domain Area Estimate (μm²)
3	108	50
3	108	70
3	108	70
6	720	260
6	720	250
6	720	270
2	80	40
6	360	80

Figure 7.17. The comparison of estimated domain sizes for GMR sandwiches patterned into different sizes. The size of Barkhausen jumps is clearly correlated to the pattern area. The NiFeCo(40 Å)/ CoFe(15 Å)/Cu(40 Å)/CoFe(15 Å)/NiFeCo(40 Å)/Ta(200 Å)/CrSi(100 Å) sandwich sensors were grown, patterned and packaged by NVE Inc. of Eden Prairie, MN.

DOMAIN DYNAMICS AND MAGNETIC NOISE

(a)

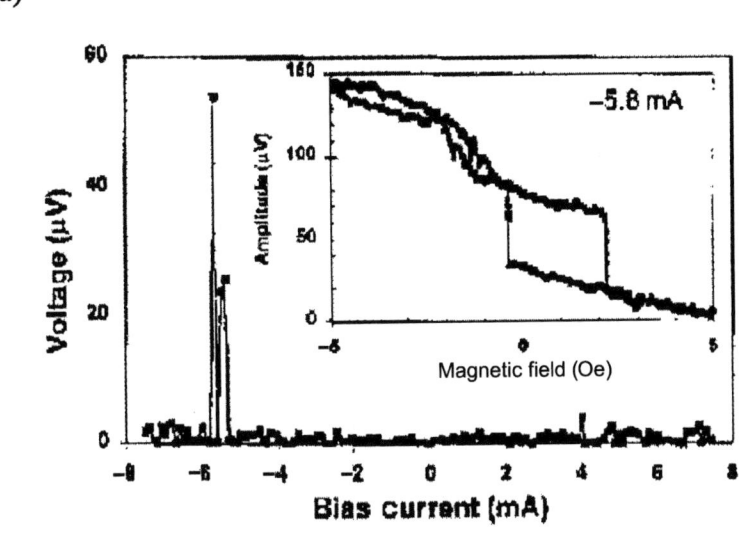

(b)

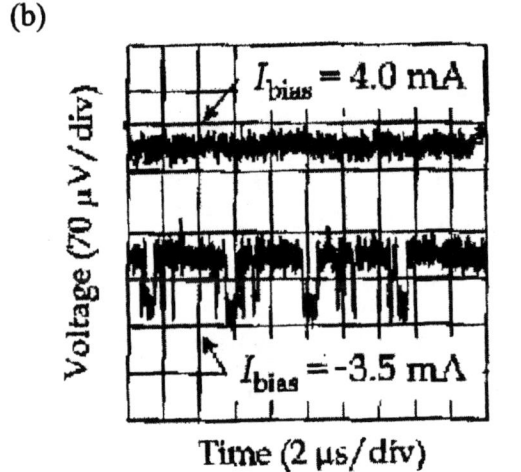

Figure 7.18. (a) MR curve voltage versus bias current for a head which shows voltage fluctuations at −5.8 mA. Inset shows low field MR curve at −5.8 mA. (b) Voltage versus time for a head at two different bias currents. Voltage fluctuations are seen at the bias current of −3.5 mA. Reprinted with permission from A. Wallash, *IEEE Trans. Magn.*, **34**, 1450 (1998). © 1998 IEEE.

field at the same bias current where the Barkhausen noise was observed. An example of these bias-current-dependent steps for a different head is shown in Fig. 7.18(b). Switching between discrete levels like this is often referred to as telegraph noise.

Two-level telegraph noise is the simplest example of noise arising from a non-Gaussian distribution of fluctuators. The study of such a signal in an MR film can be very interesting as it amounts to the study of the activity of a single domain or domain wall as the device switches between two domain configurations. If the switching is Markovian (the transition probability depends only on the current state and not on the history of the system) the power spectrum of these fluctuations will be a Lorentzian at a frequency equal to the sum of the transition rates between the two states. The switching times as a function of magnetic field and temperature for the two states and, in particular, the range of magnetic fields over which the fluctuations occur can provide some understanding of the domain dynamics including domain size estimates. Consider the case of thermally activated transitions between two states separated by an energy barrier—e.g., two positions for a domain wall. The energy barrier will have an $\mathbf{M}\cdot\mathbf{H}$ term—one of the states will involve a region of the film having a better alignment with the external field than the another state. At some applied field, the system will have an equal probability of being in either states. Adjusting the magnetic field away from this point changes the size of the energy barrier by changing this term, favoring one or another of the states. When the barrier is large enough, one state dominates completely. A comparison of the field range over which the fluctuations are active with the thermal energy available allows an estimate of the difference in magnetization between the two states. The dependence of the lifetimes of the two states, up and down, on magnetic field can be subjected to some simple analysis based on a thermally activated transition between the two states as follows:

$$1/t_{up} = 1/t_0 \exp(-\mathbf{H}\cdot\mathbf{M}_{up}/kT) \quad 1/t_{down} = 1/t_0 \exp(-\mathbf{H}\cdot\mathbf{M}_{down}/kT) \quad (7.5.6)$$

where $t_{up/down}$ is the average lifetime of the up/down state at a given field and t_0 is the average lifetime when $t_{up} = t_{down}$, $\mathbf{H}$ is the applied field measured relative to the point where $t_{up} = t_{down}$ and $\mathbf{M}_{up/down}$ is the magnetization for the state in question. Assuming that the transition is a 180° change in magnetization gives a minimum domain size estimate for a GMR material as in the case of Eq. (7.5.5). Making this assumption, we get a domain size estimate in the form of an estimate for the change in magnetization associated with the transition:

$$\Delta \mathbf{M} = (kT/H \ln t_{50})(\ln t_{up} - \ln t_{down}) \quad (7.5.7)$$

In this analysis, we only get a unique $\Delta \mathbf{M}$ if the natural log of the lifetimes are linear in $\mathbf{H}$. This type of experiment has been done using small devices made from permalloy–silver discontinuous multilayers.[21] Domain size estimates suggested that the magnetic grains did not act as uncorrelated individual domains. The authors noted that they did see behavior that was not explained by this simple model such as cases where one of the transition rates was field dependent while the other was not. Note that this type of experiment probes a different set of domains than a Barkhausen experiment. Here, we are only observing domains that happen to be thermally active at a particular magnetic field and temperature. When the field is varied from saturation in one direction to the opposite direction in a Barkhausen experiment every part of the sample must reorient so the entire reversal process can be observed including domains too large to be active at any fixed field.

Another magnetic system that has generated a great deal of recent interest is the magnetic tunnel junction. In this structure, electrons tunnel across a barrier (typically alumina) between two transition metal electrodes. The transition probability depends on a spin-dependent density of states so that the tunneling current depends on the relative magnetic orientation of the electrode layers. This relative orientation can be controlled by the same means used in various GMR spin valve structures such as pinning one layer to an antiferromagnetic layer, making the layers two different thicknesses or two different materials, etc. Magnetic tunnel junctions are getting a lot of attention as candidates for MRAM devices. Noise characterization has been performed on a magnetic tunnel junction comprised a micrometer-sized stack of NiFe/Co/Alumina/Co/NiFe.[22] $1/f$ noise is observed in the voltage at low frequencies and relatively low junction bias. When the junction bias is great enough non-magnetic telegraph noise is observed. This is attributed to charge trapping and de-trapping in the alumina layer. At lower bias voltages magnetic field dependent telegraph noise can be observed in particular magnetic configurations. The researchers attribute this to reversible domain wall motion or domain reversal in the magnetic layers. Because the device is very small a single fluctuator can sometimes be observed at just the right magnetic field just as we saw in the NiFe/Ag discontinuous multilayers we just considered.

Telegraph noise signals can often be observed in materials that exhibit Gaussian $1/f$ noise by the fabrication of a very small structure containing only a small number of fluctuators. Alternatively, the conduction process in a material can be inhomogeneous causing non-Gaussian noise in macroscopic devices. This has been observed in one of the rare-earth manganites that exhibit the colossal magnetoresistance effect (CMR). These materials were discussed in an earlier chapter. While no practical CMR devices have been demonstrated yet, the magnetics industry is keeping an eye on CMR materials since the MR response is so large. It unfortunately turns out that these materials also exhibit colossal $1/f$ noise.[23-25] Further, the noise is very non-Gaussian with telegraph switching on the order of 1/1000th of the total sample voltage observable in samples with areas in the square millimeters.[24] The main interest in this noise study was the use of noise measurements to understand the mechanism for the conduction in these materials. In the case of GMR materials, we were able to understand the $1/f$ noise in terms of a relationship between R and $\mathbf{M}$ and the existence of thermal fluctuations in $\mathbf{M}$ without needing to know the physics behind the MR. Conversely, the GMR $1/f$ noise measurements do not reveal anything about the underlying details of the spin-dependent scattering in those materials since the same $1/f$ noise will result as long as the system is described by Eq. (7.5.1) regardless of how that equation arose. Conduction in CMR materials is not well understood but the details of the resistance noise demonstrate that magnetic inhomogeneities play an important role in the mechanism for the CMR effect and support a theory of polaronic hopping conduction.[24,25] The reader is referred to the references for details including an example of the measurement of the second spectrum as a study of non-Gaussian noise without discrete telegraph steps. Of course both the large size of the $1/f$ noise and the presence of extremely large switching suggest that it may be very difficult to make stable magnetic sensors from magnetic lanthanides even if the range of magnetic field and temperature over which the MR effect occurs becomes practical.

A final example of noise from magnetic instability manifesting itself in applications is baseline popping in magnetic recording. Instead of experiencing domain activity while the head is "reading" as in Barkhausen noise or noise while the field is fixed as in $1/f$ noise or telegraph noise, the signal baseline fails to repeatably return to the same value after each transition. Each time the head reads a transition

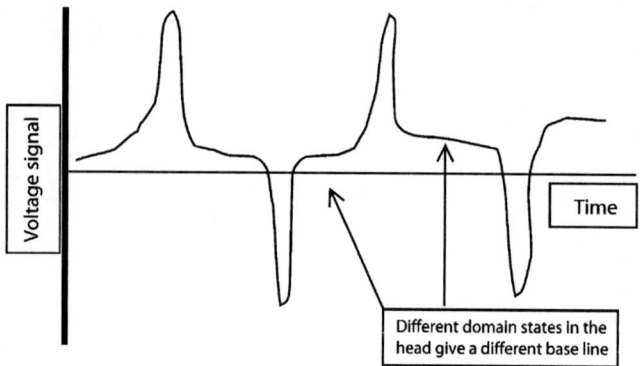

Figure 7.19. Illustration of baseline popping.

it experiences a cycling of magnetic field. Baseline popping occurs when the domain state can be different after each field cycle. The effect is illustrated schematically in Fig. 7.19. A similar change in baseline can occur after a write pulse from the head. Note that in this case the domains do not need to fluctuate to cause trouble. Each state can be stable at a fixed field. The existence of multiple quasi-equilibrium domain configurations allows the baseline to vary from one bit to the next.

We have seen that the magnetic instability tends to be a catastrophic problem rather than a contribution to background noise that can limit SNR. A great deal of effort is put into designing and fabricating devices that are magnetically stable and free of these noise problems. In Section 7.6, we will look at some examples of how that is achieved.

7.6. STABILIZATION OF MAGNETIC SENSORS

The ideal sensor would consist of a single magnetic domain. The magnetization would rotate smoothly and freely in response to the fields to be detected. In this way, the resistance would be sensitively and repeatably correlated to the external magnetic field. There would be no irreversible magnetic processes to cause the Barkhausen noise or hysteresis. Unfortunately, a single-domain state in a rectangular sensor element is energetically unfavorable. As discussed earlier, demagnetization fields at the ends of the device favor the formation of a closure domain structure. The demagnetization fields must be eliminated or the inevitable domains in the end region must somehow be stabilized. This must be done in a way that still allows relatively free rotation of the magnetization so the sensor can operate.

One possibility is to have a long narrow sensor but to pattern the electrical contacts to only drive current through the center so that the signal does not come from the very ends of the sensor. Shape anisotropy in the long strip helps to keep the magnetization along the length of the device and the end region is not active so domain activity in closure domains there does not create noise. This is shown in Fig. 7.20(a).

Figure 7.20(b) shows two more sophisticated schemes employing hard magnet stabilization. In both cases, the idea is to have permanent magnet material on the ends of the sensor. By carefully matching the flux, the poles at the ends of the

DOMAIN DYNAMICS AND MAGNETIC NOISE

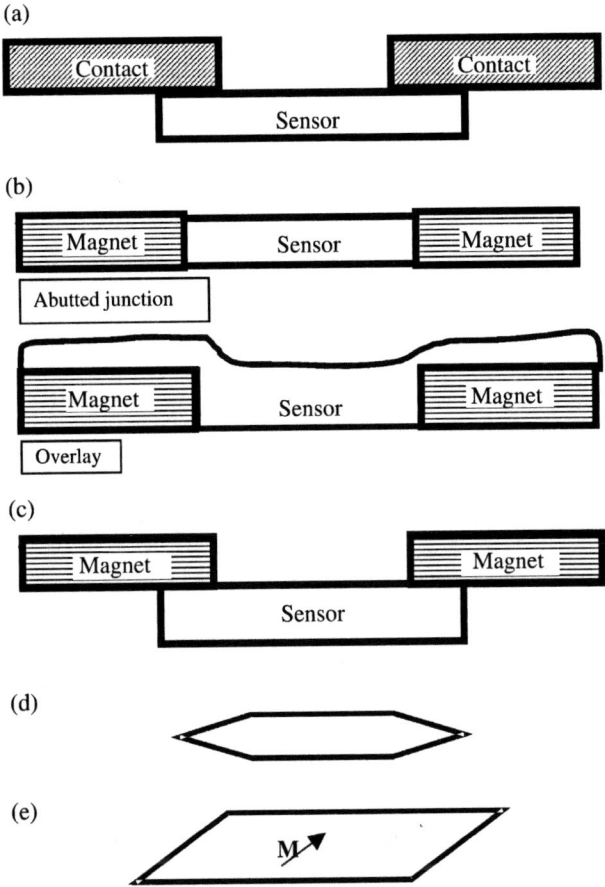

Figure 7.20. Various domain stabilization schemes. (a) Shape stabilization. (b) Hard magnet stabilization. Abutted junction and overlay methods. (c) Exchange stabilization. (d) Shaped ends—top view. (e) Rhombic sensor—top view.

sensor material can be eliminated thus eliminating the demagnetization fields. The first scheme is an abutted junction where the sensor is defined and then permanent magnet material is deposited next to it. In the second scheme, the magnets are defined and the sensor material is deposited over the entire structure.[26] Contact metal is deposited on either sides in each case and in the overlay scheme it defines the width of the sensor. Thin film MR recording heads are generally stabilized by using hard magnet stabilization. The processing required to achieve these structures is not trivial. A recent micromagnetic simulation showed that having more than 0.2 μm of overlap of the permanent magnet material onto the sensor element in an abutted junction structure can promote hysteresis and a multidomain state—creating magnetic instability instead of controlling it.[27]

Exchange coupling to a ferromagnetic or antiferromagnetic layer can also be used to pin the ends of the sensor as illustrated in Fig. 7.20(c). This type of stabilization relies on direct exchange coupling at the interface between the sensor and the magnet layer. The coupling is very sensitive to the properties of the interface, requiring atomic contact between the layers. High volume manufacturing of heads using this scheme will require a very robust process for providing good interface quality.

The sensor shape can be manipulated to enhance stability as well. A simple example of shaping the ends of an element to reduce demag fields is shown in Fig. 7.20(d). The idea is to reduce the component of magnetization that is normal to the edge to reduce the poles that create demag fields. Tapered ends are a popular way to produce better switching characteristics in small MRAM elements.[28] For MRAM as for sensors the preferred mode of magnetization reversal is magnetization rotation rather than domain wall motion. When a rectangular element has end domains the magnetization reversal will occur by the sweeping of those walls through the element as we saw earlier. The tapered ends can eliminate those walls causing magnetization rotation to be energetically more favorable. A scheme involving an elliptical active sensor region as part of a permanent-magnet-biased MR head has been patented.[29] This similarly takes advantage of the lower demag field for an ellipse. Another example is a patented rhombic sensor,[30] sketched in Fig. 7.20(e), which takes advantage of the transverse biasing in an AMR head. The magnetization in an AMR head is generally biased to lie at about a 45° angle to the length of the sensor so that the head will operate in its most linear regime. The rhombic sensor arranges for the magnetization to lie along the sensor ends instead of perpendicular to them.

Consider a rectangular MR sensor whose easy axis lies along its length. Suppose the field that the sensor will detect is along the sensor width—the hard axis of the film. This is a typical recording head configuration. We saw in Fig. 7.7(b), the domain configuration that occurs in this case when the hard-axis field is reduced from saturation. Low-angle domain walls formed because either rotation direction was equally favorable. It can be a good idea to avoid this type of ambiguity. Rotating the easy axis with respect to the length of the sensor and thus providing a preferred rotation direction has been demonstrated to reduce magnetic switching noise.[31]

A less recent example comes from permalloy thin films used for inductive read heads. In this case, a single-domain state is desired in a narrow permalloy strip with the easy axis transverse to the length. Now closure domains form along the edge instead of at the end and cause the Barkhausen noise. In this case, the Barkhausen noise is not MR. Instead, we have the more traditional case where magnetization jumps are picked up inductively by a coil. Elimination of domain walls in the core material was achieved by laminating the permalloy film—creating a sandwich or multilayer of permalloy layers separated by non-magnetic insulating spacer layers.[32] The idea is that the flux-closure can be achieved between the two single-domain layers instead of through a closure domain structure.

These examples all illustrate the effects of a good design on stability but there is also a substantial amount of materials and process development work aimed at controlling domain structure. The film property most closely tied to magnetic instability is the magnetostriction effect that was discussed in an earlier chapter. The saturation magnetostriction is the strain, or fractional change in dimension, $\Delta L/L$, a material experiences in a particular direction when magnetized to saturation along that direction. Like crystalline anisotropy, magnetostriction arises from spin–orbit coupling. This interaction causes the otherwise symmetrical charge distribution

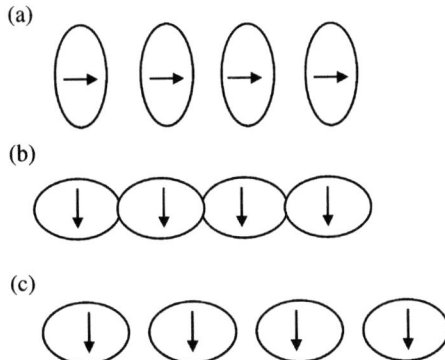

Figure 7.21. (a) The spins of magnetic ions in a lattice and the electron probability distribution distorted by spin–orbit couplilng. (b) Changes in the overlap of the electron cloud with spin angle. (c) Distortion of the lattice to reduce the overlap of the electrons.

around the magnetic ions in the lattice to be distorted along the spin axis. When the spins rotate as the magnetization direction changes the amount of overlap of the electron clouds changes giving rise to anisotropy energy and distortion of the lattice to minimize the energy. Not only does the direct magnetostriction effect produce changes in sample dimension with applied field, but also an inverse effect can cause changes in the magnetics of a material under stress. Changes in lattice spacing can cause the spins to rotate in order to minimize the overlap of the electron distributions. Figure 7.21 gives a schematic idea of the effect but highly exaggerated. Saturation magnetostriction values on the order of 10^{-6} can be a problem for magnetic stability. The effect is a magnetoelastic anisotropy proportional to the product of saturation magnetostriction and stress—magnetization will prefer to align along a particular direction relative to applied stress. This anisotropy term will be additive with the other types of anisotropy we have already considered.

If magnetostriction is not carefully controlled in a thin film device, the inevitable stresses can produce conditions favorable for domain formation and magnetic instability. This can be understood in terms of our earlier discussion on domain formation and reversal. We saw that the domain wall energy included both exchange energy and anisotropy energy. Altering the total anisotropy energy by adding an appropriate magnetoelastic term can make it either easier or harder to form domain walls. Likewise, domain wall width and mobility and the preferred domain structure will all be affected since the local anisotropy plays a role in each of these. If the magnetoelastic anisotropy tends to counteract a preferred magnetization direction that was intentionally setup through one of the stabilizing techniques already discussed the device can be destabilized. During device operation, changes in stress (e.g., due to heating) will make a particular domain orientation more favorable. Those favored domains will grow through domain wall motion just like they would in response to an applied field, possibly causing the Barkhausen-like noise.

The effect of the interaction between stress and magnetostriction on domain structure was nicely demonstrated in another study on inductive recording heads.[33] The permalloy film is the pole in an inductive reader. In the device the permalloy will be surrounded by insulating materials, notably alumina and hard-baked photoresist.

In this study, various shapes of plated permalloy were encapsulated in different combinations of alumina and photoresist layers. The Young modulus of photoresist is over two orders of magnitude less than that of permalloy or alumina so the stress experienced by permalloy in the different encapsulation structures should be very different. Changes in domain structure after annealing were observed by the Kerr microscopy, demonstrating that the anneal produced stress relaxation and changes in anisotropy. The authors were careful to rule out changes in grain size from the anneal process which was done at relatively low temperature. It was found that the samples where photoresist completely protected the permalloy from contact with alumina had much more stable domain structure. This was attributed to the redistribution of the stress from the alumina overcoat by the photoresist buffer layer.

Stress can also be changed by adjusting thin film deposition parameters. However, it is difficult to control the overall stress in a device throughout device fabrication and during operation. For this reason, generally, the soft magnetic materials in a device should have minimal magnetostriction. This includes shield materials, biasing layers, poles, etc. as well as MR sensor elements. A signal instability related to magnetostriction in the shield layers of MR heads has recently been demonstrated.[34] In MR heads, the sense element is surrounded first by insulating layers and then by soft magnetic shielding layers. Such layers are crudely depicted in Fig. 7.1. Flux from the magnetic transitions surrounding the one the head is trying to read is channeled into the shield layers so that the head only sees the flux from one transition at a time. Heads with shields made from materials with different levels of saturation magnetostriction were compared. The higher magnetostriction head exhibited two discrete output signal levels in repeated tests of alternately reading and writing. In this head spin, SEM imaging revealed a domain wall in the shield very near the sensor. The domain wall could be moved by the application of current to the write coil of the head.

Luckily, in many useful soft magnetic alloys magnetostriction can be tuned by adjusting the composition. This is the case for permalloy which has a zero magnetostriction composition ($Ni_{81}Fe_{19}$) that is near the composition which maximizes the AMR. This material is the mainstay of AMR sensors; other materials are known to have greater AMR but not at zero magnetostriction. As an added complication, for a given composition the magnetostriction can vary with other parameters that might be adjusted in order to improve the properties of the film. As one example, MR has been shown to be enhanced by increasing the (111) texture in permalloy films. The texture can be increased by growing the permalloy film on an underlayer. Both increased texture and a correlated large change in magnetostriction have been recently demonstrated for permalloy films grown on TaN and Si_3N_4.[35] In very thin films and multilayers, surface and interface contributions to the magnetoelastic anisotropy come into play producing changes in magnetostriction with film thickness and the number of interfaces.[36-38]

Of course, the entire range of processing conditions used to control materials properties can have effects on domain structure in magnetic materials. Some specific examples are grain size, texture, stress, composition of an alloy and thus, magnetostriction, and density of impurities or defects which might pin domain walls. Generally speaking, lower base pressure and higher deposition rate reduce impurities. Composition of an alloy can be altered with bias voltage as well as target

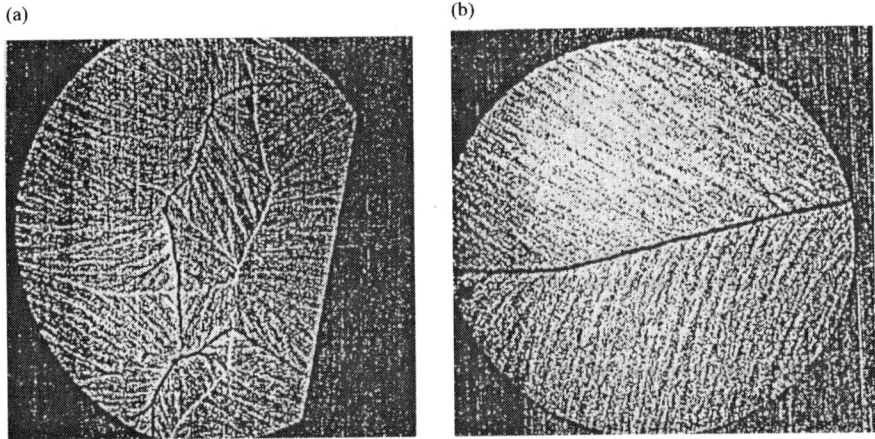

Figure 7.22. (a) The Lorentz TEM image of 300 Å thick $Ni_8[Fe]_9$ film. (b) Similar film deposited on 50 Å Ta underlayer. Both images are 28 μm in diameter. Reprinted with permission from Ahkter, M. A., D.J. Mapps, Y. Q. Ma, A. K. Petford-Long, and R. Doole, *IEEE. Trans. Magn.* **34**, 1147 (1998). ©1998 IEEE.

composition. Stress can be affected by all of the parameters especially bias voltage as well as overlayers and underlayers. Grain size and texture can be enhanced by deposition temperature and sputtering power (both of which give a higher energy process) as well as the use of post-deposition annealing and seed layers. An example is shown in Fig. 7.22 in which transmission electron microscopy (TEM) images reveal dramatic differences in domain structure in permalloy films grown with and without a Ta underlayer.[39] A complex cross-tie wall structure is cleaned up dramatically when the film is deposited on a Ta layer. The same paper also describes a reduction in the Barkhausen noise for films deposited on a Ta underlayer and for films deposited at elevated temperature.

The authors speculate that changes in grain size alter the domain structure. Changes in domain structure during annealing were observed in Fe/Cr multilayers without changes in microstructure.[40] This experiment involved *in situ* annealing of the multilayers while imaging them using the Lorentz microscopy. Increased magnetic contrast in one sample was attributed to depinning of domain walls allowing them to align through the thickness of the multilayer and also to a sharpening of the Fe/Cr interface with annealing. A second sample did not exhibit any magnetic contrast as deposited but developed large domains after annealing. This was attributed to weak antiferromagnetic coupling between the layers in the as-deposited state changing to ferromagnetic coupling after annealing. Even without an intentional anneal step, many devices will experience elevated temperatures during device fabrication and clearly that can affect domain structure through a variety of mechanisms.

Finally, the physical process of patterning the material deserves mention. Roughness at the patterned edges of the sensor can affect the domain structure and can cause pinning of domain walls. Simulations of magnetic reversal in polycrystalline Co elements comparing smooth and rough edges have demonstrated this.[41] Vortex structures were observed to form at the edge irregularities causing irreversible switching behavior while smooth-edged elements reversed by the motion of an edge domain wall through the sample at a higher switching field. Lorentz TEM images of elements with artificially rough edges have revealed a remanent state with domain walls at the irregular edge features giving a decreased switching field as shown in Fig. 7.23.[42] In addition, the process of patterning the elements can involve redeposition of material onto surfaces of the sensor. This can thwart some of the stabilization schemes discussed above where coupling between the sensor and various stabilizing magnets is crucial. Although edge smoothness can be limited by

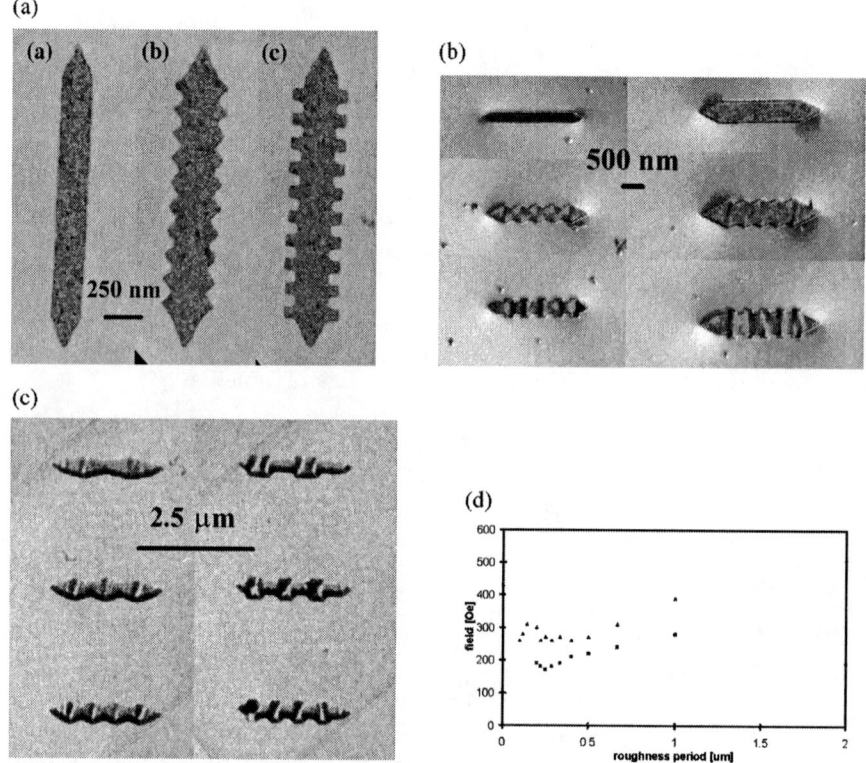

Figure 7.23. Magnetic elements were patterned with variously shaped artificial roughness (a). TEM images show magnetic structure associated with the edge irregularities (b) and (c). This generally corresponds to a reduction in switching field compared with the smooth elements (d). Reprinted with permission from M. Herrmann, S. McVitie, and J. N. Chapman, Abstracts of the 43rd Annual Conference on Magnetism and Magnetic Materials HC-12, p. 310 (1998). (a) Bright field images of magnetic elements with (a) no artificial roughness, (b) triangular and (c) rectangular roughness features. (b) Foucault images of elements showing magnetic microstructure. The mean direction of magnetization lies parallel to element length with ripple-like contrast present in elements with rough edges. (c) Fresnel images showing wall contrast. The images were taken during a magnetization experiment and show the micromagnetic states prior to switching. The applied field component is in the plane of the elements and parallel to their length. Domain walls are clearly visible. (d) Diagram of field values necessary to switch the magnetization of small magnetic elements showing the dependence on roughness type and period. Triangles are used to represent data from elements with triangular roughness features and squares for data from elements with rectangular roughness features.

the grain size, photolithographic, etch, and milling processes can all be optimized to give better-defined structures.

7.7. MEDIA NOISE IN MAGNETIC RECORDING

Media generally makes the dominant contribution to the noise background in magnetic recording. The trend towards higher and higher area density demands continual innovation in media and thermal fluctuations in the magnetic grains will place a limit on the recording density attainable in thin film media, predicted to be somewhere near 100 Gb/in.^{2} (Ref. 43). MR sensors are soft magnetic films with great sensitivity to small fields. In contrast, media must have high coercivity in order to store information. A bit is set by a high field in a particular direction and it must not be possible to unintentionally reverse the bit without a large field in the other direction. In longitudinal recording, the recording head detects the fringing field from the poles at the transition between the horizontally magnetized bits as illustrated in Fig. 7.1. The media needs to have high remanence in order to provide a large enough field to be read even when the bits are very small.

A fundamental source of media noise is the inherent granularity of the material—the fringing field comes from the superposition of fields produced by a collection of individual particles or grains. Early recording media was particulate, consisting of small single-domain magnetic particles bonded to a tape or disk. The most common material used for this is γ-Fe_2O_3. The advancement to sputtered thin film media for hard disk drives brought several advantages. Notably, the smoother surface allows smaller head–media separation and going from individual particles to grains in a film allowed higher density. While particulate media predominantly exhibit amplitude modulation noise due to non-uniformity in the magnetization of the bit or variations in effective head–media separation, the noise from thin film media is dominated by transition noise. This was illustrated by the dependence of the noise on recording density[7] as shown in Fig. 7.24. For thin film media the noise is greatest when transitions completely fill the media without overlapping. Transition noise comes from fluctuations in the geometry of the transition between the bits causing the detection of the signal to vary around the expected time.[44] Thin film media is the state of the art for longitudinal magnetic recording and the examples of recent work that we will consider are all concerned with the reduction of transition noise in thin film media.

It has been shown that the highest SNR comes from media with the greatest density of particles.[45] Maintaining SNR at higher recording density requires media with finer grain structure. Further, to reduce transition noise, the magnetic coupling between the grains must be minimized so that the transitions can be cleanly and repeatably defined. There are two types of intergranular coupling, the exchange coupling and the magnetostatic coupling. Imagine trying to draw a straight thin magnetic boundary through a thin film composed of many small magnetic grains. The magnetization in an individual grain will be aligned in a single direction and the transition line will have to bend around the grain boundaries. This is illustrated in Fig. 7.25, which shows a comparison of a high noise and a low noise media with noise power and magnetic force microscopy (MFM) images of transition written on the two media of the grain sizes.[46] Fluctuations in the magnetization of the grains

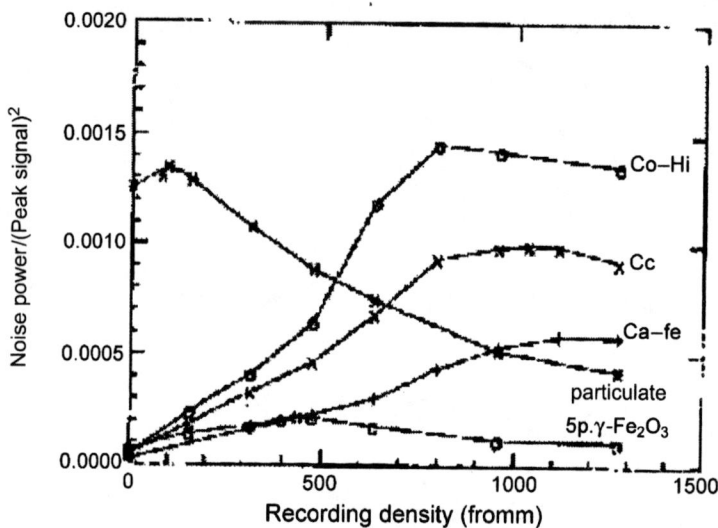

Figure 7.24. Noise power as a function of recording density in a variety of thin film media compared to particulate media. Total normalized medium noise power $\overline{N^2}$ versus recording density for various material systems. Reprinted with permission from R. A. Baugh, E. S. Murdock, and B. R. Nataranjan, *IEEE Trans. Magn.* 1724 (1983). © 1983 IEEE.

will produce fluctuations in the transition geometry and coupling between grains will effectively create larger obstacles for the boundary to detour around and greater instability of the grains near the boundary. The ideal media would have small, completely independent, densely packed particles with high coercivity. Unfortunately, the requirement for high coercivity is often at odds with fine grain structure and minimal coupling. Often, the best magnetic properties are found in films with greater crystalline perfection and larger grains. In the development of sensor materials, better epitaxy and larger grain structure are often sought for that reason. In media development many different methods are used to achieve good magnetic properties yet control grain size and intergrain coupling.

The most fundamental example of this is the universal use of underlayers. The sputtered magnetic layer will tend to reproduce the microstructure of the underlayer through epitaxy. Underlayers are used to promote texturing and control grain size. Many different underlayers have been studied. Cr and Cr alloys have been the most successful so far.[47,48] Of course, in the search for better alloys the active layer and the underlayers must be considered together. Various Co alloys have been studied with CoCrX alloys being the most popular, especially CoCrPt, CoCrTa, and CoCrPtTa. The addition of a third element to CoCr can expand the lattice; hence, the addition of a second element to the Cr underlayer in order to adjust the lattice spacing for better lattice matching. An example is a study of CrMn as an underlayer for CoCrPt.[49] The authors did observe improved texture by x-ray diffraction but also a need to preheat the substrates to achieve high coercivity. They speculate that diffusion of Mn into the magnetic layer is involved in the improved performance and further that this should reduce exchange coupling between grains giving lower noise. Sometimes, an intermediate layer is used when the active layer has a particularly large lattice

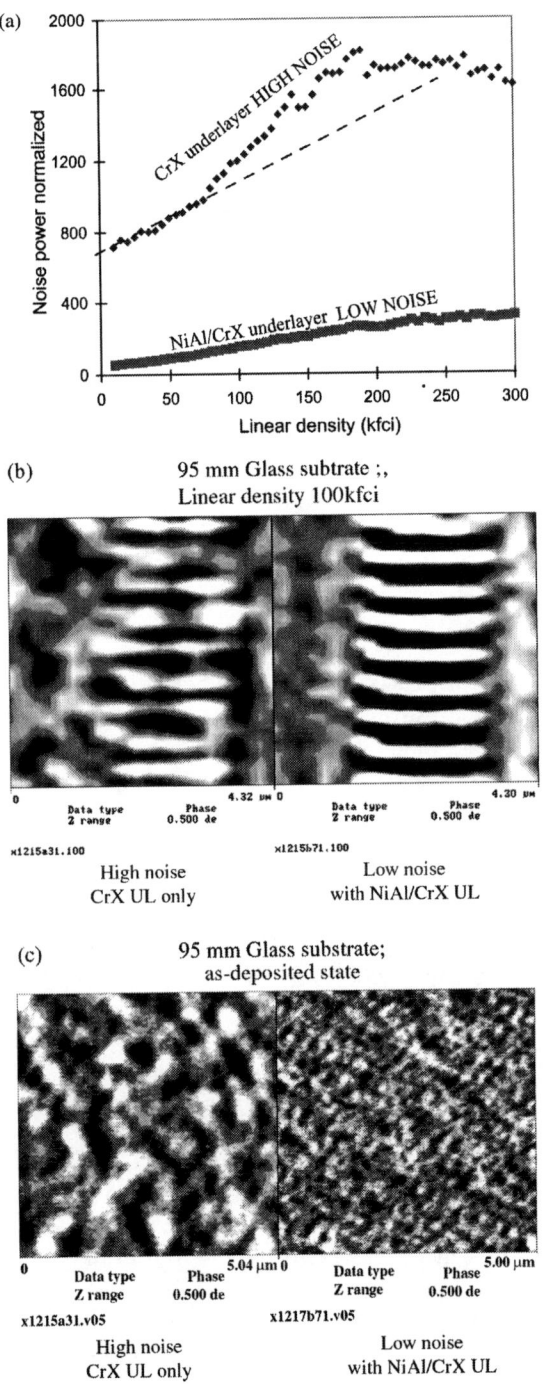

Figure 7.25 (a) Noise power for a low noise and a high noise media. (b) MFM images of transitions on the two media. (c) As-deposited MFM image showing the magnetic grain sizes for the two media. The lower noise media has NiAl under a CrX underlayer while the high noise media has only a CrX underlayer on glass. E. Yen and R. Ranjan of Seagate Technology, personal communication, 1998.

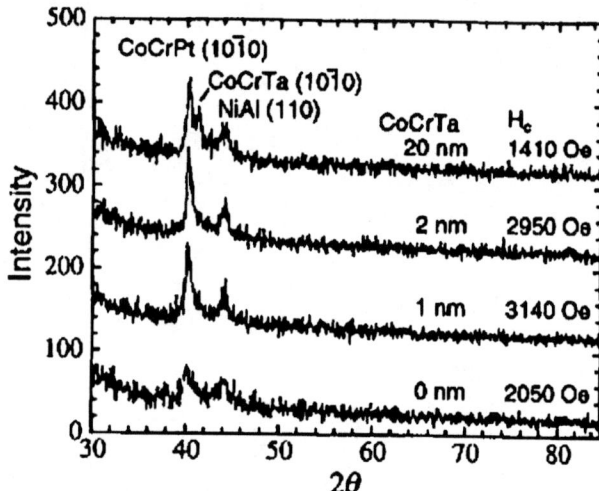

Figure 7.26. X-ray diffraction spectra showing the effect of a CoCrTa intermediate layer on CoCrPt. Reprinted with permission from J. Zou, D. E. Laughlin, and D. N. Lambeth, *IEEE Trans. Magn.* **34**, 1582 (1998). © 1998 IEEE.

spacing, as does CoCrPt with large amounts of Pt. For example, CoCrTa has been investigated as an intermediate layer between Cr and CoCrPt and better texturing and higher coercivity were achieved.[50] Figure 7.26 shows how CoCrTa improves the texturing of CoCrPt when used as an intermediate layer on a NiAl underlayer.[51] In this study, a 1000 Oe improvement in coercivity was observed for a thin CoCrTa layer.

These Co alloys can exhibit even better SNR at high recording density if a bicrystal cluster structure is achieved. In this case, the film consists of clusters of Co alloy grains which have their c-axes perpendicular to one another as shown in Fig. 7.27.[52] Computer modeling suggests that the normally oriented grains behave less collectively, reducing interactions across a transition boundary and giving an SNR enhancement similar to the effect of reducing the coupling between grains.[53] A 5 dB improvement in SNR has been reported for CoCrPtTa films with bicrystal structure compared to films with the same grain size but without bicrystal structure.[54]

As always in thin film development, the deposition conditions play an important role. Optimized deposition conditions can have a significant effect on grain size, texturing, and even intergrain coupling. One such approach was the use of high argon pressure during deposition of the magnetic films. Although this resulted in physical isolation between the magnetic grains, the grain size distribution was relatively non-uniform resulting in poor squareness of the **M–H** loop.[55,56] In recent years, the approach has been to use high-Cr containing magnetic alloys. A Japanese group has developed an "ultra-clean" sputtering process for CoNiCr and CoCrTa thin film media which promotes Cr segregation.[57] The magnetic grains have less Cr in them increasing the anisotropy while the Cr segregates to the grain boundaries reducing intergranular exchange coupling and increasing SNR. The process involves very low base pressure in the sputtering system and very clean Ar gas.

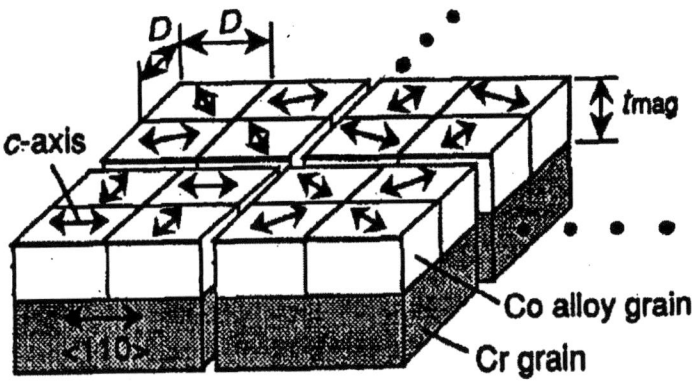

Figure 7.27. A model of the bicrystal structure in a Co alloy. Four clusters are shown. Within each cluster are multiple grains with normally oriented easy axes. Reprinted with permission from Y. Hosoe, Y .Yahisa, R. Tsuchiyama, A. Ishikawa, K. Yoshida, M. Igarishi, and Y. Shiroishi, *IEEE Trans. Magn.* **31**, 2824 (1995). © 1995 IEEE.

In another example of material diffusing into grain boundaries to reduce intergrain coupling, a recent study looks at the effect of carbon diffusion into CoCrTa from an overcoat layer.[58] The authors propose that carbon diffusion into the grain boundaries reducing intergrain coupling is responsible for the lower noise they observe in thinner films. Carbon atoms can diffuse more readily than Cr since they are smaller.

In the CoCrTa intermediate layer study described earlier the effect of substrate bias was also considered.[52] The application of substrate bias during the CoCrPt deposition increased the concentration of Pt and thus the lattice spacing. This correlated to higher coercivity. The effect of substrate bias on the CoCrTa interlayer deposition seemed to be a degradation of the NiAl underlayer surface and a poor interface between the two layers.

Finally, just as we saw in the section on magnetic sensor stability, there is much activity aimed at the design of the structure as well as at materials development. Multilayered media is an example. The active layer is made up of multiple thin layers separated by interlayer. The interlayer prevent large grains from forming and prevent grains in neighboring layers from interacting as well as serving as seed layers for the active layers. Very thin layers can be grown with a high degree of crystal perfection yielding excellent magnetic properties.[59]

On the other hand, modeling has suggested that multilayered media will not be advantageous at the upper limits of recording density (where $KV/kT < 100$ or so) due to increased thermal degradation for a double layer film compared with a single layer film of the same total thickness.[60] Thermal instability increases media noise and the same study calculates that decreasing magnetostatic coupling between grains improves thermal stability while decreasing exchange coupling between grains degrades thermal stability. This can be understood qualitatively as follows. Exchange coupling causes the grains to want to align together and act more collectively which is loosely like an effective increase in KV. On the other hand, magnetostatic coupling will produce an effective field at a grain that will tend to disturb the magnetization. Unfortunately, we have already seen that increased exchange coupling decreases SNR and many of the examples we looked at involved schemes to decrease exchange coupling. We have also seen that we want to decrease grain volume in order to improve SNR, so improved thermal stability at high densities will require large anisotropy. That is fine, unless the coercivity of the media becomes too high for the write fields that can be produced by the head. We have seen that there are many trade-offs involved in developing high-performance media. The correct balance of properties for future longitudinal recording will depend on the recording density and the available head performance.

BIBLIOGRAPHY

Magnetic Domains and Magnetization Reversal

D. Jiles, *Introduction to Magnetism and Magnetic Materials* (Chapman and Hall, London, 1991).
B. D. Cullity, *Introduction to Magnetic Materials* (Addison-Wesley, Reading, MA, 1972).

Noise

Detailed discussion of the noise formalism and the sources of electronic noise in condensed matter:
Sh. Kogan, *Electronic Noise and Fluctuations in Solids* (Cambridge University Press, Cambridge, 1996);
M. B. Weissman, *Rev. Mod. Phys.* **60**, 537 (1988).
Review on the use of noise to study materials:
M. B. Weissman, *Annu. Rev. Mater. Sci.* **26**, 395 (1996).
Derivation of the FD relation with several examples:
H. B. Callen and T. A. Welton, *Phys. Rev.* **83**, 34 (1951)

Magnetoresistive Noise

More examples of the use of noise to study various AMR, GMR, and CMR materials along with further experimental details:
H. T. Hardner, Resistance noise in giant MR materials, PhD thesis (UMI) 1996.

Stability

Description of the use of micromagnetic modeling to study domain instability:
J.-G. Zhu and D. J. O'Connor, *J. Appl. Phys.* **81**, 4891 (1997).
First principles determination of magnetostriction in bulk transition metals and thin films:
R. Wu, L. Chen, and A. J. Freeman, *J. Magn. Magn. Mater.* **170**, 103 (1997).

Media Noise

Theory and formalism:
H. N. Bertram, *Theory of Magnetic Recording* (Cambridge University Press, Cambridge, 1994). Chapters 10–12 give a detailed theoretical discussion of media noise.
J. C. Mallinson, *IEEE Trans. Magn.* **5**, 182 (1969). A clear and instructive derivation of the SNR of a tape recorder.
Overview of media requirements for 10 Gb/in^2. longitudinal recording:
D. N. Lambeth, E. M. T. Velu, G. H. Bellesis, L. L. Lee, and D. E. Laughlin, *J. Appl. Phys.* **79**, 4496 (1996).

REFERENCES

1. H. Fujiwara, *IEEE Trans. Magn.* **29**, 2557 (1993).
2. M. F. Gillies, J. N. Chapman, and J. C. S. Kools, *J. Appl. Phys.* **78**, 5554 (1995).
3. H. Cho, C. Hou, M. Sun, and H. Fujiwara, *J. Appl. Phys.* **85**, 5160 (1999).
4. C. P. Bean and J. D. Livingston, *J. Appl. Phys.* **30**, 120S (1959).
5. S. E. Russek, J. O. Oti, S. Kaka, and E. Chen, *J. Appl. Phys.* **85**, 4773 (1999).
6. R. A. Baugh, E. S. Murdock, and B. R. Nataranjan, *IEEE Trans. Magn.* MAG-19, 1724 (1983).
7. M. B. Weissman, *Annu. Rev. Mater. Sci.* **26**, 395 (1996).
8. H. B. Callen and T. A. Welton, *Phys. Rev.* **83**, 34 (1951).
9. H. T. Hardner, M. B. Weissman, M. B. Salamon, and S. S. P. Parkin, *Phys. Rev. B* **48**, 16156 (1993).
10. J. Shi, S. S. P. Parkin, L. Xing, and M. B. Salamon, *J. Magn. Magn. Mater.* **125**, L251 (1993).
11. S. Zhang and P. M. Levy, *Phys. Rev. B* **43**, 11048 (1991).
12. M. B. Weissman, *Rev. Mod. Phys.* **60**, 537 (1988).
13. H. Wan, M. M. Bohlinger, M. Jenson, and A. Hurst, *IEEE Trans. Magn.* **33**, 3409 (1997).
14. H. T. Hardner, S. S. P. Parkin, M. B. Weissman, M. B. Salamon, and E. Kita, *J. Appl. Phys.* **75**, 6531 (1994).
15. L. S. Kirschenbaum, C. T. Rogers, S. E. Russek, and Y. K. Kim, *IEEE Trans. Magn.* **33**, 3586 (1997).
16. R. J. M. Van de Veerdonk, P. J. L. Beliën, K. M. Schep, J. C. S. Kools, M. C. de Noojier, M. A. M. Gijs, R. Coehoorn, and W. J. M. de Jonge, *J. Appl. Phys.* **82**, 6152 (1997).
17. H. T. Hardner, M. B. Weissman, B. Miller, R. Loloee, and S. S. P. Parkin, *J. Appl. Phys.* **79**, 7751 (1996). The sample shown here was grown by B. Miller at the University of Minnesota.
18. H. Barkhausen, *Z. Phys.*, **29**, 401 (1919).
19. S. S. P. Parkin, R. Bhadra, and K. P. Roche, *Phys. Rev. Lett.* **66**, 2152 (1991).
20. A. Wallash, *IEEE Trans. Magn.*, **34**, 1450 (1998).
21. L. S. Kirschenbaum, C. T. Rogers, S. E. Russek, and S. C. Sanders, *IEEE Trans. Magn.* **31**, 3943 (1995).
22. S. Ingvarsson, G. Xiao, R. A. Wanner, P. Trouilloud, Y. Lu, W. J. Gallagher, A. Marley, K. P. Roche, and S. S. P. Parkin, *J. Appl. Phys.* **85**, 5270 (1999).
23. H. T. Hardner, M. B. Weissman, M. Jaime, R. E. Treece, P. C. Dorsey, J. S. Horwitz, and D.B. Chrisey, *J. Appl. Phys.* **81**, 272, (1997).
24. M. Wang, H. Yi, and S. Yan, *Solid State Commun.* **98**, 235 (1996).
25. G. B. Alers, A. P. Ramirez, and S. Jin, *Appl. Phys. Lett.* **68**, 3644 (1996).
26. J. J. Fernandez-de-Castro, Russell J. Machelski, L. G. Swanson, and G. S. Mowry, *IEEE Trans. Magn.* **32**, 3386 (1996).
27. C. Mitsumata, K. Kikuchi, and T. Kobayashi, *IEEE Trans. Magn.* **34**, 1453 (1998).
28. J. Gadbois, J.-G. Zhu, W. Vavra, and A. Hurst, *IEEE Trans. Magn.* **34**, 1066 (1998).
29. J. Nix, US Patent 5485334, (1996).
30. N. Smith, US Patent 4956736, (1990).
31. T. Kira, *IEEE Trans. Magn.* **31**, 910 (1995).
32. J. -P. Lazzari and I. Melnick, *IEEE Trans. Magn.* **7**, 146 (1971).
33. D. D. Tang and P. Kasiraj, *IEEE Trans. Magn.* **30**, 5073 (1994).
34. K. Nakamoto, S. Narumi, H. Fukui, T. Kawabe, and T. Kobayashi, *J. Appl. Phys.* **85**, 5846 (1999).
35. T. Yeh, J. M. Sivertsen, and C.-L. Lin, *IEEE Trans. Magn.* **34**, 870 (1998).
36. O. Song, C. A. Ballentine, and R. C. O'Handley, *Appl. Phys. Lett.* **64**, 2593 (1994).
37. H. J. Hatton and M. R. J. Gibbs, *J. Magn. Magn. Mater.* **156**, 67 (1996).
38. Y. K. Kim and T. J. Silva, *Appl. Phys. Lett.* **68**, 2885 (1996)
39. M. A. Ahkter, D. J. Mapps, Y. Q. Ma, A. K. Petford-Long, and R. Doole, *IEEE Trans. Magn.* **34**, 1147 (1998).
40. A. K. Petford-Long, R. C. Doole, A. Cerezo, J. S. Conyers, and J. P. Jakubovics, *J. Magn. Magn. Mater.* **126**, 117 (1993).
41. T. Schrefl, J. Fidler, K. J. Kirk, *J. Appl. Phys.* **85**, 6169 (1999).
42. M. Hermann, S. McVitie, and J. N. Chapman, Abstracts of the 43rd Annual Conference on Magnetism and Magnetic Materials HC-12, p. 310 (1998).
43. B. H. Zhou and R. Gustafson, *IEEE Trans. Magn.* **34**, 1845 (1998).
44. N. R. Belk, P. K. George, and G. S. Mowry, *IEEE Trans. Magn.* **21**, 1350 (1985).

45. J. C. Mallinson, *IEEE Trans. Magn.* **5**, 182 (1969).
46. E. Yen and R. Ranjan of Seagate Technology, personal communication.
47. J. A. Christner and R. Ranjan et al., *J. Appl. Phys.* **63**, 3260 (1990).
48. T. P. Nolan, R. Sinclair, R. Ranjan, and T. Yamashita, *IEEE Trans. Magn.* **29**, 292 (1993).
49. L. -L. Lee, D. E. Laughin, and D. N. Lambeth, *IEEE Trans. Magn.* **34**, 1561 (1998).
50. L. Fang and D. N. Lambeth, *Appl. Phys. Lett.* **65**, 3137 (1994).
51. J. Zou, D. E. Laughlin and D. N. Lambeth, *IEEE Trans. Magn.* **34**, 1582 (1998).
52. Y. Hosoe, Y. Yahisa, R. Tsuchiyama, A. Ishikawa, K. Yoshida, M. Igarishi, and Y. Shiroishi, *IEEE Trans. Magn.* **31**, 2824 (1995).
53. J.-G. Zhu, X.-G. Ye, and T. Arnoldussen, *IEEE Trans. Magn.* **29**, 324 (1993).
54. Q. Chen, J. J. K. Chang, G.-L. Chen, and R. Sinclair, *IEEE Trans. Magn.* **32**, 3599 (1996).
55. R. Ranjan et al. *IEEE Trans. Magn.* **26**, 322 (1990).
56. T. Yogi et al. IEEE *Trans. Magn.* **26**, 1578 (1990).
57. M. Takahishi et al. *IEEE Trans. Magn.* **33**, 2939 (1997).
58. J. J. -K. Chang and K. Johnson, *IEEE Trans. Magn.* **33**, 1567 (1998).
59. D. N. Lambeth, E. M. T. Velu, G. H. Bellesis, L. L. Lee, and D. E. Laughlin, *J. Appl. Phys.* **79**, 4496 (1996).
60. P. -L. Lu and S. H. Charap, *IEEE Trans. Magn.* **30**, 4230 (1994).

8

Deposition Techniques for Magnetic Thin Films and Multilayers

David Keavney*† and Charles Falco*

8.1. INTRODUCTION

In principle, any technique that can deposit thin films of metals may be used to fabricate spin-valve and magnetic multilayer structures. However, because of the device applications which are the subject of this volume, and the compatibility with existing processes that they require, certain techniques have received much more attention than others. Generally, the physical vapor deposition (PVD) methods, rather than chemical vapor deposition (CVD) and liquid phase epitaxy (LPE), have been used the most for metal films. Consequently, the PVD methods will be the primary focus of this chapter. In this we include molecular beam epitaxy (MBE), DC, RF, and ion beam sputtering (IBS), and pulsed laser deposition (PLD).

The first observation of giant magnetoresistance (GMR) was on MBE-grown films,[1] and since then the contribution of MBE to research in magnetoresistive devices has mainly been towards the understanding of the basic materials physics of these layered materials. Both MBE and sputtering were used extensively in studies of the antiferromagnetic coupling effect in thin films on which GMR was originally based.[2-5] By contrast, there have been few reports of MBE-grown spin-valves,[6] with most of the work in that area being done by sputtering. For device production, sputtering in its various forms has also found the most use to date due to the high achievable throughput and scalability to large wafer processes. MBE and evaporation typically offer cleaner vacuums and more comprehensive characterization tools, however, due to the lower achievable growth rates, they have not yet found many applications in production. PLD is still a relatively new technique, but has shown promise for deposition of some materials that are problematic with MBE and sputtering, such as complex compounds. In Fig. 8.1, we show the primary components of a system required to deposit thin metal films. These include one or more deposition sources, a sample holder with temperature control, and some means of determining the rate of deposition.

This chapter is intended to provide an overview of the technologies available, with an emphasis on those which are important for magnetic materials. For more detailed information, there are reviews and textbooks available which specialize in

---

*Optical Sciences Center, The University of Arizona, Tucson, Arizona.
†Present address: Advanced Proton Sources, Argonne National Laboratory, Argonne, Illinois.

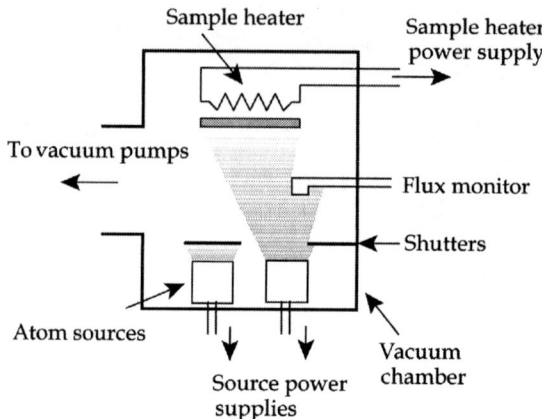

Figure 8.1. The essential components of a system needed to deposit metal thin films or multilayers: two or more sources of material, a sample holder with heater, shutters, and a vacuum system.

MBE[7–10] and sputtering,[11,12] as well as the non-PVD deposition techniques, which we do not treat here.

In the following three sections, we explain the principles of each PVD method along with the advantages and disadvantages of each one. We begin with the simplest method, thermal evaporation, and discuss it mainly in the context of MBE. Then, in Section 8.3, we move on to sputtering and discuss the most commonly used means of generating the energetic ions needed for this process. In Section 8.4, we briefly introduce PLD, and discuss some of the materials that it has been successful at depositing. Since another chapter in this volume addresses characterization techniques in more detail, in Sections 8.5 and 8.6 we briefly discuss the methods for characterizing the thickness and structure of thin films that are available for each of the deposition techniques.

8.2. THERMAL EVAPORATION

Evaporation refers simply to the production of a neutral atomic or molecular beam by raising the temperature of the material until the thermal energy of the surface is sufficient to allow a useable flux to escape. Typically, this takes place in at least a high vacuum ($\sim 10^{-7}$ Torr or better), or in an ultrahigh vacuum (UHV) ($\sim 10^{-9}$ Torr or better) to reduce contamination from the residual gases in the vacuum chamber. What a useable flux is depends on the geometry of the evaporation chamber, the total film thickness required, and the background contamination level. Since some materials sublime with sufficient flux well below their melting temperatures, while many others do not, the temperature at which evaporation occurs can best be predicted from the vapor pressure curve for the material.[13] The net atomic flux from a solid or liquid charge in atoms/s at a given temperature T is given by the evaporation rate minus the return flux due to the ambient pressure,[14]

$$\Gamma_e = a_v A_c (P_{vapor} - P_{ambient}) \left(\frac{N_A}{2\pi M k_B T} \right)^{1/2} \qquad (8.2.1)$$

where A_c is the surface area of the charge, M is the molecular weight of the evaporant, $p_{ambient}$ is the vacuum chamber pressure, and a_V is a factor which reflects the fact that not all of the incident atoms from the gas phase contribute to the return flux. The theoretical maximum Γ_e occurs when $a_V = 1$. Note that the rate is strongly dependent on the temperature through p_{vapor} and through the $T^{-1/2}$ term.

The material may be heated a number of ways, depending on the temperature required. For materials that evaporate at sufficiently high rates at temperatures of approximately 1500°C or less (e.g., Cu, Ag, and Au), direct resistive heating is often used, while for more refractory materials (e.g., W, Ta, Cr, and Fe) electron beam (e-beam) heating is often required. Both of these methods can be used alone or in the context of MBE. Further we discuss each heating method separately, giving examples of the basic technique followed by a discussion of practical implementations. Finally, we discuss these techniques as they are used in MBE.

8.2.1. Resistive Heating

Provided a material has a high enough vapor pressure at a low enough temperature, it may be heated using Joule I^2R heating by passing a sufficiently large current through a resistor in physical contact with the material. The simplest way to accomplish this is to place the material charge in a refractory metal "boat" which acts as both the heater and the crucible, as in Fig. 8.2. In this case the material flux will evaporate outward in all directions onto anything with a direct line of sight to the boat, including the substrate. Although this method meets all the requirements for producing an atomic flux, there are several features which make it unsuitable for depositing ultra thin films and multilayers of many metals. First, the rates obtained from simple evaporation sources of this type are not sufficiently stable and reproducible. This is because the rate is dependent upon the surface area of the charge, which decreases continuously as material evaporates, or worse, may change suddenly if the shape of the charge changes due to convection or boiling. Secondly, the total amount of material that can be evaporated this way is small due to the large power requirements necessitated by passing a high current through a metallic, low resistance source. This requires that the source be replenished too often to be

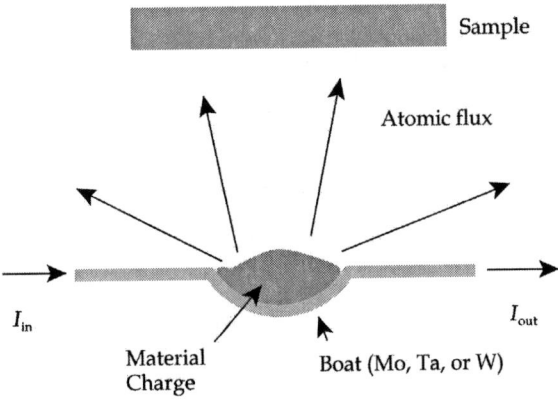

Figure 8.2. Simple thermal evaporation of material from a boat by resistive heating.

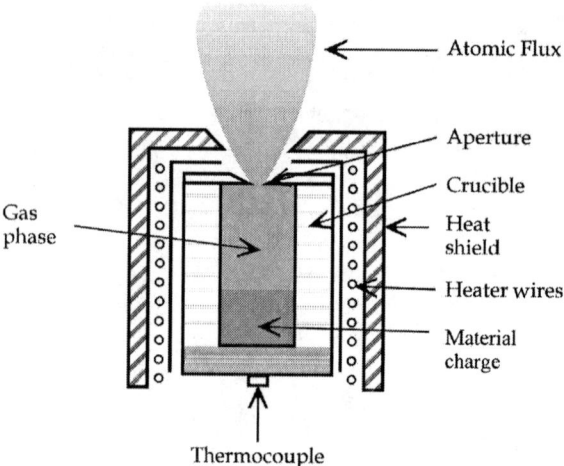

Figure 8.3. Schematic diagram of a k-cell. The material is heated indirectly by a resistive element surrounding the crucible and evaporates through the orifice at the end of the crucible. The aperture area is much smaller than the surface area of the charge, and the gas phase is in equilibrium with the charge.

practical for many applications. Thirdly, the evaporant may react with the boat, forming compounds of relatively high vapor pressure, resulting in contamination of the source material and the sample.

A practical resistive source which provides excellent rate stability and allows large amounts of material to be deposited is a modification of the above approach known as a thermal effusion cell, Knudsen cell, or K-cell. A generalized K-cell is shown in Fig. 8.3. In a K-cell, the charge is placed in an electrically insulated crucible and heated by a resistive element surrounding the crucible. This allows the crucible to be shaped, thereby directing the flux towards the substrate and producing more material deposition at the substrate per unit mass evaporated from the source. The entire assembly is typically surrounded by heat shielding to reduce the required input power for a given crucible temperature. This also allows for larger crucibles than otherwise would be possible.

An ideal K-cell is one in which the material charge is in equilibrium with the gas phase above it. In this case the gas phase pressure would be equal to the material vapor pressure at the evaporation temperature. This could be accomplished by shaping the crucible in such a way as to restrict the flux as it leaves the crucible, resulting in a higher pressure inside the crucible than outside in the chamber. For example, a long cylindrical crucible with a small circular orifice at one end would restrict the flux from the crucible.[15] If the area of the orifice A_e is small compared to the surface area of the charge, the gas phase in the crucible will be in approximate equilibrium with the solid charge. In this case, the orifice can be modeled as an evaporating disk with evaporant vapor pressure equal to the equilibrium pressure in the crucible, p_{eq}, and the rate of evaporation, Eq. (8.2.1), becomes

$$\Gamma_e \cong A_0 p_{eq} \left(\frac{N_A}{2\pi M k_B T} \right)^{1/2} \quad (8.2.2)$$

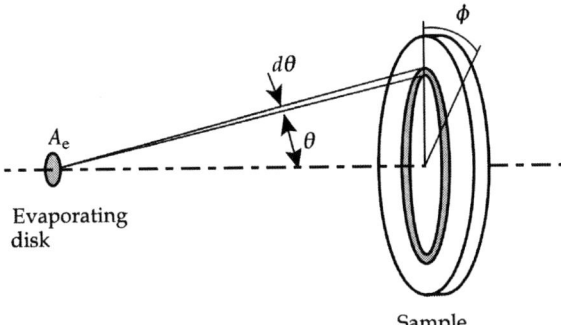

Figure 8.4. Geometry for the determination of angular flux distribution from an ideal K-cell. The evaporating disk represents the aperture of an ideal K-cell.

where we have neglected the vacuum chamber pressure because in most cases, the evaporant vapor pressure is several orders of magnitude higher than the vacuum pressure, so the latter does not have much effect on the effusion rate. Also, we have set $a_V = 1$, because any atom from the gas phase outside the crucible incident on the orifice will be returned to the crucible.

Because the aim of evaporation is to produce films with as uniform of a thickness as possible over large areas, it is necessary to consider the distribution of flux from an effusion cell as one moves off the axis of the crucible. For the ideal K-cell, the rate is not dependent on the crucible shape or the charge shape, so for a circular orifice the flux does not depend on azimuthal angle, ϕ. Therefore, consider the differential flux into the ring of angular thickness $d\theta$ as shown in Fig. 8.4. The probability that a molecule enters the orifice at an angle between θ and $\theta + d\theta$ is the ratio of the solid angle subtended by the ring to the total solid angle into which molecules can be effused. In the limit of zero wall thickness, the latter is 2π. Furthermore, the effective orifice area seen by the molecule is $A_e \cos\theta$. Combining these factors, the differential flux into the ring at θ is

$$d\Gamma_\theta = \frac{\Gamma_e}{\pi} \cos\theta \, d\omega \qquad (8.2.3)$$

where $d\omega$ is the solid angle subtended at the orifice by the ring. It is possible then to calculate the flux per unit area for an off-axis point as $I(\theta) = d\Gamma_\theta/dS$, with the area element $dS = r^2 d\omega$. This gives a flux distribution of

$$I(\theta) = I(0) \cos^4\theta \qquad (8.2.4)$$

In general, one is interested in depositing several materials sequentially, so it is desirable to have more than one effusion cell in an evaporation system. In this case they must be arranged such that the axis of any given cell is not perpendicular to the film plane, but forms some angle α, as in Fig. 8.5. Again using Eq. (8.2.3), the flux can be found as a function of θ,

$$I(\theta) = I(0)\left(\frac{r}{r_e}\right)^2 \cos\theta \cos(\theta + \alpha) \qquad (8.2.5)$$

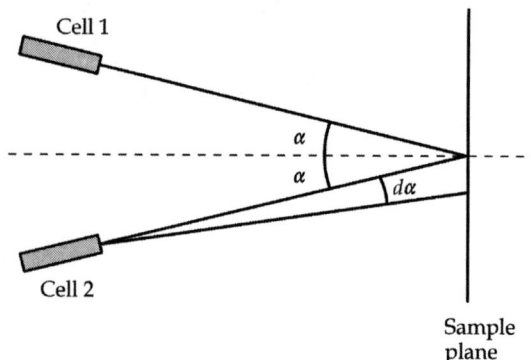

Figure 8.5. K-cell geometry in a multiple cell configuration. This requires all the cells to be offset from normal incidence by an angle α, which causes an anisotropy in the flux. This anisotropy often requires sample rotation to remove the resulting non-uniformity.

where r and r_e are the distances from the orifice to the center and edge of the sample, respectively. Using some values from the geometry of a typical MBE system, the above equation would predict a drop in flux (and consequently the thickness) of ~20% from the center to the edge of a sample 6 cm in diameter assuming an ideal K-cell located 20 cm from the sample. This amount of nonuniformity is usually not acceptable. However, by careful design of the deposition system, it is possible to overcome the sample nonuniformity by introducing substrate rotation during growth. In essence, rotation of the sample during growth averages the film thickness across the wafer by alternately moving a given point on the substrate between locations of maximum and minimum flux. With sample rotation, it is easy to achieve a thickness variation of 0.3%, even with the nonaxial source arrangement of Fig. 8.5.

For many materials, particularly metals, an ideal K-cell is simply not practical. Due to the radiative cooling, it is difficult to keep the orifice at the same temperature as the crucible, resulting in deposition of metal at the orifice. This quickly closes the crucible completely. A more practical cell is one with an open-ended cylindrical or even and outwardly tapering crucible. In these cases, the gas phase will not be in equilibrium with the material charge, and it is much more probable that molecules will sublime from the charge surface and move directly to the substrate without thermalizing. However, the mass of the charge is usually large enough that excellent rate stability still can be achieved. For many applications, active rate control is not even necessary provided the cell temperature is controlled accurately (to ±0.5°C) and the surface area of the charge does not change significantly during the deposition. Crucibles of this type also produce a more collimated beam of evaporant. This allows good film uniformity over much larger substrates than the ideal K-cell.

8.2.2. Direct Electron Beam Heating

For many materials, the vapor pressure at temperatures attainable by resistive heating is insufficient to generate a useable flux. These include Ta, W, and even some 3d metals, depending on the deposition rate required. For these applications, direct heating by electron bombardment is often used.

In this method, an e-beam is generated by thermionic emission of electrons from a hot W filament at high potential (5–10 kV) relative to the material charge to be evaporated. This potential difference causes the emitted electrons to accelerate towards the material charge, and the electrons energy is transferred directly to the charge. The crucible, typically Cu, holding the charge is water-cooled to prevent it from melting. Therefore, a temperature gradient is set up between the crucible and the point where the e-beam hits the charge. In this way, very high temperatures may be attained in a small area near the e-beam spot, even if most of the charge is not hot enough to evaporate. As with resistive heating, all this takes place in a high vacuum, or preferably, UHV to minimize atmospheric contamination of the sample. Electron beam heating is capable of evaporating even W, the lowest vapor pressure metal. However, this method of evaporation is far from equilibrium, which makes the use of active rate control essential.

In the simplest conception of an e-beam heater, the beam simply would be formed electrostatically and allowed to hit the material charge. Practically, it is necessary to direct and focus the beam, since the flux must have an unobstructed path to the sample, and since the filament cannot be in direct line of sight to the evaporating material. Manipulation of the e-beam is usually done with a magnetic field, either from a permanent magnet or a sealed coil, or both. A typical e-beam source which uses a 270° beam deflection is shown in Fig. 8.6. Many sources also exist that use 180° deflection, however 270° deflection is more desirable, since this allows the sample to be better shielded from the hot filament. This reduces the radiant heat load to the sample, and prevents contamination from the small amount of W that thermally evaporates from the filament. Another common feature found in e-beam sources are sweep coils. These are small deflection coils which sweep the e-beam over the charge to heat it uniformly. This increases the thermal mass of material involved in the evaporation process, hence improving the rate stability.

Another type of e-beam source, the rod-fed source, is shown in Fig. 8.7. In this configuration, the tip of a bar of the desired material is heated by electron bombardment from a filament which surrounds the end of the bar. All the beam

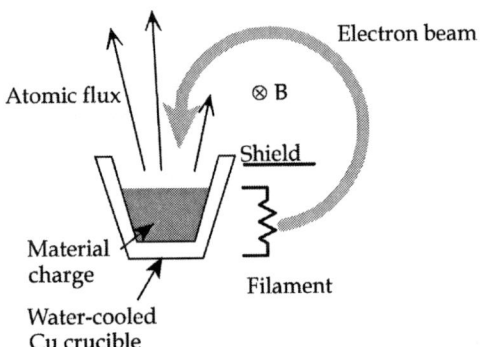

Figure 8.6. Schematic diagram of an electron-beam deposition source, in this case with 270° beam deflection. Thermally emitted electrons from the filament are accelerated electrostatically by a large negative filament bias and directed to the charge by a magnetic field. Focusing and sweep coils (not shown) are also used to provide control over the beam size and position.

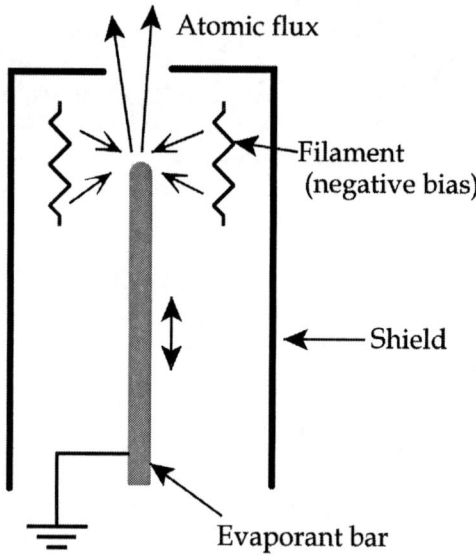

Figure 8.7. Rod-fed electron beam deposition source. Electrons are emitted thermionically from a negatively biased filament and are attracted to the tip of the grounded bar of material to be evaporated, heating it in the same way as in the crucible-base e-beam source. The bar can be advanced as material is evaporated.

direction and focusing is done by the large electric fields that develop near the tip of the bar, so that focus and sweep coils are not needed. This type of source is therefore somewhat simpler and cheaper than the crucible based source. The disadvantages are that it can only evaporate materials with good electrical conductivity that sublime at high rates well below their melting temperatures, and that the deposition rate is typically even more unstable than the crucible-based source.

As we have previously mentioned, e-beam heating is the only way to evaporate refractory metals such as Ta, Mo, and W. However, it is also often used for metals which could be evaporated from a K-cell, such as Fe and Co. It is particularly useful for rare earths, even though high temperatures are not required to evaporate them. In this case the use of a cooled crucible provides an advantage over Knudsen cells, as these metals are reactive enough to attack ceramic crucibles at their evaporation temperatures.

Electron beam heating encounters problems with evaporating alloys and compounds. In alloys, there may be differential evaporation rates for the constituents, leading to unknown and possibly irreproducible film compositions. Also, compounds may dissociate when bombarded with high energy (5–10 kV) electrons. This could have two bad effects. First, reactive gases may be released into the vacuum chamber, depending on the compound. This could degrade the performance as well as the lifetime of equipment, such as the electron gun itself or analytical instruments, and could contaminate other deposition materials in the chamber. UHV processes would be susceptible to this problem. Second, the stoichiometry of the resulting film may be affected, and may not be reproducible. For example, both of these effects have been observed in the e-beam evaporation of MgO.[16] In addition, many compounds lack sufficient electrical conductivity at the evaporation temperatures to carry the e-beam current. Then, they build up a charge which deflects the e-beam away. However, some insulating compounds are suitable for e-beam deposition, e.g., MgF_2.

8.2.3. Molecular Beam Epitaxy

The technique of MBE is essentially a combination of the different atomic flux generation techniques described earlier and extensive *in situ* structural characterization and surface analysis techniques, some of which we describe in Section 8.6. All this takes place in an ultraclean, UHV environment. The UHV environment is necessary for MBE because the growth rates are often deliberately made very low, on the order of monolayers per minute, in order to promote epitaxial growth. Therefore, to avoid impurities from the residual gases in the chamber becoming incorporated into the sample during growth, MBE systems operate with a background pressure in the high 10^{-11} to low 10^{-10} Torr range, and pressure during deposition in the 10^{-10}–10^{-9} Torr range. A typical MBE system consists of several UHV chambers: one for deposition, including several K-cells and e-guns, one for analysis, and a load lock chamber to allow for introduction and withdrawal of samples without having to vent either of the UHV chambers. All of these chambers must be interconnected, requiring mechanisms for sample transfer between them under UHV conditions. An example of an MBE system dedicated to the deposition of magnetic thin films and multilayers is shown in Fig. 8.8.

Later, we will briefly discuss some of the standard surface and structural analysis techniques that are commonly available with many of the PVD techniques described here. Also, another chapter in this volume will treat this subject in more detail. However, one such technique is so essential for MBE and is not often found outside of MBE systems, so it is worth a more detailed treatment here. Reflection high energy electron diffraction (RHEED) is an *in situ* technique in which a collimated beam of monochromatic electrons with energy in the range 5–35 keV is reflected off the sample surface at grazing incidence (1–2° from the sample plane) and the resulting diffraction pattern observed. The typical RHEED geometry is shown in Fig. 8.9. A collimated beam of high-energy electrons is incident on the sample at a very small angle (~2°), and is elastically scattered onto a phosphor screen on the other side of the sample. In this configuration, incident electrons do not have large momentum perpendicular to the film plane, even though the electron energy may be as high as 35 keV, and therefore do not penetrate the surface beyond a few monolayers. This makes RHEED scattering extremely sensitive to the surface structure and flatness, and almost completely insensitive to the out-of-plane lattice spacing. Consequently, features in the RHEED pattern are very sensitive to the surface structure, flatness, and in-plane coherence length. The RHEED geometry also allows for the pattern to be observed in real time during deposition, making it possible to detect structural changes with sub-monolayer sensitivity during growth. This makes RHEED an indispensable tool for determining epitaxy of MBE-grown films.

A RHEED pattern provides qualitative information about the surface crystallinity through the elastic scattering. When the electrons are incident on

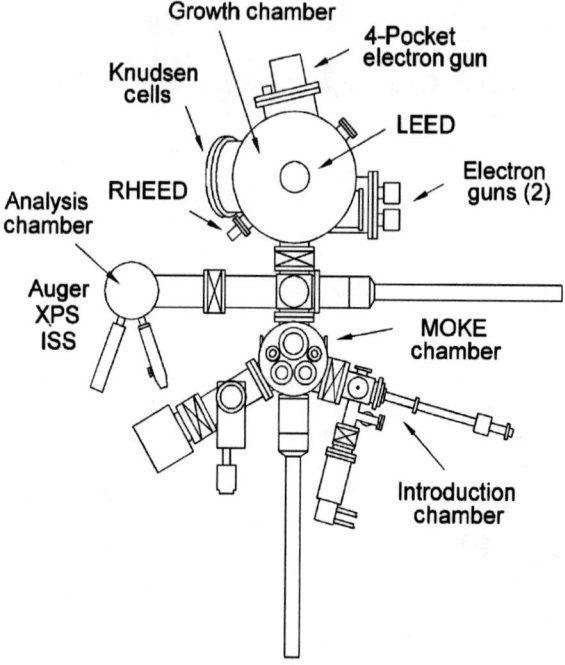

Figure 8.8. Diagram of a dedicated magnetic material MBE system, showing separate chambers for growth, structural characterization, and sample introduction. This system, which is in use at the University of Arizona, also has a separate chamber for *in situ* magnetic characterization by the magneto-optic Kerr effect.

the surface, they are subjected to the periodic potential of the surface atoms, which for most elemental solids has a periodicity comparable to the electron wavelength at the beam energy. Therefore, the elastically scattered electrons will be diffracted in directions uniquely determined by the surface lattice and its orientation with respect to the e-beam. This results in a spot or streak pattern on the phosphor screen used to observe the pattern. The angular spread of the spots or streaks depends on the lateral size of grains in the film. An example of a RHEED pattern obtained from a clean Si(111) surface prior to the deposition of a metal film is shown in Fig. 8.10.

RHEED is sensitive to the film flatness through the diffuse scattering. As with many scattering techniques, roughness tends to increase the amount of diffuse scattering. In this case, surface roughness results in a less intense elastic pattern and a higher diffuse background. This effect is so sensitive that it can easily detect the

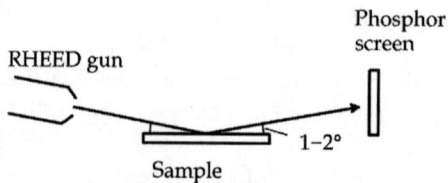

Figure 8.9. RHEED sample geometry. The electron gun and phosphor screen are both in the vacuum chamber with a UHV viewport to allow the RHEED pattern to be viewed during deposition.

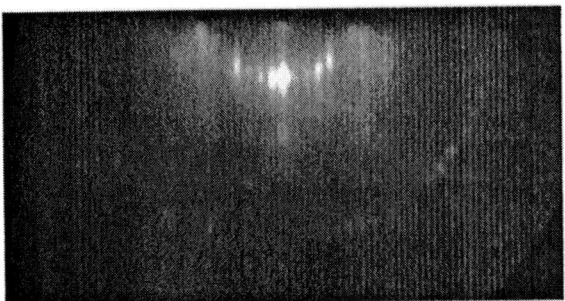

Figure 8.10. Example of a RHEED pattern, in this case from a Si(111) substrate prior to deposition of metals. Two rows of sharp streaks are visible. The fainter streaks between the bright ones arise from the 7 × 7 surface reconstruction which occurs for ultraclean Si(111).

surface roughening and smoothening transitions associated with monolayer formation, or layer-by-layer growth. As a single monolayer is formed, the surface changes from atomically flat to one with some islands and back to atomically flat as the monolayer completes formation. This results in oscillations in the diffuse scattering which are observable as oscillations in the intensity of the elastic part of the RHEED pattern with a period of exactly one monolayer. These so-called RHEED oscillations can, therefore, be used to calibrate and control the growth of films of certain materials in the monolayer thickness range.

Consider first an ideal, perfectly flat surface lattice such as in Fig. 8.11(a) with no defects, also, assume that the incident electrons interact only with the first monolayer of the surface, so that the incident electrons will see an effectively two-dimensional crystal lattice. This lattice can be described by the real space lattice vectors **R**,

$$\mathbf{R} = l\mathbf{a}_1 + m\mathbf{a}_2 + \delta_{n0} n \mathbf{z} \tag{8.2.6}$$

where $\mathbf{a}_i$ are the in-plane primitive vectors, the z-direction is out-of-plane, and l, m, n are integers. The δ_{n0} on the z term restricts the lattice vectors to two dimensions and causes the Fourier transform along the z-direction to be a constant. The in-plane reciprocal lattice vectors are found by the transformation,

$$\mathbf{b}_1 = 2\pi \frac{\mathbf{a}_2 \times \hat{\mathbf{z}}}{\mathbf{a}_1 \cdot (\mathbf{a}_2 \times \hat{\mathbf{z}})}$$

$$\mathbf{b}_2 = 2\pi \frac{\hat{\mathbf{z}} \times \mathbf{a}_1}{\mathbf{a}_1 \cdot (\mathbf{a}_2 \times \hat{\mathbf{z}})} \tag{8.2.7}$$

In momentum space the reciprocal lattice is then an array of infinitely long rods perpendicular to the sample,

$$\mathbf{G} = k_1 \mathbf{b}_1 + k_2 \mathbf{b}_2 + k_3 \hat{\mathbf{z}} \tag{8.2.8}$$

where k_1 and k_2 are integers, and k_3 may take any value [see Fig. 8.11(b)]. In practice, the space lattice is not strictly two-dimensional, and the reciprocal lattice rods are finite in length, although still very much longer than their width. Drawing the Ewald

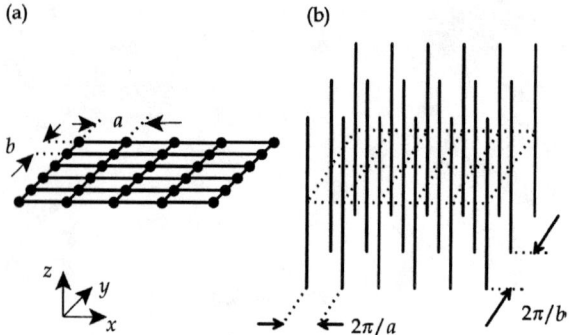

Figure 8.11. Two-dimensional rectangular lattice in: (a) real space and (b) reciprocal space. The rods in (b) represent a continuum of **G** vectors in the z-direction.

sphere in this arrangement, as in Fig. 8.12, the diffracted beam directions are given by the intersection of these rods and the Ewald sphere. This results in diffracted beams that lie on a half-cone with vertex at the origin in k-space. In the real space picture, the intersection of the half-cone with the flat phosphor screen becomes a semicircle, called the Ewald circle. Therefore, the rays that lie on the half-cone give rise to an arc of spots on the screen. If the diffracted beams are projected on to the plane of the film, they will form some angle θ with the incident beam which is dependent on the spacing of the reciprocal lattice rods perpendicular to the incident beam direction.

As an example consider the rectangular lattice with $a_1 \perp k$. For k large compared to $2\pi/a_1$, the deviation angle θ is given by

$$\sin\theta = \frac{2\pi/a_1}{k} \qquad (8.2.9)$$

as can be seen from the geometry of Fig. 8.12. This shows that the angular spacing of RHEED beams is dependent only on the real space lattice spacing perpendicular to

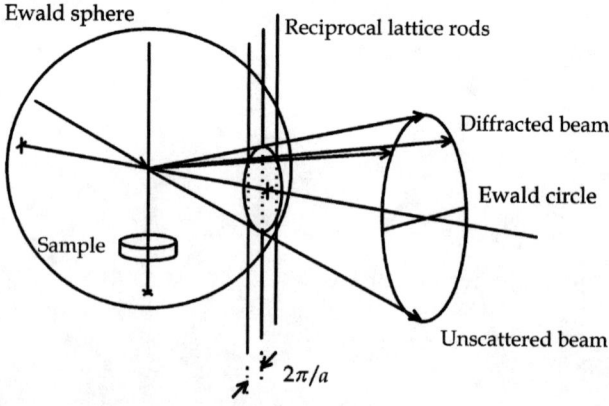

Figure 8.12. The Ewald sphere and diffraction geometry for RHEED scattering from a planar sample. The disk shows the sample geometry in real space, while the screen is coincident with the Ewald circle.

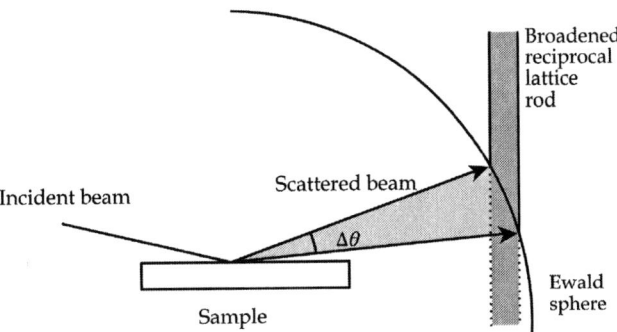

Figure 8.13. Effect of broadening of the reciprocal lattice rods on the allowed scattering vectors in RHEED. In this view, the scattering vectors, the Ewald sphere, and reciprocal lattice are projected onto the plane of incidence. In this case, the intersection of the Ewald sphere now subtends a finite angle ($\Delta\theta$) at the sample, resulting in a range of scattering directions, which broadens the diffracted spot into a streak.

the incident e-beam and the beam energy. Similar arguments can be made for oblique surface nets such as the (111) face of the face centered cubic (fcc) lattice. Therefore, it is possible to measure the in-plane lattice spacing as well as the in-plane symmetry, provided the sample can be rotated to allow the incident e-beam direction to be varied with respect to the lattice. Note that for 9.0 keV electrons, $k = 48.6 \text{ Å}^{-1}$, while a 2 Å lattice parameter corresponds to 3.14 Å^{-1}, justifying the assumption of large k.

Now consider some roughening of the surface, as in the case of a randomly stepped surface. Coherent scattering is then degraded somewhat, and the previously idealized reciprocal lattice rods may acquire some finite width. Their intersection with the Ewald sphere then allows a range of scattered electron k-vectors, as in Fig. 8.13. Note that due to the geometry, the broadening of the reciprocal lattice rods causes an anisotropic change in the diffraction pattern. Since the rods are intersected by the Ewald sphere, the range of allowed k-vectors in the direction normal to the film plane is greater than the azimuthal spread. This means that the diffracted spots on the arc will be elongated along the film normal, and the arc then becomes a set of streaks with intensity maxima at the previous positions of the spots. In practice the actual z-position of the intensity maxima may be difficult to determine due to the severe broadening in the z-direction, and the diffraction pattern may look simply like a set of lines or streaks. The angular positions of the intensity maxima will still be observable, and will occur at the same angular positions as in Eq. (8.2.9).

If the surface is roughened even further, resulting in large hills through which the incident beam may have to pass to get to the screen, or if the beam is able to penetrate the surface more than a few monolayers deep, the diffracting crystal appears more three-dimensional. The reciprocal lattice then takes on the more familiar three-dimensional form of an array of points. Its intersection with the Ewald sphere now selects the z-component of the scattered momentum according to the same rules as the in-plane components. The result is a three-dimensional Laue pattern similar to the pattern for diffraction from a bulk crystal. From these considerations of the degree of roughening, it is possible to estimate the roughness based on the features observed in the RHEED pattern. In general, the highest quality attainable from a metal surface results in the streak pattern discussed earlier for the slightly roughened surface net. Several workers, however, have observed the arc of spots from extremely high quality Fe(100) whiskers after considerable surface preparation,[17-19] but even these high quality diffraction patterns are not close to what can be obtained from semiconductor surfaces.

If the in-plane structural coherence length of the sample is less than that of the e-beam, i.e., if the sample is polycrystalline, the incident beam will encounter diffracting arrays of atoms with a continuum of orientations. The result is that the scattered beam may now lie anywhere on a cone, and the diffraction pattern becomes a continuous arc on the Ewald circle. This type of diffraction pattern is clearly distinguishable from a crystalline pattern, and its absence is evidence that the coherence length of the sample is greater than that of the electron source. For a typical thermal emission source common to many RHEED systems, this coherence length is about 0.5 µm.

As in any diffraction technique, a quantitative theory directly relating the film surface structure to the observed diffraction patterns is extremely useful. Unfortunately, a truly quantitative theory of RHEED currently does not exist, at least not at the level of the theory of low energy electron diffraction (LEED) or x-ray diffraction. Therefore we are often limited to making semi-qualitative statements about the film surface structure and quality based on features of the RHEED patterns. Despite these restrictions, it is possible to obtain the in-plane lattice parameters and the in-plane symmetry using the arguments above.

8.2.4. Microstructure of Evaporated Films

The microstructural evolution and morphology of evaporated films are both sensitively dependent on the energetics of the surfaces and interfaces involved, and on the conditions at the substrate during deposition, these include the substrate temperature, deposition rates, surface cleanliness, and background vacuum conditions. These factors can strongly affect the processes of surface diffusion, grain nucleation and growth, and interdiffusion. Consequently, these parameters allow considerable control over the film flatness, crystallinity, and purity. Below we discuss the general principles that should be considered in film deposition, but note that in many cases the optimal growth conditions for the desired result must be determined empirically.

Ignoring dynamical effects for the moment, the behavior of an overlayer will, to first order, be driven by a competition between the free energies of the substrate and overlayer surfaces, and the energy required to form the interface. To illustrate this, consider an overlayer of material B on a flat surface of material A. Associated with the formation of any surface is a surface free energy σ, defined simply as the energy per unit area required to form the surface from the bulk. Also, there is an interfacial energy σ_{AB} associated with the interface, the energy required to form the interface from the two separated systems. The interface energy, unlike the surface energies, may be positive or negative. Here we assume that the overlayer material has been allowed to diffuse over the surface and come to thermal equilibrium. We discuss these diffusion processes in more detail below. The shape of the surface of B can be determined by the minimization of the total free energy,

$$\text{Free energy} = A(\sigma_A + \sigma_{AB}) + A'\sigma_B \tag{8.2.10}$$

where A is the substrate area covered by material B, and A' is the area of the material B surface. If the substrate A has a higher surface energy than the overlayer B, the minimization of surface energy will require that the material B cover as large an area as possible, resulting in a flat, uniform film. The interface energy may reinforce this tendency or work against it, depending on its sign. It is also interesting to consider what happens if the surface energy of material B is much higher than that of the substrate. In this case the minimization of Eq. (8.2.10) would require the formation of droplets of material B with very large contact angles, which would result in very rough films. However, if the two materials are allowed to interdiffuse, and σ_{AB} is not large, the free energy can be minimized even further by the process of surface segregation, in which material B diffuses into the substrate and a layer of substrate material is formed over it.

The model discussed earlier is obviously very simplistic, as it is neither microscopic (the droplet must contain at least several thousand atoms for Eq. (8.2.10) to have any validity) nor does it take into account the dynamics of growth. It is through these dynamics that the growth morphology can be controlled. The thermodynamic model does, however, illustrate the basic processes of interface formation and some of the constraints that the properties of the materials being deposited may place on the growth.

This simple model makes the assumption of equilibrium deposition—that all of the overlayer adatoms have an opportunity to minimize their energy during the deposition. This is most often not the case. In fact, many multilayer and spin-valve structures are themselves metastable. The degree to which the adatoms can equilibrate depends on their mobility on the surface and on the time they have to diffuse before they are buried by subsequent atoms incident on the substrate. The diffusion of atoms on the surface is driven by the substrate temperature and by the binding energy of the atoms to the substrate. The total distance that atoms may travel over the surface is ultimately limited by the rate of monolayer formation. As atoms become buried by the arrival of additional material, they are then more tightly bound, and the diffusion rate is much slower. In some cases, it is possible to reduce the effects of surface diffusion by increasing the deposition rate, a condition known as saturation. The amount of diffusion will then be determined by the competition between the level of saturation and the mobility of adatoms, and can be controlled through the substrate temperature and deposition rates. The desired amount of surface diffusion will depend on the tendency of the overlayer to either form a flat film or to agglomerate, according to the surface and interface energies. When a high surface energy material is deposited onto a lower surface energy material, the tendency of the overlayer to minimize its surface energy by agglomerating must be suppressed, either by using a high deposition rate or low substrate temperature. Conversely, if wetting of the substrate is energetically favorable, it is usually better to *promote* surface diffusion, although it should be kept in mind that elevated temperatures also increase bulk diffusion, which can degrade interfaces.

Surface diffusion processes are the primary means by which adatoms can find other adatoms or lattices sites where they can be incorporated into the growing film. It is clear, then, that the final growth morphology and overall film quality will be strongly dependent on the amount of surface diffusion. Therefore, control over the growth requires control over the amount of thermal energy present on the surface, and on the time allowed for diffusion. What follows is a discussion of the processes which can be enabled by the presence of surface diffusion.

The first, and most important, is cluster formation by the aggregation of two or more adatoms. As atoms diffuse over the surface, it is possible for them to come together. The stability of the resulting clusters may be predicted using arguments similar to those for cluster nucleation and growth in the freezing transition. Consider an idealized cluster n atoms square and h atoms high on the surface of the substrate. There is a reduction in the Gibb's free energy from bringing together of these $h \times n \times n$ atoms, but an increase from the formation of the interface, top and sides of the cluster. Writing the net change in free energy,

$$\Delta G(h, n) = -\Delta\mu h n^2 - \sigma_A a^2 n^2 + \sigma_B a^2 n^2 + 4\sigma_{BP} a^2 hn \qquad (8.2.11)$$

where $\Delta\mu$ is the change in chemical potential resulting from the addition of one adatom to the cluster, and σ_{BP} is the surface tension associated with the periphery of the cluster. Equation (8.2.11) illustrates the competition between the bulk and surface terms that contribute to the free energy. The first two terms are the energy gains from bringing together hn^2 atoms and the covering of an $n \times n$ area of substrate. The three other terms represent the energy cost involved in creating the interface, top and sides of the cluster. Writing $\Delta\sigma = \sigma_A - \sigma_B - \sigma_{AB}$, Eq. (8.2.11) can be rewritten as

$$\Delta G_0(h, n) = -\Delta\mu h n^2 - \Delta\sigma a^2 n^2 + 4\sigma_{BP} a^2 hn \qquad (8.2.12)$$

From Eq. (8.2.12), it is easily seen that there can exist different regimes in which the cluster may be stable or unstable. If $\Delta\sigma > 0$, or if $\Delta\mu$ is large enough, there will be a certain critical size at which the free energy has a maximum. When the cluster is below that size, reducing the size reduces the free energy, and therefore the cluster will not be stable. However, above the critical size, it is energetically favorable to *increase* the cluster size, and the cluster becomes a stable island.

Once a cluster becomes a stable island, it can grow by the process of lattice incorporation. This occurs when an adatom diffusing over the surface encounters a site where it is more strongly bound than at any of the surface sites. If the temperature is not high enough to overcome the higher binding energy at this site, the atom will stay there and become part of the surface unless another process moves it. The most common type of site which can satisfy these conditions is a step edge,[20] such as occurs at the periphery of the island discussed above. These sites can bind adatoms more strongly because the effective coordination number is higher than at a surface site. Step edges may be numerous on the surface of the substrate to begin with, and this can play a very important role in determining the initial growth mode. Other types of sites where an adatom may stick are surface impurities, dislocations, and other crystal defects.[21] Often these types of sites can have a slightly deeper potential well than the regular surface sites. This can trap adatoms and possibly cause the nucleation of an island without the requirement of attaining the critical size first.

Earlier it was assumed that the substrate is a rigid lattice, and that substrate atoms do not participate in surface diffusion processes, except to provide the potential for the adatoms. In fact many *interdiffusion* processes are also possible, and these of course can also have a strong impact on the quality of the growth. Most common among these is simple bulk diffusion of adatoms into the substrate. The rate at which this takes place depends on the substrate temperature and the tendency for the two atomic species involved to form bulk alloys. In addition, if the overlayer surface free energy is high, it can be energetically favorable for adatoms to diffuse into the substrate in a process known as surface segregation. An even more complex diffusion process is exchange diffusion,[22] in which an adatom and a substrate atom exchange places. The adatom becomes incorporated into the lattice, freeing the substrate atom to diffuse over the surface and possibly become reincorporated somewhere else.

In the preceding discussion we have not considered epitaxy; none of the arguments we have made depend on the crystallinity or the orientation of either the substrate or the overlayer. In epitaxial growth of one material onto a different one, also known as heteroepitaxy, growth can proceed in three distinct modes, depending on the surface and interface energies involved. These are Frank–van der Merwe (FM) or so-called layer-by-layer growth, Volmer–Weber (VW) or island growth, and Stranski–Krastanov (SK) growth.[23] As indicated by the descriptive names, FM growth is characterized by complete wetting of the substrate surface by the overlayer, VW growth by agglomeration of the overlayer material into islands, and SK growth by a combination of these two behaviors. These three growth modes are illustrated in Fig. 8.14. Each mode can be associated with a regime of the quantity $\Delta\sigma$ as introduced in Section 8.2.3.[24] When $\Delta\sigma < 0$, wetting of the substrate is favored and FM growth is expected. When $\Delta\sigma < 0$, the overlayer surface tension is high and agglomeration is expected, resulting in VW growth. When $\Delta\sigma \sim 0$ for the initial growth, it may actually change sign for the later homoepitaxial growth, and SK growth is expected.

Epitaxial strain can also contribute to the energy balance which determines the structure and morphology of the overlayer. Most of the time, the lattices of the substrate and the overlayer will not be commensurate, even if they are isostructural. This results in a tendency for the overlayer to strain to match the substrate and minimize the bond energies between overlayer atoms and substrate atoms. However, this costs energy by forcing the bond lengths between overlayer atoms to take a non-equilibrium value, and the actual behavior will be determined by the competition between these two effects. A common result is that initially the overlayer will strain to match the substrate, but when it reaches a critical thickness at which the cumulative strain energy becomes too great, the film will begin to relax to its bulk lattice parameter. The critical thickness is dependent on the amount of strain and the bond strength, and can be as thin as a few monolayers. The film may relax by reverting in its entirety to the bulk structure or by introducing misfit dislocations. In addition to forcing materials in thin films to take expanded or contracted lattice parameters, some materials can be forced into non-equilibrium crystal structures which often display magnetic behavior very different from the bulk material. Examples of structures that have been stabilized by epitaxial growth include body centered cubic (bcc)[25] and fcc Co[26] on diamond, and fcc Fe.[27,28]

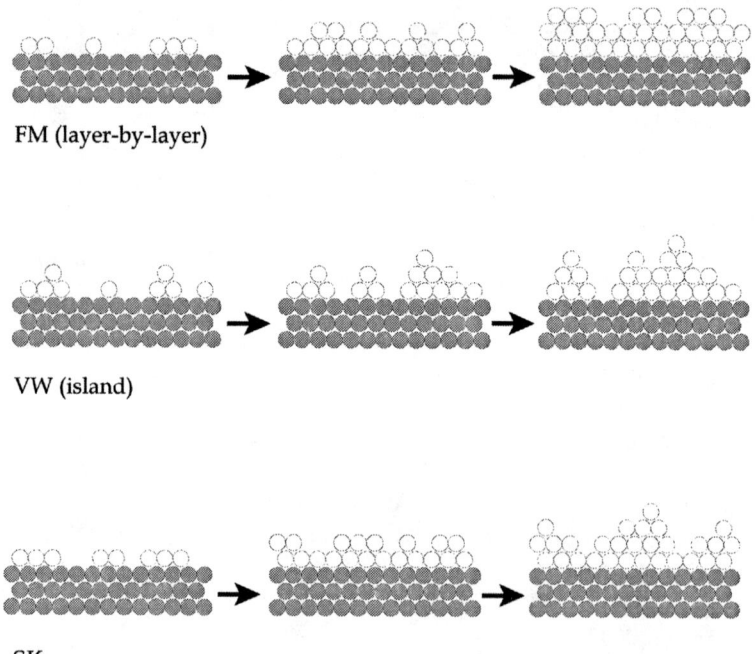

Figure 8.14. Three distinct epitaxial growth modes expected for different combinations of material surface energies. Top: Frank–van der Merve (layer-by-layer); middle: Volmer–Weber (island); bottom: Stranski–Krastanov (layer-by-layer followed by island).

The earlier discussion describes some of the factors which can affect film quality and growth morphology. In general, the relative importance of each of these factors will depend strongly on the conditions of growth, in particular growth temperature and growth rate. In principle, it is possible to predict the best conditions for any combination of substrate and overlayer. However, in practice these predictions are often not reliable, and it is up to the experimenter to determine the optimal growth conditions for a particular combination of materials.

8.3. SPUTTERING

Although almost all elemental materials can be thermally evaporated using one of the techniques discussed earlier, the obtainable rates are sometimes not sufficiently high. This is particularly true for refractory metals such as W and Ta. Also, depositing alloys and compounds often presents problems due to differential evaporation rates of the constituents or chemical dissociation of compounds. Sputter deposition can get around these problems for many materials.

Sputtering refers to the ejection of atoms or molecules from any surface by bombardment of the surface with energetic ions. It is also a common process for applications in addition to deposition, such as, sputter etching, ion milling, and substrate cleaning. The phenomenon of sputtering has been known since the previous century, but the development of it as a film deposition technology took place in the 1950s,[29,30] and several reviews have been published since then.[31-33] In sputter deposition, atoms are ejected from the surface of a target made of the appropriate material by collisions with energetic ions, and the substrate exposed to the ejected material. This results in deposition of a film with the composition of the target material.

The mechanism of momentum transfer between incident ions and surface atoms which results in ejection of the atoms is not as simple as it may initially appear. In most sputtering geometries, the incident ions are moving primarily normal to the target surface. Therefore, ejected atoms cannot be from primary collisions, since conservation of momentum requires these must be pushed into the target surface. Any ejected atoms must involve at least secondary collisions, and may come from buried layers a short distance beneath the surface of the target. Some of the events which can result in ejection of atoms from a surface are shown schematically in Fig. 8.15.

The probability of an incident ion ejecting a surface atom or atoms depends greatly on the mass of the surface atoms, the mass of the incident ions, and the energy of the incident ions. This quantity is usually expressed as the number of atoms ejected per ion, also called the sputtering yield. In general, if the masses of the incident ions and the target atoms are close to each other, the momentum transfer will be improved and the sputtering yield will be higher. Most sputtering is done with noble gas ions because of their nonreactivity, and of these argon is the most commonly used because its intermediate mass makes it a good compromise for a large part of the periodic table. Although the sputtering yield with argon is not optimal for many heavy target materials, sufficient yields for many purposes can still obtained, making argon the most useful choice of sputtering gas for a wide range of materials. The dependence of sputtering yield on the species of target material is more complex than the billiard ball physics as we discussed earlier. The binding energy of the target atoms to the lattice plays an important role. In this case, a periodic dependence of sputtering yield which is correlated to the filling of electronic

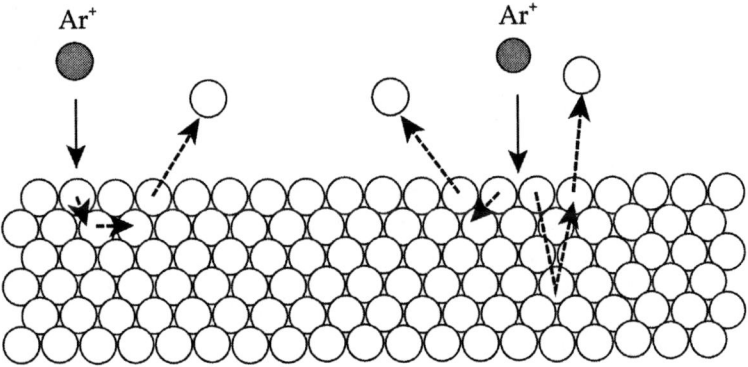

Figure 8.15. Sputtering by energetic ions at a surface, showing some of the scattering events which can result in ejection of atoms from the surface.

levels is observed. Within each shell, the yield increases with increasing atomic number, sometimes with anomalies at half-filling.

One important consideration for sputtering is the dependence of the sputtering yield on the incident ion energy. There is a threshold energy below which the yield is zero, which, for most metals, is 70–150 V. That this energy is considerably greater than the binding energy for most materials is further evidence that the sputtering process involves secondary collisions.[34] Above the threshold, the yield may be nearly linear up to a few hundred electron volts. At higher energies, the yield deviates significantly from a linear dependence. Another feature of the sputtering yield is that nearly all materials, at least metals, have yields within an order of magnitude of each other. As with any other technique, the sputter deposition rate usually must be determined empirically, but this feature does make the rate calibration and control easier. If we contrast this with thermal evaporation, we see that in the latter case the rates are much more sensitive to the relevant controlling parameter, the temperature, and to the species of evaporant.

Because sputtering yields of most materials are similar, unlike in thermal evaporation, the composition of the film sputtered from a composite target will usually be the same or very close to the composition of the target. This is particularly true of single-phase metal alloys, such as permalloy, and compounds in which none of the components are reactive gases. It is also possible in many cases to deposit binary alloy films simply by placing wires or chips of one constituent at the top surface of an elemental target of the other constituent. However, this method should probably be limited to use for doping small amounts of one material into another, and the resulting composition should be very well characterized after deposition. Some problems with preferential sputtering or deposition rates can occur with oxides, such as MgO, because oxygen which dissociates during sputtering may be pumped away or react with other components in the chamber, reducing its presence in the film. Also, since sputtering yields are not identical, if an alloy target is not single phase, there may be preferential sputtering of one phase which is rich in one of the alloy components.[35-37] However, some of these difficulties can be overcome with reactive sputtering, discussed later, and in spite of them sputtering is still useful for many composite materials.

The method of generating incident energetic ions and the exact source geometry varies widely between applications. The most common method is to ignite a plasma of the sputtering gas and electrostatically accelerate positive ions from the plasma onto a target by biasing the target at a negative potential. This method allows for easy generation of large ion current densities over very large targets, a feature needed for fast deposition of films over large wafers. The existence of the plasma can have negative consequences for the resulting film, though, and there are several variations on this basic approach that are used to minimize some of these effects by reducing the pressure of the sputtering gas. In Section 8.3.1, we discuss the simplest method of plasma generation, glow discharge in a DC diode, and how it can be used to deposit a film. In the following three sections, we then discuss some of the variations used to reduce the sputtering gas pressure and control the energetics of the sputtered material. We then discuss a very different approach to the generation of the needed energetic ions, IBS, and finally, reactive sputtering.

8.3.1. DC Diode

Although plasmas can be generated in many configurations, the requirement that the plasma source be used for film deposition limits the geometries available. The most simple sputtering device that could be used for material deposition is the planar DC diode, or cold cathode, source. In this device, the sputtering target serves as the source of ionizing electrons as well as the source of deposited material, as in Fig. 8.16.

The sputtering gas, usually Ar, is let into the chamber at a preset pressure. When the cathode is biased at a high negative voltage (-200 to -3000 V), electrons will move from the cathode into the sputtering gas and be accelerated towards the anode. If the mean free path of the electrons in the sputtering gas is long enough, they can obtain sufficient energy from the electric field to ionize gas atoms that they collide with. This requires that the pressure be somewhat less than atmospheric, but sputtering can still occur at 1000 mTorr or higher. With sufficient ionization, a plasma is generated above the negatively biased target, which then attracts Ar ions from the plasma due to its bias. When these ions strike the target they can sputter atoms in the way we have described, provided that they have opportunity to gain sufficient energy from the electric field without colliding with other Ar atoms. About 10% of the time, the sputtering process also generates secondary electrons from the target surface which accelerate into the plasma.[38] These electrons collide with and ionize additional Ar atoms, thus adding to the ionization of the plasma. If the proper conditions exist, eventually the plasma will fill the available space and a stable steady state will be reached.

The plasma potentials are governed by the well-known behavior of glow discharges. Because of the negative voltage applied to the cathode, electrons from it are accelerated away and electrons traveling from the plasma toward the cathode are reflected. This creates a region near the cathode where there are no energetic electrons, and hence no ionization. This region, the size of which depends on the cathode voltage and ion density, appears dark compared to the bright emission from the plasma, and is known as the cathode dark space or cathode ion sheath. In a typical planar diode source, the cathode sheath may exist 1–4 cm from the cathode.[39] In general, any surface at a negative voltage with respect to the anode will form an ion sheath. Most of the potential drop between the anode and other surfaces in the

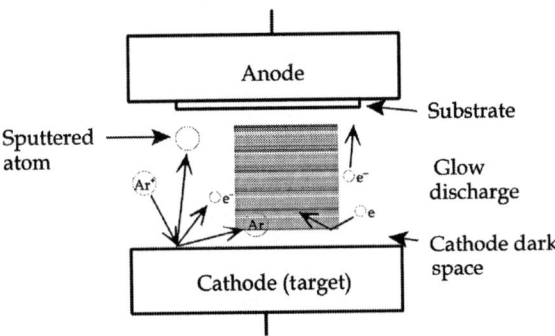

Figure 8.16. Schematic of a DC diode sputtering source.

chamber occurs across the ion sheaths. All positive ions that encounter the cathode sheath are accelerated onto it to produce sputtered atoms and secondary electrons.

The requirement for plasma stability is that sufficient electrons must be generated by the sputtering process, and that these electrons generate enough ions to replace those that strike the target. What this translates practically is that the sputtering gas pressure must be high, at least 50–100 mTorr, and that the electric field must be strong enough to accelerate electrons to the ionization energy. However, the high pressures present problems for film deposition, because the mean free path of a sputtered atom is very short at these pressures. This will reduce the energy of the atoms before they reach the substrate.[40] At 75 mTorr, sputtered atoms have suffered enough collisions to reduce their energy to the thermal energy before traveling 1 cm.[41] Consequently, in most practical sputtering geometries, the flux undergoes considerable diffusion before reaching the substrate. This deflects a portion of the flux away from the substrate, and the deposition rate drops as a result of this interaction between the working gas and the flux. Also, it may be desirable to control the energy of the incident atoms at the substrate by varying the sputtering gas pressure, but since this high gas pressure thermalizes the sputtered atoms so quickly, this is not possible unless the substrate is only a few millimeters from the target. Even if the chamber geometry permitted this, it still would not work because the sample would be within the cathode ion sheath. Finally, even though the sputtering gas is usually Ar or some other noble gas, there is a probability that it can become incorporated in the film by being buried during the film deposition. Also, impurity gasses liberated in the chamber during sputtering can be incorporated in the film. The probability of this happening is clearly dependent on the gas pressure. Therefore, although it is possible to sputter films at the pressures required for the DC diode, it is usually desirable to use the lowest possible sputtering gas pressure to improve the film purity.

8.3.2. DC Triode

One way to get around the high sputtering gas pressure requirements of the diode source is to provide an additional source of ionizing electrons independent of the target, as in the triode source.[42] Here, the geometry is similar to the diode source, with a planar target, but the cathode is a hot filament emitting electrons thermionically into the plasma such that they pass over the target. The target is biased at a negative voltage independent of the cathode, forming an ion sheath around it and accelerating ions onto it (Fig. 8.17). The additional source of ionization lowers the sputtering gas pressure requirement for plasma stability, allowing this type of source to be operated with high deposition rates at much lower pressures, in some cases as low as 1 mTorr. There are, of course, disadvantages as well. The existence of a hot filament means that at least a high vacuum is required to operate the source without damage to the filament, although in most processes a clean vacuum is needed to reduce sample contamination anyway. The filament is not generally damaged by operation in a few millitorr of noble gases, but reactive sputtering, discussed below, may not be suitable for a triode source. The filament is itself a potential source of contamination, and must be sufficiently cleaned prior to deposition, and also increases the maintenance that will need to be performed on the source.

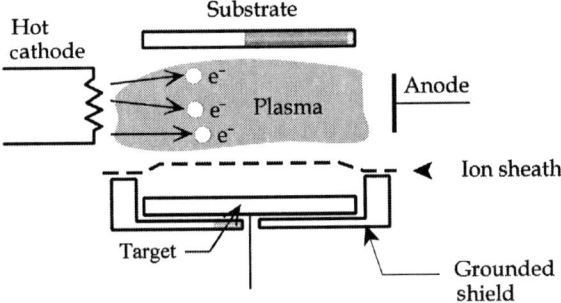

Figure 8.17. Schematic of a DC triode sputtering source.

8.3.3. Magnetron

Magnetron sputtering is another approach to reducing the required sputtering gas pressure. Although the DC triode source overcomes the sputtering gas pressure limitations associated with the diode, it is still not efficient at converting electron kinetic energy into ions. Many electrons reach the anode with considerable energy, which is then unavailable for ionization. In a magnetron source, the amount of ionization is increased not by simply increasing the supply of electrons as in the triode source, but by increasing the efficiency with which the electrons from the cathode ionize the sputtering gas.[43,44] This is accomplished by confining the electrons in a magnetic field produced by a set of permanent magnets, thereby increasing the amount of time they spend in the plasma before being lost to the anode. The magnetic field is configured in such a way as to produce a closed electron path. Therefore, each electron spends more time in the plasma, and generates several times more ions than it would if it were not confined. Magnetron sources can be configured a variety of ways, from cylindrical to planar. The planar magnetron is the most widely used magnetron device for film deposition. This is essentially a cold-cathode device as we discussed earlier, but with the electrons magnetically confined to a toroidal path (in the case of a disk-shaped target, as in Fig. 8.18) just above the target. Magnetron guns are capable of high deposition rates at very low pressures. Thus they combine the advantages of the triode source with the simplicity of operation of the cold cathode source. For this reason, magnetrons have become one of the more commonly used types of sputtering sources. Magnetrons can sputter any material that the diode and triode sources can, but there are special considerations

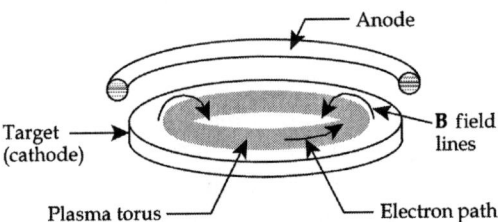

Figure 8.18. Schematic of a magnetron source, showing the confinement of the electrons by the magnetic field. The field is usually generated by permanent magnets (not shown).

for ferromagnetic materials. In this case the confinement may be disrupted by the target magnetic field, so the source magnets must be strong enough to overcome the effect of the target.[45]

Although true magnetrons are always diode sources, it is possible to apply a similar magnetic confinement approach to the triode source as well. One example of this is to apply a DC field above the target perpendicular to the electron path. This does not result in a closed electron path, but it does increase the time the electrons spend in the plasma before they strike the anode. As in the magnetron gun, this increases the efficiency with which they ionize the plasma, allows the source to be operated at lower pressures than in the absence of the magnetic field. These magnetically confined plasma sources can operate in the low millitorr range, similar to magnetrons.

8.3.4. RF Sputtering

When a positive Ar ion hits the target in a sputtering process, it normally removes electrons from the target and is reflected as a neutral atom. It can also generate secondary electrons that leave the surface and move into the plasma. In the DC techniques we described earlier, we made an implicit assumption that the target has high enough conductivity that electrons can move to the surface to replace those that are removed by sputtering. While this is true for metals, insulating materials do not satisfy this requirement. If an insulating target were placed in a DC source and negatively biased as is usually the case, it would very quickly develop a positive charge on its surface, creating a potential barrier and preventing further ions from hitting the target. The result would be that sputtering would occur only during the time the charge is building up, or about 100 ns. However, if this charged target were then biased positively, electrons would be attracted from the plasma, quickly neutralizing the surface charge. Consequently, sputtering from insulators can be achieved by using an oscillatory target voltage. In this way, a burst of atoms or molecules is sputtered each time the voltage goes negative, and the resulting surface charge is neutralized in the subsequent positive swing. Since only one 100 ns burst is obtained for each cycle of the target voltage, a high frequency is needed to obtain sufficient rates. Practical systems use an RF, usually 13.56 MHz, as this is a frequency set aside by international agreement for use in industrial processes that take place without completely shielding against RF emissions.

The behavior of the plasma potentials in an RF system is slightly different from a DC system. Because of the oscillating voltage, there is no anode and cathode in the DC sense. Rather, the two electrodes operate in much the same way, both with ion sheaths, while the plasma cloud oscillates in between. At the frequencies usually used for sputtering, the ions in the plasma do not have time to move very far in each voltage cycle, while the electrons do. Therefore, the RF plasma can be considered a gas of electrons moving through a cloud of relatively immobile ions.

Because the plasma sheaths at the various biased surfaces (target, substrate, etc.) represent loads that are not fully resistive, the implementation of an RF system is more complex than that of a DC system. To deliver RF power to the target, an impedance matching network must be used. Otherwise the load may be out of phase with the power supply, reducing the efficiency of power transfer to the target.

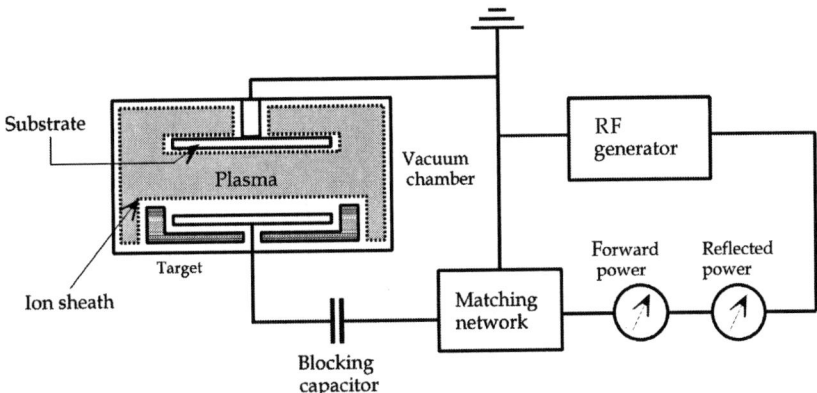

Figure 8.19. RF diode sputtering system, showing the power circuit and matching network which are required to deliver power to the target. The two power meters for forward and reflected power are used to monitor the phase matching of the RF generator to the load. For conductive targets, a blocking capacitor is needed to avoid shorting the target to ground.

This matching network is usually variable, to allow compensation for different target materials that might introduce different phase shifts, and for fine tuning the power transmitted to the load. In Fig. 8.19, we show schematically a commonly used scheme for RF sputtering, including the power circuit.

The RF technique can be used in either diode or triode sources, but most commercial system are diodes. This is because the extra ionization from the hot cathode is not needed for the source to operate at desirable sputtering pressures. At RF frequencies, the electrons in the plasma oscillate over a short distance between the electrodes, and therefore spend more time in the plasma than they would in a DC system. Hence, they expend more of their energy creating ions before they are lost to the plasma. For this reason, RF sputtering is also very useful for depositing conductive materials, and not just insulators.

The three types of sputtering discussed earlier all possess a considerable amount of flexibility in the target geometry which allows deposition of films on a wide variety of substrates. As we have mentioned before, the trend in commercial processes has always been towards larger wafers with more devices on them, as this is generally more cost effective. This requires the generation of uniform fluxes over larger and larger areas. With planar sputtering targets, the uniform area of deposition is limited primarily by the size of the target, so scaling up to a larger wafer essentially only involves using larger targets. In this volume we are mainly concerned with films, so most of our remarks have been about planar sources. However, sputtering targets may also be shaped in other geometries for deposition onto more complex shapes.

8.3.5. Ion Beam Sputtering

In IBS, instead of using a plasma to provide the energetic ions, a collimated beam of ions is directed onto the target from an ion source. Early ion beam sources were essentially cold-cathode devices configured such that a collimated beam could be extracted and directed onto a target. Since this beam is generally of similar energies to the ions in plasma sputtering, the same type of sputtering processes occur

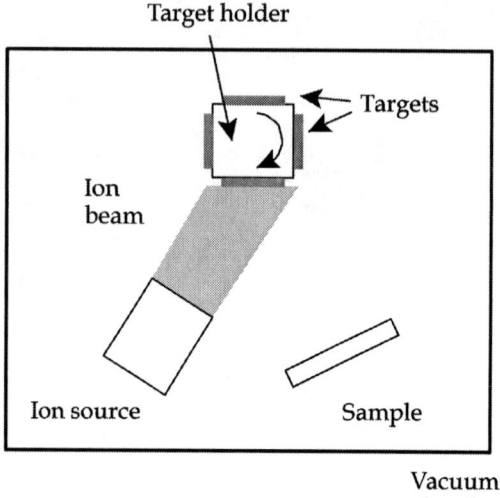

Figure 8.20. Schematic drawing of an ion beam deposition system.

when the beam is incident on the target. A substrate exposed to this target will therefore be coated with the target material, as shown in Fig. 8.20. However, IBS has the advantages that it allows independent control over the ion energy and current and reduces substrate heating from the plasma.[46] Also, it can provide high sputtering rates at very low working gas pressures. The early sources were of limited use for practical films because of the small (~1 cm) beam diameter and resulting film nonuniformity. The scalability of ion beam sources is more difficult than with plasma based sources, since it requires an increase in the diameter of the ion beam, not just the size of the target. Recently, larger size beam generators rivaling the uniformity of plasma sputtering sources have become available, and the microstructural properties of the films deposited have begun to be more thoroughly explored. Consequently, a body of work has begun to appear on the magnetic and electronic properties of ion beam sputtered spin valves and magnetic multilayers.[47–50]

8.3.6. Reactive Sputtering

As we have mentioned earlier, sputtering and evaporation of certain compounds often leads to problems with the stoichiometry of the resulting films because many compounds dissociate when bombarded with energetic electrons or ions. This is especially true when one constituent of the compound is a reactive gas, which can be pumped out of the chamber after dissociating. For example, there is evidence that deposition of MgO by both e-beam evaporation and Ar ion sputtering results in slightly oxygen-poor films. Other compounds exhibit similar behavior. This may or may not have any effect on the film structure, but for spin-valves and magnetic tunnel junctions, the electronic properties of any deposited material are of primary importance, so small stoichiometry changes are unacceptable.

In some cases, especially for oxides and nitrides, reactive sputtering can correct the stoichiometry of the deposited film. This process is similar to the plasma sputtering techniques as discussed earlier, except that a small amount of the required gas (e.g., N_2 or O_2) is mixed with the noble sputtering gas. When the plasma is lit, this gas will ionize along with the Ar. The substrate will then be exposed to this ionized reactive gas, which unlike the Ar ions will tend to react with the film. If there are bonding sites left unoccupied by the stoichiometry deficiency, the ions from the plasma will fill them. Usually, the concentration of reactive gas needed to correct the stoichiometry but not result in excess oxygen or nitrogen must be determined empirically.

Although reactive sputtering can deposit films that other techniques cannot, it is of limited use for device fabrication. This is because it is difficult to prevent the reactive gas from damaging other components of the device that have previously been deposited.

8.3.7. Microstructure in Sputtered Films

In general, the considerations which we outlined above for determining the optimal growth parameters for the desired film morphology in MBE also apply to sputtering. It is a common misconception that sputtered films are always polycrystalline and MBE-grown films are always epitaxial. In fact, sputtered epitaxial magnetic multilayers[51] and even complex intermetallics[52] have been grown. As with evaporated films, it is possible, through careful control of parameters such as substrate temperature and deposition rate, to obtain films with widely varying morphologies. However, there are some features of sputtering which introduce new considerations unique to the process. The two most important of these are the effect of the sputtering gas and the ability to bias the substrate.

In our discussion of the DC diode source, we noted that the high sputtering gas pressure required had the effect of reducing the energy of the sputtered atoms. In that case, the atoms were thermalized in less than 1 cm after leaving the target. For the more practical sources we later discussed, the sputtering gas pressures are typically 1–10 mTorr, which allows the atoms to travel several centimeters before thermalizing. For target to sample distance of 10 cm, which is typical of many sputtering systems, the atoms still have a considerable fraction of their original kinetic energy. Therefore, the gas pressure can be used as a way to control the energy of the atoms that reach the substrate.

A negative substrate bias introduces the ability to sputter material from the substrate as well as the target. This can be used to clean the substrate prior to deposition, and to backsputter material during deposition. Sputter cleaning is often an important step in sample preparation, but usually requires annealing after the cleaning to remove ion damage. Backsputtering during deposition at a rate less than the rate at which material arrives from the target can improve film microstructure and can remove some of the impurities that become incorporated in the sample from the background gas.

8.4. PULSED LASER DEPOSITION

Although sputtering has proved very useful for depositing alloys and for production applications, there are still some materials that are not suitable for sputter deposition. These include complex compounds that easily dissociate upon

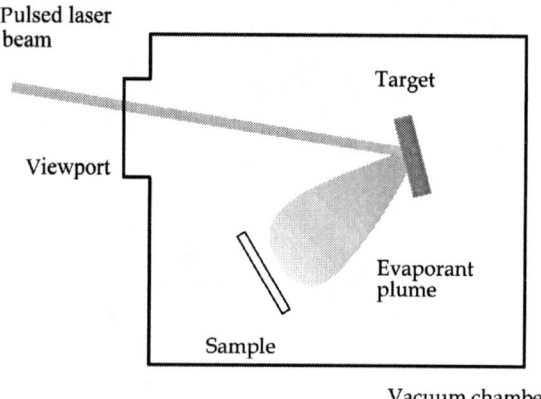

Figure 8.21. Schematic drawing of a PLD system.

bombardment with energetic particles, such as the perovskites [high temperature superconductors (HTSC) and colossal magnetoresistive (CMR) materials]. PLD, also called laser ablation, is used extensively as a means of depositing high-quality films of both HTSC and CMR materials.[53–56] It is worth pointing out, however, that high-quality films of these materials have also been grown by MBE.[57,58]

PLD has some similarities to sputtering in that the material to be deposited is removed from the surface of a target to which the sample is exposed, thereby creating a film, as in Fig. 8.21. However, as the name implies, the source of the energy needed to remove atoms from the target surface is the photons from an appropriate laser. Typically, a pulsed excimer laser is used for this purpose. The rate at which material is removed from the target also has an analogy to sputtering. Below a certain threshold pulse energy, there is no flux generated, while above this energy the rate increases nonlinearly with pulse energy. The threshold is dependent on the material.

One feature of PLD is that the laser pulses sometimes do not completely atomize what comes off the target surface. Rather, the flux may consist of particulates as well as neutral atoms and ions. The amount of particulates is dependent on the pulse energy. The particulates present a problem for depositing high-quality, flat films of many materials, especially metals, since large clusters will typically just stick to the substrate surface where they happen to land. This results in very rough films, and may also reduce the lateral grain size. To minimize this problem, the pulse energy is usually tuned to just above the threshold energy. In some cases, a method of selecting the cluster size, such as a time-of-flight spectrometer may be needed. Although the particulate problem is a disadvantage for metal deposition, it is an advantage for complex compounds. Since there is some control over the cluster mass, the optimal cluster size for any compound can be selected, and products resulting from dissociation can be rejected.

PLD is also a relatively new technology, mainly because the development of suitable lasers did not occur until the early 1980s.[59] Consequently, the process is not as well characterized as many of the other deposition techniques that we have discussed in this chapter. Also, it currently suffers from some of the same problems as early IBS work, such as difficulty in scaling up to large targets.

8.5. FILM THICKNESS MEASUREMENT AND CONTROL TECHNIQUES

In any of the above deposition techniques, it is extremely important to know accurately the thickness of the films being deposited. This requires precise knowledge of the deposition rates. Although in our discussion of the various methods of deposition we have also treated the theoretical deposition rates, in practice the actual rates obtained are dependent on factors that are difficult to predict. For example, the exact source geometry and its position with respect to the sample plays a crucial role, but differs significantly from one deposition system to another. Also, there may be a significant re-evaporation rate from the substrate, which may be difficult or impossible to predict. Therefore, some form of *in situ* rate monitoring, or at least accurate rate calibration, is necessary. The monitoring devices we discuss here do not provide absolute rate information, and must themselves be calibrated. The most common approach to this is to deposit a single film under conditions similar to those to be used for the spin valve or multilayer deposition, measure its thickness *ex situ*, then correct the calibration factor of the monitor. This calibration has to be repeated for each source. The appropriate way to determine the exact film thickness depends on the thickness range required and the materials involved. Low-angle x-ray diffraction, x-ray fluorescence, Rutherford backscattering, and profilometry are common ways to do this.

The most common device for rate monitoring is the quartz crystal monitor (QCM), also called a quartz microbalance. This device measures the resonant frequency of a quartz crystal oscillator which is exposed to the flux from the deposition source. As a film grows on the quartz crystal, its mass changes, and therefore, its resonant frequency changes. Most of the crystals available have a broad range where the mass-frequency dependence is linear, and total depositions of about 100,000 Å are possible before the crystal needs to be changed. Many of the controllers available are capable of resolving a frequency shift equivalent to ~ 1 Å of deposition, depending on the geometrical factors and the material density.

The QCM has the advantage that although it is compatible with UHV, it does not require any vacuum to operate, so that it can be used for all of the PVD techniques that we have discussed here. This, along with its wide dynamic range, has made the QCM the most popular choice for rate monitoring. The QCM does suffer from some disadvantages which limits its usefulness in some applications. The resonant frequency and the mass-frequency calibration are also very sensitive to changes in temperature. This requires that the crystal be held at a constant temperature during deposition. This is usually accomplished by water cooling, which can eliminate most day-to-day temperature variations. Unfortunately, it is less successful at removing transients that can result when the QCM is suddenly exposed to a hot source, as when a shutter opens over a K-cell. The sudden increase in temperature causes an increase in the frequency and an apparent drop in thickness, as if material has evaporated from the crystal. The effect quickly reverses as more material is deposited and the crystal temperature stabilizes again, but it can last for several seconds; long enough to render it useless for depositing a layer in the tens of Ångstrom thickness range. One way around this problem, if the source and chamber geometry allows it, is to position the crystal so that it is always exposed to the source, even when the shutter is closed. This way, the crystal will reach a stable temperature over time, and will not be affected by the shutter opening.

An alternative to the QCM, at least for some applications, is to determine the density of evaporant atoms in the flux. The evaporant density is directly related to the deposition rate. Therefore, by integrating this quantity over the time the shutter is open, the deposited thickness can be calculated. The density can be measured using a mass spectrometer placed in the flux, or by some form of optical spectroscopy. Mass spectrometers are common equipment on UHV systems for analyzing the background gases. To use a spectrometer as a flux monitor only requires that the probe be positioned such that the flux can get into the ionization region of the spectrometer. Also, optical devices are available which ionize or excite the flux atoms by electron bombardment from a hot cathode, and analyze the intensity of the resulting ultraviolet fluorescence with narrow-pass filters or spectrometers. This works well, provided the evaporant has a strong emission line within the range of the spectrometer. These methods have the advantage that they are insensitive to temperature variations, therefore the transient effects discussed for the QCM are not a consideration. They are also element specific, which allows the monitoring of alloys produced by co-deposition. Most of the spectrometers do use a hot cathode to ionize the flux, though, and thus require at least a high vacuum to operate. Also, they are poor choices for sputtering applications, because the partial pressure of the flux atoms is generally very small compared to the sputtering gas pressure. Therefore, these rate monitoring methods are mainly used in MBE applications, where the evaporant density is high compared to the background gases.

By using the signal from the rate monitor as the input to a proportional–integral–derivative (PID) controller, any of these rate monitoring methods can be configured for full closed-loop rate control. Most sputtering, e-beam, and K-cell power supplies are externally programmable by a DC input voltage, which can be supplied by the PID controller.

8.6. CHARACTERIZATION TECHNIQUES

Although a detailed description of all the available structural characterization techniques is beyond the scope of this chapter, and many are covered in a separate chapter of this volume, it is worthwhile to provide a brief overview of them here. The techniques used vary widely between laboratories depending on what is available. In this section we describe some of the most common techniques available *in situ*. Many of these techniques require UHV, and so are more commonly found in MBE systems.

8.6.1. Structural Characterization

In addition to the RHEED technique, which we described in detail above, another electron diffraction method, LEED, is also commonly found in MBE systems. LEED is also sometimes used in UHV sputtering systems. LEED differs from RHEED in that the monochromatic e-beam energy is lower (100–1000 eV), and incidence is normal to the film plane. The diffraction of backscattered electrons is observed on a spherical phosphor screen with its center at the e-beam spot (see Fig. 8.22). In this geometry, the scattering wave vector is now primarily normal to the surface. This makes LEED more sensitive than RHEED to the crystalline ordering perpendicular to the surface, and not as sensitive to surface flatness and in-plane coherence length. However, even with normal incidence, the electron penetration depth is still limited to a few tens of Ångstroms, so LEED is still a surface characterization tool. An example of a LEED pattern, again from a reconstructed Si(111) surface is shown in Fig. 8.23.

The information obtained from LEED includes the crystallinity, symmetry, and the stacking of atomic planes in the first few monolayers below the surface. The symmetry is easily observable in the diffraction pattern symmetry, and the observation of any pattern depends upon the in-plane coherence length of the sample being larger than the e-beam source coherence length, as with RHEED. The stacking information is obtained from the dependence of diffracted spot intensities with the beam energy, or the so-called LEED I–V curves. This essentially probes the periodicity of the lattice perpendicular to the surface. For LEED, it is possible to perform kinematic calculations which can predict diffraction patterns based on structural models of the crystal. These can then be used to modify atomic positions in the model to reproduce features in the I–V curves. This technique can be used to observe lattice relaxations that occur at various surfaces. In addition, surface reconstructions are easily detectable in LEED patterns.

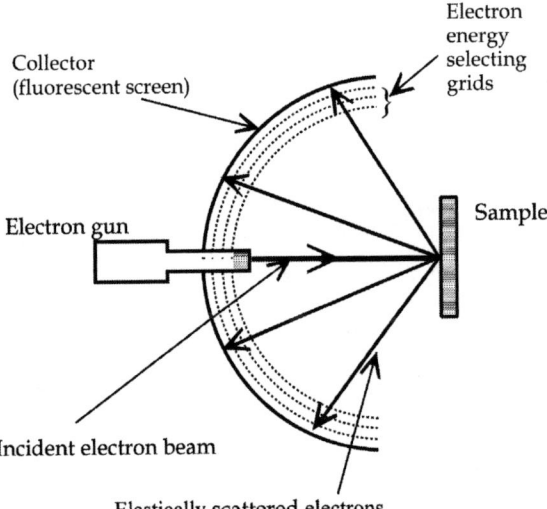

Figure 8.22. Electron optics for LEED. The diffraction pattern is viewed from behind the screen.

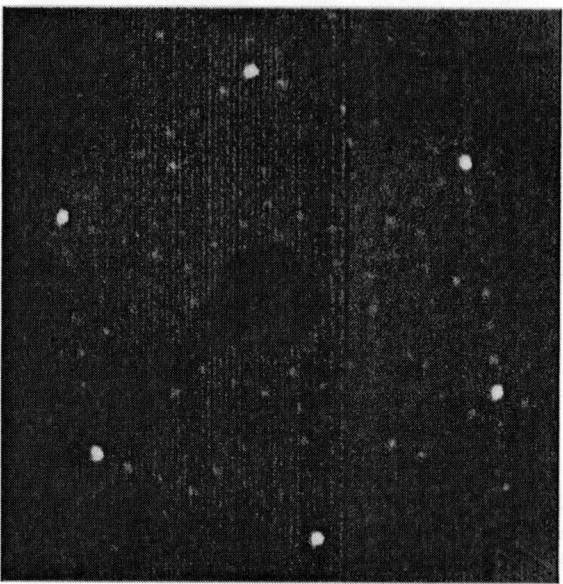

Figure 8.23. An example of a LEED pattern, in this case from a reconstructed Si(111) surface, similar to the surface for which a RHEED pattern is shown in Fig. 8.10. The bright spots are from diffraction by the bulk crystal lattice, whereas the fainter ones are from the 7×7 reconstruction.

8.6.2. Chemical Characterization

The two most common forms of chemical characterization found in MBE and UHV sputtering systems are x-ray photoelectron spectroscopy (XPS) and Auger electron spectroscopy (AES). These two are often installed together as they can both use the same electron spectrometer. In both techniques, the electronic environment of atoms in the sample is probed by measuring the emission of electron with energies characteristic of the elements. This allows quantitative analysis of the composition, including impurity levels, and determination of the chemical state of each atomic species in the sample. In XPS, the electrons come directly from the absorption of x-ray photons by core levels. In AES, the process is slightly more complicated: core levels are excited by an incident e-beam and the atom de-excites by ejecting one or more valence level electrons. The ejected electrons can have energies in the range of 20–1000 eV, which gives them an escape depth not much longer than the LEED penetration depth. Therefore, the chemical information obtained is surface sensitive.

Because of the sensitivity of these spectroscopy techniques to the chemical state of the surface, they are very useful for determining whether deposited films are oxidized or otherwise contaminated. In some cases, the peak shifts from the formation of intermetallic compounds can also be detected. This can then be used to make an estimate of the amount of interdiffusion. Another powerful way in which these electron spectroscopies can be used to obtain structural information is by angle-resolved photoelectron or Augerelectron diffraction. In this technique, the direction as well as the energy of the ejected electrons is determined by using a detector with a small acceptance angle. The angular dependence of the photoelectron intensity contains information about the location of atoms neighboring the electron emitting atom. For example, if the emitting atom is at the surface, the electrons would be emitted isotropically. On the other hand, if the emitting atom is buried below the surface, the emitted electrons must pass through the overlayers and are diffracted by them, resulting in peaks in the intensity according to the positions of the overlayer atoms. Photoelectron and Auger electron diffraction have been used to observe interdiffusion, as well as tetragonal distortions, in thin epitaxial metal overlayers.

8.7. SUMMARY

In this chapter, we described several PVD techniques, paying particular attention to those that are most commonly used for depositing magnetic films and multilayers, both for device production and for research. The two general approaches taken are evaporation, including MBE, and sputtering in its various forms.

Because the goal in MBE is usually epitaxial growth, deposition by MBE is often characterized by low growth rates, allowing sufficient time for thermal diffusion to promote crystalline growth, and extensive *in situ* characterization. The growth aspects, and most of the surface characterization instrumentation, require ultraclean vacuum conditions.

To date, the use of MBE for magnetic device production has been limited, as is the case for semiconductor MBE. This is because of the low sample throughput typically available in MBE systems, and the difficulty of obtaining uniform samples over the size wafers that industry uses (200 mm diameter as of this writing, but certain to increase). Despite this, due to its ability to grow very high quality and well-characterized films, MBE has proved to be an invaluable basic research tool, and has contributed significantly to the understanding of the physics of these magnetic materials.

Because sputtering does not suffer from the uniformity and rate limitations of MBE, it has found more use in fabrication. However, it has also has been a very valuable basic research tool for some time. DC and RF magnetron sputtering are probably the most commonly used types of sputtering, although IBS is gaining popularity due to the advantages it offers. Historically, the approach in sputtering, has emphasized fast deposition to obtain flat layers and sharp interfaces, with less emphasis on crystallinity and epitaxy. This is mainly because this type of deposition is well-suited for sputtering. However, it is a common misconception that epitaxial films cannot be made by sputtering. In fact, a wide range of growth morphologies is possible in sputtering, as well as in MBE.

ACKNOWLEDGMENTS

The authors gratefully acknowledge Motorola, DARPA, and DOE grant # DE-FG02-93ER45488 for support of various aspects of our studies of magnetic thin films. It is also a pleasure to acknowledge the contributions of Drs. James

Eickmann, Brad Engel, Gerd Fischer, Akihiro Murayama, Jon Slaughter, Uli Hiller, Kyoko Hyomi, Sung-Kyun Park, and Kirstin Petersen.

REFERENCES

1. M. N. Baibich, J. M. Broto, A. Fert, F. Nguyen van Dau, F. Petroff, P. Eitenne, G. Creuzet, A. Friedrich, and J. Chazelas, *Phys. Rev. Lett.* **61**, 2472 (1988).
2. P. Grünberg, R. Schreiber, Y. Pang, M. B. Brodsky, and H. Sowers, *Phys. Rev. Lett.* **57**, 2442 (1986).
3. S. S. P. Parkin, R. Bhadra, and K. P. Roche, *Phys. Rev. Lett.* **66**, 2152 (1991).
4. M. T. Johnson, R. Coehoorn, J. J. de Vries, N. W. E. McGee, J. aan de Stegge, and P. J. H. Bloemen, *Phys. Rev. Lett.* **69**, 969 (1992).
5. W. F. Egelhoff, Jr. and M. T. Kief, *Phys. Rev. B* **45**, 7795 (1992).
6. K. -M. H. Lenssen, J. C. S. Kools, A. E. M. De Veirman, J. J. T. M. Donkers, M. T. Johnson, A. Reinders, and R. Coehoorn, *J. Magn. Magn. Mater.* **156**, 63 (1996).
7. M. A. Herman and H. Sitter, *Molecular Beam Epitaxy* (Springer, Berlin, 1989).
8. E. H. C. Parker, (Ed.) *The Technology and Physics of Molecular Beam Epitaxy* (Plenum, New York, 1985).
9. R. F. C. Farrow, (Ed.) *Molecular Beam Epitaxy: Application to Key Materials* (Noyes, Park Ridge, NJ, 1995).
10. K. Ploog and K. Graf, *Molecular Beam Epitaxy of III–V Compounds: A Comprehensive Bibliography 1958–1983* (Springer, Berlin, 1984).
11. R. F. Bunshah, (Ed.) *Deposition Technologies for Films and Coatings* (Noyes, Park Ridge, NJ, 1982).
12. R. V. Stuart, *Vacuum Technology, Thin Films, and Sputtering* (Academic Press, New York, 1983).
13. R. E. Honig and D. E. Kramer, *Vapor Pressure Curves of the Elements* (RCA Laboratories, Princeton, NJ, 1968); J. Margrave (Ed.) *The Characterization of High Temperature Vapors* (Wiley, New York, 1967).
14. M. A. Herman and H. Sitter, *Molecular Beam Epitaxy* (Springer, Berlin, 1989), p. 31.
15. M. Knudsen, *Ann. Phys.* **29**, 179 (1909).
16. D. J. Keavney, E. E. Fullerton, and S. D. Bader, *J. Appl. Phys.* **81**, 795 (1997).
17. J. Unguris, R. J. Celotta, and D. T. Pierce, *Phys. Rev. Lett.* **67**, 140 (1991).
18. M. R. Scheinfein, J. Unguris, M. H. Kelly, D. T. Pierce, and R. J. Cellota, *Rev. Sci. Instrum.* **61**, 2501 (1991).
19. J. Unguris, R. J. Celotta, and D. T. Pierce, *J. Magn. Magn. Mater.* **127**, 205 (1993).
20. L. Eckertova, *Physics of Thin Films* (Plenum, New York, 1986), p. 96.
21. J. A. Venables, G. D. T. Spiller, and M. Handbucken, *Rep. Prog. Phys.* **47**, 399 (1984).
22. W. F. Egelhoff, Jr., in: *DOE Workshop on Surface Diffusion and the Growth of Materials*, edited by P. J. Feibelman, G. L. Kellogg, and T. A. Michalske (Santa Fe, NM, 1992).
23. A. Zangwill, *Physics at Surfaces* (Cambridge University Press, Cambridge, 1988), p. 428.
24. E. Bauer, *Z. Kristallogr.* **110**, 372 (1958).
25. G. A. Prinz, *Phys. Rev. Lett.* **54**, 1051 (1985).
26. J. A. Wolf, J. J. Krebs, G. A. Prinz, *Appl. Phys. Lett.* **65**, 1057 (1994).
27. W. A. Jesser and J. W. Matthews, *Philos. Mag.* **15**, 1097 (1967); P. Ehrhart, B. Schönfeld, H. H. Ettwig, and W. Pepperhoff, *J. Magn. Magn. Mater.* **22**, 79 (1980).
28. D. Li, M. Freitag, J. Pearson, Z. Q. Qiu, and S. D. Bader, *Phys. Rev. Lett.* **72**, 3112 (1994).
29. G. K. Wehner, *Adv. Elec. Elec. Phys.* **7**, 239 (1955).
30. E. Kay, *Adv. Elec. Elec. Phys.* **17**, 245 (1962).
31. J. A. Thornton, *Deposition Technologies for Films and Coatings* (Noyes, Park Ridge, NJ, 1982), Chapter 5.
32. G. K. Wehner and G. S. Anderson, *IN: Handbook of Thin Film Technology*, edited by L. Maissel and R. Glang (McGraw-Hill, New York, 1970), p. 3-1.
33. B. Chapman, *Glow Discharge Processes; Sputtering and Plasma Etching* (Wiley, New York, 1980).
34. R. V. Stuart and G. K. Wehner, *J. Appl. Phys.* **33**, 235 (1962).
35. M. H. Witcomb, *J. Mater. Sci.* **9**, 551 (1974).
36. J. E. Green, B. R. Natarjan, and F. Sequeda-Osorio, *J. Appl. Phys.* **49**, 417 (1978).

37. M. L. Tang and G. K. Wehner, *J. Appl. Phys.* **43**, 2268 (1972).
38. E. S. McDaniel, *Collision Phenomena in Ionized Gases* (Wiley, New York, 1964).
39. W. D. Westwood and R. Boynton, *J. Appl. Phys.* **43**, 2691 (1972).
40. W. D. Westwood, *Prog. Surf. Sci.* **7**, 71 (1976).
41. W. D. Westwood, *J. Vac. Sci. Technol.* **15**, 1 (1978).
42. J. A. Thornton, *SAE Trans.* **82**, 1787 (1974).
43. J. A. Thornton, *Metal Finish.* **77**, 45 (1979).
44. R. K. Waits, *J. Vac. Sci. Technol.* **15**, 179 (1978).
45. J. A. Thornton and A. S. Penfold, In: *Thin Film Processes*, edited by J. L. Vossen and W. Kern (Academic Press, New York, 1978), p. 75.
46. H. R. Kaufman, *J. Vac. Sci. Technol.* **15**, 272 (1978).
47. Y. Saito, K. Inomata, K. Yusu, A. Goto, and H Yasuoka, *Phys. Rev. B* **52**, 6500 (1995).
48. L. Tang, D.E. Laughlin, and S. Gangopadhyay, *J. Appl. Phys.* **81**, 4906 (1997).
49. W. E. Bailey, D. Guarisco, and S. X. Wang, *IEEE Trans. Magn.* **34**, 957 (1998).
50. T. J. Minvielle, R. L. White, and R. J. Wilson, *J. Appl. Phys.* **79**, 5116 (1996).
51. E. E. Fullerton, M. J. Conover, J. E. Mattson, C. H. Sowers, and S. D. Bader, *Appl. Phys. Lett.* **63**, 1699 (1993).
52. E. E. Fullerton, C. H. Sowers, J. E. Pearson, S. D. Bader, J. B. Patel, X. Z. Wu, and D. Lederman, *J. Appl. Phys.* **81**, 5637 (1997).
53. D. Dijkkamp, T. Venkatesan, X. D. Wu, S. A. Shaheen, N. Jiswari, Y. H. Min-Lee, W. L. McLean, and M. Croft, *Appl. Phys. Lett.* **51**, 619 (1987).
54. S. B. Ogale, D. Dijkkamp, T. Venkatesan, X. D. Wu, and A. Inam, *Phys. Rev. B* **36**, 7210 (1987).
55. J. Narayan, N. Biunno, R. Singh, O. W. Holland, and O. Auciello, *Appl. Phys. Lett.* **51**, 1845 (1987).
56. T. S. Baller, G. N. A. van Veen, and H. A. M. van Hal, *Appl. Phys. A* **46**, 215 (1988).
57. V. S. Achutharaman, K. M. Beauchamp, N. Chandrasekhar, G. C. Spalding, B. R. Johnson, and A. M. Goldman, *Thin Solid Films.* **216**, 14 (1992).
58. V. S. Achutharaman, P. A. Kraus, V. A. Vasko, C. A. Nordman, and A. M. Goldman, *Appl. Phys. Lett.* **67**, 1019 (1995).
59. J. Dielemann, E. van de Riet, and J. C. S. Kools, *Jpn. J. Appl. Phys.* **31**, 1964 (1992).

9

Magnetic Sensors

Jim Daughton* and Carl Smith*

9.1. INTRODUCTION

Ten years after the discovery of giant magnetoresistance (GMR), commercialization of the technology is evidenced by product introductions in magnetic field sensors and read heads for hard drives. This comparatively short introduction time was facilitated by the prior existence of similar products using anisotropic magnetoresistance (AMR) and other materials. In 1995, the first GMR magnetic field sensors were sold by Nonvolatile Electronics, Inc. (NVE), and in 1997, NVE introduced a line of GMR sensors, where GMR multilayer materials were integrated on-chip with silicon integrated circuits. Siemens introduced a GMR magnetic field sensor in 1997.

As explained in Section 9.2, nearly all of the applications for magnetic sensors have the ultimate purpose of measuring something besides magnetic fields *per se*. In most sensor applications, the strength of the magnetic field is used to indicate the position and speed of a ferrous body, or to indicate the amount of current flowing through a wire. In the speed and position sensing, a permanent magnet is usually used to provide a magnetic field, which is then perturbed by the position of the ferromagnetic body.

Section 9.3 explains some competing and more established magnetic sensor technologies, such as variable reluctance sensors, reed switches, Hall effect sensors, flux gates, and indium antimonide magnetoresistors. This is followed by a description of GMR materials, which could be used for sensing magnetic fields and a discussion of their properties. While the primary driving force behind the industrial introduction of GMR materials is their higher magnetoresistance, other material properties of GMR materials are extremely important to applications, and they are discussed in this section.

Section 9.5 discusses magnetoresistance sensor elements in general, Section 9.6 discusses AMR designs, and Section 9.7 discusses GMR sensor designs.

Section 9.8 describes a design concept using spin-dependent tunneling (SDT) elements. This design is illustrative of the potential for these elements, and also shows how the limitations of SDT must be accounted for in the design.

---

*NVE Corporation, Eden Prairie, Minnesota.

9.2. WHY WE SENSE MAGNETIC FIELDS

The sensing of magnetic fields has been important to commerce for over 2000 years. Compasses with naturally occurring lodestone were used for navigation by finding the direction of the Earth's magnetic field. This invention allowed mariners to confidently sail beyond the sight of shore. Magnetic compasses are still used for navigation on sea and on land. Magnetometers which measure the direction and strength of magnetic fields are used both in various fields of research and in industry wherever magnetic materials are used. However, the most extensive uses of magnetic sensors are those in which a magnetic field is sensed as a means of measuring something else. Magnetic sensors are used to sense the presence of ferromagnetic objects ranging from magnetic ink on bank notes and data on computer hard drives to automobiles and trucks. The combination of magnetic sensors and magnetic materials allows the measurement of rotational speed and position in a myriad of automotive applications ranging from ignition timing to wheel speed sensing for anti-skid breaking. Magnetic sensors with permanent magnets are used in linear position sensing of pneumatic cylinders for machinery. Electrical currents in wires are detected and measured with isolation and without physical contact by measuring their associated magnetic fields. Magnetic fields extend over a distance from their source. This fact provides the basis for a variety of non-contacting sensors which indirectly measure the property of interest.

Several books on sensors include sections on magnetic sensors.[1,2] Most commercially used magnetic field sensors can be divided into two categories: sensors which in one way or another depend upon induced voltages in coils caused by changing magnetic fields, and sensors in which the magnetic field changes a directly measurable property of a bulk sensing element. Sensors in the first class, which includes variable reluctance sensors and fluxgate magnetometers, require coils of some variety or another. Sensors in the second class, which includes the Hall effect sensors, magnetotransistors, magnetodiodes, and magnetoresistive sensors, are classified as solid-state sensors. These sensors share the characteristic that they respond directly to the magnetic field rather than by interpreting a voltage induced by a changing magnetic field. Several recent articles review various types of magnetic field sensors and the physical principles involved.[3-6]

The world market for magnetic sensors is estimated to be $1.8 billion dollars in 1998 with an annual growth rate of 3.6%.* The actual sensor elements themselves, rather than the sensor systems are about 1/3 of the market. The magnetic sensor business is a fragmented business both at the source and the end user, since many differing technologies are used in magnetic sensors, and they are used in a wide variety of industries. However, automotive applications dominate with greater than 50% of the entire market. Emerging industry trends which affect the magnetic sensor market include the switch from electro-mechanical control devices to solid-state devices and the increase of local intelligence with signals processed by integrated circuits. Both of these trends point towards smart magnetic sensors with the sensing element combined with integrated circuits in the same package.

---

*Market estimate by Nonvolatile Electronics, Inc.

MAGNETIC SENSORS

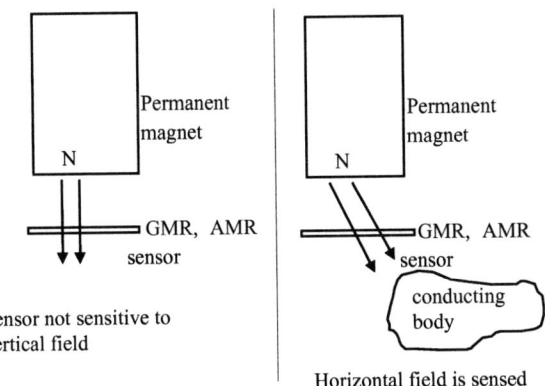

Figure 9.1. Configuration of biasing magnet and thin-film magnetoresistive sensor.

In the applications of magnetic field sensors, a magnetic field sensor can directly sense a magnetic field from permanent magnetics, electromagnets, or currents. To sense a non-magnetized ferrous object, a biasing magnet is often used in conjunction with the sensor. The biasing magnet magnetizes the ferromagnetic object such as a gear tooth, and the sensor detects the combined magnetic fields from the magnetized object and the biasing magnet. A biasing magnet is affixed to the sensor in a position such that its direct influence on the sensor is minimal. This effect can be achieved because many sensors are only sensitive to the component of magnetic field along their sensitive axis. Thin film magnetoresistive sensors are insensitive perpendicular to the film due to shape anisotropy. Only the component of the field perpendicular to slab causes a voltage in a Hall sensor. The biasing magnet can be mounted on the top of the sensor with its magnetic axis perpendicular to the sensitive axis of the sensor. The biasing magnet is then centered such that there is little or no field in the sensitive direction of the sensor. In this way, a reasonable large biasing magnet can be used which will magnetize a ferrous target at a wide gap from the sensor. Occasionally, a spacer is used between the sensor and the magnet to reduce the field at the sensor and, therefore, reduce how critically the magnet must be positioned. Figure 9.1 illustrates the configuration of a biasing permanent magnet and a thin film magnetoresistor. Figure 9.2 shows

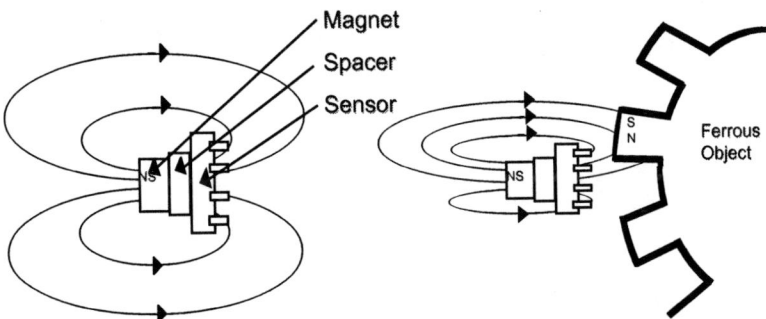

Figure 9.2. Biasing magnet and magnetic sensor for gear-tooth sensing of a ferromagnetic gear. The sensitive axis of the sensor is vertical in the plane of the figure.

the relative positions of a sensor and a biasing magnet for gear-tooth sensing. The magnetic field lines are shown both in the absence and in the presence of a ferromagnetic object. Note the induced magnetic moment in the ferromagnetic object. The technique of using a biasing magnet can only be used if the ferrous object is in close proximity to the magnet. If it is too far away, the magnet will have little influence in magnetizing the object. In some applications, such as detection of motor vehicles, the Earth's field acts as a biasing magnet resulting in magnetic dipole moments along the length of the motor vehicle and along the axles.

Magnetic sensors are used in munitions fusing to avoid premature detonation. Spin count sensors can determine that projectiles emerging from rifled barrels have traveled a minimum distance down range before they are armed. A magnetic sensor detects the Earth's magnetic field, which will pass through a maximum each revolution. A circuit counts to a preset number of revolutions before arming the projectile. For smooth-bore gun tubes, a magnetic sensor based system detects when

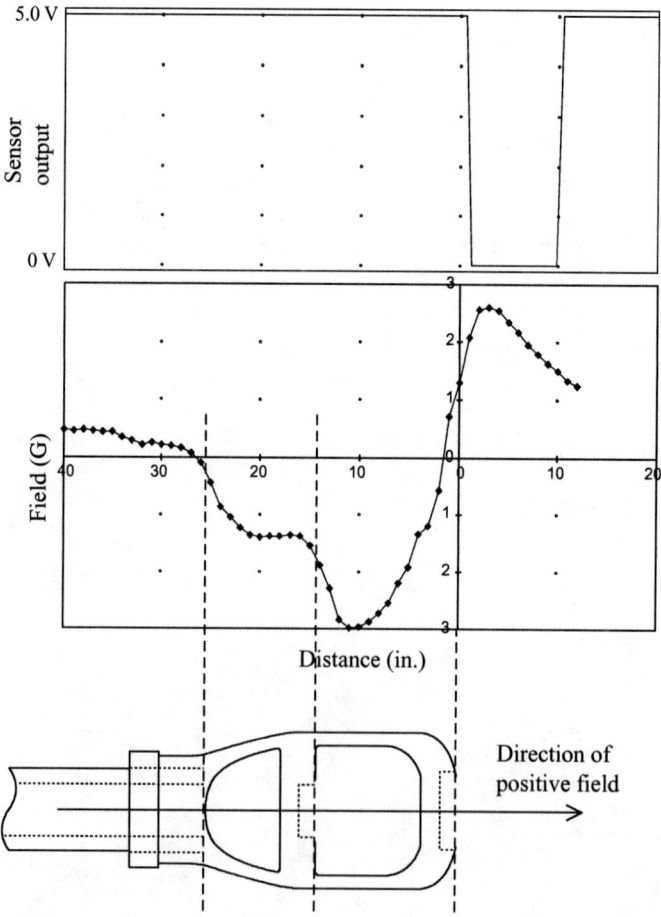

Figure 9.3. Simulated results of a magnetic field gradient sensor in a projectile exiting a gun tube at 5000 feet per second (1500 m/s).

MAGNETIC SENSORS

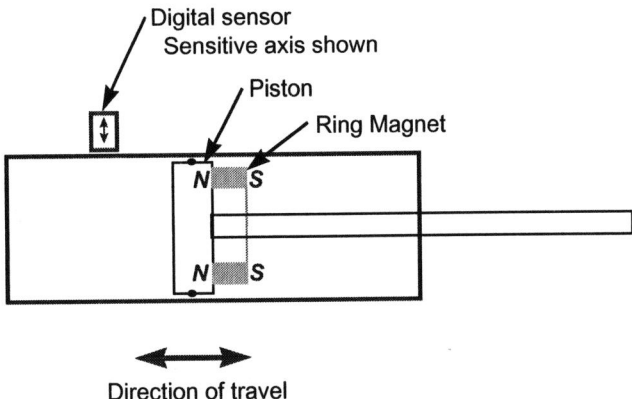

Figure 9.4. Sensing the position of a hydraulic sensor by using a permanent magnet on the piston and a magnetic sensor outside the cylinder.

the projectile leaves the ferromagnetic gun tube by detecting the magnetic field gradient. Figure 9.3 shows simulated results of sensor leaving a gun tube at 5000 feet per second (1500 m/s).

Accurate control of machinery requires knowledge of when actuators such as pneumatic or hydraulic cylinders reach their desired position. A permanent magnetic located on the piston can be sensed by a magnetic sensor outside the cylinder. Integrated magnetic sensors with comparators and output circuitry can directly interface with machine control circuitry when a preset position is reached. Figure 9.4 illustrates cylinder position sensing using magnetic sensors and permanent magnets. A variation on this application is shown in Fig. 9.5 in which the position of a permanent magnet is continuously monitored. This smart shock absorber for high-end bicycles uses a magnetic sensor to obtain an output which varies with the magnet's distance from the sensor. The control electronics include a microprocessor

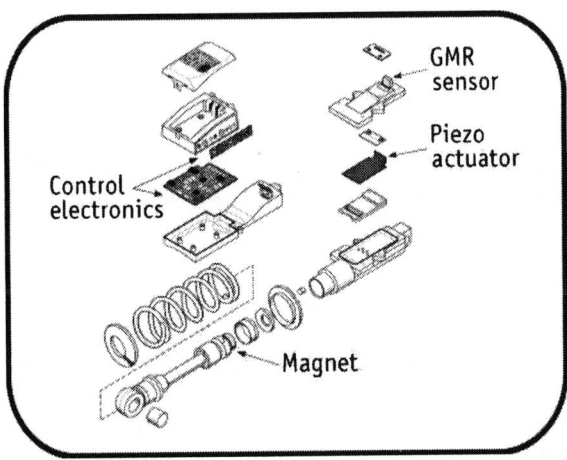

Figure 9.5. Magnetic position/speed sensor incorporated in a smart shock absorber for bicycles. Manufactured by ACX for K2.

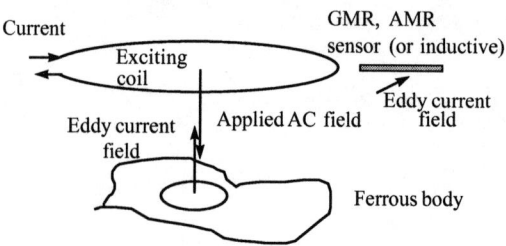

Figure 9.6. Detecting non-ferrous objects by using induced eddy currents and magnetic sensors.

which calculates the position and velocity of the piston in the shock absorber. A piezo actuator is used to adjust the damping rate by covering or uncovering one of the bleed holes in the piston and thus achieving optimum contact with the ground and control of the bicycle.

Magnetic sensors can be used to sense the presence of non-ferrous objects as well as ferrous objects. The ac magnetic fields generated by an exciting coil generate eddy currents in any conducting material. These eddy currents generate magnetic fields which oppose the applied magnetic field. A magnetic sensor will sense the change in field resulting from these eddy currents. Eddy current sensors using a variety of magnetic field sensing techniques are used in position sensing, metal detection, and crack detection in nondestructive testing. Figure 9.6 shows the principle of detecting a non-ferrous body using induced eddy currents.

The common thread of many applications of magnetic sensors is the need for non-contact sensing of some physical property or some dimension. The principle of action-at-a-distance for magnetic fields allows for sensor systems without contact wear, and systems which are quite immune to dirt, smoke, and variation of sensor gap. Non-contacting sensors also allow for electrical isolation between the sensing system and the system which is monitored. Often, the real challenge in applying magnetic sensors is the understanding of magnetic materials and the interaction between magnetic fields and magnetic materials.

9.3. MAGNETIC SENSING TECHNOLOGIES

A number of magnetic sensing technologies have been developed and used in industry over the years. This section will discuss some of these technologies that will compete with the new electron spin-based magnetoresistive sensor technologies.

9.3.1. Reed Switch Sensors

Possibly, the simplest magnetic sensor that produces a usable output for industrial control is the reed switch. It consists of a pair of flexible, ferromagnetic contacts hermetically sealed in an inert gas filled container, often glass. The magnetic field along the long axis of the contacts magnetizes the contacts causing them to attract one another closing the circuit. There is usually considerable hysteresis between the closing and releasing fields so they are quite immune to small fluctuations in the field. Closing fields for reed switches are usually in the several tens

MAGNETIC SENSORS

of oerstead (several kiloamperes per meter) range. Low cost, simplicity, reliability, and zero power consumption make reed switches popular in many applications. A reed switch together with separate small permanent magnets make a simple proximity switch often used in security systems to monitor the opening of doors or windows. The magnet, affixed to the moveable part, activates the reed switch when it comes close enough. The desire to sense almost everything in cars is increasing number of reed switch sensing applications in the automotive industry.

9.3.2. Inductive Sensors

Inductive sensors work on the principle that a changing magnetic field in a circuit induces a voltage in that circuit. They range from ac driven variable reluctance sensors to passive coils used in automotive speed sensors, but they all involve coils of wire. Their sensitivity is related to the number of turns, and therefore, their size. Passive devices which depend upon a moving magnet or moving magnetized gear-tooth are relatively insensitive at low speeds and have no output at all at zero speed. However, they are presently the industry workhorse for automotive anti-skid breaking systems. Their output is proportional to the rate of change of the magnetic field in their coil. A magnetic core is often placed in the coil in active devices. The presence of external ferromagnetic material changes the reluctance of the magnetic circuit and the inductance of the coil. A recently developed variety of inductive sensors called magnetoinductors, places a coil with a magnetic core in a resonant circuit. An external field changes the permeability of the core and hence its inductance causing a shift in the resonant frequency.

9.3.3. Hall Effect Sensors

The Lorentz force accelerates moving charge carriers in a magnetic field in a direction perpendicular to both their direction of motion and to the magnetic field. For a current in a slab of semiconductor material, the Hall effect results in a Hall voltage proportional to the magnetic field across the slab perpendicular to the current and the magnetic field. Figure 9.7 illustrates the geometry of a Hall sensor. These devices predominantly use n-type silicon when cost is of primary importance and GaAs for higher temperature capability. In addition, InAs, InSb, and other semiconductor materials are gaining popularity due to their high carrier mobilities which result in greater sensitivity and in frequency response capabilities above the 10–20 kHz typical of Si Hall sensors. Compatibility of the Hall sensor material with semiconductor substrates is important since Hall sensors are often used in integrated devices which include other semiconductor structures.

The Hall voltage increases linearly with applied field to several teslas (10 s of kilogauss); however, it is relatively small. A typical non-amplified Hall sensor will have an output of about 10 mV/kG with 10 mA of current. The temperature dependence of the Hall voltage and the input resistance of Hall sensors are governed by the temperature dependence of the carrier mobility and that of the Hall coefficient. Different materials and different doping levels result in trade-offs between sensitivity and temperature dependence.

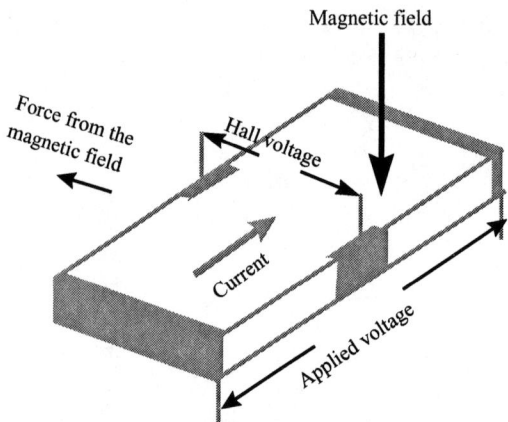

Figure 9.7. The geometry for a semiconductor used as a Hall sensor showing the directions of the applied voltage, current, magnetic field, and the Hall voltage.

9.3.4. Integrated Hall Sensors

Hall devices are often combined with semiconductor elements to make integrated sensors. By adding comparators and output devices to a Hall element, manufactures provide unipolar and bipolar digital switches. Adding an amplifier increases the relatively low-voltage signals from a Hall device to produce ratiometric linear Hall sensors with an output centered on one-half the supply voltage. Power usage can even be reduced to extremely low levels by using a low duty cycle.[7]

9.3.5. Fluxgate Sensors

Fluxgate sensors are used in many magnetic sensors used for low level magnetic fields such as in compassing, geomagnetic surveying, ferromagnetic metal detection, detection of automobiles, and even submarine hunting. The operation of a fluxgate magnetometer is based on the very large change between the saturated permeability and the unsaturated permeability of soft ferromagnetic materials with square B–H loops. If one coil is used to cyclically excite a ferromagnetic core into saturation, a second coil around the core will sense an induced voltage proportional to the number of turns, the rate of change of the magnetic field in the core, and the cross-sectional area of the core. With a core material with a square B–H loop, these changes will occur within a small fraction of each cycle, and the induced voltage will be an alternating series of short, symmetrical pulses, equally spaced in time. These pulses, although rich in harmonics, will not have even harmonics due to odd (as opposed to even) mathematical symmetry of the pulses. If an external field is superimposed on the core, the timing of the pulses will shift since it is now easier to saturate the core in the direction of the external field than in opposition to it. Even harmonics will now be present in the induced pulses due to the breaking of the symmetry. A signal proportional to the external field can be obtained by using a filter to extract the second harmonic signal at twice the drive frequency. Fluxgate technology has been developed to maximize sensitivity and to minimize direct coupling from the exciting coils to the pick up coil. Only magnetic fields with frequencies below the exciting frequency can be detected. Therefore, the upper limit is usually a few kilohertz due to the power required to drive a core at higher frequencies.

9.3.6. InSb Magnetoresistors

Semiconductor magnetoresistors are another type of sensors which utilize the effect of the Lorentz force on charge carriers. Magnetoresistors use semiconductors such as InSb and InAs with high room-temperature carrier mobility. If a voltage is applied along the length of a thin slab of semiconductor material, a current will flow and a resistance can be measured. When a magnetic field is applied perpendicular to the slab, the charge carriers will be deflected by the Lorentz force. If the width of the slab is greater than the length, the charge carriers will cross the slab without a significant number of them collecting along the sides. The effect of the magnetic field is to increase the length of their path and, therefore, the resistance. An increase in resistance of several hundred percent is possible in large fields. In order to produce sensors with hundreds to thousands of ohms of resistance, long, narrow semi conductor stripes a few micrometer wide are produced using photolithography. The required length to width ratio is accomplished by forming periodic low resistance metal shorting bars across the traces. Each shorting bar produces an equipotential across the semiconductor stripe. The result is, in effect, a series of short elements with the proper length to width ratio. A second method used in commercial devices manufactured by Siemens uses lapped wafers cut from boules which have needle shaped low resistance precipitates of NiSb in a matrix of InSb. These precipitates serve as the shorting bars.[8]

Magnetoresistors formed from InSb are relatively insensitive at low fields but exhibit large changes in resistance at high fields. The resistance changes almost as the square of the field. They are sensitive only to the component of the magnetic field perpendicular to the slab and are not sensitive to whether the field is positive or negative. The large temperature coefficients of resistivity are caused by the change in mobility of the charge carriers with temperature. Sensors are made with either single resistors or pairs of spaced resistors. The second type is used to measure field gradients and is usually combined with external resistors to form a Wheatstone bridge. A permanent magnet is often incorporated in the field gradient sensor to bias the magnetoresistors up to a more sensitive part of their characteristic curve. Figure 9.8 shows the change of resistance of a typical InSb sensor with field and temperature. Different doping levels of the semiconductor material account for the differences in characteristics as well as differences in conductivity—200 $\Omega^{-1}cm^{-1}$ for D material and 550 $\Omega^{-1}cm^{-1}$ for L material.

9.4. SPIN-DEPENDENT MAGNETORESISTIVE MATERIALS

Magnetoresistance (the percent change in resistivity with respect to its lowest value under application of a range of magnetic fields) and the magnetic field required to achieve the full range of the material's resistance (saturation field) are two of the most important material properties for GMR material applications. The rate of change of resistance with magnetic field (sensitivity) determines the signal generated by a magnetic field when the field is less than the saturation field.

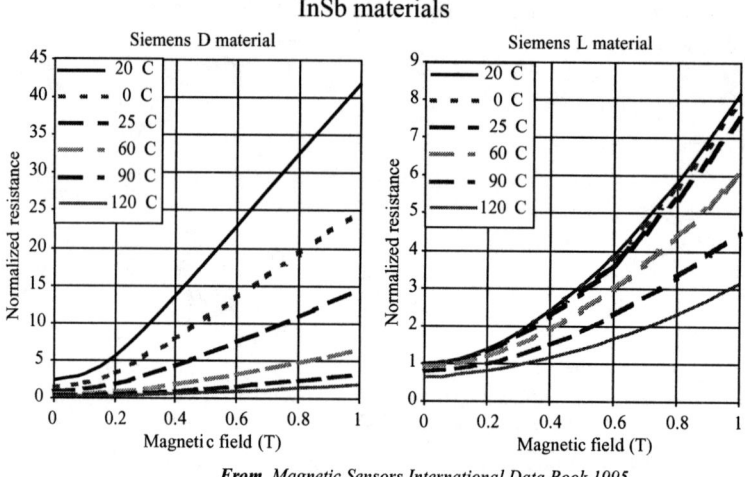

From Magnetic Sensors International Data Book 1995, (Siemens Aktiengesellschaft, Munchen, 1994).

Figure 9.8. Resistance versus field for an InSb magnetoresistor at different temperatures. Resistance is normalized to the resistance at zero field.

Table 9.1 compares some typical magnetoresistance, saturation field, and sensitivity values for a range of GMR materials. A magnetic sandwich[9] is a spin valve[10] without a pinning layer. Granular films[11] are thicker films made of immiscible magnetic and nonmagnetic conductors. SDT structures[12] use a different conduction mechanism from those materials commonly called GMR materials, but for purposes of discussion they are treated as GMR materials in this paper.

Comparison between multilayers and granular structures shows an advantage in GMR and sensitivity for multilayers. It could be postulated that the granular structure is easier to fabricate, but no serious obstacles have been encountered in making multilayers, especially at the so-called "second peak" thickness for the nonmagnetic conductor in the multi-layers.[13] With computer-controlled sputtering equipment designed for good thickness uniformity, multilayers are relatively easy to manufacture, and hence there is no obvious need for the granular structure at this point in time.

Colossal magnetoresistance (CMR) can give very large changes in resistance with applied magnetic fields,[14] but would probably be difficult to apply because

Table 9.1. Comparison of Typical Properties of GMR Materials

	% Magneto resistance	Saturation Field (Oe)	Sensitivity (% Oe^{-1})	Comments
AMR	2	5–20	0.4	
Multilayer	10–80	100–2000	0.1	Hysteresis
Granular	8–40	800–8000	0.01	Hysteresis
Spin valve	5–10	5–50	1.0	Thermal
Sandwich	5–8	10–40	0.5	
CMR	100	1000	0.1	High TCR
Tunneling	10–25	5–25	2.0	High R

of operating temperature (generally, well below room temperature), and even more so because of an extremely high temperature coefficient of resistivity (TCR). This latter property would make it difficult to compensate for temperature shifts so as to distinguish between temperature and magnetic field in a practical application. With CMR properties as they now stand, the material would not find widespread use.

The other materials listed in Table 9.1 are finding use in practical devices. AMR materials (permalloy thin films) have good sensitivity at fields below about 10 Oe, but will yield a smaller signal than spin valve, sandwich, or tunneling materials at fields higher than about 5 Oe. The sensitivity of multilayers is not as high as for the other three, but there are applications which use several hundred Oerstead field, and in these the other GMR structures and the AMR structure would be saturated, making the multilayer a better choice based on output in higher field applications.

There are several important thermal characteristics of magnetoresistive materials. TCR has already been mentioned. (GMR and SDT devices have satisfactory TCRs on the order of 1500 and -700 ppm/°C, respectively.) Short-term (a few hours) thermal stability is important in fabrication of devices, where modern processes frequently exceed 200°C. Where integrated circuits are combined with the GMR materials, the temperatures can be even higher. Long-term (thousands of hours) stability is essential to product reliability. Except for automotive and certain industrial applications, operating temperatures rarely exceed 125°C. The magnetic sandwich is stable to operating temperatures above 200°C. Thermal properties of spin valves have not been as desirable, but have recently improved greatly with the use of iridium manganese[15] and nickel manganese[16] as antiferromagnetic pinning layers, and hence operating temperature is less of a concern for spin valves than it was a few years ago.

Magnetostriction has not been reported as a problem in GMR materials, but there is potential for problems, particularly in devices with small features. Most reported alloys are nominally non-magnetostrictive in bulk, but magnetostriction in thinner layers may present problems with shifting magnetic properties. This area is worthy of more research than has been reported to date.

The resistance of a device is also an important parameter. In GMR devices, the resistance is a function of the length and width of the device, with the sheet resistivity usually in the 2–20 Ω/square range, where its number of squares is its ratio of the length to width. For very small devices which are constrained to be only a few squares (as in memory and read heads), the resistance will be on the order of 10–20 Ω. Normally, magnetic field sensors do not have this size constraint. SDT devices, on the other hand, have a resistance inversely proportional to device area. Typical resistances vary considerably depending on processing parameters, and values ranging from 10^4 to 10^9 Ω μm^2 are common. Even resistances at the lower limits of this range will result in a square micron device with 10^4 Ω resistance, and several orders of magnitude higher are easily attainable. Thus, high resistance, low power applications could be good ones for SDT devices.

SDT devices have two potential problems for sensor applications: (1) magnetoresistance declines for voltages across the device exceeding approximately 0.1 V, with an irreversible breakdown which occurs at about 2 V and (2) uncertainty in the ruggedness of very thin barriers with respect to elevated temperatures.

9.5. MAGNETORESISTIVE SENSOR DESIGN

Magnetoresistors are used in various configurations to produce sensors. The specific configuration used depends upon the characteristics of the magnetoresistive material used and the requirements of the application. For some materials, the small size of the change in resistance is most important, for others compensation for the effects of temperature.

9.5.1. Resistors

The simplest magnetoresistor configuration is a single resistor. A constant current is passed through the resistor, and the voltage across the resistor is measured. As the resistance changes due to magnetic field, so does the voltage. This scheme works best for magnetoresistors with a large change in resistance such as semiconductor magnetoresistors for which the change in resistance can be several hundred percent. For most spin-dependent magnetoresistors, the total change in resistance is 10% or less. For such materials, the base voltage must be subtracted to get a meaningful change. Most alloys have positive thermal coefficients of resistance. As the temperature increases so does the resistance. In order to differentiate between changes in voltage due to temperature and those due to magnetic fields, temperature compensation must be added to the circuit. One GMR sensor on the market is a single resistor whose resistance changes with the angle between an applied field and the axis of the resistor.[17] The data sheet, however, recommends using the sensor in a half-bridge or full-bridge configuration.

9.5.2. Resistor Pairs

There are various ways to use pairs of resistors to solve the problem of a small change in resistance. A differential pair of resistors uses two equal resistors with two equal current sources. Only one of the resistors is a magnetoresistor. A differential amplifier produces an output which represents the change in resistance of the magnetoresistor. Temperature compensation may still be necessary unless the temperature coefficients of resistance are the same for the two different resistors. A variation on this approach which reduces the sensitivity to temperature is to use two identical magnetoresistors and shield one from the magnetic field.

9.5.3. Half-Bridge

Semiconductor magnetoresistors are often used in a half-bridge configuration with the two resistors located a fixed distance apart. This configuration is used when non-uniform fields are anticipated such as in gear-tooth sensing. With a fixed voltage source, the bridge center voltage is either greater than half the applied voltage or less than half the applied voltage depending on whether the upper resistor experiences a lower field than the lower resistor or vice versa. Both resistors change with temperature so the voltage stays centered about one-half the supply voltage. However, the sensitivity will change with temperature.

9.5.4. Wheatstone Bridge

Although originally designed to be used in a balanced configuration with changes in resistance nulled out, the Wheatstone bridge has been extended by using it as an unbalanced bridge with an output proportional to the imbalance. This configuration is useful for sensor resistors with small changes in resistance such as strain gages so that the bridge is never very far out of balance. For magnetoresistors, a bridge can be fabricated by using two active magnetoresistors in diagonally opposite legs and two shielded magnetoresistors in the other two legs. Good temperature compensation is achieved, especially with a constant current source. The output as a fraction of the applied voltage is approximately one-half the fractional change in the magnetoresistors. A bridge with an output which is the same fraction of the input voltage as the change in resistance can be fabricated in instances in which two of the magnetoresistors on diagonally opposite legs of the bridge increase in resistance at the same time that the other two resistors are decreasing in resistance. This fortuitous circumstance is not always possible.

9.5.5. Anderson Loop

A recent innovation for resistive sensors is known as the Anderson loop.[18] The Anderson loop uses a current source to pass a current through a sensing resistor and a reference resistor in series. The voltage across each resistor is connected to a four-terminal dual differential subtractor. This subtractor provides an output proportional to the difference between the voltages across the two resistors. Different gains for the two channels can be used in the subtractor. With magnetoresistors, the Anderson loop could be implemented by using a shielded magnetoresistor as the reference resistor with a matched temperature coefficient.

9.6. ANISOTROPIC MAGNETORESISTIVE SENSORS

9.6.1. AMR Sensor Design

The nonlinear curve of resistance versus applied field for an AMR magnetoresistor shown in Fig. 9.9 can be linearized and made bipolar in several ways. The element can be biased by using permanent magnet or the magnetic field from a current. The response curve near the center of the slope is fairly linear—especially for small excursions. A far more clever method is to design a full sensitivity Wheatstone bridge in which all four elements are active. The elements are arranged so that at zero applied field, the current flows at an angle to the easy direction of magnetization—preferable about 45°. This arrangement places the operating point about half way down the slope of the curve shown in Fig. 9.9. An applied field in one direction increases the angle between the current and the magnetic moment decreasing the resistance. A field in the opposite direction decreases the angle increasing the resistance.

A non-zero angle between the current and the magnetic moment can be achieved with thin-film structures produced by lithography. Thin film slanted conductors called "barber poles" serve as equipotentials and cause the current to flow at a 45° angle to the stripe. This structure is shown in Fig. 9.10. A long serpentine or meander pattern

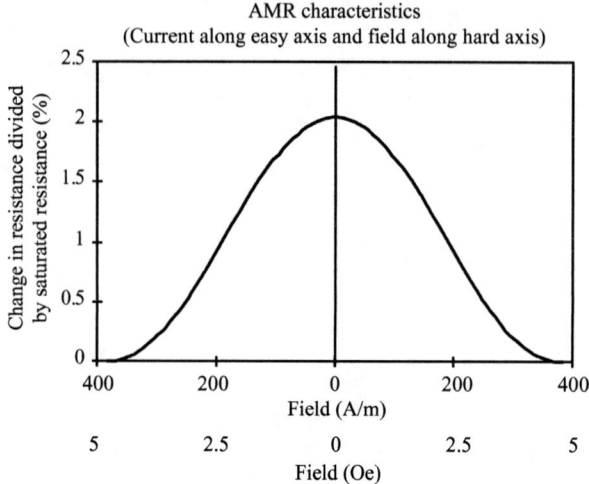

Figure 9.9. Change in resistance normalized by saturated resistance for 20 nm thick permalloy film versus field applied along the hard axis. The current is applied along easy axis.

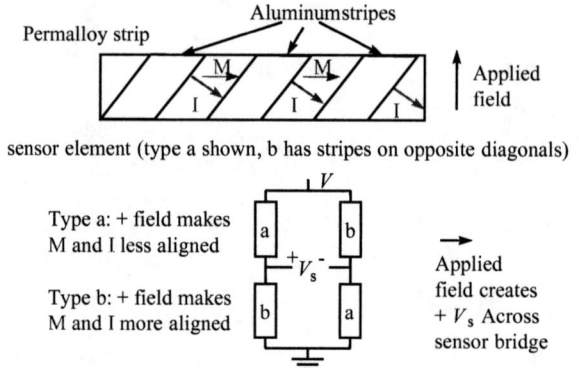

Figure 9.10. AMR sensor design by using plated barber pole conductors to force the current to flow at a 45° angle to the magnetization vector.

of permalloy film is used to achieve AMR sensor elements of sufficient resistance. Half-bridge sensors consist of two sensor elements biased in opposite direction so that the resistance of one element increases while the another decreases thereby increasing the sensitivity of the device. Full Wheatstone bridges with four sensitive elements properly biased give maximum sensitivity and zero differential output at zero field.

9.6.2. AMR Sensor Characteristics

The output of a Wheatstone bridge sensor using AMR elements biased as described earlier is shown in Fig. 9.11. Due to the inherent bi-directional

MAGNETIC SENSORS

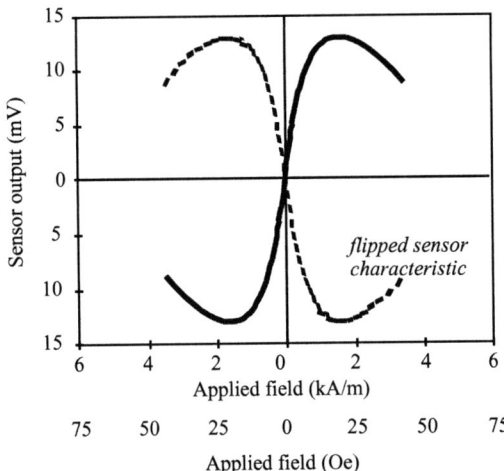

Figure 9.11. Output characteristics of an AMR Wheatstone bridge sensor with current direction 45° to the stripe and the field perpendicular to the stripe. Both the normal and flipped curves are shown.

characteristic of the magnetization of ferromagnetic materials, the possibility exists that a strong field can reverse any bias field which was built into the sensor. The reversed characteristic curve for a "flipped" sensor is shown as a dashed curve. To overcome the flipping problem, many sensors include a flipping coil which is periodically energized to ensure that the sensor is on its proper characteristic curve.

AMR sensors have good sensitivity at low fields. They are limited to maximum fields of a few kiloamperes per meter (a few tens of Oersted). Since they are thin magnetic films, the frequency response is excellent—up to 1 MHz. Reorientation of the direction of magnetization in a metal is a much faster process than moving charge carriers in a semiconductor. The thin film also gives the sensor immunity to fields perpendicular to the film. The large demagnetization factor associated with a thin film prevents any significant field in that direction. The direction of the applied field is usually across the relatively wide permalloy stripes necessary for the 45° orientation. This orientation allow the sensor to measure higher fields since relative small fields parallel to the easy direction of magnetization will saturate the stripe. However, this orientation allows the sensor to be influenced by transverse fields. Transverse fields can both affect the sensitivity and can flip the sensor's response curve.

9.7. GIANT MAGNETORESISTIVE SENSOR DESIGN

9.7.1. GMR Sensor Design

To date, the best utilization of GMR materials for magnetic field sensors has been in Wheatstone bridge configurations, although simple GMR resistors and GMR half-bridges can also be fabricated. A sensitive bridge can be fabricated from four photolithographically patterned GMR resistors, two of which are active elements. These resistors can be as narrow as 2 μm allowing a serpentine 10 kΩ

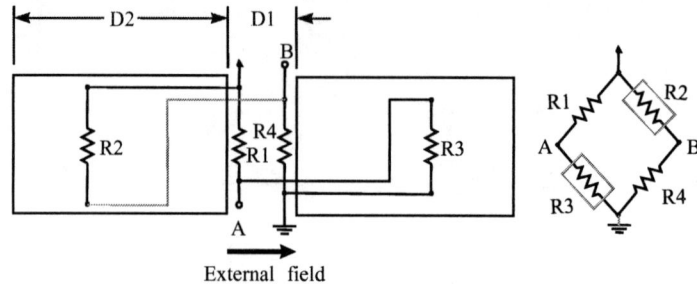

External field
Flux concentration in gap is approximately D2/D1

Figure 9.12. Configuration of GMR resistors in a Wheatstone bridge sensor. Flux concentrators are shown: D1 is the length of the gap between the flux concentrators, and D2 is the length of one flux concentrator.

resistor to be patterned in an area as small as 100 μm × 100 μm. The very narrow width also makes the resistors sensitive only to the component of magnetic field along their long dimension. Small magnetic shields are plated over two of the four equal resistors in a Wheatstone bridge protecting these resistors from the applied field and allowing them to act as reference resistors. Since they are fabricated from the same material, they have the same temperature coefficient as the active resistors. The two remaining GMR resistors are both exposed to the external field. The bridge output is therefore twice the output from a bridge with only one active resistor. The bridge output for a 10% change in these resistors is approximately 5% of the voltage applied to the bridge.

Additional permalloy structures are plated onto the substrate to act as flux concentrators. The active resistors are placed in the gap between two flux concentrators as is shown in Fig. 9.12. These resistors experience a field which is larger than the applied field by approximately the ratio of the gap between the flux concentrators, D1, to the length of one of the flux concentrators, D2. In some sensors, the flux concentrators are also used as shields by placing two resistors beneath them as is shown for R2 and R3. The sensitivity of a GMR bridge sensor

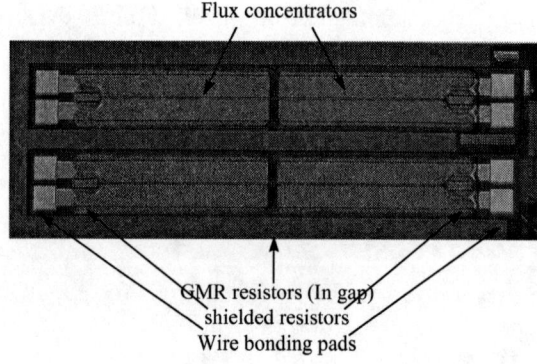

Figure 9.13. Photomicrograph of two GMR bridge sensors with flux concentration ratios of 16:1. Sensors are 3.4 mm × 0.44 mm.

MAGNETIC SENSORS

can be adjusted in design by changing the lengths of the flux concentrators and the gap between them. In this way, a GMR material that saturates at approximately 300 Oe can be used to build different sensors that saturate at 15, 50, and 100 Oe. A photomicrograph of two bridge sensors with flux concentration ratio of 16:1 is shown in Fig. 9.13. The size of each sensor is 3.4 mm × 0.44 mm and they can be packaged in 8-pin SIOC packages.

9.7.2. GMR Sensor Characteristics

An example of the output from a GMR bridge sensor is shown in Fig. 9.14. The curves in the left figure represent the output of the bridge at various temperatures with a constant voltage applied to the bridge. The curves on the right figure represent the output of the same bridge when a constant current supply is used. The maximum output still decreases with temperature, but not by a large amount since the voltage across the bridge automatically increases as the bridge resistance increases with temperature. The output at any constant field which is less than saturation remains almost constant. These curves also show the hysteresis, especially near zero when the sensor is swept to saturation in both directions. The hysteresis is reduced when magnetic field excursions are limited to one side of the origin.

9.7.3. Integrated GMR Sensors

Smart sensors with sensing elements and associated electronics are the latest trend in modern sensors. GMR materials are deposited on wafers with sputtering systems and can, therefore, be directly integrated with semiconductor processes. The small size sensing elements fit well with the other semiconductor structures and are applied after most of the semiconductor fabrication operations are complete. Due to the topography introduced by the many layers of polysilicon, metal, and oxides over the transistors, areas must be reserved with no underlying transistors or connections. These areas will be occupied by the GMR resistors. The GMR

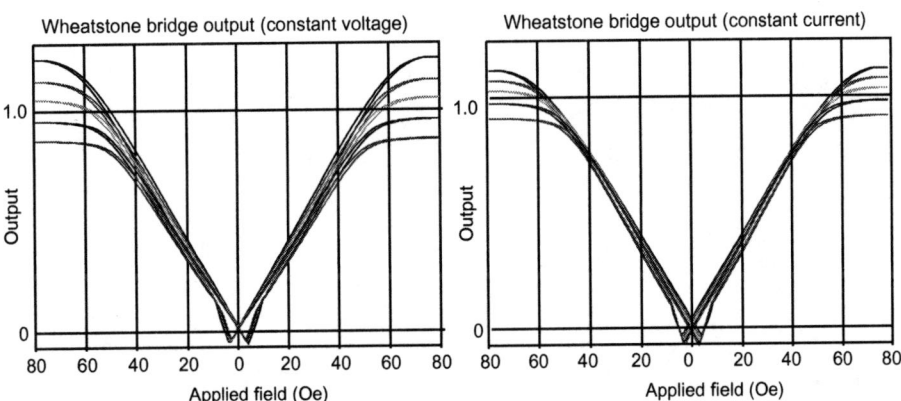

Figure 9.14. Output of a GMR Wheatstone bridge versus external field at various temperatures. The left figure represents a bridge driven with a constant voltage, the right figure, a bridge driven with a constant current.

materials are actually deposited over the entire wafer, but the etched sensor elements remain only on these reserved, smooth areas on the wafers.[19] Functions included in an integrated sensor include regulated voltage or current supplies for the sensor elements; threshold detection to provide a switched output when a preset field is reached, amplifiers, logic functions including divide-by-2 circuits; and various options for outputs. Using such elements, a two-wire sensor can be designed which has two current levels—a low current level when the field is below a threshold and a high current level when the field is above the threshold.

On-board sensor electronics can increase signal levels to significant voltages with the least pickup of interference. It is always best to amplify low-level signals close to where they are generated. Converting analog signals to digital (switched) outputs within the sensor is another method of minimizing electronic noise. The use of comparators and digital outputs makes the nonlinearity in the output of sandwich GMR materials of less concern. Even the hysteresis in such materials can be useful, since some hysteresis is usually built into comparators to avoid multiple triggering of the output due to noise.

GMR materials have been successfully integrated with both BiCMOS semiconductor underlayers and bipolar semiconductor underlayers. The wafers are processed with all but the final layer of connections made. GMR material is deposited on the surface and patterned followed by a passivation layer. Windows are cut through the passivation layer to allow contact to both the upper metal layer in the semiconductor wafer and to the GMR resistors. The final layer of metal is deposited and patterned to interconnect the GMR sensor elements and to connect them to the semiconductor underlayers. The final layer of metal also forms the pads to which wires will be bonded during packaging. A final passivation layer is deposited, magnetic shields and flux concentrators are plated and patterned, and windows are etched through to the pads.

An example of an integrated sensor and its output are shown in Fig. 9.15. The figure shows a photomicrograph of an integrated GMR sensor which serves as a digital switch. The switch output as a function of applied field is shown on the right. Such a digital sensor can be used for hydraulic or pneumatic cylinder position sensing.

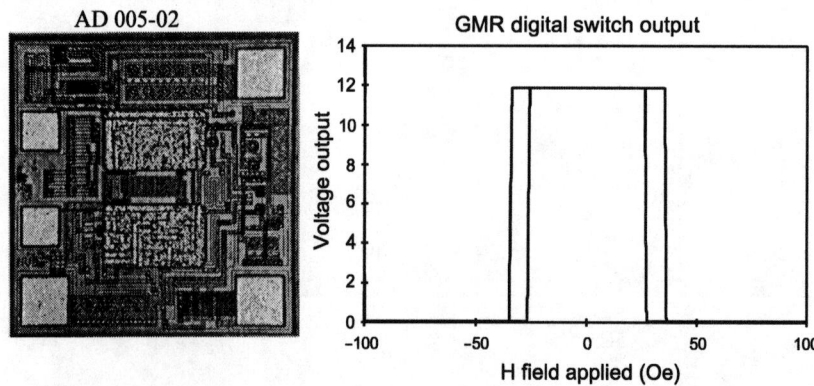

Figure 9.15. Photomicrograph of an integrated GMR sensor that serves as a digital switch. The switch output as a function of applied field is shown on the right.

MAGNETIC SENSORS 467

Digital trim sites on the chip allow the chip to be laser trimmed after manufacture for very accurate setting of the switching points. The plated soft magnetic material which serve both as flux concentrators and shields for the reference resistors are visible in the center of the chip with the sensing resistors between them.

9.7.4. Integrated Coil/GMR Sensor Concepts

Combining GMR material with semiconductors and conductor paths opens the possibility of devices utilizing coils and GMR material. As was mentioned earlier, GMR sensors can detect the magnetic field from a nearby conductor. A current in a wire over a SIOC package produces a field at the sensor of about 1 mOe/mA. However, if the current passes through a conductor on the chip immediately over the GMR sensor, the field at the sensor is increased to 200 mOe/mA. Multi-turn planar coils can produce over 1 Oe/mA at the sensor element. The geometry of an integrated coil/GMR sensor is shown in Fig. 9.16. The plated NiFe shield not only protects the sensor from external fields, but also enhances the field at the sensor produced by the planar coil winding.

It is worth noting that the direction of sensitivity of magnetoresistive materials gives them a substantial advantage over Hall devices with respect to integrating coils on-chip. The configuration shown in Fig. 9.16 would not be compatible with a Hall sensor. Generally, when Hall sensors are used as current sensors, they are placed in a flux gap in a high permeability toroid, which has windings around or a wire through the toroid to carry current and produce magnetic field in the gap.

The high-speed magnetic response of thin-film GMR material, together with the integrated coil/GMR concept, suggests a method of creating a high-speed, linear isolator.[20] A current passing through planar input coils creates a magnetic field experienced by the GMR resistors. Plated shields provided noise immunity from external fields. Over 2.5 kV of isolation has been achieved between the input and output. The small size of these devices is attractive for multi-channel isolation applications and direct fabrication with micro-electronic devices such as analog to digital converters. This concept can be extended to incorporate high-speed relay replacements.

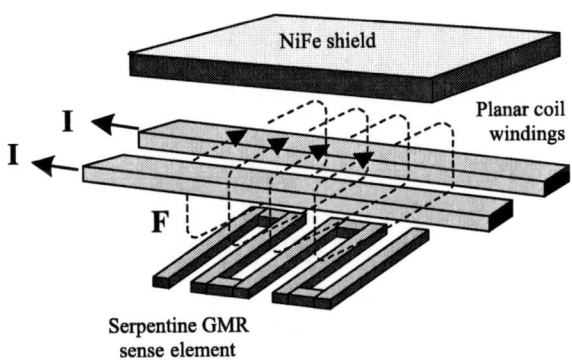

Figure 9.16. Geometry for using integrated coil to produce a field at the GMR sense element.

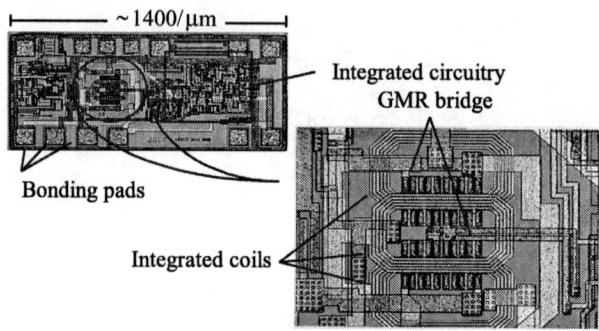

Figure 9.17. Photomicrograph of a chip designed as a linear field sensor by using planar integrated coils for feedback.

9.7.5. Integrated Linear Field Sensor with Feedback

The planar, on-chip coils used to produce a field at the sensor element can also be used in a feedback circuit to provide an integrated linear field sensor. The feedback circuit keeps the GMR material at a constant operating point thereby minimizing hysteresis and non-linearity. The feedback current is a measure of the applied field. Figure 9.17 shows a photomicrograph of a chip designed for integrated linear field sensing by using on-board planar coils. The coils are driven by the on-chip integrated circuit. The characteristics of the sensor are shown in Fig. 9.18. The sensitivity is 1.5 V/Oe with an output rms noise of 5 mV giving a field precision of 3 mOe.

9.8. SPIN-DEPENDENT TUNNELING SENSOR DESIGN

There are both military and commercial applications for magnetic field sensors which can detect magnetic fields at or below the micro-Gauss range. By virtue of

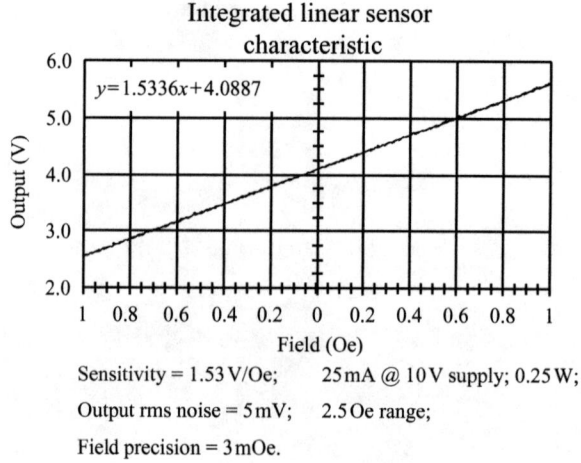

Figure 9.18. Output characteristics of the linear field sensor shown in Fig. 9.17.

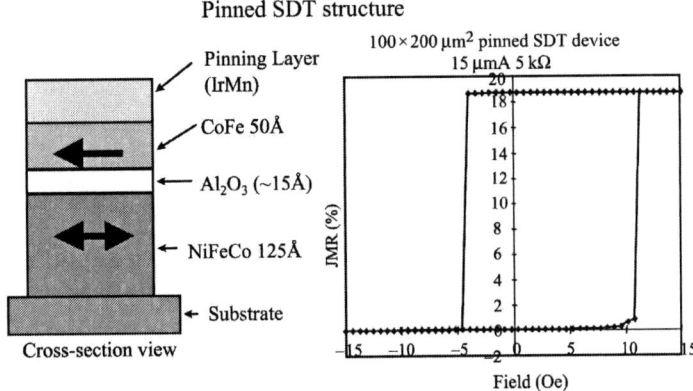

Figure 9.19. Cross-section and JMR of a pinned SDT device.

their sensitivity, SDT devices have the potential to satisfy these application needs. Potential applications include perimeter or intrusion detection, vehicle detection, and nondestructive evaluation. Currently, the market is served by squid and flux gate sensors, but the SDT sensor could be superior in size, power, and cost.

The primary advantage of SDT over other spin-dependent transport materials is sensitivity. A magnetoresistance of 20% or higher with a saturation field of less than 10 Oe has been achieved by a number of workers. Figure 9.19 shows a cross-section and Junction magnetoresistance (JMR) fabricated at NVE. The soft layer is a 125 Å NiFeCo layer, aluminum oxide is the barrier, and a 50 Å layer of CoFe is pinned by an antiferromagnetic layer of IrMn. The minor loop of JMR as a function of magnetic field shows a relatively square loop, offset by approximately 4 Oe Neel (or roughness) coupling of the soft layer to the pinned layer.

It is probably not practical to use the steep slope of this square loop as a low field sensor because the operating point would probably not be stable relative to the magnitude of the low field to be measured. A linear mode of operation is preferred. Later in this section, such a biasing scheme is described which uses on-chip coils. Finding an appropriate mode of operation is only one of the design obstacles which have to be overcome by using the SDT device as a low field sensor.

A second problem is to operate at a higher voltage level (5–10 V) in order to get high magnetoresistance. It is widely found that above about 0.1 V across a tunneling junction, the effective magnetoresistance declines. This can be overcome by making a series connection of a number of SDT devices, allowing about 0.1 V per junction. This technique is also described later in this section.

As was discussed in Section 9.5, there are many advantages to a Wheatstone bridge configuration. The design which is described next will also use that configuration, i.e., four elements wired as a Wheatstone bridge, with each element consisting of a series connection of SDT junctions.

9.8.1. Orthogonal Bias

A low dispersion (uniform easy direction and uniform anisotropy) film is very sensitive to magnetic fields in the easy direction when a hard direction bias field

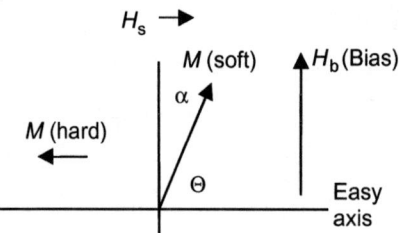

Figure 9.20. Bias field configuration for a high sensitivity SDT magnetic field sensor.

of approximately the anisotropy field is applied because under that bias condition, the anisotropy is approximately cancelled. Further, when the bias field is slightly greater than the anisotropy field, the film is essentially saturated, and noise associated with domain walls can be largely eliminated. The biasing gives the situation that is shown in Fig. 9.20.

The bias H_b is intended to be slightly higher than H_k, and if the torque equation is written in terms of the angle α, the angle that the magnetization makes with respect to the bias field, then

$$H_s * \cos \alpha - H_b * \sin \alpha + H_k * \sin \alpha \cos \alpha = 0$$

Now let $H_b = H_k (1 + \delta)$, and let $H_s/H_k = h_s$. Then,

$$h_s * \cos \alpha - \sin \alpha [1 + \delta - \cos \alpha], \text{ or}$$

$$h_s = \sin \alpha [1 + \delta - \cos \alpha]/\cos \alpha$$

Let α and δ be small, then

$$h_s = \alpha \delta, \text{ or } \alpha = h_s/\delta$$

Now since the effective resistance varies as $\sin \alpha$,

$$(\Delta R/R)(100\%) = (JMR/2)\sin \alpha \sim (JMR/2) \alpha = (JMR/2) h_s/\delta$$

This is the ideal expression for the operation of the sensor. For example, if $H_k = 15$ Oe, $JMR = 20\%$, and $\delta = 0.1$, then the ratio $\Delta R/H_s$ would be about 7% Oe^{-1}, a respectable sensitivity. This mathematical model would suggest that an infinite sensitivity as δ approaches zero. Because of non-ideal behavior of the magnetic film, an infinite sensitivity would not be possible. Two non-ideal characteristics are angular dispersion and anisotropy dispersion. Angular dispersion of the easy axis is typically from about 0.5° to about 2° in "good" uniaxial magnetic films. This is interpreted as meaning that the local easy axes in roughly 90 % of the film is within plus or minus the stated value. Anisotropy constant dispersion (or H_k dispersion) is difficult to measure, and is really probably very small (a few percent). However, a non-uniform bias field can have the same kinds of effect as H_k dispersion, and can be much larger in equivalence unless care is taken to make it small.

Figure 9.21 shows a measured characteristic of a biased junction having similar characteristics as the tunnel junction shown in Fig. 9.19. The slope of the characteristic is roughly 3% Oe^{-1} through the origin, and the hysteresis appears to be small.

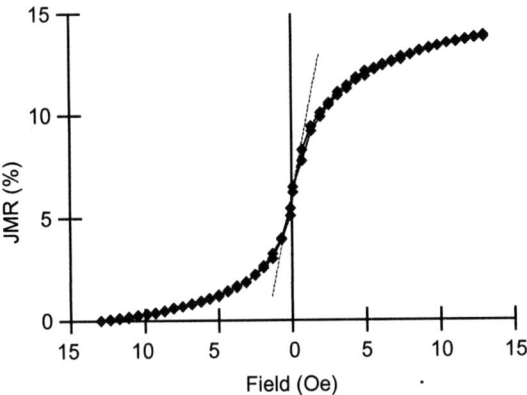

Figure 9.21. JMR as a function of magnetic field.

9.8.2. Series Connection of SDT Elements—Bridge Configuration

Because the best operating voltage for an SDT device is about 100 mV, it is probably necessary to connect a number (N) of these devices together in series. The noise should increase with the square root of N (e.g., with the square root of resistance), while the signal increase is directly proportional to N, and the signal-to-noise ratio will improve as the square root of N. With 16 devices in series, the operating voltage can be brought up to 1.6 V, and with a JMR of 20%, the maximum voltage swing across an SDT device would be 320 mV, and the full swing of bridge voltage with four active bridge elements would be 640 mV. The room temperature noise floor, or Johnson noise limit, is given by

$$v_n = 1.26 \times 10^{-10} (\Delta f * R)^{1/2}$$

where v_n is in rms Volts, the band width Δf is in s^{-1}, and R is in Ohms. Because the signal-to-noise ratio has to be larger than 1 at 10^{-8} Oe, this places a limit on how large R may be for a given signal level. From the previous analysis of sensitivity, with 10^{-8} Oe field, and without flux concentrators, the signal from a biased bridge sensor should be approximately $(0.07 \times 3.2 \times 10^{-8}) = 2.2$ nV. If the bandwidth is 1 Hz, then the resistance value for a signal-to-noise ratio of 1 can be found to be approximately 400 Ω. If that is from 16 devices in series, then the resistance of each element must be less than 25 Ω. Table 9.2 gives the size of a single element and the

Table 9.2. Area of Bridge Elements for 10^{-8} Oe Sensitivity Assuming 7% Oe^{-1} Signal Level

Resistance (MΩ μm²)	Area of Element (μm²)	Area of 16 Element String (μm²)
1	40,000	768,000 or (880)²
0.1	4000	76,800 or (277)²
0.01	400	7680 or (88)²

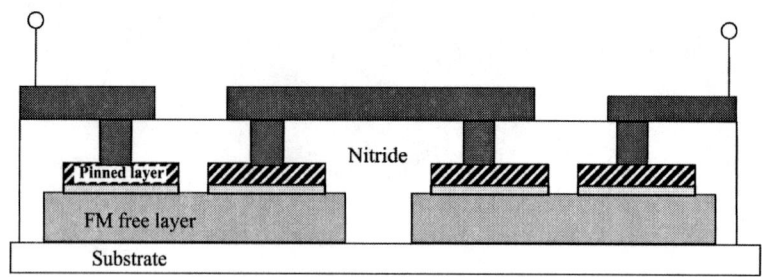

Figure 9.22. Cross-section of series-connected SDT junctions. Note that the "free layer" makes electrical connection between two junctions.

size of a 16 element string (with 20% wiring overhead) assuming several different assumed SDT resistances.

Note that with 1.0 MΩ μm^2 resistance, the area is relatively large for an integrated chip—about 35 mils on a side for each element without allowance for wiring. But for the lower values of resistance it is acceptable, and even quite good for the smallest value (3.5 mils on a side). This points us in the direction of reducing our resistance values to 0.1 MΩ mm^2 or less. There are data in the literature showing that to be feasible.

In actually connecting SDT junctions in series, it is area-efficient to place a pair of hard magnetic contact areas (pinned in the easy direction of the soft layer) in a single soft layer (which is biased in the hard direction). Then, two junctions can be connected in series using the soft layer to make the connection between the junctions, as shown in Fig. 9.22. Then, these pairs can be further connected together to give 16 or 32 junctions in series.

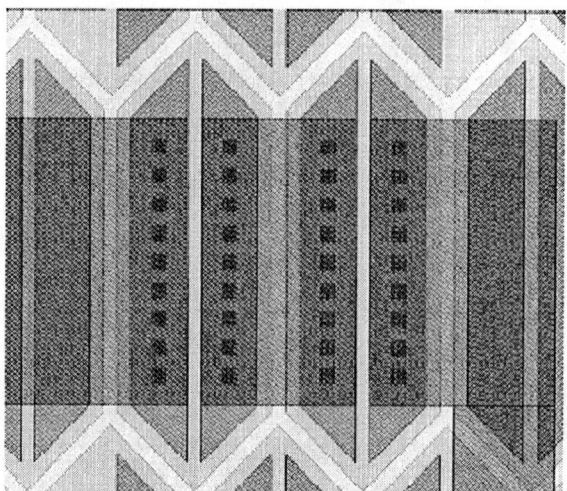

Figure 9.23. Top view of a layout of series connected SDT junctions. There are nine contact to each of the four top pinned layers that are connected in series.

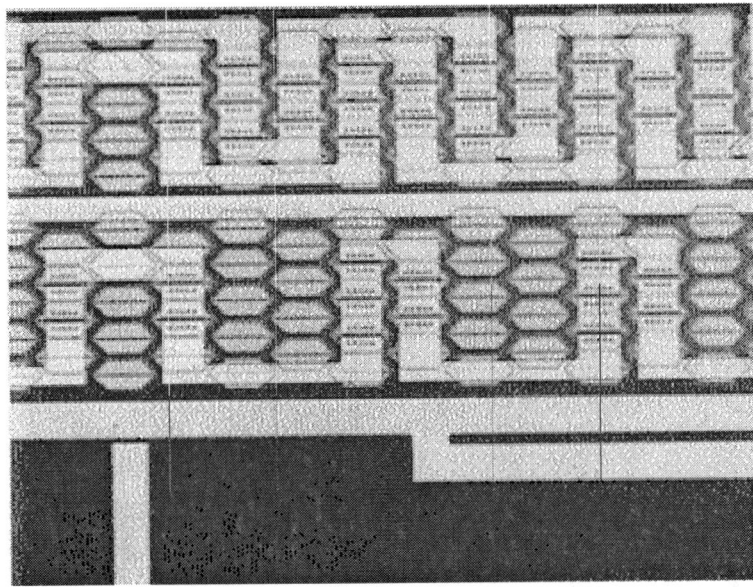

Figure 9.24. Photomicrograph of a portion of a bridge of series-connected SDT junctions.

One method of doing this is shown in Fig. 9.23. By packing neighboring elements in a close-packed arrays, the demagnetizing fields can be reduced.

Figure 9.24 shows a picture of a partially completed and wired bridge at the wafer level. In this design, all four elements of the bridge would contribute to signal.

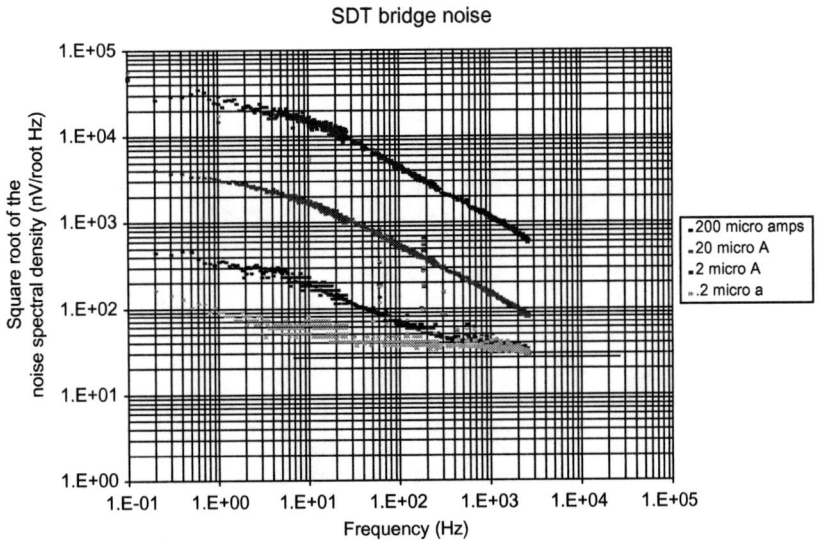

Figure 9.25. Noise spectral density measured for an SDT bridge.

The size of the devices will depend on the level of impedance achieved, but the outlook for achieving the sensor performance in a reasonable area is good.

Figure 9.25 shows rms noise as a function of frequency for a bridge of SDT devices. As the current approaches zero, the noise generated by the SDT elements approach Johnson noise. The remaining noise is $1/f$ noise, which has been demonstrated to be independent of applied magnetic field in SDT bridges of this type. By filtering, the field independent noise can be effectively eliminated, and the noise generated by the bridge should be limited to approximately Johnson noise levels, as was assumed in an earlier estimate of noise in this section

When bias fields are properly adjusted, the sensitivity of the biased bridges should improve from measured values of about 3% Oe^{-1} to about 10% Oe^{-1}. This should make the basic sensor resolution approach 1 pT/rt Hz. If additional improvements are required, then flux concentrator techniques as described in Section 9.7 can be used.

9.9. CONCLUSIONS

Sensors using spin-dependent transport have advantages over other magnetic field sensors (such as Hall effect, variable reluctance, etc.). The resulting sensors are as small as Hall effect sensors with much higher signals, particularly in the field range of 1–50 Oe. They are smaller and more reliable than either reed switches or inductive sensors, and can be combined with integrated circuits to make "smart" sensors. GMR sensors now available in the marketplace should compete well.

The direction of field sensitivity is in the plane of the thin films constituting the GMR or SDT structures, and this gives an inherent advantage for current sensing. Sensor chips with integrated drive coils or packaged chips over traces on printed circuit cards thus see relatively higher usable fields than Hall effect sensors, which are sensitive to fields vertical to the chip's surface. Sensors with integrated feedback currents are feasible with spin-dependent transport, which extend the range and accuracy of the sensor.

SDT could be the technology of choice for sensing very low fields (down to 1 pT) at room temperature. Ongoing development in this area should establish the value of SDT over the next year or two.

As a relatively new technology, there is still a lot of room for innovation and further developments in spin-dependent transport phenomena and related magnetic field sensors.

REFERENCES

1. J. Fraden, *AIP Handbook of Modern Sensors: Physics Designs and Applications* (American Institute of Physics, New York, 1993).
2. S. M. Sze (Ed.) *Semiconductor Sensors* (Wiley, New York, 1994).
3. J. Lenz, Magnetic sensors, *Proc. IEEE* **78**, 973–989 (1990).
4. J. Heremans, Solid state magnetic field sensors and applications, *J. Phys. D: Appl. Phys.* **26**, 1149–1168 (1993).
5. M. J. Caruso, T. Bratland, C. H. Smith, and R. Schneider, A new perspective on magnetic field sensing, *Proceedings Sensors Expo San Jose* (Helmers, Peterborough, NH, 1998) pp. 351–370.

6. D. R. Krahn, A Brief study of magnetic sensors, *Proceedings Sensors Expo Boston* (Helmers, Peterborough, NH, 1995) pp. 75–82.
7. P. Emerald, Low duty cycle operation of Hall effect sensors for circuit power conservation, *Sensors*, Advenstar Communications, Inc., Duluth, MN, **15**, 38 (March 1998).
8. *Magnetic Sensors International Data Book 1995* (Siemans Aktiengesellschaft, Munich).
9. J. Daughton, P. Bade, M. Jenson, and M. Rahmati, *IEEE Trans. Magn.* **28**, 2488 (1992).
10. B. Dieny, V. Speriosu, S. Metin, S. Parkin, B. Guerney, P. Baumgart, and D. Wilhoit, *J. Appl. Phys.* **69**, 4774 (1991).
11. A. Berkowitz, J. Mitchell, M. Carey, A. Young, S. Zhang, F. Spada, F. Parker, A. Hutten, and G. Thomas, *J. Appl. Phys.* **73**, 5320 (1993).
12. J. S. Moodera, L. R. Kinder, T. M. Wong, and R. Meservey, *Phys. Rev. Lett.* **74**, 3273 (1995).
13. S. Parkin, A. Li, and D. Smith, *App. Phys. Lett.* **58**, 2710 (1991).
14. L. Chen, T. Tiefel, S. Jin, T. Palstra, R. Ramesh, and C. Kwon, *IEEE Trans. Magn.* **32**, 4692 (1996).
15. H. Yoda, H. Iwasaki, T. Kobayashi, A. Tsutai, and M. Sahashi, *IEEE Trans. Magn.* **32**, 3363 (1996).
16. S. Mao, N. Amin, and E. Murdock, *J. Appl. Phys.* **83**, 6807 (1998).
17. *Giant Magneto Resistive Position Sensor, Preliminary Data S4*, (Seimens Aktiengesellschaft, Munich, 1998).
18. K. F. Anderson, *The Constant Current Loop: A New Paradigm for Resistance Signal Conditioning*, NASA TM-104260, October, 1992.
19. J. L. Brown, High sensitivity magnetic field sensor using GMR materials with integrated electronics, *Proceedings of the Symposium on Circuits and Systems*, Seattle, WA, January, 1995.
20. T. M. Herman, W. C. Black and S. Hui, Magnetically coupled linear isolator, *IEEE Trans. Magn.* **33**, 4029–4031 (1997).

10

High Speed Magnetoresistive Memories

Arthur V. Pohm*

10.1. INTRODUCTION

In order for a new memory technology to be successful in the general commercial market, it needs to not only provide additional features, but it also must be cost effective. For the present day computer hardware environment, this requires that the technology exploit the billions of dollars that have been invested in silicon technology. It also requires that the new technology have the capability of scaling as the minimum feature size in production lithography continues to diminish. Table 10.1 shows the continuing reduction in feature size for integrated circuits and the anticipated reduction in size in the future as projected by the Semiconductor Research Corporation (SRC).

It is also true that the most cost effective memory systems consist of a hierarchy made of processor latches, cache memory, main memory, disk cache, disks, and tapes, CDs, and other removable storage. As the memories get slower, they get bigger and cheaper. Figure 10.1 shows that in general the price per byte for memory increases as the square root of the speed. Figure 10.1 also indicates that in the past, because of increasing densities, memory cost have diminished 20–30% per year; this trend almost certainly will continue in the future and a successful new technology must be able to scale because a successful memory technology must fall on or below this price performance curve. Fortunately, magnetoresistive, random access, nonvolatile, memories (MRAM) can scale to minimum dimensions of less than 0.1 μm and require few masks.

High speed magnetoresistive memories (MRAM) have a number of desirable properties besides their scaling potential that make them attractive in comparison to other competitive memories such as dynamic RAMs, FLASH, or EPROMS. Table 10.2 lists these desirable features. In addition, Table 10.3 lists more detailed requirements that are necessary for MRAMs of various chip sizes to be successful assuming a fully developed technology. At this time MRAM is not a fully developed technology but could be with time and investment.

In the following material, in order to give a basic understanding of memory fundamentals and the things that limit cells as they are reduced in size, a number of factors will be examined. These will include thermal noise, thermal stability, planar film and plated wire memories, and the effect of scaling on signal level. This material will provide a basis for examining the design of magnetoresistive memories.

---

*Emeritus Professor, Iowa State University and NVE Corporation, Eden Prairie, Minnesota.

Table 10.1. DRAM Technology and Projections

Year of introduction	1988	1993	1996	1999	2002
DRAM size	4M	16M	64M	256M	1024M
Vdd	5	5/3.3	3.3	3.3	3.3
Min feature	0.8 um	0.5	0.35	0.25–0.18	0.15
Metal pitch	2.5 um	1.5	1.0	0.8	0.78
Cell size (um^2)	9	4.5	2	0.9	0.4
Min. feature^2	14	18	22	15	18
Gate oxide (Angstroms)	200	150	100	85	75
Die size (mm^2)	78	98	190	400	415
Mask levels	18	22	25	28	32

10.2. MEMORY FUNDAMENTALS

10.2.1. Noise Limits

In terms of room temperature operation, memory cells are limited by thermal noise and thermal stability as they are reduced in size. Other things such as quantum mechanical uncertainty and chip dissipation usually provide less stringent limits.

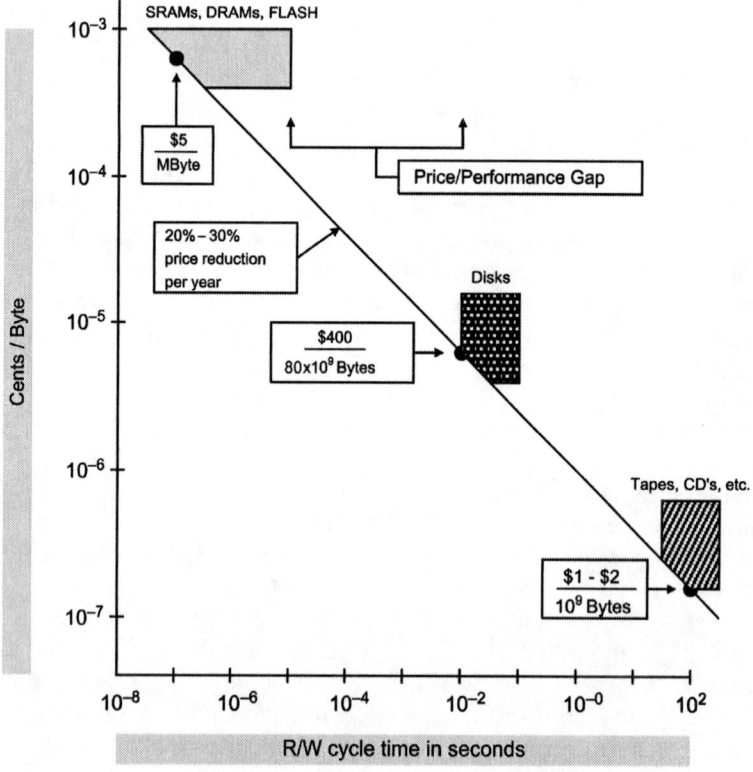

Figure 10.1. Speed vs. cost of memories.

Table 10.2. Key Magnetoresistive Memory Features

1. No wearout
2. Fast read and write (as little as 2–10 nanoseconds)
3. Very dense cells; as small as 6–8 λ^2
4. Only two or three masks beyond standard cmos
5. Fully integrated
6. Nonvolatile
7. Scales to less than 0.1µm

Table 10.3. For a Successful MRAM Memory Technology

1. Must be cost/performance effective
   A. Must be compatible with semiconductor processing
   B. Few additional masks
   C. High element yield
   D. Simple testing
   E. Easy use of spares

2. For large MRAM memories
   A. Dense element arrays; > DRAM, = FLASH
   B. Must scale to at least 0.1µm
   C. Easy testing and sparing

3. For small MRAM memories
   A. Low energy and power consumption
   B. Pad limited die
   C. High performance for energy expenditure

These limits are indicated in Fig. 10.2. In Fig. 10.2 it is noted that as the memory cycle time is reduced, the bandwidth of the sense amplifier circuit must be increased, and as a consequence the thermal noise increases. If the noise voltage increases, then the sense signal power must also increase.

Figure 10.3 shows that for band limited white noise, the noise voltage can be represented as a number (2 × bandwidth) of samples of the form of sin $(x)/x$ which are normally distributed. As noted in Fig. 10.4, if the noise is sufficiently large, it will give rise to an error during the reading process. Shown in the figure is the requirement for the signal voltage to be about 9.15 times the rms noise voltage if the chance of a noise induced error for continuous operation at 10 MHz for one year is to have only a 10^{-5} chance of occurring. (This error rate was picked as about 1/10 the chance that an integrated memory circuit would fail in a year of operation.) From the figure, it is noted that to cause an error, the noise sample must have an amplitude greater than the bipolar signal. The chance of this happening per read operation if the signal is 9.15 times larger than the rms noise voltage is just the complimentary error function of 9.15. As a practical matter to account for unwanted pickup from logic signals, sense signals are usually designed to be 20 times the rms noise voltage.

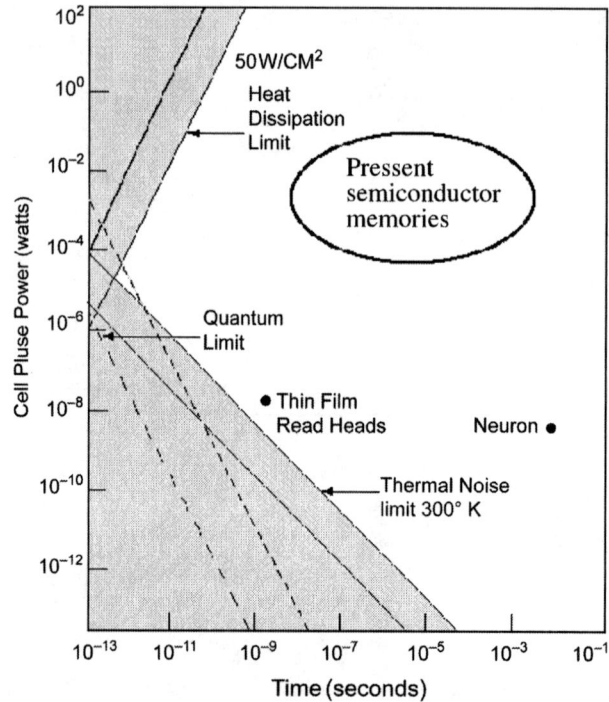

Figure 10.2. Fundamental limits.

10.2.2. Stability Limits

Another factor to consider as the size of the memory elements is diminished is thermal stability. This requires that the each memory state corresponds to an energy well of a certain minimum depth as illustrated in Fig. 10.5. If a coincidence technique in which the cell provides a logical "AND" function is used to write a cell, the well depth for the half selected cells is reduced to about 1/4 of that for a unselected cell. If the well depth for 500 half selected cells during memory operation is 55 KT where K is Boltzmann's constant and T is the absolute temperature, the chance of a

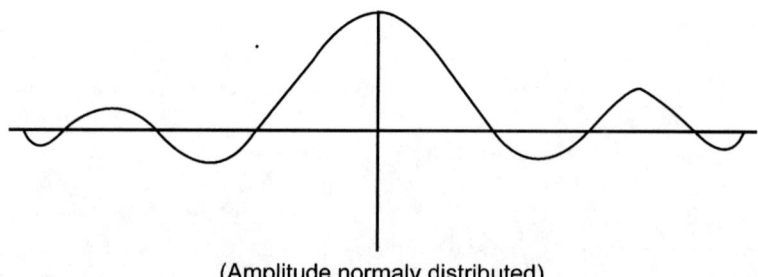

(Amplitude normaly distributed)

Figure 10.3. Thermal noise. For band limited system: Boltzmann's constant K with bandwidth ΔF, Temperature T (°K), and Resistance R (ohms). $\bar{V}_n = \sqrt{4KT\Delta FR}$ (Johnson Noise), no of samples/sec = $(1/2\Delta F)$ and sample of form = $(\sin x/x)$.

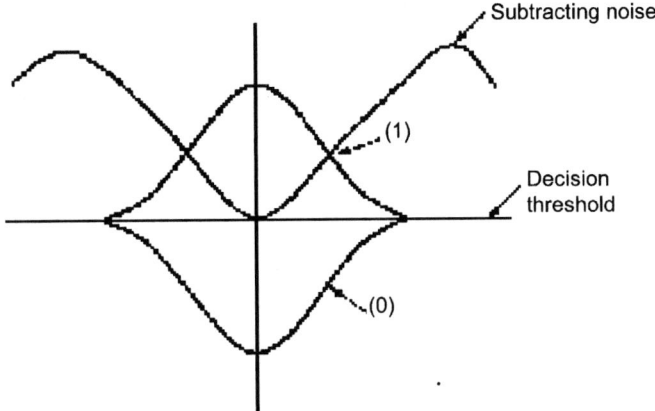

Figure 10.4. Noise induced errors.

$$p\{\text{error}\} = \int_{V_{\text{sig}}}^{\infty} \frac{1}{\sqrt{2\pi}} \exp[-(V^2/2V_n^2)] dV$$

$$= \text{erfc}\left(\frac{V_{\text{sig}}}{V_n}\right)$$

For an error probability of 10^{-19}, $V_{\text{sig}}/V_n \approx 9.1$.

thermally induced failure for a year of continuous operation at 10 MHz is about 10^{-5} assuming a relaxation time of 0.5 nanoseconds and half selection occurring 1/3 of the time. This value is less than the chance of an integrated circuit failure. The well depth threshold is a sharp one though; reducing the well depth by a value of 5 KT increases the failure rate by 150 to an unacceptable level.

10.3. PLATED WIRE AND PLANAR FILM MEMORIES, PRECURSORS

10.3.1. Plated Wire Memory

Plated wire[1] and planar film[2] memories have many magnetic characteristics in common with magnetoresistive memories, particularly the use of the coincidence of a

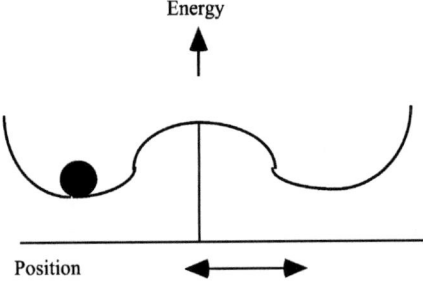

Figure 10.5. Two state mechanical storage. NEED: Half select well depth of 55KT: K = Boltzmann's constant and T = absolute temperature.

word and sense field to write a selected element. These memories used inductive sensing, and as will be shown in the following material, this sensing technique puts a very severe limit on scaling so that magnetoresistive sensing becomes a necessity. Because the fundamental magnetic behavior of a magnetoresistive memory is very similar to that of a plated wire memory, the operation of a plated wire memory will be examined. It is also worth noting that magnetoresistive sensing was first used extensively in bubble[3] and cross-tie wall memories.[4]

As noted in Fig. 10.6, plated wire memories were made by plating at first 125 µm and later 50 µm wire with a 0.5 µm permalloy layer with the easy axis of the permalloy around the wire. The two storage states corresponded to the magnetization wrapping around the wire in one direction or the other. The length of the element was established by the width of the field from the word line. As indicated in Fig. 10.6B, the memory was read nondestructively by applying the word field alone. This field rotated the magnetization in a direction along the wire and changed the flux linkages so that a positive or negative voltage of about 10 mV was induced; this indicated the state of the element. When the word field was removed, the magnetization returned to its original state. A sense line typically might hold 256 elements and a word line typically would wrap around about 200 sense lines in the 2D memory array.

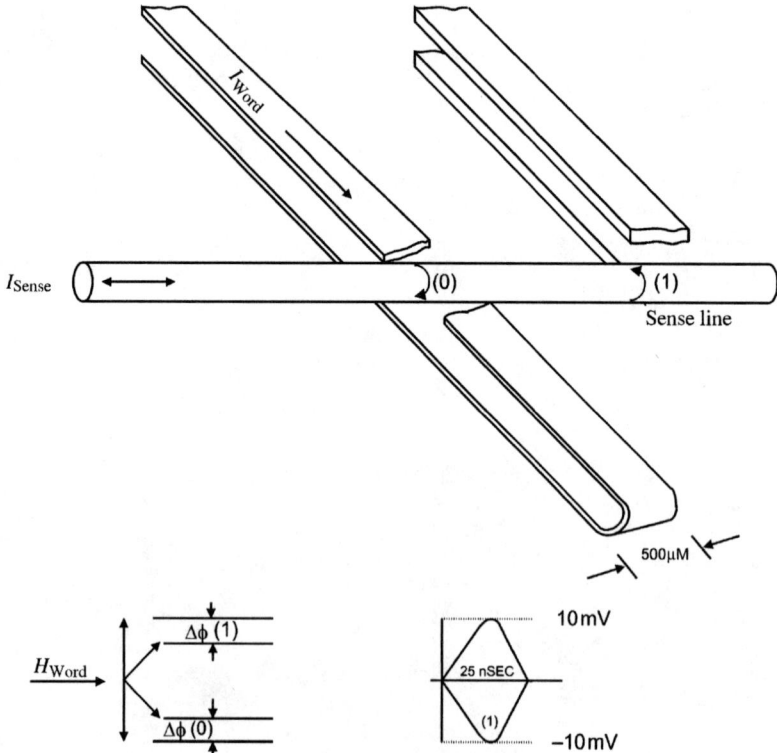

Figure 10.6. Plated wire element.

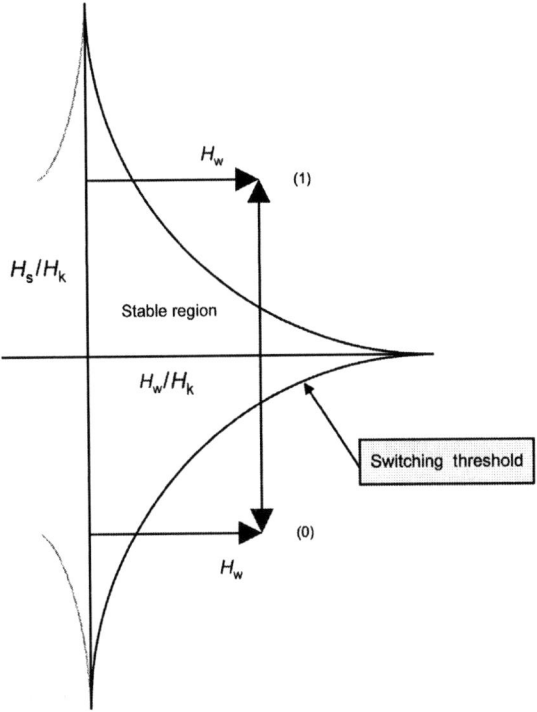

Figureee 10.7. Stoner–Wohlfarth asteroid curve. $H_s H_k = \cos^3 \theta_c$; and $H_w H_k = \sin^3 \theta_c$, where θ_c is the angle at which the magnetization switches. H_k = anisotropy constant, H_s = sense field, and H_w = word field.

Figure 10.7 shows the Stoner–Wohlfarth threshold curve that characterized the writing process. In the figure, H_k is the uniaxial anisotropy field, H_s is the sense field from the write current in the sense line, and H_w is the unipolar word field generated by the current in the word strap. Neither field alone is large enough to exceed the switching threshold, but a combination of both is enough to switch the element into the desired state. The element performs the logical "AND" function necessary for selection. The threshold curve shown is an idealized one; because of dispersion in real plated wire memory elements, a certain minimum sense field is required to write the element even if the word field is equal to the anisotropy field.

If one tried to scale plated wire elements with inductive sensing to the dimensions that are currently used in integrated memories, the signal would be much too low for adequate sensing at an acceptable speed. As an example, consider scaling the wire to 1 μm in diameter with a 0.1 μm coating and with a bit length of 20 μm. This would be a very large element by today's standards. The flux linkages and signal level would be diminished by a factor of $1/(5 \times 25)$ from the ones that are described in Fig. 10.6 and would yield an output of about 80–100 mV. To achieve an adequate signal-to-noise ratio, the memory read cycle time would be limited to about a microsecond. If the elements were shrunk further, the signal level would be totally inadequate.

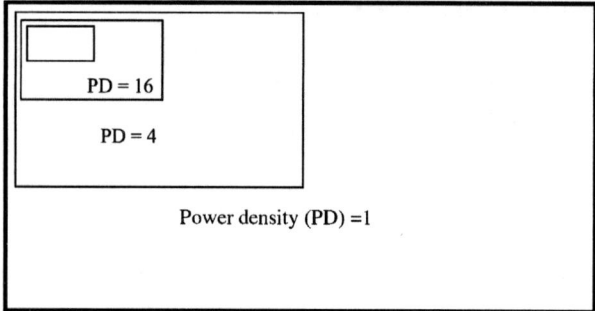

30 ohm resistors: 15 ohms/square

Figure 10.8. Impact of scaling M-R resistors (30 ohm resistors 15 ohms/square).

10.3.2. Magnetoresistive Sensing

Magnetoresistive sensing provides a way of circumventing the problem of diminishing signal level encountered by inductive sensing as elements are shrunk. Figure 10.8 shows that with magnetoresistive sensing, elements can be scaled without loss of signal until power dissipation per unit area becomes a limiting factor. Figure 10.8 shows that if a material has a particular ohms/square value and the dimensions are shrunk, the resistance value stays the same if the dimensions are reduced equally. If the current is maintained at the same level, the voltage drop across the magnetoresistive resistor that forms a memory element stays the same and consequently the sense output remains the same to the first order. As noted in the figure, the power dissipation per unit area increases as the inverse of the scaling factor. For realistic memory designs, however, other factors impose more stringent limits and these only becoming limiting for cells with widths of less than 0.1 µm.

10.4. MAGNETORESISTIVE MEMORY DESIGN FACTORS

10.4.1. M-R Materials

The first magnetoresistive memories have been made from ordinary magnetoresistive material in a sandwich structure as indicated in Fig. 10.9 with 150 Å ternary magnetic layers. The change in resistance in the material depends on the angle between the magnetization and the current and this effect is labeled anisotropic magneto-resistance (AMR). If one attempts to increase the sheet resistance by reducing the thickness of the magnetic layers, surface scattering increases the resistivity and diminishes the M-R coefficient so it is not possible to increase the sheet resistance very much; values are limited to 10–15 ohms per square and the M-R coefficient is limited to about 2%. The fact that the M-R response depends on the direction of the current imposes a serious limitation on the usable memory modes and the magnitude of the output. These factors limit the densities of AMR memories to something less than that can be achieved with typical semiconductor memories.

Giant magnetoresistive sandwich materials (GMR) are made with magnetoresistive coefficients of typically 5–10% but have been made with coefficients as high

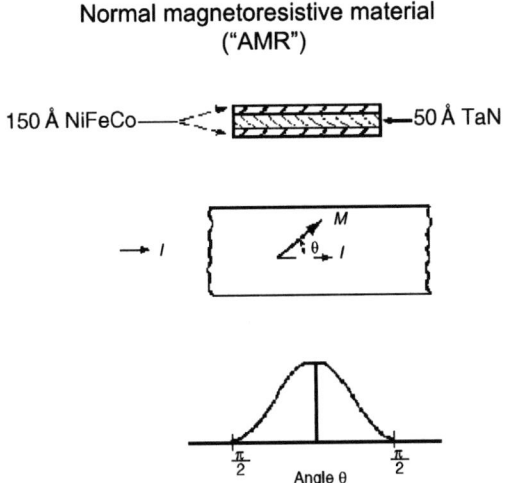

Figure 10.9. AMR sandwich material. $\Delta R/R = C \sin^2\theta$; and $C = 2\text{–}3\%$.

as 17%. As noted in Fig. 10.10, the resistance change depends only on the angles of the magnetizations in the two layers. This property of GMR materials allows for much more efficient memory elements. The magnetic layers in GMR materials can be made much thinner than the layers in AMR material without loss of the M-R coefficient.

Because of the desirable properties of GMR material, GMR memory elements can be made as dense or even more dense than semiconductor DRAM cells. Figure 10.11 shows that in terms of a figure of merit consisting of the product of the M-R coefficient, the ohm per square, and the fraction of the M-R response that generates a sense output, 6% GMR material is 36 times more effective than 2% AMR material. As shown in the figure, this results in a dramatic reduction in cell size for a given performance. A factor of 3 is gained from the M-R coefficient, a factor of 2 from the ohms per square, and a factor of 6 from the effectiveness in generating a sense output.

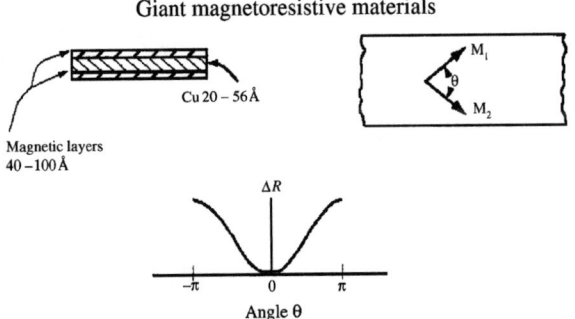

Figure 10.10. GMR sandwich material. $\Delta R/R = C \sin \theta/2$; and $C = 5\text{–}15\%$.

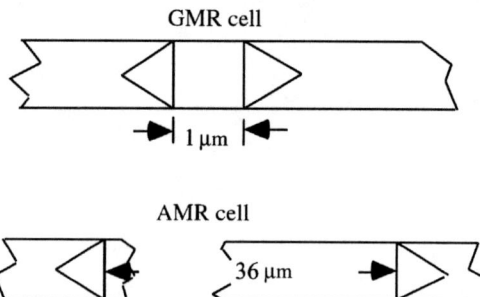

Figure 10.11. Material figure of merit. Definitions: MR = Magnetoresistive coefficient, OPS = Ohms per square, and ZOF = Fraction of MR response used in read. GMR materials (two magnetic layer): MR × OPS × ZOF ($0.06 \times 20 \times 2 = 2.4$). AMR materials: MR × OPS × ZOF ($0.02 \times 10 \times 0.33 = 0.066$); $2.4/0.066 = 36/1$.

10.4.2. Organization

In the design of various GMR memories, compromises can be made between signal level and element size as illustrated in Fig. 10.12. In a memory for minimum noise, about 1/2 of the available voltage is allocated to the string of memory elements along a sense line and the other half of the available voltage is allocated to the current source that provide the sense current. As shown in the figure for a 3.3 V device, 64 elements can be placed on the sense line with an output of ± 1.4 mV or as an extreme, 1 element could be placed on a sense line with an output of 90 mV. The actual number of elements placed on a sense line then becomes a design choice with regard to density and speed.

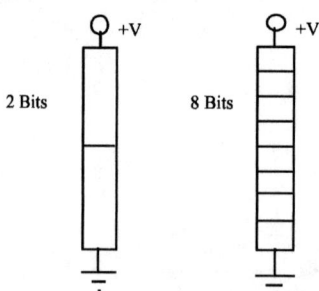

Figure 10.12. Output-cell size tradeoffs. For a 3.3 V system, about 1.5 V can be placed across the sense line. The following calculations show an output/(bit size) tradeoff. A 6% GMR is assumed and a pseudo spin valve memory cell.

$$0.06 \times 1500/64 = 1.4 \text{ mV}$$
$$0.06 \times 1500/32 = 2.8 \text{ mV}$$
$$0.06 \times 1500/16 = 5.6 \text{ mV}$$

$$0.06 \times 1500/1 = 90 \text{ mV}$$

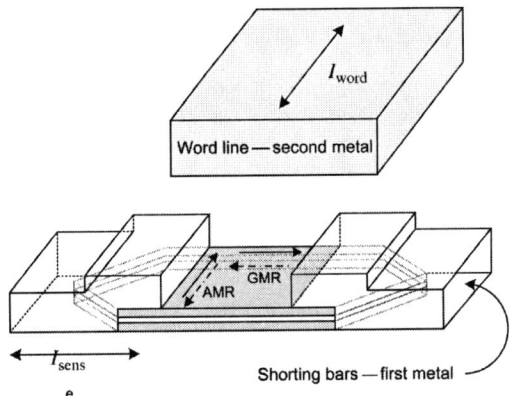

Figure 10.13. Word and sense line arrangements.

Magnetoresistive memories are generally organized as two dimensional (2D) arrays with elements strung along a sense line with a word line insulated from the sense line and perpendicular to it. The word line then provides a field along the sense line. The sense current in the sense line provides a field that scissors the magnetizations in the two layers of the sandwich. As shown in Fig. 10.13 shorting metal connects one element to the next along a sense line. As noted in the figure, it is necessary to appropriately shape the ends of the elements to get satisfactory operation. The rest direction of the magnetization may be along or across the sense line depending on the mode of operation. Figure 10.14A shows that in some memory modes, a winding under the sense line can be used to provide a field across the sense line. Fig. 10.14B shows that in a low density high performance application, it may be desirable to use semiconductor devices for selection, multiple, folded, elements to increase the signal level, and a multi-turn word line to reduce the word currents.

10.4.3. Sensing in M-R Memories

Both AMR and GMR materials have temperature coefficients in the range of 0.1–0.2% per degree centigrade. Typically, when the sense current is turned on, element temperature increases 10–20°. Thus line resistance can increase from 1 to 4% and mask the sense output. To compensate for this, the usual procedure is to have a matching array in which the sense current is turned on but no word current is applied as shown in Fig. 10.15. By use of a scheme that senses differentially, the temperature transient is canceled out to the first order. By interleaving the two arrays, the unwanted capacitive pickup from the active word line to the unselected sense lines can also be canceled out as well.

As indicated earlier, a signal-to-noise voltage ratio of 20:1 is a typical design requirement for a magnetoresistive memory. The noise in the frequency band of interest comes primarily from the memory element sense lines, from the current sources, and from the first stage of the differential amplifier. Figure 10.16 shows a calculation to estimate the signal needed to get a signal-to-noise ratio of 20 for

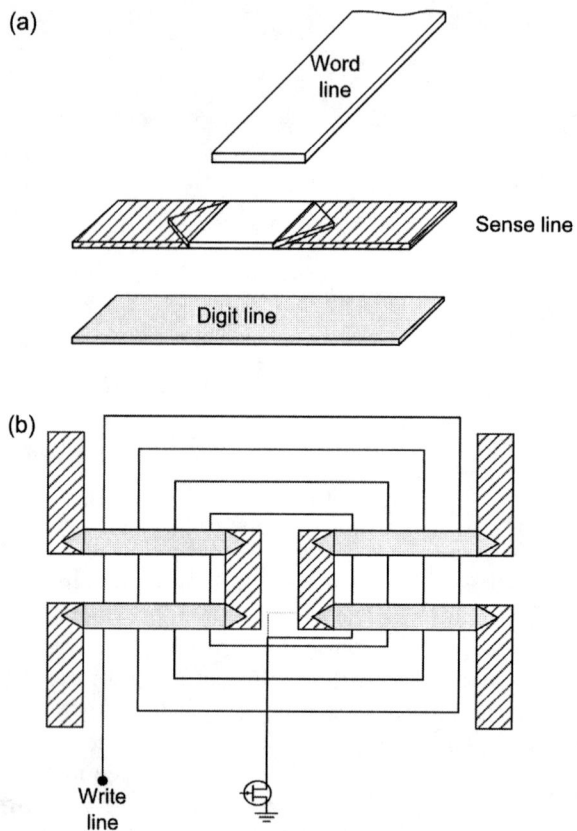

Figure 10.14. Drive arrangements.

a memory system with about a 100 nanosecond read cycle time. In the example, 1500 Ω sense lines are assumed and additional noise contribution from the current sources equivalent to 1500 Ω are assumed. The first stage of the sense amplifier and the multiplexing transistors add an additional 2000 Ω. The Johnson noise for the 20 MHz required bandwidth is 56.6 µV. Two samples are required so that the noise value must be multiplied by a factor of the square root of 2. As indicated, the required output is then calculated to be about 1.6 mV. When the details of a design are carried out, one finds as a practical matter that the required sense bandwidth in Hz is about 2–2.5 times 1/(cycle time).

10.5. HONEYWELL'S ORIGINAL MRAM

10.5.1. Memory Array Structure

The first MRAM memory[5] was produced by Honeywell Inc. and used AMR material; it was conceived before the discovery of GMR material. Figure 10.17 shows the basic memory array structure involves a type of flat wire structure that is

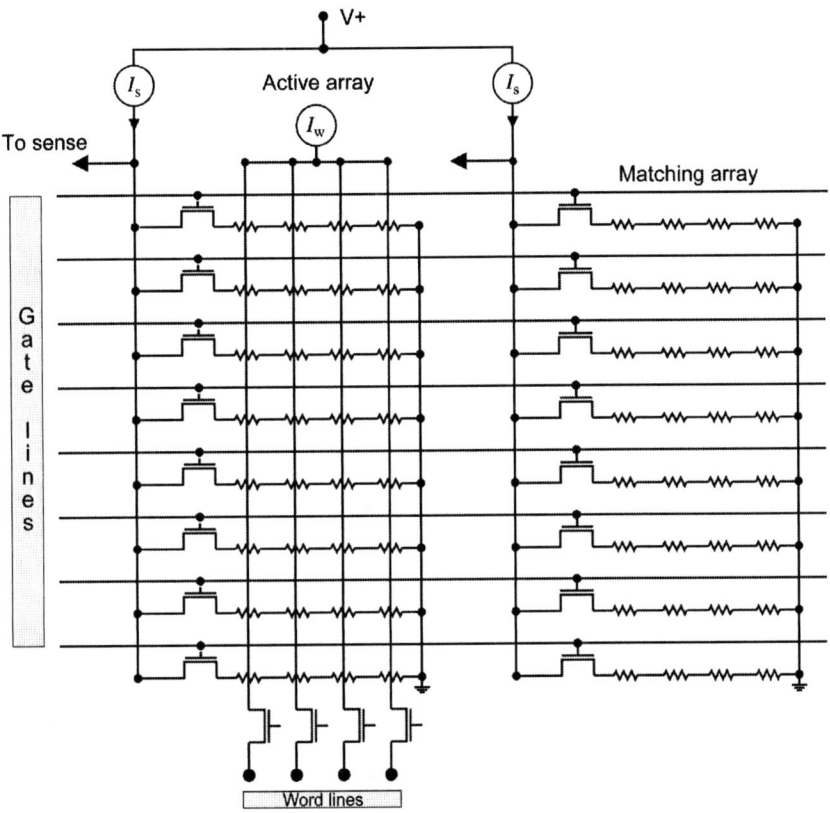

Figure 10.15. Balancing sense arrays. 512 × 32 array: $C_{\text{isolaton}} \approx 2.4$ pFd; $C_{\text{w-s}} \approx 1.6$ pFd.

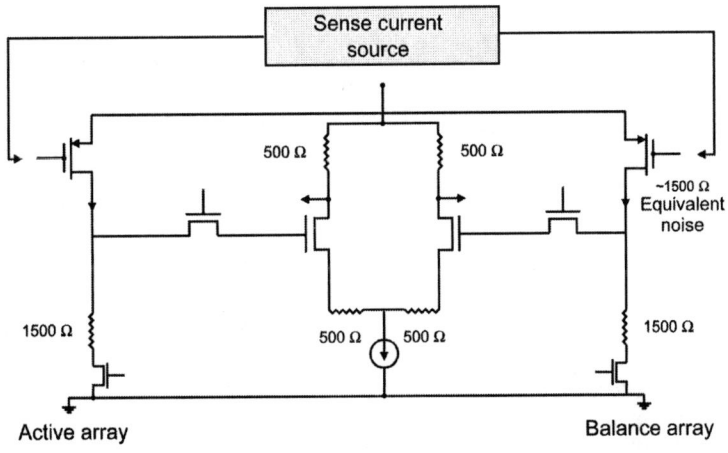

Figure 10.16. Signal-to-noise estimate. $\overline{V_n} = (4KT\Delta FR)^{1/2}$; $V_n = (2 \times 10^{-20} \times 20 \times 10^6 \times (5000 + 2000))^{1/2} = 56.6$ μV; $(2)^{1/2} \times 56.6 \times 20 = \pm 1.6$ mV; Cycle time: $1/\Delta F \times 2$ @ 100 nanosecond.

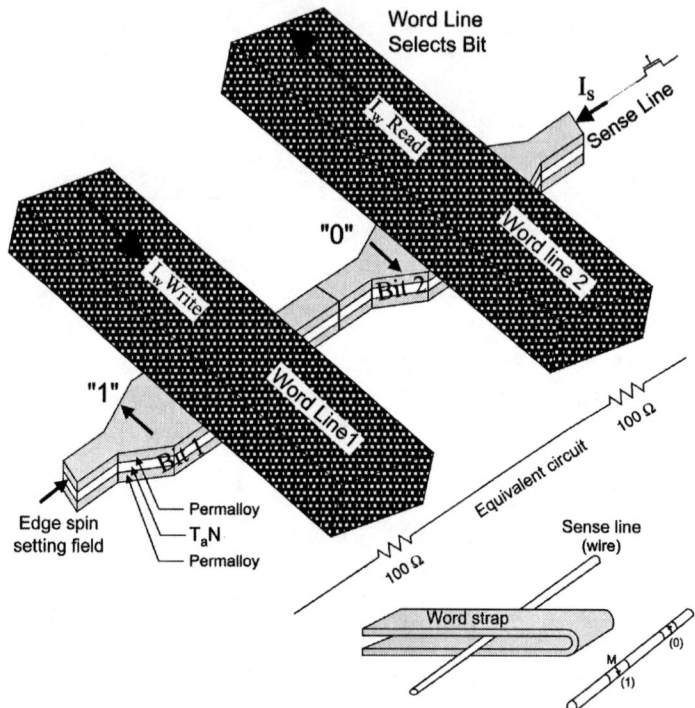

Figure 10.17. MRAM—storage element is magnetic.

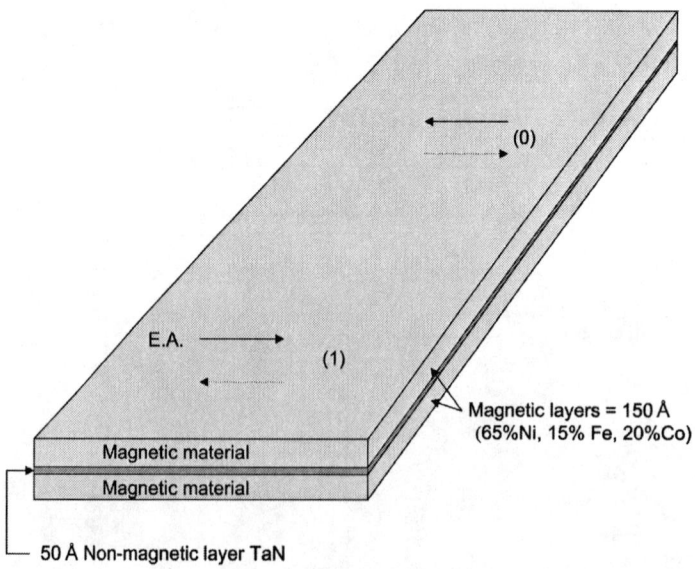

Figure 10.18. Flat wire structure.

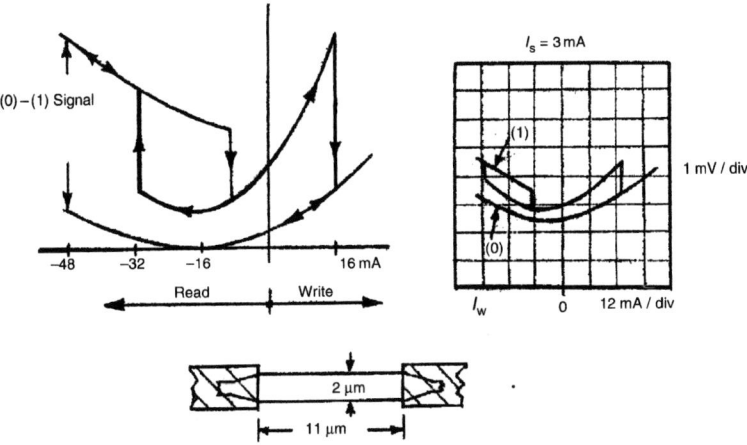

Figure 10.19. AM cell response.

formed with two ternary magnetic layers separated by resistive TaN as shown in Fig. 10.18. Rather than have discrete elements, individual elements were formed by necking the magnetic material and placing shorting metal over the necked region. The active regions of the elements were 2 µm wide and 11 µm long. The cells typically operated with 3–4 mA of sense current. The magnetization along the edges of the cells is set with a large field in the appropriate direction to establish proper operation. The necked arrangement was used to reduce demagnetizing fields.

The cell used one polarity of word current for writing and a reversed polarity for reading as shown in Fig. 10.19. The use of a reverse read current provided a larger signal and better margins than if the read and write word currents were in the same direction as in plated wire memories. As shown in the figure, the cells produced a "one"–"zero" difference of about 1 mV. The write word currents were about +24 mA and the read word currents were about −45 mA. As shown in Fig. 10.20, the cells were characterized by a modified Stoner–Wohlfarth threshold. The thresholds were reduced because of the curling of the magnetization at the edges because of the large local demangetization field. Figure 10.21 indicates that for the successful fabrication of a memory, the write thresholds must be fairly uniform so that a word or sense field alone would not disturb the state of the element. Honeywell Inc. demonstrated that this could be done.

10.5.2. Magnetization Configuration

Figure 10.22 shows that the non-destructive read operation occurs when reading a "one" by winding up the magnetization at the edges. When the read fields are removed, the energy expended in winding up the edge magnetization, returns the magnetization to its original position. Figure 10.23 shows the local direction of the magnetization across an element for a read "one", read "zero," and the two quiescent states. The pinning of the magnetization along the edges can be noted along with the increased winding during the read "one" operation. In Fig. 10.19

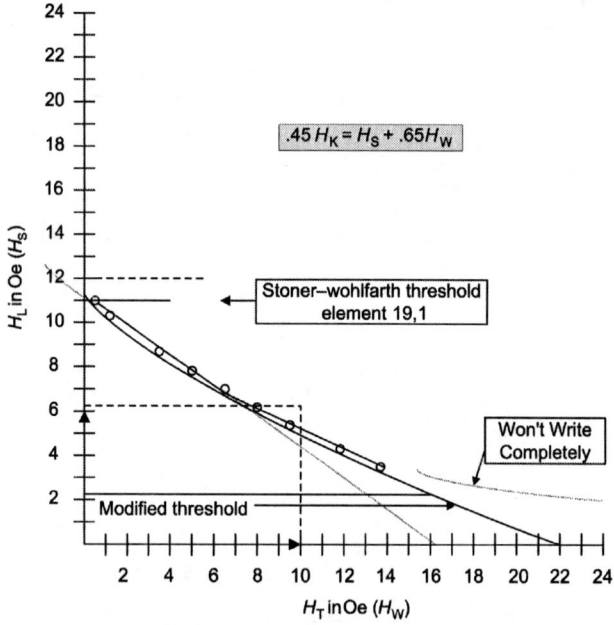

Figure 10.20. Modified Stoner–Wohlfarth threshhold.

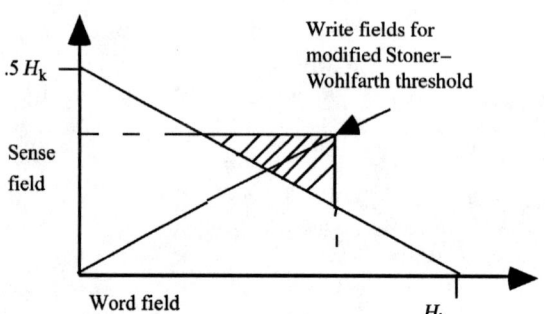

Figure 10.21. Threshold uniformity requirements.

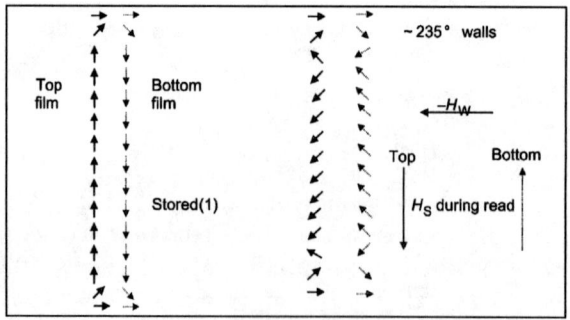

Figure 10.22. Magnetization direction during Read "1".

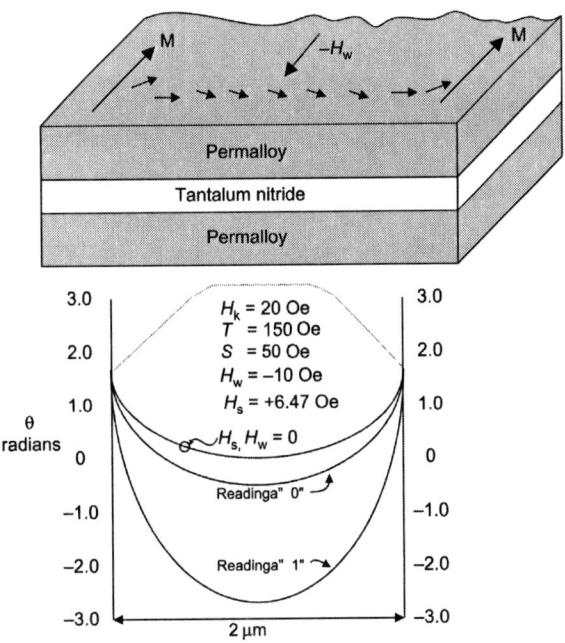

Figure 10.23. M distributions across elements.

switching in the reverse direction that occurs during the reading of a "one" is indicated by the sharpie break in the response as the field is increased in the reverse direction. If the sense field is too small, there is insufficient torque to wind up the magnetization and the sharp reverse transition does not occur. This is illustrated for a few 1.5 × 5.0 µm elements in Fig. 10.24.

The National Institute for Standards and Technology (NIST) made measurements on Honeywell M-R memory elements using a modified scanning electron microscope in which the polarization of the back scattered electrons was used to determine the magnetization direction in an element (SEMPA instrument). The bottom exposures of Fig. 10.25 show the results at a magnification of 10,000 from measuring the X and Y components (across and along the element) of the magnetization from two elements in which the edges were set and one in which an attempt was made to demagnetize the element. The Y component measurement for the set elements clearly shows the edge magnetization being set parallel to the edges even through the necked region. This edge curling region is seen to be about 0.25 µm wide. The top exposures show the normal SEM pictures of the middle and necked regions of the elements.

The middle lower middle exposure shows the formation of double Neel walls in a complicated reverse domain containing both 90 and 180° walls. The 180° walls are seen to be about 0.5 µm wide; it is also seen that walls slant at a 60° angle rather than go straight across the element. In a double Neel wall involving two anti-parallel layers, minimum energy is achieved in a wall if it slants at an angle for narrow stripes.

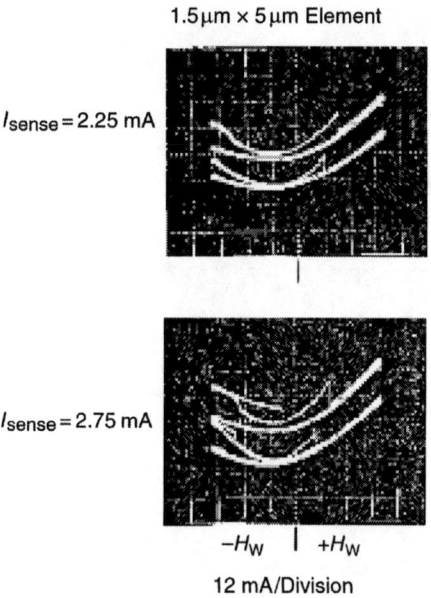

Figure 10.24. Reverse switching thresold with sense field.

10.5.3. Switching Behavior

Figure 10.26 illustrates the nature of the switching process in these AMR, necked, memory elements. The switching starts in the middle and finishes the switching by sweeping towards the ends in typically less than 2 or 3 nanoseconds.

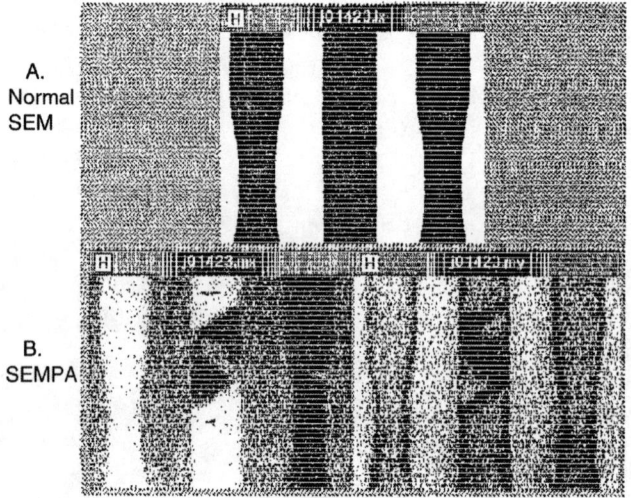

Figure 10.25. Magnetization direction measurement.

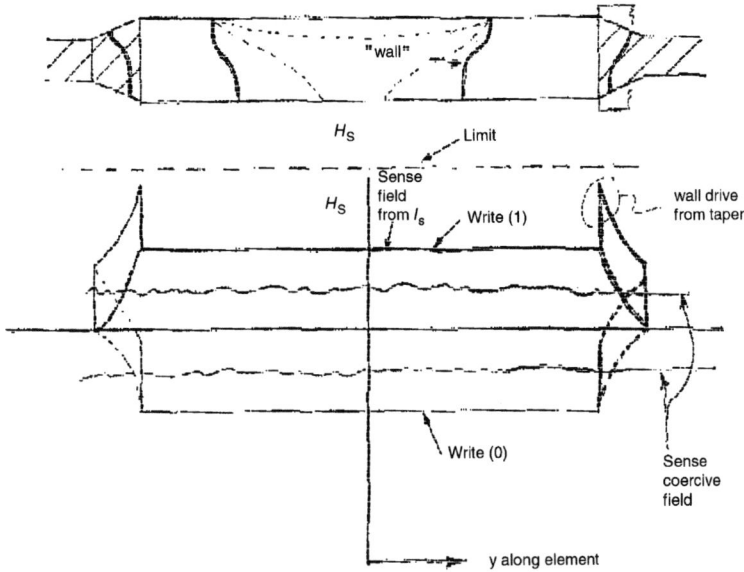

Figure 10.26. Necked element switching.

The tapered ends play a crucial role in element stability as indicated. When the wall resulting from the switching reaches the tapered region, its length diminishes as it enters the tapered region and in doing so diminishes its length and energy. The shorterning metal extracts the sense current so there is a diminishing sense field in the tapered region. This tapering effect provides a driving field that drives the wall into the necked region and prevents it from coming out when the fields are reversed as indicated.

10.5.4. Modeling of Sandwich Quasi-Static Behavior

Because, ideally the magnetizations in the top and bottom layers move with equal magnitude in opposite directions in the top and bottom layers, the demagnetizing field across the element (X demag field) can be described in a simple form. Also because of the lesser thickness of the magnetic layers, the magnetization direction can change only slightly from the top to bottom of a layer and can be characterized by the average value at the middle of the layer. Every element of magnetization in a double layer must statically have zero net torque on it if it is to be in equilibrium. Figure 10.27 indicates the torque equation[6] for the magnetization in the two layers in terms of the angle of rotation of the magnetization away from the uniaxial anisotropy axis. Note that there is a small correction to the separation of the two layers to account for the small internal z field inside the magnetic layers. The Y component of the demagnetizing field typically is approximated by assuming an ellipsoidal shape and calculating the net demagnetizing field from the magnetization in the Y direction.

Figure 10.28 shows the calculated response for a necked, AMR, element for average values of cell parameters and operating currents. Note that the calculated

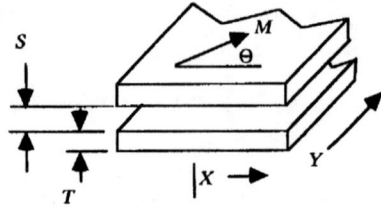

Figure 10.27. Torque equation. Torque $= 0 = M_s H_k \sin\theta \cos\theta + M_s H_s \sin\theta + M_s H_y \cos\theta$ (anisotropy) (sense Field) (word Field) (&Y demag) $+ 2Ad^2\theta/dx^2 - 6.26 M_s\, T\, S'\, (d[d(M_s \cos\theta)/dx]/dx) \sin\theta$ (exchange) (X directed demagnetizing field) $+ \cdots$ small effects. For thinner films : $s' = s + T/4$ $T < 125$ Å; and for thicker films : $s' = s + \tfrac{2}{\beta} T \geq 150$ Å.

$$\left[\frac{\cosh(\beta T/2) - 1}{\beta T \cosh(\beta\, T/2) - \sinh(\beta\, T/2)} \right] \qquad \beta = \left(\frac{4\pi\, M^2}{2A} \right)^{1/2}.$$

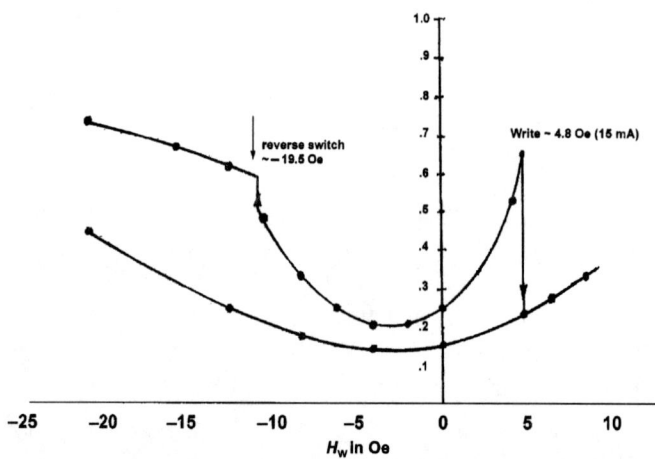

Figure 10.28. Calculated cell response: $w = 2.0\,\mu\text{m}$;, $s = 50\,\text{Å}$, $H_s = \pm H_s = +5\,\text{Oe} - 3.2\,\text{mA}$, $A = 1.2 \times 10^{-6}$, and $M_s = 1000\,\text{EMU}$.

response of an element is very close to that which is experimentally observed. The forward and reverse thresholds are accurately predicted[7] as well as the magnitude of the sense response. In the example shown, the edge magnetization was assumed to be pinned parallel to the edges by a large demagnetizing field. When sandwich layers get very narrow and fields get large, it is necessary to account for the rotation of the magnetization at the edge. This can be done by calculating the short range, large, pinning demagnetizing field arising from the equivalent north and south magnetic poles arising from the normal components of the magnetizations at the edges.

Calculations with the model show that as the width of an element is reduced for a necked element with a transverse easy direction, the fraction of the M-R response, which gives a useful output, diminishes. Figure 10.29 shows that a 2 μm element gives a zero–one difference of about 1/3 the total M-R response. If the element width

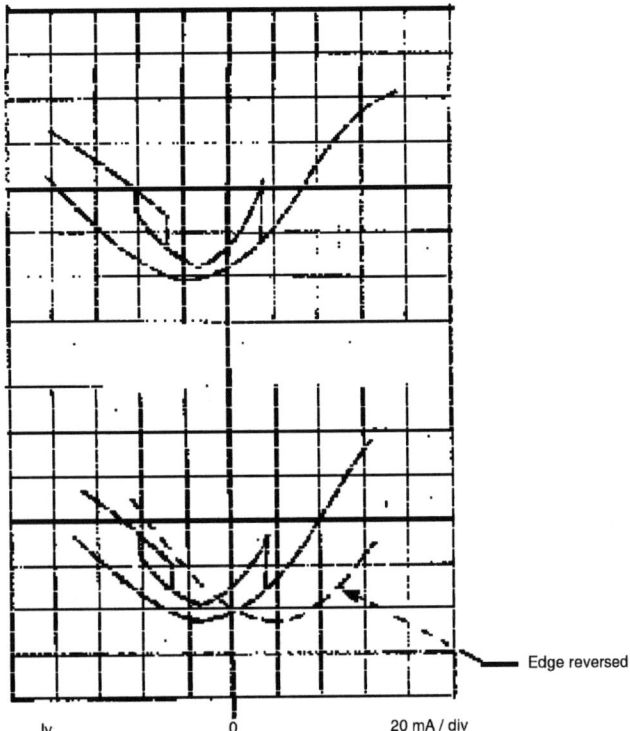

Figure 10.29. Fraction of total MR response use. Test Chip 6061-314 1-13-23; Sense #2, Word #2: $I_{sense} = 3\,\text{mA}$. (1) − (0) Difference $= 0.33 \times \Delta R/R$.

is reduced to 1 µm, this response is reduced to 1/6. As a practical matter, this mode of operation cannot be used for elements with width less than 1 µm. Incidentally, the figure shows that if the edge setting is reversed, the minimum shifts from the negative side of the word field drive to the positive side.

10.6. PSEUDO SPIN VALVE, GIANT MAGNETORESISTIVE MEMORIES

10.6.1. Pseudo Spin Valve Cell Operation

Many more operating modes that have higher efficiency can be generated for GMR sandwich materials than for AMR materials because the sense current can flow in any selected direction. For very dense, sub-micron, memory cells, pseudo-spin valve (PSV) elements are the most attractive and consequently will be examined in detail. In the pseudo spin valve structure the sandwich is made from magnetic layers of two different thicknesses.[8] The thicker layer requires larger field to switch and serves as the storage layer and the thinner layer is used for sensing. Shape anisotropy is used to form an easy axis along the element; for very large elements

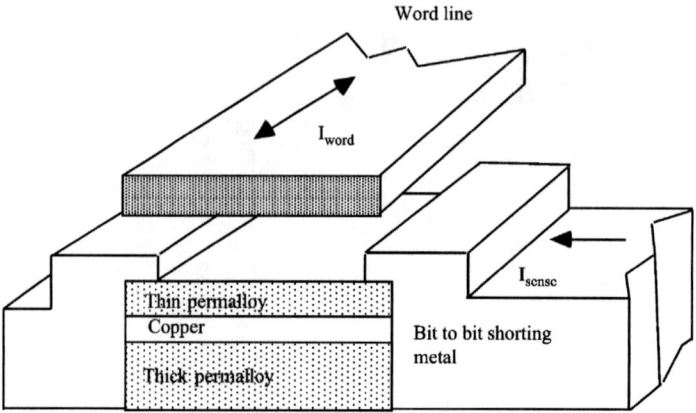

Figure 10.30. Pseudo spin valve memory cell.

with small shape anisotropy different materials can be used.[9] If the magnetizations in the two layers are parallel, the element is in a low resistance state; if they are antiparallel, the element is in a high resistance state. Figure 10.30 shows the general layout for a pseudo spin valve memory element with the word line providing a bipolar digit field along the elements and the sense current providing a unipolar hard axis scissoring field.

Figure 10.31 illustrates the threshold characteristics and memory operation for a pseudo spin valve cell. The fields required to switch the thinner of the two layers

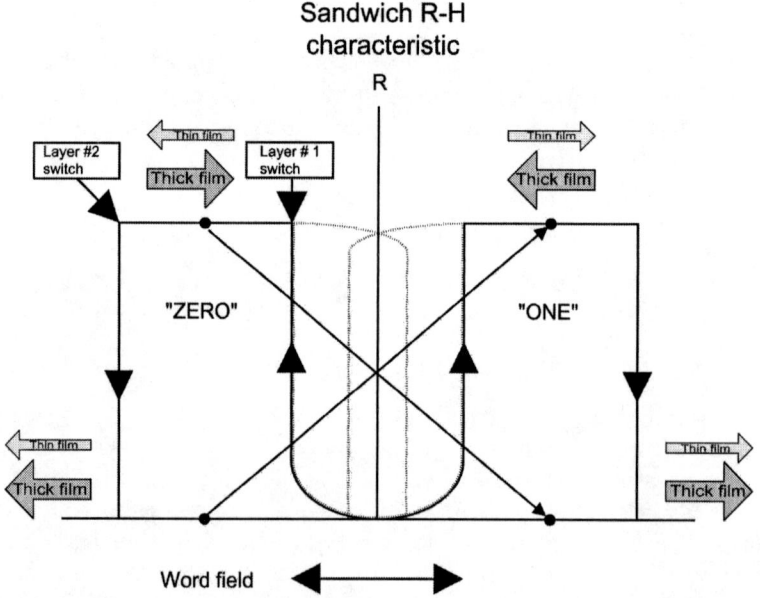

Figure 10.31. GMR memory operation. $(1)-(0)$ Difference $= 2 \times \Delta R/R$.

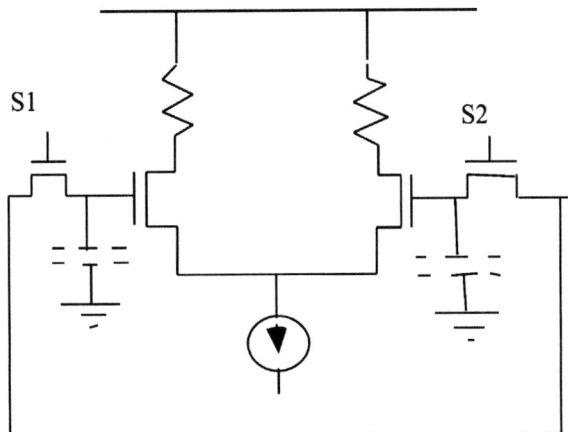

Figure 10.32. Two sample second stage amplifier.

(sensing layer) are usually only 15–30% of that required to switch the thicker layer (storage layer). As shown in the figure, for reading a negative word field is applied sufficient to switch the thinner layer but not sufficient to switch the thicker layer. With this word field applied, the sensing circuit is a nulled to a zero output. The nulling circuit is shut off and the word field then is reversed. As indicated by the arrows in the figure, the resistance of the element changes up or down by the full M-R value depending on the state of the thicker storage layer. Thus, the "zero"–"one" difference is twice the M-R coefficient as shown in the figure; this is six times more effective in the use of the M-R change than the AMR type cell. Figure 10.32 illustrates how the two sample values are stored on the gate capacitances of an output stage thus giving an output proportional to the difference.

10.6.2. Sub-Micron PSV Memory Cells

A very important property of the pseudo spin valve type of memory elements is that they can scale to sub-micron dimensions without loss of performance. A number of experimental studies have been carried on demonstrating this property.[8] Figure 10.33 shows the nominal dimensions for a number of experimental pseudo spin valve memory elements prepared by e-beam lithography at the Cornell Nanofabrication Facility.[11] Because of over etching, the widths of the elements were measured to be about 0.3 µm and 0.15 µm. The experimental elements were made from sandwich material with 60 and 80 Å thick magnetic layers with a 30 Å copper middle layer. The bulk of the magnetic layers was 65%Ni, 15%Fe, 20% Co with a 15 Å thick 95% Co, 5% Fe layer next to the copper. The M-R coefficient was about 6% and the ohms per square was about 13.

Figure 10.34 shows the major loop response for a 0.3 µm × 1.2 µm active area element. The thinner layer thresholds are about 20 Oe and the thicker layer thresholds are at about 125 Oe. Figure 10.35 shows one of the minor loops for the cell with one transition at 24 Oe and the other at 16 Oe. Tests were performed with a sense current of 0.5 mA, which is about half of the nominal 1 mA operating current.

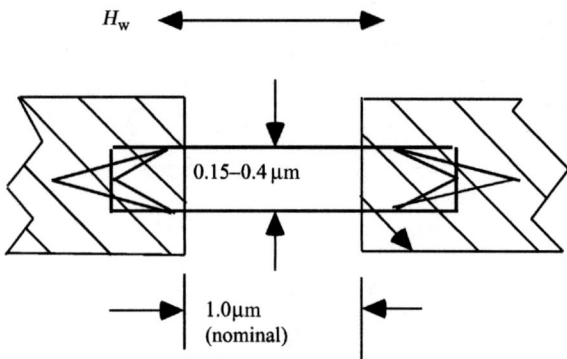

Figure 10.33. Experimental sense line structure.

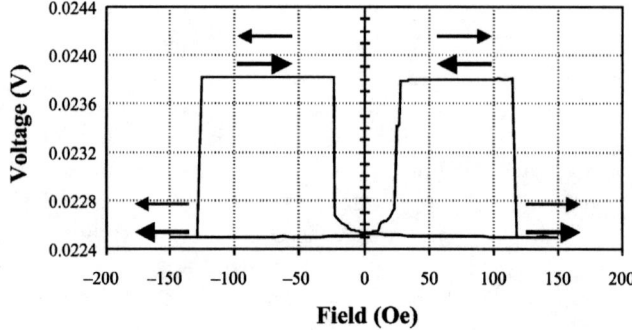

Figure 10.34. PSV element major loop. ⟶ Thin layer; 65 + 15 Å, 30 Å, 45 + 15 Å; ⟶ Thick layer; Easy axis along element 0.3 × (1.2 + contacts)μm² element.

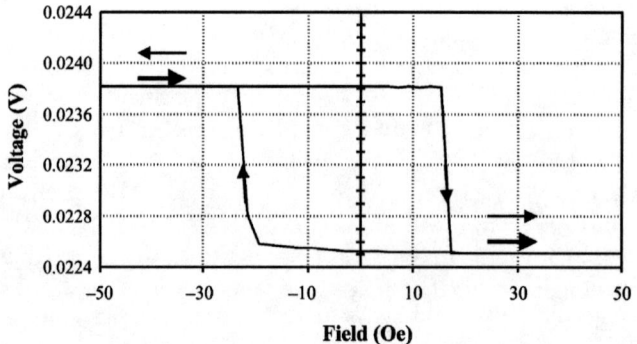

Figure 10.35. PSV element minor loop (0.3 × 1.2 μm² PSV cell).

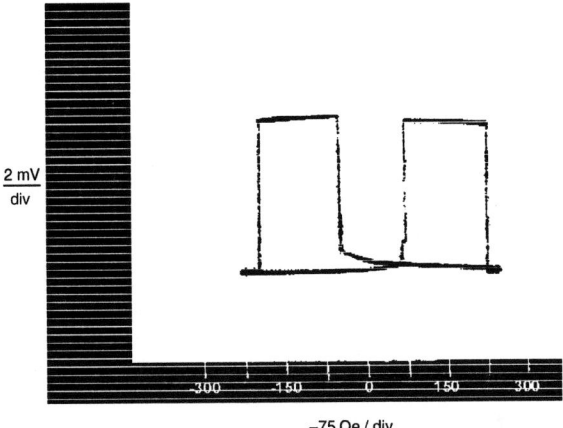

Figure 10.36. PSV memory element major loop 0.15 × 1.2 µm; 1 mA I_{sense} Slant ends; storage layer thresholds.

At nominal sense currents, cell outputs are ±2.5 mV. Figure 10.36 shows the major loop response for 0.15 µm wide element and Fig. 10.37 shows the minor loop response. With 1 mA of sense current, sense outputs are ±8 mV.

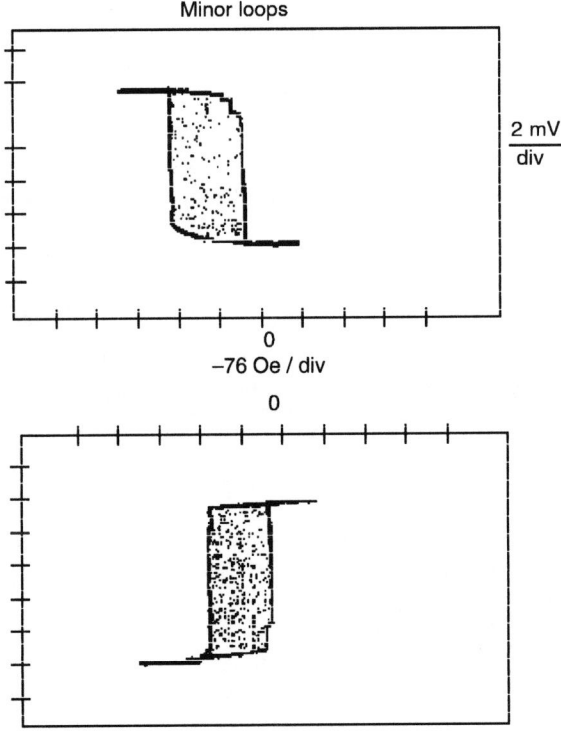

Figure 10.37. PSV memory element.

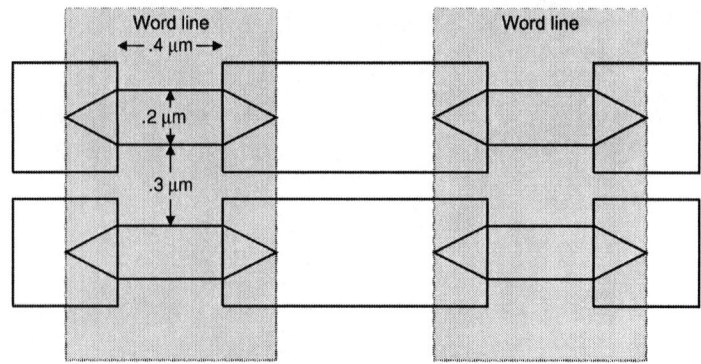

Figure 10.38. PSV memory cell. 35 Ω Resistance, 25 Ω Active, 10 Ω in Contacts; ±1.75 mV at 1 mA Sense current (with 7% GMR); Array cell area = 0.5×1.0 μm^2.

10.6.3. PSV, High Density, Memory Layout

Experience has shown that in MRAM and other high density memories, about one third to one half of the chip area is taken up by the circuitry, wiring, and spares. The rest of the area is taken up by the memory array. Therefore, if one were to design a 64 Megabit memory using 3.3 V CMOS circuitry with a chip area of less than 1 cm^2, the array area of an individual element would have to be between 1/2 and 1 μm. Figure 10.38 shows an array design where each element occupies an array area of 0.5 μm^2. Element width is 0.2 μm; element active length is 0.4 μm with 0.2 μm contacts and a 0.2 μm element to element spacing. The area of the cell is 12.5 times the minimum feature squared. Contact requirements often limit spacings and accounts for the larger than minimum line to line spacing.

In a 3.3 V design with 1 mA sense currents, 32 elements conveniently could be placed on the sense line. With a sense current of 1 mA, output signals would be ±1.8–2.0 mV, which would be adequate for a 100 nanosecond cycle time memory. To generate a 125 Oe word write field from the 0.8 μm wide word line, word currents of approximately 10 mA would be required assuming a high permeability backing (keeper) to the word line. Because of their relatively low resistance and favorable form factor, word lines can span from 512 to 1024 sense lines.

10.6.4. PSV Memory Element Modeling

As pseudo spin valve elements are made smaller, exchange coupling makes each layer act more like a single rigid domain. By assuming this single rigid behavior, one can model the quasi-static threshold behavior of a pseudo spin valve memory element with relatively simple models involving the parameters listed in Table 10.4. Figure 10.39A shows the coordinate system for the magnetization in the thinner and thicker layers in which angles of rotation of the magnetizations are measured from the same axis. The sense factors Sf1 and Sf2 take into account the fact that the magnitude of the average sense field in each layer is different because of the different thickness. Because the magnetic layers are considerably longer than they are wide,

Table 10.4. Cell Parameters

H_k = Anisotropy constant
H_w = Word field along the element
H_s = Maximum sense field at outside edge
Sf1, Sf2 = Sense factors: Average/maximum
H_{cp} = Roughness coupling field between layers
H_{d1}, H_{d2} = Demagnetizing fields across layers
R_t = Ratio of demag field along to across element
H_b = Bias field across element
θ_1, θ_2 = Local angles of the magnetization

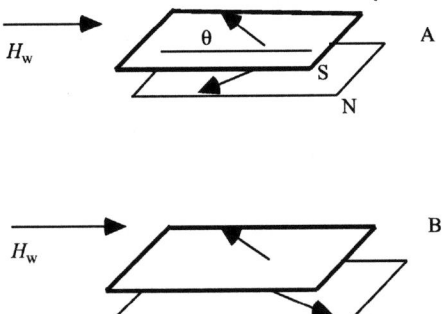

Figure 10.39. Inhibiting effect of switched thin layer on thick layer threshold.

the demagnetizing factor in the long dimension is much smaller than the one across a layer; the ratio of the two demagnetizing factors is given by R_t. Table 10.5 gives the torque equilibrium equations for the two layers when treated as single domains. Note that the Neel coupling field term (H_{cp}) is referenced to one layer and then is scaled to the other by the ratio of the thicknesses.

With the torque equations one can calculate the M-R response and the switching thresholds. Figure 10.40 shows calculated response for memory elements

Table 10.5. Torque Equilibrium Equation for Each Layer

$$T(\theta_1) = - H_k M \sin\theta_1 \cos\theta_1 + H_w M \sin\theta_1$$
$$+ H_s S_{f1} M \cos\theta_1 - H_{cp} M \sin(\theta_2 - \theta_1)$$
$$- M H_{d1} \cos\theta_1 - M H_{d2} \cos\theta_2$$
$$+ M H_{d1} R_t \sin\theta_1 + M H_{d2} R_t \sin\theta_2$$
$$+ M H_b \cos\theta_1 = 0$$
$$T(\theta_2) = - H_k M \sin\theta_2 \cos\theta_2 + H_w M \sin\theta_2$$
$$+ H_s M S_{f2} \cos\theta_2 - H_{d2} M \cos\theta_2$$
$$- H_{d1} M \cos\theta_1 + M H_{d1} R_t \sin\theta_1$$
$$+ H_{d2} M R_t \sin\theta_2 + M H_b \cos\theta_2$$
$$- H_{cp} M \sin(\theta_2 - \theta_1) H_{d1}/H_{d2} = 0$$

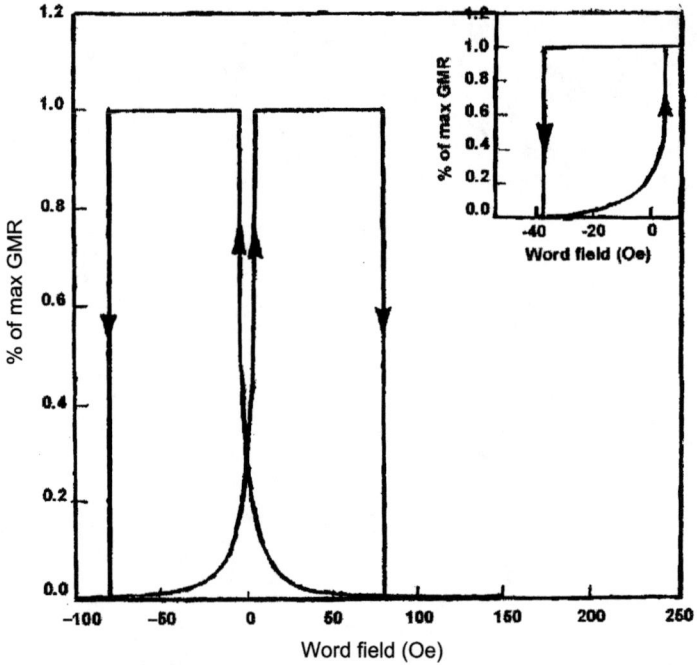

Figure 10.40. Calculated response. $H_k = 20$, $H_s = 15$, $H_p = 10$, no bias, SF1 = 0.65, SF2 = 0.58, T1 = 60, T2 = 80, $H_{d1a} = 232$, $H_{d1b} = 8.14$, $H_{d2a} = 307$, $H_{d2b} = 10.8$, 0.3×2.6 μm² bit.

with approximately the same parameters as the experimental elements. Figure 10.41 shows the dependence of the thicker and thinner layer thresholds on the sense and word fields. The two thresholds for the thinner layer correspond to the cases where the two layers are switching from parallel to anti-parallel and from anti-parallel to parallel. In one case the demagnetizing fields along the layers aid switching and in the other case they retard switching. A single threshold is shown for the thicker layer under the assumption that the thinner layer has already switched.

By taking into account that the very large demagnetizing factor limits magnetization rotation to be very nearly in the plane for thin layers, the dynamic equations describing switching[13] can be written in the simple form given in Table 10.6. Note that to be consistent with the symbols in Table 10.5, the notation for the angles is different than the traditional polar coordinates. Figure 10.42 illustrates the calculated nanosecond switching of the thinner layer when both layers start with the magnetization pointing in the same general direction. As illustrated in Fig. 10.39B, if magnetizations in both layers start out in the same general direction and rotate in opposite directions, the demagnetizing fields tend to cancel and reduce the threshold for the thicker layer switching as indicated in Fig. 10.43. However, if the risetime of the switching field is more than 2 nanoseconds, the thinner layer switches first and inhibits the switching of the thicker layer. This is a possible way of reducing the write threshold.

HIGH SPEED MAGNETORESISTIVE MEMORIES

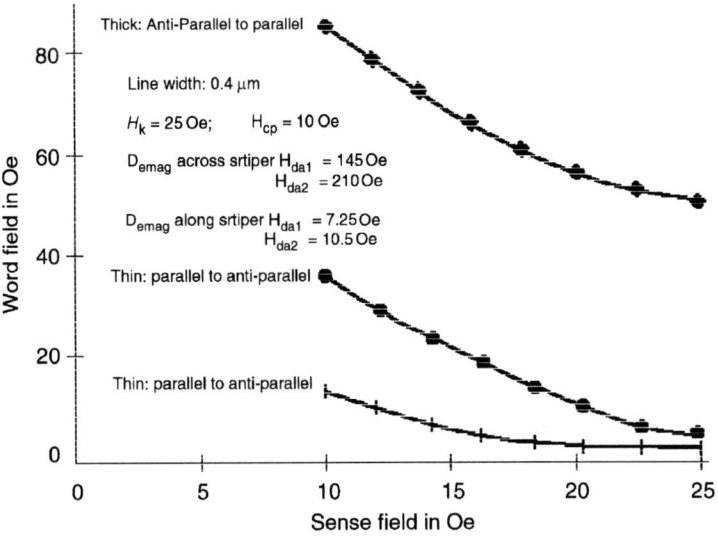

Figure 10.41. PSV cell thresholds.

Table 10.6. Gilbert's dynamic equation and coordinates

(Gilbert's equation)	$\dot{\vec{M}} = -|v|\vec{M} \times \left(\vec{H}_{\text{torque}} - \dfrac{\alpha}{|v|m}\dot{\vec{M}}\right)$
Thin film approximation:	$\dot{\theta} = \dfrac{-M|v|\phi}{(1+\alpha^2)} + \dfrac{\alpha|v|}{\mu_0(1+\alpha^2)m}T(\theta)$
	$\dot{\phi} = \dfrac{-M\alpha|v|\phi}{(1+\alpha^2)} - \dfrac{|v|}{\mu_0(1+\alpha^2)m}T(\theta)$
	$\phi \ll 1$

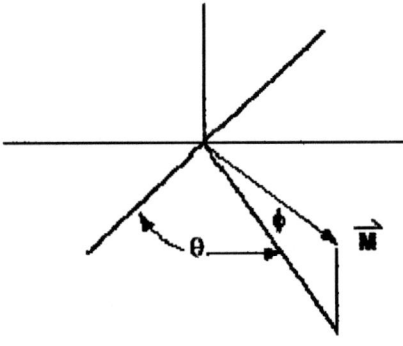

$\vec{M}$ = magnetization
$|v|$ = gyromagnetic ratio
α = damping constant

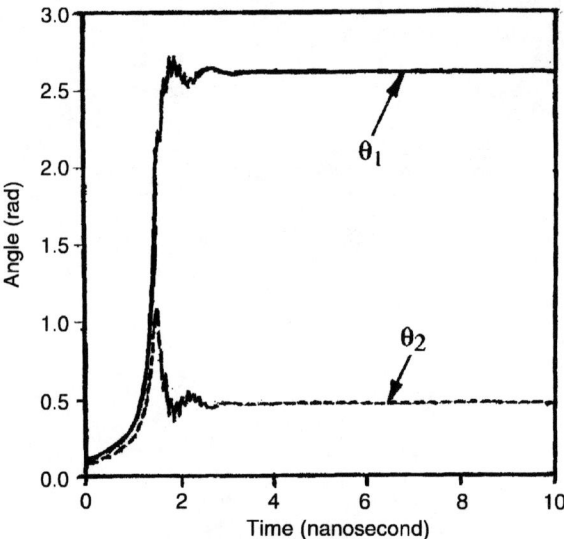

Figure 10.42. Time dependence of a 0.15 mm pseudo spin valve bit with magnetic layer thickness of 60 and 80 Å, respectively. Starting at time $t = 0$, the word field is ramped from -70 to 70 Oe over a 2 nanoseconds period. The storage layer (θ^2, dashed) does not switch along with the read-out layer (θ_1, solid), although it exhibits overshoot around 1.5 nanoseconds.

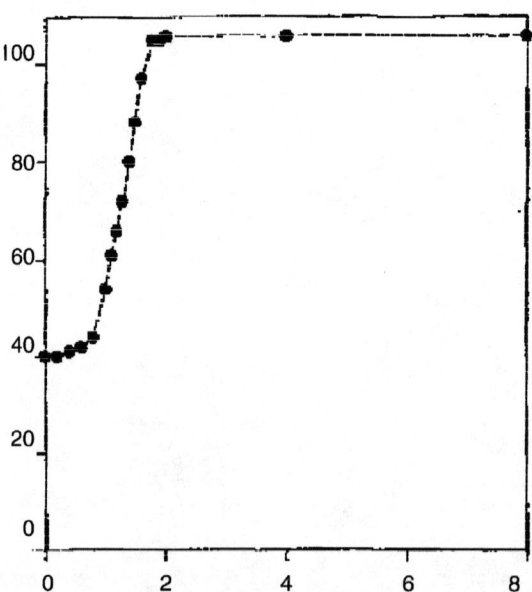

Figure 10.43. Write field vs. word field rise time calculated for a 0.15 μm × 1.8 μm pseudo spin valve bit with 50 and 80 Å thick magnetic layers. Below 2 nanoseconds, write field magnitudes decrease significantly.

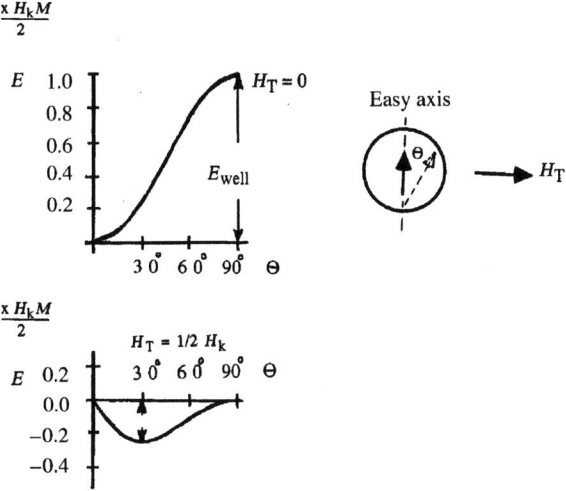

Figure 10.44. Well depth and half select. $E = (H_k M/2)\sin^2\theta - H_T M \sin\theta$.

10.6.5. Scaling Potential

Pseudo spin valve memory cells can be scaled to array densities of 10^9 bits per square centimeter with cell widths of 0.1 µm and active lengths of 0.2 µm and total lengths of 0.4 µm. Cells made with magnetic layers of 25 and 40 Å thicknesses would still have an energy well depth slightly greater than the $55 \times KT$ required for adequate stability when half selected. Figure 10.44 shows that when half selected, a typical magnetic memory cell with a Stoner–Wohlfarth threshold has its well depth reduced to 1/4 the quiescent value. A cell with the dimensions given above would have a well depth that was just adequate. As noted in Table 10.7, the 7 mA word currents and 1 mA sense currents needed to write the elements would cause temperature raise of 12.5 and 40°C. To maintain the same energy well depth if the dimensions were scaled to 3/4, shape anisotropy would have to increase by a factor of 1.78 as well as the drive fields to account for the reduced volume. Assuming the dielectric thickness is also scaled, the temperature rise from the sense current would

Table 10.7. 0.1 × 0.4 µm PSV cell parameters

Word line: 0.25 Ohms/bit
   $I_w = 7$ mA
   Temp. rise $= 12.5°$C
Sense line: 40 Ohms/element (30 + 10)
   $I_s = 1$ mA
   $V_{sign} = \pm 2.8$ mV (7% GMR)
   Temp. Rise: 40°C
Half select energy well depth just adequate!
(Well depth must be maintained; therefore, for smaller sized elements, the drive fields and anisotropy must increase.)

10.7. MEMORIES WITH SPIN VALVE MATERIALS

10.7.1. Spin Valve Materials

In spin valve materials, one of the two magnetic layers in a sandwich is pinned by an antiferromagnetic layer. Figure 10.45 shows the cross section for a spin valve material in which the pinning material is FeMn and the substrate on which the material is deposited is a silicon wafer coated with SiN. In this case the pinning layer is on top. Figure 10.46 shows that the material has an M-R coefficient of nearly 8% and that a field of over 200 Oe is required to switch the pinned layer against the pinning field. The free layer can be switched with fields of less than 50 Oe and follows a Stoner–Wohlfarth like threshold.[14] The M-R response can be further enhanced by making a double spin valve in which the free layer is located between two pinned layers as shown in Fig. 10.47.

10.7.2. Spin Valve Memory Modes

One way of employing spin valve material in a memory mode is the destructive readout one shown in Fig. 10.48. As shown in the figure, the information is stored in the direction of the unpinned layer. This layer is coincidentally selected by a word and sense field and is switched giving a change in resistance or not switched giving a change in resistance. These memory cells and mode of operation are not as attractive as pseudo spin valve memory cells for several reasons. First this mode of operation yields a destructively read, unipolar output, which reduces the effective output by a factor of two; second, because one layer is pinned, the free layer encounters a much larger

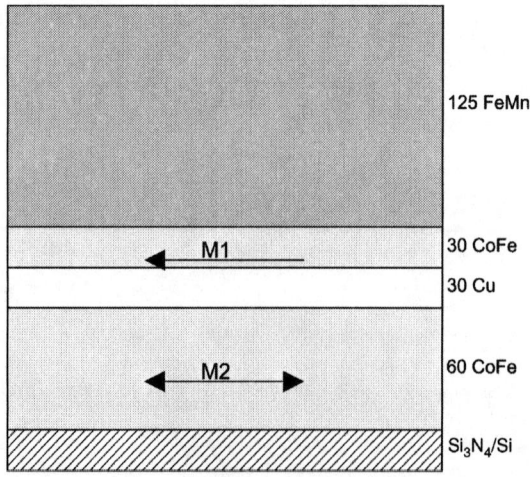

Figure 10.45. Simple FeMn pinned spin valve.

HIGH SPEED MAGNETORESISTIVE MEMORIES

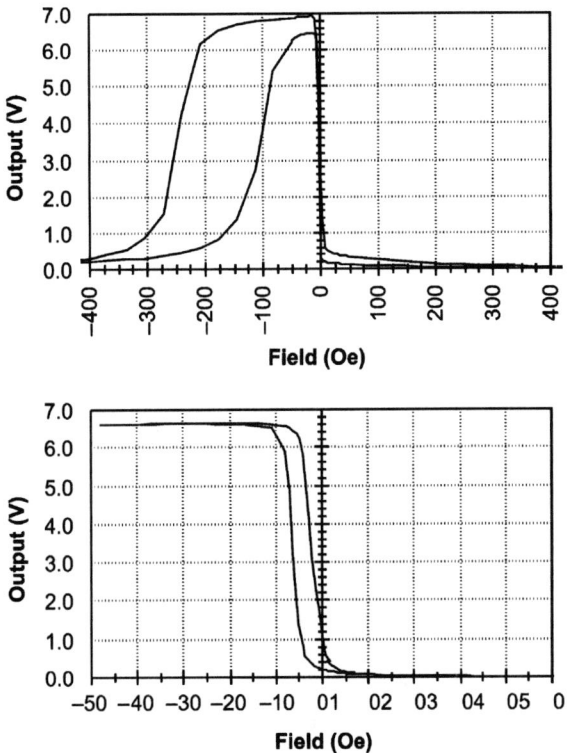

Figure 10.46. Spin valve material characteristics.

demagnetizing field for given element size when compared to the pseudo spin valve elements and additional techniques are required to reduce this demagnetizing field.

10.7.3. Spin Valve Latch Memory Cells

For small, very high speed, nonvolatile memories with non destructive readout for which array area per cell is not critical, spin valve cells appear very attractive;

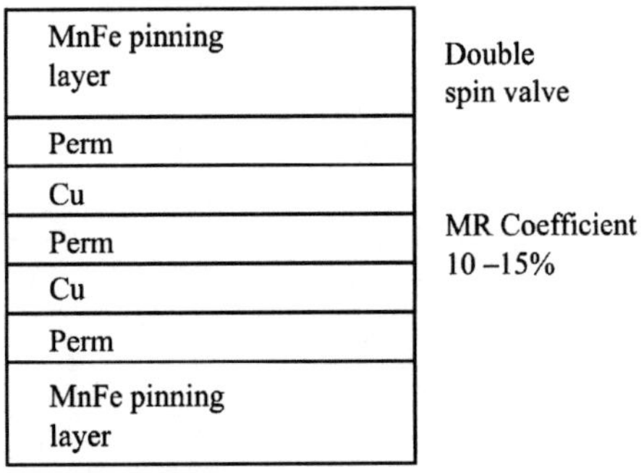

MnFe pinning layer	Double spin valve
Perm	
Cu	
Perm	MR Coefficient 10–15%
Cu	
Perm	
MnFe pinning layer	

Figure 10.47. Double spin valve material.

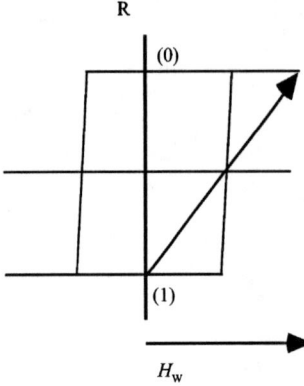

Figure 10.48. Destructive readout operation. READ (1): Switch detected; reverse word field and write. READ (0): No switch detected; No writing necessary. WRITING: Apply word pulse of desired polarity.

memories with access times of a few nanoseconds are possible. Figure 10.49 illustrates the circuit diagram for just such a memory using 0.8 μm CMOS circuitry. In this organization, the information is stored on two 700–800 Ω elements with one element being in the high resistance state and the other in the low resistance state. As shown in the figure, selection is done with semi-conductor switches. A flip-flop circuit is used to measure the resistance difference between the two elements and toggle one way or the other depending on the resistance values. Operation consists of turning on the flip-flop shorting gates for a few nanoseconds so that currents will flow in the selected elements; the gates then are opened and the flip-flop toggles. Once toggled, no sustaining current is required. A write coil selected with a semiconductor switch (not shown) is used to write the two elements at the same time. Because all the selection is done in semiconductor devices, the magnetic material can have very large margins.

10.8. SPIN DEPENDENT TUNNELING MEMORY CELLS

10.8.1. Spin Dependent Tunneling Materials

Although very similar to spin valve materials magnetically, spin dependent tunneling materials typically have M-R coefficients that are 3 or 4 times larger than spin valve materials and have much higher resistances. In spin dependent tunneling material, the copper middle layer in a sandwich is replaced by a very thin Al_2O_3 insulating layer (13–15 Å); this dielectric layer is so thin that significant conduction takes place by tunneling through the dielectric. Figure 10.50 shows the cross-section for such a device with the pinning layer on top and the free layer on the bottom. At the present time, it is possible to make cells reliably with resistances of 1000 to 100,000 Ω per square micron or greater. Figure 10.51 shows a typical response from a spin dependent cell; the M-R coefficient is 18–45%; about 100 mV can be placed across a cell before the M-R behavior degrades.

A destructive readout memory mode equivalent to the one described for spin valve memory cells can be used with spin dependent tunneling cells with 8–16 elements on a sense line. These cells yield 20 mV signals and because of the high

HIGH SPEED MAGNETORESISTIVE MEMORIES

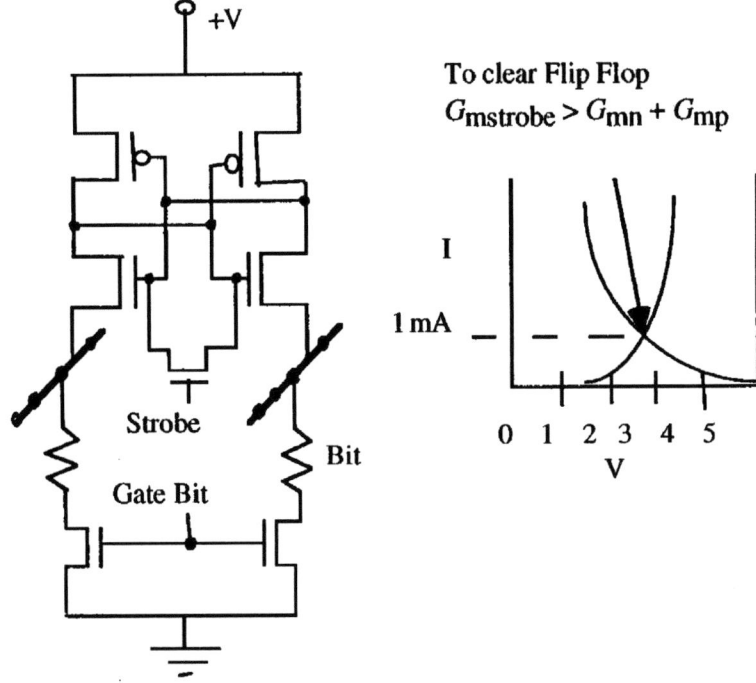

Figure 10.49. Spin valve latch type memory operation.

resistance, sense currents are very small. However, the thin dielectric provides for a large capacitance per unit area, 25×10^{-3} pfd per micron squared. As a consequence, for high ohms/square material, the intrinsic read time constant for these cells is 2.5 nanoseconds and imposes a limit on access time.

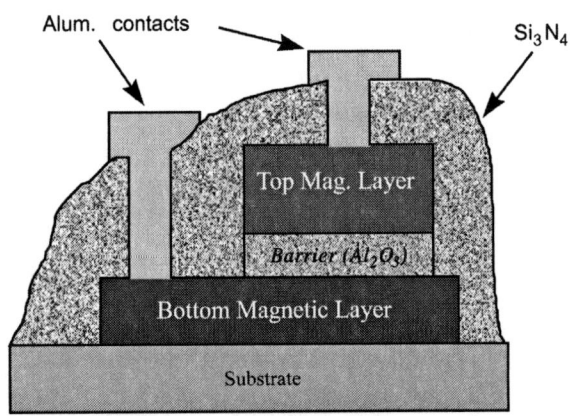

Figure 10.50. Patterned SDT device—side view.

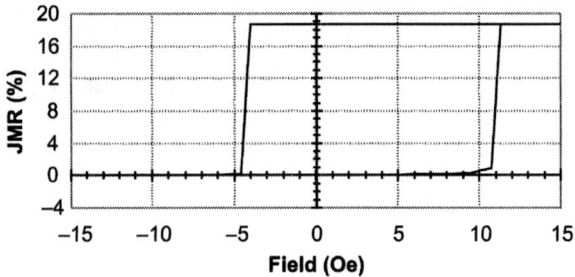

Figure 10.51. Pinned SDT Device (100 × 200 mm pinned SDT device 15 µA, 5 kΩ).

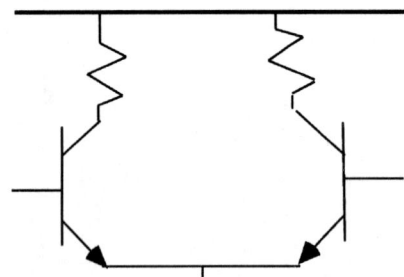

Figure 10.52. Circuit factors in static memory cells (Bipolar differential amplifier; offset is the order of 1 mV).

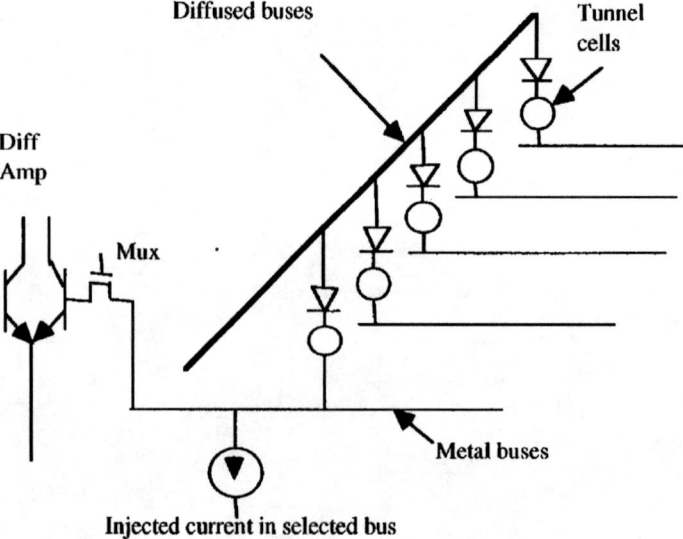

Figure 10.53. Diode switched, spin dependent tunneling cell array.

Table 10.8. Summary of MRAM technology

---

1. For a new high speed memory technology to be generally accepted, it must be cost competitive.
   A. Large, cheap batches must be possible.
   B. Easy testing and packaging must be possible.
   C. It must exploit semiconductor technology.
   D. It must exploit the scaling in semiconductor processing.

2. MRAM technology has the potential of providing superior memories to DRAM, FLASH, or EPROMs.
   A. For this to happen, processing for large uniform batches must be developed.
   B. Large investments are necessary for process development.

3. The discovery of GMR has provided the M-R properties necessary for changing MRAM from a niche technology to one with general potential.
   A. A large variety of MRAM memories can be made.
   B. Existing GMR materials are adequate for most memory needs.

Non destructive readout memories can be constructed from the spin dependent tunneling cells rather easily because of their large 20 mV signal. Figure 10.52 shows that in a BiCMOS process in which bipolar transistors are available, the offset in a differential amplifier pair typically has an rms value of less than 1 mV. Thus by comparing to a matched reference, the state of the cell can be determined by simply measuring its resistance. Figure 10.53 illustrates such a memory organization using diodes to provide selection and isolation.

10.9. SUMMARY

In a technological development such as developing a new type of memory, the costs of the various phases increase as production phase is approached. As a consequence, the risk of each phase must diminish and the probability of success must approach unity if the effort is to receive continued funding. In general, the development of magnetoresistive memories has progressed to the last stages of development with encouraging results. However, the expensive pre-production phase, which develops acceptable yields, is yet to be done in most instances.

In terms of an overview, magnetoresistive memory development can be summarized as in Table 10.8.

REFERENCES

1. F. E. Laborsky, W. R. Barber, Nondestructive readout in plated wires, *IEEE Trans. Magn.* **7** (3), 490–493 (Sept. 1971).
2. C. D. Olson, A. V. Pohm, Flux Reversal in Thin Films of 82% Nickel 18% Iron, *J. Appl. Phys.* **29**, 274–277 (March 1958).
3. A. H. Bobeck, P. I. Bonyhard, J. E. Geusic, Magnetic Bubbles, an emerging new memory technology, *Proc. IEEE* **63**, 1176–1195 (August 1975).
4. L. J. Schwee, H. R. Irons, W. E. Anderson, The cross tie memory, *IEEE Trans. Magn.* **12**, 608–613 (Sept. 1976).

5. A. V. Pohm, J. M. Daughton, C. S. Comstock, H. Y. Yoo, J. Hur, Threshold properties of 1, 2, and 4 micron multi-layer M-R memory cells, *IEEE Trans. Magn.* **23**(5), 2575–2577 (Sept. 1987).
6. H. Y. Yoo, A. V. Pohm, C. S. Comstock, 2-Dimensional analysis of laminated thin film elements, *IEEE Trans. Magn.* **24**, 2377–2379 (Nov. 1988).
7. A. V. Pohm, C. S. Comstock, The reverse word switching threshold for M-R memory elements, *IEEE Trans. Magn.* **26**, 2529–2531 (Sept. 1990).
8. B. A. Everitt, A. V. Pohm, R. S. Beech, A. Fink, J. M. Daughton, Pseudo spin valve MRAM cells with sub-micrometer critical dimensions, 3M-Intermag Conference, Jan. 1998, Paper EB-02.
9. K. Matsuyama, H. Asada, S. Ikeda, K. Taniguchi, Low current magnetic RAM memory operation with a high sensitivity spin valve material, *IEEE Trans. Magn.* **33**(5), 3283–3285 (Sept. 1997).
10. H. Asada, K. Matsuyama, K. Taniguchi, Micromagnetic study on write operation in sub-micron magnetic random access memory cell, *J. Appl. phys.* **79**, 6646–6648 (March 1996).
11. R. S. Beech, A. V. Pohm, J. M. Daughton, Simulation of sub-micron GMR memory cells, *IEEE Trans. Magn.* **31**, 3203–3205 (Nov. 1995).
12. Y. Zheng, J. Zhu, Micromagnetics of spin valve memory cells, Paper GE-12, 3M Conf. Nov. 1996, Atlanta, GA.
13. D. D. Tang, P. K. Wang, V. S. Speriosu, S. Le, K. K. Kung, Spin valve RAM cell, *IEEE Trans. Magn.* **31**(6), 3206–3208 (Nov. 1995).
14. H. Y. Yoo, C. S. Comstock, J. H. Hur, S. W. Kenkare, A. V. Pohm, Dynamic switching process of sandwich structured M-R elements, *IEEE Trans. Magn.* **25**(6), 4269–4271 (Sept. 1989).

11

Hybrid Devices

Mark Johnson*

11.1. INTRODUCTION

The majority of *Spintronic* devices involve ferromagnetic metal–nonmagnetic metal (F–N) trilayers or multilayers and use magnetoresistive properties. Most applications of these structures involve sensors, and an example is a read head that senses the magnetization state of magnetic domains associated with analog or digital information recorded in magnetic media. Recent efforts have aimed at using magnetoresistive elements in integrated circuits as nonvolatile memory cells for digital data. These all-metal devices have low impedance, typically draw high currents, and are not readily interfaced with standard silicon integrated circuitry.

Almost all digital electronics is based on silicon technology, and an important recent research trend involves the integration of a ferromagnetic element with a semiconductor device structure. This new field of semiconductor spintronics faces many interesting issues of basic and applied research. In parallel, research in digital superconducting electronics continues, fueled by interest in an alternative technology for certain high-speed applications. The motivation for extending magnetoelectronics into *hybrid device* families, composed of combinations of ferromagnetic elements with superconductors or semiconductors, is to smoothly interface the desirable properties of ferromagnetic materials, nonvolatility and speed, with superconductor and semiconductor technology.

This chapter contains sections on hybrid ferromagnet–superconductor devices and hybrid ferromagnet–semiconductor devices. Each section describes one device family based on the injection of spin-polarized carriers from a ferromagnetic component into a superconductor or semiconductor, and one device family that uses the locally strong magnetic fringe fields of a ferromagnetic component to modulate the transport properties of a superconductor or semiconductor. As background to the former sets of device families, we begin with a brief section that reviews the phenomenology of the *spin injection* technique which was originally demonstrated with an all-metal system[1-4] and which is presently being applied to hybrid systems.

11.1.1. Review of Spin Injection

11.1.1.1. Pedagogical Introduction. After a heuristic introduction, basic concepts of spin injection and spin accumulation will be reviewed more formally. This review of spin transport in metals will develop a foundation that will

---

*Naval Research Laboratory, Washington, DC.

support an understanding of basic issues of spin transport in superconductors and semiconductors.

The technique is presented conceptually in Fig. 11.1. We consider a pedagogical model of an unconventional, three-terminal device, shown in cross-section in Fig. 11.1(a). A non-magnetic metal film N is sandwiched between two ferromagnetic films, F1 and F2. Each ferromagnetic film is a single domain and is thin enough such that its axis of magnetization is constrained to lie in the plane of the film. A dc current is driven through F1 into N and returned to the current source from the bottom of N. A single voltage probe is attached to F2 (its ground will be discussed below), and we consider the case where the thickness d of N is less than a *spin depth*:

$$\delta_s = (DT_2)^{1/2} \approx \left[(v_F^2 \tau /3) T_2\right]^{1/2} \qquad (11.1.1)$$

the distance that spin-polarized electrons can diffuse in N without losing their spin orientation. Here, D is the electronic diffusion constant, v_F the Fermi velocity, τ a Drude (resistivity) scattering time, and T_2 the transverse spin relaxation time. An understanding of steady-state transport in this system is aided by considering a microscopic transport model described by the density of state diagrams of Fig. 11.1(b), where F1 and F2 are represented as transition metal ferromagnets in a simplified band model, N is represented as a free electron metal, and for simplicity the interfacial resistance between the films has been neglected.[1,2]

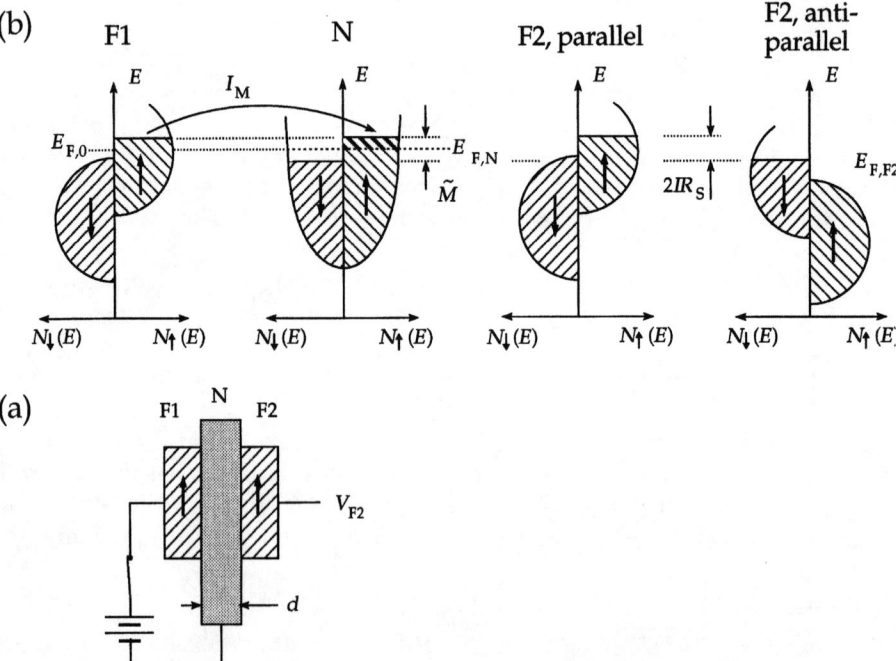

Figure 11.1. (a) Pedagogical model of three-terminal device. Arrows in F1 and F2 refer to magnetization orientation as determined by majority spin subband. (b) Diagrams of the densities of state, $N(E)$, of the ferromagnet–nonmagnetic metal–ferromagnet system depicted in (a).

In the absence of an imposed current, the Fermi levels E_F of all three films align at $E_{F,0}$. When a current is driven from F1 into N, only one spin subband in F1 is available to carry the current because transport involves only electrons within the energy range $k_B T$ of E_F. It follows that a current of magnetization, $I_M = \eta_1 \beta I_e/e$,[5] is associated with the electric current I_e. Here, β is the Bohr magneton (the magnetic moment carried by each electron), e, the electronic charge, and η, the phenomenological factor[1,2] which describes the efficiency of interfacial spin injection, $\eta = (J_\uparrow - J_\downarrow)/(J_\uparrow + J_\downarrow)$, where $J_{\uparrow,\downarrow}$ are current densities for each spin subband. In our pedagogical model, $\eta_1 = 1$, but more generally $|\eta| \leq 1$.

In the steady-state, polarized spins enter N at a rate I_M and polarization is lost, by random relaxation, at a rate $1/T_2$ (in metals, the longitudinal and transverse times are the same, $T_1 = T_2$). The result is a nonequilibrium magnetization, equivalently known as a *spin accumulation*, given by[1,2]

$$\tilde{M} = I_M T_2/A \cdot d = \eta_1 \frac{\beta}{e} \frac{IT_2}{A \cdot d} \tag{11.1.2}$$

where A is the area of electrode F1 and $A \cdot d$ is the volume occupied by the spins. The depiction of $\tilde{M}$ in Fig. 11.1(b) conserves charge neutrality. In the steady-state, the accumulation of up-spins is matched by a depletion of down-spins so that the total number of electrons is held constant. The electric current driven through F1 acts as a "spin pump" which drives a nonequilibrium density of spins into N. In turn, this nonequilibrium magnetization in N has a back effect on F1: The chemical potential of F1 rises so that the chemical potential of its up-spin subband aligns with that of the up-spin subband of N. This effect has been called a spin bottleneck[6,7] because a thermodynamic force associated with the presence of the nonequilibrium spins[8] impedes the flow of spins into N. Because charge and spin are transported by the same carrier, it follows that the electrical impedance of the F1–N interface is increased by the presence of $\tilde{M}$. A detailed discussion of charge and spin flow at the F–N interface is presented in Section 11.1.1.3.

When the magnetization of F2 is parallel with that of F1, its chemical potential will also rise so that the chemical potential of its up-spin subband aligns with that of the up-spin subband of N. This change of chemical potential has been derived[1,2,9] to be

$$eV_s = \eta_2 \beta \tilde{M}/\chi \tag{11.1.3}$$

where χ is the Pauli paramagnetic susceptibility of N. Some physical insight to this result can be gained by noting that $\tilde{M}/\chi$ has the units of magnetic field and can be thought of as the effective magnetic field (also called the *magnetization potential*, H^8) associated with the nonequilibrium spins. Then, $\beta \tilde{M}/\chi$ is the Zeeman energy of a spin-polarized electron in the presence of the field associated with all of the nonequilibrium polarized spins.[10] If the magnetization of F2 is antiparallel with that of F1, then its chemical potential lowers [$E_{F,F2}$ in Fig. 11.1(b)] so that the chemical potential of its down-spin subband aligns with that of the down-spin subband of N. Combining the above expressions for I_M, $\tilde{M}$, and V_s, and using a free electron expression for the susceptibility, $\chi = \beta^2 N(E_F) = \beta^2 3n/2E_F$, where n is the density of conduction electrons, gives the result[11]

$$R_s \equiv \frac{V_s}{I_e} = \frac{\eta_1\eta_2}{e^2}\frac{4T_2\,E_F}{3\,n\,A\cdot d} = \eta_1\eta_2\frac{\rho\delta_s^2}{Ad} \qquad (11.1.4)$$

where the second form results from using an Einstein relation for the electrical resistivity, $\rho = 1/[e^2 DN(E_F)]$.

We note that V_s is a linear function of current so that R_s has units of impedance, and that the magnitude of R_s is inversely proportional to the sample dimensions A and d. This follows from the fact that magnetization has dimensions of magnetic dipoles per unit volume, so that a constant number of nonequilibrium spins will result in a larger value of $\tilde{M}$ when the volume is diminished. We further note that Eqs. (11.1.2) and (11.1.4) were derived for the *thin limit* case, $d < \delta_s$. The density of nonequilibrium spins decreases exponentially as a function of thickness, $\tilde{M} \propto \exp(-d/\delta_s)$, so that a measurement of the diminished value of V_{F2} at thicknesses $d > \delta_s$ gives a direct measurement of δ_s. Conceptually, F1 acts as a conduction electron spin polarizer, a source of spin-polarized current, and F2 is a spin-sensitive potentiometer.

11.1.1.2. Johnson–Silsbee Theory for Charge and Spin Currents.

Johnson and Silsbee[8] applied the methods of nonequilibrium thermodynamics to study the transport of charge, heat and nonequilibrium magnetization in discrete and continuous systems. They extended the linear dynamic equations of thermoelectricity to describe interface thermoelectric and thermomagnetoelectric (spin-charge) effects.

In the thermoelectric system, the flows of charge and heat are driven by an electric field and a gradient of temperature. Classically, the dynamic laws are derived for continuous systems (e.g., bulk metals) and thermoelectric effects are described in terms of bulk properties of conductors, such as the absolute thermopower ε, the thermal conductivity κ, and the electrical conductivity σ. In the classical treatment of thermoelectric effects, the junctions between two conductors are considered to be in complete thermal equilibrium; no temperature or voltage difference appears across the junctions. This need not be the case, however, and Johnson and Silsbee used standard thermodynamic arguments to develop equations describing a number of transport effects specific to the behavior of a junction between two conductors. Of particular interest is the treatment of magnetization transport. As described with the microscopic transport model above, gradients of nonequilibrium magnetization $\tilde{M}$ can drive currents of spin *and* charge. The classical thermoelectric system was therefore generalized to include thermomagnetoelectric effects, both in bulk across metal–metal interfaces. In order to include currents J_M of nonequilibrium spin magnetization, the magnetization potential H^* is identified as the effective field associated with the nonequilibrium spin population, $-H^* = \tilde{M}/\chi - H$, where H explicitly allows for the possible presence of an external magnetic field, but $H = 0$ for the discussion herein.

The formal approach is an entropy production calculation, where *fluxes* J_N of thermodynamic parameter N (i.e., charge, heat, or nonequilibrium spin magnetization) are associated with a generalized force (or affinity) F_N (i.e., gradients of voltage, temperature, magnetization potential). Each flux can, in general, be driven by each of the generalized forces, so that each J_N can be expanded in powers of the affinities, and the coefficients are known as the kinetic coefficients L_{mn}.

For the discrete case of two metals separated by an interface with intrinsic electrical conductance G, electronic transport across the interface is given by the following linear dynamic transport equations:[8]

$$\begin{pmatrix} I_q \\ I_Q \\ I_M \end{pmatrix} = -G \begin{pmatrix} 1 & \dfrac{k_B^2 T}{e\varepsilon} & \dfrac{\eta\beta}{e} \\ \dfrac{k_B^2 T^2}{e\varepsilon} & \dfrac{ak_B^2 T}{e^2} & \eta'\dfrac{\beta}{\varepsilon}\left[\dfrac{k_B T}{e}\right]^2 \\ \dfrac{\eta\beta}{e} & \eta'\dfrac{\beta T}{\varepsilon}\left[\dfrac{k_B}{e}\right]^2 & \zeta\dfrac{\beta^2}{e^2} \end{pmatrix} \begin{pmatrix} \Delta V \\ \Delta T \\ \Delta - H^* \end{pmatrix} \quad (11.1.5)$$

The kinetic coefficients $L'_{1,3} = L'_{3,1}$ are identical with those derived from the microscopic transport model. The phenomenological parameter η' is comparable with η but is measured under conditions associated with heat flow, and ε is an energy-dependent transport parameter.[8] Finally, the $L'_{3,3}$ term describes the self-diffusion of nonequilibrium spin populations, and $\zeta \approx 1$ is an excellent approximation.

The same approach can be followed for a continuous medium, and electronic transport inside a bulk conductor is given by the following linear dynamic transport equations:[8]

$$\begin{pmatrix} J_q \\ J_Q \\ J_M \end{pmatrix} = -\sigma \begin{pmatrix} 1 & \dfrac{a''k_B^2 T}{eE_F} & \dfrac{p\beta}{e} \\ \dfrac{a''k_B^2 T^2}{eE_F} & \dfrac{a'k_B^2 T}{e^2} & p'\dfrac{\beta}{E_F}\left[\dfrac{k_B T}{e}\right]^2 \\ \dfrac{p\beta}{e} & p'\dfrac{\beta T}{E_F}\left[\dfrac{k_B}{e}\right]^2 & \zeta\dfrac{\beta^2}{e^2} \end{pmatrix} \begin{pmatrix} \nabla V \\ \nabla T \\ \nabla - H^* \end{pmatrix} \quad (11.1.6)$$

The kinetic coefficients $L_{i,j}$ are similar to those derived for interfacial transport. For example, $L_{1,3} = L_{3,1} = p(\beta/e)$ describes the flow of a magnetization current associated with an electric current, with fractional polarization p. In a nonmagnetic material, $p_n = 0$, and in a ferromagnetic material, $p_f \approx 0.05-0.45$, according to experimental estimates.[12] The fractional polarization constant p' would be associated with thermal gradients and has not yet been experimentally measured. Constants a' and a'' can be related to the thermal conductivity κ and thermopower ε with a free electron model.[8] Finally, $L_{3,3} = \zeta(\beta/e)^2$ describes self-diffusion of nonequilibrium spins, and $\zeta \approx 1$ is an excellent approximation.

11.1.1.3. Detailed Model of an F–N Interface.

A detailed description of charge and spin transport across a F–N interface can now be provided. Referring to Fig. 11.2(a), a ferromagnetic metal F and nonmagnetic material N are in interfacial contact. Although Eqs. (11.1.5) and (11.1.6) are three dimensional, for simplicity, the cross-sectional area of the interface is taken to be unity and, furthermore, isothermal flow is considered. A useful boundary condition is that a constant current J_q is imposed (a constant current source), and Johnson and Silsbee solve for the resultant magnetization current J_M. There are three regions of electron transport,

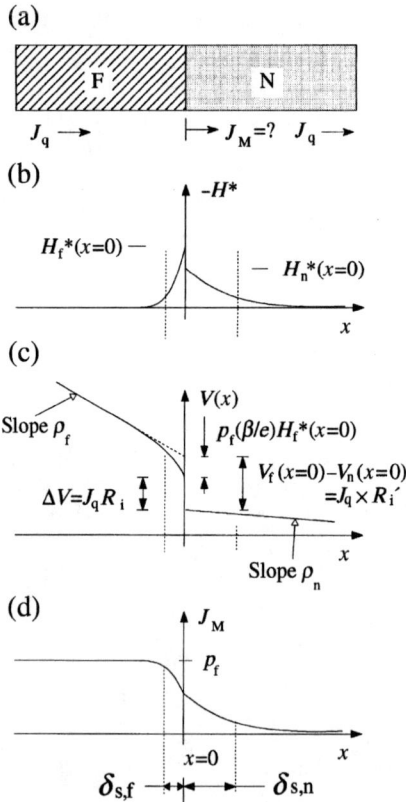

Figure 11.2. (a) Model for flow of charge and spin currents, J_q and J_M, at the interface between a ferromagnetic metal and nonmagnetic material. $x = 0$ at the interface, (b) magnetization potential, (c) voltage, and (d) current of spin magnetization.

the bulk of F and of N, and the F–N interface. Equation (11.1.6) is used to relate currents to potential gradients and to describe steady-state flows in each of the materials F and N. Equation (11.1.5) is used to relate interfacial currents with differences of potential across the interface. Under the assumption of no interfacial spin relaxation, boundary conditions demand that the magnetization currents of all three regions are equal at the interface $(x = 0)$, $J_{M,f} = J_M = J_{M,n}$ (where J_M is the interfacial magnetization current), and that the electric currents are also equal, $J_{q,f} = J_q = J_{q,n}$.

In the general case, the flow of spin-polarized current J_M into N generates a nonequilibrium magnetization (equivalently known as a spin accumulation) $-H^* = \tilde{M}/\chi$ in N [Fig. 11.2(b)]. Because of spin diffusion in N, $-H^*(x)$ decreases as x increases away from the F–N interface, and the decay length is the classical spin diffusion length, $\delta_{s,n}$. In bulk metals at low temperature, $\delta_{s,n}$ can be the order of 0.5 mm,[1,3] and in metal films $\delta_{s,n}$ is roughly 1.0 μm.[13] The nonequilibrium spin population can also diffuse *backwards*, along $-x$, going *back across the F–N interface* and into F. The nonequilibrium population $-H^*$ in F is not constrained to match

HYBRID DEVICES

that in N, $-H_f^*(x = 0) \neq -H_n^*(x = 0)$ [Fig. 11.2(a)] because, e.g., the susceptibilities can be quite different.

The interface has some intrinsic resistance $R_i = 1/G$, and the backflow of diffusing spins acts as an additional, effective interface resistance [Fig. 11.2(c)]. The apparent resistance R_i' measured at the interface, $J_q R_i' = V_f(x = 0) - V_n(x = 0)$, is larger than the intrinsic resistance, R_i, and can be measured by using an appropriate geometry.[7] The spatial extent of the backflow is described by the spin diffusion length in F, $\delta_{s,f}$. There are few reliable experimental measurements for $\delta_{s,f}$, but an estimate for transition metal ferromagnetic films is $\delta_{s,f} \approx 5$ nm.[14]

The backflow of polarized spins near the F–N interface effectively diminishes $J_{M,f}$, and the result is that the fractional polarization of the magnetization current that reaches and crosses the interface, J_M, is reduced relative to the bulk value, $J_M < J_{M,f}$ [Fig. 11.2(d)].

After appropriate algebraic manipulations, a general form for the interfacial magnetization current is found to be:[8]

$$J_M = \frac{\eta \beta}{e} J_q \left[\frac{1 + G(p_f/\eta) \, r_f \, (\xi - \eta^2)/(\zeta_f - p_f^2)}{1 + G(\xi - \eta^2)[(r_n/\zeta_n) + r_f/(\zeta_f - p_f^2)]} \right] \quad (11.1.7)$$

where $r_f = \delta_f \rho_f = \delta_f/\sigma_f$, $r_n = \delta_n \rho_n = \delta_n/\sigma_n$, $G = 1/R_i$ and recall ξ, ζ_f, $\zeta_n \approx 1$. It is important to note that spin transport is governed by the relative values of the intrinsic interface resistance $1/G$, the resistance of a length of normal material equal to a spin depth, r_n, and the resistance of a length of ferromagnetic material equal to a spin depth r_f. Typical values of these resistances are easily estimated. A thin transition metal ferromagnetic film has resistivity and spin depth of roughly 20 $\mu\Omega$ cm and 5 nm,[14] respectively, giving a value $r_f \sim 10^{-11}$ Ω cm^2, i.e., nearly temperature independent. Similarly, a nonmagentic metal film has resistivity and spin depth of roughly 2 $\mu\Omega$ cm and 1 μm,[13] respectively, at low temperature, giving a value $r_f \sim 2 \times 10^{-10}$ Ω cm^2. This value is somewhat smaller at room temperature. A typical value for R_i can be found from the contact resistance R_c measured between two copper layers making contact in a lithographically patterned structure, $R_c \approx 10^{-9}$ Ω cm^2 [15]. Since all of the typical characteristic values fall within a range of two decades, all of these terms are expected to be important for the general case.

11.1.1.4. Resistance Mismatch at an F–N Interface.

One limiting case of interest is that of high interfacial conductance $G \to \infty$ (equivalently, low interfacial resistance, $R_i \to 0$). An appropriate experimental system is a multilayer sample grown in UHV. In this case,[15] $R_i \approx 3 \times 10^{-12}$ Ω cm^2 $\ll r_f$ may justify the high conductance approximation. Equation (11.1.7) reduces to the simpler form:[16]

$$J_M = p_f \frac{\beta}{e} J_q \left[\frac{1}{1 + (r_n/r_f)(\zeta_f - p_f^2)/\zeta_n} \right] \quad (11.1.8)$$

Noting the good approximations $\zeta_f \approx 1$, $\zeta_n \approx 1$, Eq. (11.1.8) can be rewritten as

$$J_M = p_f \frac{\beta}{e} J_q \left[\frac{1}{1 + (r_n/r_f)(1 - p_f^2)} \right] \quad (11.1.9)$$

and we see that the polarization of the injected current is reduced from that in the bulk ferromagnet by the *resistance mismatch* factor $M' = [1 + (r_n/r_f)(1 - p_f^2)]^{-1}$. Using the above estimates for r_f and r_n, the mismatch factor can be expected to be as large as $M' \sim 1/20$. This case is schematically described in Fig. 11.3. With no interface resistance, the spin accumulation in N readily flows back across the interface because of self-diffusion [Fig. 11.3(b)]. The nonequilibrium spin populations in N and F are closely coupled, and there are two measurable consequences. First, spin accumulation acts as a "bottleneck" for spin *and charge* flow and there is an apparent interface resistance, even though the intrinsic resistance is assumed to be negligible [Fig. 11.3(c)]. Second, the backflow of polarized spins reduces the fractional polarization of current crossing the interface [Fig. 11.3(d)].

These effects depend directly on the transport properties of carriers in N. If spin relaxation and carrier diffusion are rapid, $r_n \leq r_f$, then spin accumulation

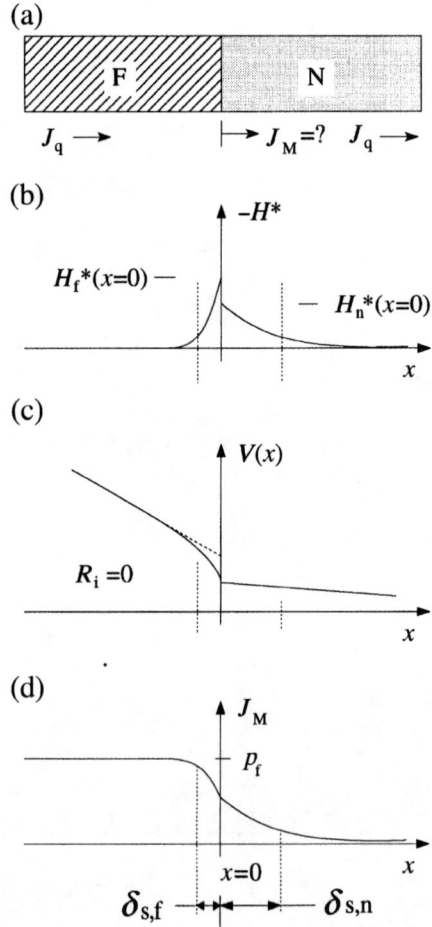

Figure 11.3. (a) Flow of charge and spin currents at an F–N interface, for the case of zero intrinsic interface resistance, $R_i = 0$, (b) magnetization potential, (c) voltage and (d) current of spin magnetization.

11.1.1.5. Transport at F–N Interfaces with Resistive Barriers.

Another important limiting case, that of high interfacial resistance, is depicted in Fig. 11.4. Spin accumulation in N can be large [Fig. 11.4(b)], but the interfacial resistive barrier prevents back diffusion. The nonequilibrium spin population in F remains small, and the voltage drop across the interface is almost entirely due to R_i. [Figure 11.4(c); note the change of scale from Figs. 11.2 and 11.3 such that the slopes $V(x)$ in F and N appear more flat.] The interfacial magnetization current is now given by

$$J_M = \eta \frac{\beta}{e} J_q \qquad (11.1.10)$$

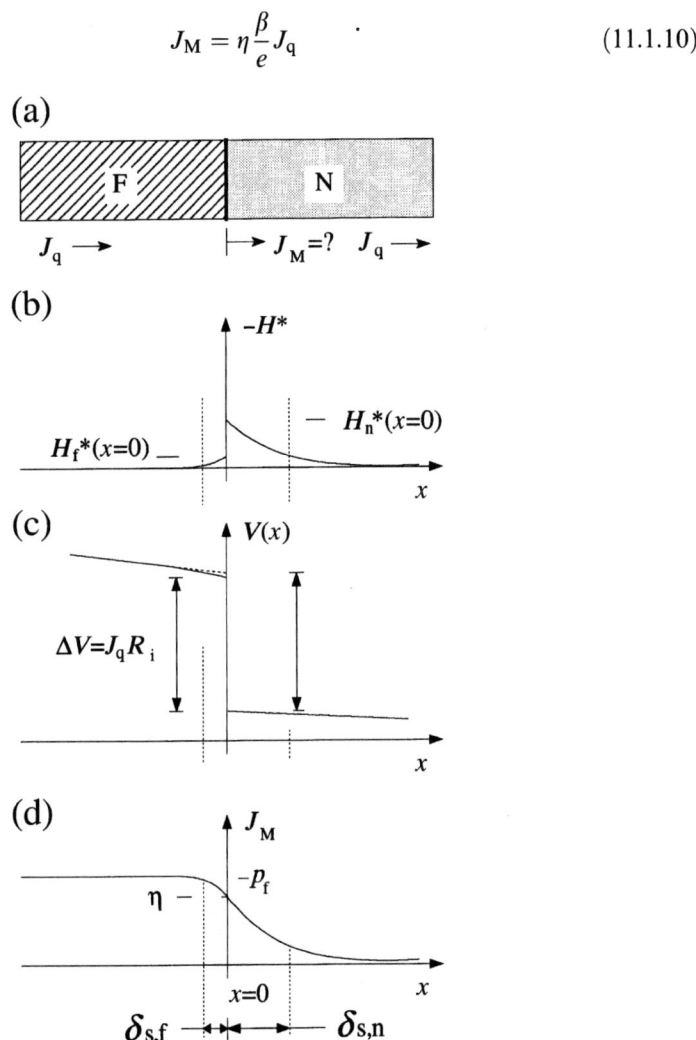

Figure 11.4. (a) Flow of charge and spin currents at an F–N interface, for the case of high intrinsic interface resistance. This case is relevant, e.g., to cases of nonzero contact resistance, R_c, or barrier resistance R_b. (b) magnetization potential, (c) voltage, and (d) current of spin magnetization.

and the fractional polarization, $P = \eta$, is dominated by the interface parameter η. The resistive barrier may be asymmetric with spin. For example, spin-up (or spin-down) electrons may have preferential transmission. The reflection of polarized carriers can, in general, diminish the polarization of current in F as it approaches the interface, and the limit $\eta \leqslant p_f$ is imposed [Fig. 11.4(d)].

The result described by Eq. (11.1.10) is appropriate for many experimental conditions. In ferromagnetic-nonmagnetic *metal* systems, the contact resistance is typically large enough to dominate transport, $R_c \gg r_f$. In ferromagnetic metal–semiconductor systems, an intrinsic Schottky barrier, or a fabricated tunnel barrier, is always present and dominates transport, $R_b \gg r_f$.

11.1.1.6. Spin Diffusion. The concepts of spin injection and spin diffusion have been introduced and discussed. Before reviewing the experimental measurements of "spin injection," the principles of spin diffusion are presented. From Eq. (11.1.5), a voltage drop across a F–N interface is seen to produce an interfacial magnetization current

$$I_M = -G(\eta\beta/e)\,\Delta V \qquad (11.1.11)$$

This is a consequence of *charge–spin coupling*: charge and spin both reside on, and are carried by, the same electron, and it follows that the motion of each electron will contribute to currents of charge and spin. This coupling, of course, only exists in N for the duration of an electron mean-free-path, ℓ. Once electrons begin to scatter among themselves, transport is diffusive, fluxes of charge and spin are statistical properties of the ensemble of particles, and currents of charge and spin become uncoupled. Equation (11.1.11) is valid because transport across the F–N interface is ballistic, and the interface is defined to be spatially restricted to a length less than ℓ.[8,17]

The equations for charge and spin transport in a bulk, nonmagnetic material are given by Eq. (11.1.6). Driven by a gradient of voltage, the currents of charge and of polarized spins are

$$\begin{aligned} J_q &= -\sigma\nabla V \\ J_M &= -\sigma(p_n\beta/e)\,\nabla V = 0 \end{aligned} \qquad (11.1.12)$$

where $p_n = 0$ in a nonmagnetic material. While there is a spin-polarized current associated with an electric current in a ferromagnetic material, $p_f = 0.05\text{–}0.5$ for transition metal ferromagnets, there is no current of polarized electrons associated with an electric current in a nonmagnetic material N. This does not mean, however, that there are no currents of polarized electrons in N. To the contrary, the currents of magnetization are described by the $L'_{3,3}$ term of Eq. (11.1.6):

$$J_M = -\sigma\beta^2/e^2\nabla(-H^*) \qquad (11.1.13)$$

where $\zeta = 1$ is an excellent approximation.[8] This term describes the self-diffusion of polarized electrons.

Spin injection and diffusion can now be described with some rigor. Spin–charge coupling occurs on the length scale of an electron mean-free-path, and an interfacial current of polarized electrons, I_M, generates a nonequilibrium distribution of polarized carriers in N, near the F–N interface. The spatial dependence of $\tilde{M}$ in N is thereafter determined by self-diffusion of the spins.

Recalling from Section 11.1.1.1 that ferromagnetic film F2 will be used as a potentiometer that selectively senses the up- and down-spin subband electrochemical potentials, it is useful to define electrochemical potentials μ_q and μ_s for charge and spin populations in N:

$$\mu_q = (E_{F,N,\uparrow} + E_{F,N,\downarrow})/2$$
$$\mu_s = (E_{F,N,\uparrow} - E_{F,N,\downarrow})/2 \qquad (11.1.14)$$

where $E_{F,N,\uparrow}$ and $E_{F,N,\downarrow}$ have already been defined as up- and down-spin subband electrochemical potentials and we introduce μ to simplify notation. It is readily shown[8] that charge flow is governed by the Laplace equation:

$$\nabla^2 \mu_q = 0 \qquad (11.1.15)$$

and the nonequilibrium spin population is governed by the diffusion equation

$$\nabla^2 \mu_s = \frac{\mu_s}{\delta_s^2} \qquad (11.1.16)$$

For any experimental configuration, Eqs. (11.1.15) and (11.1.16) are applied, along with appropriate boundary conditions.

11.1.1.7. Spin Injection Experiments.

It will be seen in Sections 11.2 and 11.3 that *spin injection* geometries devised for studies of metals have been adapted to *spin injection* studies of hybrid systems. These geometries and experimental methodologies merit a brief review.

The first experimental *spin injection* geometry consisted of a "wire" of bulk, high purity aluminum about 100 µm wide and 50 µm thick.[1,3,4] An array of ferromagnetic pads, about 15 µm wide by 45 µm long, were fabricated by photolithography and liftoff as electrodes on the top surface, with interprobe spacings x in multiples of 50 µm. The F electrodes were e-beam deposited from a single source of $Ni_{0.8}Fe_{0.2}$ in a pressure of 10^{-6} Torr after cleansing the Al surface with an Ar ion mill.

This *nonlocal geometry* can be discussed with the aid of Fig. 11.5. The quasi-one-dimensional wire extends along the $\hat{x}$ axis. A narrow ferromagnetic electrode F1 spans the width of the wire near its center, at $x = 0$. When bias current I is injected at F1 and grounded at the left end of the wire, $x = -b$, the boundary conditions for Eq. (11.1.15) are $V = V_0$ across the line at $x = 0$, and $V(x = -b) = 0$. The familiar solution is simply described as a linear voltage drop from $x = 0$ to $-b$. This is depicted by the regularly spaced equipotential (dotted) lines in Fig. 11.5. Note that the equipotential lines meet the edges of the wire with right angles because current cannot flow out the edges of the wire. However, there is no net current flow in the region $x > 0$ and the wire is at a constant potential from $x = 0$ to $x = b$. A voltage measurement between the end of the wire and a narrow electrode that spans the wire at $x = L_x$ is necessarily a null measurement, $V = 0$.

Spin-polarized electrons are injected at F1, and the boundary conditions for Eq. (11.1.16) are a steady-state source along the line at $x = 0$ and isotropic spin relaxation at a constant rate $1/T_2$.[1,3] Thus, the nonequilibrium spins diffuse equally

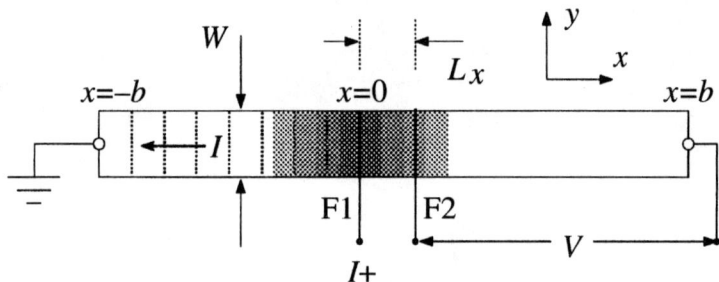

Figure 11.5. Schematic top view of nonlocal, quasi-one-dimensional geometry used in original *spin injection* experiment.[1,3] Dotted lines represent equipotentials characterizing electrical current flow. Gray shading represents diffusing population of nonequilibrum spin-polarized electrons injected at $x = 0$, with darker shades corresponding to higher density of polarized electrons.

along $\pm \hat{x}$ and self-diffusion is symmetric. The density of diffusing spin-polarized electrons is depicted in Fig. 11.5 by the shaded region, with darker shades representing higher density. When the electrode at $x = L_x$ is also a ferromagnetic film, F2, a potentiometric measurement V records a spin-dependent voltage that is relatively high (low) when the magnetization orientation $\vec{M}_2$ is parallel (antiparallel) with $\vec{M}_1$, as discussed in Section 11.1.1.1. Since $V = 0$ in the absence of nonequilibrium spin effects, this is a truly *nonlocal* measurement that uniquely discriminates against any background voltages. The ideal geometry of Fig. 11.5 is achieved when the electrodes have negligible width, W_{F1}, $W_{F2} \ll W$, and have uniform contact across the width of the wire. Departures from ideality result in a small, spin independent baseline voltage $V \neq 0$.

For an effective measurement, it is necessary to choose F1 and F2 to have slightly different coercivities, $H_{c,1} \neq H_{c,2}$, e.g., by using different materials or by using the same material but slightly different shapes for the two electrodes. Thus, in any given measurement, the detected voltage V as a function of externally applied field H should be positive whenever the magnetizations of F1 and F2 are aligned and negative over a field range $H_{c,2} - H_{c,1}$, when antialigned. For the geometry of Fig. 11.5, spin diffusion lengths can be determined by measuring V_{F2} as a function of injector–detector separation, L_x.

Figure 11.6 shows magnetotransport data for a sample with $L_x = 300$ μm at $T = 4$ K. An in-plane field $H = H_y$ is used to control the relative magnetization orientations $\vec{M}_1$ and $\vec{M}_2$ of F1 and F2. The field H_y is swept along the easy axis of F1 and F2 from positive to negative values. The region -80 Oe $< H < 30$ Oe represents the region where magnetizations $\vec{M}_1$ and $\vec{M}_2$ are reorienting between parallel and antiparallel and the detected voltage V_s drops from positive to negative. In the region -200 Oe $< H < -80$ Oe, the orientations $\vec{M}_1$ and $\vec{M}_2$ return to parallel and the original, positive voltage is regained. A field sweep from negative to positive values shows the same feature in the field range -80 Oe $< H < 200$ Oe, as expected for the hysteresis of the ferromagnetic films. The spin accumulation R_s is identified as half the difference of the dip, and measurements of $R_s(L_x)$ resulted in a determination of the low spin diffusion length (at 4 K), $\delta_s = 0.5$ mm.[1,3]

HYBRID DEVICES

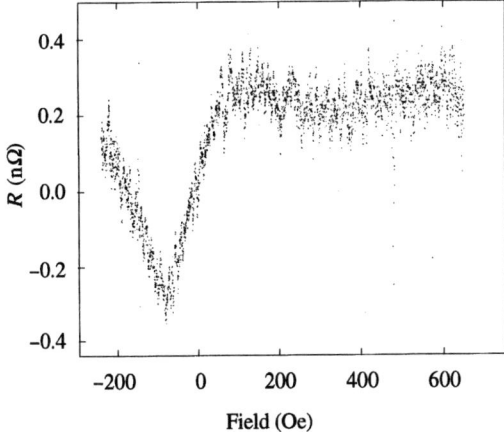

Figure 11.6. Example of data from bulk Al wire sample.[1,3] External field is applied along $\hat{y}$ axis, in the plane of F1 and F2, starting at $H_y = 600$ Oe and sweeping down. The dip occurs when $\vec{M}_1$ and $\vec{M}_2$ change their relative orientation from parallel to antiparallel. $L_x = 300$ μm, $T = 4.3$ K.

The thin film sandwich geometry of Fig. 11.7 was developed for spin injection studies in metal films and the structure is called the "bipolar spin switch."[18] Spin-polarized carriers are injected through two identical "window" junctions F1 into the sample N of thickness d, and they diffuse across the sample thickness.

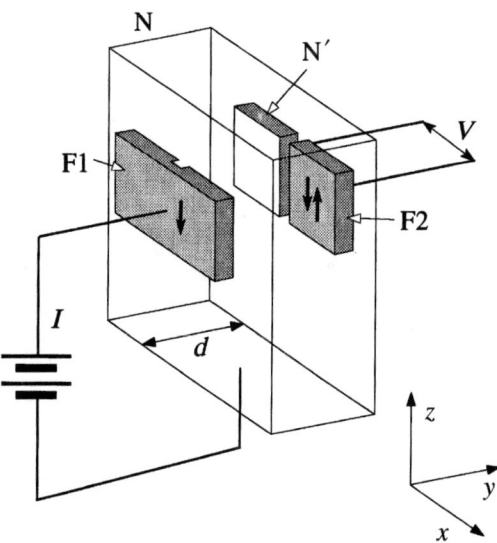

Figure 11.7. Perspective view of sample schematic for thin film geometry known as "bipolar spin switch".[18] Spin-polarized electrons are injected from F1 through two parallel "windows" and diffuse across the sample volume. Detecting film F2 measures the spin subband electrochemical potentials of the "spin accumulation." For the Hanle effect, $\vec{M}_1$ and $\vec{M}_2$ are fixed in the x–z plane and external field H_y is applied along the $\hat{y}$ axis.

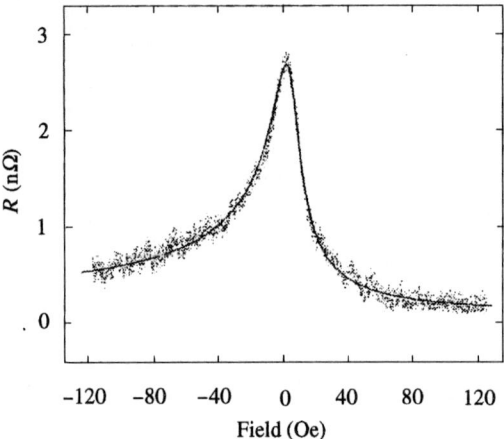

Figure 11.8. Example of Hanle data from bulk Al wire sample,[1,3] presented in units of resistance as $R = V/I$. This is an example of an "absorptive" lineshape, with small admixture of "dispersive" character. $L_x = 50$ µm, $T = 21$ K. Solid line is fit: $T_2 = 7.0$ ns, $P = 0.075$.

The electrochemical potential of electrode F2 measures the nonequilibrium electrochemical potential in N, $E_{F,n,\downarrow}$ and the voltmeter V compares this with the average electrochemical potential $E_{F,n}$ detected by electrode N' (refer also to Fig. 11.1). Spin diffusion lengths can be determined by measuring V_{F2} as a function of sample thickness d, $V(d)$, for a number of samples with varying thickness d.

An alternative methodology for measuring spin relaxation by the spin injection technique is to apply a small transverse magnetic field. In the geometry of Fig. 11.5, e.g., a field applied along the $\hat{x}$ axis or $\hat{z}$ axis is perpendicular to the orientation of the injected spins and causes them to precess. Since the spins are moving diffusively, those that enter the detector film at any given time have a distribution of arrival times and, because of spin precession, a distribution of accumulated phase angle. In the limit of zero external field, all spins have the same phase (zero) and the signal has maximum value. At a sufficiently large magnitude of field, the distribution of accumulated phase angle will span 2π radians, and the detected signal is reduced to zero. This is the same physics as transmission electron spin resonance (TESR), but performed at zero frequency, and the half-width at half-maximum (HWHM) of the observed Lorentzian signal is a measure of the transverse spin relaxation time, $\Delta H = (\gamma T_2)^{-1}$, where γ is the gyromagnetic ratio. Known as the Hanle effect, these kinds of data were taken on the wire samples of the original spin injection experiment.[1,3] An example of the Hanle effect in one of these samples is shown with the data of Fig. 11.8. The external field was applied along $\hat{z}$. The uniaxial magnetization anisotropy axes of F1 and F2 were not perfectly parallel causing the small symmetry deviation from a Lorentzian line shape. The relaxation times T_2 measured by fitting the Hanle data were in excellent agreement with the values deduced from the spin diffusion length, δ_s.

An example of a Hanle measurement on a metal thin film is shown with the data of Fig. 11.9[11] on a Nb sample, using the geometry of Fig. 11.7. The Nb film was 900 nm thick with a resistivity at 10 K of $\rho = 4$ µΩ cm, and F1 and F2 were

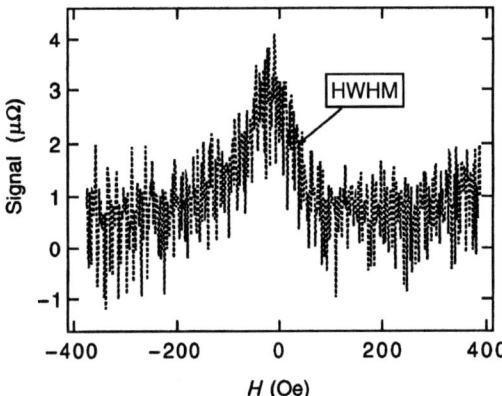

Figure 11.9. Spin relaxation measured in a Nb film by using spin injection and a small transverse magnetic field. The spin dephasing is known as a "Hanle" effect.

both composed of Permalloy. Equation (11.1.4) has two free parameters, the spin relaxation time T_2 and the fractional spin polarization, $\eta_1 = \eta_2 = \eta$. The measured HWHM results in an estimate for the spin relaxation time, $T_2 \approx 0.7$ ns. From Eq. (11.1.4) and the measured result $R_s A d = 2.5 \times 10^{-14}$ Ω cm^3, one finds $\eta \approx 0.34 = 34\%$. A fractional polarization of 34% is in the range of accepted values for Permalloy.[12,19] The data of Fig. 11.9 represent the first measurement of the transverse spin relaxation time in a thin film sample, and thereby demonstrate the utility of the spin injection technique.

One application of the *spin injection* technique in basic research is the study of spin relaxation in metals. By measuring the electron spin relaxation time T_2 and the electron mean (Drude or resistivity) scattering time τ, the fractional probability a_s of spin scattering per scattering event is deduced, $a_s \equiv \tau/T_2$. This is a useful figure of merit for discussions of spin transport. For metal films, a value $a_s \approx 0.002$ was deduced from spin injection studies[13] in the temperature range 4 K $< T <$ 100 K. A similar value, $a_s = 0.0005$ at 295 K, was deduced from a magneto-optic measurement.[20] These values correspond to spin diffusion lengths $\delta_{s,n}$ of roughly 1μm, and $\delta_{s,n}$ has relatively little temperature variation because the resistivity of disordered metal films [and therefore $\tau(T)$] has weak temperature dependence. For the high purity, bulk Al wire samples, the value $a_s = 0.001$ was deduced.[1,3] Since aluminum has low spin–orbit scattering, a smaller value of a_s was expected. This surprising result has been explained by noting that there are "hot spots" on the Fermi surface of Al where spin scattering is extremely rapid.[21]

The quasi-one-dimensional geometry of Fig. 11.5 has many advantages. The detecting F film offers a nonlocal measurement, and the noneqilibrium spins are confined to a small volume. Spin injection experiments in a Van der Pauw geometry have been proposed, but this approach has several problems.[22,23] Because spin diffusion is isotropic, the nonequilibrium spin population dissipates throughout the four arms of the cross, and the density of polarized spins in the center of the cross is relatively low. Furthermore, the intrinsic resistive background of the Van der

Pauw measurement is typically large compared with the spin-dependent shifts of electrochemical potential.

In summary, *spin injection* has become an important tool for the study of spin relaxation in metals. Conceptually, the concepts of using a ferromagnetic film as a source of spin-polarized electrons, and of using another ferromagnetic film as a spin-dependent potentiometer, were demonstrated first in metals and these concepts are now being generalized to *hybrid systems*.

11.2. HYBRID FERROMAGNET–SUPERCONDUCTING DEVICES

Rapid superconducting flux quantum (RSFQ) logic has become the basis of digital superconducting electronics. With prototype circuits operating at 100 GHz, digital superconducting electronics may fulfill requirements for some broad bandwidth, low power, high performance applications. RSFQ uses low temperature superconducting (LTS) Josephson junctions as switching elements, with characteristic speeds of order 10–100 ps, and operates with currents of order 100 µA. While RSFQ has been successful for high speed digital signal processing, there is no satisfactory design for a memory cell. RSFQ device applications, e.g., as broad bandwidth analog to digital converters (ADC), typically use off-chip semiconductor memory, and are hindered by the delay times associated with transmitting data from the cold superconducting chip to a room temperature semiconductor memory chip. Furthermore, Josepshon junctions are not capable of achieving current gain. Thus, one important need of the digital superconducting electronics industry is a superconducting device capable of current gain, and a second important need is a superconducting memory cell. The hybrid ferromagnet–superconducting devices discussed in this section have the potential to satisfy both needs. Equally important, these device structures address fundamental issues of basic research in nonequilibrium superconductivity.

11.2.1. Spin-Polarized Quasiparticle Injection Device (SP QPID)

11.2.1.1. Introduction and Concept.
In the last few years, a large effort has been devoted to the development of three terminal high-T_c superconductor (HTS) devices with current gain, and one viable approach is the quasiparticle injection device (QPID).[24–29] The injection of quasiparticles into a superconductor (S) creates a local nonequilibrium state which suppresses the superconducting order parameter and depresses the critical current density J_c.[30] This concept is briefly reviewed with the aid of Fig. 11.10. The dispersion relation for a thermal equilibrium distribution of quasiparticles (i.e., single particle excitations) is schematically depicted in Fig. 11.10(a). Particles with positive kinetic energy ε_k occupy the *electron* branch and particles with negative values of ε_k occupy the *hole* branch of the quasiparticle spectrum. There are no excitations (particles) with energies in the gap, Δ, and the total energy of a quasiparticle is $E_k = (\Delta^2 + \varepsilon_k^2)^{1/2}$. The thermal equilibrium distribution, which is an *even* mode distribution, is described by a Fermi function.

Figure 11.10(b) depicts an *odd* mode distribution which can arise, e.g., by the injection of electron-like quasiparticles at energies above the gap. A normal metal electrode [Fig. 11.10(c)], when biased with a voltage greater than Δ, drives a current

HYBRID DEVICES

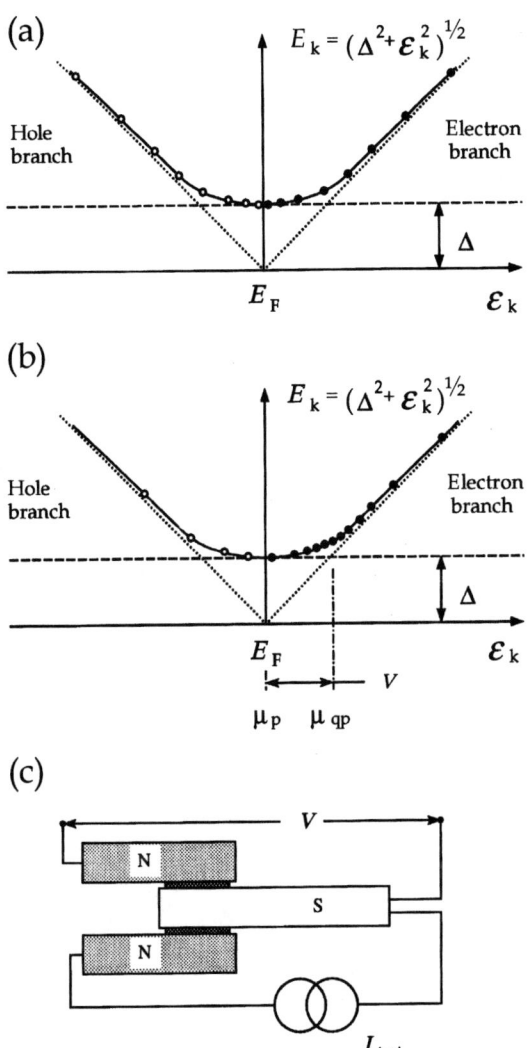

Figure 11.10. Schematic depiction of quasiparticle energy spectrum for BCS superconductors. (a) Thermal equilibrium, (b) odd mode excitation, and (c) experimental geometry for measuring the branch imbalance of (b).

I_{inj} of electron-like quasiparticles into a superconducting thin film strip. Typically, tunnel barriers at the normal metal–superconducting interface are used to inhibit diffusion and confine the injected quasiparticles to the superconductor. The high energy quasiparticles eventually recombine and rejoin the superconducting condensate, but selection rules for recombination require relaxation to the gap edge before recombination events are likely. In the steady state, the bottleneck in the recombination process results in a nonequilibrium distribution of quasiparticles, which can be detected as a *branch imbalance* by using a second normal metal detector electrode [Fig. 11.10(c)] to measure the chemical potential of electron-like excitations.

The detected voltage measures the shift of quasiparticle chemical potential of the odd mode nonequilibrium distribution from the even mode, equilibrium value [Fig. 11.10(b)], and is a measure of the density of nonequilibrium quasiparticles.[31]

The presence of a nonequilibrium quasiparticle population has an effect on the superconductor. The order parameter of S is reduced, and when the nonequilibrium density δN exceeds a critical value N_c, the gap is suppressed to zero and the sample is driven into the normal state. The value of N_c is subject to debate, with predicted values varying in the range $N_c/N_T = 0.15$–0.4,[30,32] where $N_T(T)$ is the thermal equilibrium density of quasiparticles. This effect, the reduction of the order parameter and a resulting decrease of critical current J_c, is the basis of the controlled weak link (CLINK),[33] more generally known as a QPID and schematically depicted in Fig. 11.11. A cross-strip of Pb is fabricated over a superconducting bridge of Sn, separated by a thin tunnel barrier. The operating temperature is chosen so that the Pb is in the normal state, and the Pb electrode acts as a source of electron-like quasiparticles that are injected into the Sn. The injected quasiparticles diffuse in the Sn with a characteristic length scale is of the order 10 μm. For sufficiently large values of injected current, I_{inj}, the density of nonequilibrium quasiparticles exceeds the critical value over a local, finite region and the order parameter of the Sn is suppressed to zero.

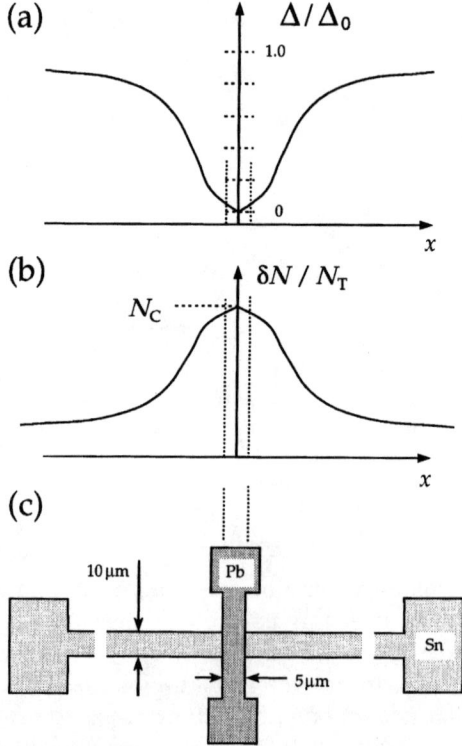

Figure 11.11. Operating principles of the CLINK, a quasiparticle injection device. (a) Suppression of the order parameter and (b) distribution of nonequilibrium density δN that results from quasiparticle injection. (c) Top view of CLINK.

The critical current of the Sn is substantially reduced in this *weak link* region. The appearance of Shapiro steps in the suppressed state[33] confirms that the Josephson phase coherence is maintained across the weak link. Since the existence and spatial extent of the weak link is controlled by the magnitude of injected current, the device was called a "controlled weak link." Current gain in the device is defined as the ratio of the critical current to the magnitude of the injected (control) current, I_c/I_{inj}.

Low-T_c superconductor (LTS) QPIDs[34] have been extensively investigated and a current gain of 4 has been achieved.[35] Substantial research has been devoted to using the CLINK concept with HTS. HTS have a lower carrier density and should be more sensitive to quasiparticle injection, but the gain of HTS QPIDs has been low. Nevertheless, these devices are attractive because of their speed, when limited by the effective quasiparticle relaxation time, may be on the order of 100 GHz.[26,36] In all prior work (both LTS and HTS), the quasiparticles were injected from normal metal electrodes and carried no spin polarization, using a geometry such as depicted in Fig. 11.12(a).

However, as discussed in Section 11.1, a ferromagnetic metal film can inject spin-polarized electrons into a metal, and a ferromagnetic metal (F) electrode can similarly be used to inject spin-polarized quasiparticles into S.[19] Recently, the approach of Fig. 11.12(a) has been modified to create a spin-polarized QPID.[37] An F layer of $La_{2/3}Sr_{1/3}MnO_3$ was used in a bilayer F–S geometry where S was a 60 nm thick $DyBa_2Cu_3O_7$ (DBCO) patterned superconducting channel. Spin-polarized carriers were presumed to be drawn from F into S and a depression of J_c was observed in S. The bilayer geometry was not designed for device applications, the T_c

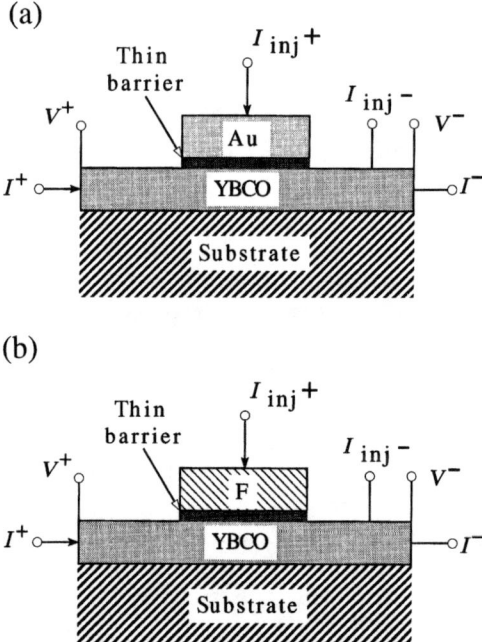

Figure 11.12. Cross-section of (a) HTS QPID and (b) HTS QPID with spin-polarized injection source.

of DBCO was rather low (71 K), and the deduced gain was about unity, i.e., comparable with the gain of unpolarized QPIDs.

However, the experiment suggested that a new mechanism of pair breaking, associated with a nonequilibrium population of spin-polarized quasiparticles in S, may be responsible for the depression of J_c.[38] The concept is described with the quasiparticle dispersion sketches of Fig. 11.13. Spin-up and spin-down quasiparticles do not intermix because spin-flip interactions, which require a torque applied to the spin magnetic moment, are rare. Thus, the spin degree of freedom of quasiparticles must be described with separate branches. Figure 11.13(a) presents a thermal equilibrium distribution of spin-up and spin-down quasiparticles. When spin-polarized electron-like quasiparticles are injected into S from a ferromagnetic electrode [Fig. 11.13(c)], the resulting nonequilibrium distribution can be described as an imbalance of the spin-up and spin-down branches as well as an imbalance of the electron and hole branches. Because quasiparticle spin relaxation is believed to be slow, the creation of a spin imbalance adds another bottleneck to the quasiparticle relaxation process and, therefore, results in a larger population of nonequilibrium quasiparticles. This larger population causes greater gap suppression per unit injected current, and the injection of spin polarized should be more effective than unpolarized quasiparticles for the creation of a depressed critical current. An improved QPID is therefore represented by Fig. 11.13(b), with a ferromagnetic injector to provide spin-polarized quasiparticles [Fig. 11.13(c)].

11.2.1.2. Spin-Polarized QPID Prototype Devices.

A recent set of experiments[39] introduced the device of Fig. 11.12(b), using a standard QPID geometry and a ferromagnetic layer, $Nd_{0.7}Sr_{0.3}MnO_3$ (NSMO), as an injection source of spin-polarized quasiparticles. In this geometry, the quasiparticles were injected fully and directly into a superconducting channel so that the effect could be studied quantitatively. Furthermore, two different types of samples were studied in order to directly compare the injection effect of spin-polarized and unpolarized quasiparticles. One structure was $YBa_2Cu_3O_7$ (YBCO)/$LaAlO_3$ (LAO)/NSMO (superconductor–insulator–ferromagnet; S–I–F) and the other was YBCO/LAO/$LaNiO_3$ (LNO) (superconductor–insulator–nonmagnetic metal; S–I–N), where the LNO is a nonmagnetic metal. In both structures the LAO, with thicknesses of 4–8 nm, is an electrically insulating layer (I) which inhibits a proximity effect between S and F. The layer thicknesses were comparable for both devices: 100 nm for the top YBCO layer and 200 nm for the bottom layer (NSMO or LNO). The advantage of a comparison between two such samples in contrast to comparing the manganate gate with a conventional metallic gate[37] is that the density of carriers, their resistivity, and the natural barrier at the interface are similar for both samples. It is a good approximation that the sole difference between sample types is that one (S–I–N) has an unpolarized quasiparticle source and the other (S–I–F) has a spin-polarized injection source. As discussed further, it was found that the S–I–N structure had characteristics comparable with other unpolarized QPIDs,[24–29] but the S–I–F device (SP QPID) had substantially larger gain.

YBCO/LAO/NSMO (or LNO) heterostructures were fabricated on a (100) LAO substrate by using pulsed laser deposition.[40] Device patterning involved photolithography, selective wet etching, and Ar ion milling and did not affect the T_c of the YBCO

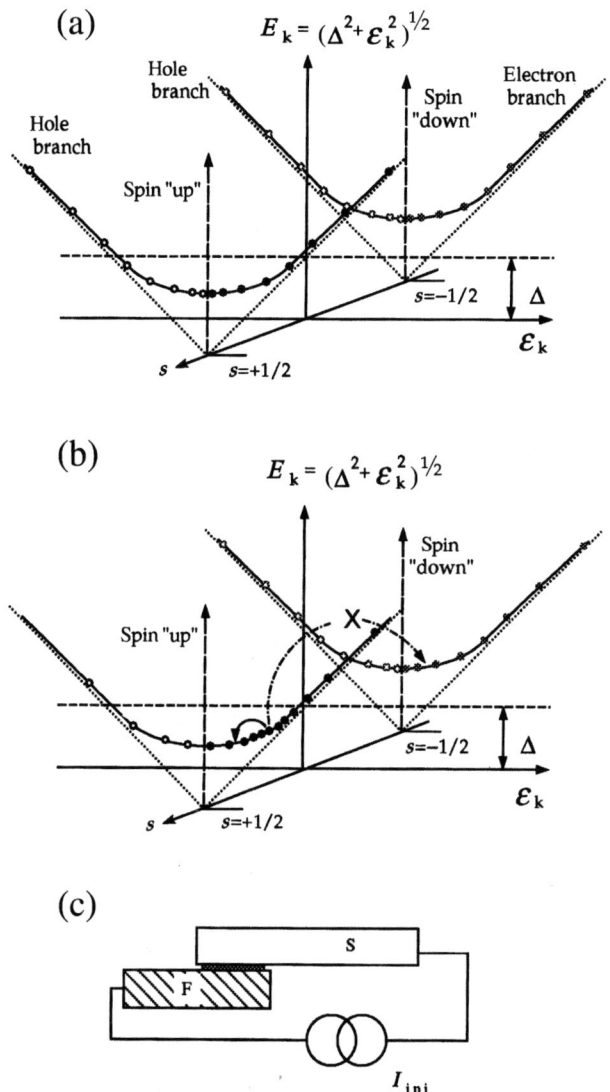

Figure 11.13. Depiction of quasiparticle energy spectrum for BCS superconductor, explicitly accounting for spin-up and spin-down branches. (a) Thermal equilibrium, (b) spin and branch imbalance that result from injection of spin-polarized quasiparticles, and (c) ferromagnetic injection source in contact with superconductor.

nor the properties of the bottom layer (NSMO or LNO). Temperature-dependent resistivity measurements of isolated layers showed that the ferromagnetic transition occurred at ~170 K for the NSMO layer, and the resistivities of the NSMO and LNO layers were ~1000 and ~300 μΩ cm at 100 K, respectively. Vertical transport measurements indicated that the resistivity of the LAO barrier layer was $\leq 10^3$ Ω cm, resulting in an area resistance $R_1 A$ on the order of 10^{-4} Ω cm^2. Typically, the T_c of the YBCO over the NSMO layer varied from 80 to 89 K.

Current–voltage curves (I–V) for a number of devices were recorded over the temperature range of 67 – 87 K for a variety of injection currents. For $I_{inj} = 0$, the I–V trace is symmetric about $I = 0$ and a $+2$ µV criterion was used to define I_{c1} and a -2 µV criterion was used to define I_{c2}. The width of the zero voltage plateau is $I_{c1} - I_{c2}$ and the center was defined to be $I_{shift} = (I_{c1} - I_{c2})/2$. Thus, for $I_{inj} = 0$ one has $I_c = |I_{c1}| = |I_{c2}|$ and $I_{shift} = 0$. As I_{inj} increases in magnitude (for both polarities), one finds $I_{shift} = -I_{inj}$ and I_c decreases slightly. The observation $I_{shift} = -I_{inj}$ confirms that all the injected current enters the superconducting channel. The gain G of the QPID is defined as $G = \Delta I_c / I_{inj}$ where ΔI_c is the reduction of critical current. Gains for the S–I–N QPIDs were of the order 0.01–0.1 and increased with decreasing temperature, consistent with previous reports.[24–29]

The family of I–V curves for a YBCO/LAO/NSMO (S–I–F) device are qualitatively similar to those of the S–I–N QPID, and one again observes $I_{shift} = -I_{inj}$. However, I_c is substantially diminished for relatively small values of I_{inj}. Gains for the S–I–F QPIDs were of order one or larger, and increased with decreasing temperature.[39] The values of current gain in the S–I–F devices varied with temperature and barrier thickness for a number of samples, and a gain as large as 5.0 at 74 K ($t = 0.87$) was observed. In all cases, for comparable devices (same barrier thickness) and temperatures, the gain of each S–I–F QPID was 10–30 times larger than the gain of the comparable S–I–N QPID.

11.2.1.3. Discussion of SP QPID Prototypes.

After considering and eliminating possible spurious mechanisms, the suppression of J_c can be modeled as the result of spin injection. It is not plausible that joule heating is responsible for critical current suppression in these devices. For S–I–N QPIDs, the maximum injection power density, $P_{inj} = J_{inj}^2 R_1 A$ was ~ 2 W/cm^2, much less than the value 100–1000 W/cm^2 required to raise the YBCO film temperature by a sufficient amount (several K) to depress I_c.[39] Since each pair of S–I–F and S–I–N devices was epitaxially grown under identical conditions, and the N and F materials are structurally quite similar, the thermal properties (heat capacities, thermal conductivities, thermal coupling to the substrate and to the bath) of the S–I–N and S–I–F devices are expected to be nearly the same. However, the S–I–F QPIDs had a lower injection power density, $P_{inj} \approx 1$ W/cm^2, and therefore the larger depression of I_c cannot possibly be related to thermal effects.

A semiquantitative nonequilibrium thermodynamic analysis[28] can be presented to facilitate a comparison of these SP QPID results with those of other QPIDs. Quasiparticles are injected into S with energies P_{inj}/J_{inj} of the order 10 mV, comparable with (or larger than) the superconducting energy gap. The energetic quasiparticles can break pairs before relaxation and recombination occurs with a lifetime τ_r. Nonequilibrium quasiparticles can cause gap suppression when the nonequilibrium density, $\delta N = [J_{inj}\tau_r(T)/et]$, with t the thickness of S, is the same order as the critical density, $\delta N \sim N_c \approx N(0)\Delta(0)$, where $N(0)$ and $\Delta(0)$ are the zero temperature density of states and energy gap, respectively. The gain of the S–I–N QPIDs is comparable with that of previously studied devices. Using the comparable values of $\tau_r \sim 10$ ps, $N(0) \sim 10^{21}$ cm^{-3} eV^{-1}, $\Delta(0) \sim 10$ meV and the assumption[28] that the equilibrium density of pairs is reduced by the ratio of the calculated thermodynamic critical field to the observed value, one finds that δN ($\sim 10^{17}$ cm^{-3}) is sufficiently large to suppress I_c if the gate current is injected into S in a region with a relatively narrow width of order 1–10 μm. The temperature dependence of the gain, which increases with decreasing temperature, is consistent with the measured temperature dependence of $\tau_r(T)$.

In conclusion, the observation of significantly larger gains for S–I–F QPIDs strongly suggests that spin-polarized quasiparticles generate a substantially larger nonequilibrium population. Evidence for the existence of a spin injected nonequilibrium quasiparticle distribution has been observed in Nb thin film samples.[41] Recently, small superconducting aluminum "islands" were subjected to spin injection, and the suppression of the order parameter was measured directly.[42]

11.2.2. Mesoscopic Magnetoquenched Superconducting Valve

11.2.2.1. Introduction and Concept. Interest in superconducting digital electronics derives from the prospect of low power, high density and high speed integrated device arrays. As discussed in the introduction to Section 11.2, the approach is usually based on microfabricated Josephson junction devices, or on superconducting quantum interference devices (SQUIDs) which incorporate two or more Josephson junctions.[43] A competitive standard for density and bandwidth is 10^7 gates GHz. At higher densities and/or speeds, the low power dissipation and negligible interconnect delays afforded by superconducting electronics may outweigh the disadvantage imposed by refrigeration costs. Despite continuing progress that has resulted in 80 GHz bandwidth operation, the technology has not achieved widespread success.

The quasiparticle injection approach described in Section 11.2.1 is a nonequilibrium approach and requires steady-state power to maintain one of the two binary states. The *mesoscopic magnetoquenched superconducting valve* (MMSV) is a low power alternative in which the device is always in thermodynamic equilibrium.[44–46] In a simple, ferromagnet–superconductor bilayer geometry, locally strong magnetic fringe fields generated by the ferromagnet create a weak link in the superconductor. The weak link can be turned ON or OFF by changing the orientation of the ferromagnet's magnetization, and the ON/OFF state is maintained without power because of the magnet's remanence. The MMSV is a tunable device, whose properties can be modified post-fabrication.

The magnetoquenched superconducting valve is schematically depicted in Fig. 11.14(a) along with integration components that will be discussed further. The basic device comprised a simple bilayer of a thin superconducting strip S and a microstructured ferromagnetic film F that overlaps the width of S. A thin layer of insulator inhibits any proximity effect between F and S. For digital operation, F is fabricated with magnetic anisotropies along the $\hat{x}$ and $\hat{y}$ axes and its function in the device is to provide a magnetic fringe field $\vec{B}$. When the magnetization $\vec{M}$ is oriented along $\hat{x}$, the dipolar field, sketched in Fig. 11.14(a), has a large magnitude near the edges of F that are parallel with $\hat{y}$. In the vicinity of S, the fringe field has a weak magnitude and a direction approximately in the substrate plane. In this *unquenched state*, there is negligible effect on superconductivity in S. When $\vec{M}$ is oriented along $\hat{y}$ [Fig. 11.14(b)], the large magnitude dipolar fringe fields at the edges of F that are parallel with $\hat{x}$ are directly above S and the perpendicular component

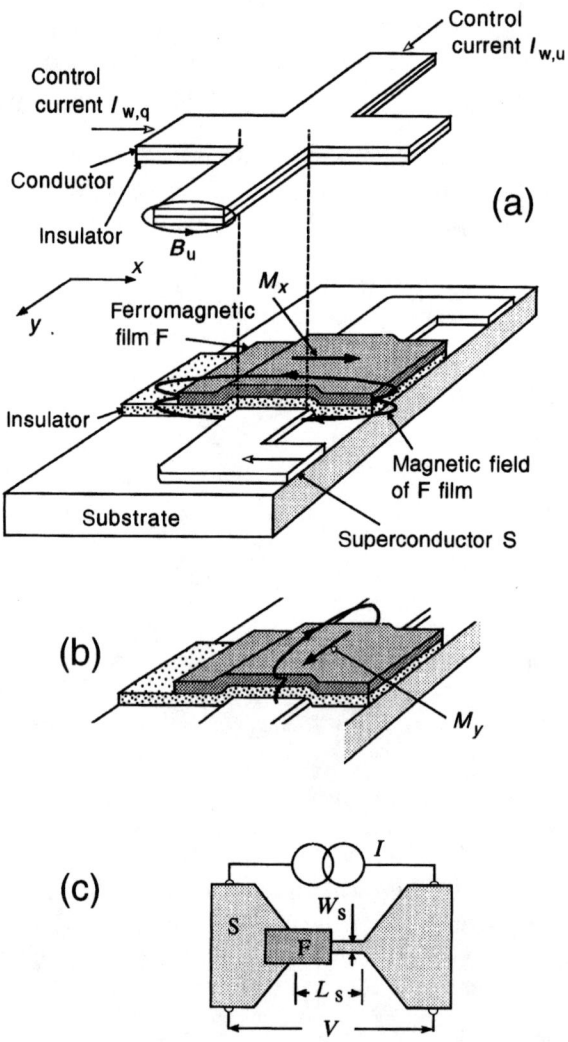

Figure 11.14. Schematic of MMSV device. (a) Unquenched state, $\vec{M} \parallel \hat{x}$. Exploded view shows "write wires" that can be used for integrated operation. (b) Quenched state, $\vec{M} \parallel \hat{y}$. (c) Top view scehmatic of device with single weak link at F film edge.

B_z is substantial. In this *quenched state*, the fringe field penetrates S and forms a "weak link" with depressed critical current I_c. As discussed below, this is a true Josephson weak link that supports phase coherence of the superconducting order parameter. Because the remanence of F maintains the magnetization orientation of F, the device is inherently nonvolatile: energy is required only to set the orientation to be along $\hat{x}$ or $\hat{y}$, after which the device state is maintained with zero power.

Figure 11.14(c) shows the geometry of device structures discussed in this section, with the F film fabricated such that a single edge of F is used to create a single weak link in a strip of width $W_s = 1, 2,$ and 5 μm. For initial prototypes, the S

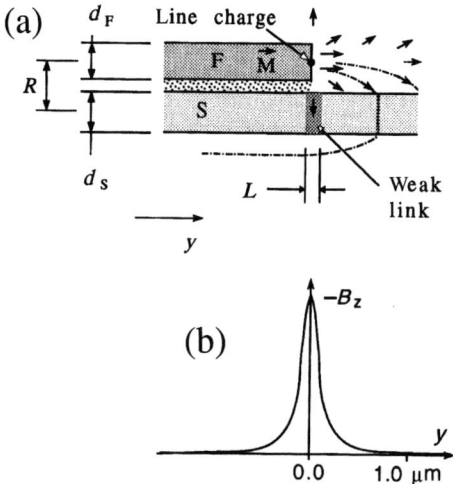

Figure 11.15. (a) Cross-section view of MMSV, showing fringe field generated at edge of F. (b) Spatial profile of B_z, from Eq. (11.2.1).

strip was Pb, thermally deposited in a vacuum of 10^{-6} Torr, oxidized by glow discharge in a 10^{-1} Torr partial pressure of oxygen for 60 s at 500 V to form a thin insulating barrier, and patterned by photolithography and an Ar ion mill. The thickness 40 nm $< d_s <$ 120 nm is the order of the bulk London penetration depth, $\lambda \approx 40$ nm. For this thickness range, Pb is effectively type II.[47] The F film is Permalloy, e-beam deposited from a single charge of $Ni_{0.8}Fe_{0.2}$ in a vacuum of 10^{-6} Torr with a thickness 120 nm $< d_F <$ 210 nm, and patterned by optical lithography and liftoff. Standard dc four-probe techniques were used to measure the transport properties of the strip and weak link.

Details of the device can be understood with a simple magnetostatic model sketched in Fig. 11.15. When $\vec{M}$ is saturated along $\hat{y}$, the dipolar fringe field $\vec{B}$* (depicted in Fig. 11.15(a) with arrows and/or dashed field lines) can be modeled as the result of a surface density of magnetic "poles" of density M_s, and $\vec{B}$ is largely shielded by S at positions remote from the edge of F. There may be small intermediate regions formed in the superconducting strip to allow the penetration of flux, but most of the flux is shielded and/or deflected to the side of S. Very near the edge of F, however, the fringe field can be modeled as the result of an equivalent *line charge* of magnetic poles of density $\lambda_s = M_s d_F$. Directly beneath the edge of F, the perpendicular component B_z is approximately given by[44]

$$B_z = \frac{2M_s d_F R}{(y^2 + R^2)} \quad (11.2.1)$$

with R the center-to-center separation of S and F. The component B_z has a large magnitude and a spatial profile that decays on the order of a fraction of a 1μm

---

*We adopt the convention that $\vec{B}$ refers to magnetic fields generated by F and H refers to externally applied magnetic fields, and use the units Oe and G interchangeably.

[Fig. 11.15(b)]. For Permalloy, with $4\pi M_s = 11,000$ Oe, and for typical values of d_F and d_S the peak magnitude of B_z is about 3000 Oe at the top surface of S and 2000 Oe at the bottom surface. Lead films typically have critical field values $H_{c2}(T=0) \approx H_c(0) = 800$ Oe, the thermodynamic value.[47] The large, highly localized magnetic field B_z generated at the edge of F is much larger than H_{c2} and creates a weak link of length L in the Pb strip.

11.2.2.2. Device Operation.

Creation of a weak link by the magnetoquenched valve mechanism is demonstrated by comparing the critical current density J_c of S for the two remanent orientations, $\vec{M} = M_x \hat{x}$ and $\vec{M} = M_y \hat{y}$. For a Pb sample with $W_s = 5$ μm, $d_F = 210$ nm, and $d_s = 120$ nm, this effect is demonstrated with the data of Fig. 11.16. External field pulses $H(\Delta t)$ are applied to set the magnetization orientation of F along $\hat{x}$ or $\hat{y}$, and I–V curves are taken in the remanent state $(H = 0)$. At $T = 7.0$ K [Fig. 11.16(a)], near T_c, the critical current in the unquenched state $I_{c,u}$ is about $I_{c,u} = 0.7$ mA (using a 2 μV criterion). In

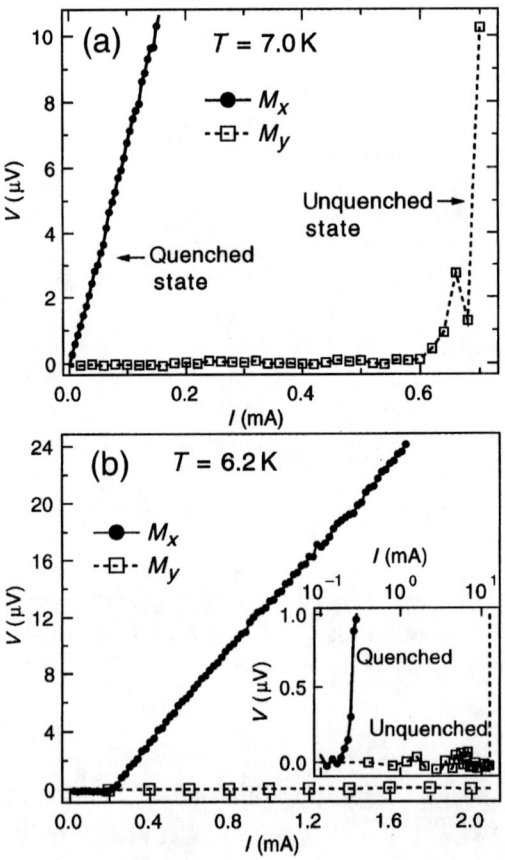

Figure 11.16. Examples of measurement of I_{cx} and I_{cy} in the quenched and unquenched states. (a) $T = 7.0$ K, (b) $T = 6.2$ K, and inset: semilog I–V plot that identifies I_{cy} as 15 mA.

the quenched state, $I_{c,q}$ is suppressed to zero. At a lower temperature [Fig. 11.16(b)], $I_{c,q}$ has a nonzero value $I_{c,q} = 0.3$ mA and the linear I–V characteristic for larger values of I is indicative of weak link behavior. As seen in the inset of Fig. 11.16(b), $I_{c,u} \approx 15$ mA is about 50 times larger. In other words, the magnetoquench effect suppresses the critical current by a factor of 50.

Figure 11.17(b) presents a semilog plot of the temperature dependence of J_{cx} ($= J_{c,u}$) and J_{cy} ($= J_{c,q}$) for the unquenched and quenched states, respectively, for the sample studied with the data of Fig. 11.16. At $T/T_c = t = 0.58$ ($T = 4.2$ K), the difference between J_{cx} and J_{cy} is nearly fivefold. As temperature increases, the gap between the two states widens and diverges near T_c, with T_c suppressed by about 0.8 K in the quenched state. Near T_c, J_{cx} (open squares) follows the Ginzburg–Landau form for a thin film, $J_{cx} \propto (1 - t)^{3/2}$ (dashed line) over a temperature range, $0.8 < t < 1$, greater than normally applied to Ginzburg–Landau theory. This confirms that F has no effect on S in the unquenched state. On the other hand, J_{cy} is

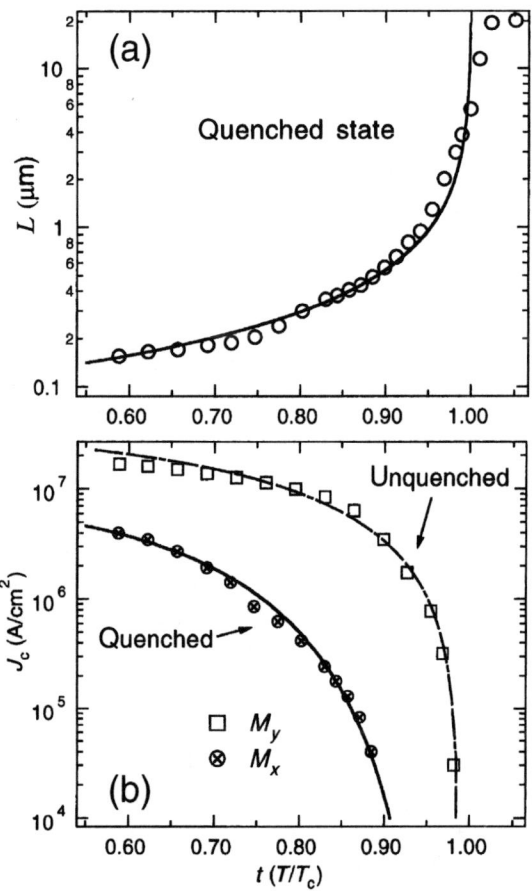

Figure 11.17. (a) $L(T)$, determined by slope of I–V and resistivity of Pb at 10 K. (b) Semilog plot of $J_c(T)$ for quenched and unquenched states. J_{cy} is suppressed by two orders of magnitude for a large range of t.

described by $J_{cy} \propto (1-t)^{1.3}$ over a similar temperature range, $0.83 < t_y < 1$, with $t_y \equiv T/T_{cy}$ and $T_{cy} = 6.5$ K. To understand the behavior of J_{cy}, it is useful to consider the product $I_{cy}(T)R_n(T)$ as a figure of merit of the junction, where $R_n(T)$ is taken as the slope of $V(I)$ for $I \gg I_{cy}$ and is strongly temperature dependent. Over a fairly wide temperature range, $0.8 < t_y < 1$, $I_{cy}R_n$ follows the linear relationship $I_{cy}R_n \approx (T_{cy} - T) \cdot 14$ µV/K (solid line), with the value $I_{cy}R_n \approx 50$ µV at the lowest experimental temperature, $T = 4.2$ K. The linear temperature dependence near T_c is predicted for a variety of S–I–S and S–I–N junctions, but the case of a temperature dependent R_n and the concomitant temperature dependent weak link length $L(t) \propto R_n(t)$ has not yet been treated. Thus, it is noteworthy that we observe this simple form for $I_{cy}R_n$ in our unusual, temperature dependent S–N–S junctions.

The slope R_n of $V(I)$ in the quenched state (refer to Fig. 11.16) and the resistivity of the Pb in the normal state can be used to deduce the length L of the weak link. Figure 11.17(a) is a semilog plot of the temperature dependence of L for the sample represented by the data of Figs. 11.16 and 17(a). The temperature dependence of L should be set by the upper critical field $H_{c2}(T)$ and the penetration depth $\lambda(T)$ of the superconducting Pb. The simple model derived from Eq. (11.2.1) and plotted in Fig. 11.15(b) must be modified to account for Meissner screening. Below T_c, we assume exponential penetration of the fringe magnetic field into S on a length scale defined by $\lambda(T)$, $B_z(y, T < T_c) \approx B_z(y=0) \exp[-|y|/\lambda(T)]$. The half-length of the weak link can be estimated as the point $y = L/2$, where $B_z(y)$ has magnitude equal to the critical field B_{c2}:

$$B_z(L/2) \approx B_z(0) \exp[-L/2\lambda(T)] = B_{c2}$$

Solving for L gives

$$L \approx -2\lambda \ln\left(\frac{B_{c2}}{B_z(0)}\right) = -2\lambda_0(1-t^4)^{-1/2}[\ln(1-t) + \alpha] \quad (11.2.2)$$

where $\lambda(t) = \lambda_0 (1-t^4)^{-1/2}$ is the usual two-fluid form.[48]

When L is the same order as the temperature dependent coherence length of the normal metal in the weak link, $L \approx \xi_n \sim 200$ nm, the superconducting order parameter should maintain phase coherence across the weak link. A unique property of a Josephson junction is that a dc voltage difference V across the junction results in an oscillating supercurrent of frequency $\omega = 2eV/\hbar$. If the junction is exposed to a microwave field of frequency ω_0, resonances occur in the junction at frequencies $\omega = n\omega_0 (n = 0, 1, 2, \ldots)$ or, equivalently, at voltages $V = n\hbar\omega_0/2e$, as Cooper pairs can absorb and emit quanta of the external field while passing the static voltage difference. This gives rise to a stepped I–V curve with steps of height $\Delta V = \hbar\omega_0/2e = (2.1 \times 10^{-15}$ V/Hz$) (\omega_0/2\pi)$(Hz), the well-known Shapiro steps associated with the ac Josephson effect.[49] Observation of Shapiro steps is the canonical test for Josephson phase coherence in a weak link.

Microwave power was applied to a magnetoquench valve sample by using a standard coax antenna.[50] In the inset of Fig. 11.18(a), data are plotted for the *unquenched* state at $t = 0.75$. The trace with closed squares and dashed lines represents the I–V curve with zero rf power. The solid line trace represents the I–V curve for identical conditions, in the presence of $(\omega_0/2\pi) = 0.75$ GHz microwave

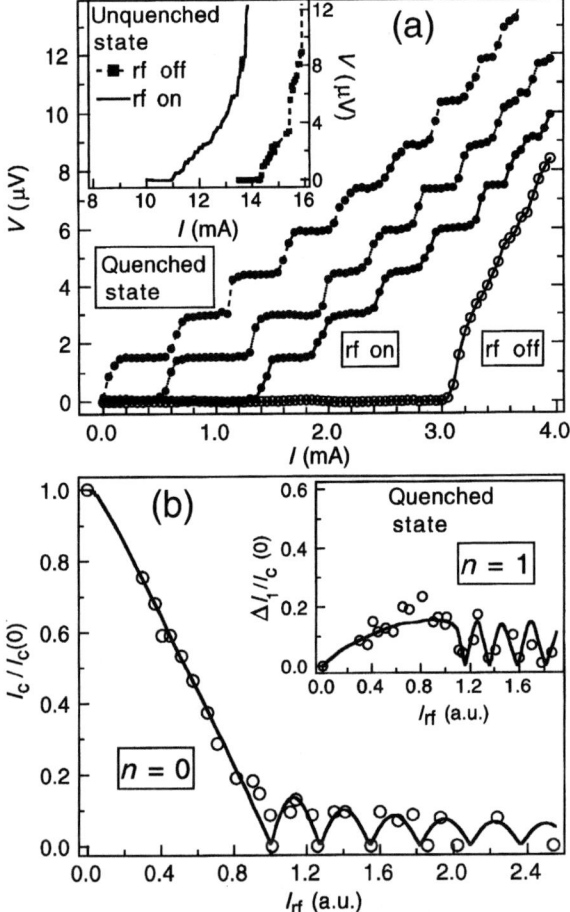

Figure 11.18. (a) I–V in the presence of microwave power. $(\omega_0/2\pi) = 0.75$ GHz, $T = 5.5$ K, quenched state. Open circles: no rf power. Close circles solid line: 2 dB m power. Closed circles dotted line: 8 dB m. Closed circles dashed line: 11 dB m. inset: unquenched state. Closed squares: no rf power. Solid line: 10 dB m. (b) Fitting the $n = 0$, Shapiro step width to an S–N–S model. inset: Fitting the $n = 1$ Shapiro step width.

radiation at an arbitrary power level of 10 dB m (order of magnitude 10 nW coupled to the sample). The critical current is reduced due to additive ac current coupled into the S strip from the microwave field. Figure 11.18(a) presents data from the same sample under the same conditions, in the *quenched* state. The data plotted with open circles and a dashed line represent the I–V curve with zero rf power, and with I_{cy} reduced by a factor of 5 relative to I_{cx}. The data plotted with closed circles and a solid line represent the same I–V curve under microwave irradiation at a power of 2 dB m, about 16% of the power of the solid line trace in the inset. The critical current is reduced, and Shapiro steps of constant height appear in the I–V curve, with $\Delta V \approx (2.0 \times 10^{-15} \text{ V/Hz})(\omega_0/2\pi)(\text{Hz})$. The two left-hand traces represent the same sample conditions with increasing microwave power, 8 dB m (or about 63%

of the power in the comparable trace in the inset) and 11 dB m (126% of the power level). The critical current is further reduced, and more steps appear in the I–V curve. In general, the width of each step varies as a function of microwave power, and the number of steps observed is limited only by the onset of "hot spot" behavior at high values of current and voltage. The power dependence of the steps follows the Bessel function form predicted for these S–N–S junction[50] for both the $n = 0$ and $n = 1$ step [inset of Fig. 11.18(b)]. The observation of Shapiros steps confirms the presence of the ac Josephson effect in these switchable weak links.

Recently, a switchable SQUID was fabricated and demonstrated interference of superconducting phase.[51] Magnetoquench weak links were fabricated on each of two arms of a loop, and a small external field H_z was applied perpendicular to the loop plane. Interference oscillations were observed when the weak links existed in the quenched state, and the oscillations disappeared in the unquenched state.

11.2.2.3. Applications of Magnetoquench Valves.

The operation of prototype devices demonstrated above used an external field to set the magnetization state $\vec{M}$ and, therefore, the device state. A variety of techniques can be used to achieve integrated operation, and a simple approach is illustrated in Fig. 11.14(a) with an exploded view of two write wires inductively coupled to F. A current pulse $I_{w,u}$ transmitted along a write wire parallel with $\hat{y}$ generates a local magnetic field $B_u \hat{x}$ oriented along $\hat{x}$, which can be used to orient $\vec{M}$ along $\hat{x}$ thereby setting the device in the unquenched state. Similarly, an appropriate current pulse $I_{w,q}$ transmitted along a write wire parallel with $\hat{x}$ can set the device to the quenched state. Because the coercivity of F is on the order of 10 Oe, current pulses with amplitudes of a few milliamperes are sufficient to control the device,[52] and thus control much larger currents in S of the order of a few hundred milliamperes. In this way, a superconducting device with current gain could be devised. While the basic principles of integrated operation have been demonstrated with prototypes,[53,*] current gain has not yet been achieved.

The control pulse needs to be applied only for a duration equal to the switching time, after which the device state is nonvolatile. Thus, the device is intrinsically a nonvolatile memory cell, with the binary values ("0", "1") stored as the magnetization state $(M\hat{x}, M\hat{y})$. When S is biased with a current of amplitude I_b intermediate between the critical current values of the quenched and unquenched states, $I_{c,y} < I_b < I_{c,x}$, the binary values also correspond to zero and nonzero voltage across the weak link, $(0, V_b)$. Since the storage function is performed by the state $\vec{M}$, readout is nondestructive and storage is truly nonvolatile: Memory is not lost when the device is warmed above the superconducting transition temperature T_c, in contradistinction with other superconducting memory cells.

The switching time τ_s of the device will be set by the times required for magnetization reorientation in F and for flux penetration in S. An upper bound for both cases is about 1 ns, and times could be as fast as 1 ps.[50] A singular advantage of superconducting electronics is low power dissipation resulting in high packing densities, short interconnects, and therefore high system speed. While SFQ devices

---

*The integrated magnetoquenched valve bears a superficial resemblance to the cryotron, an early superconducting switching device.

run on a constant clock cycle, an advantage of the switchable weak link is that it draws energy only during individual read or write processes and needs zero quiescent power. A further advantage is its simplicity. For example, a *single* magnetoquenched valve constitutes a memory cell, but four to six Josephson junctions are required to form an SFQ memory cell, which is further hindered by destructive readout.

In summary, the magnetoquenched superconducting valve is a switchable Josephson weak link. Prototype devices have been fabricated on the micrometer size scale and have demonstrated binary switching capability and coherence of the superconducting phase across the weak link. Relying only on fringe magnetic fields produced at the edge of a ferromagnetic film, the device is scalable: there is no apparent impediment to fabrication of nanometer-scale devices. Using an upper bound of $\tau_s = 1$ ns, achieving a chip of integrated devices with a density–bandwidth product greater than 10^7 gates 1 GHz seems well within reason.

11.3. HYBRID FERROMAGNET–SEMICONDUCTOR DEVICES

As magnetoelectronics moves towards greater integration with semiconductor electronics, several issues must be addressed.[54] Semiconductor device technology operates within some constraints. Supply voltages are in the order of a few volts (with a trend towards reduction to 1 V or less), device impedances are in the order of 1–10 kΩ, and typical supply currents are 0.9 mA or less. Fabrication procedures involve annealing processes at temperatures of a few hundred °C, and rapid thermal annealing requires tolerance of about 700°C for periods of a few minutes.

Regarding *Spintronic* devices, all-metal magnetoresistive devices have characteristically low impedance, of the order 100 Ω or less, and draw large currents (of the order tens of milliamperes) during readout. At high current densities, the F–N layers are susceptible to degradation by electromigration. Furthermore, these F–N layers, whose magnetic and/or magnetoresistive properties are sensitive to thickness variations of a few angstroms, show poor tolerance to high processing temperatures. Interdiffusion at the F–N interfaces results in significant and irreversible changes of device characteristics.

Magnetic tunnel junctions (MTJ) have characteristically high impedance, of the order 0.1–10 MΩ for devices of a competitive size, and also show poor tolerance to high processing temperatures. High impedance devices are susceptible to relatively high noise and relatively long RC times, where C is the device capacitance. The resistance of MTJ prototypes is sensitive to 0.01 nm variations of the tunnel barrier thickness. Prototypes have poor reproducibility of parameters (and therefore poor yield) over a full 6 in. or 8 in. wafer, and from wafer to wafer. Although there has been recent progress in making MTJs with lower impedance, these devices have poor yield and higher noise.

All magnetoresistive devices that are composed of two or more F layers, including all-metal magnetoresistors and MTJs, have magnetic properties that depend sensitively on details of interlayer coupling and of domain dynamics. They have relatively poor margins for magnetic switching characteristics.

Hybrid ferromagnet–semiconductor devices offer an alternative approach and an opportunity to avoid some of these problems. Device impedance is typically in the range of 0.1–10 kΩ, and readout currents are 1 mA or less. The hybrid devices discussed in this section use a single F film, or two F films which are in a lateral geometry and are therefore uncoupled. Interdiffusion is not a concern and hybrid devices should tolerate high processing temperatures. The margins for magnetic switching, which depend on a single F layer, are narrow, and device yield is high. This section discusses two device families: the ferromagnet–semiconductor nonvolatile gate and the spin injected field effect transistor.

11.3.1. Ferromagnet–Semiconductor Nonvolatile Gate

In this section, a hybrid device for nonvolatile memory applications is described, characteristics of a nonvolatile random access memory (NRAM) cell are discussed, and a magnetoelectronic approach to reprogrammable logic is presented. Section 11.2.2 described a superconducting device that utilized the magnetic fringe field of a microfabricated ferromagnetic element. The ferromagnet–semiconductor nonvolatile gate applies the same concept to a semiconductor Hall plate structure. In contrast with the magnetoresistive approach, this device uses a single bistable ferromagnetic layer which is electrically isolated from the Hall cross. This novel approach has several advantages. The basic device is a simple bilayer, requiring only two lithographic levels and a single alignment. It is compatible with GaAs or CMOS processing, and device isolation can be readily achieved by using Schottky diodes. The simplicity of the device places minimal requirements on fabrication of the ferromagnetic element, and evidence from prototypes suggests high yield. Signal levels are high, and the device shows inverse scalability: smaller devices have improved output characteristics.

11.3.1.1. Introduction and Concept. The fringe field, ferromagnet–semiconductor nonvolatile gate is also called the hybrid Hall effect device (HHED).[55,56] The operation of the device is based on a simple concept, presented schematically in Fig. 11.19. A thin ferromagnetic film F is fabricated above a Hall cross, electrically isolated from the carrier layer. One edge of F, referred to below as the "active edge," is positioned over a central region of the cross. F is composed of a transition metal ferromagnet, its magnetization M is constrained to the film plane, and it can be made with an easy axis along the $\hat{x}$ axis so that M is bistable with orientations $\pm \vec{M}$ along $\hat{x}$. When M is saturated along $-\hat{x}$, there is a large magnetic fringe field B near the "active edge," and this local field has a large component perpendicular to the plane of the carriers and pointing down, $-B_z$ [Fig. 11.19(b)]. The component $-B_z$ is spatially inhomogeneous but nonetheless is sufficiently large to generate a negative Hall deflection on carriers in the semiconductor [Fig. 11.19(a)] and a negative voltage is developed between terminals S1 and S2. When the magnetization orientation is reversed, so that M is saturated along $+\hat{x}$, the magnetic fringe field near the "active edge" changes sign, the perpendicular component $+B_z$ changes sign, and a positive Hall deflection is generated. Thus, reversing the magnetization of F reverses the local field at the "active edge" and reverses the polarity of the output voltage at terminals S1 and S2. The result is a magnetoelectronic device with a very simple bilayer geometry,

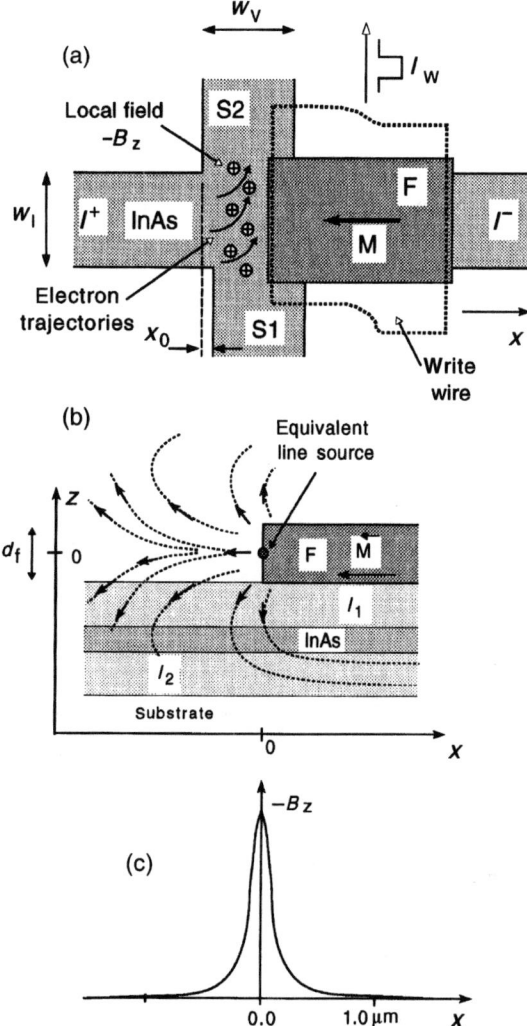

Figure 11.19. (a) Schematic diagram of HHE prototype geometry, Top view: When the magnetization is $-M\hat{x}$, the field at the Hall cross is $-B_z$ and the output voltage $V_{S1,S2}$ is negative. Dotted line represents a write wire fabricated above F. A positive (negative) "write pulse" I_w orients $\vec{M}$ to the left (right). (b) Cross-section view, showing fringe field near the "active edge" of F. (c) Profile of the spatial dependence of the perpendicular component B_z.

intrinsically bipolar output and excellent output characteristics for application as a digital storage cell.

By incorporating a small lithographic offset [x_0 in Fig. 11.19(a)], a small resistance is added to the sense measurement, $V_{S1}-V_{S2}$. In this way, the bipolar output can be shifted to levels of zero (LOW, binary "0") and $2\Delta V_H$ (HIGH, binary "1"), where ΔV_H is the Hall voltage induced by the fringe field $|B_z|$. As discussed in Section 11.2.2, B_z has a spatial decay length the order of 0.1 μm [Fig. 11.19(c)] and the HHED is characterized by *inverse scalability*: Output levels improve as the device dimensions shrink. Fabrication of the device with small dimensions w_V and w_I ensures utilization of the high field region, and output levels increase.

Integrated device operation incorporates the use of a write wire [Fig. 11.19(a)], so that the magnetization orientation $\vec{M}$ is controlled by the local magnetic fields associated with positive or negative "write" current pulses.

11.3.1.2. Prototype Fabrication and Quasistatic Characteristics.

Prototypes have been fabricated using III–V compound semiconductor heterostructures for the Hall plates.[57] In a typical prototype set, the high mobility heterostructure is grown by molecular beam epitaxy (MBE) and consists of: 3 nm InAs/25 nm $Al_{0.6}Ga_{0.4}Sb$ (I_1 in Fig. 11.1C)/15 nm InAs/200 nm $Al_{0.6}Ga_{0.4}Sb$/3 μm AlSb (I_2 in Fig. 11.1C)/semi-insulating GaAs (001) substrate. Doping of the InAs layer, the single quantum well (SQW), is achieved by an arsenic soak in the middle of the 25 nm $Al_{0.6}Ga_{0.4}Sb$ barrier.

Prototypes have been fabricated with feature sizes as small as 300 nm by using electron beam lithography. For the demonstration of integrated operation described further, a micrometer-scale Hall cross was fabricated by photolithography and a mesa etch. The F layer was then e-beam deposited from a single charge of $Fe_{0.1}Co_{0.9}$ in the presence of a growth field of about 200 Oe oriented along $\hat{x}$. The thickness was $d = 60$ nm, and the rectangular shape with transverse dimensions $a = 1.5$ μm, $b = 7.5$ μm was achieved by optical lithography and liftoff. The structure was passivated with an insulating film of silicon oxide, and a thin, patterned Au write wire was added as an additional level [refer to Fig. 11.19(a)].

The binary states of the ferromagnetic film are correlated with logical LOW and HIGH input levels by inductively coupling the magnetization to digital electric current pulses in the integrated "write wire." When a positive (negative) current I_w flows in the write wire, a positive (negative) magnetic field H_x beneath the wire and parallel to the axis of F orients the magnetization to the left (right) resulting in the binary state "1" ("0"). The magnitude of H_x is directly proportional to the amplitude of current I_w, $H = \alpha I_w$ with α an inductive coupling constant, and is weakly dependent on the distance beneath the wire. In the micrometer-scale prototype discussed herein, a current of about 100 mA in a 12 μm wide write wire resulted in a 50 Oe field.

After processing, the room temperature SQW carrier density was 1.8×10^{12} cm^{-2}, the mobility was 22,000 cm^2/V, and the sheet resistance was determined by a Van der Pauw measurement to be $R_\square = 150$ Ω/□. A small lithographic asymmetry of $x_0 \approx 70$ nm resulted in an offset of $R' = \Delta R_H$. The quasistatic output characteristic, measured with a steady-state bias current and an externally applied field H_x, is shown in Fig. 11.20(a). The coercivity of F is $H_c = 90$ Oe, with full saturation magnetization (zero slope in the $R_H(H)$ loop) achieved at about 200 Oe. The nonvolatile, nonpowered condition is $H_x = 0$, and the bistable device states correspond to the two remanent magnetization states with levels LOW $= 0$ Ω and HIGH $= 2\Delta R_H = 8.5$ Ω. The difference between LOW and HIGH is slightly less than the full output swing of 12.5 Ω. A semiquantitative magnetostatic calculation can be used to analyze the magnitude of $\Delta V_H = I \cdot \Delta R_H$, using Eq. (11.2.1). Rough agreement is achieved, with an error of about a factor of two.[55] A more sophisticated treatment[58] accounts for inhomogeneous current flow in the region of the cross, and agrees with experiment within about 10%.

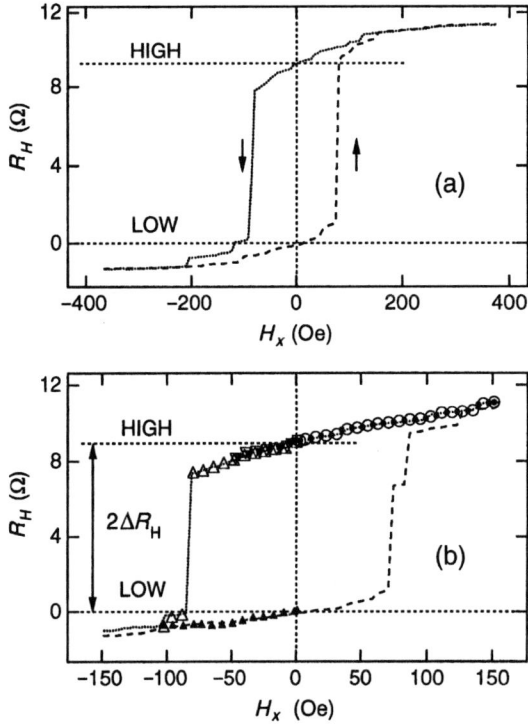

Figure 11.20. Quasistatic hysteresis loop, $R_H(H)$, of an integrated device. (a) Full hysteresis loop, as H_x is swept from -380 to $+380$ Oe and back. The two remanent states, $R_H(0)$, are marked HIGH and LOW. (b) Dotted lines, hysteresis loop measured for field sweep from -150 to $+150$ Oe and back (note change of scale from A). Symbols: response to the following quasistatic field sweeps. Starting from the HIGH remanent state, open symbols represent increasing field magnitude and closed symbols represent decreasing magnitude. Circles: 0 to $+150$ Oe and back. Triangles, down: 0 to -50 Oe and back. Triangles, up: 0 to -100 Oe and back.

Further measurements confirmed details of the $R_H(H)$ loop as shown in Fig. 11.20(b) [note change of scale from Fig. 11.20(a)]. Open symbols are used for increasing field magnitude and closed symbols for decreasing magnitude. Starting at the HIGH state, when the external field is quasistatically swept to 150 Oe and back to $H_x = 0$(circles) the $R_H(H)$ response reversibly traces the upper portion of the loop and returns to the HIGH state. Similarly, a quasistatic field sweep from $H_x = 0$ to -50 Oe and back (triangles, down) results in a reversible trace that returns to HIGH. However, when the field is swept from $H_x = 0$ to -100 Oe and back (triangles, up), a hysteretic loop is traced and the value of R_H returns to LOW.

11.3.1.3. Boolean Operations.
Since the dynamics of magnetization reversal in microstructured ferromagnetic elements occur on a timescale of nanosecond or less, the magnetic manipulations described with Fig. 11.20 could also be achieved using current pulses applied to the integrated write wire, and then the result is a magnetoelectronic Boolean logic operation. The Boolean AND

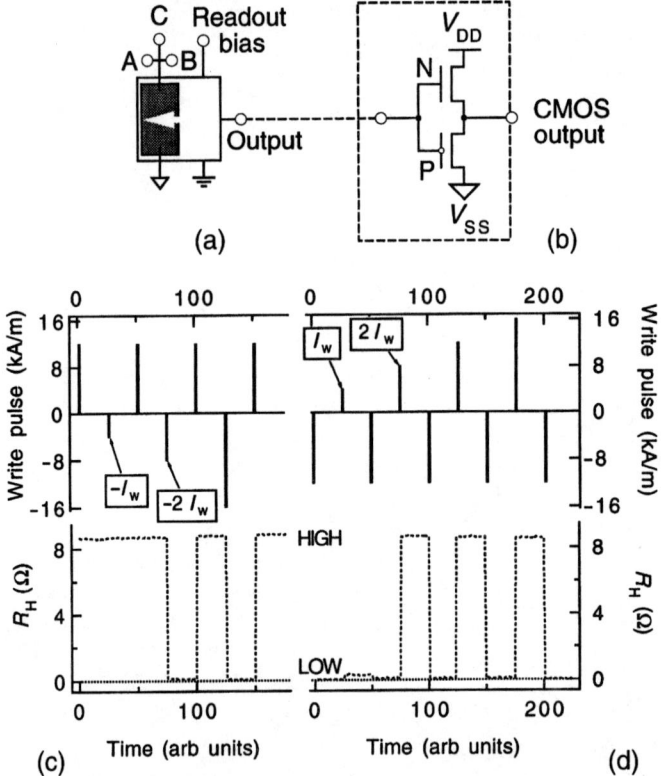

Figure 11.21 (a) Schematic representation of integrated device of Fig. 11.19, with input terminals A, B, and C, and separate terminals for bias and readout. (b) FET components that could be used to make a CMOS compatible cell. (c) and (d) steady-state response to input bias pulses that demonstrate: (c) $\overline{A \cdot B}$, $\overline{A + B}$; and (d) $A \cdot B$, $A + B$.

operation is the standard "half-select" process, i.e., utilized for nonvolatile memory applications. A two-dimensional array of memory cells is formed, and rows and columns of write wires are used to address the cells. Write pulses applied to a single row and column will have associated magnetic fields that are too small to affect cells along the row or column except at the intersection. There, the sum of the magnetic fringe fields is adequate to set the magnetization state of the target cell. More general Boolean operations are possible.

The sketch of Fig. 11.21(a) is a schematic representation of the prototype device of Fig. 11.19. Binary input pulses may be applied simultaneously to input terminals A and B or to a control terminal C and thereby transmitted down a write wire. A Boolean logic process requires only two clock steps for completion and the result is then latched, so it can be read out at any later time. This operation is demonstrated with the data of Fig. 11.21(c) using individual pulses. The Hall device is biased with a steady-state dc current (0.1 mA) and the readout voltage is recorded as current pulses are applied to the input write wire terminal [refer also to Fig. 11.19(a)].

We begin by following the same sequence of Fig. 11.20(b), a "reset" pulse of amplitude 150 Oe applied to terminal C sets the initial state of the device to HIGH. A pulse of -50 Oe is insufficient to switch the state of the device, and it remains in the HIGH state. After a reset pulse ensures that the initial device state is HIGH, a pulse of -100 Oe ($t = 75$) traverses the hysteresis loop and the device switches to LOW. By identifying a -50 Oe pulse as "unit" write current $-I_w$ associated with the binary "1," we recognize the above sequence as a demonstration of a Boolean NAND operation, $\overline{A \cdot B}$: A single unit current pulse $-I_w$ ("1") applied to either terminal A or B is insufficient to switch the device and the state remains HIGH ("1"), but two unit current pulses applied simultaneously to terminals A and B, with net magnitude $-2I_w$, successfully switch the device and the state changes to LOW ("0").

Continuing the progression, after a reset pulse sets the device state to HIGH ($t = 100$), a -200 Oe pulse ($t = 125$) also traverses the loop and the device switches to LOW. By identifying a -100 Oe pulse as a unit write current $-I'_w$, we recognize this to be a demonstration of a Boolean NOR operation, $\overline{A + B}$: A current pulse of unit amplitude I'_w applied to either terminal A or B, or to both terminals A and B, changes the device state to LOW ("0").

The inverse operations AND and OR result by changing the polarity of the input write pulses, as demonstrated in Fig. 11.21(d). After a reset pulse of -150 Oe sets the initial device state LOW ("0"), a single pulse of unit amplitude $\alpha I_w = 50$ Oe fails to change the device state, and it remains LOW ("0"). After another reset pulse ensures the initial device state to be LOW, two simultaneous pulses with net amplitude $2I_w$ ($t = 75$) traverse the hysteresis loop and set the device to HIGH ("1"). This is a Boolean AND operation, $A \cdot B$. Finally, if the unit amplitude is renormalized to $\alpha I'_w = 100$ Oe, this step shows that a single pulse I'_w applied to either terminal A or B switches the device to HIGH (1). The 200 Oe pulse ($t = 175$) shows that two pulses with net amplitude $2I'_w$ also result in setting the device to HIGH ("1"). This is a Boolean OR operation, $A + B$.

This demonstration shows that a single magnetoelectronic device can perform any of the four Boolean operations in only two clock cycles, and the function to be performed is determined instantaneously by the way the device is addressed: The normalized value of I_w has one of two values and either of two polarities.

An equivalent mode of operation is even simpler. A control pulse applied to terminal C, simultaneously with the input pulses applied to A and B, determines the function of the device. If a zero amplitude pulse is applied to C, unit pulses I_w at A and B are required to switch the device state, and it operates as an AND gate. If a unit amplitude pulse I_w is applied to C, then a single unit pulse at either A or B is necessary and sufficient to switch the device state. The 150 Oe pulse ($t = 125$) in Fig. 11.21(d) confirms that simultaneous pulses applied to A, B, and C, with a net write current of $3I_w$, also switches the device state, confirming operation as an OR gate. By changing polarity of the control pulses at terminal C, the device function can be set to be a NAND or NOR gate. Once again, a single magnetoelectronic device can perform any of four Boolean functions in only two clock cycles, and the function is programmatically controlled, instantaneously, by a single control pulse.

Figure 11.21(b) shows circuit components that could be used to make the spintronic HHE cell compatible with compensated CMOS circuits. HHE output levels of order 0.1 V are sufficient to control the threshold levels of CMOS FETs used as buffers. The buffered output levels then could be used for fanout to other CMOS circuits. The cell size represented by Fig. 11.21(a) and (b) is not larger than standard CMOS logic gate cells.

11.3.1.4. Discussion and Summary.
The fringe field, hybrid ferromagnet–semiconductor nonvolatile gate (HHE device) is a novel spintronic device with excellent operating characteristics, and with fabrication advantages over magnetoresistive devices. Because the device design is so simple, few processing steps are required for fabrication and high yields can be expected. Switching times are not yet known, but an upper bound for the magnetization reversal of a microfabricated Permalloy element is the order of a few nanoseconds. Since parasitic reactances are small, the read and write speeds should be fast, comparable with those of DRAM and Schottky diodes could be used on Si or GaAs substrates. Finally, since the storage function is performed by the ferromagnetic element, memory is radiation hard even when the cell is fabricated on Si.

Hybrid Hall effect devices may find application as reprogrammable logic cells,[54,59] or as *dynamically* reprogrammable logic cells. Since the same cell operates as nonvolatile memory, a chip with a large number of identical cells can be fabricated, and the fraction of devices used for storage and for logical processing can be dynamically apportioned. Novel kinds of "system-on-a-chip" architectures may be enabled.

11.3.2. Spin Injected Field Effect Transistor

11.3.2.1. Introduction.
Among the extraordinary properties of high mobility two-dimensional electron gases (2DEGs), one remarkable, yet relatively unexploited, characteristic is the spin splitting of the conduction band. As is well known, the high mobility carriers in the channel of a 2DEG heterostructure [Fig. 11.22(a)] are confined in a potential well with walls sufficiently steep that momentum states along $\hat{z}$ are quantized [Fig. 11.22(b)]. Typically, only states in the first quantized level are occupied. Electrons can move freely in the x–y plane and are allowed a continuum of momentum states within that plane. The Fermi wave vector, $k_F = (2\pi n_s)^{1/2}$, is of order 10^6 cm^{-1} for typical carrier densities of order $n_s \sim 10^{12}$ cm^{-2}, and the Fermi velocity, $v_F \sim 10^7$ cm/s, is weakly relativistic. When a nonzero, steady-state electric field, $\vec{E} = E_z \hat{z}$ [Fig. 11.24(b)], is perpendicular to the plane of motion, it transforms, in the frame of the carrier, as a magnetic field H' with components in the x–y plane. This "effective" field can interact with the spin magnetic moment of the carrier, μ_B, and split the conduction band into subbands separated by an effective Zeeman energy, $\Delta \propto \mu_B H'$. This section addresses the effects of spin-dependent transport when the band of carriers is split into spin subbands.

A variety of conditions can result in the presence of an appropriate field E_z. (i) A dipole field may result from inversion asymmetry in the host semiconductor bulk crystal potential, e.g., from the polar nature of III–V materials, and can be augmented by an inversion field in a heterostructure. The leading term in the spin splitting of the band is[60]

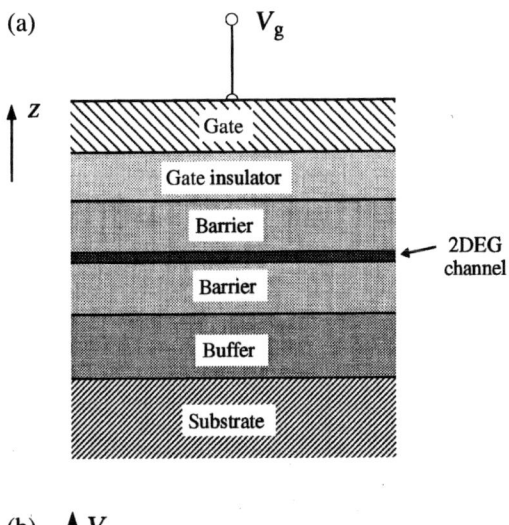

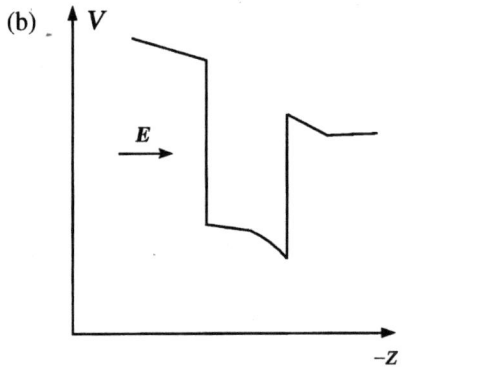

Figure 11.22. (a) Layer structure of a typical, gated 2DEG heterostructure. (b) Schematic conduction band diagram showing the potential well that confines the carriers.

$$\Delta_{\text{bulk}} \sim b\left(\frac{\pi}{d_z}\right)^2 k$$

where k is the in-plane wave vector, d_z, the width of the quantum well, and b, the material-specific constant. This bulk term, also called the "I term" related to the "DP' mechanism,"[61] is important in relatively large band gap materials, and a splitting of $\Delta_{\text{bulk}} \sim 0.01$ meV has been inferred for GaAs/AlGa As heterostructures by extrapolating ESR measurements to zero magnetic field.[62] (ii) An asymmetry of the confining well has an associated potential gradient, $\partial V/\partial z$, and is accompanied by an interfacial electric field E_z. Because this couples the electron spin and orbital terms, Rashba introduced a spin–orbit Hamiltonian:[63]

$$H_{SO} = \alpha(\sigma \times k) \cdot \hat{z} \tag{11.3.1}$$

where σ are the Pauli matrices and $\hat{z}$ is a unit vector normal to the plane of the 2DEG. The spin–orbit constant α is implicitly proportional to the magnitude of the interfacial field, E_z, and is therefore material-specific. The spin–orbit splitting,

$\Delta_{SO} \approx 2\alpha k$, is sometimes called the Rashba term and is an important effect in narrow gap semiconductors. Values of $\Delta_{SO} \approx 2$ meV[64] and $\Delta_{SO} \approx 5$ meV[65] have been measured in InGaAs/InAlAs heterostructures by analyzing the beat pattern of Shubnikov–de Haas oscillations. Using the same technique, $\Delta_{SO} \approx 3$ meV was measured for GaSb/InAs heterostructures.[66] (iii) A gate voltage V_g applied to the device [Fig. 11.22(a)] can generate a variable electric field, E_z. The authors of Ref. 65 modulated the zero-field Rashba spin splitting by about 20% by applying gate voltages of the order 1 V to their InGaAs/InAlAs heterostructures.

The discussions in this section are particularly appropriate for mechanisms (ii) and (iii). For spin splittings Δ_{SO} of a few milli electron volts, the magnitude of the effective field H', given by $\Delta = 2g\mu_B H' = [2(g/g_0)\mu_{B0}/(m^*/m_0)]\, H'$, is on the order of 1 T, where μ_0, g_0 and m_0 are the free electron values of Bohr magneton, g–value and mass, and typical values $g \sim 4g_0$ and $m_0^* \sim 0.1\, m_0$ were used. Effective fields of 1 T are quite substantial, and we will investigate novel effects on the spins of the carriers.

11.3.2.2. Novel Device Structures.

In previous discussions of spin–split subbands, the spin $\vec{s}$ of the carrier is typically assumed to have arbitrary orientation. It precesses about the vector effective field $\vec{H}'$ at the Larmor frequency Ω_L. Since the direction $\vec{k}$ of the carrier changes with each scattering event, the axis of spin precession must also change and the spin orientation is rapidly randomized. Spin dephasing has been detected by weak antilocalization measurements, and spin relaxation was estimated to be as rapid as a few picoseconds.[67,68]

A more interesting condition arises if the carriers can be *polarized* with their spin axes oriented along a common direction. Datta and Das[69] extended the Johnson–Silsbee *spin injection* concept to consider quasi-one-dimensional ballistic transport in a ferromagnet/2DEG/ferromagnet (F–2DEG–F) structure, where the carrier spins are aligned along the direction of motion. In their proposed device (Fig. 11.23), two iron contacts have magnetizations aligned along the $\hat{x}$ axis and are

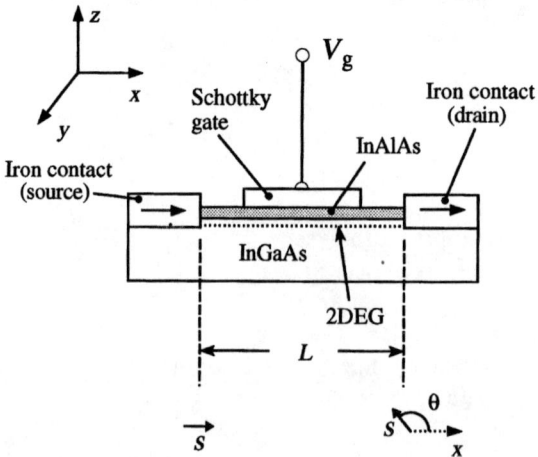

Figure 11.23. Cross-sectional diagram of Datta–Das current modulation device, with channel length L. At bottom, an electron injected at the source with its spin $\vec{s}$ aligned along $\hat{x}$ precesses during transit along its ballistic trajectory, and acquires phase angle θ by the time it reaches the drain.

used as the source and drain of a gated InGaAs/InAlAs field effect transistor (FET). One contact (e.g., the source) provides carriers to the high conductivity channel with their spins initially aligned along $\hat{x}$. An applied gate voltage causes a variable electric field E_z which transforms, in the frame of the carriers, as an effective magnetic field H'_y. The carrier spin precesses about the $\hat{y}$ axis at the Larmor frequency $\Omega_L(H')$, arriving at the second ferromagnetic contact (e.g., the drain) with a phase angle θ in proportion to the time of transit, $\theta = \Omega_L t = \Omega_L L/v_F$, and to the magnitude of effective field H'_y (refer to Fig. 11.23). The probability of transmission from the channel to the drain is proportional to the projection of the spin of the carrier on the magnetization axis of the ferromagnetic contact, $\vec{M}_d = M_d \hat{x}$.[1,69] Thus, for monatonically increasing gate voltage V_g, the source–drain conductance should vary periodically. No estimate for the amplitude of the conductance variation was made because the polarization efficiency of a ferromagnet–2DEG contact was unknown. Datta and Das also noted that the modulation amplitude should be reduced if the direction of propagation of the injected electrons is at an angle to the $\hat{x}$ axis (i.e., if there are momentum components along $\hat{y}$). If the 2DEG channel is wide (a relatively large extent along $\hat{y}$), there are many possible momentum states with $p_y \neq 0$ and the periodic conductance oscillations would be washed out. If the channel is sufficiently narrow, only a few momentum states are possible, the states are relatively far apart (i.e., discrete), and a robust effect was predicted to exist, even at room temperature.[69] The proposed device of Datta and Das has been the focus of considerable basic and applied research.

11.3.2.3. Spin Injection at a Ferromagnetic Metal–Semiconductor Interface.

The Datta–Das device requires the transmission of a spin–polarized-current at a F–N (and N–F) interface, and relies on predictions about the dynamics of spin–polarized transport in a 2DEG channel. It was shown in Section 11.1.5 that spin-polarized transport across a ferromagnetic metal–barrier–nonmagnetic material interface is characterized by the interfacial parameter η when the barrier resistance R_b is greater than the resistance of a spin depth of material in F. For the case of a ferromagnetic metal and a semiconductor, a low transmission barrier is readily employed.

Spin-polarized tunneling experiments have been performed using ferromagnetic metals and semiconductors to independently examine the issue of spin-polarized transport at a F–I–S interface. In a typical experiment, luminescence in a GaAs sample is used as a detector of polarized carriers injected across a vacuum tunnel barrier. In the reverse of optical pumping, circularly polarized light is emitted in proportion to the degree of spin polarization of the recombining minority carriers at the fundamental band gap. GaAs is an ideal material because of the relatively large spin–orbit splitting at the valence band. Room temperature experiments used nickel as the ferromagnet, and the polarization of the tunnel current was found to vary between 5% and 30% ($\eta = 0.05$–0.3).[70]

Another class of experiments has probed spin injection at a ferromagnetic metal–low transmission barrier–semiconductor interface, by convolving the exchange split density of states of F with the spin–orbit split density of states of a high mobility 2DEG.[71] In the 2DEG, the spin–orbit term of the Hamiltonian, Eq. (11.3.1), contributes to the electron energy dispersion relation:

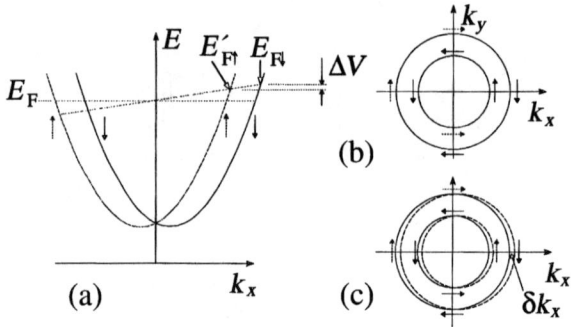

Figure 11.24. (a) Energy dispersion of 2DEG electrons along k_x, showing spin-splitting by the Rashba effect. (b) 2DEG Fermi surface in equilibrium and (c) displaced by bias current $I+$.

$$E(k) = (\hbar^2 k^2/2m^*) \pm \alpha k \qquad (11.3.2)$$

As plotted for k_x in Fig. 11.24(a), the carriers are degenerate at $k_x = 0$ and spin splitting increases linearly with $|k_x|$ up to the Fermi energy E_F. For $k_x > 0$, there are more carriers with spin down than up, and the opposite is true for $k_x < 0$. The Fermi sea is a parabaloid of revolution about the E axis, and the Fermi surface is a pair of concentric circles [Fig. 11.24(b)] with radii $k_{F,max}$ and $k_{F,min}$. Along particular directions of $\vec{k}$, spin eigenstates can be simply denoted by arrows, e.g., spin orientation along $\pm \hat{y}$ for momentum states k_x [Fig. 11.24(b)]. More generally, the circulation of spin depicted in Fig. 11.24(b) is the appropriate quantum state and spin eigenstates for particular directions of $\vec{k}$ must be appropriately constructed. A system with a circular Fermi surface is ideal for Shubnikov–de Haas measurements, and the Fermi surface depicted in Fig. 11.24(b) was first deduced from beat patterns in Shubnikov–de Haas oscillations.[66]

The total number of up-spin and down-spin carriers is equal for zero bias current ($I = 0$), but the situation changes for nonzero bias. A positive current imposed along the $\hat{x}$ axis of a 2DEG channel displaces the Fermi sea (and circles) to the right by δk_x [dotted lines of Fig. 11.24(c)]. For the branch $k_x > 0$, up-spin (down-spin) carriers are added at $k_{F,min}$ ($k_{F,max}$). The chemical potentials of both spin subbands, $E_{F,\uparrow}$ and $E_{F,\downarrow}$, are raised, but $E_{F,\downarrow}$ is raised more than that of $E_{F,\uparrow}$ because the incremental area of the Fermi sea at $k_{F,max}$ is larger than that at $k_{F,min}$. Similarly, for the branch, $k_x < 0$, down-spin (up-spin) carriers are depleted at $k_{F,min}$ ($k_{F,max}$). The result[72,73] is that the Fermi surface tilts upward towards $+k_x$ [Fig. 11.24(a)]. For negative bias current, the tilt slope reverses.

These relative differences between $E_{F,\uparrow}$ and $E_{F,\downarrow}$ represent a nonequilibrium spin magnetization $\tilde{M}$ and can be measured by appealing to the spin injection technique described in Section 11.1 for metals, and using a ferromagnetic electrode F in the *potentiometric* geometry shown in the inset of Fig. 11.25(a).[74] Here, F (terminal 5 or 6) is connected to an infinite impedance voltmeter, H_y is an externally applied magnetic field, and bias I is applied to the channel from terminals S1 to S2. The voltage difference ΔV_F measured by F, for the two magnetization states $\pm M_y$, is $\Delta V_F = \eta \Delta V$ [refer to Fig. 11.24(a)], where η is the fractional polarization of carriers

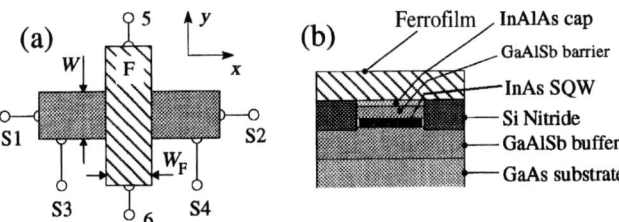

Figure 11.25. (a) Sample geometry, top view. (b) Cross-section of planarized device structure.

crossing the F–S interface. A measurement of the F–2DEG interface resistance, also called a diode measurement, will also measure the relative difference between $E_{F,\uparrow}$ and $E_{F,\downarrow}$.[71]

A recent theory[73] gives a complete and rigorous nonequilibrium calculation that averages electrochemical potential differences over the Fermi surface. The key parameters are the relative shift[73] $\eta \propto k_{F,max} - k_{F,min}$ (also proportional to α) and the fractional polarization m of carriers in F. (Here, m is equivalently called p_f in Section 11.1, and the η used by Ref. 73 should not be confused with interfacial spin transport parameter η.) It predicts that differences of resistance ΔR should be approximately the same for interface and potentiometric measurements, $\Delta R_I \approx \Delta R_P$. It further predicts that the resistance modulation should be proportional to the resistance R_λ of a length ℓ (with ℓ an electron mean free path) of the 2DEG channel, $R_\lambda = R_{2DEG} (\ell/W)$, so that the ratio $\Delta R_P/R_\lambda$ is approximately constant for device structures with similar values η and m. In the limit of low junction conductance and narrow F electrodes, $W_F \ll W$, this theory makes the quantitative prediction,

$$\frac{\Delta R_P}{R_\lambda} = |m|\,\eta \qquad (11.3.3)$$

A series of experiments were performed to test these ideas.[74,75] The samples were composed of a 2DEG mesa and a ferromagnetic metal electrode (Fig. 11.25). The mesas were fabricated from a high mobility InAs SQW structure grown with MBE. The 2DEG is confined to a 15 nm InAs layer by 20 nm above and 200 nm below the $Al_{0.6}Ga_{0.4}Sb$ layer [Fig. 11.25(b)]. The structure is capped with a 5 nm layer of $In_{0.45}Al_{0.55}As$. The carrier density in the 2DEG is electron doped from 4 ML of InAs grown near the middle of the top confining AlGaSb layer. The InAlAs cap layer is effective for reducing hole leakage current. The relatively high resistance of the F–I–2DEG interfaces, approximately constant for the temperature range 4 K $< T <$ 295 K, implies the top insulating layers form a low transmission barrier for transport.[76] The arguments of Section 11.1, and theoretical modeling for this particular experiment,[73] indicate that spin transport does not depend on details of the barrier. The carrier density, mobility, and sheet resistance at 77 K (296 K) were determined from a Hall measurement (before processing) to be 8.7×10^{11} cm^{-2} (1.3×10^{12} cm^{-2}), 75,000 cm^2 V^{-1} s^{-1} (25,000 cm^2 V^{-1} s^{-1}), and 80 $\Omega/\square$ (190 $\Omega/\square$) respectively, with weak temperature dependence below 77 K.

The 2DEG mesa is formed by optical lithography[75] and an Ar ion mill dry etch. Next, the InAs SQW is undercut with a selective wet etch [Fig. 11.25(b)]. The sample is then planarized with SiN prior to dissolving and removing the photoresist.

This planarization prevents accidental contact to the SQW at a mesa sidewall, and presents a smooth surface for the ferromagnetic films, promoting single domain behavior. Ferromagnetic electrode F is fabricated by optical lithographic patterning with a width that varies from 1.5 to 2.5 μm, followed by e-beam deposition of $Ni_{0.8}Fe_{0.2}$ or $Fe_{0.1}Co_{0.9}$, and lift-off. The planarization was accurate within the rms roughness of the sample surface, as determined by AFM. MFM images (300 K) showed continuous magnetization in the sample region in the remanent state.

Two different transport experiments were performed. In the *potentiometric measurement*,[73,74] bias current I is applied from terminal S1 to S2 [Fig. 11.25(a)] and voltage V_P is measured from terminal 5 or 6 to S4 while an external field H_y is applied to change the magnetization orientation $\vec{M}$ of F from $-M\hat{y}$ to $+M\hat{y}$. Electrode F acts as a spin sensitive potentiometer, and spin–dependent differences of voltage ΔV_P are superposed on a background voltage proportional to the interprobe spacing, $V_{5,S4}$. For the *diode* or *interface* measurement,[71,73] bias current I is applied from terminal S1 to 5 and voltage V_I is measured from terminal 6 to S2. The measured voltages are linear with I in the experimental range 10 μA ≤ I ≤ 200 μA and are presented as resistances R_P and R_I.

Typical data are shown in Fig. 11.26 for a device with a channel width $W = 15$ μm and ferromagnetic electrode width $W_F = 2.3$ μm. For the potentiometric measurement (bottom trace in Fig. 11.26), the baseline resistance $R_{5,S4} = 396$ Ω is not relevant to the data and has been subtracted. The hysteresis of R is associated with the hysteresis of F, which was measured directly and *in situ* by Kerr photometry. The observation that $\Delta R_I = \Delta R_P$ is consistent with theory.[73] Quantitatively, the measured values of ΔR can be compared with Eq. (11.3.3). Using accepted values of $m = 0.4 \pm 0.05$,[12] we deduce $\eta \leq 0.1$ which lies in the range proposed by theory.[74] Another series of experiments systematically measured ΔR_I and ΔR_P for a number of samples and over a large temperature range.[77] Experimental agreement with Eq. (11.3.3) was good, for samples with varying interfacial barrier resistance. Furthermore, there was little temperature dependence up to room temperature, consistent with this Fermi surface picture.

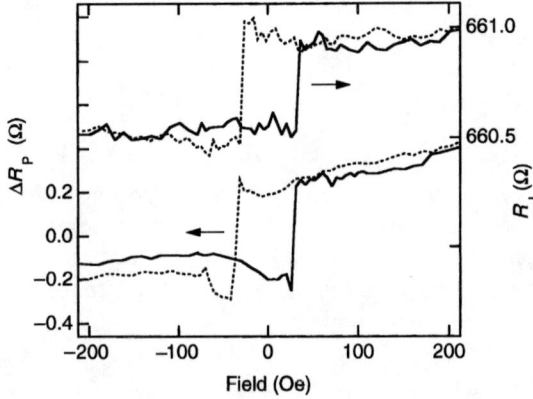

Figure 11.26. Examples of data, $T = 4$ K. Top trace, right axis: R_I. Bottom trace, left axis: ΔR_P. Solid lines: sweep field H_y up. Dotted lines: sweep H_y down. Positive bias current.

The diode geometry has the advantage of demonstrating both spin injection and detection, and the implication that spin–polarized current can be transmitted across a ferromagnetic metal–2DEG interface is important. Experimentally, it is crucial to note that the top barrier layer of the 2DEG heterostructure must be protected during device fabrication. The integrity of the barrier provides the confining potential of the 2DEG, and preserves both the Rashba effect and the high mobility of the carriers. Furthermore, it provides a tunnel barrier for transport between the ferromagnet and the 2DEG.[76]

11.3.2.4. Electrical Spin Injection and Detection in a 2DEG.

By placing two ferromagnetic electrodes of the kind represented in Fig. 11.25 on a common 2DEG channel, injection and detection experiments can be performed.[78] In this case, interprobe length scales L_x are so short as to be the order of the electron ballistic mean free path, and spin scattering is rapid.[79] The experiment cannot discriminate between ballistic and diffusive transport contributions.

The nonlocal geometry is the same as that of the original spin injection experiment, Fig. 11.5. Spin-polarized electrons are injected at F2 (Fig. 11.27), and the bias current is grounded far away at terminal S1. (Fig. 11.27 uses the convention of the authors of Ref. 79, with F2 the injecting electrode.) The portion of the 2DEG channel to the right of F2 is at a constant potential. Spin sensitive potentiometer F1 is grounded far away at terminal S2, and measures zero voltage in the absence of spin injection effects. The magnetizations of F1 and F2 are manipulated with an external field H_y, and a relatively high (low) voltage is measured with F2 when $\vec{M}_1$ and $\vec{M}_2$ are parallel (antiparallel).

The parallel and antiparallel orientations are determined by two different-coercivities, $H_{C1} \neq H_{C2}$, and the expected dependence of the detected voltage on external field, $V_S(H_y)$, is described by hysteretic dips of equal depth that occur in the field range where $\vec{M}_1$ and $\vec{M}_2$ are antiparallel.[1,78] If the remanence of $\vec{M}_1$ or $\vec{M}_2$ is not ideal, the dips will overlap. This discussion represents an idealized case where polarized electrons are injected with wave vectors k_x and reach F1 with ballistic trajectories. For diffusive transport or for a wide sample, contributions to the spin subband chemical potentials must be averaged over the Fermi surface.

The ferromagnetic probes were fabricated in the same way as described in the discussion for Fig. 11.25.[77,78] The magnetic characteristics of each electrode were measured individually using the current induced magnetization effect described

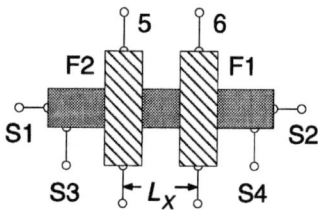

Figure 11.27. Sample geometry, top view, for spin injection and detection experiments. Following the convention of Ref. 78, the injecting electrode is labelled F2 (in contrast with Fig. 11.5).

earlier.[77] The Permalloy electrodes (F2) are not magnetostrictive and showed fairly good remanence and relatively narrow loops. The coercivity was typically $H_{C1} \approx 75$ Oe and the saturation field was $H_{S1} \approx 200$ Oe. FeCo electrodes (F2) are magnetostrictive and were characterized by relatively poor remanence and broad loops. The coercivity was typically $H_{C2} \approx 30$ Oe and the saturation field was $H_{S2} \approx 400$ Oe. Quantitative fits to theory[73,77] give a polarization parameter of $m = 40 \pm 5\%$ for both F1 and F2.

Typical data from a sample with a single $W = 15$ µm wide 2DEG channel and interelectrode spacing $L_x = 10.6$ µm are presented in Fig. 11.28(a). The baseline resistance of about $0.36 \; \Omega$ is two orders of magnitude smaller than the channel resistance $R_{ch} = 52 \; \Omega$, validating our assumptions about the nonlocal geometry. The overlapping dips that appear in the range -200 Oe $< H_y < +200$ Oe have the qualitative shape predicted by the spin injection model.

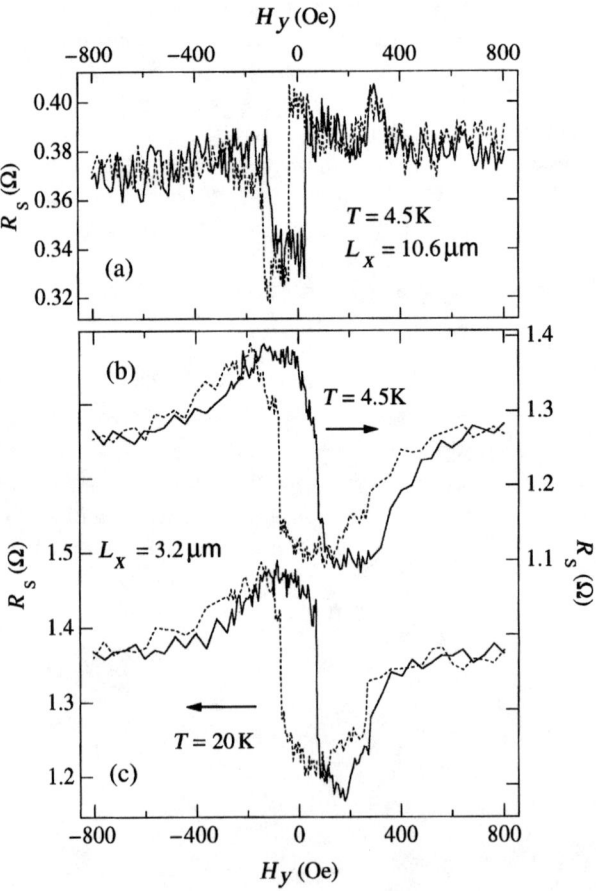

Figure 11.28. Examples of electrical spin injection/detection data.[78] (a) A sample with $L_x = 10.6$ µm. (b) and (c) A sample with $L_x = 3.2$ µm at two temperatures. Solid lines: sweep field up. Dotted lines: sweep field down.

Typical data from a sample with six $W = 0.9$ μm wide 2DEG channels in parallel and $L_x = 3.2$ μm are shown in Fig. 11.28(b) and (c) for two temperatures. The hysteretic features do not have the same detailed shape as the data of Fig. 11.28(a). Each feature has a small peak on the side of negative magnetic field followed by a dip on the side of positive field. The features occur in the field range -400 Oe $< H_y < +400$ Oe where $\vec{M}_1$ and $\vec{M}_2$ are changing their relative orientations. Hysteretic features with similar shape have been observed in spin injection studies of metal films,[11] and are associated with ferromagnetic films that have anisotropy axes that are neither parallel with each other nor with the axis of applied field $\vec{H}_y$. Because the magnetizations are not parallel when saturated, $\vec{M}_1$ and $\vec{M}_2$ can shift from parallel to antiparallel when H_y is swept from positive to negative. The voltage difference from peak to dip $(I \cdot \Delta R_S = I \cdot 0.3\ \Omega)$ corresponds to the shift of spin subband electrochemical potentials, $\Delta\mu$. The data shown in Fig. 11.28 are representative of the range of magnetotransport features $\Delta R_S(H_y)$ that were observed in these experiments.[79] As a function of temperature, the amplitude ΔR_S at 150 K was diminished to about 70% of the value at 4 K, and no modulation was observed at room temperature.

The authors model their data as the result of the ballistic transport of spin-polarized electrons. With a carrier mean free path $\ell = 1.2$ μm, only a small fraction f of the carriers injected into the $W = 900$ nm wide channels are transmitted "upstream" with ballistic or quasiballistic trajectories along $+k_x$. When the magnetization of F1 is along the $\pm\hat{y}$ axis, carriers with $\vec{k} = k_F\hat{x}$ enter the wires in spin eigenstates, maintain their spin orientation for a characteristic spin coherence length $\Lambda_{s,x} = v_F \tau_s$, and then relax and diffuse "downstream" to ground with random spin phase. $\Lambda_{s,x}$ is estimated by comparing data at $T = 4.5$ K for two nominally identical samples $(L_x = 3.2$ μm, $W = 15$ μm and $(L_x = 10.6$ μm, $W = 15$ μm), and assuming a simple exponential decay of the number of spin-polarized carriers, $N_S(x) \propto \exp[-L_x/\Lambda_{s,x}]$. $\Lambda_{s,x} = 4.6$ μm is found to be a lower bound for the spin coherent mean free path of carriers injected with spin orientation along y.

The amplitude of the voltage associated with $\Delta\mu$ is given by[79]

$$\Delta\mu = \frac{d\mu}{dn}\Big|_{E_F} \delta n = \frac{E_F}{n_s} \delta n = E_F \frac{\delta N_{\Lambda_s}}{N_{\Lambda_s}} \tag{11.3.4}$$

where

$$\delta N_{\Lambda_s} = mf\frac{I}{e}\tau_s$$

is the steady-state number of nonequilibrium, ballistic, spin-polarized electrons in a length Λ_s of the channel, and N_{Λ_s} is the number of equilibrium electrons in the same length. For a sample with $L_x = 3.2$ μm $\sim \Lambda_{s,x}$, we calculate $\delta N_{\Lambda_s} = f \times 95$ electrons and $N_{\Lambda_s} = 2.3 \times 10^5$ electrons. After simple algebraic manipulations, f is found to be[79] $f = 0.09$. This implies that roughly 9% of the electrons injected "upstream"-reach the detector with quasiballistic trajectories, a reasonable value given the dimensions of the channels and simple geometric arguments. We note that the data of Fig. 11.28(b) represent the steady-state detection, on average, of $f \times 95 \approx 9$ spin-polarized electrons, roughly one or two per channel.

These experiments are an important demonstration that spin-polarized electrons can be electrically injected into a 2DEG, and then electrically detected as a voltage. The realization of a Datta–Das, spin injected field effect transistor is now plausible, and the modulation of spin can be used as a new parameter for novel device functionality. These experiments show a remarkable sensitivity to a small number of polarized electrons, possibly suggesting applications in quantum computing, or novel kinds of single electron transistor devices.

ACKNOWLEDGMENTS

This work was supported by the Office of Naval Research.

REFERENCES

1. M. Johnson and R. H. Silsbee, *Phys. Rev. Lett.* **55**, 1790 (1985).
2. M. Johnson and R. H. Silsbee, *Phys. Rev. B* **37**, 5312 (1988).
3. M. Johnson and R. H. Silsbee, *Phys. Rev. B* **37**, 5326 (1988).
4. M. Johnson and R. H. Silsbee, *J. Appl. Phys.* **63**, 3934 (1988).
5. A. G. Aronov, *Pis'ma Zh. Eksp. Teor. Fiz.* **24**, 37 (1976) [*Sov. Phys. JETP Lett.* **24**, 32 (1976)].
6. M. Johnson, *Phys. Rev. Lett.* **67**, 3594 (1991).
7. M. Johnson, *Appl. Phys. Lett.* **63**, 1435 (1993).
8. M. Johnson and R. H. Silsbee, *Phys. Rev. B* **35**, 4959 (1987) [see the appendix for the fully generalized solution].
9. R. H. Silsbee, *Bull. Magn. Res.* **2**, 284 (1980) [see also Ref. 1].
10. M. Johnson, *Mat. Sci. Eng. B* **31**, 199 (1995).
11. M. Johnson, *J. Appl. Phys.* **75**, 6714 (1994).
12. R. Soulen et al., *Science* **282**, 5386 (1998).
13. M. Johnson, *Phys. Rev. Lett.* **70**, 2142 (1993).
14. S. Dubois et al., *Phys. Rev. B* **60**, 477 (1999).
15. K. Bussmann et al., *IEEE Trans. Mag.* **34**, 924 (1998).
16. M. Johnson and R. H. Silsbee, *Phys. Rev. Lett.* **60**, 377 (1988).
17. M. Johnson, *Phys. Rev. B* **58**, 9635 (1998).
18. M. Johnson, *Science* **260**, 320 (1993).
19. P. M. Tedrow and R. Meservey, *Phys. Rev. B* **7**, 318 (1973); J. S. Moodera, L. R. Kinder, T. M. Wong, and R. Meservey, *Phys. Rev. Lett.* **74**, 3273 (1995).
20. A. Y. Elezzabi, M. R. Freeman, and M. Johnson, *Phys. Rev. Lett.* **77**, 3220 (1996).
21. J. Fabian and S. Das Sarma, *Phys. Rev. Lett.* **81**, 5624 (1998).
22. M. Johnson, *Semicond. Sci. Technol.* (2002).
23. M. Johnson, *Nature* (2002).
24. H. Higashino, K. Mizuno, T. Matsushima, K. Setsune, and K. Wasa, *Advances in Superconductivity III*, Proceedings of the ISS'90, edited by Kajimura and Hayakawa (Springer, New York, 1991), p. 1227.
25. T. Kobayashi, K. Hashimoto, U. Kabasawa, and M. Tonouchi, *IEEE Trans. Mag.* **MAG-25**, 927 (1989).
26. C. W. Schneider, G. J. Gerritsma, and H. Rogalla, *Proceedings of the 2nd Workshop on HTS Applications and New Materials*, 8–10 May 1995, Enschede, The Netherlands, pp. 91–97.
27. Yu M. Bouguslavskij, K. Joosse, A. G. Sivakov, F. Roesthuis, G. J. Gerritsma, and H. Rogalla, *Physica C* **220**, 195 (1994).
28. I. Iguchi, K. Nukui, and K. Lee, *Phys. Rev. B* **50**, 457 (1994).

29. Q. Wang and I. Iguchi, *Physica* C **228**, 393 (1994) [It should be noted that, in prior work (Refs. 1 – 6), gains were obtained without taking into account I_{shift}].
30. For example, see J. J. Chiang and D. J. Scalapino, *J. Low Temp. Phys.* **31**, 1 (1978).
31. J. Clarke, *Phys. Rev. Lett.* **28**, 1363 (1972); J. Clarke and M. Tinkham, *Phys. Rev. Lett.* **28**, 1366 (1972).
32. C. S. Owen and D. J. Scalapino, *Phys. Rev. Lett.* **28**, 1550 (1972).
33. T.-W. Wong, J. T. C. Yeh, and D. Langenberg, *Phys. Rev. Lett.* **37**, 150 (1976).
34. For a review, see W. J. Gallagher, *IEEE Trans. Magn.* **MAG–21**, 709 (1985); R. A. Buhrman, *SQUIDs and Their Applications*, edited by H. D. Halbohm and H. Lubbig (Walter de Gruyter, Berlin, 1985), p. 171.
35. *Nonequilibrium Superconductivity, Phonons, and Kapitza Boundaries*, edited by K. E. Gray (Plenum, New York, 1981); K. E. Gray, *Appl. Phys. Lett.* **32**, 392 (1978).
36. J. Mannhart, *Semicond. Sci. Technol.* **9**, 49 (1996).
37. V. A. Vas'ko, V. A. Larkin, P. A. Kraus, K. R. Nikolaev, D. E. Grupp, C. A Nordman, and A. M. Goldman, *Phys. Rev. Lett.* **78**, 1134 (1997).
38. M. Johnson, *J. Magn. Magn. Mater.* **156**, 321 (1996).
39. Z. W. Dong et al., *Appl. Phys. Lett.* **71**, 1718 (1997).
40. Z. W. Dong, T. Boettcher, C.-H. Chen, Z. Trajanovic, I. Takeuchi, V. Talyansky, R. P. Sharma, and T. Venkatesan, *Appl. Phys. Lett.* **69**, 3432 (1996).
41. M. Johnson, *Appl. Phys. Lett.* **65**, 1460 (1994).
42. C. D. Chen et al., *Phys. Rev. Lett.* **88**, xx (2002) [Supercond Al Spin Inj].
43. See, e.g., S. T. Ruggiero and D. A. Rudman Eds. *Superconducting Devices*, (Academic Press, SD, CA, 1990).
44. T. W. Clinton and M. Johnson, *Appl. Phys. Lett.* **70**, 1170 (1997).
45. T. W. Clinton and M. Johnson, *J. Appl. Phys.* **83**, 6777 (1998).
46. T. W. Clinton and M. Johnson, *Appl. Phys. Lett.* **76**, 2116 (2000).
47. J. Dolan, *J. Low Temp. Phys.* **15**, 13 (1974).
48. See, e.g., M. Tinkham, *Introduction to Superconductivity* (McGraw-Hill, New York, 1996); [Chapter 6].
49. S. Shapiro, *Phys. Rev. Lett.* **11**, 80 (1963).
50. T. W. Clinton and M. Johnson, *J. Appl. Phys.* **85**, 1637 (1999).
51. J. Eom and M. Johnson, *Appl. Phys. Lett.* **79**, 2486 (2001).
52. V. L. Newhouse (Ed.), *Applied Superconductivity*, vol. 1 (Academic Press, New York, NY, 1975).
53. T. W. Clinton and M. Johnson, *J. Appl. Phys* **91**, 1371 (2002).
54. M. Johnson, *IEEE Spectrum Magazine* **37**(2), 33(2000).
55. M. Johnson, B. R. Bennett, M. J. Yang, M. M. Miller, and B. V. Shanabrook, *Appl. Phys. Lett.* **71**, 974 (1997).
56. M. Johnson, *J. Vac. Sci. Technol.* A **16**, 1806 (1998).
57. M. Johnson, B. R. Bennett, P. R. Hammar, and M. M. Miller, *Solid State Electron.* **44**, 1099 (2000).
58. J. Reijniers and F. M. Peters, *Appl. Phys. Lett.* **73**, 357 (1998).
59. M. Johnson, *IEEE Spectrum Magazine* **31**(5), 47 (1994).
60. J. Luo, H. Munekata, F. F. Fang, and P. J. Stiles, *Phys. Rev.* B **41**, 7685 (1990).
61. M. I. D'yakanov and V. I. Perel', *Sov. Phys. JETP* **33**, 1053 (1971).
62. D. Stein, K. von Klitzing, and G. Weimann, *Phys. Rev. Lett.* **51**, 130 (1983).
63. Yu A. Bychkov and E. I. Rashba, *J. Phys. C: Solid State Phys.* **17**, 6039 (1984).
64. B. Das et al., *Phys. Rev.* B **39**, 1411 (1989).
65. J. Niita et al., *Phys. Rev. Lett.* **78**, 1136 (1997).
66. J. Luo et al., *Phys. Rev.* B **38**, 10142 (1988).
67. G. L. Chen et al., *Phys. Rev.* B **47**, 4084 (1993).
68. P. D. Dresselhaus et al., *Phys. Rev. Lett.* **68**, 106 (1992).
69. S. Datta and B. Das, *Appl. Phys. Lett.* **56**, 665 (1990).
70. S. F. Alvarado and P. Renaud, *Phys. Rev. Lett.* **68**, 1387 (1992).
71. P. R. Hammar, B. R. Bennett, M. J. Yang, and M. Johnson, *Phys. Rev. Lett.* **83**, 203 (1999).
72. P. R. Hammar, B. R. Bennett, M. J. Yang, and M. Johnson, *Phys. Rev. Lett.* **84**, 5024 (2000).
73. R. H. Silsbee, *Phys. Rev.* B **63**, 155305 (2001).
74. P. R. Hammar and M. Johnson, *Phys. Rev.* B **61**, 7202 (2000).
75. P. R. Hammar, B. R. Bennett, M. J. Yang and M. Johnson, *J. Appl. Phys.* **87**, 4665 (2000).

76. E. I. Rashba, *Phys. Rev. B* **62**, R16,267 (2000).
77. P. R. Hammar and M. Johnson, *Appl. Phys. Lett.* **79**, 2591 (2001).
78. P. R. Hammar and M. Johnson, *Phys. Rev. Lett.* **88**, 066806 (2002).
79. W. H. Lau, J. T. Olesberg, and M. E. Flatté, *Phys. Rev. B* **64**, 161301 (2001).

Index

Activation energy E_{hop}, 59, 97, 115
   polaron trap E_{hop}^0, 73, 97, 98
   spin dependence E_{hop}^{ex}, 73, 97
Affinity, 518, 519
Aliasing, 385
AMR memory cells, 488
Analog to digital converters (ADC), 530
Anderson loop, 461
Anderson superexchange theory, 69, 73, 74
Andreev reflection, 230
Angular momentum
   orbital L, 3
   spin S, 6
   total $J = L + 2S$, 5–20
Angular variation of GMR, 250, 271
Anion coordinations, 1, 8, 10, 11, 12, 60, 86
Anisotropic magnetoresistance (AMR), 221, 224, 376, 400–402, 449, 458, 459, 485
Anisotropy, 377–384, 398–402, 408–410
   in monolayers, 168
   in superlattices, 172
   in transition metal ferromagnetics, 170
   magnetocrystalline, K_1, 39, 81
   single-ion, 33, 34
Annealing, 276
Anti-ferromagnetism, 327
Antiferromagnet, 347, 357, 370
   synthetic, xii
Antiferromagnetic coupling, 378, 388–389, 393–394, 403
Areal density, xii
Arrott plot, 326
Atomic force microscopy (AFM), 349
Aufbau principle, 31, 55
Auger analysis, 344

Ballistic multilayers, 273
Band theory, 54
   periodic potential, 94
   Fermi statistics, 105, 120
   Stoner model, 92, 94
   Overhauser spin density waves, 94
   Hartree-Fock tight-binding scheme, 94

Bandwidth, 537
Baseline popping, in recording heads, 397
Biasing magnet, 451, 459, 466
Binary applications, same as digital applications, 544, 547, 550, 551
Bipolar switch, 527–528
Bloch's law, 324
   wall, 345, 346
Blocking temperature, 277
Bohr atomic model, 3, 16
Bohr magneton, atomic m_B, 34, 77, 315
   nuclear m_n, 316
Boltzmann equation, 194
Boolean operations, 549–551
Branch imbalance, 531
Brillouin-Weiss theory, 77, 78, 80, 110, 111
   Weiss molecular-field constant, 78

Cantilever, 349
Cation energy difference U, 71–75
   Hund rule energy U_{ex}, 72
   ionization potentials (IP), 14, 15
Cation sites in anion coordinations, 10–12
Cation-cation direct exchange, 65, 69
CET (see covalent electron transfer)
Charge-coupling, 518, 524
Charge transfer integral b, definition, 57
Chemical potential, 517
CMOS compatible, 550–552
Coercivity, 325, 526, 548, 559, 560
Colossal magnetoresistance (CMR), 219, 292, 397, 458, 459
Compensated metal, 221
Compensation temperature, rare-earth garnet, 91, 92
Conductivity in
   films, 205
   layered systems, 191
   oxides, 92–125
   Holstein polarons, 100
   CET, 76, 117, 120
   CMR, 76, 107, 111, 113
Contact resistance, Rc, 521, 524

Controlled weak link (CLINK), 532
Conversion electron Mössbauer spectroscopy, 334
Correlation energy e^2/r_{ij}, 54, 55, 65, 66, 71, 73
    function, 386
Covalent bonding, 24, 27, 55
    bonding/antibonding states, 56
    nonbonding states, 61
    and bonds, 60, 61, 74–76
Coulomb blockade, 291
    energy, 12, 15, 16, 31, 54, 55, 65
Covalent electron transfer (CET), 76, 98, 117–125
Critical behavior, 324, 326
    current, I_c, 533, 536, 538, 540, 541, 542, 544
    field, H_c, 540
Crystal fields, 1, 16, 19–21, 31
Crystal-field theory, 12–33
    operator equivalents, 26
    point charge model, 12, 26
    weak field/strong field, 19
Crystallographic symmetry, 8–12
    seven crystal systems, 8
    point groups, 8
    space groups, 9
Curie law, 322
    temperature, 43, 77, 78, 83, 85, 92, 104, 106, 324
Curie-Weiss constant, 321
    paramagnetism, 321
    temperature, 321, 322
Current electrical, 222
    density, 222
    gain, 530, 532, 534, 536
Current-induced magnetic excitations, 274
Current-perpendicular-to-the-plane (CPP) MR, 208, 237, 238, 251
Current-at-an-angle-to-the-plane (CAP) MR, 252, 255
Current-in-plane (CIP) MR, 222, 237, 240
Cyclotron frequency, 220
Cylinder position sensor, 453, 466

Debye temperature, 101, 112
    frequency, 106
Demagnetization, 377, 398–399
Demagnetizing factor, 223, 318
    field, 318
Density functional theory, 177
Device scalability, 545, 546, 547
Diamagnetic susceptibility, 320
Diamagnetism, 319–320
    in superconductor, 320
Diatomic molecule, 55, 56
    homonuclear, 56
    heteronuclear, 55
Dielectric tensor, 135

Diffusion equation, 525
Digital voltmeter, 238
Diluted magnetic semiconductor (DMS), *xiv*
Dilution effect, 245
Dionne theory of magnetic dilution, 80, 89, 91
    spin canted sublattices, 81
d-orbital lobes, 22, 43, 60
Domain
    closure, 380, 398, 400
    single-domain particle, 377, 384, 395, 398, 400, 405
    size, 384, 392–394, 396
Domain wall, 341, 345, 347–352, 355, 358–359, 365
    angle, 378–380, 383, 400
    Bloch, 378–380
    chirality, 378–380
    cross-tie, 379–380
    energy, 377–378, 384, 401
    motion, 382–384, 387, 391–393, 397, 400–404
    Néel, 378–379
    nucleation, 378, 381
    scattering, 301
    thermal fluctuations, 382–383
    width, 378, 401
    360-degree, 379–381
Double exchange model, 296
Double spin-valves, 270
Drude parameters, 151
Dynamically programmable logic, 552

Easy axis, 377
Effective magnetic field, 332
    moment, 321
Einstein relation, 518
Electrochemical potential, 517, 525–528, 557, 560
Electrodeposition, 252, 254, 255, 257
Electromagnetic units for moment, emu, 314
Electromigration, 545
Electron paramagnetic resonance (EPR), 36, 49
Electron beam lithography, 548
Electron beam evaporator, 418
    rod-fed type, 420
    crucible type, 419
Electron biprism, 362
    holography, 361, 366
    energy level structures, 12–33
    mean-free-path, 524, 557, 559, 560
    spins, parallel and antiparallel, (see spontaneous magnetism)
Electrons in lattice, 92–94
    collective in metals, 92
    localized in insulators, 94
Energy product, 326
Entropy production, 518
Electron repulsion, (see correlation energy)

Evaporation, 414
  rate equation, 414
  refractory metals, 418, 420
  resistive heating, 415
  sources, 415–421
Exchange-biased spin-valve (EBSV), 243
Exchange constant J_{ij}
  direct/indirect, 63, 66
  ferromagnetic/antiferromagnetic, 67, 69, 98
Exchange coupling, 239, 242, 399–400, 405–406, 408–410
Exchange field H_{ex}, 42, 68
Extraordinary Hall effect, 221, 224, 252

Faraday effect, 135, 141, 352–353
Fermi level, 15
  statistics, applicability of, 120–122
  surface, 529, 557, 559
Ferrimagnetism, 75–92
  Néel thermomagnetization theory, 79
  Dionne refinement, 80, 89
Ferromagnetic metal (F), 219
    clusters, 275
    electrode, 525, 533, 534, 554–559
    insulator, 293
    resonance (FMR), 319
    semiconductor, *xiv,xv*
Ferromagnetism, 324
  optical and electrical control, *xv*
Fluctuation dissipation relation, 387–392
Flux closure, 380–381, 400
  concentrators, 464, 466, 471, 474
  quantum, 329

g-factor, (see spin Hamiltonian)
  iron group, 34, 35, 36, 37, 38
  rare-earth relation, 6, 7
Garnet ferrites, 86–92
Gear-tooth sensing, 450, 455, 560
Generalized forces, 519
Giant magnetoimpedance, 301
Giant magnetoresistance (GMR), 185, 206, 219, 304, 376, 378, 388–391, 393–394, 396–397, 449, 458, 466, 485
Gilbert's equation, 505
Goodenough-Kanamori rules, 74, 75, 76
Grain boundary tunneling, 296
Granular composites or alloys, 219, 235, 265, 291
  magnetoresistance, 275
Ground-state floor plans (see one-electron occupancy diagrams)
Group theory, 27–30
Growth modes
  Frank-van der Merwe, 429

  Stranski-Krastanov, 429
  Volmer-Weber, 429
Gyromagnetic ratio, atomic, 315
  nuclear, 317

Half-metallic ferromagnet, *xv*, 231, 267, 286
Half-select, 550
Hall device, 546, 548
  effect, 219, 250, 278, 300
  resistivity, 222
Hanle effect, 527, 528
  technique, 233
Hard axis, 377–384
Heat current density, 222
Heisenberg phenomenological model, 66
Heisenberg type interaction, 323
Heusler alloys, *xv*
Hexagonal ferrites (Ba,Sr M-type), 12
High mobility quantum well, 548, 552
High temperature superconductors (HTS), 530, 533
High-speed isolator, 467
Hooke's law analogy, 40
Hund's rules, 5, 13, 15–17, 31, 39, 54, 71, 72, 316
Hybrid ferromagnet-semiconductor devices, 546–562
  ferromagnet-superconductor devices, 530–545
Hybrid orbital states (see molecular orbital theory)
  systems, 516, 530
  Hall device (HHD), 546–552
  multilayer, 241
Hyperfine interaction, 331
Hysteresis, 382, 390, 394, 398–399, 549, 551, 558, 559
  loop, 325

Impurities, 225
Instability, 375–377, 394, 397–402, 406, 409
Insulator (I) or insulating layers, 219, 242
Integrated devices, 537, 544, 550
Interdiffusion, 545, 546
Interface, 244
  resistance, R_i, 520–523
Interfacial currents, 520
  doping, 249
  roughness, 165, 247
Inversion asymmetry, 552
Ising model of ferromagnetism, 78

Jahn-Teller (J-T) orbit-lattice effect, 40–42, 108, 109
  spin-orbit stabilizations, 40, 43, 44, 45, 46
Josephson junction, 530, 537, 545
Julliere model, 278
Junction bias effects, 286

Kerr effect, 135, 240, 353–355, 357, 370
    equitorial, 140
    longitudinal, 140, 353–355
    polar, 138, 139, 353
    transverse, 353–354
Kinetic coefficients, L_{ij}, 519, 524
Knudsen cell, 416
    rate equation, 416
    rate stability, 418
    flux angular distribution, 417
Kramers doublets, 35, 40
Kubo formula, 142
Kubo-Greenwood Formula, 186

Landauer formula for conductance, 211
Lande g factor, 315
Laplace equation, 525
Larmor angular frequency, 319, 554, 555
    precession, 78
Lithography, 252, 254
Localized moment, 321
Lock-in amplifier, 238
Lorentz force, 353, 358, 360
Lorentz microscopy, 383, 403–404
    differential phase contrast, 358, 360–361
    Foucault, 358–361, 366, 371
    Fresnel, 358, 359, 361, 366, 371
Low energy electron diffraction (LEED), 443–444
    Si (111) –7x7 444
Low resistance tunneling junctions, 290
Low temperature superconductors (LTS), 530, 533

Madelung energy, 31, 66, 95
Magnetic sublattices (see ferrimagnetism)
Magnetic dipole moment, 314
Magnetic domains, 341–342, 355, 361, 367, 370
Magnetic energy, 325
Magnetic field intensity, 219, 314
    flux, 314
    force microscopy, 342, 349
    forces, 352
    fringe fields, 537–539, 542, 546
    imaging, 342–343, 366, 370–372
    induction, 314
    materials, xiv
    memory, 221
    microstructure, 342, 346, 369, 372
    moment, 314
        atomic, 315
        nuclear, 316
    multilayers, 185
    multipoles, 349
    ordering, 323–328
    permeability, 319
    poles, 539
    quantum number, 318
    recording
        heads, 375–377, 385, 388, 393–395, 397–402
        media, 405–409
    sensor, 375–377, 385–388, 391, 394, 397–400, 402, 405
    design, 449, 460
        AMR, 461
        GMR, 463
        SDT, 468, 471
    eddy current, 454
    fluxgate, 450, 456
    Hall effect, 449, 455, 467, 474
    inductive, 455, 474
    integrated coil, 467, 468, 474
    integrated GMR, 449, 459, 465, 467
    market, 450
    reed switch, 449, 454, 474
    stabilization, 398–404
    susceptibility, 319
    tunnel junction (MTJ), 545
Magnetization, 219, 315
    anisotropy, 528, 537, 560
    currents, J_M, 519, 520
    dynamics, 342, 355, 372
    orientation, 517, 526, 537, 538, 546, 558, 559
    potential, H, 517, 518
    reversal, 379, 381–384, 400
    rotation, 382, 400
    switching, current induced, xvi
Magneto-optic image, 353, 355, 357
    indicator films, 357
    interaction, 352–353
    microscope, 354
Magneto-optical contrast, 357
Magneto-thermal conductivity, 219, 250
Magneto-thermopower, 219, 250
Magneto-elastic coupling, 166
Magnetoelastic ions, Mn^{3+}, Cu^{2+}, Co^{2+}, 34, 43, 44, 45, 46
Magnetoelastic properties, 87, 90
Magnetometry, 328–331
Magnetoresistance, 76, 79, 106–114, 111, 112, 113, 384, 388, 392, 395, 397, 405, 515, 545, 546
    inverse, 246, 265
Magnetoresistive random access memory (MRAM), 376, 386, 397, 400, 477, 478
Magnetoresistors, InSb, 457
Magnetostatic energy, 345, 349, 365
Magnetostatics, 539, 548
Magnetostriction, 81, 166, 376, 400–402, 459
    constants λ_{100}, λ_{111}, 81
Magnetotransport, 219
Magnons (quantized spin-waves), 221, 274

Manganate, 533, 534
Manganites, 292
Markovian process, 395
Mass magnetization, 315
Matthiessen's Rule, 225
    deviations from, 225, 226
Maxwell's equations, 223
Mean-free-path (l), 221
    effects, 275
Meissner screening, 542
Memory cell scaling, 507
Mesoscopic magnetoquenched superconducting valve (MMSV), 537–545
Metal-insulator transition, 106
Metallic oxides, 102–106
    TiO_y, 103
    CrO_2, 103, 109, 120
    double perovskites, 103, 103
Micromagnetic modeling, 342
Miller indices, 10
Molecular beam epitaxy (MBE), 548
Molecular-field coefficients, 77, 78, 79, 84, 89, 93
    intra-sublattice/inter-sublattice, 79
Molecular-orbital theory, 55–65
Multilayer, 219
    GMR, 449, 458
Multiplet structure, 18
Munitions fusing, 451
Mössbauer effect, 331

Nano-oxide layer (NOL), *xii*
Nanowires, 348
Narrow gap semiconductors, 554
Near-field optical techniques, 357
Néel coupling, 379
    model, 162, 163
    temperature, T_N, 74, 103, 327
Néel caps, 345, 366
    wall, 366
Noise, 250, 300
    Barkhausen noise, 382, 391–395, 397–398, 400, 403
    Gaussian, 387, 397
    in CMR materials, 397
    in GMR multilayers, 388–398
    in magnetic tunnel junctions, 397
    in recording heads, 393–395
    in recording media, 375–377, 384, 387, 405–410
    in spin valves, 383, 386, 391–392, 397
    Johnson noise, 376, 385, 387
    measurement, 384–391, 397
    Nyquist, 387
    non-Gaussian, 387, 395, 397
    power, 386, 405–407
    spectral density, 386–388, 390–392
    thermal, 474
    telegraph, 395, 397
    1/f, 375–377, 388–389, 391, 397, 474
Non-local conductivity, 189
Nonequilibrium (spin) magnetization, 517–525, 560
    quasiparticles, 532, 534
    superconductivity, 530
Nonlocal geometry, 525, 526, 529, 559, 560
Nonvolatile, 515, 538, 544, 546, 548
    random access memory, 546, 550
Nuclear g-factor, 316
    magnetic resonance (NMR), 335
    magnetization, 335
Nyquist theorem, 385

One-electron occupancy diagrams, 31, 32, 44, 45, 61
Orange peel coupling, 379
Orbital energy states, 12–33
    states t_{2g} and e_g, 20–30
    terms S, P, D, F, and G, 23, 24, 25, 39, 52
    overlap integral, S, 56–69, 74, 100, 110
    quenching (see crystal fields)
Ordered moment, 315
Orthogonal bias, 469
Oscillations, 242, 258, 259
Oscillatory exchange coupling, 348

Paramagnet, 384
Paramagnetic insulator, 293
    moment, 321
    state, 321
Pauli principle, 13
Pauli susceptibility, c, 517
Pauli-Landau susceptibility, 321
Periodic table, 2, 3
Perovskite B-sites, 11, 12, 95, 115
    lattices, 10, 95
    with polarized spins, 106–125
Permalloy, 221, 379, 383, 396–397, 400–403
Permanent magnet, 325
Photoemission electron microscope, 367
Photolithography, 525, 534, 539, 548, 557, 558
Photon helicity, 366, 367
Plated wire memory, 481
Point contact, 231, 274
Polarization, 343
Polarization efficiency, h, 517, 519–522, 524
Polarons, 94–106, 119, 120, 296
Polaron trap (see activation energy)
Potentiometric measurements, 526, 556–559
Predictability of CPP-MR, 265
Probing depth, 342, 344, 347, 355
Pseudo spin valve cells, 498, 499
Pseudorandom thickness variation, 274

Quantum computing, 562
Quantum well, 548, 553
Quasiparticle injection device (QPID), 530, 534–536
Quasiparticles, 530–537
Quenched orbital angular momentum, 321
Quenched state, 537–540, 543, 544

Rapid superconducting flux quantum (RSFQ) logic, 530
Rashba term, 554, 559
Read heads, magnetoresistive, 221
Recoilless nuclear gamma-ray resonance, 331
Reflection high energy electron diffraction (RHEED), 347–348, 421–426
   Ewald construction, 424
   geometry, 422
   Laue pattern, 425
   on metal surfaces, 426
   on polycrystalline surfaces, 426
   reciprocal lattice, 424
   streak pattern, 425
   Si (111) -7×7, 423
Relativistic velocity, 552
Remanence, 537, 538, 540, 548, 558, 559, 560
Remnant flux density, 325
   magnetization, 325
Reprogrammable logic, 546, 551
Resistance mismatch, 521, 522
RKKY polarization exchange, 71

Sandwich GMR material, 457, 459, 466
Saturation field, H_s, 243
Scanning electron microscopy with polarization analysis, 342–343
Scattering anisotropy, 221
Scattering-in term, 197
Schottky barrier, 524, 546
Secondary electrons, 343–345, 348, 367, 369
Semi-classical conductivity for layered systems, 192
Semi-classical limit, 191
Semiconductor, 301
   electronics, 545
   spintronics, xi, 515
   heterostructure, 548, 552, 554
SEMPA, 342–349, 350, 355, 369, 371–372
Sensing, 487
Shapiro steps, 532, 542, 543, 544
Sheet conductance, 243
Shubnikov-de Haas oscillations, 556
Signal to noise ratio (SNR), 386, 398, 405, 408, 410
Single electron transistor (SET), 562
Site preference energy, 32
   tetrahedral reduction factor (4/9)Dq, 26t, 32, 53
Slater determinant, 66
   integrals K_{ij} and J_{ij}, 66

Slater-Pauling plot, 225, 229
Spatial resolution, 344, 349, 357, 366, 369, 371
Specific resistance (AR), 261, 262
Spherical harmonics, 16, 17, 20
Spin accumulation, 515–530
   amplification by simulated emission of eadiation (SWASER), xvi
   analyzers, 345
   bottleneck, 517, 522, 531, 534
   coherent mean free path, L_s, 561
   configurations, 32, 33, 39
   dependent tunneling (SDT), 449, 458, 469, 471, 474
   dependent tunneling memory cells, 510
   dephasing, 529
   diffusion, 520, 524–525
   diffusion length (depth), d_s, 516, 518, 520, 526, 528, 529
   flip, 529, 534
   probability, 529
   Hamiltonian, 37, 52
   imbalance, 534
   injected field effect transistor (FET), 546, 552–562
   injection, 515–530, 536, 554, 555, 556, 559
     optical, xvii
   manipulation in semiconductors, electrical, xvii
     optical, xvii
   ordering, pairing and polarization, 31
   polarization, 221, 227, 231
   polarized (electron) current, 343, 344, 516, 518, 523, 524, 527, 559–561
   quasiparticles, 533, 534
   polarizer, 518
   precession, 528, 554
   relaxation time, T1, T2, 517, 525–529
   sensitive potentiometer, 518, 525–530, 559
   splitting (of conduction band), 552, 554
   superexchange, 68, 70, 74, 75, 76
   switch, 527
   transfer
     interionic energy, 72, 73
     efficiency, 99, 99, 100, 118
     probability of, 99
   transistor, 527, 546
   transport, polarized, 1, 2, 49, 73, 102, 104
   valve, 383, 386, 391–392, 397
     GMR, 457, 459
     memory cells, 508
   wave generation, xvi
Spin-dependent transport, 185
Spin-dependent tunneling, 185, 212, 227
Spin-diffusion length, 232, 235, 268
Spin-echo NMR, 336
Spin-glasses, 219
Spin-injection, 232
Spin-lattice relaxation time, 336

INDEX

Spin-memory-loss-length, l_{sf}, see spin-Diffusion Length
Spin-mixing, 259
Spin-orbit, 132, 529, 553
Spin-orbit coupling, 2, 16, 17, 33, 39, 180, 400–401
  interactions, 269
  lattice effects, 7, 46, 52, 90
Spin-relaxation at interfaces, 270
Spin-resolved photoemission, 231
Spin-spin interactions, 269
Spin-spin relaxation time, 336
Spin-valve transistor, 303
Spinel ferrites, 3, 81, 82, 83, 84, 85, 86, 87
Spintronic devices, *xvii*
  filed effect transistor (spin-FET), *xviii*
  light-emitting diode (spin-LED), *xvii*
  p-n junction, *xviii*
  resonant tunneling diode (spin-RTD), *xviii*
  spin-filters, *xviii*
Spintronics, *xi*, 515, 545
Spintronics quantum, *xix*
Spontaneous magnetism, 54, 62, 79
  ferromagnetism/antiferromagnetism, 62, 67
  ferrimagnetism, 79–81
Spontaneous magnetization, 324
Sputtering, 414, 430, 445
  yield, 431
  DC diode, 433
  DC triode, 434
  magnetron, 435
  RF, 436
  ion beam, 437, 445
  reactive, 438
Squareness, 325
SQUID magnetomer, 329
Stability limits, 480
Stationary process, 387
Stepped surface, 252
Stoner Ferromagnet, 232
Stoner-Wohlfarth threshold, 483
Superconducting digital electronics, 530, 537, 544
  order parameter, 532
  phase coherence (Josephson), 533, 538, 542
  quantum interference device (SQUID), 238, 537, 544
  gap, D, 530, 532, 534, 536
Superconductivity, 117–125
  London theory, 118
  spatial ordering condition $\nabla n_s = 0$, 118
  high-T_c, HTS, 76, 114, 117
  Pippard's ensemble wave function, 118
  coherence length, 118
  critical temperature T_c, 120–125
  normal resistivity, 121, 121f
  two-fluid model, 120–124
  real-space pair transfer, 122

$\Delta S = 0$ rule, 97, 114, 125
spin frustration $(T_N, T_c = 0)$ condition, 116, 116, 117
polaron percolation threshold, x_t, 122
Superconductor, 219, 228, 530
  perovskites, 114–125
Superparamagetism, 277, 384
Susceptibility, 388
Synchrotron radiation, 366–368
Symmetry axes, 11
  cubic, $\langle 100 \rangle$, $\langle 111 \rangle$, $\langle 110 \rangle$, 11
  tetragonal/orthorhombic $\langle 001 \rangle$, 11, 27–30, 107
  trigonal (rhombohedral) $\langle 111 \rangle$, 11f, 30, 107
System-on-a-chip, 552

Temperature gradient, 222
Thermal activation, 396
  conductivity, 222, 278, 300
Thermomagnetization (see ferrimagnetism)
Thermomagnetoelectric system, 518
Thermopower, 222, 278, 300
Thin film deposition, 402–403, 408
Thomas precession, 131
Time series, 385–387, 392–393, 395
Topographic feedthrough, 346
Topography, 342, 345–346, 349, 352, 367
Transfer curve, 385–393
Transition metal ferromagnet, 516
Transition metals, 315
  ions, 3, 5, 6, 7
Transition parameter, 387
Transmission and reflection matrices, 201
Transmission electron resonance (TESR), 528
Transmission electron microscope, 357
Transmission x-ray microscope, 369
Tunnel barrier, 524, 532, 559
  contact, *xiii*
Tunneling, 227, 228, 278
  magnetoresistance (TMR), *xx*, 219, 278
Two dimensional electron gas (2DEG), 552, 555–557, 559
Two-current series resistor (2CSR) Model, 261

Uniaxial anti-ferromagnet, 328
Units, conversion between CGS and SI, 314
  magnetic, 313

Valence bond model, 55
  Heitler-London H_2 molecule, 56
Valet-Fert Model, 261
Van der Pauw geometry, 529, 548
Vibrating sample magnetometer (VSM), 328
Vortices, 380, 382, 404

Weak link, 532, 537–542
Wolfsberg-Helmholtz model, 57, 58

Write pulses, 544, 548, 550
    wires, 544, 548, 550

X-ray dichroism, 366
    magnetic linear dichroism, 370

Zeeman effect, matrix, splittings, 31, 35, 38
Zeeman energy, 517, 552
Zero-spin polarons, 114–120, 117
    low-spin Cu^{3+}, 114–117